Balance Redox p 414

General and Inorganic Chemistry

General and Inorganic Chemistry

BY

GUNNAR HÄGG

Professor of Inorganic Chemistry
University of Uppsala, Uppsala, Sweden

Translated by

HOWARD T. EVANS, JR

United States Geological Survey
Washington, D.C., U.S.A.

ALMQVIST & WIKSELL/STOCKHOLM

© Gunnar Hägg 1969

Almqvist & Wiksell Förlag AB
Stockholm Sweden 1969

This book or any part thereof
must not be reproduced in any form
whatsoever without the written
permission of the publisher,
Almqvist & Wiksell Förlag AB.
Stockholm, Sweden

Printed in Sweden by
Almqvist & Wiksells Boktryckeri AB
Uppsala 1969

CONTENTS

Preface 7
Introduction 9

PART I. GENERAL CHEMISTRY

Chapter 1. Fundamentals, Concepts and Definitions 13
Chapter 2. Energy 50
Chapter 3. Quantum Mechanics 66
Chapter 4. The Electron Cloud of the Atom and the Periodic System 75
Chapter 5. Chemical Bonding 97
Chapter 6. The Solid State 186
Chapter 7. Structure Determination, Color, Magnetic Properties 237
Chapter 8. Reaction Kinetics in Homogeneous Systems 251
Chapter 9. The Atomic Nucleus 259
Chapter 10. The Equilibrium Equation 297
Chapter 11. Ionic Equilibria 308
Chapter 12. Acid-Base Equilibria 316
Chapter 13. Heterogeneous Equilibria 348
Chapter 14. Reaction Kinetics in Heterogeneous Systems . . . 388
Chapter 15. Surface Chemistry and Colloids 400
Chapter 16. Redox Equilibria and Electrochemistry 412
Chapter 17. Distribution of the Elements 440

PART II. SYSTEMATIC INORGANIC CHEMISTRY

Arrangement of Part II 450
Chapter 18. Hydrogen 451
Chapter 19. Group 0, Noble Gases 457

Chapter 20. Group 17, Halogens 464
Chapter 21. Group 16, Chalcogens 489
Chapter 22. Group 15, The Nitrogen Group 536
Chapter 23. Group 14, The Carbon Group 584
Chapter 24. Group 13, The Boron Group 630
Chapter 25. Group 1, Alkali Metals 647
Chapter 26. Group 2, Alkaline-Earth Metals 657
Chapter 27. The Transition Elements 668
Chapter 28. Group 3, The Scandium Group with Lanthanoids
and Actinoids 671
Chapter 29. Group 4, The Titanium Group 683
Chapter 30. Group 5, The Vanadium Group 689
Chapter 31. Group 6, The Chromium Group 693
Chapter 32. Group 7, The Manganese Group 704
Chapter 33. Survey of Groups 8, 9 and 10 711
Chapter 34. The Iron Metals Iron, Cobalt and Nickel 713
Chapter 35. The Platinum Metals 737
Chapter 36. Group 11, The Copper Group or the Coinage Metals 743
Chapter 37. Group 12, The Zinc Group 761

Appendix 1
  Symbols 776

Appendix 2
  Constants and Conversion Factors 779

Name Index and Biographical Data 780
Subject Index 784

Preface

This textbook is a translation from the fourth edition of a Swedish textbook published in 1966. The first Swedish edition appeared in 1963. It has been found appropriate to keep the book practically unchanged both as to its general character and its details. As a consequence it may show features which are unfamiliar to some of its new readers but I hope that there is no harm in its reflecting its origin in this way. However, some of these features require comment.

After consulting several English-speaking chemists I have thought it best to use a literal translation of the Swedish title although the expression "general chemistry" may have a slightly misleading meaning, at least to an American reader. In Europe this or equivalent expressions usually imply an introduction to the laws of chemistry without any specialized, descriptive material, while in America it seems to mean an elementary presentation of the entire science. An American reader may, therefore, miss a treatment of organic chemistry and biochemistry.

The Swedish edition of the book is intended to be used at the Swedish universities during the first semester of studies in chemistry which comprise general (in the European sense) and inorganic chemistry. At the start of these studies the average student is expected to rely upon knowledge in mathematics, physics and chemistry gained at the secondary school (gymnasium). The book has to take this into consideration, which means that the more rigorous parts of physical chemistry cannot be included but must be taught at a later stage. I have tried, however, to include all that I consider to be necessary for a fair and comprehensive understanding of inorganic chemistry. All this general knowledge has been brought together in Part I.

In Part II the more specialized and descriptive inorganic chemistry is treated. There is wide-spread opinion among Swedish university chemists that a fair knowledge of substances and their properties must still be obtained on the undergraduate university level and due regard has been paid to this requirement. It is to be hoped, however, that the general knowledge acquired in the study of Part I will make Part II digestible and enable the student to see most of the material in the light of general laws. Numerous cross-references have been introduced to facilitate this.

Literature references in the text are not usually found in Swedish books on this level and have not been introduced in this edition. When references appear in American elementary textbooks they commonly cover only the more recent years and thus neglect earlier work which can be equally or even more important and certainly as useful in acquiring more detailed information. According to my experience such references are very seldom made use of by the student.

The SI international unit system has been used throughout. In this book the most obvious consequence of this is that the joule is used as the general unit for energy. The calorie may be missed by some readers but I am convinced that it must be abandoned because of its irrational definition. For that reason the conversion ought to be accomplished as soon as possible. But in cases when irrational units are still used for the definition of quantities (*e.g.* atmosphere for standard state and boiling point) they must be retained.

The rules for the nomenclature of inorganic chemistry as recommended by the Nomenclature Commission of the International Union of Pure and Applied Chemistry (IUPAC) have been followed in nearly all cases. One important exception is a new notation for the groups of the periodic table which is presented for the first time in Chapter 4. The reasons for this deviation should be clear from what is said in 4–2c. The term "halogenide" and the prefix "halogeno-" are preferred by the Commission to the old "halide" and "halo-" but the latter have been retained because of their continued widespread use, especially in America. Water as a ligand is for the present still called "aquo" instead of the newly recommended "aqua". On the other hand the terms "lanthanoid" and "actinoid" recommended by the Commission have been used instead of the still common "lanthanide" and "actinide".

The American version of the rules of nomenclature of IUPAC are readily available in complete form in the Journal of the American Chemical Society, vol. 82, pp. 5523–5544 (1960). As a consequence this book does not contain any comprehensive presentation of these rules although important points are frequently referred to in the text.

I owe a lasting debt of gratitude to several friends for very valuable assistance in the production of this book. The whole manuscript of the Swedish edition was read by Professor Ingvar Lindqvist and Professor Lars Gunnar Sillén and their advice was invaluable. Professor Per-Olov Löwdin examined Chapters 3–5 and Professor Frans Wickman examined Chapter 17 and the account of radioactive dating in Chapter 9. The rest of Chapter 9 was inspected by Dr. Sigvard Eklund of the International Atomic Energy Agency (IAEA). And I am greatly indebted to Dr. Howard T. Evans, Jr. who has then presented the book in English. He has fulfilled this very difficult and heavy task with skill and attachment.

I gratefully acknowledge permission to reproduce figures and photographs as follows: To the Council of the Royal Society of London for figs. 6, 9b, 6.28 6.29; the Clarendon Press, Oxford, for figs. 5.10, 5.17, 5.18; John Wiley & Sons Inc. for fig. 14.1; Professor Inga Fischer-Hjalmars for fig. 5.20; Mr. Sten Modin for figs. 6.9a, 13.12, 34.2; Professor Ralph W. G. Wyckoff for fig. 1.2.

Uppsala, Sweden, June 1967 *Gunnar Hägg*

Introduction

"Chemistry is the science of the composition of bodies and their relations to one another." Thus wrote Jöns Jacob Berzelius in the introduction to his textbook of chemistry in 1808. For the most part, this definition still applies today. Chemistry studies the structure and properties of substances, their transformations and their reactions with one another.

But modern developments have made the drawing of boundaries between the different sciences much more difficult than in Berzelius's time. The chemist studies the structure of substances primarily to be able to draw conclusions concerning their chemical properties and how they partake in chemical processes. Nowadays, he uses for his studies a broad range of methods drawn from physics. But the physicist also is interested in the structure of substances, which he must know in order to explain their physical properties. While the chemist most often treats the atoms primarily as indivisible building units in different substances, the physicist investigates how the atoms themselves are built up of smaller parts; in addition both seek to determine how the atoms are bound to one another in different substances. That sciences in this way overlap one another is typical of our time and shows the importance of collaboration across the old boundaries.

This book is devoted to the knowledge of substances. The purely descriptive material must have a large place, but as long as it is possible with elementary methods the book will also attempt to show what is believed to be known of the reasons for the structure, properties and reactions of substances. This demands a quite extensive treatment of the general fundamentals of chemistry, which often are of a rather physical nature. It is not intended, however, to give a complete presentation of basic physical chemistry.

Chemistry can be divided into different branches according to the nature of the substances under study. The chemistry of carbon compounds is called *organic chemistry*, a designation which comes from the time when it was believed that most carbon compounds could only be formed in living organisms. The chemistry of all the other elements is called *inorganic chemistry*. The study of substances appearing in living organisms and their relationship to life processes is called nowadays *biochemistry*. Many practical reasons speak for such a division even if it cannot always be made sharp. This book deals with inorganic chemistry and only a few digressions into organic chemistry have been made. The element carbon and certain simple carbon compounds, primarily those which take part in processes important in inorganic chemistry, have mainly been considered.

General chemistry is often spoken of as being for the most part a general introduction to the whole of chemistry, wherein, not the least among other things, the fundamental physical laws must assume a central position. Since the referred-to general laws of chemistry are extensively treated in the following, the presentation departs little from what would be required of a book with the title of general chemistry. It is clear that general chemistry has much to do with *physical* or *theoretical chemistry*, which deal with the laws of chemical behavior themselves. Physical chemistry is not a separate part of chemistry, isolated from the other parts. Since the fundamental laws must be studied and applied in every branch of chemistry, physical chemistry becomes interwoven with the whole of the science of chemistry.

Nuclear chemistry is a designation which has been given in recent times to that part of chemistry which makes use of transformation of the atomic nucleus for chemical purposes (analysis, study of reactions, etc.), or studies them by chemical means (separation of the products of nuclear reactions, etc.) The boundary between nuclear chemistry and nuclear physics is very fluid; both are in rapid development. An important part of chemistry is *analytical chemistry*, which deals with the methods for the determination of the composition of substances. It thus is methodological in nature and, of course, is often used in practical work.

Even if no sharp boundaries can be drawn in the following between the presentation of general principles and the more descriptive, special treatment of the elements and their compounds, the text has for practical reasons been divided into *Part I: General Chemistry* and *Part II: Systematic Inorganic Chemistry*.

PART I
General Chemistry

CHAPTER 1

Fundamentals, Concepts, and Definitions

The various laws, empirical principles and concepts which form the general basis for the structure of chemistry are so interwoven with one another that a logically developed treatment is impossible if it is not done in a wholly historical and chronological manner. Since this would lead to a far too voluminous and indigestible presentation, this chapter gives instead a survey of a portion of the material which is needed later on. This survey is very elementary and often takes the form of a brief statement of current concepts. The purpose of this chapter also causes the discussion to be uneven; certain topics may be cursorily treated or omitted because by reason of their very importance they are more extensively dealt with later and are not needed in the introduction. The material is nevertheless introduced in logical sequence.

1-1. Some definitions

a. Properties and quantities. The *properties* of a body are either *accidental* or *essential*. Most properties of interest in chemistry are *quantities*, that is, they may be measured directly or indirectly.

Accidental properties are, for example, the quantities mass, area, volume, energy, electric charge. They are often additive and divisible and are then called *extensive quantities*. Thus, if a body is divided into several smaller parts, the sum of the masses of the parts is the same as the mass of the original body.

Essential properties are always independent of the quantity of material. It is such properties one refers to when speaking of the properties of a *substance*, and many such quantities characterize a substance. These quantities are commonly ratios of two extensive quantities, then often called *specific quantities* (for example, specific volume, specific heat).

Some properties characteristically strive to attain the same value in all parts of a region which is isolated for the property in question. If a quantity of gas is enclosed in a vessel, the pressure seeks to become uniform at all points in the gas. Inside a region thermally insulated against its surroundings the temperature gradually becomes constant at all points. Electric potential also tries to equalize itself in an electrically isolated region. Quantities such as pressure, temperature and electric potential are called *intensive quantities*.

The *chemical properties* of a substance determine how it behaves chemically under given conditions, that is, how it is transformed or reacts with other substances. Chemical properties are always the same for a given substance in the same state.

b. System. Each bounded region which is studied or discussed is called a *system*. The system may be enclosed in physical walls or boundary surfaces, but a system can also often be thought of as certain quantities of substances without specifying how they are bounded.

A system is called *homogeneous* if it has the same essential properties throughout, and *heterogeneous* if it consists of parts with different essential properties. This definition is, however, incomplete if one does not also agree on the magnitude of the region within the system whose properties are to be compared. If, for example, these regions are of atomic dimensions, the discontinuous structure of matter will naturally imply that all systems must be designated as heterogeneous. If the definition is to be applied to the treatment of chemical equilibria, the proper size of the regions is determined so that the laws of chemical equilibria are valid, which will be true only if the number of building units (atoms or molecules) of the material taking part in the process is very large. The regions must therefore be so large that they contain a sufficient number of particles to insure that statistical calculations will give a very high degree of probability. Dimensions of the order of 10^{-7} m ($=1000$ Å) are usually assumed to be adequate for this requirement. Since the distances between atoms in solid and liquid substances are of the order of 10^{-10} m ($=1$ Å), a cube of such a substance with an edge of 10^{-7} m contains a number of atoms of the order of 10^9.

This boundary drawn between homogeneous and heterogeneous systems is a practical one also because the resolution of the ordinary microscope under favorable conditions is approximately 10^{-7} m. If no heterogeneity can be detected with such a microscope, the system may be designated as homogeneous.

c. Phase. The homogeneous parts of a system with the same essential properties are considered collectively as one *phase*. Different aggregation states and different modifications of the same substance constitute different phases. The system water may thus contain the three phases ice, liquid water and water vapor. Orthorhombic and monoclinic sulfur are different phases in the sulfur system. A mixture of oil and water contains two liquid phases, an oil phase and a water phase. If such a mixture, which does not contain too much oil, is vigorously shaken, the oil is divided into small drops suspended in the continuous water phase. The oil drops nevertheless still represent the oil phase. If it is desired to indicate that it is finely divided, it is designated as a *dispersed phase* and the system as a *dispersed system*. The continuous phase, in this case the water, is called the *dispersion* *medium*. Liquid and solid phases are often denoted collectively as *condensed phases*.

If gases are mixed with each other it may happen, of course, that liquid or solid phases will separate out, but it never happens that a quantity of gas splits into several gas phases. A system can therefore never contain more than *one* gas phase.

The different phases of a system can in many cases be separated from each other by purely mechanical means. If the system is left undisturbed phases of different density are not infrequently separated under the influence of the force of gravity; this effect can be amplified by centrifugation. Liquid phases can frequently be separated from each other with the aid of a "separatory funnel", solid phases from liquid by filtration. Solid phases of different density can often be separated by stirring with a liquid of such density that some phases sink to the bottom while others rise to the surface.

1-2. Matter and energy

a. Matter and energy. In the seventeenth century Boyle began to use the balance in the study of chemical processes, and this precedent was followed in the eighteenth century by such scientists as Black and Torbern Bergman. It is clear that in this way it began to be understood, what was explicitly stated by Lavoisier in 1785, and what can now be expressed as follows: that the mass of a system remains constant under any process (chemical or physical) within the system.

Since the quality of having mass was long thought to be the characteristic property of all matter, and the quantity of matter taken to be proportional to the mass, this principle was often referred to as the law of the *conservation of matter*.

However, Einstein showed through his special theory of relativity set forth in 1905 that energy also must possess mass. This result has since been experimentally proven in many different ways. The relation between a quantity of energy E measured in joules (J) and the corresponding mass m measured in kg is determined by *Einstein's equation*:

$$E = mc^2 \qquad (1.1)$$

in which c represents the velocity of light in vacuum measured in m per s. The velocity of light is a fundamental constant of nature: $c = 2.997925 \times 10^8$ m s^{-1}.

If a system gives off energy during a process, then the mass of the system must thus be decreased. The high value of c^2 shows, however, that no change in mass can be observed under ordinary chemical processes. If 1 kg of glycerol trinitrate ("nitroglycerine") explodes, an amount of energy equal to 7.9×10^6 J is released. If this energy leaves the system, the mass of the system decreases according to Einstein's equation by $(7.9 \times 10^6)/(9 \times 10^{16}) = 8.8 \times 10^{-11}$ kg, that is, approximately 10^{-4} mg. This loss constitutes one part in 10^{10} of the mass of the system and is too small to be observable. In practice it may still be assumed that ordinary chemical reactions do not change the mass of a system.

The situation is quite different when the much greater energies which may be released in nuclear reactions are considered. Through fission of heavy atomic nuclei (9-1 b) a system of mass 1 kg can, for example, give off an amount of energy of the order of magnitude of 10^{14} J, which corresponds to a decrease of about 1 g, or one part in 10^3 of the mass of the system.

The law of the conservation of matter is thus not a strictly applicable law, but it is sufficiently accurate for all practical purposes in ordinary chemical processes.

Mainly through the work carried out in the 1840's by Mayer and Joule, the law of the *conservation of energy (energy principle)* was deduced. According to this law, energy can neither be created nor destroyed.

It is clear that the laws of the conservation of matter and energy can be combined into one absolutely exact law of the *conservation of mass*, if mass is taken to include both the mass of the matter and the mass of the energy within the system. The latter is calculated from Einstein's equation.

If the velocity of a moving body is increased, then at the same time the kinetic energy (energy of movement) is increased, which according to the theory of relativity causes an increase of mass. This becomes measurable only at very high velocities. A velocity of 3×10^5 m s^{-1}, that is 1/1000th of the velocity of light, leads to an increase in the mass of the system of only one part in 2×10^6. However, through acceleration of charged particles in an electric field (1-3a), for example, velocities can easily be obtained which cause a large increase in mass. Even under the influence of so small a potential drop as 20 kV, the velocity of an electron reaches 8.15×10^7 m s^{-1}, which produces a fractional increase in mass of 0.039, that is, nearly 4 %. Particles which are ejected from atomic nuclei through a nuclear reaction can attain very much higher velocities. In this case, when nothing further is mentioned about the mass of the particle, the mass always refers to the stationary particle, the *rest mass*.

b. Radiant energy. By radiant energy is meant electromagnetic radiation, which is considered generally as a wave motion propagated with the velocity of light, c. The wavelength λ may vary from more than 10^4 m (long radio waves) to less than 10^{-14} m (short-wave γ radiation).

In tab. 1.1 is presented a summary of wavelengths, frequencies (frequency $\nu = c/\lambda$), etc., for different types of electromagnetic radiation.

In 1900 Planck showed that when a body absorbs or gives off radiant energy of a certain frequency ν, this can only occur in the form of an even multiple of a minimum quantity of energy, which is called a *quantum*. The energy E of a quantum is proportional to the frequency of the radiation, that is,

$$E = h\nu \tag{1.2}$$

Fundamentals, Concepts and Definitions

Table 1.1. *Wavelengths, wave numbers, frequencies and energies for electromagnetic radiation.*

Type of radiation	Wavelength m	Wavelength Å	Wave number m^{-1}	Frequency s^{-1} (Hz)	Energy per quantum J	Energy per N_A quanta (einstein), kJ
radiowaves	10^4		10^{-4}	3×10^4	2×10^{-29}	1.2×10^{-8}
	10^3		10^{-3}	3×10^5	2×10^{-28}	1.2×10^{-7}
	10^2		10^{-2}	3×10^6	2×10^{-27}	1.2×10^{-6}
	10		10^{-1}	3×10^7	2×10^{-26}	1.2×10^{-5}
	1		1	3×10^8	2×10^{-25}	1.2×10^{-4}
microwaves	10^{-1}		10	3×10^9	2×10^{-24}	1.2×10^{-3}
	10^{-2}		10^2	3×10^{10}	2×10^{-23}	1.2×10^{-2}
infrared	10^{-3}		10^3	3×10^{11}	2×10^{-22}	1.2×10^{-1}
	10^{-4}		10^4	3×10^{12}	2×10^{-21}	1.2
	10^{-5}		10^5	3×10^{13}	2×10^{-20}	1.2×10
	10^{-6}	10^4	10^6	3×10^{14}	2×10^{-19}	1.2×10^2
Visible light red	8×10^{-7}	8×10^3	1.3×10^6	3.8×10^{14}	2.5×10^{-19}	1.5×10^2
red	7×10^{-7}	7×10^3	1.4×10^6	4.3×10^{14}	2.8×10^{-19}	1.7×10^2
orange	6×10^{-7}	6×10^3	1.7×10^6	5.0×10^{14}	3.3×10^{-19}	2.0×10^2
green	5×10^{-7}	5×10^3	2.0×10^6	6.0×10^{14}	4.0×10^{-19}	2.4×10^2
violet	4×10^{-7}	4×10^3	2.5×10^6	7.5×10^{14}	5.0×10^{-19}	3.0×10^2
ultraviolet	10^{-7}	10^3	10^7	3×10^{15}	2×10^{-18}	1.2×10^3
	10^{-8}	10^2	10^8	3×10^{16}	2×10^{-17}	1.2×10^4
X rays	10^{-9}	10	10^9	3×10^{17}	2×10^{-16}	1.2×10^5
	10^{-10}	1	10^{10}	3×10^{18}	2×10^{-15}	1.2×10^6
	10^{-11}	10^{-1}	10^{11}	3×10^{19}	2×10^{-14}	1.2×10^7
γ rays	10^{-12}	10^{-2}	10^{12}	3×10^{20}	2×10^{-13}	1.2×10^8
	10^{-13}	10^{-3}	10^{13}	3×10^{21}	2×10^{-12}	1.2×10^9
	10^{-14}	10^{-4}	10^{14}	3×10^{22}	2×10^{-11}	1.2×10^{10}

The constant of proportionality h is a natural constant, *Planck's constant*. If E is expressed in terms of J and ν in s^{-1}, then $h = 6.6256 \times 10^{-34}$ J s.

In 1905 Einstein found that the minimum quantity of energy $h\nu$ is not only involved in the absorption and emission of radiation, but also radiation itself must be thought of as consisting of quanta. From (1.2) it follows that the shorter the wavelength of the radiation, that is, the greater the frequency, the greater is the energy of the quanta which make up the radiation. In the visible spectrum, the energy thus increases when one goes from the red toward the violet.

Einstein found also that a quantum of radiation from a certain viewpoint can be thought of as consisting of *particles* of energy $h\nu$. Such a particle has no rest mass, but carries mass, on the other hand, just from its energy content according to Einstein's equation. (1.1) with (1.2) gives $h\nu = mc^2$, that is, $m = h\nu/c^2$. This leads to the result that a body exposed to radiation is subjected to a force, the radiation pressure. Especially when it is desired to emphasize the corpuscular properties of a quantum, it is commonly called a *photon*.

In physics the properties of a photon are usually expressed by stating that it lacks rest mass but possesses momentum. The *momentum* p of a body is the product of its mass m and its velocity v, that is, $p = mv$. A photon of mass hv/c^2 and velocity c thus has momentum

$$p = hv/c = h/\lambda \tag{1.3}$$

The wave nature of all electromagnetic radiation, on the other hand, can be demonstrated by the phenomenon of interference. The photon–light wave was the first example of the parallelism between particles and waves, which later acquired great significance and is often referred to in the following (especially in chap. 3).

Tab. 1.1 shows in the next-to-last column round energy values of quanta of different frequencies calculated according to (1.2).

Many processes occur only if definite points of attack are encountered by quanta of certain minimum energy. Only such quanta are effective and absorbed thereby. The *adding together* at a point of attack of the energies of several quanta generally does not occur. An energy-rich quantum can have an effect, therefore, which cannot be produced by any number of energy-poor quanta. Irradiation with ultraviolet light, for example, may cause a reaction which will not occur no matter how long irradiation with red light continues. Energy-rich radiation therefore most often is taken to mean radiation consisting of energy-rich quanta, and not radiation with a large total energy.

c. Energy units. In chemistry the *calorie* (cal) has been most often used as the unit of energy. The calorie cannot be expressed rationally in absolute units and has also been defined in different ways (cf. Appendix 2). Instead the rational energy unit *joule* (J) should be used.

In spectroscopy the radiant energy is often given in terms of the *wavelength* in vacuum $\lambda = c/v$ m, *wave number* $1/\lambda = v/c$ m^{-1}, or *frequency* v s^{-1} (for oscillations the unit s^{-1} is called *hertz*, Hz). None of these units have the dimension of energy, but through equation (1.2) they express the radiant energy per quantum. The wave number and frequency are directly proportional to this quantity. In table 1.1 are shown rounded wave numbers.

The unit *electron volt* (eV) was introduced originally to express the kinetic energy of a particular particle, and this unit has since been widely used in atomic physics. If a particle carrying one electrical unit of charge ($e = 1.60210 \times 10^{-19}$ C, see 1-3a), travels through a potential drop of V volts in vacuum, it acquires a kinetic energy $E = eV = 1.60210 \times 10^{-19} \times V$ J. If the potential drop $V = 1$ V, the kinetic energy becomes $E = 1.60210 \times 10^{-19}$ J. It is this quantity of energy which is called an electron volt (see also 1-4d).

Conversion factors between different energy units are given in Appendix 2.

1-3. Structure of the atoms

All matter in the ordinary sense consists of *atoms*, which in turn are built up of various kinds of *elementary particles*. These elementary particles can exist free for very short times, but the atom is the unit which retains its identity in ordinary chemical reactions. For the chemist, therefore, the atom is the real building unit of all matter. The concept of the atom in the modern sense was introduced by Dalton in 1803.

The atom contains a *nucleus*, which has a positive electric charge, and in which nearly all of the mass of the atom is concentrated. The nucleus is surrounded by an electron envelope, the *electron cloud*, containing one or several *electrons*, which have a negative electric charge. This basic feature of the structure of the atom was established by Rutherford in 1911.

a. Elementary particles. The designation "elementary particle" comes from the time when it was believed that in them had been found the most elementary building units of matter. In recent years so many different particles of a similar kind have been found that it is evident that this designation is not adequate. Nevertheless, the name continues to be used. Our knowledge in this area meanwhile is increasing so rapidly, especially with the advent of new accelerators (9-2b) that permit the study of particles of higher and higher energies, that we may hope that a more unified understanding of them will soon be evolved.

The earliest known particle is the *electron*. In 1897 J. J. Thomson showed that the so-called cathode rays, which were already observed in 1869 in connection with electric discharge in gases, consist of negatively charged particles of extremely small mass. A recently determined value of the rest mass of the electron is $m_e = 9.1091 \times 10^{-31}$ kg. The mass of other elementary particles was earlier often given in terms of the electron mass as a unit. Nowadays the rest mass of an elementary particle is usually given in terms of the corresponding amount of energy according to (1.1), expressed in MeV. If m_e in kg and c in m s^{-1} are substituted into (1.1) the energy is obtained in J. If this is converted according to 1-2c into MeV the rest mass of the electron becomes 0.51 MeV.

Every electric charge is numerically equal to the charge of the electron, or a whole multiple thereof. The electron is considered therefore to carry the *electrical unit (elementary) charge*, $e = 1.60210 \times 10^{-19}$ C. The numerical value of the charge of elementary particles, as well as of atoms and molecules, is most often given in terms of this unit. The charge of the electron thus becomes -1.

The electron is stable. It does not decompose spontaneously like most other elementary particles, but can be destroyed only by reaction with other particles.

Very soon after radioactivity became known it was found that the so-called β rays consist of electrons, which are ejected with very high velocity from nuclei.

In 1886 Goldstein observed in a gas discharge tube positively charged rays ("canal rays") which were emitted from the cathode in a direction opposite to

Table 1.2. *The "classical" elementary particles.*

	Name	Symbol Particle	Symbol Antiparticle	Rest mass MeV	Particle charge	Mean lifetime s
Baryons	Omega hyperon	Ω^-	$(\overline{\Omega}^+)^1$	1675	-1	1.3×10^{-10}
	Xi hyperons	Ξ^-	$\overline{\Xi}^+$	1321	-1	1.4×10^{-10}
		Ξ^0	$\overline{\Xi}^0$	1316	0	3.9×10^{-10}
	Sigma hyperons	Σ^-	$\overline{\Sigma}^+$	1196	-1	1.6×10^{-10}
		Σ^0	$\overline{\Sigma}^0$	1192	0	$< 0.1 \times 10^{-10}$
		Σ^+	$\overline{\Sigma}^-$	1189	$+1$	0.8×10^{-10}
	Lambda hyperon	Λ^0	$\overline{\Lambda}^0$	1115	0	2.6×10^{-10}
	Nucleons Neutron	n	$\overline{n}$	940	0	1.013×10^3
	Nucleons Proton	p	$\overline{p}$	938	$+1$	stable
Mesons	K mesons	K^0	$\overline{K}^0$	498	0	10^{-7}; 10^{-10}
		K^+	K^-	494	$+1$	1.2×10^{-8}
	Pi mesons	π^+	π^-	140	$+1$	2.5×10^{-8}
		π^0		135	0	1.1×10^{-16}
Leptons	Muon	μ^-	μ^+	106	-1	2.2×10^{-6}
	Electron	e^-	e^+	0.51	-1	stable
	Muon neutrino	ν_μ	$\overline{\nu}_\mu$	< 2.5	0	stable
	Electron neutrino	ν_e	$\overline{\nu}_e$	< 0.00025	0	stable
	Photon	γ		0	0	stable

[1] Not proved March 1966.

that of the cathode rays. Primarily through the work of J. J. Thomson at the beginning of this century it was shown that these rays consist of positively charged atoms and molecules (positive ions) derived from the gas filling of the tube. If the tube contains hydrogen the rays consist, among other ions, of hydrogen atoms with charge $+1$, that is, what later came to be considered as hydrogen nuclei. When the hydrogen nucleus proved to be one of the most important building units of the atomic nuclei, the name *proton* (Gk. πρῶτος, the first) was proposed for it.

The mass of the proton is 938 MeV and its charge is $+1$. It is, by itself, stable.

The *neutron* was discovered in 1932 by Chadwick. With a mass of 940 MeV it is not significantly heavier than the proton and, as the name indicates, is uncharged. For this reason its motion is unaffected by electric and magnetic fields. The free neutron splits spontaneously into a proton and an electron with the simultaneous emission of energy as radiation. The mean lifetime of a neutron is 16.9 min. (The mean lifetime is equal to the half-life as defined in 8a divided by $\ln 2 = 0.693$.)

Protons and neutrons, which give the atomic nucleus the major portion of its mass, are known collectively as *nucleons*.

Many types of elementary particles, among others electrons and the nucleons, have properties that would be expected if they were in continuous axial rotation. This property is customarily called *spin*.

The *photon* (1-2b) is also often included among the elementary particles.

A group of 34 elementary particles, which might be called "classical", is in many respects quite well unified. In tab. 1.2, these particles are listed in order of decreasing rest mass. Besides the photon, they are divided into *leptons* (Gk. λεπτός, thin), *mesons* (Gk. μέσος, between), and *baryons* (Gk. βαρύς, heavy). Those baryons that are heavier than the nucleons are called *hyperons* (Gk. ὑπέρ, above, on the other side). Some of the particles in tab. 1.2 are inherently stable; the others decay spontaneously into other particles. The mean lifetimes of the latter have been given in the table.

For nearly every type of particle there corresponds an *antiparticle* with the same mass, spin and half-life. The only exceptions are the photon and the π^0 meson, which are their own antiparticles. For example, to the proton there corresponds an antiproton, and to the neutron an antineutron. If a particle is charged, its antiparticle has a charge of the same absolute value but opposite sign. The antielectron is positive and is also called *positron*. The antiparticles are not normal constituents of our part of the universe. They may appear in particle collisions, and they can all arise through so-called *pair formation*. This signifies the creation of matter in the form of an elementary particle plus the corresponding antiparticle from radiation, and occurs when sufficiently energy-rich radiation comes into the influence of the electric field near an atomic nucleus. The more energy-rich the quanta, the heavier the particle pairs that are created. When the positron was discovered, energy-rich quanta, caused by cosmic radiation, were transformed into an electron plus a positron. Positrons are also formed in certain nuclear reactions (9-2a). Upon collision of an elementary particle with the corresponding antiparticle, annihilation occurs very quickly with the emission of the equivalent amount of energy in the form of radiation (*annihilation radiation*).

The possibility that somewhere in the universe matter may be made up entirely of antiparticles cannot be ruled out. Such is probably not the case, however, within our galaxy.

Since 1952 it has been found that in many cases, when strong forces arise between classical elementary particles, particles or complexes are formed that are much more short-lived than the least stable elementary particles. Mean lifetimes as low as 10^{-23} s have been found for these particles, which are called *resonance particles*. To date, about 80 resonance particles have been found. Their nature is still very obscure, but probably continued study of them will eventually throw more light on the whole area of elementary particles.

b. The atomic nucleus. The properties of the atomic nuclei are most easily explained if the nuclei are considered to be made up of protons and neutrons, that is, nucleons. To what extent nucleons occur as such in the nucleus and how the components of the nucleus are bound together is still not completely known.

The mass of an atomic nucleus is always somewhat less than the sum of the masses of the component nucleons. The difference is small, however, and never exceeds 1 %. This results from the large amount of energy which is released when the nucleons are bound together into a nucleus. According to Einstein's equation (1.1) this corresponds to a loss of mass. Since this loss is small and the proton and neutron have nearly the same mass, the mass of the nucleus is nearly proportional to the number of nucleons in the nucleus. This number is therefore called the *mass number* (A) of the nucleus.

The nucleus is always positively charged, that is, contains protons. The number of charge units indicates the *nuclear charge* or *atomic number* (Z), and is the same as the number of protons in the nucleus. If the number of neutrons in the nucleus is N, then

$$A = Z + N \tag{1.4}$$

In all atoms with the same nuclear charge and thus the same number of protons in the nucleus, for neutrality, the nucleus must be surrounded by as many electrons, which are held to it by its positive charge. Under similar conditions the electron cloud has the same structure for all such atoms. Since the chemical properties of the atoms depend almost exclusively on the structure of the electron cloud, these atoms have practically the same chemical properties. *All atoms with the same nuclear charge (atomic number) are said to belong to the same* **element.** An element accordingly contains only atoms of the same nuclear charge (atomic number).

The atoms in a single element can nevertheless contain different numbers of neutrons, that is, have different mass numbers and masses. These atoms form different *isotopes* of the element in question. The name is derived from the fact that all isotopes of an element have the same position (Gk. ἴσος, same; τόπος, place) in the periodic system of the elements (see 4-2). This position is determined exclusively by the atomic number. *Isotopes of an element therefore have nuclei with the same number of protons (Z = constant) but different numbers of neutrons.*

An atomic species with a certain nuclear composition, that is, with the same mass number and nuclear charge, and with a certain nuclear energy, is called a **nuclide.** Nuclei with the same composition can have different energies and thus according to this definition belong to different nuclides. From the chemical standpoint, however, it is not necessary to make any distinction between such nuclides (see 9-1a).

Nuclides with the same nuclear charge are therefore isotopes.

The word isotope is often used improperly when the word nuclide should be used instead. The use of "isotope" for "nuclide" has been compared with the use of "brother" for "people". Thus, for example, ^{110}Ag and ^{60}Co are two nuclides (radionuclides since they are radioactive), possibly a silver isotope and a cobalt isotope, but they are not two isotopes.

Two or several nuclides may have the same mass number but different nuclear charge and so belong to different elements. The nuclei contain the same number of nucleons (A = constant) but the number of protons is different. Such nuclides are said to be *isobars* (Gk. ἴσος, same; βάρος, weight).

c. The electron cloud. In a neutral atom the nucleus is surrounded by an electron cloud with the same number of electrons as the nuclear charge. If the number of electrons differs from the nuclear charge, the atom becomes charged, positive with electron deficiency and negative with electron excess. The positive nucleus, however, always holds together the negative electrons in the cloud. A charged atom is often called an *ion* (Gk. ἰόν, pres. part. of the verb, to go; an ion moves like a charged particle in an electric field). A positively charged ion is frequently called a *cation* and a negatively charged ion an *anion* because the former moves toward the negative electrode, the cathode, and the latter toward the positive electrode, the anode.

In all ordinary chemical reactions only the electron cloud of the atom is involved, while the nucleus remains unchanged. The structure of the electron cloud therefore determines almost entirely the chemical properties of the atom.

The electrons in the cloud constantly move extraordinarily rapidly about the nucleus. No definite paths can be assigned to them nor can it be said where a certain electron is at a certain instant. If the time of observation is taken so long that each electron meanwhile moves a very long distance around the nucleus, the designation *electron cloud* becomes entirely appropriate. In a free atom it is possible at each point in space outside the nucleus to state the probability that any electron whatever at a certain instant will be at that point. This probability can also be considered as a measure of the density of the electron cloud at the point in question. Even in an atom such as the hydrogen atom, where the nucleus is surrounded by only one electron, one can thus speak in this sense of an electron cloud.

It may be useful at this point to describe the size relationships in an atom in terms of a concrete model. In a hydrogen atom the nucleus consists of one proton, in which nearly the whole mass of the atom is concentrated. Outside the proton the electron moves in a spherical region. The probability of encountering the electron has dropped to a very low value at a distance of approximately 1 Å (10^{-10} m) from the nucleus. This implies that the density of the electron cloud at this distance is very low and that practically speaking the cloud does not extend beyond this point. If the nucleus is represented by a ball of diameter 1 mm, the distance 1 Å corresponds approximately to 50 m. Somewhere within a sphere of diameter 100 m one may therefore expect to find the electron, which on the same scale will have a diameter of the order of .1 cm.

The size relationships are only slightly changed for heavier atoms. In the most common calcium isotope with mass number 40, the nucleus contains 20 protons and 20 neutrons. The nucleons are very densely packed in the nucleus, so that on the scale just suggested the nucleus has a diameter of approximately 4 mm. Outside the nucleus 20 electrons form a cloud which in our model has a diameter of about 200 m.

It is seen therefore that the building units of the atom take up a vanishingly small fraction of the volume of the atom if it is assumed to be represented by the electron cloud. Further, nearly all of the mass of the atom is concentrated in the nucleus, whose diameter may never reach one part in 10^4 of the diameter of a given electron cloud. It is the force field within the electron cloud and the forces between the electron clouds of different atoms which prevent solid and liquid substances, in which the electron clouds are in constant contact with each other, from being compresssed to smaller volume. If the pressure is raised to the order of 10^8 bar, this resistance is, however, broken down. The separate atoms are disintegrated and it may be said that the atomic nuclei float about in a gas of electrons, which may be compressed to a substantially higher density than is possible under earthly conditions. Such high pressures are found in the interior of the stars. In the center of the sun the density is of the order of 10^2, in white dwarf stars 10^5 g cm^{-3}—and yet still higher densities may be imagined; in the atomic nuclei themselves densities exist of the order of 10^{12} g cm^{-3}.

d. Elements. In his textbook of inorganic chemistry which appeared in 1872, the Swedish chemist P. T. Cleve wrote about "Elements, fundamental substances or simple bodies, which cannot by any means known to us be decomposed into different substances." This definition of the concept of element was for the most part the same as that used by Boyle in 1661 and which to some extent was refined by Lavoisier in 1789. It was generally accepted at the end of the 19th century, and when limited to "ordinary" chemical processes, is applicable to this day.

The transformations of atomic nuclei, which first became known through natural radioactive processes, and then brought about through numerous artificially induced nuclear reactions, later made necessary the definition given in 1-3 b, that *an element contains only atoms of the same nuclear charge (atomic number)*.

Nowadays something over 100 elements are known. Ten or so of these have not been found in nature, but have been synthesized from other atoms through nuclear reactions. All the elements are represented by several isotopes. Many of these, however, have only been obtained through nuclear reactions. Even the appearance of isotopes makes the old definition of element inadequate.

Because of the similarity of the electron cloud, the chemical properties of the various isotopes of the same element are practically the same. Isotopes in general, therefore, cannot be separated from each other through ordinary chemical reactions. This has led to the result that an element in general consists of isotopes mixed in practically the same proportion wherever it occurs in the earth. The isotopes were able to be mixed uniformly before the earth mass solidified and afterward only very minor variations could occur. The most important of these variations result through the formation or decay of isotopes through nuclear reactions (in general spontaneous and associated with "natural radioactivity"), after the earth solidified.

The chemist in earlier times had always to work with the natural isotope mixtures. Even long after it was found that naturally occurring elements may consist of several isotopes, no method was known for isolating these in weighable quantities or changing their relative amounts. One still works in ordinary chemical practice with the natural isotope mixtures, which for the most part remain practically unchanged. The properties that were earlier ascribed to the supposed inalterable element therefore have great significance, even though they are average properties applicable to the ordinary isotope mixtures.

e. Symbols. In 1813 Berzelius proposed that the elements be indicated by letters and after some decades this method became common. In general the initial of the Latin or Latinized name of the element has been used, plus an additional letter from the name if it was necessary for distinction. For new symbols two letters are always used nowadays.

By means of numerical indices attached to the symbol, the nuclear charge (atomic number) and mass number may be indicated. The number of atoms which, for example, make up an atom group (molecule), and the net charge of charged atoms (ions) may also be shown by numerical indices. By international agreement, the net charge should be written in chemical literature as $n+$ or $n-$, not $+n$ or $-n$ (the sign thus indicates in the recommended mode of writing a charge unit and not a mathematical operator). Further, the following arrangement of indices has been agreed upon.[1]

$$\begin{smallmatrix}\text{mass number}\\\text{nuclear charge}\end{smallmatrix}\text{Symbol}\begin{smallmatrix}\text{net charge}\\\text{number}\end{smallmatrix}$$

The complete symbol for the most common hydrogen atom with a nucleus consisting of one proton and thus of mass number 1, is ^1_1H (this atom is sometimes called protium). A heavier hydrogen isotope with a nucleus containing one proton and one neutron is symbolized by ^2_1H (the atom is often called deuterium and not uncommonly indicated also by D, or in complete form ^2_1D). A third hydrogen isotope with an additional neutron in the nucleus becomes ^3_1H (often called tritium and indicated also by T or ^3_1T). If, as is usual in chemistry, only H is written, this means the ordinary natural isotope mixture and does not refer to any particular isotope.

The nucleus of the hydrogen atom ^1_1H, that is, the proton, is completely symbolized by $^1_1H^+$. In the same way the deuterium nucleus (the so-called deuteron, consisting of one proton and one neutron) is indicated by $^2_1H^+$.

Since the letter symbol for the element uniquely determines the nuclear charge, the charge is often not written out. With the complete symbols, however, one has the advantage of being able easily to check whether the sums of the mass numbers and the nuclear charges are the same on both sides of an equation for a nuclear reaction. In such reactions the electron cloud has no significance, hence the net charge is rarely written out. A proton then is written ^1_1H. In ordinary chemical reactions in which the elements are often isotope mixtures, the mass number is meaningless and it is unnecessary to write out the nuclear charge. On the other hand the electron cloud and the net charge become important. In such cases the proton is therefore written H^+.

Often when it is desired only to emphasize the nuclear charge (the atomic number), this is written on the same line as the letter symbol, for example, 12 Mg.

To distinguish special isotopes a symbol such as C-14 often is used instead of ^{14}C. In speech an isotope is generally expressed with the mass number following the name of the element, for example, "carbon-fourteen".

The neutron is customarily denoted by n, and in physics literature the proton often by p. More complete are the symbols ^1_0n and ^1_1p. The deuteron is frequently symbolized d, or more completely ^2_1d. The electron and positron are indicated most often by e^- and e^+ and in nuclear physics also by β^- and β^+. If the symbols $_{-1}^0e$ and ^0_1e are used the lower index denotes the charge of the particle.

---

[1] Notice that often, especially in physics literature, the following index arrangement is used:

$$\text{nuclear charge}\,\text{Symbol}\,^{\text{mass number}}$$

f. Atomic weight. The absolute mass of an atom, the *atomic mass*, seldom appears in strictly chemical calculations. On the other hand relative atomic mass values are used to a great extent, which are referred to the mass of a certain atom as a standard. A somewhat inaccurate designation for the *relative atomic mass* is the *atomic weight*. The carbon isotope of mass number 12, ^{12}C, has been chosen as a standard. *The mass of an atom ^{12}C has accordingly been set equal to* 12. The atomic weight is thus dimensionless; it is a number.

Absolute atomic masses can be expressed in units of $u = 1/12$ of the atomic mass of ^{12}C (the *unified atomic mass unit*). Using Avogadro's number N_A as defined in 1-4c, we have $u = 1/N_A \text{ g} = 1.66043 \times 10^{-27}$ kg. Expressed in this unit, the absolute atomic mass is the same as the atomic weight but has the dimension of mass. Accordingly, the absolute atomic mass for $^{12}C = 12$ u, for $^{19}F = 18.9984$ u, etc. The reason why an atomic weight is not defined as the absolute atomic mass expressed in the unit u is the uncertainty in the value of N_A and, therefore, in u.

In this way, the atomic weight, that is, the relative mass, is completely defined for a certain nuclide. But the natural elements are most usually mixtures of several isotopic nuclides. In general it is with these mixtures that the chemist works, and it is necessary for him to assume mean atomic weights appropriate to these mixtures. That these mean weights shall be effectively constant requires that the natural isotope mixtures have a sufficiently constant composition. This condition is generally so well fulfilled that the mean atomic weights can be used even for precise chemical calculations. In chemistry, when the atomic weight of an element is mentioned without qualification, it is the mean atomic weight of the natural isotope mixture which is referred to. Only if the element is represented in nature by a single nuclide is its atomic weight a relative nuclide mass.

The proportions of the component nuclides and their relative masses in a sample are determined with a *mass spectrograph*. In this instrument, positive ions are produced from the sample by heating. The ions are accelerated with an electric potential and projected in the classical type of spectrograph as a fine beam into an arrangement of electric and magnetic fields. In these fields the ions are deflected so that all ions with the same ratio M/z (M = mass of the ion, z = ionic charge expressed in charge units) reach a target surface in a single line. If the target surface is a photographic plate, there is obtained a mass spectrum in which each spectral line corresponds to a particular value of M/z. The ionic charge z can generally be easily ascertained, so that the position of the line gives M. By measuring the distance between two lines, a mass ratio M_1/M_2 can be determined with particular accuracy. In this way one can obtain the mass of a certain nuclide relative to the mass of ^{12}C, that is, find the relative mass according to the definition given above. By means of intensity measurements, the proportion of the different types of ions in the mixture can also be measured. Through a knowledge of the relative masses and amounts of the isotopes in the natural mixture of an element, the mean atomic weight of that element can be calculated. The relative masses can be determined with greater accuracy than the amounts, which besides cannot be considered to be absolutely constant from sample to sample. Therefore, in general, the atomic weights of elements which consist of only one nuclide are more accurate than those of other elements. The precision of the mass spectrograph is

Fundamentals, Concepts and Definitions 1-3 f · 27

nowadays very high, and most modern atomic weights have been determined by this means.

J. J. Thomson in 1913, with a very primitive mass spectrograph, found the neon isotopes ^{20}Ne and ^{22}Ne. The instrument was developed later by Aston (after 1919) among others, and has revealed a large number of stable isotopes.

The concept of atomic weight was introduced at the beginning of the 19th century by Dalton and Berzelius. It was not known then that the elements consisted of isotope mixtures and consequently only mean atomic weights were determined. The standard for the atomic weight scale was for a long time the element oxygen as it is found in nature. Oxygen was chosen because it combines with so many elements, and it was considered that a large number of atomic weights could be directly determined by analysis of their oxygen compounds. The atomic weight of oxygen was set equal to 16, a value which had the advantage that the atomic weight of the lightest element hydrogen became nearly 1, or more exactly, 1.0080.

When it was later found that natural oxygen consists of three isotopes in proportions which are not absolutely constant (the composition is, however, always very near to 99.76% ^{16}O, 0.04% ^{17}O and 0.20% ^{18}O), this oxygen could not be considered as a proper standard for the atomic weight scale. In physics, when atomic weights of particular nuclides were referred to, the isotope ^{16}O was therefore preferred as a standard, whose atomic weight was then set equal to 16. This new atomic weight scale was designated "physical" in contrast with the old "chemical" scale. The difference between the two scales is quite small since the isotope ^{16}O occurs in such predominant amount in natural oxygen.

It was thought desirable meanwhile to establish a universal scale. An adoption of the physical scale would, however, inconveniently bring about large changes in the enormous number of existing data which were based on chemical atomic weights. Following the proposal of Nier and Ölander, it was therefore decided by international agreement in 1960 and 1961 to base the scale on ^{12}C = 12. The relative values of the atomic weights and masses of the different scales are shown in Appendix 2. It is seen there that the change of the chemical atomic weights now is only 43 parts in 10^6 against 275 parts in 10^6 if the physical scale were adopted. ^{12}C as a standard also has the advantage that this nuclide is especially suitable for use as a direct reference in the mass spectrograph method.

Tab. 1.3 shows the atomic weights of the elements according to the international agreement of 1967. When the isotope compositions of the natural elements are reasonably constant they are shown in the table. Several elements, because of the instability of their nuclei, do not occur in nature, or appear with isotope mixtures of variable composition. A mean atomic weight is meaningless in this case, and for orientation purposes, only the mass number of the most stable isotope, which in most cases is the best known, is given in the table for those elements.

Tab. 1.4 shows the relative masses of elementary particles and lighter nuclides. The table presents the stable nuclides and their proportions in the natural elements, and also the most important of the unstable (radioactive) nuclides synthesized to the present. The two right-hand columns are touched upon in chap. 9.

From tab. 1.4 it is apparent that the relative nuclide masses depart little from the mass number (the difference is never as high as 1%). The reason is that

Table 1.3. *1967 atomic weights, isotope contents and mass numbers for elements with reasonably constant isotope composition.*

The Latin name forms are used for chemical terms derived from the name of the element, for example, aurate, ferrate. For some compounds of sulfur, nitrogen and antimony, derivatives of the Greek name ϑεῖον (e.g. thiosulfate), the French azote (e.g. azide) and the Latin stibium (e.g. stibine), are used. Variation in the atomic weight resulting from variations in isotope composition is indicated by [a]. Experimental uncertainty in the value of the atomic weight is indicated by [b]. Nuclear instability that has been definitely proved is indicated by*. For elements that do not have a constant isotope composition the mass number of the most stable (generally, the best known) isotope is listed with its half-life (8a). Isotope proportions are expressed in atomic percent.

Z	Symbol	Name	Atomic Weight	Mass number and percent content of isotopes				
1	H	hydrogen	1.00797 ±0.00001[a]	1 99.985	2 0.015			
2	He	helium	4.0026	3 10^{-4}	4 ~100			
3	Li	lithium	6.939	6 7.42	7 92.58			
4	Be	beryllium	9.0122	9 100				
5	B	boron	10.811 ±0.003[a]	10 ~20	11 ~80			
6	C	carbon	12.01115 ±0.00005[a]	12 98.89	13 1.11			
7	N	nitrogen	14.0067	14 99.64	15 0.36			
8	O	oxygen	15.9994 ±0.0001[a]	16 99.76	17 0.04	18 0.20		
9	F	fluorine	18.9984	19 100				
10	Ne	neon	20.179 ±0.003[b]	20 90.92	21 0.26	22 8.82		
11	Na	sodium (natrium)	22.9898	23 100				
12	Mg	magnesium	24.305	24 78.60	25 10.11	26 11.29		
13	Al	aluminum	26.9815	27 100				
14	Si	silicon	28.086 ±0.001[a]	28 92.2	29 4.7	30 3.1		
15	P	phosphorus	30.9738	31 100				
16	S	sulfur	32.064 ±0.003[a]	32 ~95.0	33 0.76	34 ~4.2	36 0.015	
17	Cl	chlorine	35.453 ±0.001[b]	35 75.53	37 24.47			
18	Ar	argon	39.948	36 0.34	38 0.06	40 99.60		
19	K	potassium (kalium)	39.102	39 93.08	40* 0.01	41 6.91		
20	Ca	calcium	40.08	40 96.97	42 0.64	43 0.145	44 2.06	46 0.003	48 0.185
21	Sc	scandium	44.956	45 100				

Z	Symbol	Name	Atomic Weight	Mass number and percent content of isotopes					
22	Ti	titanium	47.90	46 7.98	47 7.32	48 73.99	49 5.46	50 5.25	
23	V	vanadium	50.942	50* 0.25	51 99.75				
24	Cr	chromium	51.996	50 4.31	52 83.76	53 9.55	54 2.38		
25	Mn	manganese	54.9380	55 100					
26	Fe	iron (ferrum)	55.847 $\pm 0.003^b$	54 5.84	56 91.68	57 2.17	58 0.31		
27	Co	cobalt	58.9332	59 100					
28	Ni	nickel	58.71	58 67.76	60 26.16	61 1.25	62 3.66	64 1.16	
29	Cu	copper (cuprum)	63.546 $\pm 0.001^b$	63 69.1	65 30.9				
30	Zn	zinc	65.37	64 48.89	66 27.81	67 4.11	68 18.56	70 0.62	
31	Ga	gallium	69.72	69 60.2	71 39.8				
32	Ge	germanium	72.59	70 20.55	72 27.37	73 7.67	74 36.74	76 7.67	
33	As	arsenic	74.9216	75 100					
34	Se	selenium	78.96	74 0.87	76 9.02	77 7.58	78 23.52	80 49.82	82 9.19
35	Br	bromine	79.904 $\pm 0.001^b$	79 50.52	81 49.48				
36	Kr	krypton	83.80	78 0.35	80 2.27	82 11.56	83 11.55	84 56.90	86 17.37
37	Rb	rubidium	85.47	85 72.15	87* 27.85				
38	Sr	strontium	87.62	84 0.56	86 9.86	87 7.02	88 82.56		
39	Y	yttrium	88.905	89 100					
40	Zr	zirconium	91.22	90 51.46	91 11.23	92 17.11	94 17.40	96 2.80	
41	Nb	niobium	92.906	93 100					
42	Mo	molybdenum	95.94	92 15.86	94 9.12	95 15.70	96 16.50	97 9.45	98 23.75	100 9.62
43	Tc	technetium		99* 2.1×10^5y					
44	Ru	ruthenium	101.07	96 5.6	98 1.9	99 12.7	100 12.6	101 17.1	102 31.6	104 18.5
45	Rh	rhodium	102.905	103 100					
46	Pd	palladium	106.4	102 0.96	104 11.0	105 22.2	106 27.3	108 26.7	110 11.8
47	Ag	silver (argentum)	107.868 $\pm 0.001^b$	107 51.35	109 48.65				
48	Cd	cadmium	112.40	106 1.22	108 0.88	110 12.39	111 12.75	112 24.07	113 12.26
				114 28.86	116 7.58				

Z	Symbol	Name	Atomic Weight	Mass number and percent content of isotopes					
49	In	indium	114.82	113 4.33	115* 95.67				
50	Sn	tin (stannum)	118.69	112 0.95 120 32.97	114 0.65 122 4.71	115 0.34 124 5.98	116 14.24	117 7.57	118 24.01	119 8.58
51	Sb	antimony (stibium)	121.75	121 57.25	123 42.75				
52	Te	tellurium	127.60	120 0.09 130 34.49	122 2.46	123* 0.87	124 4.61	125 6.99	126 18.71	128 31.79
53	I	iodine	126.9044	127 100					
54	Xe	xenon	131.30	124 0.096 134 10.44	126 0.090 136 8.87	128 1.92	129 26.44	130 4.08	131 21.18	132 26.89
55	Cs	cesium	132.905	133 100					
56	Ba	barium	137.34	130 0.101	132 0.097	134 2.42	135 6.59	136 7.81	137 11.32	138 71.66
57	La	lanthanum	138.91	138* 0.09	139 99.91				
58	Ce	cerium	140.12	136 0.19	138 0.25	140 88.48	142 11.07		
59	Pr	praseodymium	140.907	141 100					
60	Nd	neodymium	144.24	142 27.13	143 12.20	144* 23.87	145 8.30	146 17.18	148 5.72	150 5.60
61	Pm	promethium		145* 18y					
62	Sm	samarium	150.35	144 3.16	147* 15.07	148 11.27	149 13.84	150 7.47	152 26.63	154 22.53
63	Eu	europium	151.96	151 47.77	153 52.23				
64	Gd	gadolinium	157.25	152* 0.20	154 2.15	155 14.7	156 20.47	157 15.68	158 24.9	160 21.9
65	Tb	terbium	158.924	159 100					
66	Dy	dysprosium	162.50	156 0.05	158 0.09	160 2.29	161 18.88	162 25.53	163 24.97	164 28.18
67	Ho	holmium	164.930	165 100					
68	Er	erbium	167.26	162 0.14	164 1.56	166 33.41	167 22.94	168 27.07	170 14.88
69	Tm	thulium	168.934	169 100					
70	Yb	ytterbium	173.04	168 0.14	170 3.03	171 14.31	172 21.82	173 16.13	174 31.84	176 12.73
71	Lu	lutetium	174.97	175 97.40	176* 2.60				
72	Hf	hafnium	178.49	174* 0.16	176 5.21	177 18.56	178 27.1	179 13.75	180 35.22

Z	Symbol	Name	Atomic Weight	Mass number and percent content of isotopes					
73	Ta	tantalum	180.948	180* 0.012	181 99.988				
74	W	tungsten wolfram	183.85	180 0.14	182 26.4	183 14.4	184 30.6	186 28.4	
75	Re	rhenium	186.2	185 37.07	187* 62.93				
76	Os	osmium	190.2	184 0.02	186 1.59	187 1.64	188 13.3	189 16.1	190 26.4	192 41.0
77	Ir	iridium	192.2	191 38.5	193 61.5				
78	Pt	platinum	195.09	190* 0.013	192 0.78	194 32.9	195 33.8	196 25.2	198 7.19
79	Au	gold (aurum)	196.967	197 100					
80	Hg	mercury	200.59	196 0.146	198 10.02	199 16.84	200 23.13	201 13.22	202 29.80	204 6.85
81	Tl	thallium	204.37	203 29.50	205 70.50				
82	Pb	lead (plumbum)	207.19	204 1.40	206 25.2	207 21.7	208 51.7		
83	Bi	bismuth	208.980	209 100					
84	Po	polonium		210*	138 d				
85	At	astatine		210*	8.3 h				
86	Rn	radon		222*	3.82 d				
87	Fr	francium		223*	22 min				
88	Ra	radium		226*	1617 y				
89	Ac	actinium		227*	21.2 y				
90	Th	thorium	232.038	232* 100	1.4×10^{10} y				
91	Pa	protactinium		231*	3.2×10^4 y				
92	U	uranium	238.03	234* 0.006	235* 0.720	238* 99.274			
93	Np	neptunium		237*	2.1×10^6 y				
94	Pu	plutonium		244*	8×10^7 y				
95	Am	americium		243*	7.65×10^3 y				
96	Cm	curium		247*	1.6×10^7 y				
97	Bk	berkelium		247*	1400 y				
98	Cf	californium		251*	800 y				
99	Es	einsteinium		254*	276 d				
100	Fm	fermium		257*	~80 d				
101	Md	mendelevium		257*	~3 h				
102	No	nobelium		255*	3 min				
103	Lr	lawrencium		256*	45 s				
104		not yet named		260*	0.3 s				

Table 1.4. *More important nuclides of elements up to 10 Ne, and masses of some elementary particles.*

Z	Symbol	A	Mass, u ($^{12}C=12$)	Stable nuclides Percent content in natural element (atomic percent)	Unstable nuclides Half-life	Nuclear reaction
0	e	0	0.000548			
0	n	1	1.008665			
1	p	1	1.007277			
1	H	1	1.007825	99.985–99.986		
		2	2.014102	0.014–0.015		
		3	3.016050		12.262 y	β^-
2	He	3	3.016030	$\sim 10^{-4}$		
		4	4.002604	~ 100		
		6	6.0189		0.85 s	β^-
3	Li	6	6.015125	7.42		
		7	7.016004	92.58		
		8	8.02249		0.841 s	$\beta^-, 2\alpha$
4	Be	7	7.016929		53.1 d	EC
		9	9.012186	100		
		10	10.01354		2.5×10^6 y	β^-
5	B	8	8.02461		0.77 s	$\beta^+, 2\alpha$
		10	10.012939	19.6–20.2		
		11	11.009305	79.8–80.4		
		12	12.014353		0.02 s	$\beta^-, 3\alpha$
6	C	10	10.0168		19.3 s	β^+
		11	11.01143		20.4 min	β^+, EC
		12	12.000000	98.89		
		13	13.003354	1.11		
		14	14.003242		5.730 y	β^-
7	N	13	13.005739		9.96 min	β^+
		14	14.003074	99.635		
		15	15.000108	0.365		
		16	16.00609		7.16 s	β^-
		17	17.009		4.16 s	β^-, n
8	O	14	14.00860		71.3 s	β^+
		15	15.00307		124 s	β^+
		16	15.994915	99.76		
		17	16.999133	0.04		
		18	17.999160	0.20		
		19	19.00358		29.4 s	β^-
9	F	17	17.00210		66 s	β^+
		18	18.00095		110 min	β^+, EC
		19	18.998405	100		
		20	19.99999		11.4 s	β^-
10	Ne	19	19.00189		17.4 s	β^+
		20	19.992440	90.92		
		21	20.99385	0.257		
		22	21.991385	8.82		
		23	22.99447		37.5 s	β^-

the masses of the neutron and the proton are nearly 1 on the relative scale and that the mass loss with the binding of nucleons into nuclei referred to in 1-3b is small in percentage.

In 1815 Prout put forth the view that all atomic weights should be multiples of that of hydrogen, which would suggest that all elements are built up of hydrogen atoms. Prout's hypothesis was discarded when considerable departures from the rule were found in accurate tests. Most decisive was the finding that the atomic weight of chlorine lay quite close to 35.5. Tab. 1.4 shows that the reason for Prout's observation is that a good number of elements in nature consist of either only one nuclide or of one nuclide in a highly preponderant amount. The mean atomic weight for such an element therefore is close to the mass number.

Knowledge of the atomic weights of the elements is necessary for all calculations of the proportions in which elements and compounds occur in matter, that is, for what is called *stoichiometry* (Gk. στοιχεῖον meaning among other things element, so that stoichiometry therefore most closely signifies determination of elements.)

1-4. Forms of matter

a. Atoms in matter. Free atoms can actually occur in very rarified gases of certain substances (noble gases, most metal gases.) The distribution of atoms in such a gas is shown schematically in fig. 1.1 a. In all other cases an atom is more or less strongly bound to other atoms. If the atomic gas is compressed the forces between neighboring atoms quickly become appreciable, and when the gas is condensed to a liquid (b), these forces become rather strong. They are stronger still in an ordered structure which forms a crystal (c or d). The attractive forces are of different kinds and can have very different strengths. For a given type of binding force, the stronger that force is, the closer the atoms are drawn together.

Often two or several atoms are bound together in a definite way into a group by forces of a different kind than those which occur between atoms of different groups, that is, between groups. An atomic group not unusually is more or less free-moving as a discrete unit and can often exist in different aggregation states of the substance in question. Such a group is usually called a *molecule*. In iodine, for example, the I atoms are normally combined into diatomic molecules I_2. Figs. 1.1 e, f, g show how molecules can occur in different aggregation states of a substance.

A molecule need not be electrically neutral, but may possess a net charge, positive or negative. Like a charged atom, a charged molecule is often called an *ion*.

In crystals with structures such as c and d, no discrete, finite molecules can be distinguished. Crystallized noble gases and most metallic crystals (fig. 6.12b, c) are built in principle in this way. Similarly, a large number of nonmetallic crystals are so constructed, for example, diamond (fig. 5.16) and the crystals in fig. 6.16. Often, for example in diamond, the binding forces are of the same kind as those within typical molecules. In these cases one may justifiably consider the whole crystal to be a *giant* or "*infinite*" *molecule*.

In certain crystals the atoms are strongly bound together into chains or layers which extend through the whole crystal, while the forces between the chains or layers are weaker. Chains of this type are encountered, for example, in the gray modification of selenium. Graphite (fig. 5.21) shows an example of a layer structure,

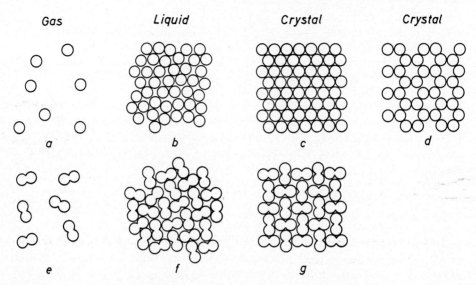

Fig. 1.1. Schematic representation of different states of aggregation: (a)–(d) of single atoms; (e)–(g) of diatomic molecules.

Also here one may speak of giant molecules, or denote chains and layers as molecules (often ions) with "unlimited" or "infinite" extent in one or two dimensions. Gray selenium contains one-dimensional and graphite two-dimensional, infinite molecules. In the same way a diamond crystal may be said to consist of one three-dimensional, infinite molecule, a *framework molecule*.

Even though NaCl molecules do not exist in a sodium chloride crystal (fig. 6.16a), the formula NaCl is used to indicate the substance and its composition. It is also practical to denote the *collection of atoms which make up a given formula* as a *formula unit*, whether or not it corresponds to an actual molecule. One formula unit of NaCl is thus 1 atom of Na + 1 atom of Cl.

If a molecule (ion) can be built up directly from free atoms or molecules, it was formerly often called a *complex molecule*, or shorter, a *complex*. Such a complex molecule can thus, in general, be more or less easily broken up into its components. Nowadays, however, the word "complex" is used mostly for any polyatomic molecule.

The geometrical organization of a group of atoms is called the *structure* of the group. The word is used for both molecules (*molecular structure*) and for the whole region of a homogeneous phase (*crystal structure*, *liquid structure*).

Two molecules which have the same composition but different structures are said to be *isomers*, or exhibit *isomerism* (Gk. ἴσος, same; μέρος, part). If molecules of a certain kind are combined into double, triple or multiple molecules, these are said to be *dimers*, *trimers* or *polymers*. If the polymerization is very great (often practically to an infinite molecule), the molecule is said to be a *high polymer*.

Atoms and molecules are constantly in rapid motion of different kinds, so-called *heat motion* (1-4f, g, h). Heat motion increases if the temperature is raised and decreases if it is lowered.

b. Pure substances and solutions. Any substance which can exist as a phase of constant composition, even if the temperature and pressure and the composition of other phases in the system change within certain limits, is called a *pure substance*. The concept includes the elements as well as compounds of the elements. By constant composition of a compound is meant here that the elements of which it consists are present in constant relative numbers of atoms (the same mole ratio, 1-4e). The composition expressed in weight percent may in fact vary in this case if the isotopic composition is changed.

A phase whose composition can be changed continuously within certain limits without destroying its homogeneity, is called a *solution*. The region of composition within these limits is called the *region of homogeneity*. The state of aggregation is unimportant; there can be gaseous, liquid, and solid solutions.

The composition of a solution of potassium nitrate in water can be continuously changed, for example, by addition of salt or water. If a sufficiently great amount of potassium nitrate is added, however, a portion remains undissolved, and constitutes a new, solid phase. If the temperature is now changed, the composition of the liquid phase (the solution) and the amount of the solid phase are altered. The composition of the solid phase remains unchanged, on the other hand. Potassium nitrate is therefore a pure substance. If one starts with a sufficiently dilute potassium nitrate solution and cools it, at first another solid phase, ice, separates. If the temperature is lowered further, more ice is formed, so that the solution becomes more concentrated. The composition of the ice is not varied, however; that is, ice is a pure substance.

A solution is thus a mixture of two or several pure substances, *components*. A sulfuric acid solution may be prepared from the pure substances, H_2SO_4 and water. One can also begin with SO_3 and water, or mix H_2SO_4, SO_3 and water The number and kinds of components is therefore to a certain degree arbitrary, but it is always possible to refer to a *minimum* number of components from which the solution can be prepared. In this example the minimum number is two. This does not mean that there are not a large number of different types of molecules in the solution: H_2O, H_3O^+, H_2SO_4, HSO_4^-, SO_4^{2-}, and so on. The quantities of these molecules, however, is determined (under equilibrium) by the amounts of the components.

If the minimum number of components is two, the solution is *binary*; if it is three, *ternary*, etc.

The component which is present in the largest amount is commonly called the *solvent* (often indicated by the index 1) and the other components *solutes* (indicated by indices 2, 3, etc.)

A *mixture* of different substances may sometimes consist of a single homogeneous phase and thus be a solution (a mixture of oil and gasoline; air is a mixture of several gases), or sometimes of several phases (a mixture of oil and water; a cement mixture).

c. Avogadro's number. If a sample 1 contains N atoms of a nuclide with relative mass M_1, and a sample 2 contains the same number N atoms of a nuclide with relative mass M_2, and so on, the masses of these samples will be in the ratio $M_1:M_2:...$. Conversely, samples with masses in the ratio $M_1:M_2:...$ contain an equal number of atoms. *The number of atoms in 12 g of the nuclide ^{12}C is Avogadro's number* $N_A = 6.02252 \times 10^{23}$. Thus, according to the values of relative mass in tab. 1.4 there are N_A atoms in 9.012186 g of ^{9}Be, for example, and in 7.016005 g of ^{7}Li. In the natural isotope mixture of an element there is the same total number of the different isotopes in an amount that weighs as many grams as the mean atomic weight (tab. 1.3). 24.312 g of natural magnesium thus contains N_A atoms ^{24}Mg + ^{25}Mg + ^{26}Mg. It is then also clear that if 22.990 g Na containing N_A Na atoms and 35.453 g Cl containing N_A Cl atoms are combined to form 58.443 g NaCl, the product contains N_A formula units NaCl.

It is difficult to appreciate the size of Avogadro's number. If an amount of sand containing N_A grains having a diameter of 0.5 mm were spread in an even layer over the state of Texas, the layer would be approximately 100 m thick. The same number of water molecules are contained in 18 g water.

d. Molecular weight, formula weight, mole. The mass of a single molecule should be called the *molecular mass*. A relative molecular mass referred to $^{12}C = 12$ is called the *molecular weight*. The molecular weight of a substance is thus the sum of the atomic weights of the atoms contained in the molecule. Molecular weight, like atomic weight, is a pure number.

The sum of the atomic weights of the atoms that make up a formula unit is called the *formula weight*. Formula weight is also a pure number.

One mole (symbol: mol) of a pure substance of given formula unit is a characteristic unit of quantity of that substance, containing the same number of formula units as there are atoms in 12 g of ^{12}C, that is, a number equal to Avogadro's number N_A.

The formula unit need not be a molecule, but may be an atom or ion or any group of atoms; the important point is that it should be clearly stated. Thus, for example, 1 mole of Mg weighs 24.312 g; 1 mole of N_2, 28.0134 g; 1 mole of H_2O, 18.0153 g; 1 mole of NaCl, 58.443 g; 1 mole of SO_4^{2-}, 96.062 g. It will be noticed that, depending on circumstances, different formula units may be given for the same substance: 1 mol O weighs 15.9994 g, 1 mol O_2, 31.9988 g; 1 mol NO_2 weighs 46.0055 g, 1 mol N_2O_4, 92.0110 g.

The concept of the mole thus replaces the formerly used terms "gram-molecule", "gram-atom" and "gram-ion".

Thus, Avogadro's number may be written:

$$N_A = 6.02252 \times 10^{23} \text{ formula-units mol}^{-1} \tag{1.5}$$

The above definition of mole is clearly equivalent to the following: *one mole of a pure substance with a given formula unit is a characteristic unit of quantity of the substance, whose mass in grams is numerically equal to the formula weight.*

The term mole is also given another meaning which, however, leads to the same amount of substance as the definitions just stated. According to this definition, *one mole is a number equal to Avogadro's number N_A*. In this case, 1 mole of Mg signifies 1 mole of Mg atoms, that is, N_A atoms of Mg. This number of Mg atoms weighs 24.312 g. 2 mol NaCl signifies 2 moles of NaCl formula units, that is, $2N_A$ formula units of NaCl. It is this definition which is applied when one speaks, for example, of 1 mole of bonds or as below, 1 mole of electrons.

N_A electrons are called 1 *mole of electrons*. The accumulated charge of this quantity of electrons is called 1 *faraday* (F) and is $N_A e = 96\ 487$ C. This charge is carried by 1 mole of univalent ions, that is, ions which each have a net charge e.

Quantities of a substance are sometimes given also in units of another type, namely, *equivalents*. One equivalent is the amount of a substance which in a certain chemical process corresponds to 1 mole of a certain other substance. "Correspondence" implies that the equivalent in each case is given off or absorbed, exchanged or in some other way reacted with one mole of the other substance. The magnitude of the equivalent thus depends on the process and is therefore not unambiguous. Frequently, one equivalent is the amount of a substance which corresponds to 1 mol H. One *electrochemical equivalent* (often merely *equivalent*) is the quantity of an ion which carries 1 faraday of charge. 1 mol Na^+, $\frac{1}{2}$ mol Ca^{2+}, $\frac{1}{3}$ mol La^{3+}, $\frac{1}{2}$ mol SO_4^{2-} represent electrochemical equivalents. In connection with a specific process or process type, the equivalent may be a practical unit, but when it is introduced, its meaning must be exactly defined.

If each of 1 mole of particles (elementary particles, atoms, molecules) has the energy 1 eV $= 1.60210 \times 10^{-19}$ J (1-2c), the total energy of all N_A particles in this mole $= 1.60210\ N_A \times 10^{-19} = 96\ 487$ J.

N_A quanta are called 1 *einstein* and can be considered as 1 mole of quanta. These quanta carry according to (1.2) the energy $N_A h\nu = 3.990\ \nu \times 10^{-10}$ J $= 0.1196/\lambda$ J (λ in m).

e. Expressions for proportion. To show the composition of a phase, the *proportion* of the component substances can be expressed in many different ways. The following are the most common in scientific work.

The *concentration* of a substance in a certain phase is given as the number of moles of the substance *per liter of the phase*. The unit of concentration is thus mol l^{-1}. The concept of concentration, however, is used in two different senses.

In some cases, reference is made to the concentration of a substance as it actually occurs as such in the phase. The magnitude of such concentration is generally indicated by c, and for a certain type of atom, molecule or ion A, by c_A or [A]. When NaCl is dissolved in water, a complete dissociation takes place into Na^+ and Cl^-. Thus, in 1 liter of solution containing 1 mol NaCl, [NaCl] $= 0$

and $[Na^+]=[Cl^-]=1$ mol l^{-1}. If a gas vessel contains 2 mol O_2 per l, $[O_2]=$ 2 mol l^{-1}.

But also, in other cases, for substances which do not dissolve simply as such in the solution, concentration is given as the quantity dissolved, without consideration of the fact that the substance itself may actually have quite another concentration, or perhaps does not exist in solution at all. This concentration related to the quantity of substance introduced is called the *total concentration* of the substance. Its magnitude is generally indicated by C. In the NaCl solution mentioned above the total concentration of NaCl $C=1$ mol l^{-1}. On the other hand, one may not write $[NaCl]=1$ mol l^{-1}.

The concentration of a substance expressed in mol l^{-1} is called its *molarity*. The designation is used both for c and C, but this hardly need be a source of error. The NaCl solution is said to be 1-molar with respect to NaCl and is designated 1 M NaCl, which implies that the total concentration unit mol l^{-1} is indicated by M. An HCl solution with $C=0.1$ mol l^{-1} (M) is indicated as 0.1 M HCl.

The quantity of a substance in a phase is also expressed as *molality*, which states the number of moles of the substance per 1 *kilogram of solvent*. (The molality is therefore also called *weight molarity*, as distinguished from *volume molarity*.) The unit of molality is indicated by m. Molality, in contrast to molarity, is independent of temperature. The difference between c and m becomes insignificant when dilute water solutions are considered, and in such cases often can be neglected.

The composition of a phase is most rationally expressed in terms of *mole fractions*. If a phase contains n_1 moles of the substance A_1, n_2 moles of A_2, and in general n_i moles of A_i, then the mole fraction of A_i is

$$x_i = n_i/(n_1+n_2+\ldots) = n_i/\Sigma n_i \tag{1.6}$$

From this definition it follows that

$$\Sigma x_i = 1 \tag{1.7}$$

The composition is also expressed in *mole percent* (if the formula units are simple atoms often called *atomic percent*), which for any substance is 100 times its mole fraction.

For a dissolved substance "i", c_i and m_i are nearly proportional to x_i, the more closely the more dilute the solution is. For any dilute *water* solution, $x_i = 18.015 c_i/1000 = 18.015 m_i/1000$ (molecular weight of water = 18.015), that is, $c_i = m_i = 55.51 x_i$.

Often the concentration is given as its negative logarithm, indicated by the operator symbol p. For example (the charge of an ion is often omitted after p if misunderstanding does not result):

$$pAg = -\log[Ag^+] \tag{1.8}$$

Fundamentals, Concepts and Definitions

As will be shown later (mainly 12-2 b, d), the operator symbol p is used also to indicate the negative logarithm of the activity of an atomic or molecular species.

f. Gases. A gas consists of units of matter, atoms or groups of atoms, which are not bound to each other, but move freely within a volume which is large relative to the volume of the units. Between these units there are forces of attraction, which, however, are so weak that they often can be neglected at the distances ordinarily occurring between the units. The free units in gases have since early times been called *molecules*, whether they are single atoms or groups of atoms. Whatever their structure, they are in fact equivalent in many respects; for example in their contribution to the gas pressure. Mercury vapor is thus said to consist of monatomic molecules.

The freedom of the molecules in a gas implies that heat motion manifests itself, among other ways, through a constant shifting of the center of gravity of each molecule (*translation*). With this motion there continually occur collisions between the molecules and with the walls of the vessel. In addition, a molecule may *rotate* around one or more axes. In a multiatomic molecule the atoms undergo *oscillation* (*vibration*) with respect to each other about a mean position, somewhat like spheres joined by helical springs.

The mobility of the molecules causes the gas mass always to fill every part of the vessel in which it is contained. The large mean distance between the molecules also allows the gas volume to be changed easily by a change in pressure. Within the gas the molecules are randomly distributed. It was probably through a sense of this random state that van Helmont at the beginning of the 17th century coined the word "gas" to suggest the word "chaos". Fig. 1.1 e shows schematically how the I_2 molecules in iodine vapor might be disposed at a certain moment. The illustration depicts approximately the density in relation to the molecular dimensions which the molecules would have at room temperature and 1 atm pressure.

One often speaks of an *ideal gas*, by which is meant a hypothetical gas where no forces exist between the molecules and in which these molecules occupy points with no volume. Real gases may frequently approach quite closely this ideal state. At low pressure the molecular volume becomes negligible compared to the total volume, and the large intermolecular distances cause the forces between molecules to become very small. If the pressure is raised, the departure from the ideal state increases and becomes very great as condensation is approached, when the attractive forces between the molecules naturally become strong. Difficultly condensed gases such as hydrogen, oxygen, nitrogen and the noble gases at room temperature and pressure below 10 atm, behave very nearly like ideal gases.

If n moles (here mole must be referred to the gas molecules as formula units) of an *ideal gas* occupy the volume V at pressure P and absolute temperature T, they obey the *general equation of state of gases*.

$$PV = nRT \qquad (1.9)$$

R is a constant, the so-called *gas constant*. Sometimes the gas constant is expressed for a single formula unit (molecule) and is then called *Boltzmann's constant* k. Thus, $k = R/N_A$. Values of R, RT and k are given in Appendix 2.

If the ideal gas equation is written $n = PV/RT$, it is immediately apparent that equal volumes of ideal gases at the same pressure and temperature contain the same number of moles, that is, the same number of molecules (*Avogadro's law*). The number of molecules is nN_A, where N_A is Avogadro's number.

From this principle it also follows that the volumes of gases involved in gaseous reactions are in ratios of small whole numbers (*Gay-Lussac's volume law* of 1808, which was the starting point for Avogadro when he formulated his law in 1811).

The volume of 1 mole of ideal gas is $V = RT/P$. At 0°C ($T = 273.15$°K) and $P = 1$ atm this volume $= 2.2414 \times 10^{-2}$ m^3 $= 22.414$ l (cf. Appendix 2).

The ideal gas equation (1.9) may be written $PV = \dfrac{m}{M} RT$, where m is the mass of the gas and M is the mass of one mole of the gas (numerically equal to its molecular weight). From this we get $M = \dfrac{m}{V} \cdot \dfrac{RT}{P} = \varrho \dfrac{RT}{P}$, where ϱ is the gas density $\dfrac{m}{V}$. Thus for two gases 1 and 2 at the same temperature

$$\frac{M_1}{M_2} = \frac{\varrho_1}{\varrho_2} \frac{P_2}{P_1} \qquad (1.9\,\text{a})$$

and if in addition the pressures are equal

$$\frac{M_1}{M_2} = \frac{\varrho_1}{\varrho_2} \qquad (1.9\,\text{b})$$

In this way, *the molecular weights of different, ideal, pure gases at the same temperature and pressure are proportional to their densities.*

(1.9a) is used for molecular weight determination. This is best done by controlling the pressures P_1 and P_2 so that $\varrho_1 = \varrho_2$, and measuring the two pressures. The gas densities are made equal by ensuring that the bouyancy of the same object is the same in both gases. Then, $M_1/M_2 = P_2/P_1$. If M_2 is known, then M_1 is obtained. (1.9b) presupposes that the gases are ideal, but by measuring several different pairs of pressures P_1 and P_2 each of which makes $\varrho_1 = \varrho_2$, M_1/M_2 may be extrapolated to zero pressure, where all gases are ideal.

Carried out in this way, the gas density method is an accurate method for determination of molecular weights of light gases, and as such has been of great importance. It is required, of course, that the gas consist of only one kind of molecule. The molecules thus may not be dissociated or associated (on the other hand, dissociation and association may be studied through determination of gas density). The atomic weight of a particular atom may be obtained from the molecular weight if the gas consists of only these atoms or if the formula of the gas molecule and the atomic weights of the other kinds of atoms in the molecule are known.

In a gas mixture, which contains a gas "i" in the mole fraction x_i, a *partial pressure* p_i is apportioned from the total pressure P of this gas, defined as

$$p_i = x_i P \qquad (1.10)$$

Addition of the partial pressures of all the component gases gives through (1.7)

$$P = \Sigma p_i \qquad (1.11)$$

The total pressure is thus equal to the sum of all partial pressures (*Dalton's law of partial pressures*). If the gases and the gas mixture are ideal, the partial pressure of each gas is the same as that which the total pressure would be if that gas alone were present in the vessel. Thus we have

$$p_i V = n_i R T$$

from which we have by addition

$$PV = RT\Sigma n_i \qquad (1.12)$$

The laws (1.9) and (1.12), which are exact only for ideal gases, can often be applied as approximate laws to real gases.

g. Liquids. In a liquid as in a gas there are units with quite large mobility but the forces between the units have become significantly greater. The units thus become so tightly packed that they are everywhere in contact with each other (fig. 1.1 b, f). The volume of the liquid therefore is little changed with changes in pressure; its compressibility is small.

The occurrence of diffusion in liquids shows that translation of the structural units also takes place here, although considerably more sluggishly than in a gas. (Diffusion is not to be confused with convection, which consists of macroscopic streams caused by differences in density.) Of course, rotation is also possible. Vibrations within the molecule are still present as in a gas, but also vibrations of the whole molecule with respect to its neighbors may become noticeable.

The structure of a liquid often shows a certain amount of order. Each freely moving unit (molecule or, if the liquid does not contain molecules, atoms) seeks to surround itself in a definite way with other similar units. With the high mobility, however, this order can never extend over a large distance, but becomes only a *short-range order*, which in addition can never be complete. The order in a liquid is therefore incomparably smaller than in a crystal, where there is also a *long-range order*. The short-range order in a liquid is often similar to the short-range order in the substance after it crystallizes (cf. the two schematic liquid structures in fig. 1.1 b and f with the corresponding crystal structures c and g; the relationship is illustrated still more clearly in fig. 6.22).

h. Solid substances. In solid substances the structural units in general are somewhat more tightly packed than in liquids. When solid and liquid phases are in equilibrium with each other at the melting point of the substance, the density of the solid phase is nearly always greater than that of the liquid. There are exceptions, however, for example, water (ice has a lower density than liquid water), bismuth, germanium and gallium. These cases do not result because the binding forces in the solid are less strong than in the liquid, but because the

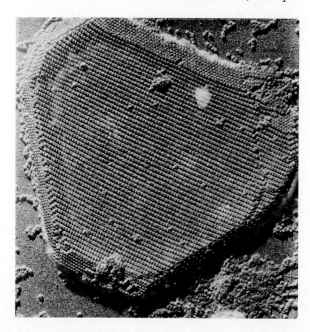

Fig. 1.2. Electron micrograph of a crystallized virus (southern bean mosaic virus). The nearly spherical molecules have a diameter of 230 Å. Magnification of the reproduction is 35 000 ×. Relief has been obtained by condensation of palladium vapor, coming from a particular direction, in a very thin layer on the crystal. Those parts which thereby lie in a "shadow" absorb electrons more weakly and therefore appear in a positive as though lying in a light shadow. (R. W. G. Wyckoff.)

crystal structures are unusually open. (If, for example, an open crystal structure such as in fig. 1.1 d melts to a liquid as in b, the density increases.)

The translation of the structural units in solid substances is much more limited than in liquids. But at higher temperatures, where all the different kinds of heat motions increase, translation may become quite apparent as diffusion. It is natural that small, free atoms may be especially easily movable. Free rotation of groups of atoms may also occur. The most common heat motions in solid substance, however, are vibrations within the groups and vibrations of the whole groups with respect to each other.

The structural units in solid substances may be disposed with quite different degrees of order. If the structure shows at the most short-range order, the substances is called *amorphous*. Homogeneous and very dense amorphous phases form *glasses*, which in a certain sense may be considered to be supercooled liquids in which the low temperature has caused the mobility to become very low. Here, thus, there can be no more than short-range order. When long-range order also begins to appear in a solid phase, the phase is said to become *crystalline*, and when the order becomes complete, an *ideal crystal* is obtained. Real crystals are sometimes nearly ideal.

Fundamentals, Concepts and Definitions

Fig. 1.2 shows an electron micrograph of a crystal made up of virus molecules. The order here is quite great.

The characteristic feature of a crystal is thus the regular internal structure and not the outer crystal form, which is only a result of the inner regularity, and which may often be lacking.

By modern research methods (primarily X-ray diffraction, 7-1) it has been found that the crystalline state in solid substances is much more common than was formerly believed.

Solid solutions may be either amorphous (many glasses are solid solutions) or crystallized.

i. Ideal solutions. In a solution (gaseous, liquid or solid), the forces between the structural units (atoms or molecules) are most often strongly dependent on the nature of the units. If the building units are A and B, the forces A–A, B–B and A–B are thus different. One may, however, imagine the ideal case in which all these forces are the same, and speak then of an *ideal solution*.

An ideal solution, like an ideal gas, is a state which is never attained in reality. Many solutions are, however, very close to ideal. This is especially true of solutions in which the different structural units are distinguished only in that they contain different isotopes of the same element, for example, H_2O and D_2O. Also, solutions between closely related substances (for example, heptane and octane, benzene and toluene) may be nearly ideal.

If a solution of A in B contains very little A, the A units in general lie so far apart that the forces A–A can be neglected in comparison the forces B–B and A–B. Such a solution is called an *ideal dilute solution*.

As with ideal gases, many laws take especially simple forms for ideal solutions.

j. Chemical compounds. A pure substance which contains two or several elements has since early times been called a *chemical compound*. That a chemical compound should be what is now called in modern nomenclature a pure substance, that is, has constant composition, became a generally accepted idea during the 18th century. A clear statement of the principle was made by Proust in 1799. Berthollet disagreed with Proust, however, and believed that the composition of a chemical compound could vary between certain limits. During the years 1799–1807 a controversy continued over the question between Proust and Berthollet. Proust apparently won out, but much later it was shown that Berthollet was right in principle. Many classical chemical compounds have, in fact, variable composition. The most important reason for Proust's victory was certainly the advent of Dalton's atomic theory, which in a convincing way was able to account for the constant composition of a compound.

The hypothesis of the constant composition of a chemical compound was mainly based on analyses of phases made up of molecules. In general, the composition of a molecule cannot be changed without obtaining appreciably different properties for the new molecule, and thus a new substance. (Only for very large molecules and a small change can the alteration of properties be negligible.) Many salts also show constant composition. An example is given by NaCl. This salt is built up of Na^+ and Cl^- ions. A change in the charges on the ions in the solution or the melt from which NaCl crystallize is not conceivable under ordinary conditions, so that the requirement of electroneutrality leads to the result that the number of Na^+ ions is equal to the number of Cl^- ions. (Variations in composition even in salts such as NaCl is discussed, however, in 6-3a.)

When solid phases were meanwhile found with variable composition, these were all designated as solid solutions, as distinguished from chemical compounds. It is not possible, however, to make such a distinction. For example, iron oxide with the composition FeO is built up of Fe^{2+} and O^{2-} in a structure of the same type as NaCl. If NaCl is a chemical compound, then obviously FeO must also be. But the same phase can be formed also after a number of Fe^{2+} has changed charge to Fe^{3+}. The increased positive charge is compensated by causing some of the positions formerly occupied by Fe^{2+} to become vacant. A homogeneous phase (solid solution) is thereby obtained within a *region of homogeneity* from the composition FeO to approximately $Fe_{0.9}O$. For the sake of simplicity however, this phase is termed the FeO phase. If it is desired to emphasize the variable composition, the sign $\sim$ (read as *circa*) may be placed before the formula, thus: $\sim$ FeO. Many similar cases are mentioned later, among other places, in 6-3.

If the concept of chemical compound is broadened beyond that of a pure substance, it becomes very difficult to define. Indeed, this does not appear to be necessary. Instead of saying that a certain system contains such and such compounds, one may consider the phases present and their regions of homogeneity. (See the example of the FeO phase.) Some phases may have constant, others variable, composition. If a phase is exclusively made up of a single kind of molecule, it will surely have a constant composition, but in many other cases it may be difficult to decide whether the phase has a constant composition or a small but finite region of homogeneity. Nevertheless, the expression "chemical compound", or more simply "compound", is still used generally in roughly its classical sense. It is frequently used in this manner in this book.

It has been suggested that phases with constant composition (for example H_2O, NaCl) be called *daltonides*, and those with variable composition (for example the FeO phase) *berthollides*.

1-5. States, processes and equilibrium

a. State. The *state* of a system can only be ascertained by studying the system, that is, by determining its properties. The properties define the state of the system and are therefore functions of the state.

From this it follows that when a property changes value from one state to another, this change depends only on the initial and final states and not on the path by which the system passed from one to the other.

b. Processes and reaction rate. When a system passes from one state to another, one speaks of a *process*. This term refers to changes of state of all kinds. *Chemical processes* imply in general a rearrangement of atoms or a transfer of atomic charges. A chemical process is often called a *chemical reaction*. But in principle, such a chemical process is not distinguished from certain more "physical" processes such as a phase change (for example fusion or vaporization) or a transfer of a substance from one phase to another (for example, solution of a substance in a solvent).

The rate with which a process takes place can vary between extraordinarily wide limits. Many chemical reactions proceed very slowly, that is, the *reaction rate* is very low. Sometimes one speaks in such cases of an *arrested reaction*. Frequently, certain substances can increase the rate of reaction without being consumed in the reaction. This phenomenon is called *catalysis* and the rate-increasing substance a *catalyst*. In a mixture of hydrogen and oxygen at room temperature, no change whatever can be observed after a very long time. When finely divided platinum is introduced into the gas mixture, however, the two elements quickly combine to form water. Platinum serves here as a catalyst.

Certain substances can reduce a reaction rate. Such a substance has formerly been called a negative catalyst (a substance which increases reaction rate is then a positive catalyst) but since it is always consumed during the course of the reaction, it is preferred now to call it by the special term *inhibitor*.

c. Equilibrium. In ordinary conversation one says that a system is *stable* and in *equilibrium* when it does not noticeably change; equilibrium is a state of rest. A system may be unvarying by having attained a state from which it has no tendency to change. This is, then, true equilibrium. But the system may be a long way from equilibrium and appear to be unchanging only because of a low reaction rate. The system hydrogen and oxygen mentioned in 1-5b, is in this way unstable even though it appears invariant at room temperature.

The expressions "equilibrium" and "stable" must be restricted to those states in which there is no *tendency* to change. In 2-3 is introduced a measure of this tendency which can demonstrate whether or not equilibrium exists.

However, it is only from a macroscopic viewpoint, in which a large number of atoms and molecules take part, that nothing seems to happen. If a gas phase containing water vapor occurs above liquid water, water molecules constantly travel in both directions between the two phases. From a certain area of water at constant temperature the same number of water molecules are always liberated per unit of time, and move into the gas phase. The number of water molecules from the gas phase which fall into the same liquid area per unit of time and are united with the liquid depends, on the other hand, on the concentration of water vapor in the gas phase. The higher the concentration, the greater is this number.

If evaporation takes place, this means that in the same time a smaller number of water molecules travel from the gas phase to the liquid than in the opposite direction. The water concentration in the gas phase then rises, and consequently the number of molecules that go from the gas to the liquid increases. Finally, this number becomes equal to the number of molecules which in the same time leave the liquid. Equilibrium has thus been reached. If subsequently, for example, the temperature is increased, the migration of molecules from the liquid increases, and only when the water vapor concentration is raised still higher, is equilibrium again achieved.

In this example equilibrium has been established between the processes $H_2O(l) \rightarrow H_2O(g)$[1] and $H_2O(l) \leftarrow H_2O(g)$. That equilibrium exists is often indicated by two arrows, thus:

$$H_2O(l) \rightleftharpoons H_2O(g).$$

The same principles apply to more "chemical" processes. If nitrogen and hydrogen are in equilibrium with ammonia, there is formed in the system ammonia molecules according to the equation $N_2 + 3H_2 \rightarrow 2NH_3$. But at the same time an equal number of ammonia molecules decompose according to the equation $N_2 + 3H_2 \leftarrow 2NH_3$. The equilibrium is thus written $N_2 + 3H_2 \rightleftharpoons 2NH_3$.

Equilibria of this kind are commonly called *dynamic* in contrast to *static*; all chemical equilibria are dynamic. By substituting a portion of the atoms in a phase or in a given kind of molecule with a radioactive isotope and then following the spread of the radioactivity to the other phases or kinds of molecules, the active exchange process can be easily demonstrated.

A catalyst hastens the process equally in both directions and therefore does not alter the position of the equilibrium.

In the same system, different processes may approach equilibrium at very different rates. If the gas mixture hydrogen + oxygen lies above water, the two gases quite rapidly dissolve in the water to establish a solution equilibrium. The water-formation equilibrium, on the other hand, is not attained. The equilibrium $N_2O_4 \rightleftharpoons 2NO_2$ is established rapidly, but both of these oxides are unstable and tend to decompose into nitrogen and oxygen. This decomposition is, however, very slow. An equilibrium that is reached for the fast processes before the slow ones have had time to proceed to any appreciable extent is called a *partial equilibrium*.

An inert state as in the mixture hydrogen + oxygen is often called a *metastable state* (Gk. μετά between, after).

---

[1] The state of aggregation is customarily indicated by s = solid, l = liquid, g = gaseous.

Fundamentals, Concepts and Definitions

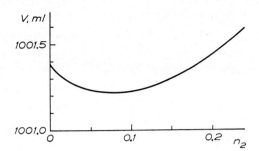

Fig. 1.3. The volume of water solutions of $MgSO_4$ at 18°C. The solutions contain n_2 moles $MgSO_4$ per 1000 g H_2O.

d. Variables of state and functions of state. In 1-5a it was said that the state of a system is defined by its properties. It can be shown that, if a system is in equilibrium, only a small number of properties need to be established for the state to be defined. Therewith all the other properties are also fixed.

For the very simple case of a system *in equilibrium*, consisting of a pure substance which forms a single phase, the values of only three properties need to be established. For liquid water in equilibrium, for example, it is only necessary to prescribe a certain pressure, a certain temperature and a certain quantity in order that the volume, density, viscosity, refractive index, dielectric constant, etc. of the water may be established.

Properties such as these may be taken as the *variables of state* (often merely *variables*) of the system, and in the example we have clearly a case with three independent variables. One may, of course, select any variables whatever as independent. All the others then become dependent on the three initial variables In a case such as that given, for practical reasons the pressure P, the temperature T, and the quantity of material n_1 (the number of moles of component 1) are often chosen as independent variables. For example, then, the volume V may be written $V = f(P, T, n_1)$. The volume is given here as a function of three independent variables, which together determine the state of the system. It is said, therefore that the volume is a *function of state*.

If the phase consists of the components 1 and 2, n_2 (the number of moles of component 2) is added as an independent variable. One obtains then, for example, $V = f(P, T, n_1, n_2)$. For k components, $V = f(P, T, n_1, n_2, \ldots n_k)$. The independent variables thus become pressure, temperature and k "variables of composition".

Essential properties (1-1a) of a phase are determined by pressure, temperature and only $(k-1)$ variables of composition. Here the total quantity of substance is in fact unimportant and only the composition of the phase is significant. To define the composition of the phase then only $(k-1)$ variables of composition are needed. As variables of composition it is here most convenient to use mole fractions x. Thus, for example, the specific volume of a phase with k components = $f(P, T, x_1, x_2, \ldots x_{k-1})$. The same principle applies to all molar quantities (1-5e).

The number of independent variables in a multiphase system is worked out in (13-2).

The independent variables of a system are also called its *degrees of freedom*. A system with, for example, three independent variables is said to have three degrees of freedom.

When a system passes from state A to state B, the *increase* in a function of state is commonly indicated by Δ. If, for example, the volumes of the system in the two states are V_A and V_B, then $\Delta V = V_B - V_A$.

If 1 mole of H_2O in the form of ice at 0°C is melted to water with no change in temperature, we may write

$$H_2O(s) \rightarrow H_2O(l); \qquad \Delta V = -2 \text{ ml } (0°C)$$

because at 0°C the volume of 1 mole of ice $V_A = 20$ ml and the volume of 1 mole of water $V_B = 18$ ml. This symbolism is also used when there is equilibrium between the two states. Δ represents then the increase of the function of state in question when the system is considered to pass *completely* from the left-hand state to the right-hand state.

e. Molar quantities. For a pure substance, the value of a certain quantity per mole is often given. This value is called a *molar quantity*. The molar volume or mole volume for water at 0°C is thus 18 ml. Frequently, a formula is given at the same time to show what a molar quantity corresponds to. As an isolated symbol for a molar quantity, a small letter is used. The mole volume is indicated thus by v.

In nearly ideal solutions (1-4i) a property is composed additively of the properties of the solution components. In such a case (for example, a solution of heptane and octane) the volume of the solution is equal to the sum of the volumes of the pure substances. If the solution consists of n_1 moles of substance 1 with mole volume v_1, n_2 moles of substance 2 with mole volume v_2, etc., the volume of the solution $V = n_1 v_1 + n_2 v_2 + \ldots$.

In nonideal solutions, on the other hand, additivity does not apply. When water and ethanol are mixed, the forces of attraction between the different types of molecules leads to a reduction of the total volume, which at a maximum amounts to nearly 10%. It is also clear that if a liquid solution contains substances which in pure form are solids, their mole volumes in the solution must be very different from the mole volumes of the pure substances. To state the contribution of a certain substance to a quantity in such cases, the following method is used.

Let us assume that the solution has the composition just described above. Then, according to 1-5d, its volume $V = f(P, T, n_1, n_2, \ldots)$. If we wish to know the contribution which, for example, substance 2 makes to V, we determine V as a function of n_2 alone while holding the other variables $(P, T, n_1, n_3, \ldots)$ constant.

Fig. 1.3 shows, for solutions of $MgSO_4$ in H_2O, how V varies with the number of moles of $MgSO_4$, n_2. The quantity of water is held constant throughout $= 1000$ g, so that n_2 in this case represents the molality of $MgSO_4$. The derivative of the curve dV/dn_2 for a certain n_2 value shows how much V increases per mole of added $MgSO_4$ for an infinitely small increment. It may also be said that the derivative at a certain point shows the increase in V when 1 mole of $MgSO_4$ is added to an infinitely large quantity of solution of the composition in question.

If a function of several independent variables is differentiated with respect to just one of them while the others are kept constant, the result is a *partial derivative*. The term *partial mole volume* has been introduced for the particular partial derivative in the example. The quantity is often given the same symbol as the mole volume, thus in this case, v_2.

Partial derivatives are written with the symbol "∂" instead of "d", so that the partial mole volume of the substance "i" becomes $v_i = \partial V/\partial n_i$. The variables which are held constant are often explicitly shown by subscripts, in this case, thus:

$$v_i = \left(\frac{\partial V}{\partial n_i}\right)_{P,T,\text{ other } n} \tag{1,13}$$

The volume of a substance is always positive, but the *partial* mole volume may well be negative. Fig. 1.3 shows that the volume of the solution decreases at the start as $MgSO_4$ is added. In this region the derivative of the curve, $\partial V/\partial n_2 = v_2$, is negative. Upon further addition of salt the volume increases again, that is, v_2 becomes positive. The negative values result from the strong forces of attraction within the solution.

If V varies linearly with n_i, then $\partial V/\partial n_i = $ constant $=$ mole volume. This happens in the ideal case when the volume is composed additively of the different mole volumes, and, of course, also for a pure substance.

If v_i indicates the partial mole volume, however, this important relation always applies:

$$V = n_1 v_1 + n_2 v_2 + \ldots. \tag{1.14}$$

Partial molar quantities for other functions of state can also be formed. They are defined analogously with (1.13) and combined to give the function of state in a manner analogous to (1.14).

CHAPTER 2

Energy

Energy plays an extraordinarily important role in the theoretical treatment of chemical problems. The prediction of the structure of a substance and the calculation of the forces which bind the atoms together are basically energy problems. The driving forces of chemical reactions can be expressed and measured energetically, from which follows the possibility of calculating chemical equilibria. The experimental determination of structures of atoms, molecules and crystals is most often based on what happens when the material absorbs or gives off radiant energy.

This chapter is concerned with the application of the energy principle to chemically important systems and the question of energy and chemical equilibrium. The special laws which are applicable to the exchange of energy in systems of particles with very small mass are referred to in chap. 3. They are then applied in several of the following chapters.

2-1. Applications of the energy principle

a. The energy principle. If any system loses energy, there is always a corresponding gain in energy by its surroundings. Moreover, when one form of energy disappears, an equivalent quantity of some other form appears in its place. This has already been stated in 1-2 as the *law of the conservation of energy* or the *energy principle*. The restrictions on the energy principle caused by the equivalence of energy and matter were also discussed in 1-2. It was pointed out that these restrictions are insignificant as long as the large quantities of energy exchanged in nuclear reactions, or the high particle velocities that can occur in nuclear reactions as well as acceleration in an electric field, are not involved.

A given quantity of heat energy cannot be completely transformed into other forms of energy, but for all other energy changes no such limitations are encountered. The special position of heat energy in this respect is related to the fact that it is an expression of the heat motion of the atoms and molecules, and thus follows statistical laws.

The science of the relationship between heat energy and other forms of energy is called *thermodynamics*. After it became possible, in the latter half of the 19th century, to study chemical equilibria through its help, thermodynamics became one of the basic foundations of chemistry. This was brought about mainly through the influence of Gibbs.

The energy principle is customarily called the *first law of thermodynamics*.

Thermodynamics deals experimentally and theoretically with systems in which the number of discrete particles (atoms or atomic groups) is very great. Thermodynamic laws are probability laws (statistical laws), but because of the enormous number of particles involved, their precision becomes so high that they operate as exact laws.

b. Forms of energy in a system. According to Einstein's equation (1.1) the total energy of a system may be set $= mc^2$ where m is the mass of the system. This total energy, however, is divided into several different forms, which will be discussed below.

If the total energy is given by Einstein's equation, the energy in a space without mass and without radiant energy $= 0$. Let us bring into this space all the atomic nuclei and electrons of a system, placed at rest, separate from each other and very thinly spread out. Energy has then been introduced, partly in the form of the rest mass of the nuclei, partly in the form of the rest mass of the electrons, and partly in the form of potential energy from the electrostatic forces between the charges on the nuclei and electrons. Of these the portion of energy in the atomic nuclei is overwhelmingly the greatest, since they are effectively the bearers of the mass of the system. This portion is called the *nuclear energy*. Now let the atomic nuclei and electrons be arranged together so that uncharged atoms are formed, still at rest and at large distances from each other. The potential energy is decreased thereby, partly through transformation to kinetic energy of the electrons when they move into the electron cloud of the atoms, and partly through loss of energy from the system. The sum of the potential and kinetic energy of the electrons is called the *electronic energy*.

The atoms may then condense into atomic groups, such as molecules or crystals. The potential energy is thus further decreased with the release of binding energy. One speaks now of the electronic energy of the whole group, resulting from the electrostatic forces between all of the electrons and atomic nuclei of the groups. The electronic energy of the group may thus be very different from the electronic energy of the free atoms that make up the group.

A part of the energy of the system is also represented by the vibration of atoms and atomic groups with respect to each other. Just as a swinging pendulum or a vibrating tuning fork possesses energy of oscillation or vibration, so also do these atoms or atomic groups. The system thus has acquired *vibrational energy*. In addition, the freely moving units of the system, atoms or atomic groups, take up energy through rotation and translation, giving the system *rotational* and *translational energy*. If the unit has mass m and velocity v, its translational energy is $mv^2/2$.

Vibrations, rotations and translations are summed up as *heat motion*, which constitutes *heat energy*. On raising the temperature, the heat motion increases, which implies that the added energy goes into vibrational, rotational and translational energy.

Electronic energy constitutes, after nuclear energy, the largest amount of energy in a system. It is principally changes in electronic energy on rearrangement of atoms which are utilized in chemical processes, and thus constitute chemical energy. Vibrational, rotational and translational energies have smaller magnitude than electronic energy.

Through what can be most nearly compared with mechanical action (for example, collision), the heat motion spreads from one atom or atomic group to another. This occurs in any transfer of heat through conduction. If the system is isolated, an equalization of the energy of heat motion takes place through such mechanical action until it is distributed according to statistical laws (Maxwell's distribution law) among the different particles. The temperature has then become constant throughout the whole system.

In 1 mole of an ideal gas which has a uniform temperature T, the collective translational energy of the molecules $= 3RT/2$. Since 1 mole contains N_A molecules, the mean value of the translational energy of 1 molecule $= 3RT/2N_A = 3kT/2$ (Boltzmann's constant $k = R/N_A$, see 1-4f). This mean value is thus proportional to T but independent of the molecular weight. How the individual molecular velocities and translational energies are distributed around the mean value is determined by Maxwell's distribution law.

Since the translational energy of 1 molecule with velocity v is $mv^2/2$, the collective translational energy of 1 mole is also $N_A m\overline{v^2}/2 = M\overline{v^2}/2$, where M is the mass of 1 mole of the gas and $\overline{v^2}$ the mean value of all the v^2 values of the individual molecules. We thus get $M\overline{v^2} = 3RT$ and $\sqrt{\overline{v^2}} = \sqrt{3RT/M}$. $\sqrt{\overline{v^2}}$ is often called the mean velocity of the molecules, but it is not quite the same as the true mean velocity $\overline{v}$. The velocity distribution according to Maxwell's law gives $\overline{v} = 0.92\sqrt{\overline{v^2}}$. Both quantities are thus inversely proportional to the square root of the molecular weight.

For example, in oxygen ($M = 32 \times 10^{-3}$ kg) at 25°C, the mean velocity of the molecules $\sqrt{\overline{v^2}} = \sqrt{3 \times 8.3143 \times 298/32 \times 10^{-3}} = 482$ m s^{-1}. In hydrogen at the same temperature, the mean velocity is 1928 m s^{-1}.

c. Internal energy. In thermodynamics it is not necessary to consider how the energy is distributed in the system. All the different kinds of energy quantities $E_1, E_2, \ldots$ are included in a sum $U = \Sigma E_i$. U is often called the *internal energy*. The prefix "internal" is not essential but is intended to show that energy terms that depend, for example, on the presence of external force fields (gravity, electric and magnetic fields) or the movement of the whole system are excluded. The internal energy of a system is a function of state (1-5d), that is, its value depends only on the present state of the system and not on its previous history. The internal energy of 1 mole of oxygen at 25°C and 1 atm is thus the same no matter how the oxygen is prepared (provided, of course, the relative amounts of the different oxygen isotopes are the same).

Energy

According to Einstein's equation (1.1), $U = mc^2$. This is of fundamental significance, but because weighings cannot be done with sufficient accuracy, the uncertainty of any determination of U according to Einstein's equation is much greater than the changes of energy which occur in ordinary chemical processes. One can never, therefore, obtain a useful absolute value for the internal energy of a system, but is obliged to work only with energy *changes*. Also, it is most often these changes which are of real interest.

If the quantities of heat q and work w (by work is meant all types of energy except heat) are introduced into a system, the increase in the internal energy of the system is

$$\Delta U = q + w \tag{2.1}$$

Negative values of q and w indicate that heat and work are removed from the system. That work is removed means that the system does work.

Through (2.1) internal energy is defined at the same time that the energy principle is expressed.

Pressure-volume work and electrical work are forms of work that are often important in chemical processes. Pressure-volume work occurs as soon as the system changes volume under an external pressure. For the sake of simplicity we may imagine that the system which is exchanging work with its surroundings is contained in a cylinder closed with a movable piston. At the beginning the pressure P is the same on both sides of the piston so that equilibrium is maintained. If the volume V of the system is increased by the amount ΔV, the piston is pushed outward. At the end of the process the same pressure is again obtained on both sides of the piston. A common and simple case is that in which the pressure at the end of the process, P, is the same as at the beginning. The system has thus increased its volume by ΔV under constant pressure P. If the area of the piston is a, the force operating on it is Pa. The increase in volume ΔV corresponds to the piston displacement $\Delta V/a$. The system has thereby done the work $Pa\Delta V/a = P\Delta V$, which means that the corresponding quantity of work is removed from the system. We thus obtain

$$w = -P\Delta V \tag{2.2}$$

If the volume is decreased, ΔV is negative and w becomes positive. Work has then been added to the system.

When the value of w according to (2.2) is introduced into (2.1), we get

$$\Delta U = q - P\Delta V \tag{2.3}$$

which thus applies if the only exchange of work in the system consists of pressure-volume work at constant pressure.

For a process without exchange of work, $w = 0$ and thus $\Delta U = q$. The change in internal energy corresponds then to the quantity of heat transferred. If no work other than pressure-volume work is exchanged (which is in general understood unless otherwise stated) the absence of work exchange requires only that the volume of the system remain constant. If under this condition the quantity of heat added is indicated by q_v, we have thus

$$\Delta U = q_v \quad \text{konst. vol} \Rightarrow P\Delta V = 0 \quad (2.4)$$

In the measurement of heat exchange in processes where gases are consumed or formed, it is often necessary to keep the entire system in a closed vessel in order to maintain the gases ("bomb calorimeter"). The volume of the system is constant in this case and (2.4) is applicable.

d. Enthalpy. In ordinary calorimetric measurements carried out at atmospheric pressure in an open calorimeter, the volume is seldom constant. Here the pressure is constant but through a change in volume work is exchanged. It is then useful to introduce a new function of state H, *heat content* or *enthalpy* (Gk. ϑάλπος, heat), which is defined as

$$H = U + PV \quad (2.5)$$

It is apparent that H is a function of state since it is a function only of other functions of state. From (2.5) it is seen also that H has the dimension of energy. The absolute value of the enthalpy of a system is not known with any greater accuracy than the absolute value of the internal energy.

For a system in two different states A and B we have

$$H_A = U_A + P_A V_A \quad \text{and} \quad H_B = U_B + P_B V_B$$

If the system goes from state A to state B, the increase in H becomes

$$\Delta H = H_B - H_A = (U_B - U_A) + (P_B V_B - P_A V_A) = \Delta U + (P_B V_B - P_A V_A)$$

If the pressure is the same at the beginning and at the end of the process, that is, $P_A = P_B = P$, then

$$\Delta H = \Delta U + P\Delta V \quad (2.6)$$

From (2.3) and (2.6) we obtain $\Delta H = q$. To remind us that this equality is true only if the pressure is the same at the beginning and at the end of the process (which in general, but not necessarily, implies that the pressure is constant during the whole process), the added heat quantity may be indicated by q_p, so that

$$\Delta H = q_p \quad (2.7)$$

For processes where gases are not consumed or formed, in general the change in volume and thus the pressure-volume work $P\Delta V$ is small. From (2.6) we see that the difference between ΔU and ΔH then also becomes small, and in such cases may often be neglected. Such is also the case when gases take part in the process, but the total number of moles of gas remains unchanged; for example, $C(s) + O_2(g) = CO_2(g)$. In this example the volume of the system does change through the consumption of the carbon, but the volume of the carbon can be neglected in comparison with the volume of the system, which is nearly all taken up by the gases.

Energy

If the number of moles of ideal gas in the system is increased by Δn and the volume thereby increased by ΔV, according to (1.9) the pressure-volume work is $P\Delta V = \Delta n RT$. (2.6) becomes in this case $\Delta H = \Delta U + \Delta n RT$. The values of RT for 0° and 25°C are given in Appendix 2.

2-2. Thermochemistry

a. Energy of reaction and transition. The branch of chemistry which is concerned with the exchange of heat in chemical reactions is called *thermochemistry*. There is, however, no reason to distinguish heat exchange associated with chemical reactions from that associated with, for example, phase transitions and transport of substances from one phase to another.

Since early times the quantity of heat absorbed or given off by a system on the transformation of a given quantity of substance has been called the "heat" of the process. The *heat of reaction* (and special forms, for example *heat of combustion*, *heat of formation*, *heat of solution*) was defined as the amount of heat *evolved* by the system in the reaction. For a reaction which absorbs heat the heat of reaction would then be negative. But on the other hand *heat of fusion* and *heat of vaporization* have been defined as the heat *absorbed* during the process.

Most processes are studied under constant pressure and the stated values of heats of reactions, etc. signify then a change in H. If the experiment takes place at constant volume so that the value signifies a change in U, it is often so stated, but unfortunately, not always.

A process in which the system gives off heat has been called *exothermic* and one in which the system absorbs heat *endothermic*. Most often ΔU and ΔH have the same sign for a given process (always if the values of ΔU and ΔH are so large that the difference cannot lead to a change in sign) but if this is not the case these terms are ambiguous.

If the method of expressing changes in functions of state described in 1-5d is used, all ambiguity caused by old conventions and terms is avoided. Energy quantities should not in general be written into reaction equations, because then it is not clear what functions of state they represent.

It is becoming more and more usual to let the terms *reaction energy* and *reaction enthalpy* indicate the quantity of heat *absorbed* at constant volume and constant pressure, respectively. Therefore, reaction energy = ΔU and reaction enthalpy = ΔH. Since the terms in both cases involve exchange of energy and also because the difference between ΔU and ΔH is usually small, both quantities are many times referred to loosely as "energy."

In the recommended method of writing a reaction, the quantity of energy transferred in a process refers to the quantity of substance that corresponds to the given reaction equation. The magnitude of the energy shows whether the formula symbols represent individual atoms and groups of atoms, or moles (in the latter case the energy is N_A times as great as in the former). Complete transformation of the substances on the left side of the reaction equation to those on the right side is assumed. Furthermore, if they are not obvious, the pressure, temperature, state of aggregation, modification or possible presence in solution (stated with the solvent—water is often indicated by aq—and concentration) of each substance should be stated. Two simple examples are given below.

The molar heat of transition for the transformation of monoclinic to orthorhombic sulfur at 25°C and 1 atm is 0.30 kJ. This is more clearly expressed

$$S(\text{monoclinic}) \rightarrow S(\text{orthorhombic}); \quad \Delta H = -0.30 \text{ kJ } (25°C, 1 \text{ atm}).$$

In this case, as in all processes involving single solid phases, the change in volume is so small that in general we may set $\Delta U = \Delta H$.

The molar heat of vaporization of water at 100°C and a constant pressure of 1 atm is 40.64 kJ:

$$H_2O(l) \rightarrow H_2O(g); \quad \Delta H = 40.64 \text{ kJ } (100°C, 1 \text{ atm}).$$

From a knowledge of the molar volumes of water and water vapor at 100°C and 1 atm $(=1.01325 \times 10^5 \text{ Nm}^{-2})$, ΔU may be calculated. We get first $\Delta V = 0.0306 \text{ m}^3$, which gives the pressure-volume work $P\Delta V = 3.10 \text{ kJ}$. According to (2.6), then, $\Delta U = 37.54 \text{ kJ}$.

The quantity of energy *absorbed* on the formation of one mole of a substance is called its *energy of formation* $(=\Delta U)$ and *enthalpy of formation* $(=\Delta H)$. In general, formation from the elements in their natural state is assumed, but to avoid misunderstanding, the reaction equation should be given.

A decomposition of a substance into simpler atomic groups or atoms is often called *dissociation*. The amount of energy which must be *absorbed* for the dissociation of one mole of a substance is called its *dissociation energy* $(=\Delta U)$ and *dissociation enthalpy* $(=\Delta H)$. Since dissociation often can be considered to occur in different ways, the dissociation products must be stated, preferably by formula.

In discussion of chemical bonding (see for example 5-3k) a substance is often considered to be formed from, or decomposed into free atoms. The amount of energy absorbed in the latter case (which is equal to the energy of formation from free atoms) is called *atomization energy (enthalpy)*. For a diatomic molecule the atomization energy and the dissociation energy are identical.

From the energy principle it follows that ΔU and ΔH for a certain process is independent of the path the system follows between the initial and final states. Whatever states occur in between are unimportant. It may also be said that this results from the fact that U and H are functions of state. It is worth noting that this condition was realized by Hess as early as 1840, before the energy principle was formulated (*Hess's law*).

As a result, ΔU and ΔH can be calculated for many processes which cannot be carried out for direct measurement. Heats of combustion, which are determined directly by calorimetry, are frequently used, for example, for calculation of enthalpies of formation. As an example, the enthalpy of formation of methane CH_4 is obtained from the following three combustion processes (the combustion of methane is written in reverse):

$$CO_2(g) + 2H_2O(l) \rightarrow CH_4(g) + 2O_2(g); \quad \Delta H = 890.8 \text{ kJ (1 atm, 25°C)}$$
$$C(graphite) + O_2(g) \rightarrow CO_2(g); \quad \Delta H = -393.7 \text{ kJ (1 atm, 25°C)}$$
$$2H_2(g) + O_2(g) \rightarrow 2H_2O(l); \quad \Delta H = -572.0 \text{ kJ (1 atm, 25°C)}$$

$$\overline{C(graphite) + 2H_2(g) \rightarrow CH_4(g) \quad \quad \Delta H = -74.9 \text{ kJ (1 atm, 25°C)}}$$

Addition of the three reaction equations and their respective ΔH values thus gives the desired result. The addition is permitted, of course, only if each substance takes part in the different processes in the same state (modification, pressure, temperature).

From the formation enthalpies the atomization enthalpies may be calculated if the atomization enthalpies of the elements in the state which they have in the process of formation are known. For methane knowledge of the atomization enthalpy of graphite (=heat of sublimation of graphite) and hydrogen (=dissociation enthalpy of hydrogen) is required. We then have:

$$CH_4(g) \rightarrow C(graphite) + 2H_2(g); \quad \Delta H = 75 \text{ kJ (1 atm, 25°C)}$$
$$C(graphite) \rightarrow C(g); \quad \Delta H = 715 \text{ kJ (1 atm, 25°C)}$$
$$2H_2(g) \rightarrow 4H(g); \quad \Delta H = 872 \text{ kJ (1 atm, 25°C)}$$

$$\overline{CH_4(g) \rightarrow C(g) + 4H(g) \quad \quad \Delta H = 1662 \text{ kJ (1 atm, 25°C)}}$$

In order to facilitate calculations of this kind thermochemical data are usually referred to the same *standard state*. The state at 25°C (298.15°K) and 1 atm is usually chosen as standard. The enthalpy increase in a process where all disappearing and formed substances are in their standard states is given the symbol ΔH^0 (possibly with an index showing the absolute temperature, e.g. $\Delta H^0{}_{298}$). Further, the zero point in an enthalpy scale is defined by assuming the enthalpies of the elements in their stable, natural forms at 25°C and 1 atm to be =0. On this scale the enthalpy of a compound is equal to the enthalpy increase when the compound is formed from the elements, that is, its enthalpy of formation.

Molar enthalpies of formation are tabulated in such reference books as W. M. Latimer, Oxidation Potentials (2nd Ed., Prentice-Hall, Inc., New York, 1952); O. Kubaschewski and E. Ll. Evans, Metallurgical Thermochemistry (Pergamon Press, London 1958); and Selected Values of Chemical Thermodynamic Properties (U.S. National Bureau of Standards Circular No. 500, 1952).

As an example we take $SO_2(g)$ for which the enthalpy of formation is given as $\Delta H^0{}_{298} = -296.1$ kJ. As the standard states of the elements refer to S(orthorhombic) and $O_2(g)$ this enthalpy of formation is the ΔH value for the process S(orthorhombic) + $O_2(g) \rightarrow SO_2(g)$ at 25°C and 1 atm. The enthalpy of a system is equal to the sum of the formation enthalpies of the included substances in the same enthalpy scale. Thus for the process $SO_2(g) + \frac{1}{2}O_2(g) \rightarrow SO_3(g)$ we have $\Delta H = \Delta H^0(SO_3) - \Delta H^0(SO_2) - \frac{1}{2}\Delta H^0(O_2) = -395.2 + 296.1 - 0 = -99.1$ kJ.

The validity of the additions, which were made above in the application of Hess's law, is easily understood if one expresses the observed ΔH values in enthalpies of formation (for example $\Delta H = \Delta H^0(CH_4) + 2\Delta H^0(O_2) - \Delta H^0(CO_2) - 2\Delta H^0(H_2O)$ etc.) and adds these expressions.

The experimental determination of fundamental thermochemical values is carried out sometimes by calorimetry, and sometimes by optical or electrical methods. Dissociation energies are often determined spectroscopically (7-2) or by a study of electron collisions.

b. Heat capacity. If a quantity of heat q_v is supplied to a system, then according to (2.4), $q_v = \Delta U$. If this added heat causes a rise in temperature ΔT, the quotient $\Delta U/\Delta T$ is called the *mean heat capacity* of the system within the temperature interval T to $T + \Delta T$ with heating at constant volume. If the amount of added heat is allowed to approach 0, the difference quotient becomes the derivative dU/dT, which is the *heat capacity* of the system at temperature T and constant volume. If this heat capacity is denoted by C_v, then $C_v = dU/dT$.

If the system is heated at constant pressure, according to (2.7) $q_p = \Delta H$, which in an analogous manner gives the heat capacity of the system at constant pressure $C_p = dH/dT$.

That the volume or the pressure remains constant on heating is customarily emphasized in the definitions of C_v and C_p by writing the partial derivatives thus (1–5e):

$$C_v = \left(\frac{\partial U}{\partial T}\right)_V \quad \text{and} \quad C_p = \left(\frac{\partial H}{\partial T}\right)_P$$

The heat capacities of solid bodies are most easily determined in an open calorimeter at constant pressure, giving in such cases C_p, which is the quantity most used in practical calculations. The difference between C_v and C_p is normally greater for liquids than for solid substances, and is greatest for gases.

The heat capacity of one gram is called the *specific heat*. In general, this refers to heating at constant pressure. The heat capacity of one mole is called *molar heat* (if the formula unit is one atom, often called *atomic heat*) and is indicated by c_v or c_p.

Heat capacity and molar heat have no meaning at transition points. Here U and H always change suddenly because of the heat of transition. Measurements of heat capacities must therefore be made within a temperature interval which does not include any phase transition.

The heat energy which is supplied to a body at constant volume without the occurrence of a phase transition, goes mainly into an increase in the heat motion. The greater the possibility that a substance has to utilize the energy for heat motion, the greater will be the molar heat c_v. Its magnitude at different temperatures can therefore provide important information about the nature of the heat motion. If the structure of the body is known and from this the heat motion can be judged, then conversely the molar heat may be estimated.

Theoretical calculations show that starting from absolute zero the molar heat should increase with rising temperature and in general approach a limiting value. The magnitude of the limiting value depends on the structure of the substance, and the approach to it occurs with varying rapidity for different substances.

For crystallized elements made up of atoms which vibrate freely about their equilibrium positions in the crystal, the limiting molar heat should be $c_v = 3R = 24.9$ J deg^{-1} mol^{-1}. The condition that the vibrations shall be free is most nearly fulfilled if the forces between the atoms are weak and the temperature is near the melting point. The substance is then often quite soft. For soft elements (metals) with comparatively low melting point, c_v thus already lies near to $3R$ at room temperature. For high melting substances, for example carbon (both as diamond and graphite), boron, and silicon, the forces between the atoms at room temperature are very strong and c_v then lies far below $3R$. At sufficiently high temperature c_v may approach $3R$ also for such substances.

In this way we may account for the rule proposed in 1819 by Dulong and Petit, that the atomic heat of a solid element at room temperature often is about 6.2 cal deg^{-1} mol^{-1} = 26 J deg^{-1} mol^{-1}. Since Dulong and Petit referred to the c_p value, and $c_p - c_v$ for solid bodies on the average is 0.8 J deg^{-1} mol^{-1}, the rule implies that in round numbers $c_v = 25$ J deg^{-1} mol^{-1}. Thus, the rule agrees well with the limiting value $3R$.

2-3. Chemical equilibrium

a. The driving force of a process. Because many processes proceed at negligible velocity (1-5 b), it is very important to be able to decide whether an apparently unchanging system is in equilibrium, or has a tendency to go over to another state. This would be possible if the driving force of the process were known.

The fact that many systems that are known to be unstable because of a low rate of reaction do not change noticeably, shows that this rate cannot be a measure of the driving force. In the mid-19th century it was believed that such a measure, on the other hand, was to be found in the heat of reaction. A large development of heat during a reaction, thus a high heat of reaction, should signify a strong tendency to react. It was generally thought that a spontaneous reaction was possible if heat was evolved, and impossible if it required absorption of heat. Expressed in language of a later time, this means that at constant pressure, the tendency to go to a state with lower enthalpy H would be the driving force. A process would thus be possible if it implied $\Delta H < 0$. For $\Delta H > 0$ the process would be impossible, but of course, in that case the reverse reaction would be possible. Equilibrium would then be achieved when $\Delta H = 0$.

This interpretation, however, cannot be correct. Many processes occur spontaneously at constant pressure and temperature with the absorption of heat, thus with $\Delta H > 0$. Such, for example, are the melting of solid bodies, many dissolution processes, evaporation of liquids, and many other reactions in which gases are formed (for example, the decompositions $2HgO(s) \rightarrow 2Hg(l) + O_2(g)$ and $CaCO_3(s) \rightarrow CaO(s) + CO_2(g)$).

Gibbs found that ΔH certainly is included as a term in the expression for the driving force, but that this expression also contains a term which depends on the difference in *probability* between the initial and final states. The probability of a state is closely related to its *disorder*. In a crystal the building units of the substance (atoms or atomic groups) are distributed with a high degree of order. When the crystal melts, the disorder increases, and if vaporization occurs, the disorder becomes still greater. It is obvious that the distribution of the building units is the more probable, the greater the disorder is.

If one looks only at the positions of the building units, an immense number of such positions must be defined in order that a particular crystalline state shall be realized. In a gas on the other hand, the molecules have the freedom to assume a multitude of different positions without altering the state of the gas. Naturally, we must consider here a system large enough so that statistical laws apply. Expressed in another way, the building units find it enormously easier to locate themselves to form a particular gaseous state, than to form a particular crystalline state. The gas may be compared to a large, disordered crowd of people in which the individuals may be distributed in innumerable ways without changing the character of the crowd as a whole. The crystal, on the other hand, is like a troop of soldiers arranged in ordered formation, in which every place must be maintained. The fixed formation is also much more difficult to form than the disordered crowd.

If a gas vessel is divided into two halves by a septum and a gas is contained in one half, the gas immediately spreads through the whole vessel if the septum is removed. This state is tremendously more probable than the more ordered state in which all the molecules remain in one half. Here it is especially easy to see that it is the great number of molecules in a real system which ensures that deviations from a uniform distribution do not become measurable. If the system contains only 2 molecules, the probability that both molecules in a certain instant would be found in one half is $(\frac{1}{2})^2 = \frac{1}{4}$. If the system contains 20 gas molecules, the corresponding probability is $(\frac{1}{2})^{20}$, that is, somewhat less than $1/10^6$. If it contains 3×10^{19} gas molecules, which corresponds to 1 cm³ of gas at 0°C and 1 atm, the probability that the number of molecules in the two halves should be different by 1% is so small that on the average, this condition would be encountered at intervals of $10^{10^{10}}$ years. All processes which involve evening out of pressure and temperature, and all diffusion processes, are examples of transitions to more probable states.

The probability (disorder) of a system is defined by the function of state *entropy* S (Gk. τροπή, turning, conversion; the entropy defines the possibility of converting

Energy

heat to work). The greater the probability and disorder is, the greater is the entropy. This should give a quite clear picture of the concept of entropy. Entropy has the dimension energy/temperature (for example, J deg^{-1}).

b. Gibbs free energy. Gibbs introduced with the help of entropy a further function of state, which directly determines the direction of a process. This function of state is called nowadays the *Gibbs free energy* and is indicated by G. It is defined by (the last part is obtained from (2.5)):

$$G = H - TS = U + PV - TS \qquad (2.8)$$

Since G is composed exclusively of functions of state, it is itself a function of state. It is seen from (2.8) that G has the dimension of energy.

For processes at constant volume an analogous function of state is also used, the *Helmholtz free energy* $A = U - TS$. From (2.8) it is seen that $G = A + PV$, that is, G and A are related to each other analogously as H and U according to (2.5). However A has less practical significance than G because processes at constant volume are less common than processes at constant pressure.

The names and symbols for the two functions G and A have varied. G as well as A have been called free energy and symbolized by F. G has also been called Gibbs' function or thermodynamic potential; A has been called maximum work or work content.

For a change of state at constant pressure and temperature (2.8) gives:

$$\Delta G = \Delta H - T\Delta S = \Delta U + P\Delta V - T\Delta S. \qquad (2.9)$$

(2.9) shows that ΔG is composed of the "energy term" ΔH and the "disorder term" $T\Delta S$. The latter decreases ΔG if $\Delta S > 0$, that is, if the system during a process goes toward a state with greater disorder than the initial state.

It is the tendency to go over to a state with lower G which is the driving force for any spontaneous process at constant pressure and temperature. The sign of ΔG determines thus whether or not a process can take place. If the transition of a system from state A to state B is written as A→B, then $\Delta G = G_B - G_A$. The process may occur spontaneously if $\Delta G < 0$. If $\Delta G > 0$, on the other hand, it cannot take place spontaneously. Then instead the process A←B may occur spontaneously. At equilibrium between the two states,

$$\Delta G = 0. \qquad (2.10)$$

It is shown later (10b) how ΔG can be determined for a process.

In 1-5b,c the instability of the gas mixture hydrogen+oxygen was discussed. In this case we have at 25°C and 1 atm:

$$H_2(g) + \tfrac{1}{2}O_2(g) \rightarrow H_2O(g); \quad \Delta G = -228.7 \text{ kJ}$$

The system thus tries to go to the right and the gas mixture on the left side is therefore unstable in spite of the fact that no reaction is noticeable (low reaction

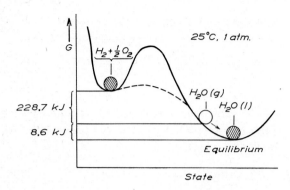

Fig. 2.1. Gibbs free energy G and equilibrium. (The figure is not to scale along the G axis.)

rate, arrested reaction). $H_2O(g)$, however, is not the most stable state for H_2O at 25°C, 1 atm. We have, in fact:

$$H_2O(g) \rightarrow H_2O(l); \quad \Delta G = -8.6 \text{ kJ}.$$

Disregarding the possibility of supersaturation, which can easily be set aside (supersaturation indeed implies an arrested reaction), the system therefore goes further to the state $H_2O(l)$.

In fig. 2.1 an attempt has been made to depict these circumstances very schematically. G is taken as a function of the state of the system, which is assumed to vary along the x-axis. The system is stable as $H_2O(l)$ at the absolute minimum of G and easily goes to this state, for example, from the unstable state $H_2O(g)$. The state $(H_2+\frac{1}{2}O_2)$ is also unstable, but corresponds to a relative minimum in G. The *energy threshold* (wall, barrier) between the two states is an expression of the arrest of reaction in this case and the system requires a certain added supply of energy to rise over the threshold. At 25°C the thermal energy of the molecules does not suffice, but if the thermal energy is increased by raising the temperature, the top of the threshold is finally reached. The system then coasts down toward the absolute minimum while the recently supplied energy is released again. At about 500°C the formation of water takes place with explosive rapidity.

A catalyst can initiate an arrested reaction by opening up a new path for the reaction, which involves a lower energy threshold (the threshold can be gone around through a pass, dashed in fig. 2.1). The catalyst may also act by giving the original reacting system the amount of energy which is required to lift it over the threshold, and then reabsorbing the same amount.

As H_2O has a lower G value than $(H_2+\frac{1}{2}O_2)$, H_2O has no tendency to decompose. For this decomposition energy must be added, for example in the form of electrical energy in the electrolysis of water.

For the first four of the following examples, the temperature is presumed to be 25°C. The pressure in all cases is 1 atm.

$$N_2(g) + O_2(g) \rightarrow 2NO(g); \quad \Delta G = 172.9 \text{ kJ}$$

Here the system tries to go to the state at the left, which implies that NO is unstable. In spite of that, NO does not decompose at a noticeable rate at 25°C.

$$C(\text{graphite}) + \tfrac{1}{2}O_2(g) \rightarrow CO(g); \quad \Delta G = -137.3 \text{ kJ}; \quad \Delta H = -110.5 \text{ kJ}$$

CO is clearly stable with respect to decomposition into the elements under the given conditions. ΔG departs quite substantially from ΔH, as a result for the most part of the doubling of the number of gas molecules in the reaction to the right. This leads to an appreciable increase in entropy. From (2.9) we find ($T = 298°K$) $\Delta S = 0.090$ kJ deg^{-1}.

$$C(\text{graphite}) + CO_2(g) \rightarrow 2CO(g); \quad \Delta G = 119.9 \text{ kJ}$$

CO is thus unstable under the same conditions with respect to decomposition into graphite and CO_2. However, the rate of decomposition at room temperature is negligible. A comparison with the previous example shows that in order to be sure that a certain state is stable, one must find out whether its G value is less than that for all other conceivable states. In a statement of stability or instability it must therefore also be stated what process is involved if it is not clear from the context.

$$C(\text{graphite}) + O_2(g) \rightarrow CO_2(g); \quad \Delta G = -394.4 \text{ kJ}; \quad \Delta H = -393.5 \text{ kJ}$$

CO_2 is thus stable with respect to decomposition into the elements at 25°C and 1 atm. Here we have $\Delta S = 0.003$ kJ deg^{-1}. This low value, which makes $\Delta G \approx \Delta H$, is related to the fact that the number of gas molecules is not changed. In such cases as this, one may also decide the direction of reaction from ΔH with sufficient certainty.

As soon as one is convinced that ΔH has the same sign as ΔG it is possible to determine the direction of the reaction from the sign of ΔH. The greater the numerical value of ΔH is, the less is the risk that subtraction of the term $T\Delta S$ will cause a change in sign. The method is especially safe if one wishes only to *compare* the behavior of closely related substances under analogous conditions. $T\Delta S$ may then be considered to be nearly the same for the different cases. $P\Delta V$ then also becomes nearly the same, so that one may also compare ΔU values (see, for example, 6-2e).

$$S(\text{orthorhombic}) \rightarrow S(\text{monoclinic}); \quad \begin{aligned} \Delta G_{298} &= 0.096 \text{ kJ} \\ \Delta G_{368.7} &= 0 \quad \text{kJ} \\ \Delta G_{383} &= -0.018 \text{ kJ} \end{aligned}$$

At $T = 368.7°K$ (95.5°C) and 1 atm orthorhombic and monoclinic sulfur are in equilibrium with each other (transition point), which means that $\Delta G = 0$. At lower temperature the orthorhombic modification is stable, as seen from the fact that $\Delta G > 0$ for the process as it is written above. At temperatures above the transition point, the monoclinic modification is stable and $\Delta G < 0$.

ΔG values for a large number of reactions are found listed in the books mentioned in 2-2a, although there they are sometimes indicated by ΔF. It is especially the ΔG values for the formation of substances from the elements in the standard state which are tabulated. Their ΔG^0 (or ΔF^0) values become the molar Gibbs free energies on an accepted G scale and can be treated analogously with what was described in 2-2a for ΔH^0 values.

The books mentioned give a compilation of a large number of thermodynamic data of various kinds.

Everything that has been said in 2-3a and b presumes that the number of building units is so great that statistical laws apply. Such is always the case for processes between phases in the sense defined in 1-1b and c. For processes between individual atoms and molecules, on the other hand, the direction of the process is determined only by the sign of the energy change ΔU (or, as it is then most often written, ΔE). Such a process between the states A and B thus goes in the direction A→B if $E_A > E_B$, that is, if $\Delta E = E_B - E_A < 0$.

c. Change of entropy with phase transition. Since $\Delta G = 0$ when two phases are in equilibrium with each other, according to (2.9) $\Delta S = \Delta H/T$ at the transition point. ΔS is thus easily calculated from the heat of transition.

For the conversion liquid → gas at the boiling point, ΔS in a great many cases lies in the vicinity of 88 J deg^{-1}mol^{-1} = 21 cal deg^{-1}mol^{-1}. This circumstance was pointed out by Trouton as early as 1884 (*Trouton's rule*) and results from the fact that the difference in disorder between the liquid phase and the gas phase at the boiling point is often approximately the same for different substances. If the freely moving units are different in the two phases, departures from this rule occur. Commonly there is strong association in the liquid phase, so that the disorder in the liquid is less than normal. The ΔS value then becomes larger than that given above. An example of this is water with $\Delta S = 109$ J deg^{-1}mol^{-1} = 26.0 cal deg^{-1}mol^{-1}.

ΔS per mole on melting is naturally less than on vaporization, but in addition, does not show the same regularity. The increase in disorder on melting may vary so much that this is not surprising. For substances which both in the crystal and the melt consist of atoms or nearly globular molecules of a single kind (Hg, Zn, Fe, W, CCl$_4$), however, melting involves a quite small and rather constant increase in disorder. Here ΔS is often close to 8.5 J deg^{-1}mol^{-1} = 2 cal deg^{-1}mol^{-1}. An irregular molecule is incorporated into a crystal structure in so special a manner that great order, that is low disorder, is involved. The increase in disorder, that is ΔS, on melting then becomes greater. The same applies to a crystal such as NaCl, in which the alternating position of the two kinds of ion in the structure implies a greater order than if the structure contained only one kind of atom.

d. Chemical potential. The equilibrium condition $\Delta G = 0$ in the last example in 2-3b implies that G for the same amount of sulfur, in this case 1 mole, is the same in the two states. The condition can therefore also be written g (orthorhombic) = g (monoclinic).

In the same way, for H$_2$O at 0°C and 1 atm, $g(s) = g(l)$. If the temperature is raised above 0°C, then $g(s) > g(l)$ and the ice transforms to water. The inverse holds for a decrease in temperature.

Equilibrium between ice, water and water vapor (at $+0.01$°C and 6.105 mbar, see 13-4a) implies the condition $g(s) = g(l) = g(g)$.

G for 1 mole, that is g, may be considered as a measure of the tendency of a substance to go from one phase to another. As soon as g is larger for one phase than for another, the substance transforms to the phase with lower g.

This applies to phases which consist of pure substances. If the substance enters into a solution, the corresponding partial molar quantity must be used. Because of its great significance this quantity has been given the special name *chemical potential*. In addition to g the symbol μ is also used. In accord with what was stated in 1-5e the chemical potential of a substance "i" is thus defined:

$$\mu_i = g_i = \left(\frac{\partial G}{\partial n_i}\right)_{P,\ T,\ \text{other}\ n} \tag{2.11}$$

The chemical potential μ_i of the substance "i" in a certain solution thus is the increase in G on addition of 1 mole "i" to an infinitely large amount of solution.

If a phase contains the mole numbers n_1, n_2, ... of the substances 1, 2, ..., G for the phase is composed analogously to (1.14) as follows:

$$G = n_1\mu_1 + n_2\mu_2 + \ldots \tag{2.12}$$

If the substance "i" is present in the phases indicated by ', '', ''', ... and in them has the chemical potentials μ_i', μ_i'', μ_i''', ..., then at equilibrium.

$$\mu_i' = \mu_i'' = \mu_i''' = \ldots \tag{2.13}$$

Whether or not one or several of the phases consists of the pure substance is unimportant. The pure substance is indeed only a limiting case for which every partial molar quantity becomes equal to the corresponding molar quantity.

As soon as a gas phase or a solid or liquid solution takes part in a process, the system may often reach equilibrium without the process having gone to completion. A simple example is the conversion of a liquid to a gas, which ceases when the concentration in the gas phase has reached a particular value (cf. the discussion in 1-5c). From an energy viewpoint, it is clear that the continuous change of composition and properties which are possible in gas phases and solutions can result in equilibrium being reached at an intermediate stage.

Examples. Water and tetrachloromethane (carbon tetrachloride, CCl_4) do not dissolve in each other. Both liquids, on the other hand, dissolve iodine, which then enters as I_2 molecules. If a water solution of iodine is shaken with tetrachloromethane, I_2 molecules pass from the water phase to the tetrachloromethane phase. Thereby, $\mu_{I_2}(H_2O$ soln.) decreases and $\mu_{I_2}(CCl_4$ soln.) increases until the two μ_{I_2} values are equal, which signifies equilibrium.

A saturated water solution of NaCl, which is thus in equilibrium with solid NaCl, is 5.3 molar at 25°C. Accordingly, at 25°C, $\mu_{NaCl}(s) = \mu_{NaCl}$ (5.3 M NaCl in H_2O). At the same temperature μ_{NaCl} is less than this value in any more dilute solution. Solid NaCl thus goes into such a solution until this value is reached.

In chapter 10 it is shown how chemical potential is used to define the activity of a substance and thereby to determine equilibrium conditions.

CHAPTER 3

Quantum mechanics

In the study of systems that consist of particles with masses as small as those of elementary particles, atoms and moderately large molecules, it is found that the energy of such systems can only have definite values, corresponding to particular *stationary states* of the systems in question. This has fundamental significance in the treatment of the exchange of energy in such systems and for the interpretation of their properties, for example, the bonding forces which appear. The theoretical foundation for the work in this area has been provided by *wave* or *quantum mechanics*.

a. Matter waves. In 1-2b was discussed the way in which electromagnetic radiation may be considered either as a wave motion characterized by the wavelength λ or frequency $\nu = c/\lambda$, or as consisting of particles with the momentum $p = h/\lambda$ (1.3). Recognizing this, it was natural to ask whether this dual particle–wave nature was universal and thus would be found for material particles, that is, particles with rest mass. L. de Broglie showed in 1924 that such is the case and found that the relation (1.3) also must apply to such particles. A particle with momentum p should therefore have the properties of a wave with wavelength

$$\lambda = h/p \tag{3.1}$$

The correctness of this expression was confirmed in 1927 by the discovery that an electron beam in striking a crystal gives rise to interference phenomena. The wavelength calculated from this effect agreed with the λ value obtained from (3.1) by setting $p = mv$, where m is the mass of the electrons and v their velocity in the experiment (if the velocity v is small in comparison with the velocity of light the rest mass may be used for m, otherwise the mass which corresponds with the velocity according to the relativity theory—see 1-2a—must be used).

Later, interference phenomena were found also in beams of other particles such as neutrons, protons and some light atoms and molecules. A wave nature is clearly connected with all matter (*matter waves*). In the macroscopic world, however, this is not observed. The large masses encountered here correspond, according to (3.1), to such small wavelengths that no wave phenomenon of this type can be experimentally demonstrated.

b. The uncertainty principle. The attempt to explain matter waves following 1925, primarily through the work of Heisenberg, Born, Schrödinger and Dirac, led to *wave* or *quantum mechanics*. A very important principle in quantum mechanics is the *uncertainty principle*, which was stated in 1927 by Heisenberg. Earlier it had been thought that, in the same way as in classical mechanics, such quantities as position, momentum and energy at a certain instant could be ascribed to the small particles of atomic physics. Heisenberg showed that this is not possible, however. Any measurement of these properties of a particle produces an interaction between the measuring apparatus and the object, and the properties of the latter become altered. This influence takes place with every measurement, but becomes of significance only when one works with the small systems of atomic physics.

Let us assume that there is a microscope that makes possible the observation of a particle, for example an electron. If it is desired to determine the position of the particle with this microscope, the particle must be illuminated, that is, struck by radiation. However, we always have an uncertainty in the determination of position that is of the same order of magnitude as the wavelength of the illuminating radiation. If the position coordinate x is measured, the uncertainty in this measurement thus becomes $\Delta x \approx \lambda$. The determination of position becomes more accurate, the shorter the wavelength of the radiation used.

But the photons in this radiation possess momentum (1-2b) and therefore change the momentum of the electron when they collide with it. According to (1.3), the momentum of a photon is $p = h/\lambda$ and the momentum of the electron should on collision be altered by a value of this order of magnitude. The momentum of an electron therefore has an uncertainty of $\Delta p \approx h/\lambda$ in the observation. Thus, the shorter the wavelength chosen to obtain a more accurate measure of position, the greater the uncertainty becomes in the value of the momentum. Multiplication of Δx and Δp gives the uncertainty relationship

$$\Delta p \Delta x \approx h \qquad (3.2)$$

For a particle of mass m, $p = mv$, and if m is constant, then $\Delta p = m\Delta v$. Substituting this in (3.2) we get

$$\Delta v \Delta x \approx h/m \qquad (3.3)$$

The small value of $h = 6.6256 \times 10^{-34}$ Js causes the product $\Delta v \Delta x$ to become very small for all ponderable bodies. For these, therefore, it has no practical significance. For a dust particle with $m = 10^{-10}$ kg ($= 0.1$ μg), $\Delta v \Delta x \approx 10^{-23}$. If the position of the dust particle can be measured so accurately that $\Delta x = 10^{-7}$ m ($= 0.1$ μm), the fundamental uncertainty in a measurement of velocity thus becomes $\Delta v \approx 10^{-16}$ m s^{-1}, an exceedingly small value.

However, for an electron with $m_e = 9.1091 \times 10^{-31}$ kg, $\Delta v \Delta x \approx 10^{-3}$. If once more $\Delta x = 10^{-7}$ m, now we find that $\Delta v = 10^4$ m s^{-1}, that is, the velocity of the electron becomes very indefinite.

If one of the two Δ values is 0, that is, the corresponding quantity is exactly known, the other Δ value becomes infinitely large, that is, the corresponding quantity is completely indeterminate. If the position of the particle were exactly known, the velocity would be wholly indeterminate, and vice versa.

Certain other pairs of quantities also are connected to each other by an uncertainty relationship analogous to (3.2) (a necessary condition, of course, is that the product of the two quantities shall have the same dimension as h, that is, $m\, l^2\, t^{-1}$). A very important such pair is energy–time, for which applies the relation

$$\Delta E \Delta t \approx h \tag{3.4}$$

Thus, if the time is defined with ever greater accuracy, the energy of the particle becomes more and more indefinite. If instead it is required that the particle have a definite energy, then time becomes indefinite. In other words, one cannot say when the particle has a particular energy value.

A very important case is met with when a particle is restricted by some force to a bounded region. A model is commonly conceived in which the particle is enclosed in a completely impenetrable box. For the sake of simplicity, it is assumed that the particle lies only on one straight path between two opposite walls of the box. The box may thus be said to be one-dimensional. The length of the path is assumed to be a and the position of the particle is given as the distance x from one wall. If it is not demanded that the position of the particle be determined more closely than that it should be found somewhere on the path a, then $\Delta x = a$, which, according to (3.3), gives $\Delta v \approx h/ma$. The particle thus cannot be at complete rest, for that would imply that the velocity is completely determined, that is, $\Delta v = 0$. The particle confined to the box must therefore have a certain motion, and this will never cease even at absolute zero. The corresponding *zero-point energy* is the smallest energy which the system can have.

For all macroscopic systems m and a are large in relation to h, so that Δv acquires so small a value that no zero point energy can be observed. On the other hand, it is appreciable (and can be indirectly demonstrated) for the various elementary particles which are bound to the nucleus and electron cloud of an atom. These may be compared with the particle in a box, but in this case, both m and a are very small, hence Δv becomes large.

The uncertainty principle acquires very great basic significance in quantum mechanics, but it might be objected that its validity always presupposes an observation. It might be possible to imagine a particle which is not observed as having both definite momentum and definite position at the same time. To this it may be answered that physics can only work with quantities which are basically measurable. To ask how matter behaves when it is not observed is meaningless.

Since in principle an electron can never be followed in its motion, the changes in state and destiny of a particular electron in a system with several electrons can never be ascertained. The electrons thus completely lose their individuality. Two electrons may change places, for example, without any possibility that the exchange can be discovered.

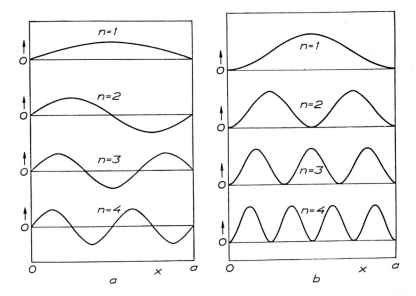

Fig. 3.1. (a) The wave function ψ_n, (b) the particle density ψ_n^2 for the "particle in a box".

c. Some basic concepts of quantum mechanics.[1] Quantum mechanics describes the energy state of a particle or a system of particles with an equation that formally is very similar to the equation for wave motion. It is therefore called the *wave equation* (also *Schrödinger equation*). In the wave equation is included the total energy of the system E and a function ψ of the space coordinates of the system (the *wave function*). The wave equation has an infinite number of solutions corresponding to different wave functions ψ_n. In order that a wave function shall correspond to a physical reality and thus be valid as a solution of the wave equation, it must be finite, unique and continuous within the entire region of space over which the system can range. These conditions are common for functions which represent physical quantities, and are sufficient in this case to give very important results. For each wave function ψ_n there corresponds a definite energy value E_n.

The model of the particle in the impenetrable box can be used again as an example. Here it is assumed in addition that the potential energy of the particle is zero for all x values within the box. The particle thus has only kinetic energy. The wave equation then becomes

$$\frac{d^2\psi}{dx^2} + \frac{8\pi^2 m}{h^2} E\psi = 0 \qquad (3.5)$$

---

[1] The equations in 3c serve only to illustrate the discussion, and are not intended to be memorized.

where m is the mass of the particle and h is Planck's constant. E is the total energy of the particle, which in this case consists of its kinetic energy. Possible solutions of (3.5) may be written

$$\psi_n = \sqrt{\frac{2}{a}} \sin \frac{n\pi x}{a} \tag{3.6}$$

The conditions given above in order that the wave function shall be valid as a solution of the wave equation are only fulfilled if n is a whole number, that is, $n = 1, 2, 3, \ldots$. The exclusion of $n = 0$ is seen to follow from the fact that this value makes $\psi_n = 0$ for all x values This is absurd since it means that the particle would then not exist (cf. below).

Fig. 3.1 a shows how ψ_n varies with x for $n = 1$ to 4. It is apparent that the fact that n is a whole number implies that the wave function has a whole number of half periods in the path length a.

The wave equation gives for each of these wave functions ψ_n a certain value for the total energy E_n, which is

$$E_n = \frac{n^2 h^2}{8ma^2}; \quad n = 1, 2, 3, \ldots \tag{3.7}$$

The total energy can thus only take on values that are determined by the whole numbers n, and which in this case are proportional to squares of whole numbers. The whole number n is called the *quantum number* of this particle system. The various energies E_n are often called *energy levels*. The particle system has the lowest energy for the smallest quantum number (here $n = 1$) and is then said to be in its *ground state* or *normal state*. At higher energies the system is *excited*. As $n \neq 0$ the expression (3.7) gives $E \neq 0$, which means that a particle must have zero point energy as was stated in 3 b.

When the particle has a certain energy E_n, determined by the quantum number n, it also has a certain velocity. This velocity corresponds, according to (3.1), to a certain wavelength, which can be shown to be equal to $2a/n$; this wavelength appears again in ψ_n in (3.6) and fig. 3.1.

From what has been said, we see how naturally quantum mechanics explains the integral quantum numbers and the energies determined by them. Schrödinger pointed out in his early work that the whole numbers enter in the same natural way as they do in connection with the standing waves in a vibrating string. The length of the string is always divided into a whole number of half wavelengths. It must be kept in mind, however, that this analogy is only a formal one.

For a definite particle position the velocity of the particle, and thus the energy, is completely indeterminate, and for a particular energy, the position is indeterminate. On the other hand, we may specify how the *probability* that a particle will be encountered in a certain energy state varies as we move along a. This probability is, in fact, proportional to ψ_n^2. The behavior of ψ_n^2 is represented in fig. 3.1*b*. While ψ_n varies between positive and negative values, ψ_n^2 is, of course, always positive. The x values for which ψ_n and $\psi_n^2 = 0$ are called *nodes*.

Since the wave function ψ_n formally has wave character (cf. fig. 3.1*a*), it is usually complex, as is generally the case with wave expressions. The above-mentioned probability is then proportional to the square of the magnitude $|\psi_n|$ of the wave function, that is, to $|\psi_n|^2$. However, we will continue to write ψ_n^2.

It is also customary to say that ψ_n^2 gives the *particle density*. The sum of the products (number) × (time interval) for all particles with a given energy that during a sufficiently long time of observation are found within a given small element of a, is an expression of the particle density within the element and is proportional to ψ_n^2. Let us assume that one could photograph the particle with a camera the instant it had a given energy E_n, but that there is no effect on the photographic emulsion at other energies. If the exposure time is sufficiently long, a blackening would appear along a which varied according to the appropriate curve ψ_n^2 in fig. 3.1*b*.

If the particle moves in a three-dimensional box, analogous but more complicated expressions are obtained. The state is then defined by three quantum numbers. It appears that as soon as the particles in a system are for some reason confined to a limited region, the system can only have particular energy values (energy levels) defined by one or several quantum numbers. In the example the walls of the box have confined the particle to a limited region. Any "quantization" of the translation energy of particles enclosed in real "boxes" or vessels, however, cannot be demonstrated experimentally because the dimensions of the vessel are too large. This is also shown by (3.7). If a is large, even for the small m values of elementary particles and atoms, the differences between neighboring E_n values become so small that transfer between them appears to be continuous. On the other hand, the quantization of energy can be demonstrated for those particle systems that are contained within the small region such as is encompassed by an atom. For the electron cloud this question is treated in more detail in chap. 4.

Two other types of energy quantization should be mentioned. If two atoms with masses m_1 and m_2 are bound to each other, they (*vibrate*) oscillate against each other about the mean position with a definite frequency ν_{vib}. The amplitude of vibration can however, only take on definite values which correspond to the particular energy values

$$E_{\text{vib}} = h\nu_{\text{vib}}(v + \tfrac{1}{2}) = \frac{h}{2\pi}\sqrt{k\frac{m_1 + m_2}{m_1 m_2}}(v + \tfrac{1}{2}); \quad v = 0, 1, 2, \ldots \tag{3.8}$$

v is the *vibrational quantum number*. k is a constant (the *force constant*) for the system in question, which is larger, the more strongly the atoms are bound together.

For *rotation* of a molecule about an axis only particular rotational velocities can occur. These correspond to the energy values

$$E_{\text{rot}} = \frac{h^2}{8\pi^2 I} j(j+1); \quad j = 0, 1, 2, \ldots \tag{3.9}$$

j is the *rotational quantum number*. I is the moment of inertia of the molecule with respect to the rotation axis in question. (If the masses $m_1, m_2, \ldots$ lie at distances $r_1, r_2, \ldots$ from the rotation axis, this moment of inertia is $I = m_1 r_1^2 + m_2 r_2^2 + \ldots$).

The expressions (3.8) and (3.9) apply to any system which vibrates (*oscillator*) or rotates (*rotator*). If the system is macroscopic, however, the large masses lead to such small differences between neighboring E values that no quantization can be detected.

The expression (3.8) remains positive, but (3.9) becomes zero for the lowest quantum number (0). The vibrations thus have zero-point energy while rotations do not. This means that rotation but not vibration may cease altogether.

Quantum mechanics explains the wave phenomena in particles through the wave character of the wave function ψ. The fact that the value of ψ^2 at a certain point is proportional to the probability that a particle will be found at this point removes the difficulty in explaining the interference phenomenon in particle rays. Earlier it was not possible to understand how two particles through interference were able in certain directions to annul each other so that the intensity became zero. According to quantum mechanics the whole phenomenon depends on the varying probability of encountering particles, for example electrons, at different places. In principle, it is possible to obtain an interference picture on a photographic plate even when the radiation intensity is so low that only one particle at a time passes the grating and reaches the plate. The interference picture shows only how the probability of incidence varies over the surface of the plate. This variation may also be calculated by quantum mechanics. On the other hand, it is never possible to predict at what point on the plate an individual particle will fall.

d. Energy exchange and spectra. If the energy of a system of particles (atoms or molecules) can only correspond to definite energy levels, energy can only be absorbed or given off in definite amounts, corresponding to the differences between energy levels. The system then takes up or gives off quanta $h\nu$, which are equal to these amounts of energy. If the system goes from a state with energy E_A to a state with energy E_B, we thus have

$$E_B - E_A = \Delta E; \quad |\Delta E| = h\nu \tag{3.10}$$

Quanta with frequency ν are *given off* if $\Delta E < 0$ and *taken up* if $\Delta E > 0$. For radiant energy one speaks mostly of *emission* in the first case, and *absorption* in the second.

If radiation is exchanged with definite frequencies (*spectral lines*), this shows the presence of energy levels. The frequencies of the spectral lines give the differences in energy between energy levels in the atomic system. These values give fundamental information about the structure of the system.

The addition of energy happens usually through electromagnetic radiation or through collisions. The collisions are caused by the heat motion of neighboring atoms (conduction of heat energy) or by particles that are emitted by a decaying atomic nucleus or accelerated in an electric field.

If the energy of an atom or molecule is to be raised from level E_A to level E_B, a quantum is required which has an energy that, according to (3.10), is at least

Quantum Mechanics

equal to ΔE. Quanta with lower frequency and energy are completely without effect even if they arrive in such numbers that their total energy exceeds the value ΔE. However, if more than one quantum occurs at the same time within a sufficiently small volume in the energy absorber, the latter can take up the sum of the energies of these quanta. Such an effect was observed in 1962 in the form of direct two-photon absorption on irradiation with an extremely high light intensity from a so-called laser.

If the addition of energy occurs through collisions, the conditions are analogous. A colliding particle must have kinetic energy of at least the magnitude of ΔE in order that any raising of energy can take place. Collisions of many particles with lower energy have no effect.

The vibrational energy of a system, which vibrates with the frequency ν_{vib}, according to (3.8), is $E_{vib} = h\nu_{vib}(v + \frac{1}{2})$. The energy difference between two states with the quantum numbers v and $(v + 1)$ is thus, in this case, $\Delta E_{vib} = h\nu_{vib}$. The frequency of the radiation that is exchanged by the vibrating system is thus equal to the vibrational frequency ν_{vib}.

Do not confuse the quantization of all radiation with the quantization of energy in a particle system! Let us assume that the vibrating system just mentioned has the energy E_{vib} and is bathed in radiation of all possible frequencies ν. This radiation thus contains quanta $h\nu$ with different energies. Because the energy of the vibrating system is quantized, however, in order to raise its energy from one energy level to the next, it can only take up from all these quanta those for which $h\nu \geqslant h\nu_{vib}$.

If the absorbed energy is held by the atomic system, this can lead to increased heat motion, that is, a rise in temperature. The energy may be released again, in the form of radiation for example, or be used in some chemical process.

When radiation with uniform intensity within a large frequency range strikes an atomic system, the system takes from the radiation those quanta with just the energy values that will raise the energy of the system to higher levels. The radiation which passes through the system is weakened at those frequencies that correspond to these quanta (*selective absorption*) and shows an *absorption spectrum*.

If the absorbed radiation is released again as radiation, this occurs in the form of quanta with energy values which signify the decrease of energy to lower levels. The corresponding frequencies form an *emission spectrum*.

Within a certain energy range the energy levels become more numerous and more dense, the more complicated the energy exchanging system is. If this consists of free atoms (a monatomic gas of not too high pressure) the energy levels are sparse. *Atomic spectra* are obtained with sharp, clearly resolved spectral lines (*line spectra*). If the gas consists of molecules *molecular spectra* are obtained with decidedly greater line density. The lines are often clustered in particularly dense groups, which since early times have been called *bands* (the spectra become *band spectra*), even though with modern spectrographs it is possible in general to resolve them into lines. (See further 7-2.)

If the gas is strongly compressed, the atoms and molecules begin to influence each other throughout the whole gas mass, and thereby as far as energy exchange is concerned takes on more the character of a single unified system. From this viewpoint it may be compared with a giant molecule. This leads to the result that the energy levels in the optical spectral range lie so densely spaced that they form bands in which individual lines cannot be separated. Spectra of liquid and solid bodies, in which the mutual influence of the atoms is greater still, also show continuous bands (6-4c). At sufficiently high temperatures this influence becomes so great that the bands run together into a continuous spectrum.

When the added energy is given off in the form of radiation of another kind than purely thermal radiation, one speaks of *luminescence*. If the energy is added as radiation and the energy of the system afterward drops directly to the initial level, the energy and frequency of the luminescent radiation become equal with those of the absorbed radiation. On the other hand, if the drop in energy takes place stepwise through intermediate energy levels, the energy and frequency of the luminescent radiation become smaller. Excited atoms and atomic systems in general give off the energy very fast, and their lifetimes are often of the order of 10^{-8} sec. The luminescence is then quenched practically in the same instant that the illuminating radiation ceases (*fluorescence*). In certain cases, the quenching may be slower so that an afterglow, *phosphorescence*, appears.

Energy released by a chemical process in the system may, of course, lead to a rise in temperature, and thereby to increased thermal radiation. But it can also happen that the radiation covers a limited frequency range, in which the radiation intensity thus becomes greater than that corresponding to the temperature of the body. The body may, for example, emit visible light at room temperature ("cold light"). This phenomenon is called *chemiluminescence*. In this way, phosphorous glows on oxidation in air (which thus, according to presently used terminology, is not phosphorescence). Many slow oxidations of organic substances result in chemiluminescence (glowing wood, glowworms, etc.).

CHAPTER 4

The Electron Cloud of the Atom and the Periodic System

4-1. State of the electrons in the electron cloud

a. Older atomic theories. Rutherford believed in 1911 that the electrons in the electron envelope were bound to the positive nucleus and revolved around it in circular paths. The similarity to a planetary system was thus striking; gravitation was replaced, however, by electrostatic forces.

Rutherford's atomic model could not explain how an atom can emit radiation of fixed energy (sharp spectral lines), nor also the stability of the atom. According to the classical laws for the electromagnetic field, the electron must send out radiation throughout its entire course of movement and thus lose energy. This would lead to a steady decrease in the radius of its orbit, so that it would finally fall into the nucleus. The shrinking of the path radius would in addition cause an ever-increasing frequency of revolution, which would lead the emitted radiation to increase its frequency continuously.

This difficulty was resolved by Bohr, who in 1913 proposed the hypothesis that the electrons in an atom can only be found in certain *stationary states*, corresponding to fixed energy levels E_1, E_2, etc. States with intermediate energies are impossible. When an electron shifts from one state to another, an amount of energy is given off or taken up (most often as a light quantum) corresponding to the energy difference; this can thus only have certain fixed values and no intermediate values. The electron cannot go below the state with the lowest energy. This portion of Bohr's hypothesis was retained in the later development.

Bohr further assumed that the stationary states corresponded to circular orbits for the electrons; he defined each orbit and its energy by a quantum number. Later, elliptical orbits were also introduced and each electronic state defined finally by four quantum numbers. However, even with this extension, Bohr's model could not be applied to the structure and spectra of more complicated atoms.

b. State of the electrons according to quantum mechanics. Bohr's fixed electron orbits conflict also with quantum-mechanical concepts. The uncertainty principle (3b) does not permit any statement about an orbit for an electron with definite energy. For a certain electron energy only the wave function can be given. This can be interpreted in a manner analogous to the case of the "particle in a box" (3c). Its square, ψ^2, at different points is thus proportional to the probability of finding an electron at these points with the energy in question. ψ^2 may also be said to give the electron density (still for an electron with the given energy). One may say that given sufficient time an electron traces out a "probability cloud" whose density ψ^2 at each point is proportional to the probability of finding the electron there. Different electron energies give probability clouds of different form and density distribution.

Even if it was necessary to modify Bohr's theory after the advent of quantum mechanics, there still remain terms relating to the picture of fixed electron orbits (for example, the terms "orbit" and "electron shell").

The state of an electron in the electron cloud is determined in the quantum mechanical treatment also by *four quantum numbers n, l, m, s*, which define a wave function, ψ_{nlms}. These quantum numbers have been called respectively *principal quantum number, azimuthal quantum number, magnetic quantum number* and *spin quantum number*. The magnitudes of the principal and azimuthal quantum numbers are represented also by letters, which for the azimuthal quantum numbers are the most common form of expression. The quantum numbers can assume the following values:

$$n = 1, 2, 3, 4, 5, 6, 7, \ldots$$
$$K, L, M, N, O, P, Q, \ldots$$
$$l = 0, 1, 2, 3, 4, 5, \ldots, (n-1)$$
$$s, p, d, f, g, h, \ldots$$
$$m = 0, \pm 1, \pm 2, \ldots, \pm l; \quad s = \pm \tfrac{1}{2}$$

The azimuthal quantum number l thus cannot be larger than $(n-1)$ and the numerical value of the magnetic quantum number m cannot become larger than l. The quantum numbers n and l are often given by symbols of the type $1s$ (which means $n=1$, $l=0$), $3d$ ($n=3$, $l=2$). An electron with $n=3$, $l=2$ is called a $3d$ electron, and so on.

Each electron in the cloud has *spin*, or at least behaves as though it has spin. If an external magnetic field is imposed, all the spin axes align themselves parallel to the lines of force of the field. If one looks along the lines of force, however, some electrons appear to rotate in one direction and some in the opposite direction. Electrons with the same spin direction are said to have *parallel spin*, and electrons with opposite spin direction, *antiparallel spin*. The two directions of rotation are distinguished by the two possible values of the spin quantum number s, that is, $+\tfrac{1}{2}$ and $-\tfrac{1}{2}$. The spin quantum number thus determines the direction of spin.

In considering the distribution of the electrons in the cloud a spin-independent wave function ψ_{nlm} is used, which depends only on the first three quantum numbers.

Table 4.1. *Electronic states, number of orbitals and electrons for* $n = 1-4$.

n	l	m	Number of orbitals $(2l+1)$	n^2	Number of electrons in the complete orbital $2(2l+1)$	$2n^2$
1 K	0 s	0	1	1	2	2
2 L	0 s	0	1	4	2	8
	1 p	−1, 0, +1	3		6	
3 M	0 s	0	1		2	
	1 p	−1, 0, +1	3	9	6	18
	2 d	−2, −1, 0, +1, +2	5		10	
4 N	0 s	0	1		2	
	1 p	−1, 0, +1,	3	16	6	32
	2 d	−2, −1, 0, +1, +2	5		10	
	3 f	−3, −2, −1, 0, +1, +2, +3	7		14	

These quantum numbers, according to Bohr's theory, determined the orbit of the electron, and ψ_{nlm} has therefore been called the "orbital wave function", and later merely *orbital*. If it is desired to emphasize particularly that the orbital refers to the state in a free atom and not to the atom involved in any combination (for example, in a molecule), one speaks of an *atomic orbital*.

It is of the utmost importance that only two electrons in an atom can have the same wave function ψ_{nlm}, or as it is said, belong to the same orbital. These two electrons thus have the same quantum numbers n, l, m and are distinguished only by the different spin quantum number s. They therefore have opposite spin directions. Two such electrons are said to form an *electron pair*. This principle may also be expressed by saying that two electrons in an atom never can have all four quantum numbers the same (*Pauli principle*).

The dependency of l and m on n, and the Pauli principle, allow for each n a limited number of states. Tab. 4.1 shows all combinations for n up to 4.

All electrons with the same n are customarily said to belong to one *electron shell*, and to indicate this letter symbols are often used for the principal quantum number. We speak then of the K shell, L shell, etc. The terminology has persisted since Bohr's theory but it is shown later that it is still justified to some extent. For a particular n value, in fact, the electrons maintain themselves preferably at a certain distance from the nucleus. This distance is larger with increasing n.

Tab. 4.1 shows that the number of orbitals possible in a shell with principal quantum number n is n^2. The shell has room then for at most $2n^2$ electrons.

Within one shell all electrons with the same l are said to belong to a *group* or *subshell*; for example, 1s group, 3d group. One group contains $(2l+1)$ orbitals and can thus contain at most $2(2l+1)$ electrons: 2 in an s group, 6 in a p group, 10 in a d group and 14 in an f group.

The orbital ψ_{nlm} determines completely the distribution of the electron, and its square ψ^2_{nlm} at a certain point is proportional to the electron density at that point.

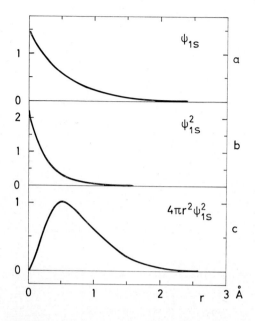

Fig. 4.1. The hydrogen atom in the ground state $1s$. (a) The wave function (orbital) ψ_{1s}. (b) The electron density ψ_{1s}^2. (c) The radial distribution function $4\pi r^2 \psi_{1s}^2$. r is the distance from the atomic nucleus.

It may be said that the size of the cloud more or less depends on the principal quantum number n. The larger n is, the more spread out is the main bulk of the cloud. The azimuthal quantum number l affects the shape of the cloud. An s cloud ($l=0$) in this way has spherical symmetry, a p cloud ($l=1$) is dumbbell-shaped, etc. The magnetic quantum number m determines the orientation of the nonspherical clouds in a magnetic field.

The energy of an electron in the cloud of the atom is generally determined by the three quantum numbers n, l and m. This is true, however, only if the atom is in a magnetic or electric field, either applied externally in the usual way, or caused by one or several neighboring atoms. In this case, there thus appears only one s level ($m=0$), but several p levels (maximum three, one for each $m=0, \pm 1$), d levels (maximum five, one for each $m=0, \pm 1, \pm 2$), etc. The energy differences between the three p levels, or the five d levels, etc., are small and may therefore be neglected in nearly all the following discussion (see however 5-3j).

If the atom is wholly free, that is, not subject to any external influence, the three p levels, and also the five d levels, come together (*degenerate*) to a common p level and a common d level. The energy of an electron is then determined only by the quantum numbers n and l. The practice just mentioned of neglecting the energy differences between the different p levels and d levels thus implies that the atoms are considered to be free.

Conversely, if we start with a free atom, on applying an external magnetic or electric field, each degenerate p level splits into three and each degenerate d level into five separate levels (the effect can be observed spectroscopically as a splitting of a normally single line and is called, in the case of a magnetic field, Zeeman effect, and for an electric field, Stark effect).

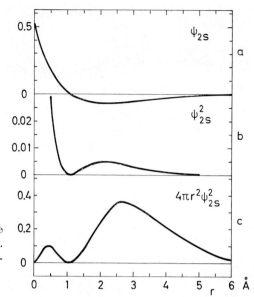

Fig. 4.2. The hydrogen atom in the state 2s. (a) Wave function (orbital). (b) Electron density. (c) Radial distribution function.

c. The hydrogen atom. The hydrogen atom with its single electron in the cloud is suitable for a discussion of the first application of the principles so far described. The electron is in its lowest energy state, the ground state, for $n=1$, $l=0$, that is, in $1s$ (this is also often called the ground state of the atom). In this state both the orbital ψ and the electron density ψ^2 are equal in all directions at the same distance r from the origin (the nucleus), that is, they have spherical symmetry. ψ and ψ^2 are shown as functions of r in fig. 4.1 a, b. They are greatest at the origin and fall monotonically with increasing r. The electron cloud thus has no sharp external limit, but beyond the distance 1 Å from the nucleus the density rapidly becomes very small. In the ground state therefore, the hydrogen atom has a quite apparent effective radius of about 1 Å.

The electron distribution is also often described by the *radial distribution function*, which, if ψ^2 has spherical symmetry, is given by $4\pi r^2 \psi^2$. Since the surface of a sphere is $4\pi r^2$, the volume of a spherical shell between the radii r and $r+dr$ is $4\pi r^2 dr$. If the probability of finding an electron in a unit of volume is proportional to ψ^2, the expression $4\pi r^2 \psi^2 dr$ gives the probability of finding the electron in the given shell. The function $4\pi r^2 \psi^2$ is thus a measure of the probability of the electron being at the distance r from the nucleus. The radial distribution function for the hydrogen electron in the state $1s$ is shown in fig. 4.1c. Since the function contains the factor r^2 it has the value 0 at the origin and must therefore pass through a maximum. It is interesting that the maximum occurs at $r=0.53$ Å which was the orbit radius Bohr obtained from his theory for the hydrogen atom in the ground state. Remember that the radial distribution function does not represent the electron *density*.

For all other s states ψ and thus ψ^2, and $4\pi r^2 \psi^2$ have spherical symmetry. Here,

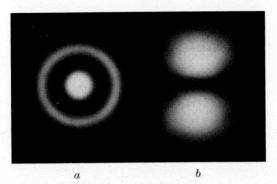

Fig. 4.3. Density distribution in (*a*) a 2*s* cloud, (*b*) a 2*p* cloud with a rotation axis of symmetry up and down in the plane of the picture.

however, there appear spherical nodes in ψ^2 with density 0, alternating with spherical density maxima. Fig. 4.2 shows the three functions for the hydrogen atom in the 2*s* state. At $r=1.06$ Å ψ changes sign and a node appears in ψ^2. The electron cloud is more spread out than for 1*s*, and since it still corresponds to one electron the density becomes much less (notice the different scales in fig. 4.1 and fig. 4.2). The radial distribution function shows that one now has the greatest chance of encountering the electron at $r=2.7$ Å.

Fig. 4.3*a* is an attempt to show the electron density in a central section through a 2*s* cloud. The variation in density with distance from the nucleus is the same as in fig. 4.2*b*.

There are always three *p* orbitals, one for each of the three *m* values $-1, 0$ and $+1$ (cf. for example, tab. 4.1). They may be described as three separate clouds of mutually similar form and rotationally symmetrical, dumbbell-shaped density distribution. Fig. 4.3*b* shows the distribution in a 2*p* cloud. The three clouds are oriented so that their rotational symmetry axes form right angles with each other. Since they can thus be placed along the coordinate axes in a rectangular coordinate system, they are often distinguished (as well as the associated electrons and orbitals) by the symbols p_x, p_y and p_z.

s and *p* clouds with other principal quantum numbers than 2 also have the same external form as 2*s* and 2*p* clouds. However, the density within the clouds varies in a different way, but in this elementary account we pass over these details. In representing the electron clouds it is often satisfactory to show just a limiting surface within which the whole electron cloud, practically speaking (for example, 90%), is found. In the top row of fig. 4.4 such surfaces are reproduced for the clouds *s*, p_x, p_y and p_z.

For $l=2$ and above (*d* electrons and following) the electron cloud takes a more complicated form.

All *d* clouds of the same *n* are mutually similar and may be said to form two cross-laid dumbbells. The rotation symmetry axes of the dumbbells coincide either with the coordinate plane diagonals (d_{xy}, d_{xz}, d_{yz} in fig. 4.4; the symbols are related to the mathematical expressions for the corresponding orbitals) or with the coordinate axes ($d_{x^2-y^2}$, $d_{x^2-z^2}$, $d_{y^2-z^2}$). The number of such orientations is six, while there are only

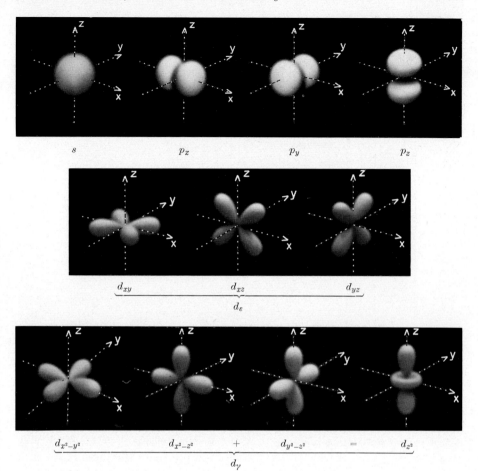

Fig. 4.4. External shapes of electron clouds for s, p and d orbitals.

five d orbitals. Two of the three clouds of the last type are, however, dependent on each other, and an arbitrary pair of them may therefore be combined into one orbital. Usually, as in fig. 4.4, $d_{x^2-z^2}$ and $d_{y^2-z^2}$ are chosen and combined into d_{z^2}. The five d orbitals then become d_{xy}, d_{xz}, d_{yz}, $d_{x^2-y^2}$ and d_{z^2}. The group of d orbitals whose clouds extend along the coordinate plane diagonals (d_{xy}, d_{xz}, d_{yz}) is designated as d_ε orbitals, and the group of d orbitals whose clouds extend along the coordinate axes ($d_{x^2-y^2}$, d_{z^2}) as d_γ orbitals.

The appearance of the electron cloud has here been discussed for the hydrogen atom. For corresponding quantum numbers, however, they have in principle the same appearance also for other atoms. The general aspect of a cloud is also independent of whether the corresponding orbital contains 1 or 2 electrons. The completely filled orbital, however, gives a cloud with greater density than the half-filled one.

6 − 689861 *Hägg*

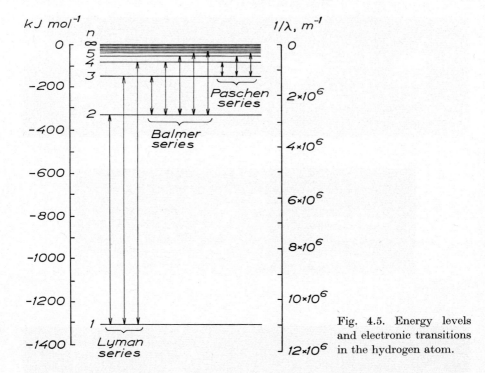

Fig. 4.5. Energy levels and electronic transitions in the hydrogen atom.

d. Energy levels in one-electron atoms. In order to state the energy of an electron in an electron cloud, a zero point must be agreed on for the energy. It is most practical to choose as a zero point the energy of the state in which the electron lies at rest at an infinite distance from the atomic nucleus, that is, wholly free. Its energy is then greater than when it lies at any finite distance. If $E = 0$ for infinite distance, then for all finite distances $E < 0$.

The expression for the energy is especially simple if the electron cloud contains a single electron. The hydrogen atom is the most important example of this, but in electric discharges, for example, atoms of other lighter elements can lose electrons so that one-electron atoms are formed. These are ions, for example, He^+. For all such cases the energy of the electron is

$$E_n = -kZ^2 \frac{1}{n^2}; \quad n = 1, 2, 3, \ldots \tag{4.1}$$

k is for this case a generally valid constant (2.1796×10^{-18} J $= 1312.8$ kJ mol^{-1}), Z is the nuclear charge and n the principal quantum number. The expression is similar in principle to (3.7), (3.8) and (3.9). For a given nuclear charge, the energy in this simple case is dependent only on the principal quantum number n.

The energy difference between two states A and B with principal quantum numbers n_A and n_B becomes

$$E_B - E_A = kZ^2 \left(\frac{1}{n_A^2} - \frac{1}{n_B^2} \right) \tag{4.2}$$

Table 4.2. *Electron distribution in the ground state for the elements* H–Ne.

	1s	2s	2p		
1 H	↑				$1s$
2 He	↑↓				$1s^2$
3 Li	↑↓	↑			$1s^2\ 2s$
4 Be	↑↓	↑↓			$1s^2\ 2s^2$
5 B	↑↓	↑↓	↑			$1s^2\ 2s^2\ 2p$
6 C	↑↓	↑↓	↑ ↑			$1s^2\ 2s^2\ 2p^2$
7 N	↑↓	↑↓	↑ ↑ ↑		$1s^2\ 2s^2\ 2p^3$
8 O	↑↓	↑↓	↑↓ ↑ ↑		$1s^2\ 2s^2\ 2p^4$
9 F	↑↓	↑↓	↑↓ ↑↓ ↑		$1s^2\ 2s^2\ 2p^5$
10 Ne	↑↓	↑↓	↑↓ ↑↓ ↑↓		$1s^2\ 2s^2\ 2p^6$

For the hydrogen atom (4.1) gives $E_n = -1313/n^2$ kJ mol^{-1}. The levels are shown in fig. 4.5. The electron has the lowest energy for $n=1$ and is then found in the orbital $1s$. In this state, the ground state, it is most strongly bound to the nucleus. An elevation of the energy to the next level ($n=2$) requires the addition of $E_2 - E_1 = 1313\ (1/1-1/4) = 985$ kJ mol^{-1}. The atom is then *excited* (an excited atom or molecule is often indicated with an asterisk, here thus H*) and the necessary, added energy is called *excitation energy*. The electron may be in either $2s$ or $2p$, which in this simple case have the same energy. Further addition of energy leads to still more excited states, corresponding to higher levels. It is seen how these levels lie more and more densely, and that the level $n = \infty$ is approached as a limiting value. Each increment of energy decreases the force with which the electron is bound to the nucleus and for $n = \infty$, the electron becomes free. The atom is then ionized. The energy which is required to ionize the atom in the ground state is called the *ionization energy* of the atom, and for the hydrogen atom this becomes

$$E_\infty - E_1 = 1313\ (1/1-0) = 1313 \text{ kJ mol}^{-1}.$$

Each increase in energy leads to a spreading out of the electron cloud (cf. fig. 4.1 with fig. 4.2).

The ratios between the different energy levels are, according to (4.1), the same for all one-electron atoms. On the other hand, all of the energy differences are proportional to Z^2. For He$^+$ they are therefore four times as great as for H.

e. Atoms with many electrons. Such atoms are said to be in the ground state when the sum of the electron energies is the lowest possible. The electron distribution in the ground state for different orbitals is therefore dependent on the energy of the orbitals. The different electrons influence each other, however, so that the energy sequence is not as simple as in the hydrogen atom. A simple rule, which was first used by Bohr, states that *the energy of the groups rises with increasing quantum number sum $n+l$ and is higher the larger n is if $n+l$ is the same*. This rule gives the energy sequence $1s<2s<2p<3s<3p<4s<3d<4p<5s<4d<5p<6s<4f<5d<6p<7s<5f<6d<7p<8s$. The higher the energy becomes, the smaller is the energy difference between consecutive groups, and this can cause overlappings which are especially marked for the adjoining groups $4f–5d$ and $5f–6d$ respectively.

All groups with $l>0$, thus groups $p, d, f, ...$, contain several orbitals (cf. tab. 4.1). All orbitals within the same group have the same energy since the atoms are considered to be free (see 4-1b). When these orbitals are filled out, the electrons do not enter in pairs in the beginning, but with one electron in each orbital. All these *unpaired* or *odd* electrons have parallel spin, that is, the same spin direction. Only after each orbital in the group has acquired one electron are electron pairs formed, which always consist of electrons with antiparallel spin, that is, opposite spin direction (*Hund's rule*).

With a knowledge of the energy sequence of the groups, the laws for the number of orbitals (4-1b) and Hund's rule, the electron distribution for all the elements in the ground state can in general be found. As an example the distribution for the elements 1 H–10 Ne is shown in tab. 4.2. The orbitals are represented by squares and the electrons in them by arrows which show the spin direction. To the right of the orbitals is shown the usual manner of symbolizing them where the superscript for a group indicates the number of electrons in the group. An extension to the whole series of elements is given below in 4-2b.

It has already been stated that the orbitals and electron clouds for all s electrons have spherical symmetry. Of the elements in tab. 4.2 this applies for H, He, Li and Be, where all electrons are s electrons. Furthermore, the orbitals and electron clouds for one group coalesce into spherical symmetry as soon as all of the orbitals contain an equal number of electrons, thus either one or two electrons each. (It can be seen immediately from fig. 4.4 that the aggregate of three similar clouds p_x, p_y, p_z must give a cloud of high symmetry.) This occurs when the group is half-filled (in tab. 4.2 for N) or completely filled (in tab. 4.2 for Ne).

For an electron cloud with spherical symmetry the radial distribution function suffices to give a good representation of the structure of the cloud. Fig. 4.6 shows this function for neon and argon. One can see the justification for continuing to speak of "electron shells". In argon, for example, the electrons with $n=1, 2$ and 3 are especially encountered in quite clearly distinct spherical shells, K, L and M shells.

Under external influences, for example, from other atoms or addition of energy, completely filled groups are less disturbed than unfilled ones. Of the latter, however, the half-filled groups are less disturbed than the neighboring stages of

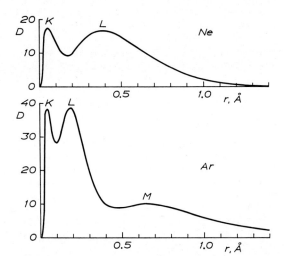

Fig. 4.6. The radial distribution function (D) for neon and argon.

orbital filling. Especially little influenced are the electron clouds of the *noble gases*, whence the name. Of the elements in tab. 4.2, He with the structure $1s^2$ and Ne with the structure $1s^2 2s^2 2p^6$ are noble gases. For the following noble gases Ar, Kr, Xe and Rn, as with Ne, the outermost shell always has the eminently stable structure of 8 electrons, $ns^2 np^6$. The $1s^2$ shell of He and the $ns^2 np^6$ shell of the following noble gases are customarily termed the *noble gas shells*, the latter also *8-shells*.

An electron distribution corresponding to the ground state is encountered in general only in free atoms. When the atoms enter into combinations (molecules, liquids, solid bodies), the distribution as a rule becomes somewhat altered. However, it is usually only the electrons in the outermost shell which are influenced, in connection with the formation of chemical bonds, for example. Since these electrons determine thereby what has been called the valence of the atom (5-2b, 5-3b), they are often called *valence electrons*. In some cases, however, the d and f electrons of the two nearest inner shells may become involved. The other electrons, on the other hand, remain for practical purposes undisturbed in all ordinary chemical processes.

4-2. Periodic system of the elements

a. History. When a considerable number of elements became known and their atomic weights determined, several scientists in the 1860's found a periodicity in the properties of the elements when they were arranged in order of increasing atomic weight. In 1862 de Chancourtois showed that at the beginning of the series elements with similar properties have differences in atomic weight of approximately 16 (this applies quite well up to Ca). Newlands in 1863 considered that similar properties recur after each interval of eight elements ("octave law").

A more elaborate system was presented in 1869 by Mendeleev in a treatise on the relation between atomic weights and properties. Mendeleev published therein a table which is similar in principle to representations of the *periodic system* which are still in use. The periodic system is associated primarily with Mendeleev's contributions, but Lothar Meyer should also be mentioned, who later in 1869, but independently of Mendeleev, arrived at approximately the same result, though not as completely developed.

What led more than anything else to the quick acceptance of Mendeleev's hypothesis was that in several cases he was able to conclude that the then accepted atomic weights were incorrect, and that these errors were later confirmed. Still more striking was Mendeleev's bold prediction in 1871 of three then unknown elements, which should fill gaps in his system. With the guidance of what he had learned about the variation of properties he predicted a large number of both physical and chemical properties of these elements. The three elements were in fact discovered later: gallium in 1875, scandium in 1879 and germanium in 1886. Their properties showed an amazing agreement with the predictions.

Nowadays we know that the elements in the periodic system should not be arranged in order of atomic weight but in order of nuclear charge (atomic number). The chemical properties follow from the structure of the electron cloud, and this in turn depends primarily on the nuclear charge. The nuclear charge is thus the essential property of the element; the number of neutrons in the nucleus and therefore the atomic weight may vary without appreciably influencing the chemical properties. The atomic number as determined by physical methods (for example, X-ray spectra according to Moseley, see 4-3 b) also gives the places of the elements in the periodic system which correspond to their properties.

That the atomic weights of the elements nearly always lie in the same sequence as the atomic numbers depends on the laws which determine how the atomic nuclei shall be composed for the highest nuclear stability (9-1). Accordingly, the ratio (number of neutrons)/(number of protons) is such that the mass and nuclear charge generally change in the same direction. Only for three pairs of elements are the atomic weights for the natural isotope mixtures reversed with respect to the atomic numbers. These are argon–potassium, cobalt–nickel and tellurium–iodine. The atomic weights of the two elements in the same pair, however, differ little from each other.

b. Stepwise building up of the electron cloud in the elemental series. In 4-1e certain general rules for the structure of the electron cloud and their application up to 10 Ne were discussed. The whole series of elements is now treated in connection with fig. 4.7. The electronic groups have here been made up of squares, each representing one orbital. These have been given arrows which indicate electrons and their spin directions. The energy rises as one goes upward in the diagram, although the energy scale could not be drawn to scale. The position of elevation of the groups shows approximately their relative electronic energy. As was said

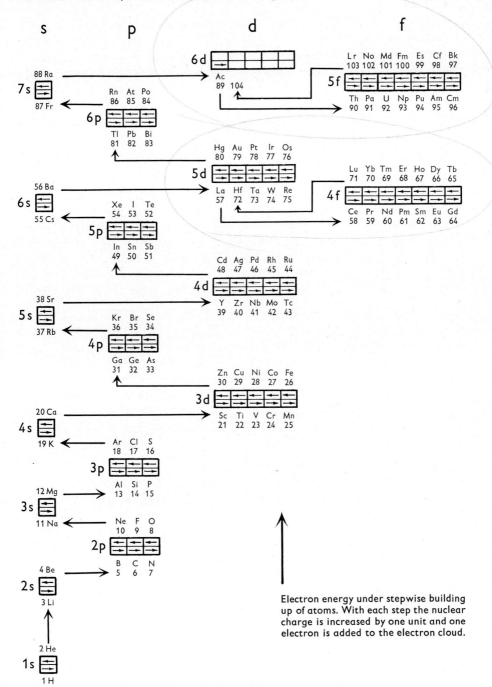

Fig. 4.7. Building up of the electron clouds of the elements.

in 4-1e the energy difference between adjoining groups becomes smaller the higher the energy is. For the groups $4f$ and $5d$ as well as $5f$ and $6d$ this leads to irregularities in the filling out of the orbitals.

It is assumed that the atoms are neutral and in going from one atom (atomic number Z) to the next (atomic number $Z+1$) one electron is thus simultaneously added to the electron cloud. The atoms are built up in the order shown by the arrows running between the groups. For each new element an electron arrow appears in the half-square lying nearest to the symbol of the element. An element is often characterized by giving the symbol of the electronic group which is added to when going to the element in question from the immediately preceding element. This becomes thus, in general, the electronic group beside which the element stands in fig. 4.7.

Following 10 Ne, the filling out of the M shell in the $3s$ group begins with 11 Na and 12 Mg and continues in the $3p$ group with 13 Al–18 Ar. The analogy with the filling out of the L shell already described in 4-1e is complete. But the M shell also includes the group $3d$, which, however, is not filled in next. Instead we continue with the N shell whose $4s$ group has lower energy than $3d$. Only when $4s$ is filled is the completing of the M shell continued by the insertion of additional electrons in $3d$ in the ten elements 21 Sc–30 Zn, which are therefore called $3d$ *elements* (belonging to the $3d$ *series*).

When $3d$ becomes complete the M shell has the structure $3s^2 3p^6 3d^{10}$ with 18 electrons. Shells of this type, thus in general $ns^2 np^6 nd^{10}$, are usually called 18-*shells*. Exposed 18-shells do not occur in neutral atoms where the s group of the next nearest outer shell is always already occupied by electrons. They can be uncovered, however, by ionization (5-2b, f) and then show great stability.

When the M shell has become complete in this way, the filling out is resumed in the N shell with the six elements 31 Ga–36 Kr, for which the new electrons occupy $4p$. With 36 Kr a noble-gas configuration is once more achieved, this time with $4s^2 4p^6$ in the outermost shell. The same building-up sequence is repeated after 36 Kr by filling the groups $5s$, $4d$, $5p$ in that order. The result is a series of $4d$ *elements*, 39 Y–48 Cd. The noble gas 54 Xe appears when $5p$ is filled.

After 54 Xe conditions become still more complicated. The group $6s$ is filled normally and the new electron for 57 La goes into $5d$, but at this point this group is left in favor of $4f$. The energies for $5d$ and $4f$ are clearly very close. The whole of $4f$ is filled through the fourteen elements 58 Ce–71 Lu, which therefore are also called the $4f$ *elements*. The outermost shells (P) and next outermost shells (O) of all the $4f$ elements have the same structure as those of 57 La, which results in the fact that they together with La form a series of elements with very similar chemical properties. The whole of this elemental series, 57 La–71 Lu, constitutes the *lanthanoids*.

After the lanthanoids the filling out of $5d$ is resumed with 72 Hf–80 Hg. These form together with 57 La the $5d$ *series*, analogous to the $3d$ and $4d$ series.

However, the position of 57 La is also occupied by the fourteen $4f$ elements, which thus, so to speak, are inserted into the $5d$ series.

After $5d$, $6p$ is filled out, which is terminated with the noble gas 86 Rn.

For the most part the same arrangement is followed hereafter as after the preceding noble gas 54 Xe. The energy differences between $6d$ and $5f$ are, however, still smaller than between $5d$ and $4f$. The new electron for 89 Ac goes into $6d$, but for at least the three following elements 90 Th, 91 Pa and 92 U the new electrons may sometimes belong to $6d$ and sometimes to $5f$. From 95 Am on it is fairly clear that the electrons are added to $5f$. For the sake of simplicity, however, in fig. 4.7 90 Th–92 U have also been placed in the $5f$ group and thus included with the other $5f$ *elements*. In this way also the analogy to the $4f$ elements is emphasized. The $5f$ elements with 89 Ac are grouped together as *actinoids*. The elements after 92 U, which with some extremely rarely occurring exceptions are not found in nature, are called *transuranium elements*.

The series of actinoids may be expected to be complete with 103 Lr after which the building up of $6d$ should continue. A nuclide of the element 104 (mass number $=260$) has been synthesized, but it is very unstable and its properties are little known.

If one or several electrons are taken from an atom, it is transformed to a positive ion. As will be described later (5-2f), in the first place those electrons are generally removed which were last added in the stepwise building up according to fig. 4.7. However, especially with d and f elements, certain reversals occur between electronic groups with close-lying energies. Fig. 4.7 shows, in fact, the change in electron energy when one electron is added (removed) while simultaneously the nuclear charge is increased (decreased) by one unit, so that the neighboring element is formed. With ionization, on the other hand, the nuclear charge is unchanged; the energy requirement thus becomes greater and the sequence of arrangement among the groups may be different.

The spectroscopically determined electronic distribution for atoms in the ground state also departs in some cases with very close-lying energy levels from the distribution according to fig. 4.7. In the d elements, sometimes 1 or 2 electrons go from the outermost ns group to the $(n-1)d$ group lying inside. In Cr, Cu, Nb–Rh, Ag, Pt and Au, 1 electron goes in, and in Pd and Ir, 2 electrons transfer. For Cr, for example, the distribution in the two outer groups thereby becomes $3d^54s$ instead of $3d^44s^2$. This is of special interest for Cu, Ag and Au where the distribution becomes $(n-1)d^{10}ns$ instead of $(n-1)d^9ns^2$. In Cu, Ag and Au the $(n-1)d$ group is therefore filled as in Zn, Cd and Hg.

In this connection the concept of *transition element* or *transition metal* should be defined. Some authors use these terms to mean the elements belonging to the d series as well as the f series inserted into the $5d$ and $6d$ series. The transition elements then include the groups 3–12 of the periodic system (for the group numbers see tab. 4.3). Since the concept is particularly intended to embrace those elements that show the characteristic properties (often referred to in the following) that depend on incomplete d groups it is better to use the terms to mean those *elements that can have an incomplete d group*. The elements in group 12 (Zn, Cd, Hg) are then excluded. On the other hand the elements in group 11 (Cu, Ag, Au), which lack incomplete d groups in the ground state of the neutral atoms (see above) and in monovalent ions, do have such groups in the multivalent ions (for example $3d^9$ in Cu^{2+}, $5d^8$ in Au^{3+}; cf. 5-2f) and are thus transition elements.

c. The periodic system. What has been said up to now shows that the quantum laws require that many features in the structure of the electron cloud shall recur periodically, although with varying length of period, when one goes from element to element in the atomic number series. This periodicity should be reflected in the properties of the elements. The filling out of electronic groups in inner shells also explains why elements immediately succeeding each other within certain series may be chemically very similar.

Fig. 4.7 leads very easily to the development of the periodic system which is shown in tab. 4.3. Each horizontal row corresponds to a *period*, which always ends with a noble gas. The first six periods contain 2, 8, 8, 18, 18, and 32 elements, respectively. The length of a period is therefore $2z^2$, where z is a whole number. The seventh is not known in its entirety, but may be expected to contain 32 elements.

With the exception of helium, the elements that have analogous positions in fig. 4.7 (immediately above one another), stand in the same column. This analogy of position leads in general to analogous chemical properties, and elements in the same column are said to belong to the same *group* (not to be confused with electronic group).

In tab. 4.3 as well as the rest of this book, the groups have been numbered without regard to the structure of the electron cloud. Other designations have long been used, of which the most common are compared below with that used here. Roman numerals and capital letters have also been used. Sometimes also, letters are not used in periods 1–3, but only in higher periods (a great disadvantage of this practice is that elements in the same column can have different group symbols; for example, Na in group 1 and K in group 1a). Notice that groups 0–7 have the same number in all three systems, and that in groups 11–17 the unit's digit is the same as the older numbers.

1	2	3	4	5	6	7	8	9	10	11	12	13	14	15	16	17	0
1a	2a	3a	4a	5a	6a	7a	8	8		1b	2b	3b	4b	5b	6b	7b	0
1a	2a	3b	4b	5b	6b	7b	8	8		1b	2b	3a	4a	5a	6a	7a	0

With the numbering of this book, with the exception of the lanthanoids and actinoids, groups with odd (even) numbers contain only elements with odd (even) atomic numbers. Even though this numbering is entirely arbitrary, the following rules of thumb are helpful. In periods 1–3 the unit's digit of the group number

Table 4.3. The periodic system.

n	ns^{1-2}		$(n-1)d^{1-10}$ with the insertion of $(n-2)f^{1-14}$ in $_{58}$Ce–$_{71}$Lu* and $_{90}$Th–$_{103}$Lr**										np^{1-6} with the exception of He ($1s^2$)					
	1	2	3	4	5	6	7	8	9	10	11	12	13	14	15	16	17	0
1	$_1$H																	$_2$He
2	$_3$Li	$_4$Be											$_5$B	$_6$C	$_7$N	$_8$O	$_9$F	$_{10}$Ne
3	$_{11}$Na	$_{12}$Mg											$_{13}$Al	$_{14}$Si	$_{15}$P	$_{16}$S	$_{17}$Cl	$_{18}$Ar
4	$_{19}$K	$_{20}$Ca	$_{21}$Sc	$_{22}$Ti	$_{23}$V	$_{24}$Cr	$_{25}$Mn	$_{26}$Fe	$_{27}$Co	$_{28}$Ni	$_{29}$Cu	$_{30}$Zn	$_{31}$Ga	$_{32}$Ge	$_{33}$As	$_{34}$Se	$_{35}$Br	$_{36}$Kr
5	$_{37}$Rb	$_{38}$Sr	$_{39}$Y	$_{40}$Zr	$_{41}$Nb	$_{42}$Mo	$_{43}$Tc	$_{44}$Ru	$_{45}$Rh	$_{46}$Pd	$_{47}$Ag	$_{48}$Cd	$_{49}$In	$_{50}$Sn	$_{51}$Sb	$_{52}$Te	$_{53}$I	$_{54}$Xe
6	$_{55}$Cs	$_{56}$Ba	$_{57}$La*	$_{72}$Hf	$_{73}$Ta	$_{74}$W	$_{75}$Re	$_{76}$Os	$_{77}$Ir	$_{78}$Pt	$_{79}$Au	$_{80}$Hg	$_{81}$Tl	$_{82}$Pb	$_{83}$Bi	$_{84}$Po	$_{85}$At	$_{86}$Rn
7	$_{87}$Fr	$_{88}$Ra	$_{89}$Ac**	104														

Lanthanoids and actinoids

*6	$4f^{0-14}$	$_{57}$La	$_{58}$Ce	$_{59}$Pr	$_{60}$Nd	$_{61}$Pm	$_{62}$Sm	$_{63}$Eu	$_{64}$Gd	$_{65}$Tb	$_{66}$Dy	$_{67}$Ho	$_{68}$Er	$_{69}$Tm	$_{70}$Yb	$_{71}$Lu
**7	$5f^{0-14}$	$_{89}$Ac	$_{90}$Th	$_{91}$Pa	$_{92}$U	$_{93}$Np	$_{94}$Pu	$_{95}$Am	$_{96}$Cm	$_{97}$Bk	$_{98}$Cf	$_{99}$Es	$_{100}$Fm	$_{101}$Md	$_{102}$No	$_{103}$Lr

Table 4.4. *Metals and nonmetals in the periodic system.*

	1	2	3	4	5	6	7	8	9	10	11	12	13	14	15	16	17	0
1	H																	He
2	Li	Be											B	C	N	O	F	Ne
3	Na	Mg											Al	Si	P	S	Cl	Ar
4	K	Ca	Sc	Ti	V	Cr	Mn	Fe	Co	Ni	Cu	Zn	Ga	Ge	As	Se	Br	Kr
5	Rb	Sr	Y	Zr	Nb	Mo	Tc	Ru	Rh	Pd	Ag	Cd	In	Sn	Sb	Te	I	Xe
6	Cs	Ba	La	Hf	Ta	W	Re	Os	Ir	Pt	Au	Hg	Tl	Pb	Bi	Po	At	Rn
7	Fr	Ra	Ac															

☐ nonmetals   ┆ ┆ semimetals   metals unenclosed

gives the number of electrons outside the outermost noble-gas shell, that is, those electrons that here act primarily as valence electrons. In the higher periods the whole group number gives the number of electrons outside the outermost 8-shell, and in addition the unit's digit of the two-digit group numbers 11–17 gives the number of electrons outside the outermost 18-shell. In these periods the unit's digit in groups 1–8 and 11–17 usually shows the number of valence electrons.

Helium is assigned to the other noble gases. On purely formal grounds hydrogen can be placed in group 1. However, most of its properties do not account for this position and it is common therefore to think of hydrogen as standing completely alone.

The column furthest to the left gives the principal quantum number n for the s and p groups that are filled within the period in the same row. n in addition shows the number of the period. In the uppermost row the electronic groups which are filled in the groups under them are indicated.

The lanthanoids and actinoids have been collected in the two lowest rows, which are completely independent of the rest of the system. The electronic group that is filled out stands here in the second column, and it is presumed that all the actinoids are $5f$ elements.

In this book, the elements are divided into *metals* and *nonmetals*. Solid and liquid metals have metallic luster and are good conductors of electricity and heat. All metals except mercury are solid at room temperature. The metallic state is discussed in more detail in 6-4. Nonmetals lack the above-mentioned properties. Several of them are gaseous at room temperature and atmospheric pressure. The regions of metals and nonmetals in the periodic system are shown in tab. 4.4. The great majority of the elements are metals. The boundary between metals and nonmetals, however, is not sharp. At least the elements within the dashed frame in fig. 4.4 show intermediate properties and have in more recent times been called *semimetals*, (another, poorer name is metalloid, that is, metal-like). Some semimetals, for example tin and arsenic, can occur in both metallic and nonmetallic modifications. It is worth mentioning that in all nonmetals except hydrogen and helium the lastly occupied orbital is a p orbital.

The term metalloid was introduced by Berzelius for what are here called nonmetals. The word was long used in this sense and only recently has it been carried over in its meaning to semimetals. It is because of this confusion of meaning that it should be avoided. When the word is encountered, one must assure himself as to what is referred to.

The elements of certain groups of the periodic system have been collected under special names. The elements in group 1 except hydrogen are called *alkali metals*, the elements in group 2 *alkaline earth* metals and the elements in group 11 *coinage metals*. The elements in group 17 are called *halogens* (=salt maker from Gk ἁλός, salt; γεννάω, I make), in group 16 *chalcogens* (=ore maker from Gk χαλκός, copper and later also generally metal and ore), in group 0 *noble gases*. In groups 8–10 one often speaks of the *iron metals* which include Fe, Co and Ni, and the *platinum metals*, which include Ru, Rh and Pd (the "light" platinum metals) and Os, Ir and Pt ("heavy" platinum metals). The groups are also called by their first elements, for example, group 5 is the vanadium group, group 16 the oxygen group.

The change of physical and chemical properties from element to element in the periodic system is discussed in many connections in the following.

d. Screening and effective nuclear charge. The electrons are retained in the electron cloud by the attraction of the positive nucleus. The force with which the nuclear charge attracts a particular electron depends on the average position of the electron in the electron cloud and the structure of the cloud. The nuclear charge naturally becomes shielded or *screened* to some degree from an electron by the portions of the negative electron cloud which lie inside the electron. It is known that a charge cloud with spherical symmetry exerts the same force on a charge lying wholly outside the cloud as it would if the entire cloud charge were concentrated in the center of the sphere. All the electrons in the groups lying inside a given electron therefore screen the nuclear charge from it by practically their whole charge, that is, by one charge unit per electron. Other electrons in the same group as the given electron screen it less effectively; this screening effect may be considered to amount to approximately 0.5 charge unit per such electron. The electrons in groups outside practically speaking exert no screening effect. The sum of all charges that screen a certain electron is called the *screening constant* S. If Z=the true nuclear charge of the atom, the electron in question is therefore acted upon by an *effective nuclear charge* $=(Z-S)$.

According to Coulomb's law, the force that holds the electron to the atom is directly proportional to the effective nuclear charge acting on it, and inversely proportional to the square of the mean distance of the electron from the nucleus.

In going from an element to the element with next higher atomic number the nuclear charge is increased by one unit at the same time that a new electron is added to the electron cloud. The effective nuclear charge thereby increases practically by one unit for the entire portion of the electron cloud lying inside the new electron. For the portions of the cloud that may lie outside, on the other hand, the effective nuclear charge is practically unchanged, because the increase in nuclear charge is compensated by the screening effect of the new electron. The result is that the electron cloud contracts, the more so the greater the portion of the cloud that is not screened by the new electron, that is, lies inside it.

If the newly added electron initiates a new shell outside those already existing, these latter do indeed contract through the simultaneous increase in nuclear charge, but the space required for the new shell is so large that the total extent of the cloud increases (cf. Ne and Ar in fig. 4.6 where this change, however, is more marked for the reason that the nuclear charge and number of electrons both increase by as much as 8 units).

4-3. Atomic spectra

Spectra were considered generally and very briefly in 3d. Here will be added an abbreviated review of atomic spectra, the most important source for everything that has already been said concerning the structure of the atomic cloud in the free atom. Atomic spectra are experimentally studied mainly by means of emission spectra (important absorption spectra of free atoms are, however, represented by spectra of the sun and fixed stars, where the continuous radiation from the denser, inner parts is selectively absorbed in the thinner, outer stellar atmosphere).

a. Spectra of one-electron atoms. All spectral lines must have energies that correspond to (4.2). Since, according to (3.10), $E_B - E_A = h\nu$, and further $1/\lambda = \nu/c$ (1-2c), the wave number of a spectral line is

$$\frac{1}{\lambda} = \frac{k}{hc} Z^2 \left(\frac{1}{n_A^2} - \frac{1}{n_B^2} \right) = k_R Z^2 \left(\frac{1}{n_A^2} - \frac{1}{n_B^2} \right) \qquad (4.3)$$

k/hc is replaced here by the constant k_R. The law of (4.3) was demonstrated for the spectrum of hydrogen in 1890 by J. Rydberg; k_R, which is therefore called Rydberg's constant, has the value 1.09733×10^7 m^{-1}.

All electronic transitions from higher levels to a particular level are said to belong to the same spectral series. In fig. 4.5 some transitions to the levels $n = 1, 2$ and 3 have been indicated by arrows. The corresponding spectral series have been named for their discoverers. The approximate wave numbers of these transitions may be read from the scale at the right. The order of magnitude of the wave numbers are on the average 10^7, 2×10^6 and 10^6 m^{-1} for the Lyman, Balmer and Paschen series, respectively. Reference to tab. 1.1 shows that these series lie in the ultraviolet, visible and infrared regions, respectively.

Within each series (n_A = constant) the wave number converges with increasing n_B toward a limit (the "series limit"), which is reached when $n_B = n_\infty$. The energy corresponding to the wave number of the series limit is $E_\infty - E_A$, that is, the energy that must be added to the atom in state A to ionize it. From the wave numbers of any series, when n_A and n_B are ascertained, $E_\infty - E_1$ may be calculated, that is, the ionization energy for the atom in the ground state (4-1d).

On raising Z in one-electron atoms (for example, to 2 in He$^+$), the wave number of the corresponding spectral lines increase, according to (4.3), in proportion to Z^2.

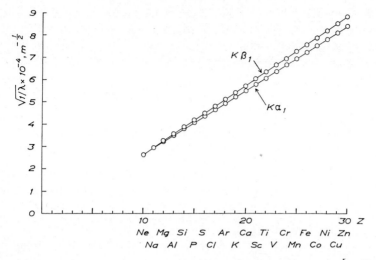

Fig. 4.8. Moseley's law for the X-ray lines $K\alpha_1$ and $K\beta_1$ for the elements 10 Ne–30 Zn.

b. Spectra of many-electron atoms. For many-electron atoms the simple expressions (4.1), (4.2) and (4.3) no longer apply, partly because the energy of the electron does not depend only on the principal quantum number, and partly because it is not proportional to Z^2 but to $(Z-S)^2$ (see 4-2d). The spectra become much richer in lines, and their interpretation more complicated. The expression (4.3) may, however, serve as a starting point for an explanation of certain general features. If for different atoms only those spectral lines are compared that are caused by electronic transitions between corresponding energy levels (4.3) may be simplified (n_A and n_B are then constant, so that the quantities in parentheses are constant) to:

$$\frac{1}{\lambda} = aZ^2 \quad \text{or} \quad \sqrt{\frac{1}{\lambda}} = bZ \tag{4.4}$$

where a and b are constants.

In many-electron atoms the innermost electrons are acted on by nearly the whole nuclear charge. For such atoms (4.4) therefore applies quite well in the comparison of spectral lines caused by electronic transitions between inner energy levels. (4.4) shows that the wave numbers for these "inner" spectra increase rapidly as Z increases. As soon as the lowermost Z values are left, the long-wavelength X-ray region (tab. 1.1) is entered and thereafter the inner spectra of the elements lie in the X-ray range.

In the year 1913 Moseley found for equivalent X-ray spectral lines from different elements a relationship between wave number (and naturally also frequency) and the atomic number of the element which nearly corresponded to (4.4). The relation has been called *Moseley's law*. This is shown in fig. 4.8 where the square root of the wave numbers for two X-ray lines from the so-called K series are presented as functions of Z (K-series spectral lines correspond to transitions to $n=1$, which was the reason that the electronic shell with $n=1$ later was called the K shell; L-series lines correspond to transitions to $n=2$, etc.).

Moseley's law immediately acquired great significance because it gave the first opportunity to determine uniquely the atomic number of an element, and thereby its place in the periodic system.

The X rays have high energy and the excitation energies must therefore also be large. In general, X rays are obtained by bombardment with fast electrons. In order that they be produced only by thermal processes a temperature is required of the order of magnitude of 10^6 °K.

Spectra caused by electronic transitions between higher energy levels, on the other hand, are little displaced when Z is changed. The electrons that give rise to these "outer" spectra are strongly screened from the nucleus, and therefore for them the effective nuclear charge $(Z-S)$ never becomes significantly greater than 1. According to (4.3) one would thus expect wave numbers of approximately the same order of magnitude as in the spectrum of hydrogen. Such is indeed the case; the outer atomic spectra of all the elements lie in the visible region and in the immediately adjacent portions of the infrared and ultraviolet regions.

CHAPTER 5

Chemical Bonding

5-1. Introduction

a. Outline. Nowadays, the concept of chemical bonding embraces all the forces that in one way or another hold atoms together. The concept thus has a much broader meaning than when it began to emerge in the 19th century in connection with the bonds between atoms in a molecule. The forces that build a crystal or hold a liquid together, for example, were thought to be principally of another kind and of little interest to the chemist. On the other hand, substances such as sodium chloride were thought to consist of molecules, and the bond Na–Cl was considered in the beginning as chemical bonding of the same kind as the bonding in CCl_4. Thus, in such cases, these ideas were extended far beyond typical molecular bonding, leading to difficulties when unified explanations were sought for chemical binding forces.

If two atoms approach each other, their electron clouds eventually come in contact, and then more and more penetrate each other. This *overlapping* of electron clouds becomes appreciable only when rather dense portions of the clouds are encountered, that is, when the interatomic distance becomes quite small (some Å). Then the interaction between the outer electrons in the two clouds may lead to attractive forces that draw the two atoms together. On the other hand, repulsive forces between the electrons in the inner, filled orbitals of the clouds always arise. As the interatomic distance decreases, the repulsive forces increase more and more rapidly, and finally balance the attractive forces. If the atoms remain at this equilibrium distance, they are bound to each other.

Both the attractive and repulsive forces are caused by electrostatic forces in the multiparticle system, which consists of the two atomic nuclei and the electrons in the clouds. The theoretical treatment of this system must be carried out by quantum mechanical methods, and even for very simple systems becomes so difficult that simplifying assumptions and approximate methods must be introduced. In spite of this, quantum mechanics has brought about a very great forward stride in our understanding of chemical bonding.

While the quantum mechanical treatment in principle can be carried out for all kinds of bonding, it leads to such different types of bonding conditions that it is practical to distinguish among certain main types of bonds. This is most true of an elementary presentation. It must be remembered, however, that the main types are not sharply delimited, but fuse into each other. They also seldom appear in pure form.

The attraction between two atoms is especially easy to understand when one atom more or less completely draws to itself one or more electrons from the other atom. The first atom then acquires an excess negative charge and the other an excess positive charge. The atoms have become ions and the attraction can easily be thought of as caused by the electrostatic attraction between the oppositely charged ions as separate units (*ionic bond*; other but less proper names are polar or heteropolar bond). In certain cases, neutral molecules with irregular charge distribution are involved, which thereby become electric *dipoles*. An ion may be bound to the part of the dipole molecule that has an excess of charge of sign opposite to that of the ion (*dipole bond*, in particular, *ion-dipole bond*). Often these types of bond, which can be fairly well interpreted with classical electrostatics, are grouped together under the designation *electrostatic bond*.

A completely pure ionic bond presupposes for each atom wholly separate electron clouds, which thus do not overlap each other. No electrons belong to more than one atom, but are all *localized* to particular atoms.

In overlapping clouds, electrons in their outer portions may become common to the two clouds. These electrons become *delocalized*, giving rise to bonding of the type commonly called *covalent bond*. Covalent bonding can only be interpreted by quantum mechanics. It is also called molecular bond, or, since it is most often brought about by electron pairs, electron-pair bond (other, older names such as nonpolar or homopolar bond are improper).

The character of a bond thus depends on the degree of delocalization of the electrons in the electron clouds of the participating atoms. *A bond becomes more covalent with increasing electron delocalization*.

In polyatomic molecules delocalization may extend over more than two atoms. In a condensed phase, for example a crystal constituting a three-dimensional, framework molecule, it may even extend over all the atoms within a homogeneous region of the phase. Under certain conditions, which are described in 6-4 (cf. also 5-6a), the delocalized electrons in such cases may move in an electric field with metallic conductivity as a result. *Metallic bonding* is therefore, in principle, covalent.

A special type of bond, in which a hydrogen atom under certain conditions binds together other atoms, is customarily considered as a separate type of bond in itself, the *hydrogen bond*.

Also, in the absence of all the types of bond mentioned, there occur rather weak bonding forces between atoms and molecules. These are called *van der Waals forces*.

Berzelius as early as 1819 advanced the theory that chemical bonding was caused by electrostatic attraction between the atoms; a bond formed if one atom was positively and the other negatively charged. With this theory Berzelius succeeded in explaining the bonding in a large number of inorganic substances, but when the science of organic compounds developed, difficulties were encountered. For example, Berzelius was obliged to assume that the hydrogen atom was always positive and the chlorine atom always negative. Then it was found that in organic molecules a chlorine atom can often be exchanged for a hydrogen atom, which would be impossible if they had different charges. Difficulties arose also in inorganic chemistry. For example, it was not possible to explain how a molecule could be formed from two atoms of the same kind. With increased experience those cases that could not be explained by Berzelius's theory accumulated more and more, so that it gradually fell into disrepute.

At the end of the 19th century several different kinds of chemical bond began to be recognized. Berzelius's interpretation to a large degree was correct for what is now called ionic bond, and after Bohr set forth his atomic theory, Kossel in 1916 was able with its help to give a reasonably satisfactory explanation of the ionic bond.

On the other hand, Berzelius's theory ran aground on the covalent bond, and this later also proved not to be interpretable by classical methods. However, it should be mentioned that in 1916 G. N. Lewis with great intuition presented the hypothesis that in this type of bonding the bond was formed by electrons—two in general—which belonged simultaneously to the atoms that were bound together. Nevertheless, a basically satisfactory explanation of the covalent bond first became possible only after the problem could be attacked by quantum mechanical methods. The way to this development was opened when Heitler and London in 1927 accounted for the bonding in the hydrogen molecule. Later important contributions have been made by Mulliken, Pauling, Slater and others.

b. Some general concepts. Certain concepts can be applied to several different types of bonds and should therefore be made clear before special treatments of the individual types are presented.

When an atom participates in combination with other atoms, either in a molecule or in a crystal, in general a number of neighboring atoms may be distinguished to which the primary atom is directly and strongly bound. The distance from the center of the primary atom to the center of such a bound atom is called the *bond length* for the bond in question. The bond length is of great importance in the estimation of the strength of the bond. The bond length between two atoms in a certain element pair is less the stronger the bond is.

Viewed with respect to the primary atom, a bound atom or group of atoms is called a *ligand* (Lat. ligare, to bind). The directly bound atom, which either itself forms the ligand or is part of a polyatomic ligand and therefore contributes bonding to it, is called the *ligand atom*.

It is said that the primary atom *coordinates* its ligands (= arranges them, places them in position). If the geometric distribution around the primary atom is considered as a whole, one often speaks of *coordination*. For example, the coordination around a certain atom may be octahedral. The number of ligands around a primary atom is called its *coordination number*. Octahedral coordination implies the coordination number 6 (a case of "6-coordination").

In many molecules an atom occurs at the center of the molecule and has a more or less symmetrical environment of ligands. This atom is customarily called the *central atom* (possibly a *central ion*) or *nucleus*. Some examples of how an atom

M functions as a central atom or nucleus, surrounded by ligands X, shown in a plane, are:

$$\text{X—M—X} \qquad \begin{matrix} & X & \\ & | & \\ & M & \\ \diagup & & \diagdown \\ X & & X \end{matrix} \qquad \begin{matrix} X \\ | \\ \text{X—M—X} \\ | \\ X \end{matrix}$$

Sometimes several nuclei can be distinguished; the molecule is then said to be *polynuclear*. The following are *dinuclear* molecules:

$$\begin{matrix} X & & X \\ | & & | \\ \text{X—M—X—M—X} \\ | & & | \\ X & & X \end{matrix} \qquad \begin{matrix} X & X & X \\ \diagdown \diagup \diagdown \diagup \\ M & & M \\ \diagup \diagdown \diagup \diagdown \\ X & X & X \end{matrix}$$

A ligand which is united to two or more nuclei is said to be a *bridging ligand*. In the right-hand molecule there are two bridging ligands. Unlimited, "infinite" molecules (1-4a), for example,

$$\begin{matrix} X & X & X & X \\ | & | & | & | \\ \text{—X—M—X—M—X—M—X—M—} \\ | & | & | & | \\ X & X & X & X \end{matrix}$$

are said to be *infinitely nuclear*. In complexes having no geometrically distinguishable central atom, for example, ClO⁻, one often speaks of a *characteristic atom*, in this case Cl.

In *formulas* the symbol of the central atom or the characteristic atom (atoms) is placed first. Then follow in sequence the anionic, neutral and cationic ligands. Often it is necessary to denote a complex by enclosing it in parentheses or brackets [ ].

In *names* the central or characteristic atom is placed after the ligands, which are stated in the sequence given above. In *complex anions* the suffix -ate is added to the name of the central or characteristic atom. (For these names see tab. 1.3.) *Complex cations* and *neutral molecules* are given no special suffix.

Anionic ligands are given the suffix -o. In systematic names sulfur as a ligand should be designated as *sulfido-*. However, in acids and anions with trivial names, *thio-* may be used to indicate the replacement of ligand oxygen by sulfur. The names of *neutral* and *cationic ligands* are used without alteration. Water and ammonia as neutral ligands are called *aquo* and *ammine*.

Examples: SO_4^{2-}, tetraoxosulfate, which, however, is in general shortened to sulfate; $[Al(OH)(H_2O)_5]^{2+}$, hydroxopentaquoaluminum ion; H_3PO_3S sulfidotrioxophosphoric acid or monothiophosphoric acid.

Many neutral complexes, however, are considered as binary compounds and are given names as described in 5-2a and 5-3f. Thus, SO_3 is called sulfur trioxide, not trioxosulfur.

c. Bond energy. Let us assume a system consisting of two atoms, A and B, which can be bonded together in a pair (often a molecule) AB. The forces occurring between A and B may be divided into attractive forces and repulsive forces. These

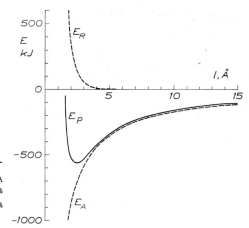

Fig. 5.1. The potential energy E_P, composed of the energy of attraction E_A and the energy of repulsion E_R, for a system of two atoms A and B with a distance between centers l.

forces give rise to potential energies, *energy of attraction* E_A and *energy of repulsion* E_R. The total potential energy becomes $E_P = E_A + E_R$. The energies must be measured from an arbitrary zero point. It is most practical to set the potential energy $=0$ when the atoms are at infinite distance from each other. The potential energy for a finite interatomic distance l is then equal to the work done by the system when the atoms are moved from the distance l to the distance ∞. This displacement takes place *against* the attractive and *with* the repulsive force. The energy of attraction then becomes negative and the energy of repulsion positive.

A typical example of how E_A, E_R and E_P change with the distance l is shown in fig. 5.1. A decrease in l always leads to an increase in both the attractive and repulsive forces. But by the definition of the potential energy, E_A hereby decreases and E_R increases. The repulsion becomes of larger significance, however, only when the l value is so small that the electron clouds of the atoms begin more appreciably to penetrate each other (overlap). This is here primarily caused by repulsion between the electrons of filled orbitals. If the atoms are compressed very near to each other, the E_R curve rises steeply. The shapes of E_A and E_R lead to the result that the total potential energy $E_P = E_A + E_R$ has a minimum. At this minimum the bonding is most stable and the atoms remain at the corresponding distance. About this distance as equilibrium position, however, they can undergo small oscillations.

The energy which is given up when the two atoms A and B approach each other from an infinite distance and are bound together, is called the *bond energy B* for the bond A–B. This is equal to the distance from the minimum in E_P to the zero line of the energy ($B = -E_P$ at the minimum). The greater the bond energy is, the stronger is the bond.

For l values greater than that corresponding to the minimum in E_P, the numerical value of E_R in general can be neglected in comparison to the numerical value of E_A (see fig. 5.1). Then we may set $E_P \approx E_A$. For rough calculations of bond energies this approximation can often be made even at the minimum, thus giving $B \approx -E_A$.

For the simple case where the attractive force is predominantly classically electrostatic in nature ("Coulomb force"), the force between two ions, which can be assumed to have the charges $+z_1$ and $-z_2$, according to Coulomb's law, is kz_1z_2/l^2. This force gives the potential energy (Coulomb energy) $E_A = -kz_1z_2/l$, which, when the above-mentioned approximation can be made, leads to $B \approx kz_1z_2/l$. If z_1, z_2 and l are known, then approximate comparisons between different bond energies can easily be made.

The bond energy must be added to the pair AB to break the bond and transport the two atoms to an infinite distance from each other. In referring to this process, one often speaks of *bond dissociation energy* for the bond A–B. For a diatomic group (molecule) the bond dissciation energy is the same as the *atomization energy* (2-2a).

These energy values may be given for reactions between individual atoms or, more commonly, for reactions between 1 mole of each kind of atom. Sometimes they are given for reaction at constant volume, sometimes for reactions at constant pressure, and expressed then as ΔU and ΔH, respectively. If fig. 5.1 corresponds to the reaction at constant volume between 1 mole of each kind of atom, we may write

$$A + B \rightarrow AB; \quad \Delta U = -560 \text{ kJ} \quad \text{or} \quad AB \rightarrow A + B; \quad \Delta U = 560 \text{ kJ}.$$

According to the circumstances, we may say that the bond energy or the dissociation energy (atomization energy) is 560 kJ.

What has here been generally said about bond energy is independent of the character of the bond, and for every bond between two atoms the shape of the energy curves in principle is the same as in fig. 5.1 (cf. for example fig. 5.9). For molecules that consist of more than two atoms, however, the conditions are no longer so simple (5-3j); and when the bond forces, as in the most common cases of ionic bond, lead to the formation of crystals, other approaches must to some extent be applied (6-2b).

d. Effective radii. The electron clouds of atoms and ions have no sharp external boundary (cf. fig. 4.1–4.3 and 4.6), but nevertheless an *effective radius* can be assigned to each type of atom which is relatively constant *for a certain type of bond* under similar conditions of coordination. The distance between the centers of two atoms that are bound to each other is equal to the sum of the effective radii of the two atoms for the situation in question. The usefulness of the concept of effective radius results from the fact that the repulsion energy E_R (fig. 5.1) climbs so sharply in a very short distance interval. Completely definite, constant effective radii would imply that the electron clouds are undeformable, which means that the E_R curve would be a perpendicular line. The slow decrease of density of the electron cloud with increasing distance from the nucleus (cf. the figures just

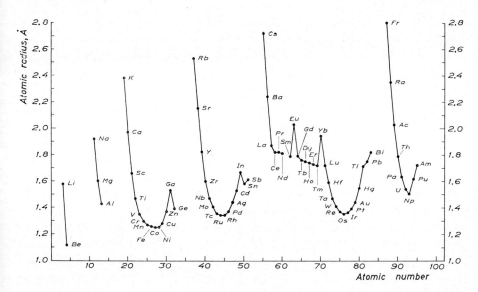

Fig. 5.2. Metallic atomic radii. All values are adjusted to correspond to coordination number 12.

mentioned) indeed also leads to the result that there is no physical basis for a completely fixed effective radius.

In different types of bond and especially in different degrees of ionization, the same kind of atom may have very different effective radii. A distinction is made therefore among ionic radii (for a certain charge), covalent radii, metallic radii and van der Waals radii. In general, the effective radii may be considered uniform in all directions and the atoms treated as spheres with radii equal to the effective radii. It must be remembered, however, that an atom, for example, may be bound to another atom with a covalent bond and thus exhibit a covalent radius toward it, but in other directions show its van der Waals radius toward atoms in neighboring molecules (5-4).

For comparison of sizes of the elements, radius types of the same kind must naturally be used. The metallic radii permit a unified comparison of the greatest number of elements and in fig. 5.2 these have therefore been plotted together. They may generally be obtained by bisecting the shortest interatomic distance in a crystal of the element. The interatomic distance is determined by X-ray diffraction methods (7-1).

The periodicity of radius values is clear in fig. 5.2, and the course of the values follows in general from the discussion in 4-2d about screening, effective nuclear charge and atomic size. In each period the alkali metal has the largest radius because it begins a new shell. As long as the subsequently added electrons are placed in the outermost shell, the radius rapidly decreases because these electrons do not screen the rest of the cloud from the nucleus. However, when the transition element series are encountered in periods 4 and 5 (third and fourth in the figure), the decrease becomes less rapid because the addition of electrons takes place more deeply inside the cloud. The new electrons shield those outside so well from the nucleus that they are little affected by the increase in nuclear charge. Toward the middle of the d series the radii become nearly constant, and even begin to increase at the end of the series. This is most probably caused by changes in the strength of the metallic bond (more on this in 6-4f).

When the addition of electrons occurs so deeply within the cloud as in the lanthanoids (the $4f$ elements), a still greater portion of the cloud escapes the effect of the increase in nuclear charge, so that the decrease in radius becomes still slower (for Eu and Yb, see below). Because of the large number of lanthanoids, however, the electron cloud at the end of the series has shrunk so much that the following $5d$ elements have approximately the same radii as the corresponding $4d$ elements. This is most apparent for the elements lying nearest to the lanthanoids and particularly so for Hf, whose radius is practically the same as for Zr. It is customarily said that this is caused by the *lanthanoid contraction*. Since the contraction is particularly small for the lanthanoids, however, the term is somewhat misleading.

Two lanthanoids, namely Eu and Yb, depart from the other lanthanoids in showing significantly larger metallic radii. With the ordinary electron distribution for the lanthanoids, these two elements and their nearest neighbors should have in the $4f$ shell and outward the distributions shown below at the left (cf. fig. 4.7):

        62 Sm: $4f^5\,5s^2\,5p^6\,5d\,6s^2$
        63 Eu: $4f^6\,5s^2\,5p^6\,5d\,6s^2$, but has instead $4f^7\,5s^2\,5p^6\,6s^2$
        64 Gd: $4f^7\,5s^2\,5p^6\,5d\,6s^2$

        69 Tm: $4f^{12}\,5s^2\,5p^6\,5d\,6s^2$
        70 Yb: $4f^{13}\,5s^2\,5p^6\,5d\,6s^2$, but has instead $4f^{14}\,5s^2\,5p^6\,6s^2$
        71 Lu: $4f^{14}\,5s^2\,5p^6\,5d\,6s^2$

In Eu and Yb one electron is lacking in the left-hand distribution to make the $4f$ group half full or completely full. The tendency to form half-filled or completely filled groups is so great, however (cf. 4-1e), that in these two cases the $5d$ electron enters the $4f$ group so that the distribution on the right is obtained. Now, the $5d$ and $6s$ electrons are involved in bonding in the lanthanoids in the metallic form (more on this in 6-4f). In Eu and Yb, thus, only $6s^2$ are bond-forming, while in the others $5d\,6s^2$ are bonding. This leads to weaker bonding and thus longer bond lengths and radii for Eu and Yb. Many other properties also deviate from the series for these two elements.

The actinoids should also be expected to show a retarded decrease in radius, similar to that of the lanthanoids. However, the decrease after Ac is still rapid and first begins to abate after U. This shows that the first actinoids in the metallic state behave as $6d$ elements (cf. 4-2b). However, after Np a radius increase sets in which can be ascribed to a transition to $5f$ character.

If elements in the same group of the periodic system are compared, it is found

Chemical Bonding

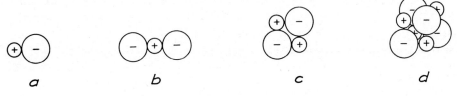

Fig. 5.3. Examples of ion pairs and ion clusters.

generally that the radius increases with the number of the period. This is indeed to be expected; for each period a new electron shell appears, with a considerable need for space. Exceptions to this rule constitute the above-mentioned 5d elements.

5-2. Electrostatic bonding

a. Ionic bonding. The most important kind of electrostatic bond is the bond between ions of opposite charge. For the sake of clarity this bond should be called *ion-ion bond*, but the most common name is *ionic bond*. Typical ionic bonds appear only in crystals built up of ions (*ionic crystals*) and in these is the most important bonding force. However, electrostatic forces of the same type occur, of course, in all systems which contain ions. If, for example, a solution or gas contains ions, as will soon be shown, pairs and groups (clusters) of oppositely charged ions are formed. The clusters may, however, contain a varying number of ions.

If ionic bonding is considered to be a result of classical electrostatic forces between ions as individual units, it is clear that no definite bonding direction can be distinguished. The pure ionic bond is thus in itself *undirected*. The occurrence of regular coordination depends on the striving for uniform distribution of charge in combination with purely geometric factors (5-2d). If beyond this any tendency for directed bonding can be detected, it depends on the presence of covalent bonding.

Ions may consist of charged atoms, *atomic ions*, or charged atomic groups, *complex ions*. In the latter the atoms are bound together by forces that in general are rather strongly covalent. Their structure will therefore be described later in connection with covalent bonding (5-3).

As a consequence ionic bonding will not be treated here in its entirety and only the particular formation, structure and size of the atomic ions can now be taken up in any detail. The description of the way ions build up crystals, one of the most important problems in ionic bonding, will also be postponed to a later chapter (6-2). Here we give only a few words in explanation of why the attraction between ions so often leads to crystal formation.

Let us assume that a Na^+ and a Cl^- ion form an ion pair Na^+Cl^- (fig. 5.3a). The pair is neutral, but the centers of its positive and negative charges do not coincide (such a system is called an electric *dipole*; see 5-2g). A second Cl⁻ ion is attracted by the positive charge of the Na⁺ ion and is repelled by the negative charge of the first Cl⁻ ion. It can therefore be attracted to the pair as in *b* because in this position the attraction is stronger than the repulsion (the Na⁺ ion lies nearer than the first Cl⁻ ion). If a completely new pair Na⁺Cl⁻ comes into the vicinity of the first pair, opposite charges are attracted and the ions arrange themselves as in the cluster *c*. Here, too, ions of opposite charges lie nearer each other than ions with the same charge. Four pairs would build up a cluster as in *d*, and it is easy to see how thereby the beginning of a NaCl crystal has been formed (cf. fig. 6.16*a*). It is clear that each of the clusters in *c* and *d* have less energy than a system which would consist of respectively two or four ion pairs separate from each other. The cluster with lowest energy and therefore the stablest that can be built up of four Na⁺ and four Cl⁻ ions should in principle look like *d*. A larger number of ions must try to build up a still larger group, which would become an extension of *d*, that is, a crystal. The energy of the crystal should be significantly lower than that of the free ions, and also lower than the energy of any distribution more dissociated or disordered than the crystal. In this way the formation and growth of well-ordered crystals is favored.

In a *formula* the electropositive component is always placed before the electronegative. When the bond is strongly ionic in nature, it is easy to decide the character of the components; when the bond is strongly covalent, the rule given in 5-3f is applied. In *systematic names* the electronegative component is given the suffix *-ide* if it is monatomic (NaCl sodium chloride) and the suffix *-ate* if it is polyatomic (CaSO₄ calcium sulfate). However, certain polyatomic anions have names that end in *-ide*, for example, OH⁻ hydroxide; NH₂⁻ amide.

b. Ionization energy and electron affinity. In 4-1d there was a discussion about the energy that must be added to ionize an atom in the ground state (that the atom is in the ground state requires, among other things, that it be free in a gas phase), called the *ionization energy*, I, of the atom. On ionizing an atom with many electrons, the most weakly bound electron (the electron in the highest energy level) is removed first. The energy necessary to remove this electron, and following that electron after electron, can be measured, usually spectroscopically. These energy values I_1, I_2, I_3, ... are called the 1st, 2nd, 3rd ... ionization energies. Then for an atom A,

$$\begin{aligned} A(g) \to A^+(g) + e^- & \quad \Delta U = \Delta H = I_1 \\ A^+(g) \to A^{2+}(g) + e^- & \quad \Delta U = \Delta H = I_2 \end{aligned} \qquad (5.1)$$

and so on.

The fact that $\Delta U = \Delta H$ means that the volume of the system may be assumed to be constant (2-1d).

To ionize A to A²⁺ the energy $\Sigma I = I_1 + I_2$ is thus required. If one speaks without further explanation of the ionization energy of a neutral atom, one always means

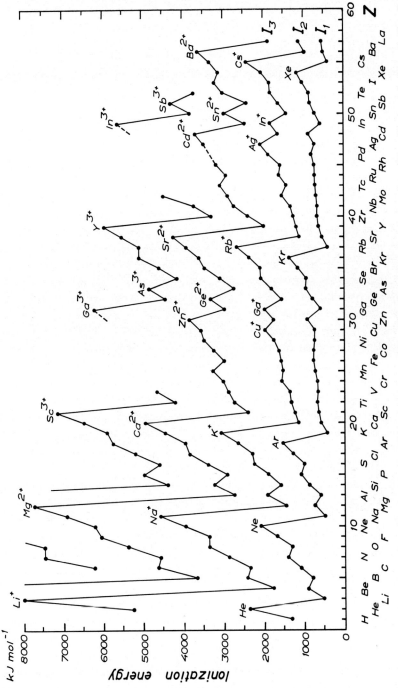

Fig. 5.4. Ionization energies for the elements 1–57. The incomplete curve segments above the I_3 curve belong to the I_4 curve. Atoms and ions with noble-gas shells and ions with 18-shells and (18+2)-shells have been labeled.

Table 5.1. *Ionization energies, I_1, kJ mol^{-1}.*

Period						
1	H	1312			He	2372
2	Li	520	F	1681	Ne	2081
3	Na	496	Cl	1255	Ar	1521
4	K	419	Br	1143	Kr	1351
5	Rb	403	I	1007	Xe	1170
6	Cs	376			Rn	1037

I_1. The ionization energy is also called ionization potential and is often given for a single atom in electron volts (1-2c).

In general, the electrons are given off in order of decreasing principal quantum number. Exceptions to this are found in the $4f$ and $5f$ elements (5-2f).

The ionization energies are very useful to show how the elements behave as positive ions. At the same time they give information about the stability of different electron configurations.

Fig. 5.4 shows the ionization energies for the elements 1–57. As will be shown in 5-2f several of the higher stages of ionization cannot be reached in chemical reactions at moderate temperatures, but only through the addition of high energy in the form of high temperature or energetic radiation. Tab. 5.1 lists some I_1 values which are used in later calculations.

If I_1 values are compared within one group in the periodic system, they appear in general to decrease when one goes to higher atomic numbers and thus to larger atomic radii. In such a comparison a simple electrostatic argument can be applied. It is the outermost, most weakly bound electrons which are given up first on ionization, and they are bound more and more weakly the further they lie from the nucleus. Among the transition elements, however, there are considerable departures from this rule.

Within one period I_1 increases for the most part from the alkali metal at the beginning of the period to the noble gas at the end. This increase can be understood mainly as a result of the decrease in atomic radius with increasing atomic number within the period. When for the alkali metal of the next period a single s electron is added outside the noble-gas shell, it is very loosely bound, resulting in a sharp decrease in I_1. Thus, the I_1 curve acquires a pronounced maximum at each noble gas (cf. also tab. 5.1).

When an additional s electron is added outside the noble-gas shell, it becomes more difficult to remove an electron from the s^2 pair thus formed than to remove a lone s electron. The s^2 pair is thus comparatively stable and is therefore sometimes called an *inert pair*.

If a lone p electron is added outside the s^2 pair, on the other hand, it will clearly be loosely bound. The I_1 curve therefore shows a maximum for the element with the outermost electrons ns^2 and a minimum for the following element with ns^2np. Compare in this way the I_1 values with the electron configurations given below for Be–B and Mg–Al (the notations "K" and "L" represent complete K and L shells inside the incomplete outer shells):

$\quad\quad$ 4 Be: $K\ 2s^2$ $\quad\quad\quad\quad$ 5 B: $K\ 2s^2 2p$
$\quad\quad$ 12 Mg: $KL\ 3s^2$ $\quad\quad\quad$ 13 Al: $KL\ 3s^2 3p$

After Ca, on the other hand, the continued building up in the $3d$ group takes place according to:

$\quad\quad$ 20 Ca: $KL\ 3s^2 3p^6 4s^2$ $\quad\quad$ 21 Sc: $KL\ 3s^2 3p^6 3d 4s^2$

which causes a steady increase in I_1 together with the decrease in radius. The same applies to corresponding regions in following periods. At the end of the $3d$ series of elements there is a maximum for Zn and a minimum for Ga, which is easily explained by the electron configurations

$\quad\quad$ 30 Zn: $KL\ 3s^2 3p^6 3d^{10} 4s^2$ $\quad\quad$ 31 Ga: $KL\ 3s^2 3p^6 3d^{10} 4s^2 4p$

The same is correspondingly true for the pairs Cd–In and Hg–Tl in following periods. Shells of the type $(n-1)s^2(n-1)p^6(n-1)d^{10}ns^2$ in Zn, Cd and Hg are called $(18+2)$-*shells*.

In some cases the I_1 values reflect the greater stability of the half-filled groups (for example, smaller maxima for N, P and As with half-filled p groups).

Cs has the lowest of the I_1 values known to date. Fr, which follows after Cs among the alkali metals, must have a very similar I_1 value but it is uncertain whether it is lower or higher.

On stepwise ionization of a given atom the ionization energy increases with each step. A rather regular increase results from an increase in the effective nuclear charge for the outermost portion of the electrons. In addition, there is a sharp jump when the next electron to be removed belongs to a shell just exposed with a lower principal quantum number, that is, a noble-gas shell or an 18-shell. See, for example, I_1, I_2 and I_3 for Mg (Mg^{2+} has a noble-gas shell) and for Zn (Zn^{2+} has an 18-shell).

It is of great interest to compare atoms and ions with the same number of electrons in the electron cloud. Such *isoelectronic* atoms and ions, in general, have the same electron distribution. This is the case, for example, with Ne, Na$^+$, Mg^{2+}, Al^{3+} which all have 10 electrons and a Ne cloud. The difference lies only in that the electron cloud is held together more strongly as the nuclear charge increases, as shown both in increased ionization energy and decreased effective radius (cf. the radius values in fig. 5.5 for the three ions mentioned). Because of the equal electron distribution in isoelectronic atoms the different I_n curves in fig. 5.4 have a very similar shape. For each stage of ionization, however, the curves are shifted one unit toward increasing Z and at the same time toward higher energy. The discussion given above on the shape of the I_1 curves therefore applies to all curves.

Table 5.2. *Electron affinities E_1 or ΣE for the transitions indicated, kJ mol⁻¹.*
H → H⁻ : 72

C → C⁴⁻ : −2960	N → N³⁻ : −2290	O → O²⁻ : −652	F → F⁻ : 349
		S → S²⁻ : −335	Cl → Cl⁻ : 369
		Se → Se²⁻ : −406	Br → Br⁻ : 341
			I → I⁻ : 312

However, for the transition elements a different electron distribution may be found for an nonionized atom on the one hand and the ions isoelectronic with it on the other. This is most striking in the nonionized atoms isoelectronic with the 18-shell ions, for which the 18-shell never occurs. Compare, for example, Ni: $KL\,3s^2\,3p^6\,3d^8\,4s^2$, with Cu⁺ = Zn²⁺ = Ga³⁺: $KL\,3s^2\,3p^6\,3d^{10}$. The clear maxima in I_2 for Cu⁺, I_3 for Zn²⁺ and I_4 for Ga³⁺, which show the difficulty in breaking up these 18-shell ions, therefore has no counterpart in I_1 for Ni.

On the other hand, Zn has the same electron configuration as Ga⁺, Ge²⁺, As³⁺, that is, the (18 + 2)-shell $KL\,3s^2\,3p^6\,3d^{10}\,4s^2$. The relatively inert s^2 pair referred to earlier therefore gives maxima in all cases.

The increase in the I_1 values toward the end of each period shows that the atoms here become more and more unwilling to give up electrons. In many cases these atoms try to *take up* electrons with the release of energy and thereby acquire an electron cloud of the same structure as the immediately following noble gas. The tendency of an atom to take up electrons is measured in terms of the *electron affinity* (Lat. affinitas, relationship) E, that is the energy which is released on the taking up of an electron. All participating atoms must be in the ground state in this case (which requires among other things that they be free in a gas phase). We have

$$X(g) + e^- \rightarrow X^-(g) \qquad \Delta U = \Delta H = -E_1$$
$$X^-(g) + e^- \rightarrow X^{2-}(g) \qquad \Delta U = \Delta H = -E_2 \qquad (5.2)$$

and so on.

In the transition X → X²⁻, therefore, the energy $\Sigma E = E_1 + E_2$ is released. The electron affinity is difficult to measure experimentally and rather few values are well known. Tab. 5.2 shows E_1 and ΣE values for the formation of more or less typical negative monatomic ions.

In the formation of halide ions and the hydride ion H⁻, energy is released. The halogens have the largest electron affinities of all the atoms because it is only necessary to add a single electron to form a noble-gas shell. Chlorine shows the highest value among the halogens.

That chlorine has a greater electron affinity than fluorine may be surprising when it is known that fluorine can oxidize chloride ions. However, here we are concerned with the process $\frac{1}{2}F_2 + Cl^- \rightarrow F^- + \frac{1}{2}Cl_2$. The ionization of the halogen molecule X₂, that is, the process $\frac{1}{2}X_2 + e^- \rightarrow X^-$, is therefore decisive. This process consists of the partial processes $\frac{1}{2}X_2 \rightarrow X$ and $X + e^- \rightarrow X^-$. In these there are released the amounts of energy $-D/2$ (where D is the enthalpy of dissociation for X₂) and E respectively, therefore $(E - D/2)$ in total. The halogen molecule for which $(E - D/2)$ is the greatest is the most easily ionized. For F_2 we have $349 - 79 = 270$ kJ and for Cl_2 $369 - 121 = 248$ kJ. The reversal thus depends on the fact that F_2 is dissociated more easily than Cl_2.

For the transitions from O, S and Se to O^{2-}, S^{2-} and Se^{2-} energy is taken up, that is, the addition of energy is required. This results from the fact that when the second electron is to be added to O^-, S^- and Se^- in order to form the noble-gas shell, it is repelled by the excess negative charge caused by the first added electron. Still greater amounts of energy are required for the formation of N^{3-} from N and C^{4-} from C.

It is clear that the elements in the lower left corner of the periodic system (Fr, Cs) most easily give up electrons and thus form positive ions. It is customary to say that these elements are the most *electropositive*. The electropositive character decreases as one goes upward and to the right in the system. In the upper right corner, thus among the nonmetals (not counting group 0), the electropositive character is least, and here are found the strongly *electronegative* elements which easily take up electrons to form negative ions.

The word *valence*, which has been used with several different meanings, has referred also to the charge of a monatomic ion. If the word is to be used in this sense it should be more precisely expressed as *electrovalence*.

c. Ionic size. The size of an ion is of extraordinarily great significance for its chemical properties. The force field on the surface of an ion depends both on the charge and the distance of the surface from the nucleus. But, in addition, the manner in which the ions are packed in a crystal depends to a large degree on the space required by the ions (6-2c). Here, purely geometrical circumstances often appear to be nearly decisive.

In general, ions with opposite charge are brought into contact with one another in a crystal, and if monatomic ions are involved, the interionic distance represents the sum of the two *ionic radii*. The interionic distance, as a rule, is easy to determine by X-ray diffraction (7-1), but the share of the two ionic radii in this distance is more difficult to decide. There are a few cases in which it is known with certainty that ions of the same kind (in all cases anions) are in contact, and then half of the interionic distance gives an ionic radius which can be used in the calculation of other radii. In this and other ways the radii of most monatomic ions have now been determined. A summary of the more typical monatomic ionic radii is given in fig. 5.5.

If a certain ion can be bonded to a varying number of ions of opposite charge, the strength of each bond will naturally be greater, the smaller the coordination number is. With greater bond strength, the oppositely charged ions penetrate more deeply into each other. The radius of an ion is therefore dependent on its coordination number. Since coordination number 6 is very common, it is often customary to give the cor-

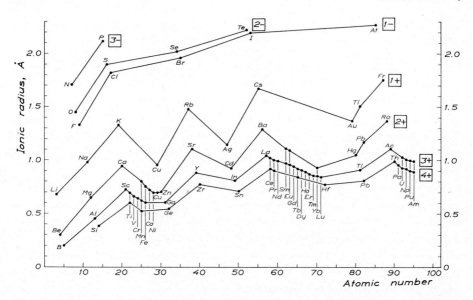

Fig. 5.5. Ionic radii. All values are referred to 6-coordination. The lines join ions of the same charge. Between many ions joined with a line may lie several other ions of the same charge. Their radii are less well known and may deviate more or less from this line.

responding radii (see also fig. 5.5). With this as the unit radius, the following approximate relative radii for other coordination numbers are obtained, which may be used for calculation:

Coord. number	12	9	8	6	5	4	3	2	1
Rel. radius	1.09	1.05	1.04	1.00	0.98	0.95	0.92	0.87	0.80

A cation is significantly smaller and an anion significantly larger than the corresponding neutral atom. For example, we have:

Na 1.92 Å   Cl 0.99 Å
Na^+ 0.98 Cl^- 1.81

In cation formation the number of electrons decreases and in addition the remaining cloud is drawn together because its outer portions are subjected to a greater effective nuclear charge (4-2d). In anion formation, on the other hand, the number of electrons increases while at the same time the outer parts of the electron cloud become more loosely held as a result of the decrease in effective nuclear charge. Only the very largest cations (Rb^+, Cs^+, Au^+, Tl^+, Fr^+, Ra^{2+}) attain a size larger than the smallest anion (F^-), and most are far smaller.

The force with which an ion attracts neighboring ions of opposite charge increases when the numerical value of the ionic charge increases. As a result the electron clouds of the ions are pressed more deeply into each other. The increased attraction therefore leads to a strengthening of the radius-decreasing effects of

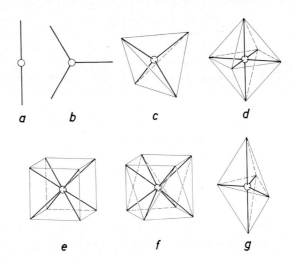

Fig. 5.6. Coordination configurations.

an increased positive charge mentioned above. On the other hand it weakens the radius-increasing effects of an increased negative charge.

Highly charged cations are therefore very small. If an ion can have varying positive charge the radius markedly decreases with increasing charge (cf. Mn^{2+}, Mn^{3+} and Mn^{4+}). The same is true of a series of isoelectronic cations (for example, Na^+, Mg^{2+}, Al^{3+}, Si^{4+}). On the other hand, the corresponding effects for anions leads to rather small differences in the radii of the isoelectronic F^- and O^{2-}, and also Cl^- and S^{2-}.

A comparison of equally charged ions shows for the most part the same progression of radius values as in fig. 5.2. Such a comparison is practicable mostly in the transition-element series. For example, we see the gradual decrease in radii for $Sc^{3+}-Fe^{3+}$, $Mn^{2+}-Zn^{2+}$ and above all for the well-represented M^{3+} ions of the lanthanoids (*lanthanoid contraction* 5-1 d; this can be seen more clearly in fig. 28.1). The amount of the lanthanoid contraction is just enough to cause the ions of the 5d elements to have approximately the same size as the corresponding ions of the 4d elements (cf. for example $Zr^{4+}-Hf^{4+}$).

Both M^{3+} and M^{4+} ions of the actinoids, particularly if the first actinoid elements are overlooked, show a gradual decrease in radius analogous with that of the lanthanoids. One may therefore speak of an *actinoid contraction*. The effect must depend on a contribution of 5f character, which for the ions begins to be noticeable earlier in the series than for atoms in the metallic state (cf. 4-2b, 5-1d and fig. 28.1).

Closely similar sizes of equally charged ions results in similarities in many physical and chemical properties. This is especially true for neighboring lanthanoids and the pair Zr–Hf, but also the similarities are great for the other 4d–5d element pairs. Also within the same series of d elements similarly charged ions of neighboring elements may resemble each other in many aspects.

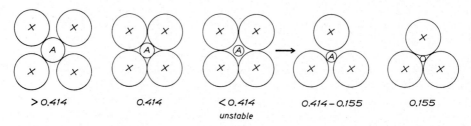

Fig. 5.7. Limiting radius values for plane coordination. The numbers indicate the ratio r_A/r_X.

It has been mentioned that ionic size sometimes is more significant than ionic charge. Ions with close-lying radius values and a difference in charge that does not exceed one unit may thus behave quite similarly, and, for example, substitute for each other in a solid solution (the charge difference must then be compensated, which can be effected in various ways; see 6-3a). Such ion pairs are Li^+–Mg^{2+}, Na^+–Ca^{2+}, O^{2-}–F^-.

The size of complex ions may vary from relatively small, discrete ions, to giant ions with "infinite" extensions in one, two or three dimensions. Even the smallest complex ions are in general larger than the largest monatomic ions. An important exception is the ion OH^-, which because of the smallness of the hydrogen atom has practically the same size and form as the O^{2-} ion. Since O^{2-} and F^- have nearly the same radius, OH^- and F^- happen to be nearly equal in size. Since they also have the same charge, they easily replace each other in solid solutions; an example is shown by the apatites (22-2a). The form of complex ions can, of course, be very irregular, but if they consist of a central atom that is uniformly surrounded by a large number of ligands of the same kind, they may attain high symmetry and sometimes become nearly spherical (for example, $PtCl_6^{2-}$ with octahedral coordination).

d. Coordination configurations. When a central ion coordinates ligand ions of one kind, for a particular coordination number the lowest energy and thus the most stable arrangement will be achieved when the bonding directions are distributed in space as evenly as possible. The electrostatic repulsion between the ligand ions is then a minimum. For coordination numbers 2, 3, 4 and 6, then, linear, equilateral-triangular, tetrahedral and octahedral coordination are obtained respectively (fig. 5.6 a–d). For coordination number 8, from a strictly electrostatic standpoint, the most stable coordination polyhedron is not a cube (fig. 5.6e), but a square antiprism (fig. 5.6f). For the rather uncommon coordination number 5, the trigonal bipyramid (fig. 5.6g) is the most stable. In this coordination, unlike the previously mentioned ones, not all the bond directions are equivalent.

A system of ions, in which certain ones function as central ions and coordinate others of opposite charge, has the lowest energy if the central ion coordinates as

Table 5.3. *Minimum values for the quotient r_A/r_X for various coordinations.*

Coordination configuration	Coordination number	Minimum value for r_A/r_X
equilateral triangle	3	0.155
tetrahedron	4	0.225
trigonal bipyramid	5	0.414
octahedron	6	0.414
square antiprism	8	0.645
cube	8	0.732

many ligand ions as possible. But the arrangement is stable only if all coordinated ligand ions lie in contact with the central ion. The coordination number can thus only be as high as is consistent with this condition. Ligand ions of a certain size are therefore coordinated in larger numbers by the central ion, the larger this ion is. A simple example in a plane in which the central ion A (radius $=r_A$) coordinates monatomic ligand ions X (radius $=r_X$), is shown in fig. 5.7. Here a square coordination becomes unstable when the quotient r_A/r_X falls below the minimum value required for contact $=\sqrt{2}-1=0.414$, and then goes over to a triangular coordination, which is stable down to $r_A/r_X=(2/\sqrt{3})-1=0.155$. For the coordinations in fig. 5.6 the minimum values for r_A/r_X given in tab. 5.3 are obtained. It can be seen how larger and larger coordination numbers become possible, the higher is the quotient r_A/r_X.

Since cations most often are smaller than anions, it is very common that the coordination number for a central cation is limited by the radius ratio. Thus we find a purely geometric cause for the occurrence of a *maximum coordination number*. On the other hand, the coordination number of anions is seldom restricted in this way.

It should not be expected that a calculated radius ratio determines conclusively the coordination configuration. This would presume that the ions were quite hard spheres with prescribed radii and that only electrostatic bonding forces were involved. Larger deviations from spherical form, which are primarily found with polyatomic ligands, require an allowance for the special case.

The tendency toward the highest coordination number which is consistent with the condition of contact provides an explanation for the fact that the coordination numbers 5 and 7 are rather uncommon. If the central ion can surround itself with five or seven ligands, then there is also room for six or eight, respectively. This is apparent for the number 5 from the fact that the trigonal bipyramid and the octahedron require the same minimum value for the radius ratio. Under such conditions the higher coordination number is preferred.

From what has been said it can be understood that it is not necessary to assume definite bonding directions in order to account for such coordination configurations as the equilateral triangle, tetrahedron and octahedron. The two latter are the most frequent coordination configurations and this can be explained by the fact that they are the ones which give the most evenly distributed bond directions for the most common radius ratios. If, on the other hand, other coordination configurations occur, these probably depend on special causes, for example contribution of covalent bonding. The plane-square coordination, which is common around the Ni^{2+} ion, for example, implies a bond distribution that is so uneven that a special explanation is necessary.

e. Polarization of ions. In any completely free atom or ion the center of positive charge (the atomic nucleus) and the center of negative charge (center of charge of the electron cloud) coincide. When ions come within a short distance from each other, however, different parts of the cloud in each ion are differently influenced by the charges of the neighboring ions. The cloud thus becomes deformed. Since this often causes the two charge centers no longer to coincide, the ion is said to be *polarized*.

The electron cloud of an anion is generally more diffuse than that of a cation, and also is not so tightly bound as in the latter with its excess positive charge. Anions are therefore, as a rule, more easily deformable and polarizable than cations. *Polarizability* increases with increasing ionic size and number of electrons in the cloud.

The *polarizing effect* of an ion on another ion is greater, the nearer it can approach the other ion, and therefore increases when the ionic radius decreases. Also, of course, it increases with increased ionic charge. Cations, which are usually significantly smaller than anions (cf. fig. 5.5), and whose charges often have a numerically higher value, therefore play the greatest role as polarizing ions. At each stage in the positive ionization of an atom, its radius decreases and its charge increases. Thus, its polarizing effect is increased.

In estimating the polarizing effect of an ion, it must be observed that this effect is clearly determined more closely by the effective nuclear charge (4-2d) in the outermost parts of the electron cloud of the ion than by the ionic charge. These two quantities most often are not the same.

It follows from the above that a strongly polarizing ion is slightly polarizable, and vice versa. The polarization phenomenon in this way becomes mainly unilateral; in general an anion is polarized by a cation.

In typical ionic bonding polarization is slight and the different ions therefore occur with fairly individual and undeformed electron clouds. If the polarization of the anions is increased, their electron clouds are drawn more and more toward the cations. The bonded atoms therefore acquire a more and more united cloud and lose their typical ionic properties.

The common electron cloud is a sign of covalent bonding (5-1a). It may be said, therefore, that increased polarization gives rise to decreased ionic bonding contribution and increased covalent contribution. This view of covalent bonding and its connection with ionic bonding arose before the quantum mechanical treatment of chemical bonding. It can still be of great use.

The very small size of the hydrogen ion, the proton H^+, causes it to be very strongly polarizing in spite of its low charge. Within each group of the periodic system, the radius of similarly charged ions increases, as a rule, as the atomic number increases (cf. fig. 5.5). At the same time the polarizing effect of the ion thus decreases. The alkali ions, because of their size and low charge, have little polarizing effect, which decreases in the series $Li^+ \rightarrow Fr^+$. Fr^+ has the smallest polarizing effect of all cations. For the cations in one period (for example, $Li^+-Be^{2+}-B^{3+}$ or $Na^+-Mg^{2+}-Al^{3+}-Si^{4+}-P^{5+}$) a rapid increase in the polarizing effect is caused by both decrease in radius and increase in charge. Be^{2+} and Al^{3+} are already quite strongly polarizing and take part in bonding with a definite covalent contribution. In bonds with Si^{4+} the covalent contribution is significant and with P^{5+} predominant. The increased polarization with rising ionic charge means that at least with charges of $+4$ and higher one can no longer speak of typical ions when atoms occur in combination with one another (a completely free ion is possible, on the other hand, but at moderate temperatures can only occur in an extremely thin gas). When in such a case the ion Cl^{7+} is mentioned, for example, this is really only a way to show the oxidation number of the atom (5-3 g).

As has been mentioned above, the polarizing effect of an atomic species increases with each stage of positive ionization. At the same time, the covalent contribution to its bonding therefore increases.

The structure of the electron cloud also has an influence on the polarizing effect of a cation. The ions of the transition elements have thus an especially great polarizing effect. For example, of the similarly charged and nearly equally large ions Na^+ and Cu^+, the latter forms bonds with significantly greater covalent contribution than the former.

The polarizability of anions is the least for the small ion F^-. It is large for large ions such as Te^{2-} and I^-. From this we may conclude that CsF, for example, should exhibit typical ionic bonding and HI a bonding with strong covalent contribution.

f. Formation of monatomic ions. When monatomic ions are formed as atoms of two different kinds come together, this implies that one kind of atom gives up electrons to the other. Thus, the former forms cations and the latter anions. In electrochemical processes (chapter 16) electron transport can take place through external conductors (electrodes and connecting wires), so that the giving off and absorption of electrons can occur in different places in the system. In principle, there is no difference between this and the direct transfer of electrons between the atoms.

Addition of (5.1) and (5.2) shows that the amount of energy $(I - E)$ must be added if electron transfer from one atom to another is to take place. In this it is presumed that the initial atoms as well as the resulting ions are free.

The energy requirement is least if I is as small and E as large as possible. The smallest value of $(I-E)$ is attained for electron transfer from Cs(g) to Cl(g) and according to tab. 5.1 and 5.2 is $376-369=7$ kJ mol^{-1}. Except for the external addition of energy, the transfer of electrons clearly would never be possible energetically if energy could not be gained from some other process. Most important is the bond energy that is released when the ions formed combine with each other or with other atoms and groups of atoms. In a solid phase crystals are most often formed (6-2) and in solution the ions combine in general with molecules of the solvent (solvation, 5-2i). The energy conditions in amorphous solid phases and in pure, fused substances are little different in principle from those in crystals. In a gas phase ion pairs and ion clusters are formed. Here, as an example, let us take the simplest of all cases, namely, the formation of an ion pair in a gas phase.

The initial substances are assumed to be Na(g) and Cl(g), which will form an ion pair, indicated by Na$^+$Cl$^-$(g). The entire process may then be written:

$$\left\{ \begin{array}{l} \text{Na(g)} \xrightarrow{+I} \text{Na}^+\text{(g)} \\ \text{Cl(g)} \xrightarrow{-E} \text{Cl}^-\text{(g)} \end{array} \right\} \xrightarrow{-B} \text{Na}^+\text{Cl}^-\text{(g)} \qquad (5.3)$$

$$\Delta U$$

Increase of energy of the system for each partial process is given above the corresponding arrow. The energy of bonding for the ion pair Na$^+$Cl$^-$ is designated B. For the total process (indicated by the long arrow), we then have:

$$\Delta U = I - E - B \qquad (5.4)$$

For the ion pair Na$^+$Cl$^-$, $B=510$ kJ mol^{-1} and we get $\Delta U = 496-369-510 = -383$ kJ mol^{-1}. The fact that the process can go in the indicated direction is in agreement with the quite strongly negative ΔU value. This depends in turn on the rather large bond energy B.

The expression (5.4) can be used to give an estimate of the possibility of different atoms forming ions by an electron transfer analogously with (5.3). A comparison of the possibility of different atoms forming *cations* can be made if an electron transfer is assumed to take place from all these atoms to the same kind of atom (for example Cl). The only variables are then I and B, and ΔU varies only with the difference $(I-B)$, or for a more highly charged cation $(\Sigma I - B)$. A great lack of accurate B values hinders quantitative calculations, but an estimate of the variation of B is obtained from the Coulomb energy according to the expression $B \approx kz_1z_2/l$ (5-1c). Since B varies in approximately the same way in the formation of other ionic combinations as in ion pairs (crystal formation, solvation) the result applies in large measure also to these more complicated but much more important cases. An example of a more realistic calculation as applied to crystal formation will be given in 6-2d.

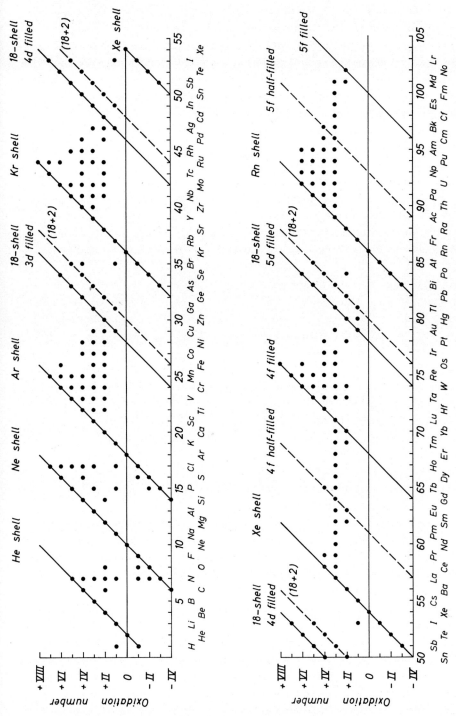

Fig. 5.8. The more important ionization stages (in the figure the concept of oxidation number set forth in 5-3g is used). The stage 0 has been indicated only for the noble gases.

With the stepwise ionization of an atom, I (and ΣI) increases. At the same time the charge increases and the ionic radius decreases, which lead together to an increase in B. The difference $(I-B)$ becomes sufficiently small if I is not too large and B is not too small. The first condition hinders the higher stages of ionization. If not before, further ionization always ceases at the point of the great increase in I that is required to break up a noble-gas shell. The second condition may make the lower stages of ionization difficult.

In the following, reference is made to the periodic system in tab. 4.3, to the electronic structure of the elements according to fig. 4.7, and to fig. 5.8.

Like Na, the other *alkali metals* easily form noble-gas ions M$^+$ by giving up the single s electron outside the noble-gas shell.

The *alkaline-earth metals* mainly form the noble-gas ions M^{2+} by giving up the s^2 pair outside the noble-gas shell. An ionization only to M$^+$ requires a significantly lower I value, but generally does not occur because the second of the two conditions stated above is not satisfied. The B value is low, partly because the ions M$^+$ have low charge and partly because they have large radii (one s electron remains outside the noble-gas shell). Starting with the system (2Ca+2Cl), for example, this transforms to the state (Ca+CaCl$_2$), in which it has a significantly lower G value than in the state 2CaCl. (Under certain conditions, however, M$^+$ ions are believed to be formed in melts.)

Among the *2p and 3p elements* only Al and possibly Si form fairly typical cations (cf. 5-2e). The noble-gas ions Al^{3+} and Si^{4+} are by far the most important. Lower stages of ionization give low B values because of low charge and large ionic radii, and are unstable (the existence of the ions Al$^+$, Al^{2+} and Si^{2+} has been recently established, however).

The outer electronic groups of the *3d and 4d elements* are usually

$$(n-1)s^2(n-1)p^6(n-1)d^x ns^2 \quad (x=1\text{--}10)$$

Also in the ions the groups $(n-1)d$ are often incomplete, characteristic of typical transition-element ions (cf. 4-2b). On ionizing, the ns electrons are always given up first since they have the highest principal quantum number (see 5-2b). Most often both the ns electrons are given up so that at first M^{2+} ions are formed. M$^+$ ions appear as a rule only when the total number of electrons permits an exposed 18-shell, that is, as in Cu$^+$ with $3s^2 3p^6 3d^{10}$ and Ag$^+$ with $4s^2 4p^6 4d^{10}$ (cf. I_2 maxima for these ions in fig. 5.4).

The removal of a third electron occurs most easily with Sc and Y (cf. the I_3 values), which are therefore ionized to Sc^{3+} and Y^{3+} with no divalent stage. These are noble-gas ions and so are the most stable ions also for some of the following 3d and 4d elements. In these, the ns^2 pair and all the $(n-1)d$ electrons are given up. For example, in the 3d series the noble-gas ions Sc^{3+}, Ti^{4+}, V^{5+}, Cr^{6+} and Mn^{7+} are known, of which the latter are bonded with very strong covalent contribution, however. Lower stages of ionization (except M$^+$) also often occur, and are the most important for the later elements of the two series. For example, for

Fe the most stable ions are Fe^{2+}: $KL\ 3s^2 3p^6 3d^6$, and Fe^{3+}: $KL\ 3s^2 3p^6 3d^5$. The very high I_3 values at the end of the two series explain the increasing difficulty in ionizing the M^{2+} ions further. Starting with Ni and Pd, M^{2+} is the most stable ion, and for Zn and Cd M^{3+} is unknown.

The two last elements in the series can form ions with 18-shells, that is, with complete d groups. In this way, from the $3d$ and $4d$ elements the iso-electronic pairs Cu^+ and Zn^{2+} with $3s^2 3p^6 3d^{10}$, and Ag^+ and Cd^{2+} with $4s^2 4p^6 4d^{10}$, are obtained. The multivalent ions of Cu and Ag (Cu^{2+} common; Cu^{3+}, Ag^{2+} and Ag^{3+} uncommon) have incomplete d groups (for example $3d^9$ for Cu^{2+}) and thus become typical transition-element ions.

In *period* 6 La is ionized to La^{3+}, analogous to Sc^{3+} and Y^{3+}. Also, all of the lanthanoids following thereafter—the *$4f$ elements*—with the outer electronic groups

$$4f^x\ 5s^2 5p^6 5d\ 6s^2 \qquad (x=1\text{--}14)$$

give up primarily the electrons $6s^2$ and $5d$ so that M^{3+} ions are formed. These have externally the stable 8-shell $5s^2 5p^6$, but are not noble-gas ions in the true sense because of the incomplete $4f^x$ group lying inside. Since only this changes, it is easy to understand that all these ions must have very similar chemical properties. It happens also that in addition one $4f$ electron is given up (Ce^{4+}, which is a true noble-gas ion; Pr^{4+}, Tb^{4+}). In some cases only $6s^2$ are given up while at the same time the electron cloud is shifted to inner levels, so that it can be said that the remaining $5d$ electron goes into $4f$ (Sm^{2+}, Eu^{2+}, Yb^{2+}). It is shown in fig. 5.8 that the formation of these ions depends on the tendency to attain a noble-gas shell or to obtain the $4f$ group either half-filled or completely filled. With these departures from the M^{3+} stage of ionization the properties are significantly altered.

With Lu the $4f$ group is full ($x=14$) and thereupon the *$5d$ elements* begin, whose outermost groups are analogous to those of the $3d$ and $4d$ elements (see above). For the most part the ions that occur are also analogous. It will be noticed that, especially when the $5d$ elements are considered, there is a displacement of ionic stability toward higher stages of ionization as one goes within a system group toward heavier elements. In group 6 the stability thus increases for the M^{6+} ions in the sequence $Cr^{6+} < Mo^{6+} < W^{6+}$, while the stability of the M^{3+} ions changes in the opposite direction. Re_2O_7 is the most stable rhenium oxide and is clearly more stable than Mn_2O_7. FeO_4 is unknown while RuO_4 and OsO_4 are well known, the latter being the most stable. In these high ionization stages, however, the bonds are predominantly covalent.

The elements in *period* 7 vacillate at the start between being $5f$ and $6d$ elements. Fig. 5.8 shows both the similarities with the lanthanoids and the departures from them. Up to U the most important ions are noble-gas ions. Thereafter there occurs a shift toward lower ionization stages. Starting with Am the analogy with the lanthanoids appears to be complete. Here the $5f$ character is obvious.

For the *$4p$, $5p$ and $6p$ elements* the outer electronic groups are

$$(n-1)s^2(n-1)p^6(n-1)d^{10}ns^2 np^x \qquad (x=1\text{--}6)$$

Here np^x and ns^2 are often given up so that the 18-shell is exposed. We then obtain, for example, Ga^{3+}, Ge^{4+}; In^{3+}, Sn^{4+}; Tl^{3+}, Pb^{4+}. Also, ions with $(18+2)$-shells are common. Then, only np^x are given up and the "inert" pair ns^2 remains. Such ions are known for the $4p$ elements (Ga^+, which however is very unstable; Ge^{2+}, As^{3+}), but play a greater role with the $5p$ elements (In^+, Sn^{2+}, Sb^{3+}), and for the $6p$ elements constitute the most stable ionic types (Tl^+, Pb^{2+}, Bi^{3+}). No breaking up of the 18-shell ever takes place. The maximum ionic charge is therefore $(2+x)$.

The possibility of different atoms forming *anions* can be compared by assuming that these atoms take up electrons from the same kind of atom. The variables in (5.4) then become only E and B, so that ΔU varies with $(-E-B)$. This expression—or for a more highly charged anion $(-\Sigma E - B)$—must thus be sufficiently small.

From tab. 5.2 it is seen that the halogens should easily be able to form anions. Among them the iodine atom is the most difficult to ionize. Because of its size the iodine atom is strongly polarized and therefore is the least typical halide ion.

The formation of O^{2-}, S^{2-} and Se^{2-} is more difficult. The electron affinity for the transition $Te \rightarrow Te^{2-}$ is not known, but bonds with an appreciable ionic contribution occur in compounds of Te with the most easily ionizable metals (groups 1 and 2).

Nitride ions N^{3-} and carbide ions C^{4-} can only be formed in compounds with metals in the groups 1 and 2. There is also a weak ionic contribution in compounds of phosphorus with these metals and one can therefore imagine a phosphide ion P^{3-}. The bonds between metals and other elements in the groups 4 and 5, on the other hand, have almost wholly metallic character. The hydride ion H^- is much more easily formed than the last-named ions, and in this respect stands nearest to the halogens.

Many compounds have long been called chlorides, sulfides, nitrides, hydrides, etc., even though the bond character is almost entirely covalent or metallic.

All of the anions mentioned up to this point have noble-gas shells. It is worth pointing out that the elements in group 16 in solid and liquid phases do not form monovalent anions of the type O^-. In such phases also, there are never found negative monatomic ions with variable charge.

No indication has been found up to now that the elements at the end of the transition element series should tend to form 18-shells by absorption of electrons.

Everything that has been said heretofore about the formation of ions applies if this takes place between one atom and another, and without external addition of energy. If high energy is imposed by collisions with energy-rich particles (produced at highly elevated temperature, acceleration in an electric field or by nuclear decay) or by radiation with energy-rich quanta, completely different possibilities for ion formation arise. If, for example, a high-voltage spark strikes through oxygen at low pressure, all the ionic species from O^+ to O^{6+} are formed (the ionization of O to O^{6+} requires 41 720 kJ mol^{-1}). When the proportion of relatively free, charged particles (positive ions and electrons) begins to become large and the gas thus becomes highly conducting, it is said to constitute a *plasma*. At a temperature of 10^6 to $10^7\,°K$, the mean energy of the heat motion is sufficient to strip off the greater part of the electron cloud, or all the electrons altogether. Nuclei and electrons become rather evenly and randomly distributed and most of the electrons no longer belong to any particular nucleus. At very high pressure (about 10^8 bar or higher) complete destruction of the electron cloud takes place, even at low temperature (cf. 1-3c and 6-4d).

g. Molecules as electric dipoles. In 5-2e polarization of atoms and monatomic ions was discussed. The concept will now be broadened and applied primarily to molecules.

When atoms occur in combination, the whole common electron cloud for the combination may have a density distribution such that its center does not coincide with the center of the nuclear charges. A neutral molecule, that is, a molecule with no net charge, acquires in this way an electric moment, and becomes an *electric dipole*. If the total positive charge = total negative charge = ne, and the distance between the charge centers = l, the *dipole moment* of the molecule is defined as

$$\mu = nel \tag{5.5}$$

If a molecule is a dipole regardless of whether or not it lies in an electric field, it is called a *permanent dipole*. If an external field is imposed the molecule seeks to position itself so that the dipole opposes the field (*orientation polarization*). On the other hand, the molecule does not move in the field since it has no net charge. Orientation polarization occurs both if the molecule lies in a field between two electrodes and if it is subjected to the electric field from an ion in its vicinity. Orientation polarization is counteracted by heat motion and therefore decreases with increasing temperature.

An external field can also cause shifts of the electron cloud in relation to the atomic nuclei, analogously to what was described for atoms in 5-2e. It can also lead to smaller changes in the relative positions of the nuclei. The effect is called *induced polarization*. Atoms as well as molecules that are not permanent dipoles then become *induced dipoles*. Different atoms and molecules are acted upon to varying degrees by the external field; they have different *polarizabilities*. The induced polarization is practically independent of temperature.

The permanent dipole moment is determined in general from measurements of the *dielectric constant* of the substance. This is defined as the capacitance of a condenser with the substance in question between the plates if the capacitance of the same condenser with no matter (vacuum) between the plates is taken as unity. In principle the measurement is carried out according to this definition. The dielectric constant by a simple calculation gives the polarization, which is the sum of the orientation polarization and the induced polarization. Each can be determined, either by measurement at different temperatures or by using alternating fields of different frequencies. With sufficiently fast reversal of the field direction only the electron cloud is able to shift while the molecules do not reorient themselves. Orientation polarization is then absent. From the value of the orientation polarization the permanent dipole moment can be directly computed. In comparing substances of approximately the same molecular weight and density (for example, water and other liquid solvents) it may be assumed that in general the dipole moment is greater, the greater the dielectric constant is. For water, both quantities are higher than for most other solvents. Water is therefore said to be a strongly *polar* solvent.

The dipole moment is a directional quantity (vector). As with the direction of an electric field, it is customarily represented by an arrow which extends from the center of positive charge toward the center of negative charge.

It has been found that the permanent dipole of a molecule often can be taken as the sum of the moments appropriate to the individual bonds. This summation is vectorial and can be carried out (analogously, for example, with the summation of forces by means of the force parallelogram) if the separate moments are represented by arrows whose lengths are proportional to the magnitudes of the moments.

If the molecule has sufficiently high symmetry, it will have no dipole moment. For example, there is no reason why the electron cloud in a molecule of HH or ClCl should be unsymmetrically distributed with respect to the nuclei. The carbon dioxide molecule also has no dipole moment and since a C–O bond can be expected to have a certain moment, this can only be explained if the molecule is linear with the appearance OCO. The two bond moments then become directed as ← → and compensate each other.

On the other hand HCl and H_2O are dipoles. The former has the moment $\overrightarrow{HCl}$. The dipole moment of H_2O can only be explained if the molecule is bent, as H^{O}H (the angle between the H–O bonds is 104.5°). Each bond moment has the direction $\overrightarrow{HO}$. With the molecular orientation given they are thus directed as ↗ ↖ and the resultant is a moment ↑.

NH_3 is also a dipole. The three H atoms lie at the corners of the base of a triangular pyramid with the N atom at the apex. Each bond moment has the direction $\overrightarrow{HN}$ and the three combine to give a moment along the altitude of the pyramid through the N atom and directed toward it.

In all three, HCl, H_2O and NH_3, the H end of the molecule thus becomes positive and the opposite side negative.

BCl_3, on the other hand, has no dipole moment. This can be explained only if the Cl atoms form an equilateral triangle, in whose center the B atom is placed. The bond moments are thus arranged in a plane as ↑↙↘ and cancel each other out. Such highly symmetrical molecules as CH_4 and CCl_4, which both have tetrahedral coordination, also lack dipole moments. The same is true of benzene C_6H_6 with its sixfold ring.

h. Dipole bonding. A molecule with a permanent dipole moment can be bound to an ion (*ion-dipole bond*) or to another molecule with a permanent dipole moment (*dipole–dipole bond*) in a manner analogous to that in which an ion pair is bound to an ion, or two ion pairs to each other, respectively (5-2a). Dipole–dipole bonding can affect the cohesion between molecules with permanent dipole moments in liquids (see 13-4b). Here, however, we consider only the stronger and more important ion-dipole bond. In this bond the dipole molecule is oriented in relation to the ion so that the end which has the charge opposite to that of the ion comes nearest to it. The center of opposite charge then lies closer to the ion than the center of like charge, and the net effect is an attraction. Since the dipole molecule is neutral, the complex formed has the same charge as the simple ion.

The bond energy for ion-dipole bonding is always much less than for ion-ion bonding. As the distance is increased the energy of attraction for the former bond type more rapidly approaches 0 (that is, the energy value for infinite distance; see 5-1c) than for the latter type.

The first dipole molecules that are bound to an ion are, of course, attracted until they come in contact with the ion. In this way an "inner sphere" of bound ligands is formed. If ligands are present in sufficient quantity, however, (for example, if the ion is in a solution in which the solvent consists of ligand molecules) they often also are bound in one or more "outer spheres". The ion then becomes surrounded by a cloud of ligands, which, at least in its outer portions, is rather ill defined.

For a given dipole molecule the bond energy is proportional to the electric field at the midpoint of the dipole. The bond energy therefore becomes larger, the higher the charge of the ion, and the nearer the dipole lies to the center of the ion. For the same charge, the bond energy thus increases for dipoles in the innermost sphere when the ionic radius decreases.

For the same field strength the bond energy for different dipole molecules is proportional to their dipole moments.

For the innermost sphere, the condition that the dipole molecule shall be in contact with the ion leads to a geometrically determined maximum coordination number, which depends on the relation between the sizes of the ion and the dipole molecule. The larger the ion is, the more molecules of a given type will have room to be in contact with its surface. What was said concerning this in 5-2d, can, of course, be applied also to ion-dipole bonding.

Ion-dipole bonding of H_2O causes ions in water solution in general to be *hydrated* (forming *aquo ions*). It follows from the discussion given above that the H_2O envelope of an ion in solution is not particularly well defined; it must be assumed that H_2O molecules are bound both in an inner sphere directly to the ion and in outer spheres.

Of the alkali-metal ions in water solution, the smallest, Li^+, is the most hydrated, and the degree of hydration decreases as the ionic radius increases within the group; the bond strength at the surface of the ion decreases and finally becomes insufficient to prevent effectively the heat motion from shaking off the H_2O molecules. Ions of the type M^{2+} are more hydrated than Li^+. Here also the decrease in degree of hydration with increasing radius is noticeable; the hydration decreases, for example, in the series $Mg^{2+} > Ca^{2+} > Ba^{2+}$. Ions of the type M^{3+} are still more hydrated. Here, however, the tendency of ions with high ionic charge to form covalent bonds begins to be apparent. H_2O is thus no longer bound by pure ion-dipole bonding.

Since anions are generally large and of low charge their hydration becomes of lesser significance than for most cations. The highest degree of hydration is found for the smallest anions present in water solution, OH^- and F^-.

The special hydration properties of the ions H^+ and OH^- are treated in more detail in chap. 11 and 12. For practical reasons, and since the degree of hydration as has been said is difficult to define, the hydration of ions is ordinarily seldom given if there is no need for it to be emphasized. Such specification may be important, however, when hydrated ions behave as acids (chap. 12).

When an ionic crystal is formed from water solution the H_2O envelope of the cations may very often at least partly be carried with it into the crystal as *water of crystallization*. The solid phase then becomes a *hydrate*. In crystals a definite coordination number is attained and in general a very regular coordination (6-2). Here also it is clearly seen how the ionic charge and ionic radius determine the bond strength and maximum coordination number.

Many cations bond NH_3 with the formation of *ammine ions* (the corresponding solid phase is called an *ammine*). The alkali-metal ions bond NH_3 more weakly than H_2O, probably because NH_3 has a smaller permanent dipole moment than H_2O. In water solutions in which NH_3 competes with H_2O for bonding, only Li^+, the smallest alkali-metal ion, forms ammine ions and for this a quite high NH_3 concentration is required. The alkaline-earth-metal ions, on the other hand, bond NH_3 somewhat more strongly than H_2O ($CaCl_2$ therefore cannot be used for drying NH_3). This has been explained by the fact that NH_3 is more polarizable than H_2O, so that a larger induced dipole moment arises in the former molecule when the field is as strong as it is at the surface of the divalent cation. But approximately the same thing is expressed when it is said that NH_3 has a greater tendency than H_2O to be covalently bound, and that this bond type begins to develop as one goes to the right in the periodic system.

As one goes to the right from group 2, that is into the transition-metal series, bonding with NH_3 becomes more and more strong and at the same time its covalent contribution increases. Here NH_3 is obviously bound more strongly than H_2O throughout. Thus, for example, the diamminesilver ion $Ag(NH_3)_2^+$ is much more stable than the aquo ions of silver. If NH_3 is added to a water solution of Ag^+ ions, the H_2O envelope about Ag^+ is supplanted by NH_3, and with sufficiently high NH_3 concentration $Ag(NH_3)_2^+$ is formed. The H_2O envelope of the Ag^+ ion can be removed by Cl^- with the formation of the difficultly soluble $AgCl$, but Cl^- can produce $AgCl$ from a strongly ammoniacal Ag^+ solution only with difficulty.

Table 5.4. *Enthalpy of hydration* $Y = -\Delta H$ *for ions in infinitely dilute water solution, kJ mol^{-1}, 25°C.*

H$^+$	1080	NH$_4^+$	300	Mg^{2+}	1840	Al^{3+}	4500	OH$^-$	502
Li$^+$	506	Ag$^+$	420	Ca^{2+}	1510	La^{3+}	2900	F$^-$	509
Na$^+$	396	Tl$^+$	300	Sr^{2+}	1360	Fe^{3+}	4350	Cl$^-$	370
K$^+$	310			Ba^{2+}	1260	Th^{4+}	6070	Br$^-$	340
Rb$^+$	290			Fe^{2+}	1880			I$^-$	300
Cs$^+$	260			Hg^{2+}	1800				

i. Solvation. Many different solvent molecules are bound to the ions of the dissolved substance just as has been shown above for H$_2$O and NH$_3$. One speaks generally of *solvation* and *solvate ions*. If the solvated molecules are taken into a solid phase, this is often called a *solvate*. The name is used independently of bond type, which, as with H$_2$O and NH$_3$, can vary from ion-dipole to covalent bonding.

With the solvation of an ion, *energy* or *enthalpy of solvation* (in water solution *energy* or *enthalpy of hydration*) is released, and it is this which gives the larger part of the energy that is required to obtain ions in dilute solution.

Tab. 5.4 presents the enthalpy of hydration $Y = -\Delta H$ for a number of ions. The enthalpy of hydration is often large, and also gives an idea of the relative extent of the hydration. It can be seen how with the same charge the enthalpy of hydration is larger the smaller the ion is. With approximately similar ionic radius (cf. the pairs Li$^+$–Mg^{2+}, Na$^+$–Ca^{2+}) the enthalpy of hydration is larger the higher the charge is.

If the hydrated ions are indicated by (aq), the following scheme for the formation of hydrated ions from Na(g) and Cl(g) is obtained, analogous to (5.3). The enthalpy of hydration for the metal ion is designated Y_M and for the halide ion Y_X. Since Y_M and Y_X are enthalpy changes, ΔH and not ΔU is obtained for the total process (for the ionization processes $\Delta U = \Delta H$; cf. (5.1) and (5.2)).

$$\begin{Bmatrix} \text{Na(g)} \xrightarrow{+I} \text{Na}^+(g) \xrightarrow{-Y_M} \text{Na}^+(\text{aq}) \\ \text{Cl(g)} \xrightarrow{-E} \text{Cl}^-(g) \xrightarrow{-Y_X} \text{Cl}^-(\text{aq}) \end{Bmatrix} \quad (5.6)$$

$$\Delta H$$

From this we obtain for the whole process

$$\Delta H = I - E - Y_M - Y_X \quad (5.7)$$

and with values from the tables, $\Delta H = 496 - 369 - 396 - 370 = -639$ kJ mol^{-1}. The large total enthalpy of hydration 766 kJ mol^{-1} thus gives a ΔH value that precludes any doubt concerning the direction of the process.

According to the energy of ionization the alkali metals become more and more electropositive in the series Li, Na, K, Rb, Cs, that is, they give up their valence electron more easily the heavier they are. This seems not to agree, however, with the observations in processes in which alkali metals give up electrons in water solution

Table 5.5. *Enthalpy of sublimation* $L = \Delta H$, kJ mol^{-1}, 25°C.

Li 155	Be 321					B 407	C (graphite) 715
Na 109	Mg 150					Al 314	Si 370
K 90	Ca 193	Sc 390	----------	Cu 341	Zn 131	Ga 276	Ge 330
Rb 86	Sr 164	Y 430	----------	Ag 289	Cd 113	In 244	Sn (white) 300
Cs 79	Ba 176	La 370	----------	Au 344		Tl 186	Pb 195

and thus are converted to hydrated ions. But in this case, the process $M(s) \rightarrow M^+(aq) + e^-$ is decisive. It can be thought of as divided into the stages (the electron is immediately taken care of, for example, by a halogen):

$$M(s) \xrightarrow{+L} M(g) \xrightarrow{+I} M^+(g) + e^- \xrightarrow{-Y_M} M^+(aq) + e^- \qquad (5.8)$$

$$\underbrace{\phantom{M(s) \xrightarrow{+L} M(g) \xrightarrow{+I} M^+(g) + e^- \xrightarrow{-Y_M}}}_{\Delta H}$$

The first of these stages is the vaporization of the solid metal, which requires the enthalpy of sublimation L. A number of L values are recorded in tab. 5.5.

For the total process, the enthalpy $\Delta H = L + I - Y_M$ thus needs to be added. From the tabulated values we get for the series Li, Na, K, Rb, Cs the ΔH values 169, 209, 199, 199, 195 kJ mol^{-1}, respectively. Li thus now requires the smallest energy for ionization, and this results from the large enthalpy of hydration for Li$^+$. It is also true that Na requires the largest energy for ionization in water solution. The standard potentials (16c and tab. 16.1) range Li, Na, K, and Rb in the same order as the above ΔH values.

5-3. Covalent bonding

The theoretical treatment of the covalent bond must be carried out by quantum mechanical methods. These have rendered extraordinarily important results, but they are still very incomplete. For a given system of atoms there are in general a small number of bonding states that have so much lower energy than all others that the choice can be limited to these few. But the energy differences between these are often less than the uncertainties in the energy calculations. Therefore, in choosing the correct state, various observed properties, such as interatomic distances, coordination behavior, dipole moments and magnetic properties, must also be taken into account. The result is, in general, that one of the remaining bonding states explains all these properties better than any of the others. This state is with high probability the right one, but is not necessarily the only possible one from a purely quantum-mechanical viewpoint. Future developments affording greater accuracy in the energy calculations may make possible more definite predictions.

a. Simple cases of covalent bonding in diatomic molecules.

When the structure of the electron cloud of free atoms was discussed in 4-1 and 4-2, we started with an atomic nucleus whose charge was increased stepwise while at the same time electrons were added outside the nucleus in one orbital after another. In the derivation of the electronic structure in a group of combined atoms—often a molecule—we may proceed in an analogous manner. We start with the atomic nuclei lying in the same positions as in the final group and then introduce electrons one at a time. The energies of the electrons are determined by orbitals that involve the whole of the final nuclear and electronic system, and which are called *molecular orbitals* (often abbreviated MO). The first electron is added to the molecular orbital with the lowest energy and one continues then in a manner similar to that of the building up of atoms. As with single atoms, the Pauli principle and Hund's rule apply here also (4-1 b, e).

The first problem at hand is to decide in what order the molecular orbitals should be filled with electrons, that is, their energy sequence, and how the various occupied orbitals influence the bonding between the atoms.

It has been found that the molecular orbitals are formed by combination of the atomic orbitals of the participating atoms. In order that an atomic orbital ψ_A of atom A shall be able to combine with an atomic orbital ψ_B of atom B, it is necessary:

1. that the electron clouds corresponding to ψ_A and ψ_B sufficiently overlap; this requirement leads to the fact that for the most part orbitals in the outermost electron shells of the two atoms are combined (atomic orbitals belonging to higher shells than the outermost in each atom cannot be used for stable bonding; electrons in these "unstable" orbitals would have too high energy);
2. that the two orbitals ψ_A and ψ_B in each atom correspond to approximately the same energy; this also leads to the result that mainly orbitals in the outermost shells are combined;
3. that ψ_A and ψ_B have the same symmetry with respect to the line of bonding A–B (*bond axis*).

The combination of ψ_A with ψ_B always gives two molecular orbitals ψ_+ and ψ_- (notice that the number of orbitals in the system thus is not changed by the combination), which with fairly good approximation may be expressed as a linear combination of atomic orbitals (LCAO) by

$$\psi_+ = c_A \psi_A + c_B \psi_B \tag{5.9}$$
and
$$\psi_- = c_A \psi_A - c_B \psi_B \tag{5.10}$$

c_A and c_B are constants for given atoms A and B, and determine the contribution of the atomic orbitals to the molecular orbitals. If A and B are identical, $c_A = c_B = 1/\sqrt{2}$.

The electrons in the molecular orbitals no longer belong only to one atom but are common to the whole atom pair. It has been said earlier in 5-1a that this can

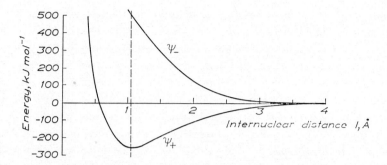

Fig. 5.9. The energy of a system of two protons and one electron in either of the molecular orbitals ψ_+ and ψ_-. The dashed line corresponds to the internuclear distance in the hydrogen molecule ion H_2^+.

be the case for the outer portions of the electron clouds that strongly overlap each other.

Atomic orbitals that do not fulfill the conditions given above do not combine, and may be considered as atomic orbitals relatively undisturbed by the neighboring atom. The electrons in these orbitals confine themselves mainly to their respective atomic nuclei.

The above principles are applied first in some very simple cases with two atoms of the same kind in period 1, which thus in the ground state make use only of the atomic orbital $1s$. Simplest of all are the ionized hydrogen molecule ("hydrogen molecule ion") H_2^+ with only one electron, and the neutral hydrogen molecule H_2 with two electrons. These systems were treated by Heitler and London in 1927.

If the $1s$ orbital of one atom is designated ψ_A and of the other atom ψ_B, these are combined according to (5.9) and (5.10) into two molecular orbitals.

$$\psi_+ = \frac{1}{\sqrt{2}}(\psi_A + \psi_B) \tag{5.11}$$

$$\psi_- = \frac{1}{\sqrt{2}}(\psi_A - \psi_B) \tag{5.12}$$

The three given conditions for combination are fulfilled.

In fig. 5.9 a hydrogen molecule ion H_2^+ is considered to be formed by the approach to each other of a hydrogen atom and a proton (cf. fig. 5.1). At the start the energy of the system is that corresponding to the electron in either ψ_A or ψ_B. When the $1s$ orbitals begin to overlap, combination can take place to give either ψ_+ or ψ_-. In the first case the energy of the system decreases and passes through a minimum. The nuclei thus are attracted to each other and assume an equilibrium distance ($l = 1.06$ Å) corresponding to the energy minimum. Only when they are compressed to a distance of about 0.5 Å does the energy become greater than when the nuclei are at an infinite distance from each other. With

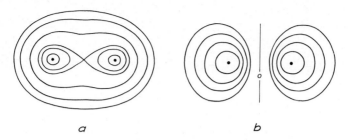

Fig. 5.10. Electron density distribution in the hydrogen molecule ion H_2^+. (a) Attractive state, bonding orbital ψ_+. (b) Repulsive state, antibonding orbital ψ_-. The plane of the picture is an arbitrary plane through the nuclei. The electron clouds have rotational symmetry about the bond axis. (C. A. Coulson.)

the combination for ψ_- the energy increases continuously. The system then never becomes stable, and the nuclei repel each other at all distances.

For these reasons ψ_+ is called a *bonding orbital* and ψ_- an *antibonding orbital*. In relation to the energy of ψ_A or ψ_B, the excess of energy for ψ_- at the equilibrium distance is greater than the deficiency of energy for ψ_+. However, the difference is quite small. Therefore it is usually assumed that the corresponding bonding and antibonding orbitals practically compensate each other if they are occupied at the same time. This compensating action is generally applicable.

The electron density in the electron clouds of the two molecular orbitals becomes

$$\psi_+^2 = \frac{1}{2}(\psi_A^2 + \psi_B^2 + 2\psi_A\psi_B) \tag{5.13}$$

$$\psi_-^2 = \frac{1}{2}(\psi_A^2 + \psi_B^2 - 2\psi_A\psi_B) \tag{5.14}$$

The electron density between the nuclei A and B is of especial interest. It can easily be given for the point midway between the nuclei where by symmetry $\psi_A = \psi_B$. Here we then obtain

$$\psi_+^2 = 2\psi_A^2; \qquad \psi_-^2 = 0$$

The electron density at the midpoint thus becomes high if the orbital ψ_+ is occupied and zero if the orbital ψ_- is occupied. For the hydrogen molecule ion H_2^+, the electron density is shown in fig. 5.10, in which for both cases the nuclei are placed at the equilibrium distance for ψ_+. It can be seen how the electron cloud for ψ_+ is concentrated between the nuclei and for ψ_- is pushed away from this region.

If both orbitals are occupied simultaneously, by addition of (5.13) and (5.14) we get the resulting electron density

$$\psi_+^2 + \psi_-^2 = \psi_A^2 + \psi_B^2 \tag{5.15}$$

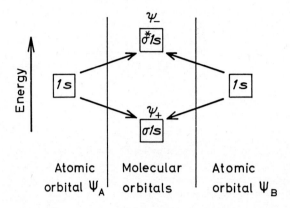

Fig. 5.11. Schematic diagram of the combination of two $1s$ orbitals into molecular orbitals.

which is practically the same electron density as that around two free atoms at the same distance. The same result can be seen schematically by imagining that fig. 5.10a and b are placed over one another with the nuclei in coincidence. This electron density compensation is related to the energy compensation between ψ_+ and ψ_- referred to above.

A very schematic picture of the combination of two $1s$ orbitals into molecular orbitals is given in fig. 5.11. Here ψ_+ is indicated by $\sigma 1s$ and ψ_- by $\sigma^* 1s$. σ is a generally accepted symbol indicating that the molecular orbital has rotational symmetry around the internuclear bond axis, and $1s$ shows that it is composed of $1s$ atomic orbitals. The antibonding orbital is indicated by an asterisk.

The filling of the two molecular orbitals $\sigma 1s$ and $\sigma^* 1s$ in the formation of diatomic molecules from elements in period 1 is shown in tab. 5.6. The orbitals are filled from the bottom up while observing the Pauli principle (4-1 b). In H_2^+ and H_2 only the bonding orbital is occupied, in H_2^+ by one electron and in H_2 by two electrons. In the latter case the two electrons enter with opposite spin directions as an electron pair. The electron density resembles that in fig. 5.10a, although the electron cloud is more dense throughout. The bond in H_2^+ is called a *one-electron bond* and may be indicated by one dot for the bonding electron, thus in this case $[H \cdot H]^+$. *Electron-pair bonding* as in H_2 is the most common covalent bond. Lewis showed the *bonding* or *shared electron pair* by two dots as in $H:H$. The dots have the same significance as the dash in the old valence-bar formula H–H. Since the dash is more visible in complicated formulas and is easier to write,

Table 5.6. *Electron distribution in molecular orbitals for molecules consisting of two atoms belonging to period 1.*

Orbitals are shown by a dash, an unpaired electron by a dot and an electron pair by two dots. Energy increases upward.

	H_2^+	H_2	He_2^+	(He_2)
$\sigma^* 1s$	—	—	$\cdot$	$:$
$\sigma 1s$	$\cdot$	$:$	$:$	$:$
bond number:	0.5	1	0.5	0

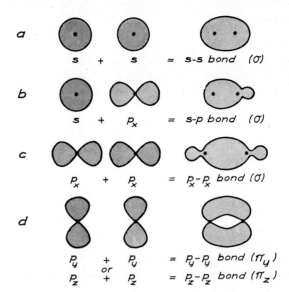

Fig. 5.12. Shape of the electron clouds for different types of bonding orbitals. The line between the two atomic nuclei (the bond axis) has been chosen as the x direction.

it has been most commonly used subsequently. It thus signifies an electron-pair bond and should not be used for other types of bonds. Sometimes the dash is omitted; if OH occurs in a formula, an electron-pair bond is generally understood between O and H.

The one-electron bond is very rare. It is weaker than the electron-pair bond. This follows from the fact that the nuclear distance and bond energy for H_2^+ are 1.06 Å and 255 kJ mol^{-1} and for H_2 0.74 Å and 435 kJ mol^{-1}, respectively. It is seen then how the common electrons draw the nuclei together. The hydrogen molecule ion cannot be isolated but can be detected spectroscopically in an electric discharge in hydrogen at low pressure. It has a short lifetime; it is stable with respect to decomposition, but easily takes up an electron and goes over to the neutral molecule.

In He_2^+, which can be detected spectroscopically in an electric discharge in helium, the antibonding orbital σ^*1s is occupied by one odd electron. This leads to a repulsion which is approximately half as strong as the attraction caused by the two electrons in $\sigma 1s$. Together the three electrons thus give a bond that is roughly half as strong as an electron-pair bond. We speak here of a *three-electron bond* and write schematically $[He \cdots He]^+$.

For He_2 σ^*1s would be filled and cause a repulsion which would be somewhat stronger than the attraction through $\sigma 1s$. The result is that a molecule He_2 is not stable.

Fig. 5.12 shows very schematically the outer contours of the electron clouds of the s and p orbitals and the bonding molecular orbitals that are formed by combination of them (the molecular orbital for the s–s bond in fig. 5.12a is thus a schematic representation of fig. 5.10a). In all cases the condition 3 stated earlier is

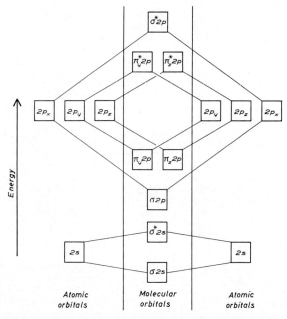

Fig. 5.13. Formation of molecular orbitals for a bond between two like atoms belonging to period 2.

fulfilled, in that the two orbitals that are combined have the same symmetry with respect to the bond axis. The three combinations $s+s$, $s+p_x$ and p_x+p_x give molecular orbitals which are rotationally symmetric about the bond axis. These molecular orbitals are all designated as σ *orbitals* and the bonds they form as σ *bonds*. The combinations p_y+p_y and p_z+p_z, on the other hand, give molecular orbitals without rotational symmetry (often likened to a pair of sausages tied together at the ends). Here one speaks of π *orbitals* and π *bonds*. Notice that mixed combinations such as p_x+p_y violate condition 3 and therefore cannot be formed.

The concentration of electrons along the rotational-symmetry axis in the cloud for the p atomic orbitals causes the overlapping at constant internuclear distance, and consequently the bond strength of the combinations, to increase in the sequence $s-s < s-p < p_x-p_x$. All these σ bonds in general have greater overlapping and thus greater bonding strength than π bonds between the same kinds of atoms.

The appearance of the clouds for antibonding molecular orbitals is of less interest. However, they are characterized by the fact that the electrons avoid the region between the nuclei.

All orbitals with the exception of the s orbitals are markedly directional (see fig. 4.4). When such orbitals of an initial atom participate in σ bonds (for example, of the type $s-p$ or p_x-p_x), the bond directions extending from this atom coincide though possibly with certain deviations, with the directions of the orbitals. In this way, the definite bond directions, which most often characterize covalent bonds, are explained. This also leads to the result that, when the bonds from a central atom to its ligands have a notably covalent contribution, this central

Table 5.7. *Electron distribution in molecular orbitals for molecules consisting of two atoms belonging to period 2.*

Symbols the same as in table 5.6.

	Li₂	(Be₂)	N₂, CO, CN⁻, NO⁺	NO	O₂	O₂⁻	F₂, O₂²⁻	Ne₂⁺	(Ne₂)
σ^*2p	—	—	—	—	—	—	—	·	:
π_y^*2p, π_z^*2p	— —	— —	— —	· —	· ·	: ·	: :	: :	: :
$\pi_y 2p, \pi_z 2p$	— —	— —	: :	: :	: :	: :	: :	: :	: :
$\sigma 2p$	—	—	:	:	:	:	:	:	:
σ^*2s	—	:	:	:	:	:	:	:	:
$\sigma 2s$	:	:	:	:	:	:	:	:	:
bond number:	1	0	3	2.5	2	1.5	1	0.5	0

atom often prefers a particular type of coordination and thus a certain *characteristic coordination number*. With large ligands, however, the lack of space may prevent the attainment of the characteristic coordination number of the central atom (5-2d).

In bonds between elements in period 2, orbitals in the L shell participate almost exclusively. The K orbitals lie so deeply inside the atom that they practically speaking remain as undisturbed atomic orbitals, *nonbonding orbitals* (do not confuse these with antibonding orbitals). The combination of the L orbitals into molecular orbitals is shown in fig. 5.13 for molecules formed from two identical atoms in period 2.

The $2s$ orbitals are combined into one bonding ($\sigma 2s$) and one antibonding (σ^*2s) molecular orbital, analogous to the $1s$ orbitals in fig. 5.11. From the $2p$ orbitals, because of the greater overlapping, the strongest bonding molecular orbital is obtained by combination of the two $2p_x$ orbitals (the bond axis is assumed to be the x direction). The corresponding electron clouds appear approximately as in fig. 5.12c. The bond is a σ bond and the molecular orbital is therefore given the symbol $\sigma 2p$. That $\sigma 2p$ is the most strongly bonding orbital implies also that it has the lowest energy. The corresponding antibonding orbital σ^*2p thereby acquires the highest energy of the $2p$ orbitals.

The atomic orbitals $2p_y$ and $2p_z$ are combined in pairs as in fig. 5.12d into the bonding π orbitals $\pi_y 2p$ and $\pi_z 2p$, which obtain mutually equal energies. The antibonding π orbitals π_y^*2p and π_z^*2p also have mutually equal energies.

The molecular orbitals in fig. 5.13 are filled with electrons from the bottom up. Besides the Pauli principle, Hund's rule (4-1e) must be applied to orbitals with the same energy. All such orbitals are thus occupied first by odd electrons with parallel spin, and thereafter electron pairs are formed. A number of important examples are considered in tab. 5.7.

Furthest to the left in the table the electron distribution in a Li₂ molecule is shown, which has altogether two L electrons. In lithium gas, as in gases of other alkali metals, in addition to single atoms, there is a small amount of diatomic

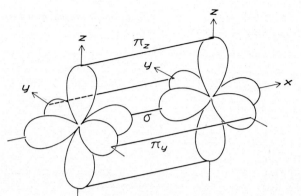

Fig. 5.14. Formation of one σ and two π bonds in N_2.

molecules. The two L electrons in Li_2 fill $\sigma 2s$. We thus have an s–s bond (fig. 5.12 a), analogous to the bond in H_2. The molecule may be written Li–Li. The spherical electron distribution in the s clouds of the atoms causes the overlapping in the combination $s+s$ to be small in general, and the s–s bond is therefore weak. The H_2 bond is an exception, however, because the $1s$ cloud is quite concentrated. The bond strength drops rapidly with increasing atomic radius. The bond energy for H_2 is 435, for Li_2 114 and for K_2 53 kJ mol^{-1}.

In a molecule of Be_2 with a total of four L electrons, the antibonding orbital σ^*2s would also be filled and cause a repulsion, which is somewhat stronger than the attraction through $\sigma 2s$. The molecule therefore cannot be stable.

In N_2 with a total of 10 L electrons, it may be assumed that the bonding $\sigma 2s$ and the antibonding σ^*2s practically compensate each other. Thus, the three bonding $\sigma 2p$, $\pi_y 2p$ and $\pi_z 2p$ remain, of which the first bonds the most strongly. These form the σ bond $2p_x$–$2p_x$ and the two π bonds $2p_y$–$2p_y$ and $2p_z$–$2p_z$. It may be imagined that the two atoms, when they are ready for bond formation, place their p-electron clouds in parallel as in fig. 5.14. As suggested by the horizontal lines, the combinations then take place in pairs resulting in clouds of the types shown in fig. 5.12 c and d. The molecule is thus held together by a *triple bond*. With the symbolic method used here, this is written $|N \equiv N|$. The three horizontal dashes show the three electron pairs that participate in the triple bond, but without showing the difference in character between the three individual bonds. All the valence electrons should be indicated and this is done here by the two other dashes, which show the remaining electron pairs in the L shell, contained in $\sigma 2s$ and σ^*2s. Since these practically compensate each other, in elementary presentations they are often considered as being wholly irrelevant from the bond standpoint, and are called *lone* or *unshared electron pairs*. Because of the symmetry of the molecule, they should be distributed equally between the two atoms, which is also expressed by the formula. In the phosphorus molecule P_2 the electrons in the M shell partake in a bond that is analogous to that in nitrogen.

It may always be expected that a triple bond is a $\sigma\pi\pi$ bond. A *double bond* in general is a $\sigma\pi$ bond and should occur, for example, if only one of the bonds $2p_y$–$2p_y$ or $2p_z$–$2p_z$ in fig. 5.14 is able to be formed. It should be observed that more than one σ bond can never arise between two atoms. A double bond thus cannot be a $\sigma\sigma$ bond.

The different character of the individual bonds incorporated into a multiple bond leads to the result that a double bond need not be twice as strong nor a triple bond three times as strong as a single bond between the same atoms (cf. 5-3 k).

Because of the rotational symmetry of the electron clouds that form the σ bond, they do not hinder the two bound atoms from turning around the bond axis. This *free rotation*, of course, may be restricted or wholly prevented by other atoms joined to the two atoms in question. Forces between these attached atoms, or purely spatial considerations, may then be decisive. With rotation about the bond axis of a π bond on the other hand, the absence of rotational symmetry rapidly leads to a reduction of the overlapping. Therefore, rotatability is lacking or at least very slight in this case.

Multiple bonds arising from one or two π bonds p–p occur mainly between elements belonging to period 2 and to a lesser degree between these elements and the elements with the smallest atomic radii in period 3 (P, S, Cl). The reason for this probably is that the overlapping between the p_y and p_z clouds rapidly diminishes with internuclear distance, and that this distance becomes too great for sufficient overlapping in other combinations of elements. The orientation of d_ε orbitals (see fig. 4.4) permits π bonds p–d and d–d at larger internuclear distances. In this way double bonds may be realized when d electrons begin to appear in period 4 and the following periods (5-3 i, j).

If a molecule is formed from two different kinds of atoms, the two arrangements of atomic orbitals have different energies, and the energy sequence of the molecular orbitals may be altered. If the two kinds of atoms lie near to each other in the same period, however, the energy sequence may be the same. In tab. 5.7, therefore, it has been possible also to include a few such molecules.

For carbon monoxide CO, the cyanide ion CN⁻, and the nitrosyl cation NO⁺, all isoelectronic with N_2, the same electron distribution in the molecular orbitals is found as in N_2. The formulas therefore become $|C\equiv O|$, $[|C\equiv N|]^-$, and $[|N\equiv O|]^+$.

In NO with a total of 11 L electrons, an electron pair lies in $\pi_y 2p$ and an odd electron in $\pi_y^* 2p$. Thus, we have here a three-electron bond. In addition there are two single bonds arising from $\sigma 2p$ and $\pi_z 2p$. The orbitals $\sigma 2s$ and $\sigma^* 2s$ compensate each other and therefore may be considered to contain lone electron pairs. The symbol becomes $|N\!\!\stackrel{..}{\cdot}\!\!O|$.

NO, like H_2^+ and He_2^+, is a molecule in which the total number of electrons in the electron cloud is odd. Such molecules are often called *odd molecules*. They must contain an electron that does not belong to an electron pair, an *unpaired* or *odd electron*. It is chiefly in odd molecules that one- and three-electron bonds are found. They most often try to come together into double molecules in which the unpaired electrons can form a pair; they are therefore uncommon. The presence of unpaired electrons may be detected by the fact that the substance in question becomes paramagnetic (7-4a, b). Substances containing odd molecules are often strongly colored.

Although O_2 has 16 electrons, the rules for filling the molecular orbitals lead to unpaired electrons in both π_y^*2p and π_z^*2p. These give with π_y2p and π_z2p two three-electron bonds. In addition there is a single bond arising from $\sigma 2p$. The oxygen molecule should therefore be written $|O\!:\!\!:\!\!:\!O|$, and the two unpaired electrons with the same spin direction cause oxygen to be paramagnetic in all its states of aggregation. The sulfur molecule S_2 is built up analogously to O_2 and is therefore also paramagnetic.

Tab. 5.7 also shows how the 13 L electrons in the hyperoxide ion O_2^- are distributed. Here we have a three-electron bond and a single bond ($\sigma 2p$). The four remaining electron pairs compensate each other and may be considered as lone pairs. The formula becomes $[\overline{O}\cdots\overline{O}]^-$, and the unpaired electron causes the ion to be paramagnetic.

In F_2 there is only a single bond, coming from $\sigma 2p$. Six electron pairs compensate each other and appear as lone pairs. The formula is $|\overline{\underline{F}}\!-\!\overline{\underline{F}}|$. The bonds are similar in other halogen molecules. For the peroxide ion O_2^{2-}, isoelectronic with F_2, the formula is analogously $[|\overline{O}\!-\!\overline{O}|]^{2-}$. The peroxide ion is diamagnetic in contrast to the hyperoxide ion O_2^-.

Ne_2^+ is an odd molecule with a three-electron bond and occurs with a short lifetime in ionized neon gas.

In a diatomic neutral molecule of neon, Ne_2, all of the molecular orbitals formed from the L shell should be filled. For each bonding orbital there then corresponds an antibonding orbital, so that attraction of the covalent type cannot occur. Therefore, no molecules are formed.

With greater differences between the two combined atoms it is more difficult to decide on the energy sequence of the molecular orbitals, and such an elementary presentation as the foregoing becomes impossible. The same is true if the number of atoms is increased beyond two. The circumstances may then also be complicated by the fact that a molecular orbital and its electron cloud in principle belong to the whole molecule. In many cases the orbital and the cloud belong practically to only two atoms (*localized orbital*). In other cases they quite clearly belong to more than two atoms, perhaps to the whole molecule (*delocalized orbital*). Several examples of these will be presented later.

b. Bond number and covalence. It is often practical to state the *bond number* of a covalent bond. This is defined as half the difference between the number of bonding and the number of antibonding electrons (this definition corresponds to a certain extent to the "bond order" of quantum mechanics). An ordinary single bond has the bond number 1, a double bond the bond number 2, and so on. Bond numbers higher than 3 do not occur. In bonds that can be written just with bond dashes, the bond number is thus equal to the number of dashes. However, it does not need to be a whole number; in the one-electron and three-electron bonds it is 0.5, and in 5-3e bond numbers are discussed that are not integral for other reasons. In tab. 5.6 and 5.7 the bond numbers for the molecules included therein are given.

If the bond number increases for the bond between the same pair of elements, the bond strength also increases. However, there is no proportionality between bond number and bond strength (cf. 5-3k).

The sum of the bond numbers from an atom to all its ligands is called the *covalence* of the atom. The covalence is thus in general equal to the number of bond dashes extending out from the atom. If all the ligands are bound by single bonds the covalence is equal to the coordination number, but if bonds with higher bond number are present, it is greater than the coordination number.

c. Valence bond. In the presentation of electron-pair bonding in 5-3a only one bonding molecular orbital per single bond was considered. The other molecular orbitals compensate each other so completely that they may be neglected from the standpoint of bonding (they are said to be occupied by lone or unshared electron pairs). It is found in general that the bonding orbital may be thought of as being formed by a combination of one atomic orbital from each of the atoms that are combined. The complete, free atoms are thus brought together and combination of their atomic orbitals is brought about, while still observing the conditions 1, 2 and 3 in 5-3a.

Conditions 1 and 2 enable an atom to use for bonding all orbitals in the valence electron shell and the orbitals available in the shell nearest inside with energies lying very close to them (in the transition elements). In this way the maximum number of atomic orbitals that can be used for bonding is determined. How many of these that are actually used then depends on other circumstances, for example the total number of electrons, the energy relationships and spatial considerations.

This method of combination of available atomic orbitals (*valence bonding*) can be generally applied to covalent bonding, and permits an elementary and graphic description of more complicated cases of bonding. It also adheres quite closely to the conceptions the chemist usually has concerning the occurrence of a bond. For these reasons it is frequently used in the following discussions.

Starting from the valence electron distributions in tab. 4.2, the formation of some of the bonds already considered may now be schematically written:

$$H \cdot + \cdot H \to H-H \qquad |\dot{N} \cdot + \cdot \dot{N}| \to |N \equiv N|$$

$$|\overline{F} \cdot + \cdot \overline{F}| \to |\overline{F}-\overline{F}| \qquad H \cdot + \cdot \overline{F}| \to H-\overline{F}|$$

The same bonds as in 5-3a are thus obtained if the half-filled orbitals of the participating atoms are combined. Analogously, we have:

$$\text{H} \cdot + \cdot \ddot{\underline{O}}| \;\;\begin{matrix}\text{H}\\ \cdot\\ +\\ \end{matrix}\;\; \rightarrow \;\; \text{H}\!-\!\begin{matrix}\text{H}\\ |\\ \underline{\text{O}}|\\ \end{matrix} \qquad\qquad \text{H} \cdot + \cdot \dot{\underline{\text{N}}}| \;\;\begin{matrix}\text{H}\\ \cdot\\ +\\ \cdot\\ \text{H}\\ \end{matrix}\;\; \rightarrow \;\; \text{H}\!-\!\begin{matrix}\text{H}\\ |\\ \underline{\text{N}}|\\ |\\ \text{H}\\ \end{matrix}$$

The covalence of an atom is determined in all these cases by the number of its unpaired electrons. However, this depends here only on the accessability of the electrons and does not constitute a rule. As an example, N has four valence electron orbitals ($2s$, $2p_x$, $2p_y$, $2p_z$; the "hybridization" of these discussed later may here be overlooked) but it cannot bond 4 H with these since the system, an uncharged group NH_4, would contain an odd electron that cannot be accommodated. If this is removed, however, all four orbitals may be used. This occurs in the ammonium ion, which may be formed according to the scheme:

$$\begin{matrix}\text{H}\\ |\\ \text{H}\!-\!\underline{\text{N}}|\\ |\\ \text{H}\\ \end{matrix} + \text{H}^+ \;\rightarrow\; \left[\begin{matrix}\text{H}\\ |\\ \text{H}\!-\!\text{N}\!-\!\text{H}\\ |\\ \text{H}\\ \end{matrix}\right]^+ \tag{5.16}$$

The two participants in this process here have not each contributed its own electron to the bonding pair, but rather NH_3 has given the whole pair. An atom or molecule that contributes one of its lone (unshared) electron pairs to a bond is called an *electron-pair donor* (in this case NH_3). The atom or molecule which receives this electron pair is called an *electron-pair acceptor* (in this case H^+). An electron-pair donor that thus likes to be bound to a hydrogen nucleus as well as to other atoms with a deficiency of electrons in some valence-electron orbital is also said to be *nucleophilic* (Gk. φίλος, friend). An electron-pair acceptor, which likes to be bound to the lone electron pair of an atom, is said to be *electrophilic*.

The valence-bond method would give for the oxygen molecule the structure $\overline{\underline{O}}\!=\!\overline{\underline{O}}$, and thus could not explain its two unpaired electrons. It is less suited to explaining one- and three-electron bonds, and its usefulness in treating delocalized orbitals is limited.

In 5-1a it was mentioned that in the quantum mechanical treatment of the chemical bond it is necessary to work with approximate methods. One of these, the *valence-bond* (VB) *method*, works with a procedure similar to that just presented. It is graphic but suffers from limitations both as regards accuracy and applicability. Another method, the *molecular-orbital* (MO) *method*, follows closely the presentation of 5-3a. It is often less graphic but more accurate and more adaptable. Depending on the type of problem, however, sometimes one, sometimes the other may be most appropriate. In general descriptions parts are often borrowed from both methods.

d. Hybridization. In the treatment of the bonds in more complicated molecules the question arises as to which atomic orbitals are to be used for bonding. It is apparent that the electron cloud of a central atom often is so strongly influenced

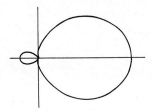

Fig. 5.15. Outer contour of the electron cloud of a tetrahedral sp^3 orbital.

by the force field from the surrounding atoms that the original orbitals are greatly altered. They are combined and transformed into new orbitals with other geometric properties than the initial orbitals have. This transformation is called *hybridization*. The new orbitals, the *hybrid orbitals*, are always equal in number to the initial orbitals. Their electron clouds in general are distributed very uniformly in space and they can be thought of as mutually equivalent when they participate in bonds to atoms of the same kind. The energetic cause of hybridization lies in the fact that the new orbitals under the actual circumstances can give stronger bonds than the original orbitals. The energy of the system can thus be decreased through hybridization.

Hybridization is very common in covalent bonds. This is first illustrated here by an especially clear-cut case, the carbon atom.

One of the basic principles of structural organic chemistry is the "tetrahedral carbon atom." By this is meant that the carbon atom can form four mutually equivalent single bonds, the axes of which are directed from the center of a tetrahedron to its corners. This is certainly difficult to reconcile with what has been said up to now about the formation of covalent bonds and the appearance of the electron clouds. The bonds of the carbon atom must be formed by the electrons in the L shell, whose three $2p$ orbitals should give bond axes at right angles to each other, while the $2s$ orbital should give a fourth, weaker bond in an arbitrary direction.

Four orbitals, ns, np_x, np_y, and np_z, can, however, be coalesced into four hybrid orbitals. In this case we speak in particular of sp^3 *hybridization*, and thereby indicate the number and kind of the initial orbitals. The electron clouds of the hybrid orbitals are tetrahedrally directed (sp^3 *orbitals*, *tetrahedral orbitals*) and the outer contours of the electron cloud of one such orbital appears approximately as in fig. 5.15. The sp^3 orbitals can only participate in σ bonds, and since each of the four clouds is highly elongated, the overlapping at constant internuclear distance, and thus the bonding ability, becomes greater than for the original orbitals with the spherical s cloud and the three dumbbell-shaped p clouds. In this way the reason for hybridization can be seen.

Simple examples of these bonds are found in methane, CH_4, and tetrachloromethane (carbon tetrachloride), CCl_4. Here the bond axes are exactly tetrahedrally directed; the angle between them, the bond angle, is equal to the "tetrahedral angle" of 109.47°. In a plane, however, we may write the structural formulas

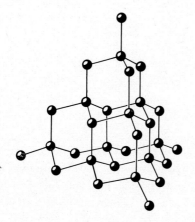

Fig. 5.16. Portion of the structure of cubic diamond.

$$\begin{array}{ccc} & \text{H} & & & |\overline{\text{Cl}}| \\ & | & & & | \\ \text{H}-\text{C}-\text{H} & \text{and} & |\overline{\text{Cl}}-\text{C}-\overline{\text{Cl}}| \\ & | & & & | \\ & \text{H} & & & |\underline{\overline{\text{Cl}}}| \end{array}$$

If the environment of the carbon atom is less symmetric, the forces between the bound atoms or groups may cause departure from the tetrahedral angle.

An important case of sp^3 hybridization is found in diamond, where each carbon atom is tetrahedrally surrounded by four other carbon atoms (fig. 5.16). The diamond crystal is held together in this way by a three-dimensional framework of covalent bonds. The whole crystal may be considered as a giant molecule of carbon, and it cannot be broken down without rupturing these strong bonds.

Because the sp^3 orbitals can only participate in σ bonds, two or three such orbitals cannot give double or triple bonds to another atom. If a carbon atom has fewer than four ligands, it thus should not be able to use all four of the sp^3 orbitals. Since the system seeks the lowest possible energy, and a bond decreases the energy, the carbon atom tries to use as many orbitals as possible for bonding. This can be made possible by other types of hybridization.

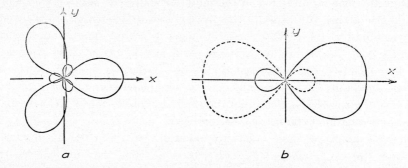

Fig. 5.17. Electron clouds for (a) the three sp^2 orbitals, and (b) the two sp orbitals. (C. A. Coulson.)

Fig. 5.18. (*a*) The ethylene carbon atoms with sp^2 clouds in bonding position. Each carbon atom has in addition a p cloud with axis of elongation parallel to the z direction (normal to the picture plane). (*b*) The acetylene carbon atoms with sp clouds in bonding position. Each carbon atom has in addition two p clouds with axes of elongation parallel to the y and z directions (the latter normal to the picture plane). (C. A. Coulson.)

The first of these is *sp² hybridization*, which makes use of only two p orbitals. The three hybrid orbitals give a cloud oriented in a plane as in fig. 5.17*a*. The sp^2 orbitals also can be used only for σ bonds. The third p orbital of the carbon atom is undisturbed by the hybridization, and the axis of elongation of its electron cloud is directed normally to the plane of the figure. This orbital can, as usual, also give π bonds.

When a carbon atom has three ligands, one can in general assume sp^2 hybridization to be present. A typical example is ethylene, C_2H_4, where the two carbon atoms are bound to each other by a σ bond as in fig. 5.18*a*. In addition, π bonding occurs between the unhybridized p orbitals. The four hydrogen atoms are bound with σ bonds by the remaining sp^2 orbitals. The formula may thus be written

$$\begin{array}{c}\text{H}\qquad\qquad\text{H}\\ \diagdown\qquad\diagup\\ \text{C}=\text{C}\\ \diagup\qquad\diagdown\\ \text{H}\qquad\qquad\text{H}\end{array}$$

All of the atoms lie in the same plane because of the π bond. The angle between the C=C bond and each C–H bond is close to 120°.

The other type is *sp hybridization*. In this only one p orbital takes part. The clouds for the two hybrid orbitals fall with their axes of elongation along the x axis in fig. 5.17*b* but extended in opposite directions. Two p orbitals are undisturbed by the hybridization, and the axes of elongation for their clouds lie along the y and z axes (the latter normal to the picture plane) in fig. 5.17*b*.

In the carbon atom, sp hybridization occurs when the atom has one or two ligands. An example is acetylene, C_2H_2, wherein the two carbon atoms are united by a σ bond as in fig. 5.18*b*. In addition the p orbitals are bound in pairs so that two π bonds are formed. The two hydrogen atoms are bound with two σ bonds by the remaining sp orbitals. The whole molecule becomes linear and the formula is H–C≡C–H. The covalence of carbon is thus still four.

sp hybridization of the carbon atom is believed to occur both in CO and CO_2. σ bonding then takes place in CO between a hybridized sp orbital (oriented along the x direction) of the carbon atom and the p_x orbital of oxygen. In addition there are two π bonds between the respective p_y and p_z orbitals. Under this supposition, we obtain a $\sigma\pi\pi$ bond and the formula $|C\equiv O|$ as was assumed in 5-3a. The lone electron pair of the carbon atom fills the remaining sp orbital. The bonds in carbon dioxide are discussed in 5-3e.

The ammonium ion NH_4^+ is isoelectronic with CH_4 and also has a tetrahedral structure. Consequently sp^3 hybridization is present here as well. This applies also to the ammonia molecule NH_3. The two molecules, which may be written in a plane

$$\begin{bmatrix} \text{H} \\ | \\ \text{H—N—H} \\ | \\ \text{H} \end{bmatrix}^+ \quad \text{and} \quad \begin{array}{c} \text{H} \\ | \\ \text{H—N}| \\ | \\ \text{H} \end{array}$$

thus have analogous bonds. In NH_3, however, one of the tetrahedrally directed hybrid orbitals is not used for bonding but is occupied by the lone electron pair (cf. CO above). The bond angle H–N–H in NH_3 is 108°, quite close to the tetrahedral angle.

The NH_3 molecule thus forms a trigonal pyramid with N at the apex and H at each corner of the base (pyramidal structure). When the molecule exists in a gas phase, it "inverts" (approximately as when an umbrella turns inside out) and returns to its initial structure 24×10^9 times per second. The nitrogen atom thus lies equally often on either side of the hydrogen atom plane. The inversion frequency of 24×10^9 Hz lies within the radio-frequency range and has therefore been used for the control of precision time keepers.

In the H_2O molecule in the gas phase the bond angle H–O–H is 104.5°. It was assumed earlier that the bonds were formed from the p orbitals of the oxygen atoms. This should give a bond angle of 90°, but since the charge distribution in the molecule (see 5-3g) confers a positive charge on the hydrogen atoms, these should repel each other, resulting in an increase in the bond angle. This assumption was believed to be supported by the fact that the bond angle in H_2S is 92.3°. Here the repulsion effect should be less, partly because of a lower positive charge on the hydrogen atoms, and partly because the larger size of the central atom holds these at a greater distance from each other.

It is probable, however, that in the H_2O molecule in the gas phase some, but not complete, sp^3 hybridization of the orbitals of the oxygen atom occurs. This leads to a bond angle that lies between 90° and the tetrahedral angle, although clearly nearer to the latter. In ice the bond angle is equal to the tetrahedral angle and here the hybridization is doubtlessly practically complete. Also for the H_2O molecules in liquid water the hybridization is presumably very nearly complete. The oxonium ion H_3O^+, isoelectronic with NH_3, probably has a similar structure, and in that case we may depict[1] analogous bonds for these molecules thus:

---

[1] If an atomic group is not linear this is often shown in the following by writing it bent, even if the actual bond angle is not represented. Sometimes this is not done for lack of space.

$$\left[\begin{array}{c} \text{H} \\ | \\ \text{H—O|} \\ | \\ \text{H} \end{array}\right]^+ \quad \text{and} \quad \begin{array}{c} \text{H} \\ | \\ \text{H—O|} \end{array}$$

In general, tetrahedrally hybridized orbitals are assumed for the oxygen atom. Complete, tetrahedral sp^3 hybridization is generally present (for exceptions see 5-3i) in covalent bonding with fourfold coordination. Examples: SiO_4^{4-}, PO_4^{3-}, SO_4^{2-}, ClO_4^-, BeF_4^{2-}, BF_4^-. But it is most often sufficient that, as in NH_3, H_3O^+ and H_2O, the total number of single bonds (σ bonds) + lone electron pairs around the central atom shall be four for sp^3 hybridization to be assumed. Beautiful examples of this are provided by the oxochlorate ions. In all of them tetrahedrally directed sp^3 orbitals are present, whether they contain shared or lone electron pairs. As a result, the structures shown schematically in fig. 5.19 are obtained. It is clear why the ion ClO_2^- is not linear but bent and why the ion ClO_3^- is not plane but pyramidal.

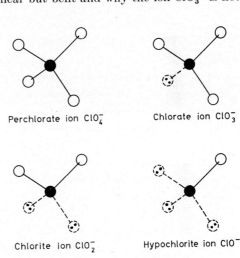

Fig. 5.19. Schematic pictures of the sp^3 hybridization in the oxochlorate ions.

e. Unsymmetrical and delocalized orbitals. If two atoms of different kinds are bound to each other, in general the electron density in the electron clouds of the bonding orbitals is unsymmetrically distributed. Fig. 5.20a shows this for the σ cloud for a bond in a molecule AB. The density distribution for a π cloud in a corresponding case is shown in fig. 5.20b. In each cloud the two electrons spend more time in the vicinity of B than in the vicinity of A. It may be said then that the bond acquires a certain degree of *polarity*; the electrons are to a certain extent displaced toward the distribution they would have in a purely ionic bond A^+B^-.

A representation of the density distribution in the electron cloud of the bonding orbitals such as in fig. 5.20 requires a very penetrating knowledge of the bond. The state of such a bond can seldom be shown with the symbols and rough formula expressions which have been developed up to this point. In such cases it has often

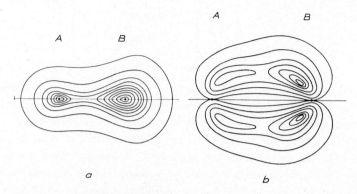

Fig. 5.20. Electron density distribution in the electron cloud of an orbital that bonds the atoms A and B. (*a*) σ bond. (*b*) π bond. (Inga Fischer-Hjalmars.)

been necessary to give the state as an intermediate stage between two or more imagined states that are easily described and in general become some sort of limiting states. In the present example, perhaps a purely covalent bond A–B and a purely ionic bond A^+B^- would be appropriate limiting states.

Especially in organic chemistry such intermediate states have been called *mesomeric states*, or it is said that *mesomerism* occurs (Gk. μέσος, between, in the middle; μέρος, part). The expression may be quite useful so long as it is remembered that the choice of the limiting states is a practical matter and therefore often arbitrary. It is also said that the intermediate state is the result of *resonance* between the limiting states. This word is less proper, however, because it has been found that it often leads to the completely erroneous idea that the state oscillates between the limiting states and that a mean value of these is observed.

After the proper limiting states have been chosen, it is often customary to give the relative "weights" in which they must be "mixed" to obtain the actual state. In this way a fairly good description of the state may be obtained and it is no doubt convenient to state, for example, that the ionic contribution in the molecules HF, HCl, HBr and HI is 43, 17, 13 and 7%, respectively. However, the very formal nature of such values must not be forgotten.

In the presence of π bonds, which are necessary for the occurrence of double and triple bonds, the bonding orbitals often belong not to just two atoms but to a greater portion of the molecule. They are thus *delocalized*. A condition for delocalization is that an atomic orbital of an atom shall be able to overlap effectively atomic orbitals of more than one neighboring atom. In an ordinary structural formula the possibility for this is most often seen when a double bond can be extended from one atom to any one of several neighboring atoms with equal justification.

Also, in the case of delocalization, for lack of a proper symbolic system, the state must often be described as a mesomerism, implying an intermediate state between structures that can be written down. A very simple example is SO_2. The orbitals of the sulfur atom are sp^2 hybridized. Two of the sp^2 orbitals are used for σ bonds with the oxygen atoms and the third is occupied by an unshared electron pair. The molecule is therefore bent, with a bond angle of about 120°. The p orbital lying normal to the sp^2 plane of the sulfur atom can interact with the parallel p orbital of either of the oxygen atoms, so that a delocalized π orbital arises. The mesomeric state may be expressed by a double-headed arrow between the limiting states, thus:

$$\overline{S} \diagup\!\!\!\diagdown \quad \leftrightarrow \quad \overline{S} \diagup\!\!\!\diagdown$$
$$|\underline{O}|\ |\underline{O}| \qquad |\underline{O}|\ |\underline{O}|$$

The electron cloud of the delocalized π orbital is spread out along the whole molecule on both sides of the molecular plane. Since this orbital contains two electrons and participates in two sulfur–oxygen bonds, the π-bond contribution per bond corresponds to the bond number 0.5 (5-3b). Each σ bond has the bond number 1 so that the total bond number in each bond becomes 1.5.

Ozone, O_3, and the nitrite ion, NO_2^- (isoelectronic with ozone) are analogous to SO_2.

In all cases of mesomerism, the limiting states may differ from each other only with respect to the electron distribution while the mutual positions of the atomic nuclei must remain unchanged.

Other examples of delocalized orbitals are found in the isoelectronic BO_3^{3-}, CO_3^{2-}, NO_3^-, BF_3 and in the analogously built SO_3. The central atom in all of these is sp^2 hybridized and thus lies in the same plane as the ligands. The p orbital lying perpendicular to the sp^2 plane has equal possibilities of combining with the parallel p orbital of any of the ligands. In this way a delocalized π orbital is formed. If this is expressed as a mesomerism, we have, for example, for NO_3^-:

$$\left[\ \begin{array}{c}|\underline{O}|\\ \|\\ \diagup N \diagdown\\ |\underline{O}|\ \ |\underline{O}|\end{array} \leftrightarrow \begin{array}{c}|\overline{\underline{O}}|\\ |\\ \diagup N \diagdown\!\!\!\diagdown\\ |\underline{O}|\ \ |\underline{O}|\end{array} \leftrightarrow \begin{array}{c}|\overline{\underline{O}}|\\ |\\ \diagup\!\!\!\diagup N \diagdown\\ |\underline{O}|\ \ |\underline{O}|\end{array}\ \right]^{-}$$

All three of the bonds from the central atom thus become equivalent and have the bond number 1.33. Structure determinations show in this case plane, equilateral molecules. All three of the central atom–ligand distances are equal and shorter than would be expected for single bonds.

Delocalized bonds are very common in organic chemistry. Benzene C_6H_6 must be mentioned as an example. The equivalence of all the carbon atoms and carbon–carbon bonds in the benzene molecule (which is shown, among other ways, by the fact that all of the carbon–carbon distances are exactly equal, 1.39 Å) has led the structure to be considered as a mesomer between a number of structures, of which a and b below are the most important, and for reasons of symmetry are equally likely:

H H
 | |
 C C
H\\ // \\ // \\H   H\\ // \\ // \\H
 C   C   C   C
 || ||
 C   C   C   C
H/ \\ // \\ //H   H/ \\ // \\ //H
 C   C   C   C
 | |
 H   H
 a   b

All of the carbon atoms are sp^2 hybridized and the sp^2 orbitals are used for σ bonds between the carbon atoms and to the hydrogen atoms. This requires that the whole molecule be plane. The six p orbitals, which stand normal to the molecular plane, are combined into delocalized π orbitals, whose electron clouds follow the carbon chain around the whole molecule on both sides of the molecular plane. These π orbitals take up the six remaining electrons.

If altogether six electrons participate in π orbitals over the six carbon–carbon bonds a π-bond contribution per carbon–carbon bond with bond number 0.5 is obtained (5-3b). Since each σ bond has the bond number 1, the total bond number for each carbon–carbon bond becomes 1.5. The same value is obtained directly if it is assumed that the structures a and b partake with equal weight in the actual structure.

The possibility that the π electrons have to move around the whole benzene ring leads to easy transfer of electrical forces from one part of the molecule to another. An atom or atomic group bound to one carbon atom can thus influence the properties of the other parts of the ring. This is typical of all ring systems

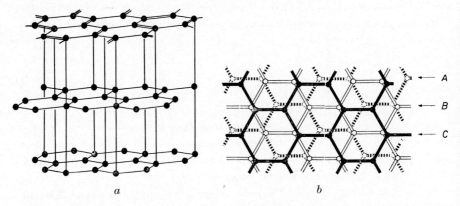

Fig. 5.21. The crystal structure of graphite. (a) The most common form, *hexagonal graphite*. In the uppermost layer a possible combination of double bond positions is shown. (b) A somewhat rarer form, *trigonal (rhombohedral) graphite*, which also contains parallel six-ring layers, which are here seen in a direction nearly perpendicular to the layers. With the letters of the figure this form is built up with the layer sequence *ABCABC* .... The hexagonal graphite is built up of layers in only two mutual positions, e.g. *A* and *B* in the sequence *ABABAB* ....

$I_A - E_B$, that is, the ionization energy I_A for A minus the electron affinity E_B for B (cf. 5-2b). In the second case the ions A⁻ and B⁺ are formed for which the energy $I_B - E_A$ is required. A and B thus have the same electronegativity if $I_A - E_B = I_B - E_A$, that is, $I_A + E_A = I_B + E_B$. If A is more electronegative than B, that is, if it is easier to form A⁻B⁺ than A⁺B⁻, then $(I_A + E_A) > (I_B + E_B)$. The quantity $(I_A + E_A)$ is thus a valid measure of the electronegativity of A.

The values of I and E must, however, apply to the atom in the energy state in which it partakes in the bond (for example, for a carbon atom in CH_4, the hybridized state sp^3). This means that the electronegativity of a given atom depends on its state of bonding. In general only one value is given corresponding to the most common bonding state. For example, for the halogens with the ground state as the bonding state, it may be ascertained from tab. 5.1 and 5.2 that the values in fig. 5.23 are nearly proportional to $(I + E)$.

Fig. 5.23 shows the electronegativity values (x) in Pauling's scale since this is most used by chemists. Hydrogen and the elements on the boundary between metals and nonmetals lie in the middle of the scale with values of about 2. The nonmetals have higher values, with fluorine ($x = 4.0$) as the most electronegative element. The metals have in general lower values than 2, with cesium and francium the least electronegative ($x = 0.7$). (An easy point to remember is that x for the elements in period 2 changes in steps of 0.5). The transition elements have a somewhat irregular distribution within the enclosed x intervals.

The electronegativity values must be treated with care and judgment, and they allow only estimates at best. They are nevertheless often very useful. For example, they can give an idea of the polarity of a bond. From comparisons of estimated values of the percent contribution of ionic bonding in different bonds, it has been found possible to say for a bond A–B that, approximately,

$$\text{percent ionic-bond contribution} = 16|x_A - x_B| + 3.5|x_A - x_B|^2 \qquad (5.17)$$

For an electronegativity difference $|x_A - x_B| = 2.1$ the ionic-bond contribution is about 50 %. It is found that the ionic-bond contribution in the alkali halides varies from 32 % for LiI to 91 % for CsF. For NaCl it is 49 %. For the hydrogen halides HF, HCl, HBr and HI, the ionic-bond contribution is found to be 43, 17, 13 and 7 %, respectively.

It can be assumed that the bonds in metal fluorides have at least 50 % ionic bond contribution. The same is true of a large number of metal oxides. Most metal oxides are considered at first hand to be ionic compounds.

For *formulation* and *nomenclature* according to 5-2a, when strong covalent bonding is present, it may be difficult to determine the charge distribution. So that in this case the formula and name shall not depend on electronegativity values that are sometimes variable and uncertain, in binary compounds of *nonmetals* the component written first is the one that appears first in the series

$$\text{B, Si, C, Sb, As, P, N, H, Te, Se, S, At, I, Br, Cl, O, F}$$

In the name, the element that comes after in the series takes the ending *-ide*. Examples: H_2S, dihydrogen sulfide; Cl_2O, dichlorine oxide; OF_2, oxygen difluoride.

In atomic groups that contain three or more elements, however, the order should as far as possible agree with the arrangement in which the atoms are bound in the group; for example, the thiocyanate ion is written SCN$^-$, not CNS$^-$.

g. Oxidation number. In the treatment of chemical processes that involve electron transfer from one atom or atomic group to another, that is, what is called in chemistry oxidation or reduction (16a), the concept *oxidation number* is of great use.

The oxidation number of an atom gives the charge of the atom under certain assumptions concerning the charge distribution in the combination (for example, molecule) in which the atom takes part. In deciding the oxidation number of an atom the following rules apply:

1. The oxidation number of a monatomic ion is equal to its charge.
2. The oxidation number of an atom in the element is zero. In typically metallic phases the oxidation numbers of all the atoms can in general be assumed to be zero.
3. In a covalent compound with known structure, the oxidation number of each atom is equal to the net charge of the atom after all the electrons shared in bonds are transferred to the most electronegative of the two atoms participating in the bonding. Electrons that are shared between atoms of the same kind are commonly divided equally between them.
4. In compounds of unknown structure the oxidation number of an atom can be determined if the oxidation numbers of the other atoms are known. The sum of the oxidation numbers of all atoms in a molecule is equal to the net charge of the molecule.

The oxidation number of an atom determines what is customarily called the *oxidation state* of the atom. The oxidation number should always be written with Roman numerals (for zero and nonintegers, however, Arabic numerals must be used).

The application of the rules given above is illustrated in the following examples, in which the oxidation number is placed as a superscript index by the symbol of the element:

$Na^{+I}Cl^{-I}$ $Mg^{+II}Cl_2^{-I}$ H_2^0 O_2^0 C^0 (in diamond or graphite)

$H^{+I}Cl^{-I}$ $Li^{+I}H^{-I}$ $O^{+II}F_2^{-I}$ $Cl^{+I}F^{-I}$ $H_2^{+I}O^{-II}$ $N^{-III}H_3^{+I}$ $[N^{-III}H_4^{+I}]^+$

$[N^{+V}O_3^{-II}]^-$ $S^{+IV}O_2^{-II}$ $S^{+VI}O_3^{-II}$ $[S^{+VI}O_4^{-II}]^{2-}$ $[Mn^{+VII}O_4^{-II}]^-$

Ordinarily the oxidation number is shown in formulas only for clarification, especially for an atom that can occur in several oxidation states. When there is no doubt that the sign is positive, this is often omitted. Thus we write $Fe^{II}Cl_2$ (or $\overset{II}{Fe}Cl_2$). The name of the compound is written iron(II) chloride (read "iron-two chloride"), and we speak of iron(II) compounds.

In *names of ions* the charge of the ion is sometimes given in a manner which must not be confused with the notation using oxidation numbers. The charge is then written with Arabic numerals within parentheses immediately after the name. In this system the ion Fe^{2+} will be called iron(2+) ion and the compound $FeCl_2$ iron(2+) chloride. The ion $Fe(CN)_6^{4-}$ can thus be called either hexacyanoferrate(II) ion or hexacyanoferrate(4−) ion.

The oxidation numbers of the elements show the following general features (cf. also fig. 5.8):

It is very common for an element with an odd (or even) group number to occur with odd (even) oxidation number. If the element possesses several characteristic oxidation numbers, the difference between them, therefore, is frequently an even number. This depends, of course, on the tendency to avoid unpaired electrons.

Fluorine, the most electronegative element, has the oxidation number −I in all compounds with other elements.

Oxygen, which next to fluorine is the most electronegative, has the oxidation number −II in compounds with all other elements except fluorine (see OF_2 above). In the peroxide ion $[|\overline{O}-\overline{O}|]^{2-}$ and in the hyperoxide ion $[\overline{O}\cdots\overline{O}]^{-}$ (5-3a) oxygen has the oxidation numbers −I and −0.5, respectively.

The mean position of hydrogen on the electronegativity scale leads to the result that it has the oxidation number +I when it is bound to nonmetals and −I when it is bound to metals.

The other nonmetals in groups 15, 16 and 17 all have several oxidation numbers. The elements in group 15 have numbers from −III to +V, the elements in group 16 (except oxygen) numbers from −II to +VI (usually even) and the elements in group 17 (except fluorine) numbers from −I to +VII (usually odd). Here the highest oxidation number is equal to the unit's digit of the group number and from there it can extend downward 8 units. At the lowest and highest oxidation numbers a noble-gas shell is attained, in the first case by filling up, in the second by removal of the valence-electron shell.

The metals in groups 1 and 2 almost always have the oxidation number +I and +II, respectively, that is, that to be expected from the number of their valence electrons.

In groups 13 and 14 the elements in periods 2 and 3 most often have the oxidation numbers +III (B, Al) and +IV (C, Si), respectively. The elements in the higher periods usually have the numbers +I and +III in group 13 (Ga, In, Tl), and the numbers +II and +IV in group 14 (Ge, Sn, Pb) (cf. 5-2f).

With the transition elements it is very common, it could almost be said typical, that every element shows several oxidation numbers (the exceptions consist mainly of certain lanthanoids).

The more electronegative the ligands of an atom are, the more strongly are electrons drawn from it and thus the greater is the possibility that it will show a high oxidation number. Especially high oxidation numbers are therefore found when the ligands are O and F.

Examples:

$S^{VI}O_3$, $S^{VI}O_4^{2-}$, $S^{VI}F_6$, $Cl_2^{VII}O_7$, $Cl^{VII}O_4^-$, $Mn_2^{VII}O_7$, $Mn^{VII}O_4^-$, $Fe^{VI}O_4^{2-}$, $Os^{VIII}O_4$, $Cu^{III}F_6^{3-}$

Since the oxidation number is based on certain conventions regarding the electron distribution, it takes on a quite formal character. In certain cases, primarily with close-lying or uncertain values of electronegativity of the bound atoms, their oxidation numbers may be quite uncertain.

Neither is it feasible to draw far-reaching conclusions concerning the charge distribution in a molecule from the oxidation numbers. The oxidation numbers correspond to the charge distribution that the molecule would have if all its bonds were purely ionic bonds. With small ionic bond contribution, therefore, the true charge distribution is quite different. Since the bonding electrons in covalent bonding belong to two or more atoms, it is quite unrealistic here to place all of the valence electrons on particular atoms.

Sometimes the *formal charge* of one or several atoms is given. This is the net charge of the atom if all electrons shared in bonding are divided equally between the two bound atoms. The sum of the formal charges of all the atoms in a molecule, of course, is equal to the net charge of the molecule. The following examples (formal charges in parenthesis) can easily be derived from structural formulas given earlier:

$H_2^{(0)}O^{(0)}$ $N^{(0)}H_3^{(0)}$ $[N^{(+)}H_4^{(0)}]^+$ $[N^{(+)}O_3^{(\frac{2}{3}-)}]^-$ $[S^{(2+)}O_4^{(-)}]^{2-}$

With bonds of largely covalent character the formal charges give a truer picture of the charge distribution than the oxidation numbers.

Structures in which neighboring atoms have formal charges of the same sign seem to be less stable (Pauling). Of the limiting structures given below, which at first hand would be expected for N_2O, isoelectronic with CO_2 (5-3e), the last one for this reason probably does not occur:

$$\begin{array}{ccc} (-) \ (+) \ (0) & (0) \ (+) \ (-) & (2-) \ (+) \ (+) \\ \underline{N}=N=\overline{O} & |N\equiv N-\overline{O}| & |\underline{N}-N\equiv O| \end{array}$$

h. The octet rule. It has already been said (rules 1 and 2 in 5-3a and 5-3c) that for covalent bonding an atom can use the orbitals in the valence-electron shell and possibly orbitals with very close-lying energy in the shell lying closest inside it (in the transition elements). In periods 1 and 2, therefore, only the orbitals in the valence-electron shell come into question, that is, in period 1 the orbital $1s$ and in period 2 the orbitals $2s$, $2p_x$, $2p_y$ and $2p_z$. The result is that the highest possible covalence is one for hydrogen and four for the elements in period 2. In period 2, even if all four orbitals are not used for bonding, the valence electrons are nevertheless in most cases distributed around the nuclei in the molecule in such a manner that each nucleus is surrounded by four electron pairs, that is, eight valence electrons. In this way, the available orbitals are used to the maximum for bonding, which leads to a decrease in energy and thus greater stability. In the molecules N_2 and NO_3^-, for example, the electron distribution therefore becomes:

Chemical Bonding

$$|N\equiv N| \text{ and } \left[\begin{array}{c} |\underline{O}| \\ \parallel \\ {}_{|\underline{O}|}\diagup^{N}\diagdown_{|\underline{O}|} \end{array}\right]^{-} \text{ but not } \overline{N}-\overline{N} \text{ and } \left[\begin{array}{c} |\overline{O}| \\ | \\ {}_{|\underline{O}|}\diagup^{N}\diagdown_{|\underline{O}|} \end{array}\right]^{-}$$

This rule, known as the *octet rule*, is illustrated by most of the formulas given earlier in this chapter. Its significance was understood as early as 1916 by Lewis.

When after period 2 other orbitals beside s and p orbitals can also be used for bonding, exceptions to the octet rule appear. If the transition elements are overlooked, it still applies very commonly, because the s and p orbitals of the valence-electron shell have lower energy than the higher orbitals (d orbitals, etc.) and therefore are preferred for bonding. In many cases in which structural formulas can be written that fulfill the octet rule, clear departures from it occur. For example, in the sulfate ion the S–O bonds very likely have some double-bond character, so that the formula usually written

$$\left[\begin{array}{c} |\underline{O}| \\ | \\ |\underline{O}-S-\underline{O}| \\ | \\ |\underline{O}| \end{array}\right]^{2-}$$

is not completely correct. A similar situation applies in a considerable number of other oxo anions, but even so, these are written with single bonds in the following.

The application of the octet rule is illustrated mainly in 5-6.

i. Participation of higher orbitals in bonding. Both nitrogen and phosphorus form trichlorides (NCl_3, PCl_3), but only phosphorus forms a pentachloride PCl_5. Here is clearly an example of a case where d orbitals begin to be used for bonding in period 3, with a departure from the octet rule as a result.

The molecules PF_5, PCl_5 and PBr_5 (PI_5 does not exist, probably because five iodine atoms, on account of their size, cannot simultaneously approach sufficiently closely to a phosphorus atom) in the gas phase have a structure in which the central phosphorus atom is surrounded by five halogen atoms at the corners of a trigonal bipyramid (fig. 5.6g). The bonding orbitals of the phosphorus atom here probably are sp^3d *hybrids*, composed of $3s\,3p^3\,3d$. This hybridization is rather unusual, owing probably to the fact that an odd number of electron pairs is for the most part uncommon, if hydrogen is excepted.

It is noteworthy that crystals of PCl_5 are composed of equal numbers of tetrahedral PCl_4^+ ions and octahedral PCl_6^- ions. Crystals of PBr_5 consist of $PBr_4^+ + Br^-$. One must be careful therefore in comparing the bonding conditions in different states of aggregation. Solutions of PCl_5 or PBr_5 in ion-free solvents also contain ions (as shown by electrical conductivity). Two molecules of the same kind have here reacted, for example, according to $2PCl_5 \rightarrow PCl_4^+ + PCl_6^-$ (*autoionization*).

In many other cases d orbitals are hybridized with s and p orbitals. The two most important types will be mentioned here. In one, one s, three p and two d orbitals, which have nearly the same energy, are hybridized. In the transition elements the d orbitals often belong very likely to the d group nearest below the valence-electron shell, but in other cases to the d group in the valence-electron shell. The sequence of the groups in the first case has led to the hybrids generally termed d^2sp^3 *hybrids*. The six hybrid orbitals have their axes directed as from the center of an octahedron toward its corners (*octahedral orbitals*).

Octahedral coordination is extremely common and occurs above all with the transition elements and the p elements immediately following them. If the difference in electronegativity is great between the central atom and the ligand atom, the bond has a more classic electrostatic nature, that is, ion–ion bonding if the ligand is charged (for example, $Fe^{III}F_6^{3-}$), and ion–dipole bonding if it is uncharged (for example, $Cr^{III}(H_2O)_6^{3+}$). As was mentioned in 5-2d, the octahedral coordination in such cases may be explained on a purely electrostatic basis. A weak contribution of covalent bonding may also be expected, however, in which the octahedral orbitals will support the octahedral coordination. As the difference in electronegativity decreases, the covalent contribution increases. For example, in $Fe^{II}(CN)_6^{4-}$ it is very strong.

Octahedral coordination is never encountered in period 2, both because the elements in this period have to small radii to make 6-coordination possible, and because they lack d orbitals in their outermost shells.

The second of the two most important types of hybrid with d orbitals is dsp^2 *hybridization*. The four hybridized orbitals give electron clouds with their axes of elongation directed as from the center of a square toward its corners (*square orbitals*). Square coordination occurs chiefly in complexes of transition element ions that have a nearly but not quite full d group. Most important are (cf. fig. 5.8) Cu(II), Ag(II) with nine d electrons, and Ni(II), Pd(II), Pt(II) and Au(III) with eight d electrons. $Cu^{II}(NH_3)_4^{2+}$ and $Cu^{II}(H_2O)_4^{2+}$ are well-known complexes with square coordination. Square coordination must always contain a covalent contribution. Otherwise, 4-coordination will be tetrahedral, either because of an even distribution of bond directions in electrostatic bonding (5-2d), or because of sp^3 hybridization in covalent bonding.

In square coordination two additional ligands are very often bound, one on each side of the plane of the square, but at a greater distance than the first four ligands. This may be thought of as a transitory form between pure square coordination and octahedral coordination, where in the latter two opposite ligands have a greater distance from the central atom than the others. Only the ligand-field theory (5-3j) gives a satisfactory explanation of this effect.

Coordination numbers greater than six, which, however, are rather uncommon with stronger covalent bond contribution, also require the participation of d orbitals. They occur especially when a highly charged and not too small central atom bonds fluoride ions. The size relationships are then favorable for the simultaneous proximity of a large number of ligands to the central atom. Examples are $Zr^{IV}F_7^{3-}$, $Nb^VF_7^{2-}$ and $Ta^VF_7^{2-}$. Particularly interesting are the octacyano complexes of molybdenum and tungsten, among others $Mo^{IV}(CN)_8^{4-}$ and $Mo^V(CN)_8^{3-}$. In a bond $M-C\equiv N|$ the three atoms lie in a line, and further, the unshared electron pair is directed away from the central atom. Therefore, the ligands can come quite close to each other without any great repulsion.

As already mentioned in 5-3a, d orbitals can also participate in π bonds and thereby cause double bonding, but space does not permit any further discussion of this. More recent views, especially concerning the function of d orbitals in complexes of transition metals, are described in 5-3j.

j. Ligand-field theory. Since about 1950 a series of problems, concerning primarily the properties of transition metal complexes, have been dealt with by a new method. Here the covalent bonds between central atom and ligands are at first left to one side, and consideration is primarily given only to the electrostatic field that the ligands set up at the central atom.

In the compounds to which the method can be applied, the ligands are either negative ions or molecules with lone electron pairs directed toward the central metal atom. In the transition metals in which a d group (sometimes also an f group) is incompletely filled by electrons, there occurs an important effect. As mentioned in 4-1b, in a completely free atom the five d levels degenerate into a common d level. In the ligand field, this is split up in different ways, depending on the nature of the ligands and their configuration around the metal atom.

To show this, let us first consider the special case of a regular octahedral complex MX_6 in which M is the transition metal and X is a ligand (for example, H_2O, NH_3 or Cl^-). The five d clouds appear approximately as in fig. 4.4, where the coordinate axes lie along the bond axes M–X. The clouds for the two d_γ orbitals ($d_{x^2-y^2}$, d_{z^2}) extend in the directions toward the ligands, while the clouds for the d_ε orbitals (d_{xy}, d_{xz}, d_{yz}) avoid them. Thereby the energy for the electrons in the former is increased through repulsion by the negative charges of the ligands, while the energy of the electrons in the latter remain on the whole unaffected. The symmetry of the complex, however, causes the energies of the three d_ε orbitals to be mutually equal, and similarly the energies of the two d_γ orbitals to be equal. The fivefold degeneracy of the d orbitals in a free M ion (fig. 5.24 b) is thus partly relaxed by the splitting of the d level into a triply degenerate d_ε level and a doubly degenerate d_γ level of higher energy (fig. 5.24c). The energy difference between the d_ε and d_γ levels, which is customarily called Δ, can be determined spectroscopically (see below) and varies with M and X. If M is a $3d$ element, Δ usually lies in the range 80–250 kJ mol^{-1}; in complexes of the transition elements in the following periods Δ is 40–80% greater than in the corresponding $3d$ complexes. For the same M it may be expected that increased ligand field will produce a greater splitting between d_ε and d_γ, and thus greater Δ. If some of the more common ligands are arranged according to increasing Δ with the same M, the following series is obtained, in which the ligand field strength increases to the right: $I^-<Br^-<Cl^-<F^-<H_2O<$ oxalate ion $<NH_3<NO_2^-<CN^-$. The ligand field strength clearly does not correspond to the action of purely electrostatic forces (for example, the Coulomb attraction between a cation and F^- is much stronger than that between the same cation and NH_3, but nevertheless F^- gives a weaker ligand field than NH_3). Covalent bonding forces are also certainly involved.

In a regular tetrahedral complex MX_4 the coordinate axes bisect the angles between

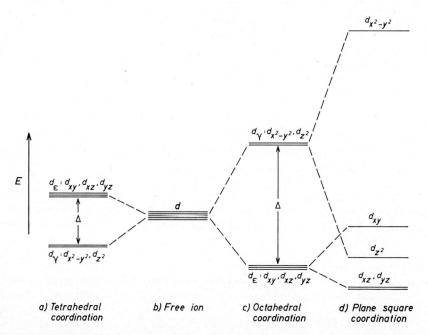

Fig. 5.24. Energies of the d orbitals in different ligand fields. The orientation of the complex is defined in (a) in that the tetrahedral edges are bisected by the coordinate axes; in (c) in that the ligands lie on the coordinate axes $\pm x$, $\pm y$, and $\pm z$; and in (d) in that the ligands lie on the coordinate axes $\pm x$ and $\pm y$, the plane of the square thus lying perpendicular to the z axis.

the bond axes, with the result that the three d_ε clouds extend more closely to the bond axes than the d_γ clouds. The two orbital groups d_ε and d_γ are therefore also separated here, but now the former has higher energy than the latter (fig. 5.24a). For the same ligand-field strength the splitting and thus the energy difference Δ is less than in the octahedral case.

In going to coordination types of lower symmetry, the orbitals become mutually still more dissimilar. The degenerate energy levels are further split up. An important example is the plane square coordination (fig. 5.24d). This may be thought of as arising from the octahedral coordination by the withdrawal of two opposite ligands from the central atom. If this is assumed to take place with the ligands on the z axis (the xy plane thus becomes the plane of the square), the d_γ level is split into a higher level for $d_{x^2-y^2}$ and a very much lower level for d_{z^2}, from whose cloud the ligands now lie at a great distance. The d_ε level also is split: d_{xy}, whose cloud lies in the plane of the square, takes on higher energy, while d_{xz} and d_{yz}, which remain equivalent, but whose clouds extend out of the plane of the square, form a doubly degenerate level with lower energy (the energies of the d_ε orbitals may, however, lie in different relationship to d_{z^2} than is shown in the figure).

In this way the plane square coordination is considered as a special case of the octahedral, and it is found that there are often intermediates between these two types. It was pointed out in 5-3i that in cases designated as square complexes, two further ligands are often found on each side of the plane of the square but at greater distance from the central atom than the square-coordinated ligands.

Table 5.8. *Various d-electron distributions.*

	Octahedral coordination						Tetrahedral coordination		
	Weak field high spin			Strong field low spin			Weak and strong field high spin		
Number of d electrons[1]	d_ε	d_γ	Form	d_ε	d_γ	Form	d_γ	d_ε	Form
d^0	---	--	reg			= weak field	--	---	reg
d^1	·--	--				= weak field	·-	---	
d^2	··-	--				= weak field	··	---	reg
d^3	···	--	reg			= weak field	··	·--	irreg
d^4	···	·-	irreg	···	--		··	··-	irreg
d^5	···	··	reg	⁞··	--		··	···	reg
d^6	⁞··	··		⁞⁞·	--	reg	⁞·	···	
d^7	⁞⁞·	··		⁞⁞⁞	·-	irreg	⁞·	···	reg
d^8	⁞⁞⁞	··	reg	⁞⁞⁞	⁞-	irreg	⁞⁞	···	irreg
				or = weak field					
d^9	⁞⁞⁞	⁞·	irreg			= weak field	⁞⁞	⁞··	irreg
d^{10}	⁞⁞⁞	⁞⁞	reg			= weak field	⁞⁞	⁞⁞⁞	reg

[1] For the d elements (group numbers 3–12), for oxidation numbers > I (for group 11 also for oxidation number I) the number of d electrons = group number – oxidation number.

We now consider how the electrons are distributed in the d_ε and d_γ levels in octahedral MX_6 complexes with various numbers of d electrons (tab. 5.8). For certain numbers of electrons a distinction must then be made between fields with weak ligand field (small Δ value) and with strong ligand field (large Δ value).

With from one to three d electrons ($d^1 - d^3$) these go with both weak and strong ligand fields into each of the three d_ε levels, but when the following d electrons are placed, two possibilities occur. If the ligand field is so weak that the energy gap Δ is sufficiently small, in d^4 the fourth electron can go into a d_γ orbital and assume the same spin direction as the first three electrons. If the ligand field is strong enough on the other hand, the gap Δ may be so great that energy is gained if the fourth electron instead goes into a d_ε orbital and there forms a pair with one of the first three electrons. The result is that the number of unpaired d electrons, all with parallel spin, is the highest possible with weak field and is less with strong field. It is therefore said that the two types of distribution are distinguished by *high spin* and *low spin*, respectively. The same applies to d^5, d^6 and d^7. For d^8 one finds with strong field either a distribution with the least possible spin or a distribution which is equivalent to that in a weak field. For d^9 and, of course, for d^{10}, the distribution is once more independent of the field strength.

Since the number of unpaired electrons determines the magnitude of the spin magnetic moment of the molecules (7-4a, b), it may be decided experimentally whether the d-electron distribution in a complex corresponds to high spin or low spin. For Fe(III) complexes (d^5) it is found that $Fe^{III}F_6^{3-}$ and $Fe^{III}(H_2O)_6^{3+}$ have five unpaired electrons and thus are high-spin complexes, while $Fe^{III}(CN)_6^{3-}$ has one unpaired electron and is thus a low-spin complex (cf. the position of these ligands in the field-strength series mentioned above). Another low-spin complex is $Co^{III}(NH_3)_6^{3+}$ (d^6).

Earlier (especially through Pauling) the magnetically determined number of unpaired electrons has been used in complexes with d^4 to d^8 to distinguish between "ionic-bond complexes" and "covalent complexes". The high-spin complexes have, in fact, the same d-electron distribution as in a free ion, in which all five of the d orbitals always have the same energy, and it was easy to assume that the central atoms in these complexes occurred as ions—even more so since the electrostatic bond contribution here is often large. On the other hand, the bonds in the low-spin complexes unquestionably have strong covalent contribution. Even if there exists a certain relationship between ligand-field strength and bond character, there is no reason to allow the limits between high spin and low spin as determined by field strength to set the limits between ionic-bond complexes and covalent complexes. It is preferred now instead to characterize the complexes as high spin and low spin only according to the designations based directly on the magnetic measurements.

As already stated, the ligand-field theory in the main does not take into account the covalent bonds, but on the other hand, it permits them to be interpreted in most cases as already described before. Tab. 5.8 shows that in all complexes up to d^3 and in low-spin complexes up to d^6 the two d_γ orbitals are free to participate in d^2sp^3 bonds. In the low-spin complex $Fe(CN)_6^{3-}$, then, $3d^24s4p^3$ bonding occurs in which the electrons are donated by the ligands. If not both d_γ orbitals are free "outer" d orbitals may be used. If d^2sp^3 bonding is present in the high-spin complex FeF_6^{3-}, it is probably $4s4p^34d^2$ bonding. It should finally be mentioned that with octahedral coordination, the d_ε electrons of the metal certainly often take part in π bonds with the ligands (the d_ε clouds in fact diverge from the bond directions and therefore can participate in π bonding). In this way double bonds appear.

Tab. 5.8 also shows the most common d-electron distribution in tetrahedral coordination. It has already been mentioned that here also a splitting into a d_γ and a d_ε level is obtained, but that the former now has the lowest energy (cf. also fig. 5.24a). The d_γ level is therefore occupied first. With the same field strength, Δ is here smaller than for octahedral coordination, with the result that, at least in complexes with a $3d$ element as central atom, no pair formation takes place even in strong fields before all the d orbitals are occupied by unpaired electrons. Therefore, high spin is most often obtained regardless of the field strength.

Here we have for the most part assumed that the ligands form a regular octahedron or tetrahedron around the central atom, that is, the complex has "cubic" symmetry (cf. 6-1e). Strictly speaking, however, this requires that the charge cloud caused by the d electrons under discussion does not have less than cubic symmetry. A lower symmetry in the cloud affects the positions of the ligands and can cause a greater or lesser decrease in symmetry of the complex. (In this way, of course, the energy levels may be displaced and formerly degenerate levels split up). A necessary (but not always sufficient) condition for the charge cloud to have cubic symmetry is that each of the three d_ε orbitals and each of the two d_γ orbitals contain an equal number (0, 1 or 2) of electrons. In tab. 5.8 these cases have been designated "reg" under the headings "Form".

With octahedral coordination different numbers of electrons in the two d_γ orbitals $d_{x^2-y^2}$ and d_{z^2} should cause especially large deviations from the regular octahedron because the clouds of the d_γ orbitals point toward the ligands. The two opposite ligands in the z direction then take on a different distance from the central atom than the other four, which still form a square in the xy plane. This case has been designated "irreg" in tab. 5.8. When the two opposite ligands lie at an extended distance a transition to square coordination is brought about, which can therefore occur with d^4 and d^9 (both high spin) and d^7 and d^8 (both low spin). Several cases of square coordination with d^9 and d^8 were mentioned in 5-3i. With d^4 it is known for Cr(II) and Mn(III).

Chemical Bonding

Different numbers of electrons in only the three d_ε orbitals should have little symmetry-decreasing effect on an octahedral complex, and in these cases the form is not indicated in the table.

With d^{10}, according to tab. 5.8, a regular coordination polyhedron would be expected. However, here sometimes for Cu(I) and Ag(I), and very often for Au(I) and Hg(I), an octahedron is found in which two opposite ligands lie much closer to the central atom than the four others. Two opposite bonds are clearly especially strong and we often speak here of 2-fold linear (or nearly linear) coordination.

The tetrahedral coordination may be expected to be markedly irregular as soon as all of the d_ε orbitals, whose clouds lie here nearest to the bond axes, do not contain an equal number of electrons ("irreg" in tab. 5.8).

Concerning the appearance of different complex types, it can be stated that octahedral coordination is the most common. Examples of *regular octahedral complexes* are: $Ti^{IV}Cl_6^{2-}$ $(d_\varepsilon^0 d_\gamma^0)$, $Cr^{III}(NH_3)_6^{3+}$ $(d_\varepsilon^3 d_\gamma^0)$, $Fe^{III}F_6^{3-}$ $(d_\varepsilon^3 d_\gamma^2)$, $Co^{III}(NH_3)_6^{3+}$ $(d_\varepsilon^6 d_\gamma^0)$, $Ni^{II}(H_2O)_6^{2+}$ $(d_\varepsilon^6 d_\gamma^2)$, $Zn^{II}(NH_3)_6^{2+}$ $(d_\varepsilon^6 d_\gamma^4)$. The conditions for square coordination have been stated above. Tetrahedral coordination is less likely than octahedral or square coordination with the exception of those cases in which it can become regular, that is, d^0, d^2, d^5, d^7, and d^{10} ("reg" in tab. 5.8). Known *regular tretrahedral complexes* are: $Ti^{IV}Cl_4$ $(d_\gamma^0 d_\varepsilon^0)$, $Mn^{VII}O_4^-$ $(d_\gamma^0 d_\varepsilon^0)$, $Fe^{VI}O_4^{2-}$ $(d_\gamma^2 d_\varepsilon^0)$, $Fe^{III}Cl_4^-$ $(d_\gamma^2 d_\varepsilon^3)$, $Co^{II}Cl_4^{2-}$ $(d_\gamma^4 d_\varepsilon^3)$, $Zn^{II}Cl_4^{2-}$ $(d_\gamma^4 d_\varepsilon^6)$.

The above also explains changes in type of coordination that occur when one goes from one metal to an adjacent one with the same oxidation number; for example:

Fe(II) (d^6)	Co(II) (d^7)	Ni(II) (d^8)
octahedral complexes	octahedral and tetrahedral complexes	octahedral and square complexes

Our knowledge is still insufficient to predict complex types safely, but the above can be useful to some degree. It should be kept in mind that when a complex occurs in a crystal, its environment can impose on it another form and other properties than what may be expected when it occurs more freely in solution.

When the d groups are incompletely filled, as is the case for all transition-metal complexes with d^1 to d^9, transfer of electrons can easily take place between the different, neighboring d levels. It has already been stated that the energy difference Δ between d_ε and d_γ for octahedral complexes of the 3d elements lies in the region 80–250 kJ mol^{-1}. The corresponding wave-number range is approximately 0.7×10^6–2.1×10^6 m^{-1}, that is, lying for the most part in the visible spectrum (cf. tab. 1.1), and this applies also to other d–d transitions. Solutions and crystals containing transition metal complexes are therefore nearly always colored. Spectroscopic determinations of the positions of absorption bands and from these energy differences between different levels have been of great value in the development of the ligand-field theory and the interpretation of bonding conditions according to it.

k. Bond energy. The bond energy between the atoms in a diatomic molecule is equal to the energy of formation of the molecule from single atoms, or to the identical energy of dissociation into single atoms (5-1c).

In molecules with more than two atoms, on the other hand, it is difficult in general to define the energy of each individual bond. The sum of all the bond energies should be equal to the atomization energy (2-2a), but as a rule, there is no indication as to how this should be divided among the different bonds. Only in the special case where all the bonds are of the same kind can the division be easily made. All bonds of the same kind should have the same energy and thus the atomization energy will be divided equally among the bonds.

As an example, methane CH_4 has the atomization energy (actually, enthalpy) 1662 kJ (2-2a), and each of the four C–H bonds has a *mean bond energy* of $1662/4 = 416$ kJ per mole of bonds. For water with atomization energy of 926 kJ, each O–H bond has a mean bond energy of 463 kJ.

Taking the example of water, however, we find that the mean bond energy of 463 kJ is not equal to the energy that is required for the dissociation $H_2O \to H + OH$, the *bond dissociation energy*. For the next dissociation step $OH \to H + O$ we find yet another value for this quantity. We have, namely:

$$\begin{array}{ll} H_2O(g) \to H(g) + OH(g); & \Delta H = 495 \text{ kJ} \\ \underline{OH(g) \to H(g) + O(g);} & \underline{\Delta H = 431 \text{ kJ}} \\ H_2O(g) \to 2H(g) + O(g); & \Delta H = 926 \text{ kJ} \end{array}$$

Summation gives the obvious result that the sum of all the bond dissociation energies is equal to the atomization energy.

The example also shows that one energy value for the bond O–H is found in OH (uniquely defined and $=431$ kJ) and another energy value in H_2O (mean bond energy $=463$ kJ). It is thus clear that a constant bond energy cannot be assumed for each bond, so that the sum of such bond energies in different molecules would always be equal to the atomization energy of the molecule. Each bond is too greatly affected by its environment for this to be possible.

Nevertheless, the variations are often so small that bond-energy values can be used with the assumption that they are additive in order to obtain a useful estimate of the atomization energy of a molecule. Of course, only those bond-energy values should be used that correspond to the bond types present. The method has been used primarily for organic molecules. It is important because of the fact that a large difference between the observed atomization energy and that calculated in this way indicates that one or more bonds are of another type than that assumed (see below).

Bonds between atoms of the same kind can be assumed to be practically completely covalent. Of these the bond H–H has the energy 435 kJ, which far exceeds the energies of other bonds between like atoms. This is because the interatomic distance is so small and also because all electrons are bonding (5-3a). In other bonds of this type the energy does not exceed 350 kJ.

On the other hand, bonds between atoms of different kind often have higher energies. This results chiefly from the advent of ionic bonding, which occurs when the difference in electronegativity increases. The bond energies are therefore greatest for the largest differences in eletronegativity. A value of 565 kJ, high for a single bond, is found for the bond H–F. In this the ionic bond contribution is probably somewhat more than 40 % (5-3f), and at least 250 kJ of the total bond energy probably results from this contribution.

If the bond number can vary for a given atom pair, the bond energy increases with increasing bond number. However, there is no direct proportionality between these two quantities, and either may increase more slowly or more rapidly than the other. This is shown by the following example:

	Bond energies, kJ per mole of bonds		
	Single bond	Double bond	Triple bond
C–C	345	615	835
N–N	180	418	942
O–O	145	495	

In going from one of the bond types given above to a neighboring one for the C–C bond, the bond energy is changed by a smaller amount than that corresponding to the bond energy of many kinds of single bonds. For this reason the bonds $C=C$ and $C\equiv C$ can be easily opened by "adding" atom pairs; they are *unsaturated*. For example, for the addition of bromine to ethylene according to the reaction

$$H_2C=CH_2 + Br_2 \rightarrow H_2C-CH_2$$
$$\;|\;\;\;|$$
$$Br\;Br$$

assuming that the bond energies are additive, it is found that $\Delta H = -95$ kJ, which indicates that the reaction will take place.

On the other hand, the nitrogen molecule shows no unsaturated character and is very stable. The same applies to the oxygen molecule, which, however, is somewhat less stable than the nitrogen molecule.

With departures from normal bonding conditions, theoretical calculations of atomization energies may lead to interesting results. This is true for molecules with delocalized orbitals, among others. If the atomization energies are calculated for the assumed limiting structures with localized orbitals, it is found that these are less than the atomization energy for the true structure with delocalized orbitals. Thus, the true structure has a lower energy and is more stable than any of the limiting structures. Calculations of this kind as a rule must be made more precisely than merely by addition of bond energies.

l. Covalent atomic radii. The effective radii of atoms in covalent bonding are in principle more easily determined than ionic radii (5-2c) because the initial values can be obtained by halving the distance between atoms of the same kind bound together. In any application of covalent atomic radii, however, it is important to remember that the radius is fairly constant only as long as the atom occurs in bonds with the same bond number.

Tab. 5.9 presents the most important covalent radii. The values for single bonds correspond in general to sp^3 orbitals ("tetrahedral radii"). On going to other bond types, however, the changes are not great as long as the bond number does not change. For a number of elements in period 2 and 3, it may be seen

Table 5.9. *Covalent atomic radii*, Å.

	H 0.30								
	Li	Be			B	C	N	O	F
single bond	1.34	1.06			0.88	0.77	0.70	0.66	0.64
double bond					0.76	0.665	0.60	0.55	
triple bond					0.68	0.60	0.55		
	Na	Mg			Al	Si	P	S	Cl
single bond	1.54	1.40			1.26	1.17	1.10	1.04	0.99
double bond						1.07	1.00	0.94	0.89
triple bond						1.00	0.93	0.87	
	K 1.96		Cu 1.35	Zn 1.31	Ga 1.26	Ge 1.22	As 1.18	Se 1.14	Br 1.11
	Rb 2.11		Ag 1.53	Cd 1.48	In 1.44	Sn 1.40	Sb 1.36	Te 1.32	I 1.28
	Cs 2.25		Au 1.50	Hg 1.48	Tl 1.47	Pb 1.46	Bi 1.46		

how sharply the radius decreases when the bond number and therefore the bond strength increases.

Within each period the radius decreases as the atomic number increases, and within each group it increases downward.

Properly, covalent radii should be given for purely covalent bonds and the sum of two such radii should then be reduced if the bond strength increases as a result of increased contribution of ionic bonding. However, it has not been possible to find any generally useful expression for such a correction. Therefore, it is most often better to use without correction the values in tab. 5.9, which, especially for strongly electronegative atoms, presuppose a certain ionic bonding contribution. Deviations of the ionic bonding contribution from this in the actual case are of minor importance.

The dependency of the radii on the bond number makes it possible to obtain important information about bonding conditions from experimentally determined interatomic distances (most often by X-ray diffraction but also spectroscopically). Tab. 5.9 shows that the bonds C–C, C=C and C≡C should have the interatomic distances ("bond lengths") 1.54, 1.33 and 1.20 Å respectively. If these values are plotted as a function of the bond numbers 1, 2 and 3 respectively, a curve is obtained which may be used to estimate the bond numbers of other carbon–carbon bonds from their interatomic distances. For example, the carbon–carbon distances of 1.39 Å in benzene and 1.42 Å in graphite give with this curve bond numbers of 1.6 and 1.4 respectively, in fairly good agreement with the values 1.5 and 1.33 given in 5-3e.

While this method seldom can be used so successfully for other atoms, in many cases it gives qualitative results. For example, in the ion NO_3^- each of the three equal N–O distances is 1.21 Å. The radius values give for the bonds N–O and N=O the distances 1.36 and 1.15 Å respectively, which shows that the bond number should lie between 1 and 2 (the simplified discussion in 5-3e gave the value 1.33).

An atom naturally does not present an effective covalent radius in all directions but only toward the atom or atoms to which it is covalently bound. The covalent "radius" therefore must not be considered as the radius of a spherical, effective surface but only as the atom's share in a covalent bond length. In this way the covalent radii are distinct from the effective radii of other bond types. Because these latter can operate in all directions (undirected bonds), an atom has a spherical effective radius in directions which diverge sufficiently from its covalent bond directions. In general, the radius of this surface is set equal to the van der Waals radius of the atom (5-4 b).

5-4. Van der Waals bonding

a. van der Waals forces. If noble-gas atoms or uncharged molecules with no permanent electric dipole moment (5-2g, h) are condensed into a liquid or solid phase (cf. for example, condensation of iodine molecules according to fig. 1.1), this most often depends on bonding forces of another type than those considered up to now. Such forces have been called *van der Waals forces*.

The fact that a free atom lacks a permanent electric dipole moment only means that the mean charge distribution measured over a sufficiently long time is symmetrical. Because of the oscillations in the electron cloud around the nucleus, however, the charge distribution varies, and at a given instant is generally unsymmetrical, so that a dipole is formed. This variable dipole acts upon neighboring atoms and induces variable dipoles in them also. The oscillations of the various dipoles are coupled to each other so that they occur in rhythm. In this way a decrease in the energy of the whole system occurs leading to attraction. An equilibrium position is attained when the atoms approach so closely to each other that the repulsive force between the electron clouds becomes equal to the attractive force (cf. 5-1 c). The atoms are then said to lie in van der Waals contact with each other.

The van der Waals forces are completely undirected and always very weak. The energy of a van der Waals bond between two atoms in contact is often of the order of magnitude of 5 kJ mol^{-1} and seldom rises above 20 kJ mol^{-1}. The strength falls off very rapidly with increasing interatomic distance. In 5-1c it was shown that the energy of electrostatic attraction between two ions at a distance l is proportional to l^{-1}. The van der Waals energy is proportional to l^{-6}. The van der Waals forces therefore play an appreciable role only in condensed phases and are effective even there mainly between very close-lying atoms. When a condensed phase is vaporized, the van der Waals forces fall to very small values.

The van der Waals force between two atoms at a certain distance is greater, the more easily their electron clouds can be displaced in an electric field, that is, the greater the polarizability of the atoms is. The polarization depends mainly on displacements in the outermost electron shell, and these occur more easily,

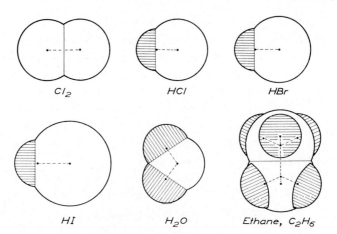

Fig. 5.25. Effective surfaces of molecules. Centers of atoms are indicated with dots. The lines joining them are thus covalent distances.

the farther this shell is from the nucleus. The polarizability of atoms with outer shells of analogous structure, that is, atoms in the same group of the periodic system, therefore increases with increasing atomic number. In consequence the van der Waals forces between such atoms increases with their atomic number. In general, this is true even if the atoms compared have different structures. An important exception to this is found in the noble gases. The high symmetry and stability of the electron clouds of these atoms results in an especially small polarizability and thereby weak van der Waals forces.

The total van der Waals force between two molecules is the sum of the van der Waals forces between the atoms facing each other in the two molecules. It therefore increases, in general, with the number of atoms and their atomic numbers, and thus often appears to increase with increasing molecular weight. The total force can, of course, become large if large areas of the molecules lie in contact with each other; for example, for parallel-oriented chain or layer molecules.

The van der Waals forces play no great role in the building up of stable molecules. On the other hand, they are of great significance for properties related to the formation and transformation of condensed phases, as well as their properties of cohesion. Some of these effects are treated in 6-5c and especially in 13-4b.

b. van der Waals radii. The effective radii shown by atoms when they are in van der Waals contact (van der Waals radii) are often difficult to establish. In general an atom is usually also bound in another direction with a covalent bond and the position of this bond in relation to the point of contact can affect the contact distance (cf. fig. 5.25). This also becomes appreciably dependent on the total attractive force between the two molecules to which the atoms belong. The van der Waals radii are therefore often quite variable.

If an atom is covalently bonded in one direction and is surrounded by an electron octet, in general it will exhibit in other directions a van der Waals radius that is close to the radius of the corresponding ion with noble-gas envelope. It is quite clear, for example, that a chlorine atom in $|\overline{\text{Cl}}\!-\!\overline{\text{Cl}}|$, in directions sufficiently far away from the covalent bond, should act with approximately the same effective radius as the ion $|\overline{\text{Cl}}|^-$ has in all directions.

On the other hand, the van der Waals radius of an atom is generally significantly larger than its covalent radius. In this way, molecules acquire van der Waals surfaces of the type shown in fig. 5.25.

Many crystals, especially in organic chemistry, are built of molecules chiefly under the action of van der Waals forces. In such cases the molecular packing is determined for the most part by the size and shape of the molecules, as defined by their van der Waals surfaces.

5-5. Hydrogen bonding

A hydrogen atom covalently bound to a strongly electronegative atom A as in A–H is often bound further to a strongly electronegative atom B (which may be = A) that has at least one lone electron pair. This *hydrogen bond* is frequently written A–H···B. Since a hydrogen atom cannot covalently bond more than one other atom (only one orbital can be used for covalent bonding), the hydrogen bond cannot be an ordinary covalent bond.

That A and B should be strongly electronegative is related to the fact that the hydrogen bond obviously must have a strong electrostatic contribution. Because of the strong electronegativity of A the hydrogen atom in A–H is already appreciably positive. This positive charge is also further strengthened through electron displacement toward B. The positive hydrogen atom, often designated as a proton, as a result holds together the negative A and B.

But the hydrogen bond also has a probably equally large covalent contribution, in that the available electrons are delocalized over all three of the participating atoms.

It is the small size of the hydrogen atom, or more accurately the proton, that makes it possible for it to form a bond of this kind. The positive atom must be buried in one electron cloud common to the three atoms. This leads also to the result that the two atoms drawn together by the proton most often come significantly closer to each other than would correspond to their ordinary contact distance, that is, the sum of their ionic or van der Waals radii (5-4). However, the distance never becomes as small as the covalent distance.

The given condition is more difficult to fulfill if more than two atoms are to be bound by one proton. The coordination number of the hydrogen atom in a hydrogen bond is therefore often, but not always, 2.

The energy of the hydrogen bond is, in general, 20–30 kJ mol^{-1}. It is thus a weak bond, but nevertheless is stronger than most van der Waals bonds. For the most part, the strength increases with increasing electronegativity of the participating atoms. Hydrogen bonds therefore occur most frequently between fluorine, oxygen and nitrogen. Chlorine, which has the same electronegativity as nitrogen, partakes less often in hydrogen bonding, presumably because of the larger atomic radius. Sulfur may sometimes take part, and in rare and exceptional cases even carbon (when a neighboring atom to the carbon atom has already drawn electrons to itself and thereby made the carbon atom more inclined to take up electrons).

The hydrogen atom most often lies close to the line joining the two atoms A and B. However, it nearly always lies nearer to one of these (for example, see ice, below). The stronger the bond is, and the closer A and B are therefore drawn together, the more central the position of the hydrogen atom becomes. The strongest hydrogen bond has been found in the hydrogen difluoride ion FHF^- (bond energy about 150 kJ mol^{-1}) and in this the hydrogen atom apparently lies surrounded completely symmetrically by the two fluorine atoms. Here the covalent bonding is probably completely delocalized.

Hydrogen bonds are found in most substances that contain hydrogen bound to fluorine, oxygen or nitrogen. Typical examples are HF, H_2O and NH_3. The number of hydrogen bonds decreases in the phase sequence solid→liquid→gas (only in HF can they play any role in the gas phase), and many of the bonds must thus be ruptured in the phase transitions solid→liquid and liquid→gas. As a result, the melting and boiling points for these hydrides are significantly higher than would be expected by comparison with the hydrides of the other elements in the same group of the periodic system (fig. 13.3 and fig. 13.4).

The crystal structure of ordinary ice is shown in fig. 5.26. The oxygen atoms of the H_2O molecules form a tetrahedral lattice and the molecules are oriented so that one hydrogen atom lies on each line joining two oxygen atoms. Of the six possible orientations of this type, however, all are found equally often and their distribution is believed to be largely random. One oxygen–hydrogen distance, 1.01 Å, is significantly shorter than the other, 1.75 Å, and corresponds to a typical, covalent O–H bond. The H_2O molecules have therefore retained their individuality. The tetrahedral angles indicate complete sp^3 hybridization (5-3d).

When ice melts, considerable numbers of hydrogen bonds are broken so that the structure becomes much more irregular and changing. However, the hydrogen bonds still play a large role and the result is that the short-range structure of water is probably not very different from that of ice (cf. 21-5c).

As stated before, the atom B in the hydrogen bond A–H$\cdots$B possesses a lone electron pair. Often, as in ice, there is one lone electron pair per hydrogen bond. In crystalline ammonia, all of the hydrogen atoms of the NH_3 molecules take part in hydrogen bonds, so that a skeleton is built up that extends in three dimensions throughout the whole crystal. The NH_3 molecule contains a single, lone electron pair which participates in three hydrogen bonds at the same time.

Crystallized hydrogen fluoride is built up of "infinite" chains as shown on the next page. In liquid and gaseous hydrogen fluoride broken segments of such chains

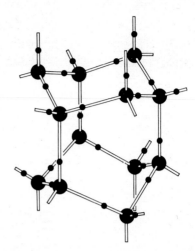

Fig. 5.26. The structure of ordinary ice. Large spheres = O, small spheres = H.

occur as well as rings, of which the ring (HF)$_6$ (at right below) is the most stable and therefore most frequently represented.

In carboxylic acids hydrogen bonding between two carboxyl groups is very common, leading to polymerization. In this way may be formed, for example, dimeric molecules with the structure

If n molecules, each with the dipole moment μ (5-2g), are bound together so that the dipole moments follow each other in a line, thus: → → → ...., the molecule so formed has the dipole moment $n\mu$. Since the dielectric constant of a phase consisting of dipoles is nearly proportional to the number of dipoles per unit of volume and the square of the dipole moment, the formation of such straight chains leads to a high dielectric constant. The hydrogen bond in HF, H$_2$O and NH$_3$ also makes itself evident in that these substances in liquid form have considerably higher dielectric constant than would correspond to the value estimated from the dipole moments of single molecules.

Hydrogen bonds play an extraordinarily great role in the building up of biological materials. Nearly all biological processes are believed in some stage to involve the formation or the rupturing of hydrogen bonds. Biologically important compounds very often contain just those elements that can partake in hydrogen bonds, and the small amounts of energy that need to be transferred in their formation or rupture may account for their prevalence.

5-6. Review of certain relationships of bond and structure

The foregoing treatment of chemical bonding is now applied in an elementary discussion of certain bonds and structures produced by them. The presentation must be very schematic and only a few simple and typical cases can be considered. Development in different directions will be carried out later in several connections. The metallic bond is also considered to some extent, but here this can be done only in a very primitive way. A more detailed treatment of this bond is presented in 6-4.

In addition to what has already been said in other connections (especially 4-2c, 5-2b, c, d, e and 5-3h, i) about the general variation of properties in the periodic system, it should first be mentioned here that the elements in period 2 deviate in many properties from those standing below them in their respective groups. The most important reasons for this are: (a) they can form double bonds through π bonding between p orbitals more easily than the later elements; (b) no d orbitals can be used for bonding; the covalence and coordination number therefore cannot exceed four, so that no double bond can be associated with an atom that has already formed four σ bonds; and (c) the elements in period 2 often have so much smaller radius than the later elements that their coordination numbers are smaller and their polarizing effect with the same charge is larger (5-2d, e).

a. The elements. The supply of valence electrons and the number of orbitals usable for bonding determines to a large extent the bond type and structure. The atoms generally show a tendency to place the valence electrons in all the orbitals that can be used for bonding. In order to achieve this end, as a rule two or more atoms must join together their orbitals and valence electrons.

We begin the discussion with the elements in period 2, for which only the four orbitals in the valence-electron shell can be considered usable for bonding, and to which thus the octet rule (5-3h) may be applied. When we go backward from Ne in the period, we see the effect of the constantly decreasing supply of valence electrons. In $|\overline{\underline{Ne}}|$ the octet is complete in the single atom, but on the other hand two atoms $|\overline{\underline{F}}\cdot$ join together with the formation of $|\overline{\underline{F}}\!-\!\overline{\underline{F}}|$. For $|\overline{O}$ the formation of $\overline{\underline{O}} = \overline{\underline{O}}$ would be expected, but a departure from the rule occurs with the structure $|O \vdots\!\vdots\!\vdots O|$, in which, however, oxygen still has the covalence 2. Next we have $|N \equiv N|$. Since a bond number of 4 is impossible (5-3b), in carbon each atom can be surrounded by a complete octet only if it is bound to more than one other atom. This occurs in diamond (fig. 5.16) and graphite (fig. 5.21). The three- and two-dimensional "infinite" molecules (giant molecules) formed here give elementary

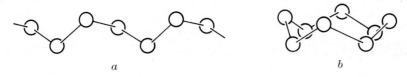

Fig. 5.27. Important structural elements in sulfur, selenium and tellurium. (*a*) Portion of the infinite chain, which is found in this regular, helical form in gray selenium and tellurium. More or less deformed fragments of such chains occur in melts of all three elements. (*b*) Eight-membered ring.

carbon a very low volatility. The light, finite molecules Ne, F_2, O_2 and N_2 on the other hand form difficultly condensed gases.

In every three-dimensional, giant molecule the outer orbitals are so delocalized (6-4) that if certain of these orbitals are incompletely filled, the electrons can easily move through the whole molecule without encountering noticeable energy thresholds. This electron mobility confers metallic properties on the substance. On the other hand, if all the orbitals are filled as in diamond a blocking of the electron mobility occurs and the substance becomes an insulator. Delocalization and thus metallic properties associated with unfilled orbitals are possible even in a liquid phase as long as there is covalent interaction among all neighbor atoms. On the other hand they cannot occur in a gas phase because of the large interatomic distances, which make such interaction impossible.

According to the above, metallic properties would be expected in solid and liquid phases of all elements before C in period 2. Li and Be are truly metallic while pure B is possibly a semiconductor (6-4a, 24-2b). In lithium gas there is besides lithium atoms a small proportion of Li–Li molecules (5-3a). No metallic properties can appear in this gas.

The above also explains why solid or liquid hydrogen does not have metallic character, although the hydrogen atom, like the alkali metal atoms, only has one s electron in the valence-electron shell. The hydrogen atom can use only one orbital ($1s$) for bonding and this is always filled. As far as electron supply is concerned, hydrogen behaves most like a halogen.

In period 3 the $3d$ orbitals have so much higher energy than the $3s$ and $3p$ orbitals that they play no large role in bonding. The octet rule therefore applies to a large extent. For the elements before Si (Na, Mg, Al), however, not all the $3s$ and $3p$ orbitals are filled, so that metallic character occurs in the solid and liquid state. As will be shortly demonstrated, also in following periods certain structural features may be understood with the help of the octet rule. Larger uncertainties concerning the supply of valence electrons and usable orbitals, however, make it more difficult to predict metallic character. All elements before group 14, however, are metals, and metallic properties appear also in the lower parts of groups 14–17 (see tab. 4.4).

The structures of the nonmetals in period 3 and higher periods are influenced

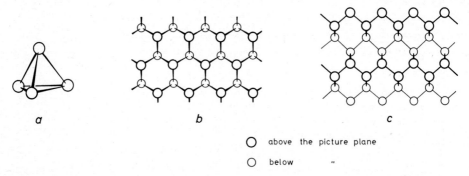

○ above the picture plane
○ below   "

Fig. 5.28. (a) P_4 molecule; (b) layer in black arsenic; (c) layer in black phosphorus. With one lone electron pair on each atom, the octet rule is fulfilled in all cases.

by the decreasing tendency for π bonding. In group 17 where only single bonds are required, there is no noticeable difference from period 2, but in groups 16 and 15 in general completely different structures are found. In group 16 the elements in the gaseous form occur as diatomic molecules (S_2, Se_2, Te_2), of which at least S_2 has a structure like O_2. But, especially in the solid and liquid forms, the octet rule is attained with only single bonds by the formation of chains, for example, $-\bar{\underline{S}}-\bar{\underline{S}}-\bar{\underline{S}}-\bar{\underline{S}}-$. These chains may be open (fig. 5.27a) as in crystalline, gray selenium and tellurium (the chains here extend throughout the whole crystal; they are "infinite" and may be said to form one-dimensional, giant molecules) and also in melts of these elements and of sulfur. They may also be closed into rings, of which the eightfold (fig. 5.27b) is the most common. Eightfold rings are found in crystalline, orthorhombic and monoclinic sulfur, crystalline red selenium, and also in melts and vapors of these elements. The bond angles are in general about 100°.

For the elements in group 15 diatomic molecules occur in gases at high temperature, and of these P_2 probably has a structure like N_2. But more common is a peculiar type of tetrahedral, four-atom molecule (fig. 5.28a), as P_4 (white phosphorus in solid form, melt and solution, and also phosphorus gas), As_4 and Sb_4 (yellow arsenic and gases of both elements). The octet around each atom here consists of three shared and one unshared electron pair. In the metallic modifications of arsenic, antimony and bismuth, the same condition is achieved by the formation of infinite layers, consisting of nonplanar, sixfold rings in which the coordination number is still three (fig. 5.28b). The structure is built up of such parallel layers so that the shortest interatomic distance between the layers is considerably longer than within the layer. The symmetry is trigonal. Black phosphorus has a related, orthorhombic layer structure, also with the coordination number three (fig. 5.28c).

In the tetrahedral molecules probably σ bonds occur between p orbitals but the bond angles of 60° must imply quite large strain and consequent decrease in stability. In the structures in both fig. 5.28b and c, on the other hand, the bond angles are about 100°. Both these structure types with their infinite layers of sixfold rings are reminiscent of the graphite structure but all bonds in the layer are single σ bonds, which rules out delocalized orbitals and plane layers as in graphite. As one goes downward in the group and the elements become more metallic, the interlayer distance decreases in relation to the interatomic distances within the layer, which shows that the bonding forces between the layers increase.

The formation of large, often infinite molecules in the elements lying below period 2 in groups 15 and 16 explains why these elements are less volatile than the halogens in the corresponding periods.

In group 14 Si, Ge and the gray modification of Sn have the diamond structure (fig. 5.16). No modifications analogous to graphite are found, probably because these require π bonds, which here are not easily formed. White Sn and Pb are metals.

b. Binary compounds. The survey is continued with binary compounds, that is, compounds consisting of two kinds of atoms.

With a large difference in electronegativity the ionic bond contribution is large (5-3f) and the compounds form ionic crystals. These will be considered later. If both kinds of atom are metallic the phase formed from them also has metallic character.

In the formation of more covalent compounds the supply of valence electrons and orbitals usable for bonding again plays a fundamental role. In pure elements the number of valence electrons per atom is, of course, constant, but as soon as several different kinds of atom interact, this can vary with change in composition. In addition, the relative sizes of the two kinds of atom can here often have great importance.

In determining the supply of valence electrons in relation to the number of orbitals usable for bonding, we may begin with the premise, according to 5-6a, that incompletely filled orbitals are present in all metals. The arrangement of these according to the supply of valence electrons cannot be predicted by elementary methods. In the nonmetals, in general, an increase in the supply of valence electrons and an increase in the electronegativity go together. An exception is hydrogen, which certainly cannot partake in bonding with more than one electron, nor have more than one electron-receptive orbital.

When more than two atoms of two different kinds are bound together, the atom which has the largest supply of valence electrons in general participates in larger numbers. This can be explained as a striving for a large number of valence electrons. In the molecule thus formed the atom with the smaller number of valence electrons can then appear as the central atom and thus to the greatest possible extent use its orbitals for bonding. In most complex molecules that do not contain hydrogen, therefore, the central atom is the least electronegative kind of atom and thus takes a positive oxidation number (CrO_4^{2-}, SO_2, NO_3^-, ClO_3^-, PCl_4^+, BrF_4^-). In this way is also explained the fact that the ions XO_2^-, XO_3^- and XO_4^- exist for $X = Cl$ but are unknown or at least are very unstable for $X = F$, which has a greater electronegativity than O (on the other hand, oxygen difluoride is formed with the formula $|\overset{-I}{F}-\overset{+II}{O}-\overset{-I}{F}|$). In hydrogen compounds hydrogen, because of its single bonding orbital, is always a ligand. If then the atom acting as central atom has a greater electronegativity than hydrogen, it takes a negative oxidation number (NH_4^+, NH_3, NH_2^-, H_2O).

How *changes in the number of electrons* can change the coordination number is illustrated here with a number of hydrogen compounds with nonmetals. If an atom of a nonmetal in groups 14–17 bonds hydrogen it can bond only as many hydrogen atoms as the number of electrons that are lacking from the next noble-gas configuration. For example, in period 2 the uncharged hydrides CH_4, NH_3, H_2O, and HF are formed. The L shell (octet) is complete in all these hydrides and further neutral hydrogen atoms cannot be bound. If the atomic system is deprived of electrons, however, often more orbitals can be used for bonding. If electrons are added instead, the L shell is completed with fewer bound hydrogen atoms. In the former case positively and in the latter negatively charged molecules result. The diagram below shows such hydrides of the elements mentioned plus B and Ne. In the columns one proton H^+ is added at each step upward. Isoelectronic molecules lie in the same horizontal row.

BH_4^-	CH_4	NH_4^+	H_4O^{2+}		
		NH_3	H_3O^+	H_3F^{2+}	
		NH_2^-	H_2O	H_2F^+	H_2Ne^{2+}
		NH^{2-}	OH^-	HF	HNe^+
		N^{3-}	O^{2-}	F^-	Ne

Only the molecules below the step-like line are known. (The ions H_2F^+ and HNe^+ are, however, very unstable. The former is found in certain solutions in liquid HF, the latter can be demonstrated spectroscopically in electric discharges in gas mixtures of hydrogen and neon.) For B, C and N the upper limit depends on the fact that all four L orbitals are used in BH_4^-, CH_4 and NH_4^+. The course of the boundary line to the right of the N hydrides depends on the decreasing stability upward in the columns and to the right in the horizontal rows. The first effect results from the fact that each newly added electron encounters ever-increasing electrostatic repulsion from the increasing positive charge of the preceding molecule. The second effect may be understood if the principal formation processes are considered, for example, $3H^+ + N^{3-} \rightarrow NH_3$, $3H^+ + O^{2-} \rightarrow H_3O^+$, $3H^+ + F^- \rightarrow H_3F^{2+}$. In all these cases $3H^+$ are added to isoelectronic noble-gas ions, whose radii are quite similar in size (fig. 5.5). Even if the final bonds have a strongly covalent contribution, the bond energy must decrease when the numerical value of the charge of the noble-gas ion decreases.

In a molecule containing n atoms there must be at least $(n-1)$ bonds. If all of these are electron-pair bonds, $2(n-1)$ electrons will participate in them. If the total number of valence electrons in the molecule is less than $2(n-1)$, not all bonds can be electron-pair bonds. Such a molecule is called an *electron-deficient molecule*. The most common electron-deficient molecules are hydrides and alkyl compounds of elements in groups 1, 2 and 13; for example, boron hydrides (of the many isolated boron hydrides the simplest neutral molecule is B_2H_6; BH_3, which would not be an electron-deficient molecule, has not been isolated) and $[Al(CH_3)_3]_2$.

The electron-saving bonds, which must be present in an electron-deficient molecule, often involve the combination in a molecular orbital of orbitals from three atoms (three-center bond). In B_2H_6 there are formed in this way *hydrogen bridges* (do not confuse these with the hydrogen bond described in 5-5). A hydrogen bridge may be very schematically written as in the formula a below. In B_2H_6 two BH_2 groups are bound together by two such bonds according to the formula b:

$$\begin{array}{cc} & H \\ B & \cdot \cdot \; B \end{array} \qquad \begin{array}{ccc} H & H & H \\ \diagdown & \cdot & \diagup \\ & B \; \cdot \cdot \; B & \\ \diagup & \cdot & \diagdown \\ H & H & H \end{array}$$

$\qquad\qquad a \qquad\qquad\qquad\qquad b$

In the tetrahydridoborate ion BH_4^-, on the other hand, the increased supply of valence electrons makes possible an entirely normal molecule, which is isoelectronic with CH_4 and NH_4^+. As would be expected, the structure is tetrahedral. The tetrahydridoaluminate ion AlH_4^- is analogous.

An example of the way the *supply of usable orbitals* determines the coordination has already been given in 5-3i, where it was pointed out that octahedral coordination is never encountered in period 2, probably because, among other things, the lack of d orbitals prevents d^2sp^3 hybridization.

Also, the *space about an atom* can be decisive for the coordination number (5-2d). The small covalent radii of carbon and nitrogen (cf. tab. 5.9) are believed to be the reason that these atoms cannot coordinate four oxygen atoms tetrahedrally. But three oxygen atoms can be coordinated in a triangle around carbon or nitrogen without coming too close to each other. Thus the isoelectronic ions CO_3^{2-} and NO_3^- arise. The central atom can nevertheless attain the covalence of 4 through a (delocalized) π bond (5-3e). Boron forms BO_3^{3-}, isoelectronic with these ions. The larger covalent radius of the boron atom, however, allows it also to coordinate four oxygen atoms. As an example, the ion $B(OH)_4^-$, among others, exists in water solutions of borates.

For Si, P, S and Cl in period 3, on the other hand, the covalent radii are so large that tetrahedral coordination of oxygen becomes normal. The isoelectronic ions SiO_4^{4-}, PO_4^{3-}, SO_4^{2-} and ClO_4^- show this. In period 5 the radii are so large that octahedral coordination of oxygen becomes possible. Examples of this are provided by the tellurate ion TeO_6^{6-} and the periodate ion IO_6^{5-}.

The striving of a central atom for a certain coordination number often leads to the formation of *polynuclear complexes* when the number of ligands per central atom is less than this coordination number. For borates this can easily be made clear in a plane.

Boron never coordinates less than three oxygen atoms. The coordination number three is found in the mononuclear BO_3^{3-}, which occurs in orthoborates (fig.

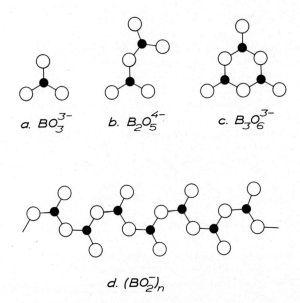

Fig. 5.29. Various borate ions, all with three oxygen atoms around each boron atom.

5.29 a). For example, magnesium orthoborate $Mg_3(BO_3)_2$ is obtained by fusing together $3MgO + B_2O_3$. The crystal is built up of the ions Mg^{2+} and BO_3^{3-}. But with the proportions $2MgO + B_2O_3$ a melt is obtained with only 2.5 oxygen atoms per boron atom. From this magnesium diborate $Mg_2B_2O_5$ crystallizes, whose structure contains Mg^{2+} ions and the dinuclear diborate ion $B_2O_5^{4-}$. In this, however, the boron atom can still coordinate three oxygen atoms because two oxygen triangles have a common corner (fig. 5.29b). The boron atoms thus use the oxygen atoms with greater economy by linking together through them, forming polynuclear polyborate ions.

The ions BO_3^{3-} and $B_2O_5^{4-}$ are isolated, *finite* ions, but with still lower oxygen ratio, an even larger fraction of the oxygen atoms must be shared with two boron atoms. "*Infinite*" ions may then occur. In the metaborates, in which the anion contains only two oxygen atoms per boron atom, the coordination number three can be attained by the formation of chains. The chains are either closed into rings as in potassium triborate $K_3B_3O_6$ (fig. 5.29c) or are open as in calcium metaborate $Ca(BO_2)_2$ (fig. 5.29d). A crystal of the latter compound is built up of straight chains which, placed parallel and with cations lying between, extend throughout the entire crystal. We thus have here infinite ions, which can be denoted by the formula $(BO_2^-)_n$, where n signifies a very large number.[1]

Polynuclear anions like $B_2O_5^{4-}$, $B_3O_6^{3-}$ and $(BO_2^-)_n$, are generally called *polyanions*. If such ions, like those just mentioned, have central atoms of only one

---

[1] The gross composition of an "infinite" molecule may be most easily obtained by finding in a schematic picture or formula of a sufficiently large portion of the molecule the smallest unit from which the molecule can be built up.

kind, they are given the special name *isopoly anions*, to distinguish them from *heteropoly anions*, in which there are different kinds of central atoms (31-6 a).

The tendency toward square coordination for Cu(II) (5-3i, j) causes crystals of anhydrous CuCl$_2$ to be built up of infinite plane chains:

$$\cdots\mathrm{Cu}(\mu\text{-Cl})_2\mathrm{Cu}(\mu\text{-Cl})_2\mathrm{Cu}(\mu\text{-Cl})_2\mathrm{Cu}\cdots$$

(The chains, however, are packed so that two chlorine atoms from neighboring chains are coordinated at a greater distance, thereby giving an irregular octahedral coordination around Cu(II).)

The metaborate ions and the chloro complex of Cu(II) just mentioned thus form one-dimensional infinite complexes. An infinite complex can also be extended as a net or layer in two dimensions through the whole crystal, or fill it up in a three-dimensional framework. In this way the central atoms can still further conserve ligands.

On melting or dissolving a crystal containing infinite complexes, these are, of course, more or less broken up and deformed. Groups that may be considered as fragments of the infinite complexes, however, often remain. A melt of Ca(BO$_2$)$_2$ thus contains chains of various lengths, whose gross composition with sufficient length lies near the formula BO$_2^-$. Since the chains have lost the support they have in the crystal, and also because of the free rotation around the single bonds, they become bent in many different ways. With solution in a solvent a high coordination number can be maintained in spite of the rupture of the infinite complex, if the molecules of the solvent enter as ligands. In the example of CuCl$_2$, fragments such as the following may be formed in water solution:

$$\left[(\mathrm{H_2O})_2\mathrm{Cu}(\mu\text{-Cl})_2\mathrm{Cu}(\mathrm{OH_2})_2\right]^{2+} \quad \left[\mathrm{Cl_2Cu}(\mu\text{-}\mathrm{H_2O})\mathrm{Cl}\right]^- \quad \mathrm{Cl}(\mathrm{H_2O})\mathrm{Cu}(\mathrm{OH_2})\mathrm{Cl}$$

Silicates, phosphates and sulfates with less than four oxygen atoms per central atom (Si, P and S) may be easily prepared from oxide melts. The tendency of these central atoms to coordinate four oxygen atoms results then in the formation of polynuclear or infinite ions. The manner in which the decreased ligand supply leads to increased complexity in the series: mononuclear complex→polynuclear, finite complexes→infinite chains→infinite layers→infinite framework, is shown most beautifully by the silicates (23-3h). Analogous polynuclear complexes can be formed by many other central atoms, however, as soon as the supply of ligands is deficient. The groups thus formed are described in more detail under the respective central-atom element in part II.

If a substance containing polynuclear complexes with oxygen ligands is dissolved in water, the complex often takes up oxygen ions from the water, so that a decomposition takes place into complexes with a lower number of nuclei. For example, if a disulfate containing the disulfate ion $S_2O_7^{2-}$ is dissolved in water there occurs the reaction:

$$S_2O_7^{2-} + H_2O \rightarrow 2SO_4^{2-} + 2H^+ \tag{5.18}$$

Complexes of this type therefore often behave as acids with water (12-1a). Decomposition is facilitated if the protons are effectively tied up through addition of base.

c. Metal complexes. Complexes with metal cations as the central atom are very common. Completely free metal cations probably never occur in liquid solutions. In any case at least all the nonalkali metals in liquid solution form complexes with molecules of the solvent or with other dissolved substances. Often the metal complex can be formed by direct addition of metal ions and free ligands, thus constituting a complex in the older sense of the word (1-4a). It is mainly such complexes which we consider here.

A good deal has already been said of the conditions of bonding in metal complexes, especially when the central atom is a transition metal (5-2d, 5-3i, j). It has been pointed out that the bonds between the central atom and the ligands are of very variable type. The covalent contribution may be strong when there is little difference in the electronegativity of the central and ligand atoms (example: $Fe(CN)_6^{3-}$), and less strong with greater electronegativity difference (examples: AlF_6^{3-} with ionic bonding or $Ca(H_2O)_6^{2+}$ with dipole bonding). If a central atom can exist in several oxidation states, the covalent contribution increases as the oxidation number increases (5-2e). It should be remembered that triangular, tetrahedral and octahedral coordination does not necessarily imply directed covalent bonding, but may be accounted for by purely electrostatic forces (5-2d).

Most frequently, hybridized bond orbitals from the metal ion participate in the covalent bonding. With the most common coordination numbers 4 and 6, mostly sp^3 tetrahedral or dsp^2 square orbitals and d^2sp^3 octahedral orbitals may be expected. As stated earlier (5-3i, j), d orbitals may also participate in π bonds and thereby give rise to double bonding.

With covalent bonding, a certain central atom as a rule exhibits a characteristic coordination number (5-3a). In a liquid solution this is often satisfied by solvent molecules taking part as ligands, thus complementing a prevailing shortage of other ligands. In water solution it may thus be assumed that metal cations are nearly always hydrated to *aquo ions* (5-2h). For example, in a water solution of $CdSO_4$ tetrahedral $Cd(H_2O)_4^{2+}$ ions are present. As CN^- ions are added to this solution, more and more of the H_2O ligands are replaced by CN^-, and at the final stage $Cd(CN)_4^{2-}$ is obtained. "Complex formation" with CN^- thus does not mean addition of CN^- but substitution of H_2O by CN^-. Occasionally ligands can be added beyond the characteristic coordination number. As an example, the ion Cu^{2+} bonds four ligands particularly strongly in plane, square coordination

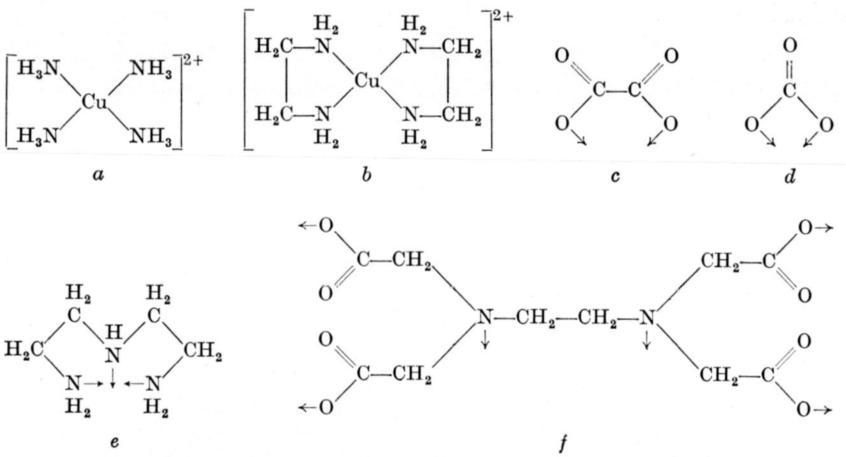

Fig. 5.30. (a) Tetramminecopper(II) ion, to be compared with (b), where two molecules of ethylenediamine through chelate bonding gives the same square coordination of N around Cu. (c–f) Other chelate ligands in which the ligand atoms are indicated by arrows.

and occurs therefore usually with coordination number 4, which is characteristic. But Cu^{2+} may sometimes bond two further ligands (5-3i, j), which, however, lie at a greater distance from the central atom than the first four, so that an irregular coordination octahedron is formed.

The properties of the bond are naturally especially dependent on the nature of the central atom and the ligand atom (5-1 b). The shared electron pairs in the covalent bonds are generally given by the ligand atoms, which thus act as electron pair donors (5-3c). The *ligand atoms* are usually composed of rather electronegative nonmetals, and the most important are halogens, oxygen, nitrogen, sulfur and carbon. They can all be ligand atoms in either inorganic or organic ligands but examples here are mainly limited to the former. Ions of *halogens* and *oxygen* particularly often occur as monatomic ligands. In addition, oxygen is the ligand atom in a great number of molecules, the foremost ligands being H_2O, OH^- and oxo anions. In organic ligands one oxygen atom in a carboxylate ion often acts according to:

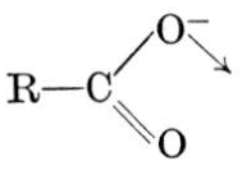

$$R-C\begin{matrix}O^-\\O\end{matrix}$$

Nitrogen is active in NH_3 and derivatives thereof, and in many organic ligands. *Sulfur* as a ligand atom is found, for example, in S^{2-}, HS^-, $S_2O_3^{2-}$, most often in SCN^- and in a considerable number of organic ligands. In water solution the only ligand occurring with *carbon* as the ligand atom is CN^-.

Outside of water solutions there is also the ligand atom *hydrogen*, for example, in AlH_4^-. Here we have in addition *carbon* as a ligand atom in CO, which gives carbonyl metals, for example, $Ni(CO)_4$.

Often different atoms in a ligand can act as ligand atoms. In NO_2^- the nitrogen atom is most often the ligand atom, but sometimes an oxygen atom acts as such instead (22-10c). In CN^- and SCN^- carbon or sulphur respectively most often are ligand atoms but also nitrogen may have this function. If a ligand can act through several ligand atoms with proper mutual positions, it may fill more than one of the coordination positions of the central atom. Ethylenediamine NH_2–CH_2–CH_2–NH_2 (often indicated by "en" in complex formulas) is coordinated by Cu^{2+} to give the complex *b* in fig. 5.30, which may be compared with the tetrammine complex *a*. The coordination is square in both cases. Ethylenediamine is said to be a *chelate ligand* (Gk. χηλή, crab's claw) and the complex *b* is called a *chelate*. The oxalate ion *c* and the carbonate ion *d* are chelate ligands in which the bonds are formed by oxygen atoms. In all these cases the ligand is *bidentate* (Lat. dens, tooth). In *e* a *terdentate* ligand is shown, and in *f* a ligand which can be as high as *sexadentate*. The latter is the anion of ethylenediamine tetraacetate ("EDTA"), which clearly must twist itself quite elegantly around the central atom in order that all six "teeth" shall be effective.

Chelate complexes are most often stronger than ordinary complexes (*chelate effect*). Five-membered rings (formed in *b, c, e* and *f*) generally give the greatest stability and after that six-membered rings. The chelate effect depends mainly on the fact that the processes producing the chelates have a higher probability than those producing nonchelate complexes (cf. 2-3b).

Let us assume that the bidentate ligand A–C–C–A is first attached to a metal atom M by only one of its ligand atoms A. The complex formed M–A–C–C–A should have approximately the same stability as a complex M–A with the ligand atom alone. If now the chelate ligand is bound to the same M with the second ligand atom also, this means simply a ring closure to

$$\begin{array}{c} C\text{—}C \\ \diagup \quad \diagdown \\ A \qquad A \\ \diagdown \quad \diagup \\ M \end{array}$$

In order to form A–M–A from M–A, on the other hand, a second free ligand A must be taken from the solution. This process has a much lower probability than ring closure, for which the second ligand atom already lies near the metal atom. The shorter the chain A–C–C–A is, the nearer the second ligand atom lies, and consequently the probability for chelate formation increases. This causes the entropy increase ΔS to become larger. On the other hand, the shorter the chain that is bent into a ring is, the more the bonds are strained, which causes the enthalpy increase ΔH to become larger. The Gibbs free energy change $\Delta G = \Delta H - T\Delta S$ (2.9) generally has a minimum for a five-membered ring, which thus because of both the high probability of formation and the low strain becomes the most stable type of ring.

EDTA, which can form five five-membered rings, gives complexes with all the transition elements, with the alkaline-earth metals and to a small extent with Li^+ and Na^+. Among the chelate ligands that are used for analysis, EDTA has therefore assumed very great importance. In cases of lead poisoning a calcium complex of EDTA is introduced intravenously. In the body calcium is replaced by the lead, following which the new complex is secreted without injury to the organs.

Metal atoms combined in biological substances generally are bound as chelates. This is true, for example, of Mg in chlorophyll and Fe in the "heme" groups of hemoglobin.

In the following section we discuss how the nature of the metal complex depends on the *central atom*. For the most part, the conditions in water solution are considered. The most important metal cations may here be properly divided into three types, all of which have been treated in 5-2f.

The first type includes *metal cations with noble-gas shells*, possibly also with a more or less filled $(n-2)f$ group deeper in the electron cloud. The strength of the complex increases rapidly as the charge of the metal ion increases, and in a series of ions with the same charge, the strength is greater the smaller the ion is. The alkali metal ions because of their low charges are very weak complex formers, but the smallest, Li^+, is the strongest among them. Also, the alkaline-earth metals form weak complexes if we exclude the smallest, Be^{2+}. When metal ions of this type form complexes in water solution, it is mainly with the two most electronegative ligand atoms, fluorine and oxygen. Oxygen occurs then chiefly in H_2O, OH^- and oxo anions. Therefore, if NH_3, S^{2-} or CN^-, for example, are added to solutions of these ions, the H_2O molecules already coordinated are displaced only to a small extent. But in the examples, where the additives are bases, protons are given up by more or less of the H_2O molecules, which are transformed to OH^-. It is also often found that fluoro complexes, but no other halo complexes, exist in water solution (for example, ZrF_6^{2-} but not $ZrCl_6^{2-}$).

The bond strength in the complexes being discussed increases with the charge of the ligand; oxygen in OH^- is bound more strongly than oxygen in H_2O. Differently charged isoelectronic oxo anions form complexes with relative strengths as follows: $CO_3^{2-} > NO_3^-$; $PO_4^{3-} > SO_4^{2-} > ClO_4^-$. The fact that the perchlorate ion ClO_4^- for the most part has an especially small tendency to act as a ligand is made use of when an experiment is to be conducted in a salt solution but complex formation with the salt ions must be avoided.

The above discussion indicates that complexes of this type have a large contribution of electrostatic bonding.

The second type embraces *metal cations with 18-shells or (18+2)-shells*. Here we observe, particularly on comparing the monovalent ions Cu^+, Ag^+ and Au^+ with alkali ions of the same charge and approximately the same size, that the bonding must be of a totally different type. In water solution, difficultly soluble halides and sulfides are in equilibrium with halo and sulfo complexes. Stable ammine and cyano complexes are also known. The stability of these complexes often varies according to the series: $Cu^+ < Ag^+ < Au^+$. For these three cations the stability for various ligands follows $F^- < Cl^- < Br^- < I^-$ and $F^- < OH^- < NH_3 < CN^-$. The bond strength increases thus as the radius of both the central atom and the ligand increases. The fact that NH_3 is bound more strongly than OH^- shows that charge does not play a decisive role. It is therefore clear that the electrostatic contribution to bonding must be small. When the stabilities of complexes of isoelectronic metal cations of different charge are compared, for the most strongly electronegative ligand atoms like F^- and OH^-, the sequence $Ag^+ < Cd^{2+} < In^{3+} < Sn^{4+}$ is found; but for less electronegative ligands an opposite order occurs.

The third type includes *transition metal ions*. According to 4-2b these have an incomplete d group, that is, they are characterized by $(n-1)d^x$ ($x = 1-9$). Here, in general, strong complex formation occurs, and if the strength of complexes with similarly charged cations in water solution are compared, for most ligands it is found that it increases as one goes to the right in the periodic system. Thus we have usually $Mn^{2+} < Fe^{2+} < Co^{2+} < Ni^{2+} < Cu^{2+}$. The ligand-field theory (5-3j) has been found to be valuable for the interpretation of complexes of the transition metal ions.

Especially with organic ligands it may happen that the bond to the central atom does not take place with any definite ligand atom. In unsaturated organic molecules (for example, ethylene, which has no lone electron pair, cf. 5-3d) the electrons in the π orbital of the double bond can participate in a bond to a transition metal ion. The ion is then bonded, so to speak, to the double bond. Such bonds can also occur with delocalized π orbitals (5-3e).

Examples of bonding to delocalized π orbitals are given by derivatives of the unsaturated hydrocarbon cyclopentadiene, C_5H_6 (*a* below). This is a very weak acid which gives the cyclopentadienide ion $C_5H_5^-$. This can be written as in *b*, where the delocalized π bonds are indicated, giving the ion an aromatic character. It bonds to ions of nontransition metals with a strong ionic contribution and without noticeable involvement of the π orbitals. On the other hand, transition metal ions are bound to these orbitals, and then so-called *sandwich complexes* are formed. The best known is the very stable iron(II) complex $Fe(C_5H_5)_2$, called *ferrocene* and shown in *c*.

If a central atom coordinates several different kinds of ligands, the various possibilities for the distribution of these may lead to isomerism of different kinds, which has been used for structure determination (6-1h, 7-5).

A metal atom may have an oxidation number in strong complexes that is unknown or corresponds to a very unstable state in weaker complexes, or when the atom is more or less "free". Thus, the ion Co^{3+} is very unstable in water solution when there are no ligands to which it can be strongly bound. On the other hand, Co(III) exists in a large number of stable complexes, for example $Co(NH_3)_6^{3+}$.

The appearance of high oxidation numbers of an atom when it bonds strongly electronegative ligands like O and F (5-3g) should be considered as an example of the stabilization of an oxidation state through complex formation. This may also be seen as a tendency for the covalent bonding of such ligands to increase as the oxidation number increases, which can be related to the general rise in polarizing effect that follows each increase in ionization (5-2e). For example, in sulfates of V(II) and V(III), the ions V^{2+} and V^{3+} exist both in crystals and in water solution (though hydrated). In the stage V(IV), on the other hand, we find the oxovanadium(IV) ion VO^{2+}. In the stage V(V) we have the ion VO_2^+ and even ions with so many oxygen ligands per vanadium(V) atom that the ion becomes negative, for example, the (ortho)vanadate(V) ion VO_4^{3-}.

The transition elements normally never have an oxidation number lower than I in their compounds, and even that is rare except for the coinage metals in group 11 (cf. 5-2f). With bonding to certain special ligands, however, such oxidation numbers as 0, −I and −II can exist (the number I also becomes more common with these ligands). Inorganic ligands of this type are CO (in carbonyl metals, 23-2f), the cyanide ion CN^- isoelectronic with CO, and PCl_3. We have, for example, $Ni^0(CO)_4$, $Fe^{-II}(CO)_4^{2-}$, $Ni^0(CN)_4^{4-}$, $Ni^0(PCl_3)_4$. It should be remembered, however, that such low oxidation numbers only have formal significance. They correspond to electron distributions that give a net charge of 0 for the CO and PCl_3 groups and −1 for the CN group. Obviously the electron cloud is more displaced toward the ligands than these net charges would indicate. All the ligands behaving this way also have orbitals free to accept π-bond electrons.

The importance of care when interpreting oxidation numbers of metal atoms in complexes can be illustrated by the case of reduction of hexacyanoferrate(III) ion $Fe^{III}(CN)_6^{3-}$ to hexacyanoferrate(II) ion $Fe^{II}(CN)_6^{4-}$. One electron is added to the iron d orbitals so that d^5 changes to d^6. The ion thus changes from paramagnetic to diamagnetic (cf. tab. 5.8, octahedral coordination and low spin) but at the same time the iron atom delocalizes some of its electron cloud (equivalent to about one electron) on to the six cyanide ligands. As a result the electron density at the iron atom is almost identical in the two ions; the oxidation numbers are therefore somewhat misleading.

d. Energy-rich molecules and free radicals. The foregoing discussion about the structure of molecules applies chiefly to molecules in the ground state. When energy is added, for example by heating, collisions with energy-rich particles or by irradiation with energy-rich quanta, the molecular energy is raised to higher levels for a larger number of molecules; the molecules become excited. In excited states the molecules may have other structures. In addition, completely new molecules which already in the ground state have high energy may now become the most stable and be formed. Therefore, at high temperatures molecules may be encountered which are not usually found at ordinary temperature. Often they have a very simple structure and at sufficiently high temperatures complicated molecules do not exist (in stellar spectra no molecules with more than two atoms are believed to have been demonstrated with certainty). Further, charged molecules (ions) often appear, and molecules with unpaired valence electrons. An atom or molecule which has one or more unpaired valence electrons, which are easily utilized in a bond, is usually called a *free radical*. Examples of free radicals are many single atoms (cf. tab. 4.2) and monatomic ions like $H\cdot$, $|\overline{F}\cdot$, $|\overline{Ne}\cdot^+$, and odd molecules (5-3a) such as the hydroxyl radical $\cdot\overline{O}$—H and the methyl radical $\cdot CH_3$. The oxygen molecule with two unpaired electrons (5-3a) is an example of a special case for which the term free radical is not used. Neither does one speak of free radicals when unpaired electrons lie in the inner electron shells as in the transition elements. Free radicals may be very stable toward decomposition but are extremely reactive and try to combine with each other or with other atoms or molecules. This can be inhibited if the temperature is rapidly lowered to a few °K where the reaction velocity is very slight. The reactive molecules thereby become "frozen in" and may be preserved for a long time. The same result may be obtained if they are formed in a solid phase or a liquid phase with high viscosity so that they have low mobility and therefore little chance to collide with neighbors.

Among the molecules that have been identified in hot gases may be mentioned the following: CH, CH_2, CH_3, NH, NH_2, OH, AlO, Al_2O, C_2, CN, CS, CF, AlF. At high temperatures solid substances such as BO, AlO, Al_2O and SiO may also be stable.

The higher the temperature, the more common the single atoms become. In an oxyhydrogen flame there are appreciable proportions of H and O. Electrons also begin to be given off so that monatomic ions are formed. The material begins to go over to a plasma (5-2f), and at temperatures of approximately 10^7 °K the ionization is complete.

Energy-rich molecules and atoms often occur also as short-lived intermediate products in chemical reactions (such, for example, are the activated complexes mentioned in 8d, f, and g, and the single atoms H, Cl and Br). It should be remembered that a molecule at the moment of reaction need not have the structure corresponding to that of the ground state in which it is generally pictured.

e. Simple rules for electron distribution and bond angles in molecules with covalent structures. If it is known which atoms in a molecule are bound to each other, it is often possible by means of simple rules to derive a quite reasonable electron distribution and approximate bond angles. These rules represent an application of what has already been said, but they may be useful as a summary.

The wide validity of the octet rule means that for all cases with a maximum coordination number of four one should try to distribute the electrons so as to fulfill the rule. In this way also the location of possible double and triple bonds is decided. If these can be arranged in different ways, then the corresponding π orbitals are delocalized.

If the total number of valence-electron pairs (shared and unshared) around an atom is at the most four, it may be assumed that all electron pairs that are not involved in π bonds lie in hybridized orbitals of the types sp^3, sp^2 or sp. Therefore, in the electron-distribution formula, at first all π bonds may be ignored, that is, in each double and triple bond only one of the bonds, the σ bond, may be considered. If the sum of the number of electron pairs in σ bonds and unshared (lone) electron pairs around a certain atom is four, then they are distributed tetrahedrally about the atom (cf. fig. 5.19); if there are three such electron pairs they lie in a plane triangle (120° apart); if there are two they lie on a straight line (180° apart). This means that the electron pairs are distributed as uniformly as

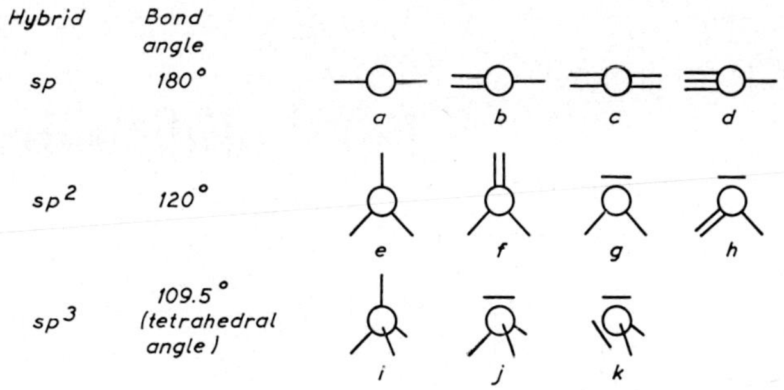

Fig. 5.31. Bond angles for various combinations of bonds about a central atom.

Hybrid	Bond angle
sp	180°
sp^2	120°
sp^3	109.5° (tetrahedral angle)

possible about the central atom, that is, with the greatest possible distance from each other. This may be expressed as a result of a mutual repulsion between the valence-electron pairs. The various possibilities are illustrated in fig. 5.31 (in cases a, b, e, and g the octet rule is not fulfilled). From these rules it immediately follows, for example, that CO_2 is linear (c and d) while SO_2 is bent (h), NO_3^- is plane (f), and ClO_3^- is pyramidal (j). The most important exception is the case with square, fourfold coordination (dsp^2 hybridization, 5-3i), which is mainly encountered in complexes with Ni(II), Pd(II), Pt(II), Cu(II), Ag(II) and Au(III).

Even if the total number of valence-electron pairs is greater than four, they are generally distributed as uniformly as possible about the central atom. In the comparatively rare cases where the number is five, therefore, a distribution such as in fig. 5.6g is most often obtained, and when the number is six the distribution is usually octahedral.

CHAPTER 6

The Solid State

The physics and chemistry of solid bodies was for a long time very incompletely understood. Nearly all chemical knowledge was based on reactions in liquids or gases, and therefore hardly anything was known about the structure of large groups of solid substances. It was, in fact, only with Laue's discovery in 1912 of interferences on the passage of X rays through a crystal, and the subsequently developed X-ray diffraction methods for the study of the structure of solid bodies (7-1) that this area was opened for investigation. The study of the solid state is also of importance because it gives fundamental information about the structure and properties of the most important technical construction materials.

6-1. Geometrical crystallography

The atoms in a solid phase may be distributed in space in widely different degrees of order (1-4h). If the structure at the most shows short-range order (1-4g), we have an *amorphous body*. Complete order, which would characterize an *ideal crystal*, would require that all of the atoms remain at rest. We know that this is not the case even at absolute zero (3b, c) and at higher temperature there is always heat motion. In what is generally designated a crystal the order is sufficiently complete that the crystals discussed in a large part of the introductory discussion may be considered as ideal crystals.

If long-range order (1-4g) is completely lacking, all of the properties of the body are independent of direction within the body. If differently oriented sample rods are removed from the body, it is found that all the rods give the same values for thermal expansion coefficient, heat conductivity, electrical conductivity, elasticity, refractive index, light absorption, etc. If a sphere is made out of the body, it has the same hardness and is affected equally rapidly by solvents at every point of its surface. The body is *isotropic* (Gk. ἴσος, like; τρόπος, direction, way).

With sufficiently high degree of long-range order, however, many—but not all —properties are different in different directions. The body is *anisotropic*. A crystal is always anisotropic with respect to a large number of properties.

a. Periodic repetition and identity. Lattices. An ideal crystal is characterized, as stated above, by a completely ordered structure. For a crystal to be homogeneous, the structure must be uniform throughout the entire crystal. This requires a

The Solid State

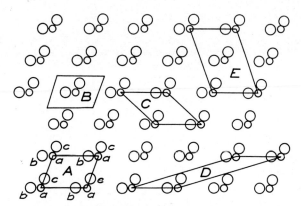

Fig. 6.1. Two-dimensional, periodic structure.

periodic repetition in three dimensions of a particular group of atoms. Fig. 6.1 shows a two-dimensional example, which may be considered to be extended very far (often said "infinitely" far) in the plane of the picture in all directions. Many illustrations are given in the following sections as two-dimensional examples, but corresponding analogies in space can easily be imagined.

In the periodic structure, each mathematical point has corresponding to it an infinite number of other points with an environment of the same form and orientation. All such points are said to be *identical*. In fig. 6.1 all "atoms" a are identical among themselves as well as all b, and all c. The atoms b and c on the other hand are not identical with each other even if they are the same kind of atom. Also, other elements (lines, planes, entire atomic groups) that have surroundings of the same form and orientation are said to be identical. Groups of a, b and c lying close together (let us say a molecule abc) are repeated thus by infinitely many identical groups, and the structure may be considered with advantage as being built up of these groups.

All lines and planes that are identical with each other respectively are parallel, and are said to form sets of lines and planes (fig. 6.2). The identical lines and planes in a set are repeated at a constant distance. Fig. 6.2 may also be thought of as a section through a three-dimensional periodic structure and the lines then also indicate traces of different sets of planes in the plane of the picture.

It should be clear that a set of planes in a crystal has greater physical meaning, the more densely its planes are populated with atoms. It follows from fig. 6.2 that the interplanar distance in a set is greater, the more densely populated the plane is.

The length of any line joining a point with another identical point without passing through any other identical point between, is called an *identity distance*. All straight lines drawn in fig. 6.1 are identity distances.

The collection of all points that are identical with each other is called a *lattice*. The two-dimensional structure in fig. 6.1 may in a purely geometrical way be thought of as being built up of atoms belonging to three congruent lattices in

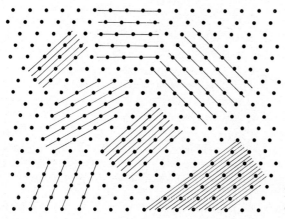

Fig. 6.2. Sets of lines in a two-dimensional, periodic structure.

parallel position, each containing only one kind of mutually identical atoms. It thus consists of an a lattice, a b lattice and a c lattice, all congruent. Every crystal that contains more than one kind of mutually identical atoms, may be considered as being built up of three-dimensional lattices in an analogous way. The complete, material crystal structure should not, if it can be avoided, be called a lattice.

b. Primitive and elementary regions. If two identity distances from one point in a two-dimensional periodic structure form two sides of a parallelogram, this is called a *primitive area* if no other points identical with the corner points belong to the parallelogram (that is, fall within it or on its sides). In fig. 6.1 the areas A–D are primitive areas. They all contain one of each of the atoms a, b and c. This is immediately seen in the area B. In A, C and D the same result is obtained when one studies how the atoms are distributed among the area in question and the adjacent areas.

The area E contains two identical points; it is not primitive, but *doubly primitive*.

All areas A–E may be designated as *elementary areas* or *unit areas*. For an area to be a unit area it must be possible to build up the two-dimensional structure by laying duplicates of the area next to each other in parallel orientation. A primitive area is always a unit area, but a unit area does not need to be primitive.

A two-dimensional periodic structure is described in terms of a coordinate system whose axes are the two different sides of a unit area. A primitive area is often chosen as a unit area, and then usually that area that has the shortest sides.

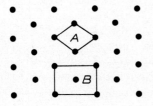

Fig. 6.3. Choice of different unit areas.

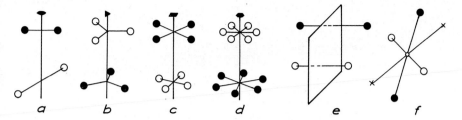

Fig. 6.4. Some important symmetry elements.

In fig. 6.1 the proper choice would be A ($= B$ if the origin is displaced). In certain cases, however, a nonprimitive unit area may be chosen for reasons of symmetry, or in order to obtain an orthogonal coordinate system. In fig. 6.3 A is the primitive area with the shortest sides, but, because of the special properties of the structure, the advantage of an orthogonal coordinate system is obtained if the doubly-primitive area B is chosen as a unit area instead. A unit area of this type is also said to be *centered*.

In a three-dimensional periodic structure the circumstances are analogous. If three identity distances starting from one point form the edges of a parallelopiped, it is primitive (a *primitive cell*) if no other points identical with the corner points belong to the parallelopiped. The structure can be built up from an *elementary cell* or *unit cell* if duplicates of it are laid next to each other in parallel orientation. The three edges of the unit cell define the coordinate system which is used here. A coordinate system is often chosen that corresponds to a primitive unit cell but for reasons analogous to those mentioned above for two-dimensional structures, it may happen that a coordinate system will be chosen that corresponds to a nonprimitive, centered unit cell (see 6-1 e).

To get the best idea of the general features of a crystal structure, it is not always best to consider only one unit cell. On the other hand, the unit cell is essential in any accurate description of the structure.

Like the unit cell, the coordinate axes are repeated periodically and form, so to say, the framework on which the crystal structure is built. The geometrical elements (points, lines and planes, especially those which are symmetry elements; see 6-1 c) that are needed for the description of the structure, and which, of course, are repeated periodically, are often also included in this framework.

The weight of the contents of the unit cell, divided by its volume, is equal to the (ideal) density of the crystal. Often it is most convenient to express the contents of the unit cell in terms of a certain formula unit. If the formula weight of this unit is M, it weighs M/N_A g, where N_A is Avogadro's number. If there are n such formula units in the cell volume V cm³, the density is then

$$\varrho = \frac{nM}{N_A V \text{ (cm}^3\text{)}} = \frac{1.6604 \, nM}{V \text{ (Å}^3\text{)}} \text{ g cm}^{-3} \tag{6.1}$$

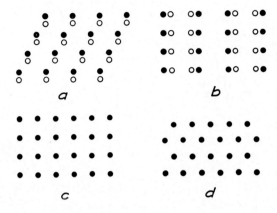

Fig. 6.5. Two-dimensional structures with various symmetries.

In the right hand expression the value of N_A is inserted and V is expressed in Å^3. Example: a unit cell of NaCl contains $n=4$ formula units NaCl ($M=58.443$) and has the volume $V=178.31$ Å^3, which gives $\varrho=2.1769$ g cm^{-3}.

After V has been measured by X-ray diffraction, which in favorable cases can be done with an accuracy of 0.1 ‰, the expression (6.1) is often used in various situations to determine ϱ, n, M or N_A, when the other three of these four quantities are known.

c. Symmetry. A very important property of a crystal is its *symmetry*. The symmetry is defined by *symmetry elements*, which describe operations (other than parallel displacements) by which the crystal structure can be made to superimpose on itself. All crystal elements (points, groups of points, lines, areas) that can be made to superimpose on themselves by symmetry operations are said to *equivalent*. (Note the difference between identity and equivalence!)

Some important examples of symmetry elements are shown in fig. 6.4, in which the symmetry operations are applied to a few selected points. Equivalent points have the same symbol.

A *rotation axis* (*a–d*) causes the structure to superimpose on itself by a rotation through the axial angle $360°/n$ or a whole multiple thereof. n is called the *order* of the axis, and in a crystal can only have the values 2, 3, 4 and 6.

A *mirror plane* (*e*) causes the structure to superimpose on itself by "reflection" in the plane. The parts of the structure lying on opposite sides of the mirror plane are related to each other as an object to its mirror image.

A rotation axis and a mirror plane establish a direction in the crystal, the former by the direction of the axis and the latter by the normal to the plane.

A *symmetry center* (*f*) implies that each direction from a point (=the symmetry center) is equivalent to the opposite direction. The structure is caused to superimpose on itself by a symmetry operation called *inversion*.

If a symmetry element passes through a lattice point, an identical element

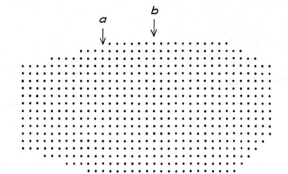

Fig. 6.6. Different possibilities for the deposition of building units in crystal growth.

must pass through every point identical with the first point. All symmetry elements thus occur in sets of identical elements.

Here it is practical to show only a few examples for two-dimensional cases (fig. 6.5). In *a* there is no symmetry element. In *b* there are 2-fold rotation axes (in the two-dimensional example, the same as symmetry centers) and mirror planes, in *c* in addition 4-fold, and in *d* 3- and 6-fold rotation axes.

The different symmetry elements in a crystal are not independent of each other; the existence of certain elements can require the presence of certain elements and exclude the presence of others. The number of possible combinations of symmetry elements therefore is limited.

Molecules can, of course, also have symmetry elements other than those mentioned. There is nothing to prevent a molecule from having a 5-fold rotation axis, for example, but if such a molecule enters into a crystal, this axis has no connection with the symmetry of the crystal, which cannot contain 5-fold axes.

d. The external symmetry and shape of crystals. The external symmetry and shape of crystals depends on the mechanism of their formation and growth. This question must be introduced here, but is developed more extensively later in chap. 14.

Fig. 6.6 shows very schematically a section of a crystal bounded by surfaces of different kinds (cf. the actual crystal in fig. 1.2). It is apparent that a surface becomes "smoother", the more densely it is populated with building units. The more sparsely populated surfaces are more uneven, often built up in steplike form. As the crystal grows, the newly added building units are deposited primarily on surfaces of the latter type where at the "steps" they are bound to the subsurface from more directions than on a densely populated surface. A new building unit thus is attached more easily in a position at the point of the arrow *a* than in a position at the point of the arrow *b*.

A further important condition is that a building unit that encounters the crystal in a position such as *b* does not readily leave the crystal surface completely when its temporary bond is broken. Instead, it jumps over to analogous, neighboring positions, and in this way wanders over the crystal surface until it finally

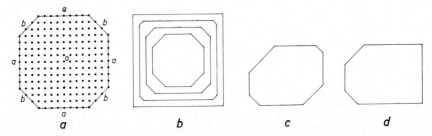

Fig. 6.7. Crystal growth. (*a*) Section of a structure with 4-fold symmetry. (*b*) Different stages in the growth of the corresponding crystal with uniform supply of material. (*c*), (*d*) Distorted crystals with uneven supply of material.

comes to an edge where it becomes permanently attached. This surface wandering of building units thus contributes to the deposition at such points as a. In this way it is possible to explain how parts of a crystal which are not directly encountered by building material from the surrounding phase can still grow. A crystal needle can thus grow into a material-free phase or a vacuum (often, however, this can also result from diffusion through the interior of the crystal; see 6-3c).

Densely populated crystal planes thus grow by extending the surface layer while the tendency to begin a new such layer by the deposition of building units on the existing plane is small. The growth velocity normal to a crystal plane therefore on the whole becomes less the more densely populated the plane is, and the crystal faces that appear therefore correspond in general to relatively dense planes.

If the conditions of growth are otherwise equal (especially with respect to supply of material), all faces that by symmetry are equivalent are equally developed. Fig. 6.7 *a* thus schematically shows the result of growth from *o* if a 4-fold rotation axis stands normal to the picture plane at this point. Here, the slowest growth has been assumed to be normal to the four equivalent surfaces *a* and the fastest growth normal to the four equivalent surfaces *b*.

On further growth the surfaces *a* increase at the expense of the surfaces *b*, and finally the latter may entirely disappear. The course of growth is shown schematically in fig. 6.7 *b*.

An uneven supply of material may cause different growth velocities in directions that are equivalent with respect to the symmetry. For the same crystal as in fig. 6.7 *a–b* the result may then be as in *c* or *d*. Crystals with their outer form more or less distorted in this way are extremely common. Nevertheless, *corresponding faces form the same angles with each other as in the symmetrically grown forms* (Niels Stensen, "Steno", 1669). By measurement of the directions of the faces, therefore, the true outer symmetry can be determined provided a sufficient number of faces is developed. The study of a large number of crystals may reduce the risk of error.

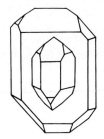

Fig. 6.8. Symmetrically grown quartz crystal drawn for comparison inside a distorted crystal with the same faces.

In fig. 6.8 a symmetrically grown crystal of quartz is compared with a distorted crystal.

If the growth velocity is particularly great in a certain direction, the crystal will naturally become elongated in this direction. For example it may become needle- or filament-shaped. If the growth velocity is particularly slow in a certain direction the crystal becomes platelike, with the surface of the plate lying normal to this direction.

Crystals formed under different conditions, for example from different solvents, may show very different relative development of faces, and thus have varying appearances (different "habits"). Sodium chloride crystals obtained from pure water solutions are bounded by cube faces, but if the solution contains 10 wt % of urea, the crystals are bounded by octahedral faces. Both types of crystals nevertheless have the same external symmetry, and the internal structure is the same in both cases. This type of behavior may depend among other things on the fact that molecules from the solution (molecules of solvent or of a foreign substance) are adsorbed in different ways on mutually nonequivalent faces, so that their growth is influenced in varying degrees.

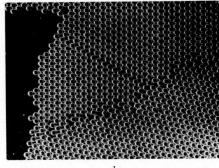

a   *b*

Fig. 6.9. (a) Copper as an example of a solid, polycrystalline body. A ground copper surface has been polished and etched with an acid iron(III) chloride solution, which attacks the differently oriented crystal grains in different ways, whereby they acquire varying reflectivities. Magnification 280 ×. (S. Modin). (b) Soap bubbles (0.3 mm dia.) floating on a liquid surface and forming a model of a polycrystalline body. (W. L. Bragg and J. F. Nye.)

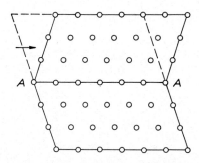

Fig. 6.10. Schematic drawing of the lattice of a twin. (The formation of a twin by mechanical deformation of an initial crystal partly outlined by a dashed line is also shown; cf. 6-5a).

The external symmetry of a regularly grown or idealized crystal naturally reflects the symmetry of the crystal structure to a certain extent. However, this symmetry can be completely determined in general only by diffraction methods (7-1).

The dissolution velocity of a crystal also is different in different directions, and the same applies to the reaction velocity in many chemical reactions. It is quite natural that a direction of high growth velocity generally should be a direction of high dissolution velocity. The corners and edges of the crystal are therefore attacked most rapidly by the solvent.

When a crystalline phase is formed from a liquid phase, the crystallization often starts at a large number of points at the same time (14-2). All of the crystals grow simultaneously, and if the liquid phase completely converts to the crystalline phase, growth in any direction ceases only on the contact of two crystals. In general, these have differently oriented lattices. The result is a *polycrystalline* body. All metallic bodies in ordinary use are polycrystalline (fig. 6.9) and consequently not single crystals.

Often a substance crystallizes in several simultaneously grown crystals so that *crystal aggregates* are formed. The different crystals in such an aggregate are rarely randomly oriented with respect to each other. Generally, they have *one* crystal direction oriented more or less in parallel, or they may even be grown together oriented completely in parallel. The first case occurs often with needle or filamentlike crystals oriented with parallel elongation axes, or with platelike crystals packed with plate surfaces parallel.

A phenomenon distinct from this is *twinning*. In this case two or more crystals contact each other with a definite mutual orientation that is other than a purely parallel position. This mutual orientation is dependent on the structure of the crystal. In fig. 6.10 a twin is shown very schematically, and it can be seen how the atoms in the contact plane AA take part in the lattices of both crystals.

Twins are formed either during crystal growth, and then often at a very early stage, or also by mechanical deformation of already-formed crystals (6-5a).

Crystal formation and crystal growth are discussed also in chap. 14.

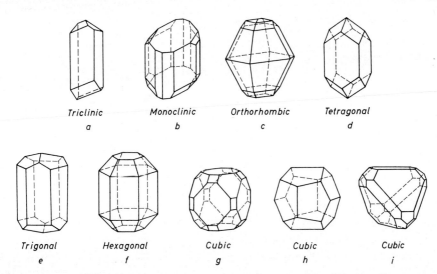

Fig. 6.11. Examples of crystals belonging to the different crystal systems. The crystals are: (a) strontium dihydrogen tartrate tetrahydrate $SrH_2(C_4H_4O_6)_2 \cdot 4H_2O$; (b) realgar As_4S_4; (c) orthorhombic sulfur; (d) zircon $ZrSiO_4$; (e) calcite $CaCO_3$; (f) beryl $Be_3Al_2Si_6O_{18}$; (g) fluorite CaF_2; (h) pyrite FeS_2; (i) tetrahedrite $Cu_{12}Sb_4S_{13}$.

e. Crystal systems. For a long time, crystals have been divided according to their external form into seven main groups, the *crystal systems*.

For nearly all cases the following features suffice to characterize the crystal systems. Some examples of the different systems are illustrated in fig. 6.11.

Triclinic system. No symmetry element, or only a symmetry center. No direction is therefore established by any symmetry element.

Monoclinic system. Either *one* 2-fold rotation axis or *one* mirror plane or both, in which case the rotation axis and the normal to the mirror plane coincide. *One* direction in the crystal is thus established by the symmetry elements.

Orthorhombic system. *Three* mutually perpendicular, *nonequivalent* directions are established by either 2-fold rotation axes, mirror-plane normals or both.

Tetragonal system. *One* 4-fold rotation axis.

Trigonal (formerly also rhombohedral) system. *One* 3-fold rotation axis.

Hexagonal system. A 6-fold rotation axis.

Cubic (*isometric*) system. *Three* mutually perpendicular, *equivalent* directions, established by either 2-fold or 4-fold rotation axes. Recognition of a crystal belonging to the cubic system is aided by the fact that in addition it always has four 3-fold rotation axes, which have the same relation to the axes just referred to as the diagonals of a cube to its edges. Compare, for example, the three cubic crystal forms, the cube, the regular octahedron and the regular tetrahedron. The three mutually perpendicular, equivalent rotation axes in the cube and octahedron are 4-fold, and in the tetrahedron 2-fold, but in all three forms the four

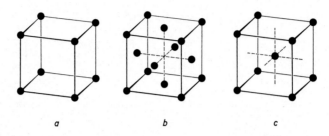

Fig. 6.12. Cubic unit cells. (*a*) Primitive or simple; (*b*) face-centered; (*c*) body-centered.

3-fold rotation axes are present. In a crystal belonging to the cubic system the velocity of light is the same in all directions, the crystal is *optically isotropic*.

The crystal system determines the most suitable type of coordinate system for the description of the crystal, and thereby often also the type of unit cell.

In crystals belonging to the cubic system the unit cell is frequently not primitive but centered, either in all cube faces, in a *face-centered cubic lattice* (quadruply primitive), or in the center of the cube, in a *body-centered cubic lattice* (doubly primitive). Fig. 6.12 shows unit cells of cubic lattices, the primitive (simple) and the two centered. For the two latter the primitive cells are oblique and do not show directly the cubic symmetry. In this case it is natural to prefer the convenient cubic coordinate system as represented by the centered cells, which in addition show the positions of the symmetry elements.

f. Close packing. If the building units of a crystal are bound together by undirected forces, the structure is determined to a large extent by spatial factors. The building units are then often distributed in such a way that the packing is as dense as possible, which, of course, gives the structure the lowest energy (6-2b). If directed, that is, typically covalent bonding forces are present between the building units, the spatial factors become less important. If the building units are charged, they must in addition be fitted together so that the charge is distributed as evenly as possible, but nevertheless this is often permitted by close packing.

Especially common are the closest possible packings that can be formed by spherical building units of equal size. These are what are customarily simply called close packings. They are best described by following the way they may be built up stepwise from close-packed layers of spheres.

Spheres of equal size packed as densely as possible in a layer form a hexagonal packing like the spheres with centers as at A in fig. 6.13 *a*. If a new layer of spheres is placed over the first, and the packing is to be as dense as possible, the centers of the spheres must lie directly over either one or the other of the positions B or C. If B is chosen, the spheres in a third layer are placed with their centers either over the third position C or again over A. Thus the sequence ABC or ABA is obtained. If these are repeated periodically the two most common close packings

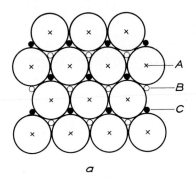
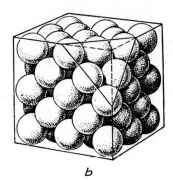

Fig. 6.13. (a) Close-packed layer of spheres and the projections A, B and C of the centers of spheres in all the layers in close packing lying above this layer. (b) Cubic close packing.

result, *cubic close packing ABCABC* ... and *hexagonal close packing ABABAB* ... The designations "cubic" and "hexagonal" indicate the crystal system to which these close packings belong. The cubic close packing has a face-centered cubic lattice in which the octahedral faces in the unit cell are parallel to the packing layers (fig. 6.13b). The relationship is apparent on comparison with fig. 6.12b.

It can be seen in fig. 6.13 that an atom in a close packing always is surrounded by twelve neighbors; an atom in one layer in *a* is surrounded by six neighbors in the same layer, plus three neighbors below and three above this layer. The co-ordination number of 12 is quite high.

Fig. 6.13b shows how a crystal structure might appear pictorially. Contacts between the effective spheres of the atoms must clearly occur in many directions. However, the atoms are more often represented as spheres of radii smaller than that corresponding to the interatomic distance. The reason for this is mainly that the figure thus becomes more "transparent" and can show more detail.

g. Isomorphism and polymorphism. Different substances that crystallize in essentially the same structure are said to *isomorphous* or to show *isomorphism* (Gk. ἴσος, alike; μορφή, form; the expressions "isotype" and "isostructural" are also used). Of course, the dimensions of the lattice may vary from substance to substance but the relative positions of the building units will remain the same. For example, crystallized argon, copper and γ iron are isomorphous; so also are diamond and gray tin; sodium chloride, silver chloride and calcium oxide; calcite and sodium nitrate.

If one substance can exist in several different crystal structures (often also called *modifications*) it is said to be *polymorphous* or to show *polymorphism*. (If it exists only in two different crystal structures it is often said to be *dimorphous* or to show *dimorphism*). Examples of polymorphous substances are carbon (which possesses at least two graphite and two diamond stuctures) and silicon dioxide (with at least six different crystal structures). The different crystal struc-

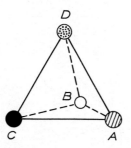 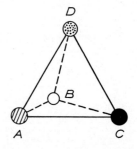

Fig. 6.14. Two enantiomorphous molecules ABCD. The enantiomorphism is retained if central atoms of the same kind are inserted into the molecules.

tures of one substance in general constitute different phases of the substance (cf. however 6-3b and 13-3c.)

Different crystal structures of the same substance are not designated as different isomers (1-4a) of the substance. The differences between graphite and diamond, for example, are more fundamental than between the two isomers normal butane and isobutane.

Allotropism (Gk. ἄλλος, other; τρόπος, way, direction) in an element may be caused either by different molecular structure, for example ordinary oxygen (O_2) and ozone (O_3), or—much more commonly—by the arrangement of the same building units in different crystal structures, for example α and γ iron, or orthorhombic and monoclinic sulfur. The term allotropism can be dispensed with and nowadays is not used as often as formerly.

h. Enantiomorphism. Two objects that in relation to each other appear as an object and its mirror image cannot always be made to coincide only by rotation and translation. A pair of shoes or a pair of gloves, for example, are objects of this type, of which there are always therefore a left-hand and a right-hand form. The necessary conditions that coincidence shall be impossible are that the objects lack both mirror planes and symmetry centers (on the other hand they may possess rotation axes). The two forms of such a pair are said to be *enantiomorphs* (Gk. ἐναντίος, opposite; μορφή, form).

A group of atoms may be arranged so that they constitute one form of an enantiomorphous pair. The group may be a finite molecule or form an entire crystal. A simple example of enantiomorphous molecules is given by tetrahedral molecules consisting of four different atoms (fig. 6.14). The two molecules in an enantiomorphous pair are said to be *antipodes* of each other. We have here a type of *stereoisomerism* (Gk. στερεός, solid, here in the sense of space). A large number of properties are the same for the two antipodes. If they occur in an environment which has sufficiently low symmetry, or take part in a process of sufficiently low symmetry, however, they have mutually different properties. In natural products often only one antipode of a pair is found.

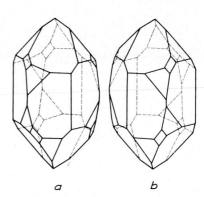

Fig. 6.15. Crystals of (a) left-hand quartz, (b) right-hand quartz.

If a crystal is built up of one antipode of an enantiomorphous molecular pair, the crystal can never have a mirror plane or a symmetry center (a number of identical left-hand gloves can never be arranged so that the group has these symmetry elements). The crystal thus represents one form of an enantiomorphous crystal pair. The other form (the mirror image) is obtained if one starts with the other antipode. Cane sugar molecules lack mirror planes and symmetry centers, and cane sugar crystals, which only contain one antipode (only one occurs in nature), also lack these elements.

But the building units of a crystal can have higher symmetry and just by their arrangement in the crystal lattice give enantiomorphous crystals (both a left-hand and a right-hand spiral staircase can be built up of parallelopiped-shaped bricks). An example is sodium chlorate. When sodium chlorate crystallizes from water solution the highly symmetrical building units of sodium ions and chlorate ions arrange themselves in such a way that a crystal with no mirror plane or symmetry center is formed. Here the chance is equally great for the formation of one group as for its mirror image. Therefore, among a sufficiently large number of crystals there will be equally many "left-hand crystals" as "right-hand crystals" (which type will be associated with one term or the other depends, of course, on convention). Quartz is another example (fig. 6.15).

A phase in which one form of an enantiomorphous pair is present, as a rule exhibits *optical activity*, that is, rotates the plane of polarization of a polarized light beam that passes through it. If one form of the pair rotates the plane of polarization in one direction, the other form under otherwise equal conditions rotates the plane an equal amount in the opposite direction. This is true whether the form in question is a molecule, which may occur in some arbitrary phase, or a crystal. If it is a crystal, whose lack of mirror planes and symmetry center depends only on the crystal structure (sodium chlorate, quartz), the optical activity disappears as soon as the crystal breaks down. A sodium chlorate solution or a melt of quartz (liquid or in solidified form as silica glass) is not optically active. On the other hand, a cane sugar solution is optically active.

6-2. Crystal structure and lattice energy

a. Types of crystals. Crystals are often divided into different types that are distinguished by the kind and distribution of their bonds.

A crystal that is built up of distinct molecules is called a *molecular crystal*. That the molecules are distinct implies that the bond forces within each molecule are substantially stronger than the bond forces between molecules. The former generally have a strong covalent contribution. The latter are often chiefly van der Waals forces but also hydrogen bonds or weak covalent forces may be present.

Most substances that occur as independent molecules of the classical type crystallize as molecular crystals with finite molecules. To these belong the majority of organic crystals. From inorganic chemistry may be mentioned the examples N_2, I_2, crystalline sulfur, CO_2. Ice may also be included although very obvious hydrogen bonds occur between the H_2O molecules (5-5).

Molecular crystals may also contain molecules that are infinite in one or two dimensions, chain or layer molecules. As examples the chain molecules in gray selenium (fig. 5.27 a) may be mentioned, and the layer molecules in graphite (fig. 5.21).

If covalent forces extend in three dimensions throughout the crystals, separate molecules can no longer be distinguished. The whole crystal becomes a giant molecule, a framework molecule. The most typical example is diamond (fig. 5.16). Quartz (fig. 23.3) can also be classed as such although the Si–O bonds here also have an appreciable ionic contribution. Crystals of this type do not have any generally accepted designation (they should not be called molecular crystals).

When the crystal is bound together by reasonably typical ionic bonds one speaks of an *ionic crystal*. The ions may, of course, be complex and built up of bonds of other types. In sodium chloride and potassium nitrate all the ions are finite. In calcium metaborate $Ca(BO_2)_2$ the anions form infinite chains (fig. 5.29 d) in mica infinite layers and in feldspar infinite frameworks are formed, in whose cavities the cations are located (23-3h).

Finally, we have *metallic crystals*, in which the bond forces to a large extent are of the same kind throughout the whole crystal (6-4).

b. Lattice energy. *Lattice energy* X (*crystal energy* would be better) denotes the quantity of energy given up per mole of crystal when the crystal is formed from discrete units that up to now in this book have been called the building units of the crystal. Since the lattice energy refers to the energy *given up* on crystal formation, it is equal to the decrease of the energy of the system under the formation process. Thus, $X = -\Delta U$.

The definition of the term "building unit" is not entirely unambiguous, but most often it refers to the units that in general are transported as such out of the adjoining phases during the growth of the crystal. For example, the building units in diamond and copper consist of the respective individual atoms; in sodium chloride of the ions

The Solid State

Na$^+$ and Cl$^-$; in potassium nitrate of the ions K$^+$ and NO$_3^-$; in iodine of I$_2$ molecules. It is unimportant whether the building units actually exist free, however; the main concern is that they be defined, and that it is understood what building units are referred to when an energy value is given.

For the decomposition of 1 mole of crystal into free building units, an amount of energy equal to the lattice energy must be added.

c. Lattice energy and crystal structure. When a phase crystallizes while the system approaches equilibrium, the building units of the crystal generally arrange themselves so that the crystal has the lowest possible energy, from which it follows that the lattice energy becomes as high as possible.

The lattice energy, of course, is the sum of the bond energies of all the bonds between the building units. The bond energy between two atoms has a maximum on "contact" between the atoms (5-1c). If the building units are neutral and there are no directed (covalent) forces between them, then the tendency toward the highest possible lattice energy causes the building units to find as many mutual contacts as possible. Therefore, they pack as densely as their shapes will permit.

Under the conditions mentioned, therefore, nearly equal-sized and spherical or nearly spherical building units very often give close packing, most often cubic and somewhat less often hexagonal (6-1f). These are thus found in a very large number of metallic crystals and in the noble-gas crystals. A considerable number of metals (for example, alkali metals and α iron) crystallize in a body-centered cubic lattice (fig. 6.12c). This also gives a high coordination number (8 as compared with 12 in the close packings) and is only slightly more loosely packed than the close packings. If the atomic radius is r, the volume per atom is $5.66r^3$ for the close packings and $6.16r^3$ for the body-centered cubic lattice. On the other hand the simple cubic lattice (fig. 6.12a) is very rare in these cases, which is certainly a result of its smaller coordination number (6) and its significantly looser packing (the volume per atom is here $8r^3$).

If the building units consist of elongated molecules they most often pack with their lengths parallel. Flat molecules usually are packed with their planes parallel.

If the building units are electrically charged (ions) or are electric dipoles, the charge distribution in addition must be as even as possible. Ionic crystals are a compromise, often difficult to predict, between the tendencies toward close packing and even charge distribution. When monatomic ions are concerned, the cations generally are smaller than the anions (fig. 5.5) and besides often have numerically higher charge than the latter. The structure then is nearly always distinguished by having a cation surrounded only by anions. Cation-cation contacts do not occur.

In many ionic crystals one type of ion, as a rule an anion, is so much larger than the other that it plays a decisive role in packing. In such cases the anions may often practically speaking be packed in close packing, in whose interstices the cations are inserted so as to obtain the most even charge distribution. The fact that this can take place shows that anion–anion contacts are possible, but each anion then has its charge compensated by one or several neighboring cations and exerts therefore only a slight electrostatic effect on its anion neighbors (see, for example, fig. 12.1b).

To obtain the highest possible lattice energy, the cations and anions must be in true contact with each other. The stability of different structure types therefore becomes determined by limiting values of the radius ratios of the same kind as those considered in 5-2d. The same reservation that was raised then about the validity of the geometrically derived conditions of stability naturally also applies to crystals. All of the observations, which primarily were examined by V. M. Goldschmidt in the 1920's, show, however, the great significance that the purely geometrical factors have for the existence of different structures.

For ionic crystals of the composition AX the *sodium-chloride type* (fig. 6.16a) and the *cesium-chloride type* (fig. 6.16b) are by far the most common structure types. Both are cubic.

In the sodium-chloride type both kinds of ion have the coordination number 6. It can be seen (cf. 6.12b) that the structure may be thought of as being built up of two interpenetrating face-centered cubic lattices placed parallel to each other. The sodium-chloride type as a whole is not close packed, but if one of the kinds of ion is clearly dominant in size (which is often the case), its lattice is practically speaking close packed. The ions of smaller size lie in the largest interstices in the close packing.

In the cesium-chloride type both types of ion have the coordination number 8. The structure may be thought of as being built up of two interpenetrating simple cubic lattices placed in parallel position.

With the exception of NH_4F all alkali and ammonium halides have the sodium-chloride or the cesium-chloride structure. The boundary between the two types, however, is not especially clearly defined by the ionic radius relationship, and this certainly depends on the fact that their lattice energies are so little different from each other. The halides of the smaller cations Li^+, Na^+ and K^+, however, all have the sodium-chloride structure. Rb halides also have this structure at room temperature and atmospheric pressure, but at other temperatures and pressures the cesium-chloride structure may appear. This shows the tendency toward higher coordination number with larger ionic radius of the cation but, since both types are in equilibrium with each other at the transition point, it also shows that their lattice energies may be equal. The same applies to the ammonium halides (the ion NH_4^+ occurs as a nearly spherical ion with an effective radius approximately as large as that of Rb^+). For CsCl, CsBr and CsI the cation is so large that the salts normally have the cesium-chloride structure, but the sodium-chloride structure may appear at more extreme temperatures and pressures.

The sodium-chloride type is represented by a large number of compounds from

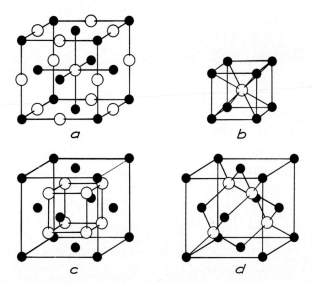

Fig. 6.16. Unit cells of (a) sodium-chloride type AX, (b) cesium-chloride type AX, (c) fluorite type AX_2, (d) sphalerite type AX. Solid circles = A, open circles = X. The cubic diamond structure is obtained from the last if the two types of atom become identical. In d the trigonal rotation axes are the body diagonals of the cube, while the fragment of the diamond structure shown in fig. 5.16 has been oriented with the trigonal rotation axes vertical.

different classes of substances, ionic compounds as well as others. In addition to the halides already referred to, we may mention oxides and sulfides of alkaline-earth metals, AgF, AgCl, AgBr, ∼FeO, ∼TiC, PbS and LiH.

The cesium-chloride type is also found among different classes of substances. It is common among alloy phases, for example, ∼CuZn (β brass).

The *fluorite type* (cubic, fig. 6.16c) is a common structure type among ionic crystals and also other compounds of the composition AX_2. The figure shows that the unit cell contains four A and eight X ions. The A ions coordinate eight X ions in a cube. The X ions coordinate four A ions in a tetrahedron. The cubic coordination about the A ions requires that the radius ratio $r_A/r_X > 0.732$ (tab. 5.3). In fact, a change to sixfold coordination takes place at about $r_A/r_X = 0.67$. Then structures of the *rutile type* often appear, named after a modification of TiO_2 (29-f).

Beside CaF_2, the compounds $SrCl_2$, ThO_2 and several alkali-metal oxides and sulfides such as Na_2O and Na_2S, for example, crystallize in the fluorite structure. In the latter the number and positions of the cations and anions have been reversed with respect to CaF_2. Therefore, it is said that CaF_2 and Na_2O are *anti-isomorphous* and that Na_2O crystallizes in an *antifluorite type* structure.

Many compounds with complex ions may also be considered as fluorite and antifluorite types. To the former belong several ammine complexes such as

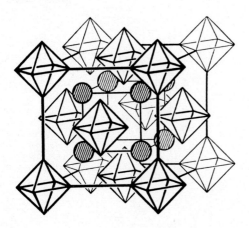

Fig. 6.17. Unit cell of K_2PtCl_6. The spheres = K^+, octahedra = $PtCl_6^{2-}$. Pt is located at the centers of the octahedra and Cl at each octahedral corner.

$Ni(NH_3)_6Cl_2$ and to the latter compounds such as K_2SnCl_6 and K_2PtCl_6 (fig. 6.17). In these the divalent monatomic ions in the simple structure are replaced by nearly spherical complex ions such as $Ni(NH_3)_6^{2+}$ and $PtCl_6^{2-}$.

If the covalent contribution to the bonding becomes significant, structures often arise that may be considered as being built up of layer molecules with unlimited extent in two dimensions. The electrostatic forces are compensated within each layer molecule and the forces between the layers are weak. Portions of layer-structure types common in compounds with the general formulas MX_2 and MX_3 are shown in fig. 6.18. The X atoms are distributed in the same way in both cases and practically speaking form a close packing. The M atoms lie in octahedral cavities. In the MX_2 layer all, and in the MX_3 layer $\frac{2}{3}$ of the cavities in the layer are occupied by M atoms. MX_2 layers of this kind are found in many halides, sulfides and hydroxides (X = OH); for example, $FeCl_2$, $CdBr_2$, CdI_2, PbI_2, SnS_2, $Ca(OH)_2$ and $Mg(OH)_2$. The MX_3 layer is found in halides, for example $CrCl_3$, $FeBr_3$ and BiI_3, and in some hydroxides, for example $Al(OH)_3$. In halides and sulfides mainly van der Waals forces occur between the layer molecules, and in some of the hydroxides hydrogen bonds are formed under the influence of the hydrogen atoms of the OH groups. The layers are nearly always laid upon one another so that the X atoms become close packed throughout the whole crystal. However, close packing is consistent with different relative positions of the layers so that the same layer type can give different crystal structures.

When the building units are bonded by forces with strong covalent contribution, their directive effect overcomes the tendency toward close packing. A typical covalent structure is the cubic *diamond structure* (fig. 5.16), in which the tetrahedral coordination of all atoms can be attained only by a very open packing (a diamond structure of spherical atoms with radius r in contact with each other has the volume $12.3r^3$ per atom). This structure is a special case of the *sphalerite type* (zinc blende; cubic, fig. 6.16d). Sphalerite is the most common modification of zinc sulfide ZnS, but this substance also exists in a hexagonal modi-

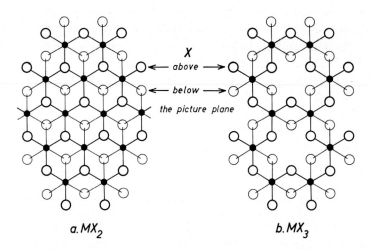

Fig. 6.18. Portion of a layer in layer structures with the composition (a) MX_2 and (b) MX_3. Solid circles = M, open circles = X. The layers have infinite extent in the picture plane.

fication called *wurtzite* of which a hexagonal diamond structure found in 1966 (23-2 b) is a special case. There are an unlimited number of possible ways to build up structures of substances AX so that all atoms have tetrahedral coordination. They are distinguished only by the relative orientation of the tetrahedra and all must have closely similar energies. The sphalerite and wurtzite types, however, have the highest symmetry and are the most common of these tetrahedral structures.

In order that there shall be one bonding electron pair per tetrahedral bond in the sphalerite and wurtzite types, four valence electrons per atom are required. This condition is fulfilled in a large number of compounds belonging to these structure types. It can be fulfilled for elements if they belong to group 14 (diamond, Si, Ge, gray Sn) and for binary compounds if both kinds of atom either belong to group 14 (SiC) or to two groups that lie symmetrically around group 14 (AlN, AlP, InSb, ZnS, HgS, CuCl, CuBr, CuI, AgI). In all these substances the covalent bond contribution is very large.

In ice (fig. 5.26) the oxygen atoms form from a geometrical viewpoint a wurtzite structure, in which they are held together by hydrogen bonds. The nitrogen and fluorine atoms in ammonium fluoride, isoelectronic with ice (notice the analogy between the formulas NH_4F and OH_4O), are bound together by hydrogen bonds, and form a completely analogous structure.

Transition metals lying to the right of group 5 often form with some *p* elements in groups 14–16 (Sn, Pb; As, Sb, Bi; S, Se, Te) metallic phases of the so-called *nickel-arsenide type* (first found for NiAs). The structure is hexagonal, the transition-metal atoms coordinate six *p*-element atoms octahedrally while the latter are surrounded by six transition-metal atoms at the corners of a trigonal prism. Interstitial solution (6-3a) with variation of the number of transition-metal atoms in the structure often causes the composition to depart from the ideal formula MX.

d. Magnitude of lattice energy. For a crystal that can be decomposed into free, and thus gaseous, building units, the lattice energy is equal to the sublimation energy. The lattice energy is therefore often easily determined for crystals that consist of finite, neutral molecules, and for metal crystals.

The lattice energy for ionic crystals, on the other hand, cannot be experimentally determined directly, since ionic crystals on vaporization usually give ion pairs and ion clusters (5-2a) and only a small number of free ions. It must therefore here be determined indirectly, in general according to a method that is called the *Born-Haber cycle*, after its originators.

The Born-Haber cycle is a hypothetical process that gives the relationship between a series of energy quantities. An example is given here for NaCl. The formation of crystalline NaCl from the free building units of the crystal corresponds to the equation $Na^+(g) + Cl^-(g) \rightarrow NaCl(s)$. The energy $-\Delta U$ given off in this process is the lattice energy X for NaCl. In the Born-Haber cycle, however, the enthalpies of sublimation L and dissociation D are included, so that all other energy quantities must also be expressed as enthalpy changes. In the process of crystal formation 2 moles of gaseous ions disappear and if these can be assumed to have formed an ideal gas, for this process according to 2-1d we may set $\Delta H = \Delta U - 2RT = -X - 2RT$. The Born-Haber cycle may then be represented by the following scheme, where the enthalpy *increase* in each partial process is shown above the respective arrows (cf. (5.3) and (5.6)):

$$\left\{ \begin{array}{l} Na(s) \xrightarrow{+L} Na(g) \xrightarrow{+I} Na^+(g) \\ \tfrac{1}{2}Cl_2(g) \xrightarrow{+D/2} Cl(g) \xrightarrow{-E} Cl^-(g) \end{array} \right\} \xrightarrow{-X-2RT} NaCl(s) \qquad (6.2)$$

$$\Delta H$$

The scheme is analogous to (5.3) but here the initial substances are $Na(s)$ and $Cl_2(g)$, and the final product is crystalline NaCl instead of free ion pairs Na^+Cl^-.

From (6.2) it follows that the enthalpy of formation is

$$\Delta H = L + D/2 + I - E - X - 2RT \qquad (6.3)$$

The quantities $-\Delta H$, I, L and D are in general directly measurable. The electron affinity E is experimentally known in rather few cases. The lattice energy X can be calculated theoretically for simple ionic crystals. The Born-Haber cycle is used both to calculate X when E is known, and vice versa. If the values for NaCl (kJ mol^{-1}; 25°C) are inserted into (6.3), we get $\Delta H = 109 + 121 + 496 - 369 - 774 - 5 = -422$ kJ mol^{-1}.

The lattice energy (774 kJ mol^{-1}) makes a very large contribution to the enthalpy of formation of NaCl(s) from Na(s) and $Cl_2(g)$. The fact that the lattice energy is significantly greater than the bond energy in the free ion pair Na^+Cl^- (510 kJ mol^{-1}; see 5-2f) means that crystals and not ion pairs are formed at ordinary temperature. These energy values also illustrate the discussion in 5-2a concerning the way the energy of the system decreases when crystals are built up of ions. Each individual bond Na^+–Cl^- is weaker in the crystal than in the free ion pair where the electrostatic attraction can be concentrated in the one bond. This is made clear by the fact that the interionic distance in the crystal is 2.81 Å

Table 6.1. *Lattice energies, kJ mol⁻¹, 25°C.*

All values except in the last group correspond to ions as building units.

LiF	1021	KF	816	CsF	724	MgO	3930
LiCl	841	KCl	716	CsCl	649	CaO	3480
LiBr	808	KBr	691	CsBr	632	SrO	3220
LiI	753	KI	653	CsI	600	BaO	3040
NaF	908	RbF	774	AgF	915	Ne	2.5
NaCl	774	RbCl	678	AgCl	860	Ar	8.50
NaBr	745	RbBr	657	AgBr	845	Kr	11.7
NaI	699	RbI	624	AgI	835	N_2	7.79
						Cl_2	31.1
				CaF_2	2630		

and in the ion pair 2.40 Å. That the total energy per mole in spite of this is greater in the crystal depends on the much greater number of bonds per mole there.

Tab. 6.1 lists a number of lattice energy values. To these may be added the values of the sublimation enthalpies in tab. 5.5, which are only slightly different from the sublimation energies and thus the lattice energies.

In most alkali halides and in the oxides of magnesium, calcium, strontium and barium, the ionic bond contribution is great. The strength of the ionic bond is clearly fully comparable to and often greater than that of the covalent bond (5-3k).

According to 5-1c the energy of electrostatic attraction between two ions is $E_A = -kz_1z_2/l$. The lattice energy of substances that all form ionic crystals of the same type should therefore increase as the ionic radii and thus the interionic distances l decrease. Further, the lattice energy should increase with increasing ionic charge. That this is so is shown by a comparison of values in tab. 6.1 for a number of substances that crystallize in more or less typical ionic lattices of the same type (NaCl type). These substances are the halides of lithium, sodium, potassium and rubidium, CsF and the oxides of magnesium, calcium, strontium and barium. The effect of size is observed in the energy decrease from lithium to rubidium, fluorine to iodine, and magnesium to barium. The effect of charge is noticeable in that the values for the oxides mentioned are much higher than for all the alkali halides.

Even if the structure type is changed we notice that the lattice energy still often increases with decreasing ionic radius and increasing ionic charge. An obvious requirement, however, is that the ionic contribution to the bonds shall be reasonably similar. CsCl, CsBr and CsI, which crystallize in another structure type (CsCl type, 6-2c), in this way fit in with the other alkali halides. Less consistent, on the other hand, are the more covalently bonded silver halides, of which yet AgF, AgCl and AgBr crystallize in the NaCl type (the ionic radius for Ag⁺ would place them between the sodium and potassium halides).

The last group in tab. 6.1 shows lattice energy values for crystals that are held together by van der Waals forces. The low strength of these forces can be seen, and the way they increase with the molecular weight, at least for similar molecules (5-4a).

As would be expected, a substance with a high lattice energy has a high melting point, and vice versa. For example, we may compare the melting points for MgO (2800°C), CaF_2(1360°C), LiF(870°C), CsI(621°C) and N_2 (−210°C), with the lattice energy values in tab. 6.1. It is obvious, however, that other factors also influence the melting point, and that there is no precise relationship between lattice energy and melting point.

e. Some applications to equilibria. In all equilibria in which crystallized phases take part, their lattice energies naturally play a large role. However, the circumstances most often are too complicated to permit the quantitative use of lattice energies, and also, they are known for only a very few crystals. What is known qualitatively about lattice energies nevertheless often suffices to permit us to draw important basic conclusions. Above all, a clearer picture is obtained of the energy balance in the processes in question. Some examples of this are given in the following.

If numerical values are available for those terms that according to (6.3) contribute to the enthalpy of formation of a substance, in general the reason for the greater or lesser stability of the substance with respect to decomposition into the elements can be understood. A discussion of the stability of alkali and silver halides will illustrate this. The similarity of the different systems means that we may here assume that the ΔH values lie in the same sequence as the ΔG values and thus are useful in estimating the relative stabilities (cf. 2-3b). For the halides of all alkali metals we find that ΔH becomes less negative as we follow the series F^-, Cl^-, Br^-, I^-. (For example, for the lithium halides the ΔH values are: LiF, −612; LiCl, −409; LiBr, −350; LiI, −271 kJ mol^{-1}.) The chief cause of this decrease is the decrease in lattice energy as the radii of the anions increase. For the halides of the same alkali metal the stability should thus decrease in going to the right in the anion sequence mentioned.

The least negative ΔH value of all the alkali halides is −271 kJ mol^{-1} for LiI. For all the silver halides, however, we find still less negative values (AgF, −203; AgCl, −127; AgBr, −100; AgI, −62 kJ mol^{-1}). This is caused mainly by the higher ionization energy of silver (fig. 5.4), but also by its higher sublimation enthalpy (tab. 5.5). The silver halides should thus be less stable than the alkali halides. Of the former, AgF should be the most stable, and this is found to be a result of its high lattice energy, which in turn is caused by the small size of the fluoride ion. AgBr and AgI should be the least stable.

These conclusions concerning the stability are in agreement with observations of the light sensitivity of the halides. Of the silver halides, AgBr and AgI are most easily decomposed by light in the visible spectrum, AgF and the alkali halides not at all. For the decomposition of the latter, more energy-rich radiation (ultraviolet or shorter wavelengths) must be applied.

It is generally true that the stability of the halides of a metal decreases with increasing atomic number of the halogen. The explanation always is that the lattice energy decreases with increasing anionic radius. Further, compounds of Cu(I) and Au(I), similarly to the compounds of Ag(I), are significantly less stable than the corresponding compounds of alkali metals. In all cases this results for the most part from the higher ionization energy of the metals in group 11. Similar conditions are also found on comparison of compounds of the metals in groups 2 and 12.

When a crystal is dissolved in a solvent, an amount of energy equal to the lattice energy must be taken up. On solvation of the building units that go into solution, enthalpy of solvation is released (5-2i).

For the solution of sodium chloride crystals in water we may write:

$$\text{NaCl(s)} \xrightarrow{X+2RT} \left\{ \begin{array}{l} \text{Na}^+(\text{g}) \xrightarrow{-Y_M} \text{Na}^+(\text{aq}) \\ \text{Cl}^-(\text{g}) \xrightarrow{-Y_X} \text{Cl}^-(\text{aq}) \end{array} \right\} \underset{\Delta H}{\longleftarrow} \qquad (6.4)$$

The enthalpies of hydration for the metal and halide ions are denoted as in 5-2i by Y_M and Y_X. The heat of solution at infinite dilution is $-\Delta H$. The scheme gives

$$\Delta H = X + 2RT - Y_M - Y_X \qquad (6.5)$$

The values in tab. 5.4 and 6.1 give for NaCl: $\Delta H = 774 + 5 - 396 - 370 = 13$ kJ mol^{-1} (at 18°C, $\Delta H = 4.9$ kJ mol^{-1} has been obtained by direct measurement). That $\Delta H > 0$, that is, that heat is absorbed on solution, means that the lattice energy of NaCl is somewhat greater than the sum of the hydration enthalpies of the ions.

On the other hand, for LiCl the lattice energy is less than this sum, that is, heat is given off on solution. The tabulated values give: $\Delta H = 841 + 5 - 506 - 370 = -30$ kJ mol^{-1}, which shows that it is the great hydration enthalpy of the Li$^+$ ion that succeeds in overcoming the otherwise quite large lattice energy.

For salts, the absorption of heat on solution in water is more common than the contrary effect.

The last examples show that ΔH for the solution process is composed of the small difference between the large terms of lattice energy and the sum of the hydration enthalpies. The error in calculated ΔH values can therefore be quite large and it may be difficult to draw any conclusions from these values. Because ΔH is numerically small, changes in entropy on solution play a large role in determining the sign of ΔG (equation (2.9)). Therefore, a discussion of solubility equilibria only from the standpoint of lattice energies and hydration enthalpies is hardly possible. However, it is found that the high lattice energy of LiF is the chief cause of the fact that this compound is the only difficultly soluble alkali halide. Also, for CaF_2, the high lattice energy is the main reason for the low solubility.

For reactions between crystallized substances only, ΔH for the process can be calculated from their lattice energies. Since the lattice energy represents the energy *decrease* on crystal formation, for a process $A + B \rightarrow C + D$, in which all the substances are crystalline, we have $\Delta H = (X_A + X_B) - (X_C + X_D)$. With values from tab. 6.1, we get, for example,

$$\text{LiBr(s)} + \text{KF(s)} \rightarrow \text{LiF(s)} + \text{KBr(s)}; \quad \Delta H = -88 \text{ kJ } (25°C)$$

For a process of this kind the change in entropy must be small, which according to (2.9) means that $\Delta G \approx \Delta H$. The calculated value for ΔH thus shows that at room temperature the state on the right side is the most stable. Reaction toward the right, however, proceeds at a noticeable rate only on raising the temperature some hundred degrees (if a mixture of LiBr and KF is ground at room temperature, a rapid double decomposition takes place, but this results from the temperature rises at the contacts between the grains; because of the hygroscopic nature of the salts all experiments must be carried out with well-dried salts and in a completely dry atmosphere).

It can be shown that in cases such as this, in which the structure type is the same (here the NaCl type) and the ionic-bond contribution is reasonably strong in all four phases, that combination is the most stable that corresponds to mutual combination of the two smallest ions (here Li$^+$ and F$^-$) and the two largest (here K$^+$ and Br$^-$).

Calculations of this last kind, for lack of lattice energy values, have until now been possible in only rather few cases.

6-3. Disorder phenomena in solid substances

Here are considered primarily those disorder phenomena that are of particular interest from a chemical standpoint.

a. Solid solutions. Solid solutions, and thus solid phases with extended homogeneity ranges (1-4b), may be either amorphous or crystalline. However, we may take crystalline solid solutions as a model. Amorphous solid solutions are formed according to the same principles and differ from the crystalline ones only by their greater disorder.

The two most important types of solid solutions are shown schematically in fig. 6.19. In types I and II we start with a crystal of the element A (*a*). In type I B atoms are dissolved, whereby they replace A atoms (*substitution solution*). The replacement often occurs randomly so that B is distributed irregularly over the original A sites (*b*). With compositions in which a simple ratio exists between the number of atoms of the different kinds, however, the distribution may become ordered (*c* for the composition AB). An ordered distribution is much less probable than a disordered one of the same composition and therefore has a lower entropy than the latter. If it nevertheless exists, this is because it has so much higher lattice energy, that is, lower energy content, that according to (2.9) $\Delta G < 0$.

The solutions in I may be formulated $A_{1-x}B_x$, which shows that the total number of atoms in the crystal is constant. An example is the solid solution of gold in copper, $Cu_{1-x}Au_x$, which extends all the way from pure copper to pure gold ($0 \leq x \leq 1$). For $x = 0.25$ (the composition Cu_3Au) and $x = 0.5$ (the composition CuAu) ordered distribution can exist. Often the solubility is limited, which can be indicated as limits for x (for example, $x \leq 0.2$).

In type II the element C is dissolved in A. Its atoms are assumed to be so small that they may be inserted into interstices in the A lattice without disrupting it. The distribution may be irregular (*b*) or ordered (*c* and *d* for the compositions A_2C and AC). This type of solution may be thought of as an *addition solution* if one goes to the right in the figure from a certain composition, and as a *subtraction solution* if one goes to the left. For example, if carbon is dissolved in α or γ iron, carbon atoms are inserted into the interstices in the iron lattice (*interstitial atoms*), which in principle corresponds to the step from *a* to *b*. We often speak in this case of an addition solution. All types of solution that belong to case II may be considered together as *interstitial solutions*.

The formulation should be AC_x if one starts with A and adds C, or AC_{1-x} if one starts with AC and subtracts C. Both formulas show that the number of A atoms is constant and the number of C atoms variable.

In type III is shown a substitution solution of D in AB, wherein D replaces B. The solution is assumed to extend up to AD. The substitution is random in *b* and ordered in *c* (the composition A_2BD). In cases of type III it is often said that

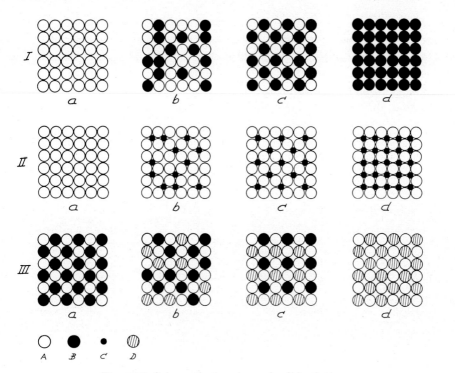

Fig. 6.19. Schematic drawings of solid solutions.

AD dissolves in AB (for example, NaBr in NaCl; double substitution may also occur, for example, when RbBr dissolves in KCl).

These solutions may be formulated $AB_{1-x}D_x$, which shows that the total number of atoms $B + D$ in the lattice is constant. A common but in complicated cases less informative formulation is obtained by placing the atoms or atomic groups that replace each other, separated by commas, inside parentheses. In this example, thus, we have $A(B,D)$.

A change in the composition of a solid solution in general results in a change in the dimensions of the lattice. Exchange of one kind of building unit by another that requires a greater volume causes an increase in dimensions. For interstitial solutions an increase in dimensions occurs when the content of interstitial atoms increases. No such changes in dimensions have been shown in fig. 6.19.

Crystalline solid solutions have also been called mixed crystals, an improper term since a solid solution is a homogeneous phase.

In fig. 6.19 atoms either replace each other or are inserted in holes. A solid solution can also occur in which molecules of different kinds replace each other, or small molecules are inserted in holes.

When greater or less order in a solid solution is spoken of, it is the order in the *distribution* of the building units over the lattice points that is referred to. The crystalline state requires that the building units lie on geometrically rather well-ordered lattice points, but they may be distributed on these in all degrees of order from completely disordered (as in I–IIIb) to completely ordered (as in I–IIIc). The laws of periodic repetition and identity (6-1a) are naturally only valid *statistically* in cases of disorder of distribution.

For a solid solution to occur, the building units that are exchanged or changed in number must fulfill certain conditions as to their size. Furthermore, the phase must remain electrically neutral.

If atoms or ions are to be able to replace each other to any great extent, the difference in radii may not exceed approximately 15 % of the smaller radius. The ions Na$^+$ ($r=0.98$ Å) and K$^+$($r=1.33$ Å) therefore cannot be exchanged for each other to any appreciable degree (NaCl and KCl are not soluble in each other at room temperature). Neither can F$^-$ ($r=1.33$ Å) and Cl$^-$ ($r=1.81$ Å). On the other hand replacement can occur between K$^+$ and Rb$^+$ ($r=1.48$ Å), and between Cl$^-$ and Br$^-$ ($r=1.96$ Å). F$^-$ and OH$^-$ (which can be considered as spherical with $r=1.45$ Å) often exchange with each other. If molecules are to be able to exchange with each other, their external effective surfaces must, of course, be quite nearly congruent.

Interstitial solution requires that the building units that vary in number be so small that they easily fit into the interstices. Small-atom nonmetals such as hydrogen, carbon and nitrogen are often inserted into lattices of transition metals. Because cations are often small in relation to anions (5-2c; cf. also fig. 5.5), most commonly it is the number of cations in a lattice that varies. The anion lattice on the other hand often forms a more rigid framework in whose openings the cations may vary both in number and kind. In general it is only for very small anions (for example F$^-$) or very large cations that this relationship may be reversed.

When the heat motion of the building units increases with rise in temperature, the space requirements generally become less restrictive. Thus, the possibilities for solution become greater and the homogeneity ranges are broadened. For NaCl and KCl the mutual solubility increases and they become completely miscible at temperatures above 400°C.

The requirement of electrical neutrality does not prevent exchange or interstitial solution by uncharged atoms or molecules. Solid solutions (extended homogeneity ranges) are therefore very common in intermetallic phases. In certain crystalline phases water enters into solid solution in variable amount ("zeolitic" water, see 13-5e2). Ions with equal charge of the same sign can exchange with each other, but if the ionic charges are unequal or the number of ions changes, two or more compensating solution processes must take place at the same time. Naturally, cations and anions are never exchanged for each other.

The compensating processes may both be substitutions. An example is the feldspar mineral (23-3h5) albite, NaAlSi$_3$O$_8$, in which Na$^+$ can be replaced by Ca^{2+} at the same time that Si^{4+} is replaced by Al^{3+}. The oxygen lattice remains unchanged.

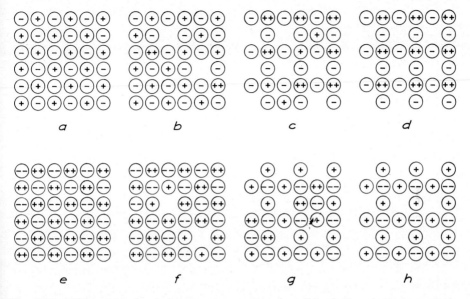

Fig. 6.20. Different kinds of charge compensation in solid solutions in ionic crystals.

The most important cases, where one of the processes leads to an interstitial solution, are shown schematically in fig. 6.20. In the progressive changes $a \rightarrow b \rightarrow c \rightarrow d$ cations are replaced by other cations of higher charge. The increase in positive charge is compensated by vacancies in a corresponding number of cation sites (subtraction solution). The anion lattice is unchanged. The exchange can occur between ions of different elements. In LiCl Li$^+$ can be replaced by Mg^{2+}. For each exchange of 1 Mg^{2+} for 1 Li$^+$, in addition 1 Li$^+$ leaves the crystal. Overall, there thus occurs an exchange of 1 Mg^{2+} for 2Li$^+$ and the solid solution can be written Li$_{2(1-x)}$Mg$_x$Cl$_2$, or (Li$_2$, Mg)Cl$_2$. This mechanism corresponds in a formal way to an increase in the content of MgCl$_2$ by the phase, and it has often been said that MgCl$_2$ dissolves in LiCl.

But also atoms of the same kind can change charge. The iron oxide Fe$_3$O$_4$ crystallizes in the cubic system with 8Fe^{2+} + 16Fe^{3+} + 32O^{2-} in the unit cell. If it is oxidized at temperatures below about 350°C the oxygen lattice remains unchanged. More and more of the Fe^{2+} ions change to Fe^{3+} ions and the increase in positive charge is compensated by cation vacancies. The final stage is (metastable) γ Fe$_2$O$_3$, which contains 21$\frac{1}{3}$Fe^{3+} per 32O^{2-}, while 2$\frac{2}{3}$ of the 24 original cation sites are empty. This type of solid solution appears frequently in compounds of the transition elements, whose ions easily change charge. If we start with the complete structure ("ideal composition", "stoichiometric composition"), the charge of the metal ions must increase. Therefore, we encounter this type of solution in compounds that in the ideal composition contain metal ions that can increase in charge. Beside Fe$_3$O$_4$ may be mentioned FeO, FeS, Cu$_2$O and CuI, whose metal content can be decreased to a point corresponding at least to the formulas $Fe_{0.88}O$, $Fe_{0.82}S$, $Cu_{1.996}O$ and $Cu_{0.995}I$.

Progressive changes in the opposite direction, thus $d \to c \to b \to a$, give addition solutions. Here cations are replaced by other cations with lower charge and the decrease in positive charge is compensated by the insertion of cations into the interstices of the original structure. The anion lattice remains unchanged. This solution type is known, for example, in ZnO and CdO, and results here from the appearance of monovalent cations Zn^+ and Cd^+. In these cases the homogeneity regions are very small.

In fig. 6.20 e–h the change in charge in the cation lattice is compensated by a change in the number of anions, which, as mentioned above, is more rare. An example of this case, although with other charge values and compositions, occurs in the system $PbF_2 - BiF_3$. The components PbF_2 (in principle corresponding to h) and BiF_3 (in principle corresponding to e) dissolve in each other in all proportions, over which the number of F^- ions in the crystal changes.

A change in charge coupled with interstitial solution permits the average charge of an atom type in a homogeneous phase to vary continuously and the composition of the phase to depart from the ideal composition that is required by constant ionic charge. Both of these results have great fundamental significance. The latter, for example, may exclude the use of a phase in quantitative analysis and stoichiometric atomic weight determinaton. As mentioned previously (1-4j), phases of this type are often, in spite of their extended homogeneity range, denoted by a formula corresponding to the ideal composition; for example, FeO phase, or $\sim$FeO.

The composition can vary even in salts such as the alkali halides. If an alkali halide crystal is exposed to X, γ or electron radiation, it becomes colored. Sodium chloride first becomes yellow but after intensive irradiation and heating to about 200°C the color becomes blue-violet. The same color is obtained when the crystal is heated in sodium vapor at the same temperature and then rapidly cooled. These colored crystals show an excess of metal and their densities are less than normal. It is believed that the crystals contain anion vacancies and that an electron is trapped in each vacancy left by one monovalent anion. In this way electrical neutrality is maintained. An anion vacancy + a trapped electron is called an F center (after the German "Farb-zentrum"). The trapped electron becomes involved with the electron clouds of the surrounding cations, and therefore an F center can be thought of very schematically as an anion vacancy + a neutral alkali atom. A crystal with F centers then is not different in principle from the solution type of fig. 6.20f. In NaCl it has been possible to obtain approximately 1 F center, that is 1 anion vacancy, per 2000 anions, which corresponds to the composition $NaCl_{0.9995}$. With stronger heating of colored crystals more and more F centers come together so that finally alkali metal particles of colloidal dimensions are formed. Such particles have been observed in the ultramicroscope.

Ion exchangers and *clathrate compounds* may also be classed as solid solutions. Ion exchangers consist of framework molecules that form a solid framework provided with tubes or channels of molecular dimensions. On their inner walls atoms or molecules are bound in varying numbers while the solid framework is undisturbed. Because these channels are in contact with the surrounding phase, the atoms or molecules can undergo exchange. Ion exchangers are considered in more detail in 15-3.

In the clathrate compounds (Lat. clatratus, enclosed by a grating) atoms or molecules ("guests") are enclosed within cavities of molecular dimensions ("cages") at the time that the crystal is formed. The cages lack openings to the surroundings that would permit the "guests" to escape. It is typical of clathrate compounds that many different kinds of atoms and molecules can be enclosed in a given cage type. They must only have enough space in the cages and exert no appreciable bonding forces on the cages other than van der Waals forces. *Gas hydrates*, which in general are clathrate compounds, sometimes crystallize from water containing gases dissolved under pressure. The water structure that forms the cages is none of the structures which have been observed for pure ice (21-5c), but is specially adapted for cage formation. A common structure has a cubic unit cell containing $46H_2O$. The enclosed guests can occupy 6 large and 2 small interstices per cell. If the guests are sufficiently small, *e.g.* Ar, Kr, Xe, O_2, H_2S, CH_4, they can be accommodated in both types of interstices so that the saturated gas hydrate obtains the composition $8X \cdot 46H_2O$. Larger guests as Cl_2, Br_2, SO_2, N_2O have enough room only in the large interstices so that the composition at saturation will be $6X \cdot 46H_2O$.

b. Disorder. Distribution disorder in solid solutions implies large lattice imperfections and it appears reasonable that similar imperfections may be found even in crystals of ideal composition (for example, "pure substances"). Thus, also in these exchanges (between fairly equally sized and generally equally charged building units) and lattice vacancies (in ionic crystals in this case there are equal numbers of cation and anion vacancies) may occur. Here and there an atom may also have left a normal site, causing a vacancy, and taken an interstitial position somewhere nearby (the atom should then be small in relation to the other building units). It has been possible to show the presence of such lattice imperfections (*defects*) in several ways. As a rule the number of defects at ideal composition is relatively small but can nevertheless be of great significance for many properties (for example, diffusion rate and conductivity, which together with X-ray diffraction and entropy measurement, are used to determine the extent of crystal defects). In exceptional cases the number is great; in ~TiO at the ideal composition TiO and at room temperature, approximately 15 % of all the cation and anion sites are empty.

In the same crystal the defects may be very different for different building units. Building units of small relative size are often more subject to disorder than the larger units. An extreme case is α AgI (stable above 140°C) in which the Ag^+ ions can be found practically speaking anywhere in the interstitial space between the I^- ions. It may be said that the cation portion of the crystal is "molten" and irregularly fills the space in the anion framework approximately like a liquid in the pores of a sponge.

Distribution disorder is normally present at equilibrium (the examples TiO and α AgI correspond to equilibria), but may be greater before equilibrium is reached. Both in crystalline solid solutions and pure substances the distribution is completely ordered only at absolute zero (if equilibrium has had time to be established). The order decreases then with rise in temperature.

The change in distribution order with increase in temperature may be illustrated by β brass, $\sim$CuZn. With the composition CuZn and complete order the structure is of the CsCl type (fig. 6.16b). In this the lattice points in a simple cubic lattice are occupied only by Cu atoms, and the lattice points in another lattice congruent with the first only by Zn atoms. We call the first kind of lattice points "Cu sites" and the second kind of lattice points "Zn sites". As long as the Cu sites throughout the whole crystal are occupied by more Cu atoms than Zn atoms (from which the opposite follows for the Zn sites) the distribution still has long-range order. The distribution has short-range order if the atoms of one kind are surrounded by more atoms of the opposite than of the same kind. Short-range order may remain even without long-range order, and probably never completely disappears.

The long-range order can properly be represented by the quotient (number of Cu in Cu sites – number of Zn in Cu sites)/(total number of Cu sites) for the whole crystal. This quotient becomes 1 for complete order and 0 for complete disorder, that is, random distribution of Cu and Zn over all Cu and Zn sites. With increasing temperature the long-range order decreases according to the schematic curve in fig. 6.21. It decreases slowly at first but then falls rapidly to 0, that is, to complete disorder. The reason that the curve falls more rapidly as the temperature rises is that the greater the disorder is, the less energy is required for another Cu atom and Zn atom to change places with each other. The temperature T_c, above which the long-range order is 0, that is, the distribution disorder is complete, for β brass is 470°C.

The above applies to distribution order. The purely geometrical order of the structure, which is required for crystalline properties, decreases with rise in temperature because of the increased thermal vibrations, but nevertheless remains to a large extent until the substance melts, when it completely or nearly completely disappears. Only a certain short-range order can then remain. Most often melting or some other phase transition occurs before the temperature T_c is reached. An entire course such as in fig. 6.21, an *order-disorder transition*, however, is not uncommon (see also 13-3c).

Order-disorder transition can also occur in other kinds of distributions than the distribution of atoms on lattice points. An important type is the *Curie transition*, which originally referred to the transition ferromagnetism-paramagnetism (7-4a). Nowadays the term is also applied to a similar dielectric transition.

A third type of order-disorder transition results from the fact that more and more of the building units in the crystal begin to rotate as the temperature rises. Above a definite temperature T_c all the building units of the same kind rotate. Rotation occurs mostly with building units that have nearly rotational symmetry, especially if they are nearly spherical. Among such building units may be mentioned the molecules CH_4, HCl and the ions NH_4^+, ClO_4^-.

Beside all these disorder phenomena that may be consistent with equilibrium, there are many kinds of disorder that imply lack of equilibrium. Most common are the imperfections in the geometrical order of the crystal structure. Often these

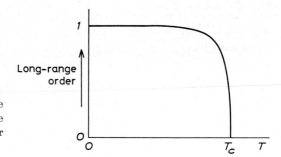

Fig. 6.21. Change in long-range order of distribution with change in temperature (order–disorder transition).

are concerned with *deformation* caused by mechanical forces (6-5a). Imperfections may also be caused by crystal growth that is too rapid for equilibrium to be attained. Later the mobility of the building units in the solid phase has been too low for subsequent correction. Examples of this type are glasses (6-3d), in which the building units have not had time to order themselves in any long-range order whatever.

c. Diffusion and ionic conduction. Diffusion in a crystal, that is migration of building units to an equilibrium position, is very common. If the crystal permits building units to occur in interstitial locations, or lattice vacancies to form, diffusion can take place, and this more easily the greater the disorder is. Often, however, temperatures higher than room temperature are necessary for the building units to have sufficient mobility.

If charged building units, ions, can migrate through the crystal, migration occurs also in an electric field. The crystal thus conducts electricity through material transport; it is an *ionic conductor*.

A building unit in an interstitial location can move from one such location to the next (interstitial migration). Another migration mechanism is based on a lattice vacancy. A neighboring building unit that will fit will jump into it. In this way a new vacancy is formed, which is filled in turn, and so on. The most obvious effect is that the vacancy migrates and its direction of migration is opposite to that of the building units. Since building units with small relative size are both most common in interstitial locations and most often cause lattice vacancies, these are the most common migrators, which anyway seems quite natural. Since defects are required for migration, it is also clear that both diffusion and ionic conduction are encouraged by an increase in the number of defects. A pure substance therefore most often has low ionic conduction, but this increases rapidly even with solution of a small amount of substances that increase the number of defect sites. At room temperature, however, the ionic mobility is usually so small that both diffusion and ionic conduction are quite insignificant. But they may nevertheless become quite appreciable at a temperature of a few hundred degrees (most alkali halides at 300°C have an ionic conductivity of the order of magnitude of 10^{-8} ohm^{-1}cm^{-1}).

A substance may transport charge both through ionic conduction and electronic conduction (6-4). Therefore, we must know how large a fraction of the current is carried by cations, anions and electrons. These fractions are called *transference numbers* and designated by n_+, n_- and n_e (thus the relation $n_+ + n_- + n_e = 1$ is true). By direct measurement of the amounts of ions transported, n_+ and n_- are determined, from which also n_e is obtained.

For KCl at 525°C, $n_+ = 0.88$, $n_- = 0.12$, $n_e = 0$, and at 600°C, $n_+ = 0.71$, $n_- = 0.29$, $n_e = 0$. The current transport thus takes place mainly through the more easily mobile cations, but when the whole structure becomes less ordered through increasing thermal motion, the share of the anions in the transport increases. In AgCl, AgBr, and AgI the mobility of the cations is so overwhelming (cf. the discussion of α AgI in 6-3b) that $n_+ = 1.00$ up to above 300°C. Cu(I) halides show only electronic conduction ($n_e = 1.00$) at room temperature, but as the ionic mobility increases with heating, the current transport is taken over more and more by the cations. Between 300° and 400°C n_+ rises to very near to 1.

Diffusion through the crystal of its own building units is best studied by the use of radioactive trace atoms (9-3e). Understanding of the diffusion mechanism has clarified a good number of difficult problems concerning the reaction of solid phases with gases. Among these problems, for example, are many cases of the oxidation of metals in oxygen. If the oxide layer thus formed is porous and brittle the oxidation continues rapidly inward. Such is the case with the alkali and alkaline earth metals with the exception of beryllium. If the metal oxide is volatile it easily dissipates and does not protect the metal. On heating molybdenum in air the oxide MoO_3 is formed, which over 500°C quickly begins to vaporize. With a dense and nonvolatile oxide layer, however, the continued action of oxygen depends on its ability to diffuse through the layer. If the oxide layer has a tendency for defect formation, diffusion takes place easily and the oxidation continues, though less rapidly than with a porous oxide layer. Examples are copper and iron at 400–500°C, where the diffusion rate begins to become sufficiently high to be effective. If the tendency of the oxide to form defects is small, the layer inhibits diffusion and protects against continued oxidation. Aluminum and chromium are examples of this type.

When the oxidation continues because of defects in the oxide, it is commonly found that the oxide layer grows outward into the gas phase and not inward into the metal. This results from the fact that the metal in the form of metal ions most often diffuses through the oxide layer much more rapidly than oxygen in the form of oxide ions. Let us assume the oxide to be Cu_2O growing on metallic copper in oxygen. At the phase boundary $Cu_2O(s)$–$O_2(g)$ new layers of the Cu_2O structure are formed by the process $2Cu^+ + 2e^- + \tfrac{1}{2}O_2(g) \rightarrow Cu_2O(s)$. For each $\tfrac{1}{2}O_2$ then $2Cu^+ + 2e^-$ are brought from the nearest underlying lattice plane. In this there is thus produced cation vacancies and electron deficiencies, which are filled from the next underlying lattice plane, and so on. In this way, a stream of Cu^+ ions and electrons is set up through the whole Cu_2O layer. This stream is ultimately supplied by the metal phase through the reaction $Cu \rightarrow Cu^+ + e^-$. The outward migration of the copper is demonstrated by the fact that a completely oxidized copper wire becomes hollow.

If the metal forms several oxide phases with different oxygen content, the course of the oxidation is more complicated, but may be similar in principle.

d. Amorphous solid phases and glasses. An amorphous solid phase is formed if the atoms, because of insufficient mobility, are arranged with no long-range

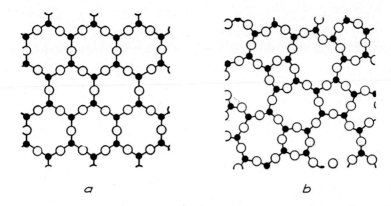

Fig. 6.22. Schematic picture of (*a*) an infinite molecule in a crystal, (*b*) a liquid or glass of the same substance with extensive short-range order remaining, but no long-range order.

order. Glasses are the most typical amorphous solid substances and are characterized by their density and homogeneity. They may be considered as liquids that are so strongly supercooled that the relative mobility of the building units is reduced to roughly the same value as in a crystal. The liquid structure with its lack of long-range order has, so to speak, been "frozen solid". Glasses, however, like liquids, may have more or less short-range order.

The crystal structure of substances that can form glasses usually explains why their melts are reluctant to crystallize. It is found in most cases that the corresponding crystals contain either infinite molecules constructed with covalent bonds (or possibly hydrogen bonds), or very large, irregularly shaped, finite molecules.

On melting a crystal that contains one- and two-dimensional, infinite molecules (neutral or charged), the covalent bonds of a large proportion of the molecules remain intact. Bond rupture certainly does occur here and there, resulting from the greater mobility and decrease in the mutual support of the molecules when the long-range order disappears. But more or less deformed fragments, which undoubtedly can be quite large, still remain. Their existence is often made evident by the fact that the melt becomes viscous. If the temperature of the melt is reduced below the melting point it becomes difficult for the large fragments to arrange themselves again in a crystal. They move sluggishly and become entangled with each other. Many bonds must be broken and reformed in order to build up the crystal. Crystallization thus proceeds so slowly that supercooling easily occurs. On further cooling of the supercooled liquid it becomes even more viscous and gradually goes over to a glass.

On the other hand, in a melt with only small building units, these need to move only short distances in order to find their places. Because of their relatively compact form, they move and turn more or less unhindered by their neighbors. Crystallization proceeds rapidly and glass formation does not take place.

If a glass is warmed, it gradually softens within what is usually called a *transformation range*. The glass thus has no melting point. The reason for this is that the disordered structure is held together by forces of widely varying magnitudes, which are successively broken on heating.

Substances that crystallize in infinite framework molecules bound together mainly by covalent bonds can also give glasses. Melting even here does not break all of the bonds, and fragments of the giant molecule remain in the melt. It is also possible that melting gives a mobile structure in which long-range order has disappeared but short-range order remains (sketched two-dimensionally in fig. 6.22b; fig. 6.22a shows the corresponding crystal structure). In both cases it is easy to see why crystallization is hindered.

Since a glass is a supercooled liquid, it is unstable and always has a tendency to crystallize. At sufficiently low temperature, however, the crystallization rate is very low. But if a glass is held long enough within the transformation range, where the building units have appreciable mobility, it often crystallizes.

Many oxides like B_2O_3, SiO_2, GeO_2, As_2O_3 (in the modification stable closest below the melting point), and P_2O_5 (in its stable modification) form framework molecules because only in this way can the central atom achieve the desired coordination number (5-6b). These oxides have a strong tendency to form glasses. If O^{2-} ions in the form of a "basic" oxide (12-1e, 21-5a), for example K_2O or CaO, are added to the oxide melt, the ligand supply is increased and the coordination number may perhaps be attained by the formation of infinite sheets. The salt melt so obtained thus still tends to form a glass, even though it is less viscous than the oxide melt. An increased addition of basic oxide may make possible the formation of infinite chains. The viscosity decreases further, but the tendency to form a glass is still great. Only when the amount of added basic oxide is so great that finite groups may form does this tendency become appreciably smaller.

Sometimes infinite molecules are built up by hydrogen bonds. Fragments of such molecules may also exist in a melt and hinder crystallization. It is for this reason that alcohols, especially polyalcohols (for example, glycerol), may form glasses. Water, in which hydrogen bonds cause a three-dimensional coalescence of the molecules (5-5), can be obtained in the form of a glass.

Glasses of elements are found primarily with carbon, sulfur, selenium and tellurium. Carbon glass, which consists of fragments of graphite layers, however, may never be completely free of long-range order, and besides is very difficult to obtain pure. Sulfur glass (amorphous or plastic sulfur) contains sulfur chains (5-6a). If crystalline sulfur is heated to just above the melting point (somewhat over 120°C) the melt contains mainly S_8 rings, is fluid and crystallizes easily. On further heating the rings are broken up and long chains are formed. The melt then becomes viscous, and the viscosity reaches a maximum at about 200°C. On rapid cooling to room temperature this melt does not have a chance to crystallize. The resulting glass is still within the transformation range, however, and crystallizes gradually. When the melt is heated above 200°C the viscosity decreases again, because of the increasing heat motion and a disintegration of the chains. Glasses of selenium and tellurium are obtained on rapid cooling of the molten gray modifications, which contain infinite chains.

Melts containing large and irregular organic molecules may also form glasses. High polymers and plastics may have structures like glasses.

Substances containing small building units, for example salts and metals, may be obtained as amorphous, solid phases. For this as a rule it is necessary to let a gas of the substance be condensed at so low a temperature that the atomic mobility is very low. With increasing temperature these amorphous phases most often crystallize even before room temperature is reached.

6-4. Solid substances with electronic conduction

a. Different types of electronic conductors. Electric currents are often carried by solid substances by movement of electrons, *electronic conduction*. This conduction mechanism, in contrast to ionic conduction, implies no material transport. Often, however, electronic and ionic conduction can occur simultaneously (6-3c).

Electronic conductors may be divided into two large classes, *metals* and *semiconductors*.

The conductivity of *metals* at 0°C is of the order of 10^4 ohm^{-1} cm^{-1} (the limits may be given by silver, 67×10^4 and bismuth, 1×10^4 ohm^{-1} cm^{-1}). The conductivity decreases rather slowly with rising temperature and also decreases with increasing proportion of dissolved foreign substances.

When the conductivity at 0°C is in the range of 10^2–10^{-9} ohm^{-1} cm^{-1}, we customarily speak of a *semiconductor*. The conductivity of a semiconductor increases rapidly with rising temperature and with the proportion of foreign substances in solid solution. In these respects it is similar to an ionic conductor (6-3c) but is different from a metal.

When the conductivity is as low as 10^{-10} ohm^{-1} cm^{-1} and below, the substance is generally said to be an *insulator*. Since no absolutely insulating substance is known, the boundary between semiconductor and insulator is a matter of convention.

A special phenomenon is that the conductivity of many metallic substances and also of certain semiconductors becomes extraordinarily high (the resistivity becomes immeasurably small) if the temperature falls below a point characteristic of the substance in the neighbourhood of absolute zero. The substance is then said to be *superconducting*.

b. The metallic state. In addition to the high conductivity of the special kind mentioned above, the metallic state is also distinguished by high thermal conductivity and metallic luster. All of these properties are caused by the presence of electrons with high mobility.

The metallic elements generally crystallize with very high coordination numbers, and this commonly applies also to other metallic phases. In the close packings (6-1f) the coordination number is 12 and in the body-centered cubic lattice 8 (6 additional atoms lie at only 15% greater distance so that sometimes the coordination number 14 is referred to here). For the same element often several such structures appear (existing within different temperature and pressure regions) and this is especially common in the transition metals.

The metallic bond and metallic properties are certainly caused by no other electrons than the valence electrons. In lithium, for example, it is certain that the $1s^2$ electron pair is too tightly bound to the nucleus to participate (even in the molecule Li_2 this orbital is practically speaking an undisturbed atomic orbial; see 5-3a). Only one valence electron ($2s$) per atom thus must take care of the bonding in the body-centered cubic structure of lithium, with its high coordination number. The bonding capacity of each electron must therefore be spread out, or "delocalized". An individual bond between two neighboring atoms thus becomes much weaker than in Li_2, where it is served by two electrons. This is apparent from the interatomic distance, which in Li_2 is 2.67 Å and in the metal 3.03 Å. In compensation, there are in the metal many more bonds per atom, which leads to an increase in the total bond energy from 57 kJ per mol Li in Li_2 to 155 kJ per mol Li in the metal. The mobility of the valence electron is related to its strong delocalization (cf. benzene and graphite in 5-3e).

The delocalization of the bonding electrons allows the metallic bond to be properly treated by using the concept of molecular orbitals. It has been shown that this method of treatment is also suitable for semiconductors.

c. Molecular orbitals in "infinite" atomic groups. If one atom after another is laid next to each other at a distance that allows the combination of their outer orbitals, one molecular orbital after another is formed. From an atomic orbital in the free atom two molecular orbitals arise when two atoms are joined, three molecular orbitals when three atoms are joined, and so on (it was mentioned earlier in 5-3a that the total number of orbitals is not changed with combination into molecular orbitals). Since the energies of the molecular orbitals are different from the energy of the orbitals from which they are formed, they are spread out over a certain energy range. Within this range the number of molecular orbitals and thus the number of energy levels becomes equal to the number of atoms in the atomic group within which the interaction takes place. Fig. 6.23 shows this schematically for each of two initial atomic orbitals. If the atomic group is "infinite" the number of levels becomes very great (for example, in a cubic crystal of lithium

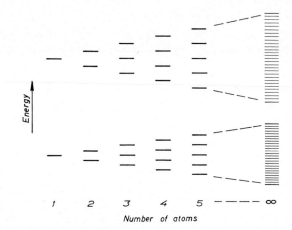

Fig. 6.23. Sharp energy levels (two such represented) in a free atom each broadened into an energy band by the addition of one atom after another.

with an edge of 10^{-3} mm the number of atoms is approximately 4×10^{10}; in the figure it is shown $= \infty$). An *energy band* is then obtained, in which the energy levels lie too densely to be resolved. If two electrons per level, that is, per atom, belong to one such energy band, it is filled.

The energy band is an expression of the delocalization of corresponding orbitals and their electrons over the whole of the infinite group. Nevertheless, delocalization often does not prevent the treatment of the bonds in the infinite groups in terms of apparently localized orbitals. This requires, however, that the number of valence electrons is sufficient for the employment of all the usable orbitals in normal covalent bonds (for example, diamond).

Fig. 5.9 showed how the energy levels for the molecular orbitals ψ_+ and ψ_- in the beginning become more and more separated as the interatomic distance decreases, that is, as overlapping increases. If an infinite group is thought of as being built up of atoms arranged first in correct relative positions but at very large distances, and this loose grouping then being closed up uniformly, the outer atomic orbitals first are combined, whereby their energy levels are spread out. With continued compression deeper and deeper energy levels are successively broadened. The process is shown for lithium in fig. 6.24a. When the nuclei arrive at their equilibrium distance (the dashed line) the outer orbitals thus are associated with energy bands, which often even overlap each other. On the other hand, many inner orbitals may still be undisturbed and correspond to definite energy values. In this way, the spectral lines in the outer spectra of condensed phases become spread out (cf. 3d), while the lines in the inner spectra (short-wave X-ray spectra) in general remain sharp.

The electrons in a band occupy it from the lowest levels upward as long as their number permits. The band thus is "filled" from below. Figures such as fig. 6.24b, which corresponds to the equilibrium distance for lithium (notice the relation to a), are often used. The sharp 1s level completely filled with 2 electrons per atom

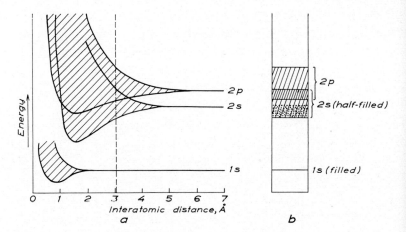

Fig. 6.24. Energy bands in metallic lithium (a) for varying interatomic distance, and (b) for the equilibrium distance corresponding to the dashed line in a.

can be seen, as well as the half-filled 2s band (1 electron per atom; occupancy denoted by stippling) and the empty 2p band which overlaps the 2s band. Usually only the valence-electron bands are shown in such figures.

d. Electronic conduction. The electrons in the valence-electron bands of an infinite atomic group thus belong to the whole group (the crystal). They move in all directions within the group, but equally many in a given direction as in the opposite direction so that no current arises. The necessary condition for a current to be able to flow when an electric field is applied across the group is that the electrons that move against the direction of the field can acquire a higher velocity, that is, higher energy, than other electrons. This condition can be fulfilled in various ways, as shown in fig. 6.25.

In a, one valence-electron band is half filled. In this there are thus a large number of electrons and a large number of empty levels, between which the energy differences are extremely small. Under the influence of the field a large number of electrons can therefore attain increased velocity in a direction opposite to the direction of the field. The conductivity as a consequence becomes high; the substance is a *metallic conductor*. The decrease in metallic conductivity with increasing temperature and with increasing proportion of foreign, dissolved atoms is explained in both cases by the decrease in the order of the crystal. The "passability" of the current-carrying electrons thereby also decreases.

Metallic conduction also occurs in cases such as represented by b. Here one valence electron band is filled so that it contains no empty levels which would allow higher electronic velocities. But the filled band is overlapped by an empty band, and the electrons can be raised to levels in this band by the addition of very small energies. Such is the case, for example, with beryllium, in which the 2s band is filled but is overlapped by the 2p band.

The Solid State

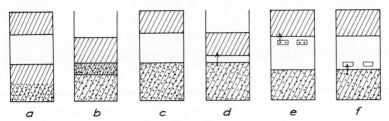

Fig. 6.25. Various cases of electronic conduction. (a) Metal, lower band half filled. (b) Metal, overlapping bands, incompletely filled. (c) Insulator, lower band filled, upper band empty. (d) Intrinsic semiconductor, depending on the small energy gap between a filled and an empty band. (e) Impurity semiconductor, excess conducting (n type). (f) Impurity semiconductor, deficiency conducting (p type). Shaded area: allowed energy bands. Dots: electrons. ▭: impurity levels.

In c, on the other hand, a filled, valence-electron band is separated from a higher, empty band by an "energy gap" with no permitted energy levels. Here the energy of electrons in the lower band must be raised to the lower boundary of the upper band in order that they may carry current by means of the small energy increments caused by the electric field. Often the energy gap is too large for enough electrons to be able to bridge the gap for any appreciable conductivity due to electronic conduction to arise. If there is no ionic conduction, we then have an *insulator*. This applies, for example, to diamond, quartz and many ionic crystals with ideal composition, at least at room temperature.

The energy gap between two bands, however, may be so narrow that with the addition of energy in the form of heat or radiation it can be bridged by sufficiently many electrons to cause a noticeable electronic conduction (d). At the same time equally many empty levels appear in the formerly filled, valence-electron band, and thus the possibility arises also for the remaining electrons in this band to carry a current. The number of electrons raised to the upper band, and thus the number of empty levels in the valence-electron band, however, is generally small, so that low conductivity results; we have a *semiconductor*. Since the semiconduction here, in contrast to that to be described further on, does not require the addition of foreign atoms, we speak of *intrinsic semiconduction*. The lower, nearly filled band, in which the small number of free levels leads to the idea that they constitute *electron holes*, gives rise to another conduction mechanism than that associated with the upper band with its good supply of free levels. In the former case an electron can be accelerated only by moving, so to say, into one of the few holes. It then leaves behind a new hole which in turn is quickly filled by another electron, and so on. It naturally is not appropriate to try to give a concrete picture of how this electron hole looks in the actual crystal. In the same way that one can think of an electron forming a cloud of negative charge, however, the electron hole may be thought of as forming a cloud more positive in charge relative to the electron cloud. Since the current-carrying electrons move in a direction opposite to the field these positive charge clouds move with the field direction. It is customary then to regard formally each electron hole as a positive current carrier, which moves with the field direction (*hole conduction*).

Under very high pressure the conductivities of solid substances may be completely different from those at room temperature. The ever greater breadth and overlapping of the orbitals with the decrease in the interatomic distance (cf. in principle fig. 6.24a) leads to the result in general that rise in pressure shifts the conductivity in the direction insulator → semiconductor → metallic conductor. Metallic conductivity has been found in selenium at pressures above 128 kbar, and in iodine, RbI and CsI at pressures over about 250 kbar. At sufficiently high pressures the electron cloud is completely separated from the atomic nucleus and then all matter has metallic conductivity (cf. 1-3c, 5-2f).

The higher the energy of heat motion is, the more easily an electron can be lifted over an energy gap. The number of conducting electrons and holes increases so rapidly with rising temperature that the associated hindered "passability" of the electrons recedes into the background. The conductivity of intrinsic semiconductors therefore increases rapidly with the temperature. Many substances that are insulators at low temperature may become intrinsic semiconductors at higher temperature. Only in a very few cases, however (for example, silicon, germanium), is the energy of heat motion at room temperature great enough for so many electrons to be able to bridge the energy gap that we may speak of semiconduction. In most cases heating to higher temperature is necessary. Sometimes the required energy can be supplied in the form of light radiation (*photosemiconduction*). A classical example of a photosemiconductor is selenium in its gray modification.

The most common cases of semiconduction, however, are illustrated by *e* and *f*. Here the crystal departs from its ideal composition (*impurity* or *extrinsic semiconduction*). As the simplest example we may take germanium (the previously mentioned intrinsic semiconduction of germanium, which at ordinary temperature gives only a very small conductivity, is not considered here). Germanium has a diamond structure, which thus in pure germanium is held together by four valence electrons per atom. Germanium can dissolve arsenic by substitution, and this impurity means that here and there in the crystal there are atoms with one valence electron more than the four of germanium. These excess electrons can carry current through the crystal. In a band scheme electrons from the impurity atoms lie in energy levels (*impurity levels*) just under the empty band (*e*). With very small addition of energy (most often from heat motion) these electrons are lifted up into the empty band and there function as conducting electrons.

Germanium atoms can also be replaced by aluminum atoms. In this way the crystal contains a number of atoms with only three valence electrons. These give rise to free impurity levels immediately above the full valence-electron band (*f*) and electrons may be lifted into these with very small addition of energy. Thus holes in the valence-electron band arise and thereby the means for hole conduction are produced.

Excess conduction in e is said to be of n type (conduction is performed by negative electrons) and deficit conduction in f to be of p type (conduction is performed by positive holes).

In crystals with more than one kind of atom a departure from the ideal composition can have the effect of an impurity although no foreign atoms are present. Impurity semiconduction in ionic crystals generally occurs with the types of solid solutions shown in fig. 6.20. It is unimportant, of course, whether the ions with deviating charge consist of a kind of atom already present at the ideal composition, or is a foreign type. In fig. 6.20 b and g cations appear with higher charge (fewer electrons) than in the ideal compositions a and h, respectively. Here we have thus electron-deficit conduction (p type). In fig. 6.20c and f cations occur with lower charge (more electrons) than in the ideal compositions d and e, respectively, which leads to electron-excess conduction (n type). If solid solution is possible over the whole range between the given limits (thus the ranges a–d and d–h, respectively), the conduction mechanism changes somewhere within the range from n type to p type.

The number of current carriers (electrons or positive holes) can at most become of the same order of magnitude as the number of impurity atoms. For this reason the conductivity of impurity semiconductors is very small in comparison to that of metals. It is also clear that the conductivity of an impurity semiconductor at constant temperature should be approximately proportional to the concentration of impurity atoms. By the precise addition of impurities semiconductors can therefore be prepared with the desired conduction mechanism and desired conductivity. However, the sensitivity to impurities is so great that the utmost purity of starting material is necessary. In these the number of impurity atoms must be reduced to 1 in 10^7–10^{10} atoms. This lies below spectroscopically detectable concentrations and the conductivity must be used to test purity. New methods of purification (13-5f2) had to be evolved in connection with the development of semiconductor techniques. A disadvantage is that changes in composition at higher temperature (oxidation, vaporization) often restrict the temperature range in which semiconductors can be used. From this viewpoint, silicon, and to a still greater degree silicon carbide SiC, which nowadays can be obtained in sufficiently pure crystals, are favorable.

The sensitivity of conductivity to impurities and changes in composition naturally makes it difficult to decide whether a substance that has been found to be a semiconductor still has this property in completely pure form and at ideal composition.

The change in conductivity with impurity content can reveal the type of the semiconductor and solid solution. An increase in the proportion of anions in an ionic-crystal semiconductor of p type (fig. 6.20b and g) means an increase in the number of "impurity atoms" and thus also in electron deficit and conductivity. In an ionic-crystal semiconductor of n type (fig. 6.20c and f) on the other hand, the number of "impurity atoms" decreases and thus also electron excess and conductivity. If the oxygen pressure over Cu_2O and ZnO is increased, the conductivity increases and decreases, respectively. Cu_2O is thus p type and ZnO n type.

At low temperature only a small fraction of the possible electrons is raised to or from the "conduction band" (in the direction of the arrows in fig. 6.25 e, f). However, the number increases very rapidly with rising temperature, so that the conductivity of impurity semiconductors similarly to intrinsic semiconductors increases rapidly with temperature.

The large temperature coefficient of conductivity is made use of in semiconductor elements, *thermistors*, whose conductance is used as a measure of temperature. The thermistor makes possible accurate temperature measurement in spite of very small dimensions, which allows small heat capacity and hence very rapid readings.

The conduction mechanism of semiconductors is also exploited in a very important class of *rectifiers*, which as a rule are built up of an n and a p conductor in contact with each other. Let us assume first that the n conductor is made positive and the p conductor negative. In the n conductor conduction electrons are then drawn away from the contact layer and cannot be replaced by conduction electrons from the other side of the contact, since they are lacking there. In the p conductor positive holes are drawn away from the contact layer, also with no possibility of replacement. The contact layer thus becomes impoverished of current carriers. Its resistivity becomes high, and it is said that a *barrier layer* is formed. If instead the n conductor is made negative and the p conductor positive, electrons and holes migrate from each side toward the contact layer. Both electrons and holes are constantly supplied from the two electrodes, and when they are brought together in the contact layer, the holes are filled by the electrons. A previously formed barrier layer is wiped out. The supply of current carriers becomes good at all places and the current can flow without extra resistance. Common semiconductors used in rectifiers are selenium, silicon and Cu_2O.

The transport of current in a contact of metal–semiconductor or semiconductor–semiconductor does not follow Ohm's law and this property is used for protection against voltage surges. In a body of sintered semiconductor powder (often silicon carbide SiC) there are a large number of such contacts and it can be made to have a large resistance at low voltages and low resistance at high voltages.

The functional components in a *transistor* are semiconductors. Up to now crystals of germanium or silicon with 1–10 foreign atoms added per 10^6 atoms of Ge or Si have been most used. Silicon carbide is also coming into use.

e. Luminescence of solid substances. Fig. 6.25 also shows the conditions for luminescence of solid substances (3d). If the added energy can raise electrons over an energy gap, they may return to the ground state with the emission of the difference in energy in the form of radiation. Insulators clearly require the addition of very high energy (highly energy-rich radiation) in order to luminesce. Luminescence is facilitated as in fig. 6.25 e and f by the introduction of foreign atoms (*activators*) or deviations from an ideal composition. Different activators can give different impurity levels and thus different luminescent colors. A host of luminescent substances are in use, but as examples we may mention sulfides (ZnS or CdS activated with Cu, Ag or Mn are excited especially intensively by electron beams and are used in TV tubes; activated CaS and BaS are classic "phosphors"), phosphates (apatite, activated with Mn and Sb in fluorescent lamps), silicates (Zn_2SiO_4, activated with Mn), alkali halides (acti-

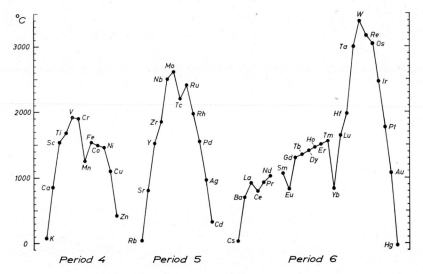

Fig. 6.26. Melting points of metals in periods 4, 5 and 6.

vated with Tl for scintillation counters) and salts of transition metals, in which impurity levels can be formed by deviations from the ideal composition and change in charge (6-3a).

f. Metallic bond. It is assumed that it is the electrons in the incompletely filled energy band (whether or not it consists of several overlapping bands) that wholly or partly participate in the metallic bond. However, the various functions of these electrons are not completely known, and can be referred to here only in very general terms.

Lithium and beryllium have already been discussed, and as with these it is generally true that in the alkali metals 1 electron per atom and in the alkaline-earth metals 2 electrons per atom participate in the overlapping ns and np bands. In the transition-element series that follow, the $(n-1)d$ electrons of the free atoms also participate, with the ns, np and $(n-1)d$ bands overlapping each other. It appears as if one bonding electron is added for each element up to group 6. The number of electrons involved in bonding in period 4 thus should be: K, 1; Ca, 2; Sc, 3; Ti, 4; V, 5; Cr, 6. After group 6 no further increase occurs. It is possible that the number of bonding electrons in period 4 remains at about 6 up to Ni and thereafter decreases. In periods 5 and 6 the decrease probably begins already with group 7. In any case the number of bonding electrons is very small in groups 11 and 12. In group 13 the d electrons probably no longer play any role in metallic bonding. Here, as with the metals of group 14, chiefly the np electrons are operating.

This variation in the number of bonding electrons within one period with a maximum in the vicinity of group 6 agrees well with many physical properties of the metals. The bond strength may be considered to increase with the number of bonding electrons, and this leads to the result, for example, that the metals within one period show a pronounced melting-point maximum in the vicinity of group 6 (fig. 6.26). The stronger the bonding is, the more difficult it is to compress the metal with pressure, and therefore the compressibility shows a minimum in the neighborhood of group 6. The increase in radius at the end of a transition element series mentioned in 5-1d (cf. fig. 5.2) probably also depends on the fact that at this point the bond strength decreases. The hardness of the metals also is greatest near group 6.

In the lanthanoids in period 6 as a rule the number of bonding electrons is three, as with Sc and Y. Fig. 6.26 shows that the bond strength rises relatively slowly within the series with the exceptions of the marked minima for Eu and Yb. This results from the fact mentioned in 5-1d that the number of bonding electrons for these elements is two instead of three (one electron has entered into the $4f$ group in order to make it half-filled and wholly filled, respectively). Besides the positions of their melting points, Eu and Yb in the metallic form in several other respects are reminiscent of Ba (relatively soft, oxidizes rather rapidly in air), which also has two bonding electrons.

g. Metallic phases. Besides the pure, metallic elements, all phases that contain only such elements (intermetallic phases) have metallic properties.

Also, many metallic phases are known containing metallic elements and nonmetals. In these, the metal must be a transition element. For the nonmetal probably all except the halogens can participate; seldom oxygen, however, because of its large electronegativity, which easily causes ionization. These conditions for metallic properties are necessary but not sufficient. Thus, the proportion of the nonmetal must not be too great. One vanadium oxide, $\sim V_4O$, is metallic while V_2O_5 lacks metallic properties. Phases of this type that are technologically very important (primarily carbides) occur in steel.

A substance that contains several elements, and in which all phases have metallic properties, is called an *alloy* (Lat. alligo, to bind to). The term alloy is used for both single- and multiple-phase substances.

The great majority of metallic elements, as mentioned previously (cf. 6-4b), have crystal structures with very high coordination number. Such is commonly the case also in other metallic phases. Near the boundary between metals and nonmetals (cf. tab. 4.4), however, there are many more open structures with lower coordination number (on the metallic side, for example, gallium, antimony and bismuth).

The absence of ions in metallic phases means that the composition of the phase is not subject to any conditions of electrical neutrality. Solid solutions, leading to extended homogeneity regions, are therefore common. However, a certain composition within or in the vicinity of the homogeneity region can often be given as characteristic of the phase in question. If the range of homogeneity is very narrow, its position most often corresponds to the characteristic composition. The formula corresponding to the characteristic composition generally has no relation to any electrovalences of the participating atoms.

If the participating elements are very different in electronegativity, transitions may occur between metallic bonding and ionic bonding. Since the metallic bond is a kind of covalent bond, such cases may be interpreted in the same way as intermediate forms between ordinary covalent bonding and ionic bonding. The shift of electrons toward the more electronegative atoms is evident, among other ways, in that the composition of the phase corresponds to the ordinary electrovalences and that the homogeneity regions become small. Examples are Mg_2Si, Mg_3Sb_2, Li_3Bi. Such phases often have notably high melting points (for example, Li_3Bi, 1145°C).

Since the electrons in the incompletely filled energy band are responsible for the metallic bonding, it is clear that the number of valence electrons per atom is of great significance for the composition and properties of the phase. In many systems of the coinage metals (group 11), among others, with elements from groups 12–15 phases with particular crystal structures are formed when the number of valence electrons per atom (the so-called *valence-electron concentration*) lies near certain definite values. The number of valence electrons is set equal to the unit's digit of the group number. In the systems Cu–Zn, Cu–Al, and Cu–Sn the so-called β phases thus occur, in which the atoms are distributed over the points of a body-centered-cubic lattice. These β phases exist around the characteristic compositions CuZn, Cu_3Al and Cu_5Sn. The valence-electron concentration for all these compositions is 3/2. The applicability of laws of this type was demonstrated in 1926 by Westgren and Phragmén, and independently by Hume-Rothery. Phases that require fulfillment of conditions of this type for their formation are called *electron compounds*.

As is always true for crystals, the relative sizes of the participating atoms are of great significance for the structure of the metallic phase. The relative sizes also play a great role in the formation of solid solutions (6-3a).

6-5. Mechanical deformation and division of solid phase

By moderate mechanical action a solid body can be *deformed* and by stronger action broken, that is *divided* into smaller pieces. In both cases the energy of the system is raised by retention of a part of the applied energy (the remainder forms heat).

a. Deformation. The *deformation energy* taken up on deformation may be considered as potential energy. However, the deformation may be of different kinds. If it disappears when the deforming forces cease, it is called *elastic*. In the elastically deformed state the environment of the atoms remains practically unchanged and has only become slightly deformed. The body therefore can, so to speak,

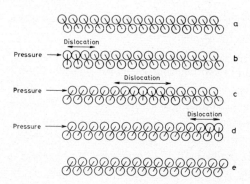

Fig. 6.27. Gliding with dislocation.

spring back to the original state, whereupon the energy of deformation is given up again. If the deformation does not disappear when the forces cease, it is called *plastic*.

Plastic deformation of crystals often involves gliding along lattice planes. Fig. 6.27 shows schematically how gliding takes place. The picture illustrates most closely gliding in a metal crystal, in which the conditions are especially simple because the building units are atoms with no strong electrostatic forces between them. In an ionic lattice the mechanism is affected by such forces but the result is the same in principle. In a molecular lattice the building units, the molecules, slide over each other, and in fig. 6.27 may be represented by the circles. Because of the importance of metals as construction material, mechanical deformation of crystals has been studied mostly in metal crystals.

The pictures in fig. 6.27 show a row of atoms on either side of a plane, the *glide plane*, along which gliding takes place. The atoms below the glide plane are thought to be fixed whereas a pressure towards the right is applied to the atoms above the glide plane. This causes the latter atoms to glide to the right parallel to the plane of the picture. In gliding, energy must be added in order to lift the atoms in the moving half out of the "holes" in the other half. The major portion of this energy is recovered, however, when the moving atoms reach new "holes". The atoms in the glide plane do not move all at the same time, but by the movement of an imperfection, a so-called *dislocation*. The dislocation is shown at its start in b and has a width approximately corresponding the length of the double-headed arrow. It extends through the whole crystal perpendicular to the picture plane. If the pressure continues, the dislocation migrates into the crystal. In c it lies in the center of the crystal, and in d the dislocation has left it partially and in e completely. A glide of the length of one interatomic distance has taken place. The process can then be repeated. Because of the relatively weak forces between densely populated lattice planes, gliding generally occurs most easily along such planes.

Other types of dislocation also occur. That described above is called more particularly *edge dislocation*, and in 14-2b *screw dislocations* are mentioned.

Sometimes the energy conditions are such that a plane slides very easily over into a new, stable position that does not, as above, correspond to gliding through an interatomic distance. The first plane then cannot glide further, but the plane above it glides an equal amount relative to the first plane, and so on (see fig. 6.10). The total displacement of a plane thus becomes proportional to the distance of the plane from the first

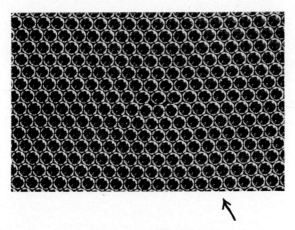

Fig. 6.28. Soap-bubble model showing a dislocation (best seen if one looks along the paper in the direction of the arrow). (W. L. Bragg and J. F. Nye.)

glide plane. In this way there is formed a mechanically deformed crystal with the same structure as the initial crystal but in twin position to it (cf. fig. 6.10).

It has been possible to observe dislocations with the electron microscope. Fig. 6.28 shows a model picture of a dislocation. It is believed nowadays that most dislocations that are used in gliding are incorporated into the crystal on crystallization. They thus represent defect locations with higher energy and can easily initiate gliding. If a crystal is constructed as in fig. 6.27c the dislocation can disappear by migrating to the crystal boundary, either to the left whereupon the structure a is formed, or to the right with formation of structure e.

Very little energy is required to *move* an already existing dislocation within a single crystal of a pure substance and thus complete a glide. The presence of dislocations in most single crystals therefore makes them much more easily deformable and thus

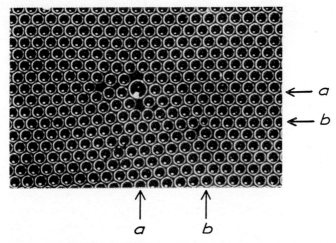

Fig. 6.29. Soap-bubble model showing lattice distortions caused by a large dissolved atom (indicated by the arrows *aa*) and a small dissolved atom (indicated by the arrows *bb*). (W. L. Bragg and J. F. Nye.)

softer than what would be expected of an ideal crystal. The deformability and softness is significantly less in a polycrystalline material in which the glides are to a large extent locked by neighboring grains, and in any case cannot extend beyond a grain boundary (cf. fig. 6.9). Foreign, dissolved atoms also lead in general to hardening. By their disparate sizes or force fields they cause distortions that prevent migration of dislocations and therefore gliding. Fig. 6.29 shows this in a soap-bubble model, in which, however, the size deviations and thus the distortions are greater than are possible in real, solid solutions.

It has been found that single crystals that have grown undisturbed out into a gas or liquid phase in the form of very fine threads ("whiskers", diam. $\sim 10^{-3}$ mm) nearly always have the mechanical properties appropriate to an ideal crystal. Plastic deformation then requires uncommonly large forces. The method of formation and form have clearly caused these whiskers to include unusually few dislocations. Because of their great strength whiskers are being used for reinforcing composite construction materials.

b. Mechanical division. On mechanical division deformation may also take place and a portion of the energy taken up may therefore be deformation energy. Beside heat energy, the remainder is *surface energy*. The surface energy results from the fact that the building units in the free surface of a homogeneous phase are not surrounded as completely by neighbors as those in the interior of the phase and therefore are more loosely bound. The total surface energy of the phase increases when its total free surface is increased and thus when the phase is divided into smaller parts. If a cube of edge 1 cm is divided into cubes of equal size with edges 10^{-n} cm, the total surface is increased by the factor 10^n. With fine division the increase therefore becomes very great. If a 1 cm cube is divided into cubes with edge 10^{-5} cm, the total surface increases from 6 cm^2 to 60 m^2. The surface energy per unit area is, however, also dependent on the curvature of the surface. The building units in convex surfaces are all more loosely bound, the more curved the surface is, that is, the smaller its radius of curvature is. Generally, it is therefor true that a smaller particle has greater solubility and vapor pressure than a larger particle of the same substance at the same temperature. When sufficiently large particle dimensions (approximately 10^{-3}–10^{-2} mm) are attained, however, the changes in solubility and vapor pressure with further increases in the dimensions become too small to be demonstrated experimentally. All larger particles therefore for practical purposes show the solubility and vapor pressure corresponding to that of a flat surface.

Every phase tries to reduce its surface energy and thereby go toward greater stability. This becomes the more noticeable the smaller are the particles of which the phase consists. If a phase consists of many small particles there are often strong cohesive forces between them, which lead to aggregation of the particles if they are sufficiently mobile (particles of the same kind, however, often become electrically charged with the same sign, whereby they are prevented from coming together; cf. 15-4). If a very small and a very large particle of the same substance lie near to each other in a phase that permits transport of the building units of the particles (gas phase if the substance has a sufficiently high vapor pressure;

liquid phase if its solubility therein is adequate), the building units go over from the small to the large particle until the former is used up. In a polycrystalline body in which disoriented crystal grains of the same substance adjoin each other in grain boundaries, the larger grains can grow without material transport. Proceeding from the grain boundaries the building units of the smaller grain are rearranged so that the lattice attains the same orientation as that of the adjoining larger grain. It may be said that the result is a *grain-boundary migration* into the smaller grain. All these processes require a temperature high enough so that the building units have sufficient mobility. However, the final equilibrium arrangement is approached so slowly that it can seldom be realized, but it would correspond to the whole phase being gathered into one homogeneous body with the highest degree of order that the composition and temperature of the phase would allow.

The cohesive forces mentioned above indicate that the outward-directed, free ("unsaturated") bonding forces from the building units on the free surface are appreciable. These can naturally also be satisfied by bonding to all types of atoms or atomic groups that are attracted from the surroundings. This tendency for bonding becomes stronger the more exposed the building units are. It therefore increases when the radius of curvature of a convex surface decreases, and becomes very strong at outward edges and above all at corners.

In a crystal a mechanical division often leads to a cleavage of the crystal along a lattice plane. The ability to cleave is greatest along densely populated lattice planes, *cleavage planes*, between which the forces are especially small. Often the cleavage along certain planes becomes even more pronounced because of a special distribution of the bonding forces. Typical cases are graphite (fig. 5.21) and mica (23-3h4).

When a fracture occurs with sufficiently high mechanical stress, the fracture surface is often a glide or cleavage plane. In a polycrystalline body, however, where no glide or cleavage planes passing through the entire body are available the fracture may be more irregular and cross the crystal grains. If the cohesion between the grains is low enough, the fracture will follow the grain boundaries (grain-boundary fracture, intercrystalline fracture).

Amorphous substances and glass lack both glide and cleavage planes. Deformation and cleavage here become wholly irregular. In glasses this leads to "conchoidal" fracture.

c. Hardness. The hardness of a substance is difficult to define. In the first place, a physically acceptable statement of hardness must apply to a single crystal or to the individual crystal grains in a polycrystalline body. Further, the hardness of the crystal is generally different in different directions. In any case, if the hardness varies appreciably with direction, the crystal surface on which the measurement of hardness was carried out must be stated. And finally, the concept of hardness includes the collected effects of several completely different properties, such as gliding ability, cleavability, and so on. These properties have varying influences on different methods of measurement, which therefore do not always give consistent results.

In the most important methods of hardness measurement, a very hard body of definite shape (diamond pyramid or steel ball) is pressed with known pressure into the surface under investigation. The hardness is determined by the size of the impression. For a more qualitative comparison of hardness, scratching is often used. A substance that scratches another, but is not scratched by it, is said to be harder than it. Especially in mineralogy, Mohs' hardness scale (1822) is used, which adopts the following degrees of hardness:

1 talc	3 calcite	6 orthoclase	9 corundum
2 gypsum or halite	4 fluorite	7 quartz	10 diamond
	5 apatite	8 topaz	

1 and 2 can be scratched with the fingernail, 3–5 with a knife, 6–10 are not scratched by a knife but scratch glass. The difference in hardness between 9 and 10 is very great.

The hardness often reflects the bonding conditions in a crystal. Molecular crystals, built up of molecules mutually held together by van der Waals forces, are generally very soft (hardness 1–2, more rarely as high as 3). This depends, of course, on the low strength of the van der Waals forces. To these belong the majority of organic crystals. From inorganic chemistry may be mentioned crystals of N_2, P_4, CO_2. If hydrogen bonds are present as in H_2O, the hardness is not increased appreciably.

Ionic crystals are generally significantly harder (most often hardness 2–6). Here the parallelism between hardness and lattice energy can often be observed. The hardness thus often increases with increasing ionic charge and decreasing ionic radius (6-2d). Fluorite is therefore harder than sodium chloride.

Crystals containing framework molecules, in which the bonds have a strong, typically covalent contribution, are often very hard (hardness 6–10). This is because of the strength of the covalent bonds but also because the strongly directed nature of the bonds makes gliding difficult. As examples quartz, silicon carbide (carborundum), and most especially diamond may be mentioned.

Metals may have very different hardnesses, from the softest metal cesium (hardness <1) to very hard metals such as chromium. The variation of hardness with the strength of the metallic bonds has been discussed in 6-4f.

Dissolved substances increase hardness, mainly by interfering with gliding (6-5a).

In technology hardness is often measured directly on polycrystalline or even multiphase material and then gives a kind of average hardness for a greater or lesser number of grains. This hardness depends strongly on mechanical working and heat treatment (14-3). In general it increases as the grain size decreases.

Deformation and mechanical properties, especially of polycrystalline, solid bodies, are treated further in 14-3c.

CHAPTER 7

Structure Determination, Color, Magnetic Properties

Structure determination, that is, the establishment of the geometrical arrangement of the atoms, in general must be accomplished by physical methods. It is only in organic chemistry, where the problem applies to molecular structures with certain generally applicable and understood principles of structure, that chemical methods of structure determination (based on chemical reactions, identification of fragments of the molecule and the synthesis of the molecule from such fragments) has played a large role. These methods, however, have never been able to give, for example, interatomic distances and accurate bond angles. As the problems become more complicated, even here physical methods are more and more brought to bear.

Accurate, complete structure determinations can only be carried out by diffraction and spectroscopic methods. These are first taken up in the following brief survey. In addition, a large number of physical properties can be used to give important, general information or solve special problems. Electric dipole properties and magnetic properties have been especially useful. The first have already been considered in 5-2g, but the latter are taken up in the following. The color of substances is also discussed.

7-1. Diffraction methods

Let us assume that a wave front encounters particles, which then themselves become the source of wave systems with the same wavelength as the incident wave. If the distances among the particles is of the same order of magnitude as the wavelength, *interference* occurs, that is, the different wave systems interact, or interfere. In certain directions the wave systems reinforce each other so that a higher intensity ("interference") occurs, and in other directions weakening takes place. The interference phenomenon or *diffraction* gives in this way a variable intensity distribution, an *interference* or *diffraction pattern*. The pattern becomes especially pronounced (sharp interference lines) if the particles that scatter the waves are numerous and arranged periodically in some way. If one or more interparticle distances are repeated many times in the system of scattering particles, notable diffraction phenomena occur even if the system has no other order. The positions and intensities of the interferences in the diffraction pattern often make possible a determination of the respective locations of the scattering particles.

Diffraction phenomena, which can be used for structure determination, occur if the applied waves have a wavelength of the same order of magnitude as the interatomic distances in the system under study, that is, approximately 1 Å. Up to now, X rays, electrons and neutrons have been most used.

X-ray diffraction was discovered in 1912 by von Laue and thereafter developed, especially by W. H. and W. L. Bragg, into unquestionably the most important method of structure determination. Primarily because of the results gained thereby concerning the structure of solid substances, it has been the main cause of the rebirth of inorganic chemistry that has come about in recent decades. Nowadays even very complicated crystal structures can be determined in detail by X-ray diffraction. It has also been used for the study of glasses, liquids and gas molecules. In these systems with no long-range order and thus no periodicity, only the most important interatomic distances can be established, without any information about their mutual directions. However, this may suffice for the determination of the nearest-neighbor coordinations in glasses and liquids, or the structures of simple molecules.

In X-ray diffraction the X radiation is scattered by the electron clouds of the atoms. The intensity of the radiation scattered by an atom is nearly proportional to the number of electrons in the cloud and the method therefore gives the electron density distribution of the system. Light atoms, particularly hydrogen, may therefore be difficult to locate, especially if the system at the same time also contains heavy atoms.

In *electron* and *neutron diffraction* the wave nature of electrons and neutrons (3a) is made use of. The charges on the electrons cause their scattering to be determined by the electric field in the atomic system impinged on by the electron beam. Light atoms thus become somewhat easier to distinguish by electron diffraction than by X-ray diffraction. Electron beams generally penetrate very little into solid bodies and electron diffraction can therefore often give information about the structure of thin surface layers (for example, oxide films). For the determination of the structure of molecules in gases electron diffraction is more appropriate than X-ray diffraction, since among other things much shorter exposure times can be used.

In neutron diffraction a beam of slow neutrons, which up to now can only be obtained in sufficient intensity from an atomic reactor (9-4a), is used. The method has therefore not yet been used so widely. One great advantage is that it generally permits the location of light atoms in the presence of heavy atoms. The neutron radiation is scattered only by the atomic nucleus (there are also special interactions with magnetic moments in the system, 7-4a) and the intensity of the scattered radiation is not necessarily less for a light atom than for a heavy one.

The purely geometrical results of a complete structure determination (coordination, bond lengths, bond angles) can also provide fundamental information for the establishment of types of bonds. Furthermore, the most advanced structure determinations go a step further by giving the electron density at each point in the unit cell.

Even without a complete interpretation of the diffraction pattern, the diffraction methods very often can give the dimensions and symmetry of the unit cell. The dimensions with the help of the expression (6.1) give the number of formula units per cell, or the molar weight of the formula unit. The symmetry may make possible conclusions concerning the symmetry of the building units. With no interpretation whatever the pattern may be used for identification and analysis. This applies especially in the case where sharp interference lines occur, that is, when crystals are involved.

7-2. Spectroscopic methods

The spectroscopic determination of structure refers mostly to molecular structure determination, and therefore implies the study of molecular spectra (3d). If we overlook the nuclear energy, the energy of each molecule consists of electronic energy, vibrational energy, rotational energy and translational energy (2-1b). A molecule at rest has only the first three of these forms of energy. It is also only these that are notably quantized (3c) and therefore give definite energy levels, which can be observed spectroscopically. Aside from the nuclear energy the energy of a molecule at rest is thus

$$E_{mol} = E_{el} + E_{vib} + E_{rot} \tag{7.1}$$

These three energy terms are determined by expressions analogous to (3.7), (3.8) and (3.9) respectively, and hence by quantum numbers of the types n, v and j. Changes in these quantum numbers give changes in the three terms, which thus contribute to changes in the molecular energy. If an energy change (positive or negative; one quantum number may increase when another decreases) is indicated by ΔE, we may then write schematically

$$\Delta E_{mol} = \Delta E_{el} + \Delta E_{vib} + \Delta E_{rot} \tag{7.2}$$

Changes in E_{mol} reveal themselves as spectral lines (cf. 3d).

The numerical value of the changes are very different for the three different forms of energy. In general it is true that

$$|\Delta E_{el}| \gg |\Delta E_{vib}| \gg |\Delta E_{rot}|$$

If only the rotational quantum number j changes, then $\Delta E_{mol} = \Delta E_{rot}$. A *rotational spectrum* is obtained which in general is relatively simple. Because of the low numerical value of ΔE_{rot} the frequencies of the spectral lines are small, that is, their wavelengths are large. Rotational spectra therefore lie in the *microwave region (microwave spectra)* and in the *far infrared*[1] (see the following diagram).

---

[1] The infrared region is customarily divided into *near infrared* (wavelengths about 2–40 $\mu m = 2 \times 10^{-6} - 4 \times 10^{-5}$ m) and *far infrared* (wavelengths about 40–1000 $\mu m = 4 \times 10^{-5} - 1 \times 10^{-3}$ m). The division is a practical one, in that the nearer region can be studied with prisms (single crystals of alkali and thallium halides). For the far region gratings must be used.

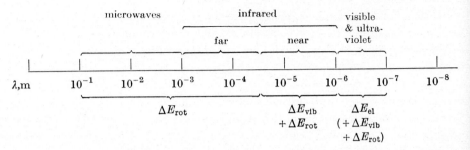

A change in the vibrational quantum number v is always accompanied by a change in j. If n is not changed at the same time, then $\Delta E_{mol} = \Delta E_{vib} + \Delta E_{rot}$. A *vibrational-rotational spectrum* is obtained. A particular change in v may be combined with a large number of changes in j and since all changes ΔE_{rot} are very small, these combinations give ΔE_{mol} values that are grouped closely around ΔE_{vib}. The corresponding spectral lines therefore form a band. The magnitude of ΔE_{vib} places the bands mainly in the *near infrared*. Since the number of possible vibrations most often is larger than the number of possible rotations, and further since a large number of combinations can occur, vibrational-rotational spectra are generally significantly more complicated than rotational spectra.

An *electronic energy change* (change in quantum numbers of the type n) is most often accompanied by changes in both v and j. All three terms in (7.2) then contribute to the change ΔE_{mol}. The spectra therefore have in general a complicated band character, which makes them difficult to interpret. ΔE_{el} is dominant in magnitude so that the band most often lies in the *visible* to *ultraviolet range*.

Microwave and infrared spectra are most often studied by absorption, spectra in visible and ultraviolet both by absorption and emission. However, only very simple molecules that tolerate the treatment required (strong heating or bombardment with particles accelerated in an electric field) can be studied by emission. Microwave spectra are only used with gaseous substances, the others with substances in all states of aggregation.

If the quantum number changes that caused the different lines in a molecular spectrum have been found, the force constants for different bonds and moments of inertia for rotation about different axes may be obtained from expressions such as (3.8) and (3.9). These data provide extremely important information about the molecule and for simpler molecules can be used for accurate calculations of bond strengths and structure. In the same way that ionization energies are calculated from series limits in atomic spectra (4-3a), very accurate values of dissociation energies may be calculated from series limits in spectral bands. Spectral data can also be used for calculation of certain thermodynamic quantities such as specific heat and entropy.

Microwave and infrared spectra are very sensitive to structural changes. Every molecule therefore gives an individual spectrum that can be used for identification and analysis. The sensitivity to structure decreases as the wavelength decreases, but instead more and more often bands are observed that are characteristic of certain atomic groups and coordinations. This sensitivity to groups and coordination is very great in the visible to ultraviolet region and has been of great importance in the analysis of both these features.

A wholly different type of molecular spectrum is the *Raman spectrum*. Here the system is irradiated by an intense beam containing one or a few frequencies and the radiation scattered to the sides is analyzed. The incident radiation must have less energy than is required to cause fluorescent radiation. Let us assume that a molecule is irradiated with radiation of frequency v_i. These quanta may be properly considered as photons (1-2b), which collide with the molecule. The photons rebound from the molecule, are "scattered" by it, either with unchanged energy ("elastic" collision) or after exchange of energy. In the latter case, if it has given up energy, the amount of energy lost is just equal to that required for an increase in the rotational or vibrational energy of the molecule, or both. Let us assume that the vibrational energy is raised by the value $\Delta E_{vib} = h v_{vib}$. The photon then leaves the molecule with the energy $h v_i - h v_{vib} = h(v_i - v_{vib}) = h v_1$. This means that the molecule scatters radiation with the frequency lowered to $v_1 = v_i - v_{vib}$. If the molecule had acquired rotational or vibrational energy before the collision, this can instead be given to the photon. The molecule then scatters radiation with increased frequency, for example $v_2 = v_i + v_{vib}$.

In the radiation scattered by the substance in this case, therefore, "Raman lines" with the frequencies $v_i - v_{vib}$ and $v_i + v_{vib}$ may be observed spectroscopically around the original frequency v_i. The frequency value v_{vib} may be detected in this way without the necessity of working with this actual frequency. Other rotational and vibrational frequencies can give rise to further Raman lines grouped around v_i. When these frequencies are measured, they can be used to a large extent in a manner analogous to the frequencies obtained from microwave and infrared spectra.

It should be observed that Raman lines are thus caused by photons whose energy has been *changed* through energy exchange with an atomic system. The lines in emission and absorption spectra, on the other hand, are produced by photons of definite energy that have been *created* or *consumed*.

7-3. Color

a. Transmission color, surface color and metallic luster. Light that passes through a body is deprived of radiation of those wavelengths that are absorbed. If the absorption occurs within the visible spectral range, the light becomes colored after passage; the body has a *transmission color*. This is the color complementary to the color of the light that is absorbed. The transmission color thus depends on the absorption spectrum of the body and can be fully understood only if the energy changes that cause the absorption spectrum can be explained.

If light scattering is neglected, a body throws back incident light partly by reflection from the outermost surface layer, and partly by total reflection. The ability to reflect light from the surface layer itself increases as the absorption capacity increases. This reflection therefore plays a relatively small role in substances with low light absorption. The overwhelming proportion of the reflected light here results from total reflection. It thus penetrates into the body, is totally reflected at a boundary with a medium of lower optical density (often air), and then goes back through the body. On this passage it suffers absorption and the reflected light therefore shows transmission color.

Let us assume that a colorless, homogeneous body (a crystal or a piece of glass) is divided into fragments, for example, by crushing. From an aggregate of fragments (for example, a powder) the light is reflected in a certain direction from more and more closely spaced points, the finer the subdivision is. With sufficiently fine division a white aggregate surface is seen in white, incident light. Colorless substances are therefore often described as white.

If the body is colored, the light absorption and therefore the color becomes stronger, the longer the path through the body the light has traveled before and after the total reflection. If the body is pulverized, the light paths are shortened, the absorption is less and the color becomes lighter. A large crystal of copper sulfate pentahydrate is deep blue both in transmitted and reflected light but gives a light blue crystalline powder. A "streak", which occurs when the surface of a substance is scratched, contains small fragments of the substance and is therefore lighter than the unscratched surface.

The color of a solid substance can also vary with the degree of structural order (6-3b). Fe_2O_3, obtained by heating iron(III) sulfate to a maximum of 700°C, has quite low structural order and a red color. If it is heated over 700°C, the order increases and the color darkens (14-3b). Deepening of the color on heating a solid substance, however, may also be caused by crystal growth. In both these cases the color is preserved on cooling, in contrast to so-called "heat color", which results from a reversible transformation and therefore reverts on cooling.

With very high light-absorption capacity the reflected light passes such a short distance through the body that it may be said to be reflected from the surface itself. The body then takes on a more or less pronounced *metallic luster*. Those wavelengths that are most absorbed are most strongly reflected, whereby a *surface color* can arise. If transmitted light can be observed in a very thin sheet of the body, it is found that the wavelengths that are diminished are just those that are dominant in the reflected light. The surface color is thus complementary to the transmission color.

Metals have a very high absorption capacity for visible light. They are therefore opaque even down to layer thicknesses of approximately 10^{-4} mm (gold leaf of this thickness is transparent). The high absorption capacity is related to the high mobility of the electrons that can easily change energy within the incompletely filled energy bands. High reflection capacity and hence metallic luster results from the high absorption capacity. The greater the electron mobility and hence also the conductivity is, the greater are the absorptivity and reflectivity.

In general the absorptivity and reflectivity of metals do not vary much within the visible spectral region. However, they always diminish as the wavelength decreases. This occurs especially rapidly in copper and gold. The lower absorption of shorter wavelengths is apparent in gold leaf from its blue-green transmission color. The dominance of long-wavelength light on reflection leads to red and yellow surface colors. In silver and aluminum, for example, the reflectivity is high and more constant within the visible spectrum (especially in aluminum, which has good reflectivity up into the ultraviolet), giving a "white" surface color.

Metals with lower conductivity have lower reflectivity and thus "darker" metallic luster (for example, iron). In semiconductors with their still lower conductivity, the reflectivity is still less. In many cases, however, these show a metallic luster, although most often dark (for example, the mineral galena, PbS). Also, certain other substances with strongly delocalized molecular orbitals and consequent electron mobility within the molecule (5-3e) may have metallic luster and surface colors (numerous organic dyes; for example, crystals of fuchsin have a metallic luster and greenish surface color but red transmission color, the latter also being the color of a fuchsin solution).

If a surface with metallic luster is covered by a film (for example, of oxide), whose thickness is of the same order of magnitude as the wavelength of light, a color arises through interference that depends on the thickness of the film (*iridescent color, tarnish color*). Iridescent color thus has no direct relation to the nature of the substance.

b. Appearance of transmission colors. In the following only the color of nonmetallic substances is discussed, that is, the transmission color. Only inorganic substances are considered.

As described in 7-2, absorption in the visible and ultraviolet spectral region is caused by changes in electronic energy (which besides often are accompanied by changes in vibrational and rotational energy). Electronic energy changes are always possible but the energy jumps are often so large that the absorption lies in the ultraviolet. Special circumstances, however, can make possible so small electronic energy jumps that absorption occurs in the visible region, and thus color appears.

If color is caused by a certain atomic or electronic configuration, it appears in all the states in which this configuration exists. The I_2 molecule thus gives approximately the same violet transmission color both in iodine vapor and iodine solutions in weakly polar solvents (for example, carbon disulfide, tetrachloromethane), in which relatively unchanged I_2 molecules occur. Iodine solutions in strongly polar solvents (for example, water, alcohols) on the other hand are brown, because the I_2 molecules form solvates, so that their electron clouds are more drastically altered.

For a substance AB that dissociates into the ions A^+ and B^-, the light absorption on complete dissociation is often combined additively from the absorptions of each ion A^+ and B^-. Let us assume, for example, that B^- does not absorb in the visible spectral region, but is "colorless", like the ions Cl^-, NO_3^-, ClO_4^-. With additivity all sufficiently dilute solutions of salts with the same A^+ and such colorless B^- ions should have the same absorption spectrum and thus the same color, that of the A^+ ion. This conception of characteristic *ion colors* and their additivity is very useful but must be applied with care. An ion is never completely free and its color may be altered by changes in the environment that can change the influence on the ion from outside. Whether or not the light absorption also of molecules or solid phases is more or less equal to the sum of the absorptions of the separate kinds of ions depends on the conditions of bonding. The additivity may often apply quite well for a typical, ionic crystal but on the other hand not if the electron clouds of the ions are strongly altered by covalent bonding.

If we except the transition-element ions (including lanthanoid and actinoid ions), in general no absorption of visible light takes place in an ion with a relatively unpolarized (undeformed) electron cloud. Such ions are thus colorless. If a substance besides colorless anions contains only ions of nontransition elements and none of these are appreciably polarized, which implies predominantly ionic bonding (5-2e), the substance is accordingly colorless. If more noticeable *polarization* of any kind of ion occurs, implying greater covalent contribution, this leads as a rule to appreciable absorption within the visible spectral region, and thereby to color. The cations with noble-gas structure of elements in groups 1 and 2 are little polarizable (5-2e) and therefore always colorless. This applies also to Al^{3+}. The cations with 18- and (18+2)-shells (5-2b, f) are also not easily polarizable, but on the other hand they exert a significantly stronger polarizing effect than the first-named cations. Together with less easily polarizable anions (see below) they therefore give colorless compounds, but cause strong polarization of easily polarizable anions, so that a large covalent contribution and color results.

The simple anions, of which the halide and chalcogenide ions are the most important, are more polarizable than the cations (5-2e). The polarizability increases with increasing size. Especially the larger, such as I^-, often vary from colorless to colored, depending on how strong the polarizing effect to which they are subjected is. While all the alkali halides are colorless, among the silver halides we find thus colorless AgCl, yellow-white AgBr and yellow AgI. The 18-shell ion Ag^+ is colorless like the alkali ions, but the former has much greater polarizing effect and this becomes noticeable in relation to Br^- and still more in relation to I^- (compare also how the increasing covalent contribution is observed in the decrease in solubility in the sequence AgCl > AgBr > AgI). The alkali metal oxides M_2O and sulfides M_2S are colorless and easily soluble, in contrast to the corresponding silver compounds which are black and difficultly soluble.

Oxo anions of the elements in groups 13-17 (for example, borate, carbonate, silicate, nitrate, phosphate, sulfate and halate ions) are nearly always colorless. Only strongly polarizing cations, for example, Ag^+, can sometimes develop color. For example, Ag_2CO_3 and Ag_3PO_4 are yellow.

In any case, when the covalent bond contribution becomes so great that the electron cloud clearly becomes common with the other bound atoms, the color is caused by the state of the whole cloud and cannot be attributed to the different atoms individually.

A completely different cause of color occurs in ions with *incomplete d or f groups*, that is, in most of the transition-element ions (5-2f). These are incorporated as a rule as central atoms in complexes (with the solvent molecules or with other ligands) and it was shown in 5-3j why complexes of the d elements so often absorb visible light and are therefore colored. The source of color is similar in the f elements (lanthanoids and actinoids). The color of a complex of a d element is often very dependent on the nature of the ligands and the type of complex. Thus a solution containing $Ni(H_2O)_6^{2+}$ is green, $Ni(NH_3)_6^{2+}$ blue and $Ni(CN)_4^{2-}$ yellow. The example shows that a definite color cannot be spoken of for the central ion common to all cases, Ni^{2+}. The sensitivity to the environment is much less for the f-element ions and for these therefore there is a rather constant ionic color. The explanation is that the color is caused by electronic transitions within the incompletely filled f group, and that the deep-lying f orbitals are practically undisturbed by the complex formation. The absorption bands of the f-element ions are also often remarkably narrow.

The atoms of the transition elements very often occur in oxidation states in which they may be considered as noble-gas ions (empty $3d$ group) and thus should be colorless. In period 4, for example, we have the series Sc(III), Ti(IV), V(V), Cr(VI), Mn(VII) with the ions Sc^{3+}, Ti^{4+}, V^{5+}, Cr^{6+}, Mn^{7+}. With increasing charge (stronger polarizing effect), however, the covalent contribution in the bonds coming from these ions becomes greater (cf. 5-2e), $3d$ orbitals take part in these bonds and color may appear. As one goes to the right in the above series, for example, the tendency to form oxo anions increases and at the same time they become more and more strongly colored. Oxo anions are known beginning with Ti(IV) and color begins to appear with V(V). The strongly colored oxo anions in chromate(VI) and permanganate(VII) are well known.

A third cause of color is the presence of *unpaired electrons* (odd molecules, 5-3a). Such is the case, for example, with NO_2, ClO_2, O_2^- (hyperoxide ion) as well as many organic "free radicals". An exception is nitrogen oxide NO, which is colorless despite its unpaired electron.

The most intensively colored, inorganic substances are generally found, however when a molecule or crystal contains one kind of atom in several oxidation states. The high absorption often occurs over the whole visible spectrum so that the color becomes dark or even black. $CsAuCl_3$, which should be written $Cs_2[Au^ICl_2][Au^{III}Cl_4]$, forms coal-black crystals, while $Cs[Au^ICl_2]$ is colorless and $Cs[Au^{III}Cl_4]$ is golden yellow. Another example is Prussian blue $KFe^{II}Fe^{III}(CN)_6$ with a dark blue color. In many dark minerals, for example magnetite $Fe^{II}(Fe^{III})_2O_4$, the same cause for strong light absorption is found. If metallic copper is added to a light-green solution of copper(II) chloride $CuCl_2$, the solution becomes brownish-black before all of the Cu(II) is reduced to Cu(I), when it becomes colorless. Often the color appears even with a very small proportion of atoms in another oxidation state. The phenomenon is not understood, but is certainly related to the possibility for electronic transfer between atoms with different oxidation number (cf. the high light absorption of metals, 7-3a).

In many cases color arises from other causes than those mentioned above, but space does not permit discussion of them. Among these, for example, is the color of the I_2 molecule, although this must be related to the fact that the great deformability of the I atom has become strongly involved. Transmission colors in solid bodies caused by the so-called F centers are described in 6-3a.

7-4. Magnetism

a. Different kinds of magnetism. All substances show magnetic phenomena in the general sense, and the reason for this is the mobile electric charges that exist in all matter. Here we treat only those magnetic phenomena that depend on the movements of the electrons in the electron clouds. Rotation (spin) of the atomic nucleus also causes magnetic effects, but these are small in relation to those just mentioned and therefore can often be neglected. They are utilized, however, in the methods of nuclear magnetic resonance ("NMR") developed in 1946, which have gained great significance in the treatment of problems in structural chemistry.

Every magnet is characterized by a *magnetic moment M*, which determines the torque exerted on the magnet in a homogeneous magnetic field. The magnetic moment is defined in a magnetic bar with pole strength p and length l as $M = pl$. The magnetic moment is a directed quantity (vector) and is indicated by an arrow.

If a metal wire loop carrying no current is moved into a magnetic field, a current is induced in the loop, whereby it acquires a magnetic moment. The current quickly stops because of the resistance in the loop, but if this were zero, the current would continue until the loop was moved out of the field (when a current is induced in the opposite direction which cancels the first current). The direction of the current in the loop is such that its magnetic field (magnetic moment) acts against the external field.

If the loop already carries a current at the start, it has an original (permanent) magnetic moment. If it were freely movable it would orient itself so that its field and the external field reinforce each other. When it is introduced into the field, however, a pulse of inductive current arises, which acts against the external field.

In every substance the electrons behave in their movements around the atomic nucleus like circular currents with no resistance. The atoms thereby generally acquire magnetic moments, caused by the "orbital movement" of the electrons (*orbital magnetic moment*). However, s electrons give no orbital magnetic moment. The "current strength" that results from the orbital movement of an electron, according to elementary principles, is proportional to its number of revolutions per unit of time.

The spin of an electron leads in addition to a *spin magnetic moment*.

The *permanent magnetic moments* mentioned above combine in a definite way to a total permanent moment. Within an electron pair the permanent moments of the two electrons compensate each other so that the permanent moment of the pair becomes zero. The permanent moment of an atom as well as a molecule is thus caused only by its unpaired electrons.

In an external field the total permanent moment tries to orient itself so that the field is reinforced. In this way *paramagnetism* arises. This is thus associated with the unpaired electrons. The strength of paramagnetism falls off with increase in temperature. The heat motion hinders the orientation of the permanent moment in the field direction.

The external field also changes the frequency of revolution of the electrons in a way to oppose the field. The change in velocity corresponds to the inductive pulse in the wire loop but in the atom the altered velocity is maintained until the body is taken out of the field. The weakening of the field implies *diamagnetism*. Diamagnetism is independent of temperature.

Paramagnetism and diamagnetism thus have opposite effects on the magnetic field; the former reinforces it and the latter weakens it. Every substance must possess diamagnetism. If it in addition has paramagnetism, this generally overrides in strength so much that the result is a strengthening of the field. The substance is then said to be paramagnetic.

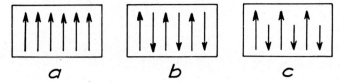

Fig. 7.1. Schematic picture of the magnetic moments of individual atoms (with no heat motion) within one domain in a crystal that is (a) ferromagnetic, (b) antiferromagnetic, (c) ferrimagnetic.

In crystals with sufficiently high order sometimes an exchange action takes place between the electron clouds that causes the permanent magnetic moments of the atoms, even in the absence of an external field, to try to orient themselves in parallel and in the same direction. This leads to *ferromagnetism*. Alignment is complete, however, only within particular volume elements, *magnetic domains*, that are smaller than the crystal (the form of the domains is often very regular, and their dimensions may vary from a few tens of Å to about 0.1 mm; larger domains can be observed microscopically by their effect on colloidal, ferromagnetic particles). The permanent moments of different atoms within one domain would thus lie as shown very schematically in fig. 7.1a. Each domain thereby acquires a strong *spontaneous moment* and this prefers a number of "favorable" directions within the crystal lattice. But the distribution of the domains and the mutual orientation of their spontaneous moments are such that the total moment becomes zero or very small (fig. 7.2a). If an increasing external field is applied, the domains that have nearly the same direction of moment as the external field grow at the expense of the others (fig. 7.2b). Gradually the moments of the latter turn to another favorable direction that is more closely aligned with the field direction (often a shift of 180° occurs) than the original one (fig. 7.1c) Finally, all moments are turned out of the favorable directions to attain even greater parallelism with the field (fig. 7.1d). All these processes lead to an increase in the total moment of the crystal and finally this reaches a limiting value (*magnetic saturation*) that is far greater (often 10^6 times) than the largest total moment of a paramagnetic body of the same volume. In a body consisting of several crystals, the same processes take place in each crystal and the total result is the same.

If the strength of the external field is again decreased, the shifts in the domains take place in the opposite direction. However there is then observed, as with every change in field strength, a pronounced lag (*hysteresis*). If the external field is reduced to zero there thus remains a portion of the realignments and of the total magnetic moment due to them (*remanence*). For demagnetization to zero moment the application of an external field in a direction opposite to the first is required (the strength required for the new field is called the *coercive force* of the substance).

It has been customary to distinguish between *soft* and *hard* magnetic material. The former has small, the latter large coercive force. To reduce the effect losses at continual remagnetizations, for example, of poles in electrical machines and in transformers, a small hysteresis is required, that is, a soft material. Pure iron is such a material, but still smaller losses may be obtained with certain alloy steels (34-3e) and oxides. Permanent magnets, on the other hand, should be magnetically hard, and a large coercive force is necessary. Carbon steel was formerly used for permanent magnets but is now replaced by special alloy steels or other alloys.

The alignment is complete within each domain at absolute zero. With increasing temperature the alignment decreases. The action of temperature is an order–disorder

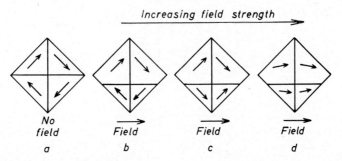

Fig. 7.2. Magnetic domains in a crystal. (a) With no external field, (b)–(d) with increasing strength of an external field. The arrow within a domain shows the direction of the total moment of the domain.

transformation (6-3b) and follows a curve of the same type as fig. 6.21. When the temperature approaches T_c (the *Curie point*) the order (alignment) drops rapidly and above this temperature the moment directions are completely disordered if no external field is applied. The spontaneous moments of the domains have thus completely disappeared and the substance shows only ordinary paramagnetism. Of course, as with all paramagnetic substances, a certain alignment of the permanent moments of the atoms occurs if an external field is applied.

Ferromagnetism requires in the first place that the atoms have permanent magnetic moments, that is, be "paramagnetic". They must thus have unpaired electrons. The alignment of permanent moments within a domain leading to ferromagnetism is only possible in the solid state and with ordered atomic distribution, and thus only in crystals. Ferromagnetism has only been detected in transition elements and phases containing them. The only known pure, ferromagnetic elements are iron, nickel, cobalt and gadolinium (Curie temperature +16°C). Of nonelement substances, particularly metallic phases and semiconductors containing these elements and also manganese may be ferromagnetic.

It has been found in recent years that the exchange action between close-lying atoms may also cause the permanent moments of neighboring atoms to be directed in exactly opposite directions (to be "antiparallel"). If the moments pointed in opposite directions are equally large (fig. 7.1b) the total moment at absolute zero becomes zero (*antiferromagnetism*). If the moments pointed in one direction are greater than those pointed in the other (fig. 7.1c), a total moment arises that appears as a more or less weakened ferromagnetism. For the last case the term *ferrimagnetism* has been proposed.

Paramagnetic ions, especially of transition elements, often orient their permanent moments in antiparallelism if they are bound together by ions such as F^-, O^{2-}, S^{2-}, Se^{2-}. In this way, MnF_2, Cr_2O_3, MnO, MnO_2, FeO, MnS and $MnSe$ are antiferromagnetic. Magnetite Fe_3O_4 (34-5b) and pyrrhotite (34-5a) are ferrimagnetic. Among pure elements, antiferromagnetism has been demonstrated in chromium and manganese.

The most direct information about the permanent dipoles of atoms in a crystal has been provided by the method of neutron diffraction. Since neutrons have a magnetic moment themselves, they are scattered by the permanent magnetic moments of the atoms as well as by their nuclei, so that the former strongly influence the diffraction effects caused by crystals of ferromagnetic, antiferromagnetic and ferrimagnetic substances. Thus, it has been possible to deduce in detail the distribution and orientation of permanent magnetic moments in such crystals by the analysis of neutron diffraction patterns.

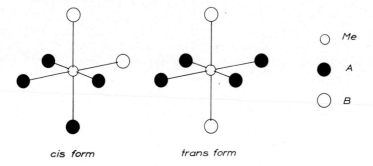

Fig. 7.3. Cis-trans isomerism in octahedral groups MeA_4B_2.

b. Magnetism and bonding. In 7-4a it was stated that paramagnetism shows the existence of unpaired electrons. The value of the spin moment per atom or molecule gives direct information about the *number* of unpaired electrons in the atom or molecule. A calculation of the spin moment from magnetic measurements is therefore very valuable, especially in bond studies.

The diamagnetism of a substance is composed nearly additively of the diamagnetisms of the component atoms. Its magnitude is thus little affected by the nature of the bonds and can often be calculated. The magnetic measurements can therefore often be corrected for the diamagnetic contribution to get the permanent magnetic moment. A more general method for obtaining this is to make measurements at different temperatures and utilize the temperature dependency of paramagnetism. That portion of the permanent moment that constitutes the orbital moment of the electrons can generally in turn be calculated and the spin moment obtained as the remainder.

More recently it has been possible to set up a resonance between an applied alternating, external field and the precession motion that the spin axes of the electrons undergo in a magnetic field (electron spin resonance, "ESR"). This method has become very valuable for the study of electron spin in paramagnetic substances.

In 5-3a a considerable number of molecules with unpaired electrons, which are thus paramagnetic, were listed, and in 5-3j transition-metal complexes with unpaired electrons were discussed. One further example may be mentioned: solutions of compounds of monovalent mercury are diamagnetic. Therefore, they contain not the ion Hg^+, which must have one unpaired electron, but Hg_2^{2+}, in which the two odd electrons form a pair.

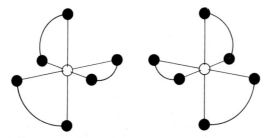

Fig. 7.4. Enantiomorphism in octahedral coordination of three chelate ligands (each indicated by a pair of black spheres joined by an arc).

7-5. Stereochemical methods

In the 1890's Werner began to develop new theories for the structure of inorganic molecules which, as far as the general principles of structure are concerned, have been largely confirmed subsequently. No physical methods of structure determination were then known, and Werner based a number of his structural hypotheses on stereochemical considerations.

He was able thus to prove that the six-coordinated, metal complexes studied by him had octahedral coordination. Complexes of the type MeA_4B_2 had two, but never more than two isomers (the ion $Co(NH_3)_4Cl_2^+$ thus exists in a violet and a green form). Octahedral coordination here allows two isomers, a cis form and a trans form (Lat. cis, on this side; trans, on the other side), as shown in fig. 7.3. With the two other most easily imagined types of sixfold coordination—plane-hexagonal and trigonal-prismatic—three and four (of which two are enantiamorphic) isomers, respectively, should appear. Neither did complexes of the type MeA_3B_3 show more than two isomers, which also spoke for octahedral coordination.

The same is the case with the enantiomorphic structures, which Werner succeeded in preparing with chelate ligands (5-6c). In one type of these, three molecules of ethylenediamine $NH_2CH_2CH_2NH_2$ were octahedrally coordinated about a Co^{3+} ion as in fig. 7.4.

Werner assumed that four-coordination was generally tetrahedral. However, in diammineplatinum(II) chloride $Pt(NH_3)_2Cl_2$, he found two isomers, which is not consistent with tetrahedral coordination. He therefore interpreted these two isomers as cis and trans forms of a plane, square coordination. A later structure determination by X-ray diffraction proved that he was correct.

CHAPTER 8

Reaction Kinetics in Homogeneous Systems

Reaction kinetics is concerned with the rate and mechanism of chemical reactions.

The *reaction rate* is expressed as the amount of a participating substance that is consumed or created per unit of time. The quantity of the substance in the system may be expressed as the number of molecules of the substance within the system or, most commonly, in concentration measure. Since the reaction rate most often changes with time, it must be given as the time derivative of the number of molecules or the concentration. If the substance A is consumed and the substance B formed, in the infinitesimal time dt the change in concentration of A becomes $-d[A]$ and of B, $d[B]$. The reaction rate may thus be given as $-d[A]/dt$ or $d[B]/dt$.

The distinction is made between *homogeneous reactions*, in which the participating substances are in the same homogeneous phase, and *heterogeneous reactions* which take place at the boundary surface between two phases in a heterogeneous system. In this chapter only homogeneous reactions are considered. Heterogeneous reactions will be taken up after heterogeneous equilibria have been covered (chap. 14).

a. Reactions of the first order. Let us assume that a molecule A has a tendency to decompose in some way and that the probability that this decomposition will occur is completely independent of the presence of other molecules. The number of molecules that decompose per unit of time and unit of volume then is proportional to the number of molecules A present per unit of volume, that is, to the concentration [A]. The reaction velocity thus becomes

$$-\frac{d[A]}{dt} = k[A] \tag{8.1}$$

The constant of proportionality k is called the *rate constant* or in this special case the *decay constant*. A reaction for which the rate can be expressed by a formula of the type (8.1) is said to be of the *first order*.

By integration of (8.1) we obtain

$$kt = \ln\frac{[A]_0}{[A]} = 2.303 \log\frac{[A]_0}{[A]} \quad \text{or} \quad [A] = [A]_0\, e^{-kt} \tag{8.2}$$

where $[A]_0$ signifies the initial concentration of A, that is, [A] at $t=0$.

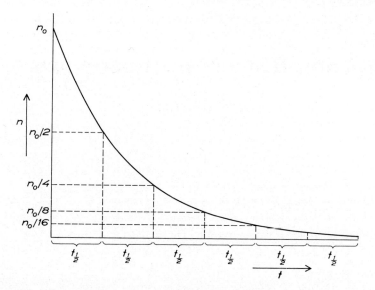

Fig. 8.1. Reaction of the first order. Each succeeding halving of the number of molecules and associated half-life time is shown by the dashed lines.

In (8.1) and (8.2) [A] and $[A]_0$ may be replaced by the *number* of molecules of A in the system, n at time t and n_0 at time $t=0$.

Often the *half-life* $t_{\frac{1}{2}}$ is given, that is, the time after which half of the molecules A have disappeared. Then $[A]=[A]_0/2$. From (8.2) it follows that

$$t_{\frac{1}{2}} = \frac{\ln 2}{k} = \frac{0.693}{k} \tag{8.3}$$

The more rapidly the decomposition takes place, the larger k is and the smaller is the half-life $t_{\frac{1}{2}}$. Both k and $t_{\frac{1}{2}}$ are independent of the initial concentration of the decomposing substance. Fig. 8.1 shows the decomposition curve for a reaction of the first order.

A reaction in which a molecule decomposes independently of other molecules is called a *monomolecular reaction*. A monomolecular reaction is always of the first order. On the other hand many first-order reactions are not monomolecular but more complicated, although this is not apparent from the course of the reaction rate. As an example, the decomposition of dinitrogen pentoxide N_2O_5 may be mentioned. The final products of decomposition are N_2O_4 and O_2, which gives the overall reaction

$$2N_2O_5 \rightarrow 2N_2O_4 + O_2$$

However, the reaction probably takes place in the following steps:

a. $N_2O_5 \rightarrow N_2O_3 + O_2$ b. $N_2O_3 \rightarrow NO_2 + NO$
c. $NO + N_2O_5 \rightarrow 3NO_2$ d. $4NO_2 \rightarrow 2N_2O_4$

The monomolecular reaction *a* goes rather slowly but the reactions *b–d* are nearly instantaneous. Obviously it is the slowest reaction that chiefly determines the rate of reaction. The reaction is thus of the first order.

The decomposition of N_2O_5 is also an example of the common and very natural case where the overall reaction in no way shows the true process.

One type of reaction that is strictly monomolecular and therefore always first order is the decomposition of radioactive atomic nuclei (9-2a). This is determined only by the unstable nucleus and generally is not affected by its environment. The course of the decomposition (decay) therefore also is independent of whether the system is homogeneous or heterogeneous.

b. Reactions of higher order. If two types of molecule A and B react with each other according to the reaction $A+B \rightarrow C$, reaction can only take place if one molecule A collides with one molecule B. The probability that a particular molecule A will collide with any molecule B is to a first approximation proportional to [B]. The probability that any molecule A at all will collide with a molecule B then becomes proportional to both the concentrations [A] and [B]. The reaction rate therefore is

$$-\frac{d[A]}{dt} = -\frac{d[B]}{dt} = k[A][B] \tag{8.4}$$

A reaction with this rate behavior is said to be of the *second order*. For the special case $A=B$, that is, the reaction $A+A \rightarrow A_2$, the rate $=k[A]^2$. In both the cases given two molecules react with each other and the reactions are *bimolecular*.

A reaction $A+B+C \rightarrow D$ has the reaction rate $=k[A][B][C]$ and is said to be of the *third order*.

Fig. 8.2 shows for reactions of different orders how the concentration of one of the participating kinds of molecule drops with time with the assumption of the same initial rate. The conversion thus slows down more rapidly, the higher the order is.

If the initial rate remained constant the concentration would follow the dashed line in fig. 8.2. Reactions with constant rate do occur and depend on the fact that other factors than the concentration limit the rate. The rate of a photochemical reaction may be limited by the number of incoming quanta and the rate of an electrolysis is limited by the current strength. If these factors are constant, the rate becomes constant. Sometimes it is said that such reactions are of *zero order*.

It is apparent from fig. 8.2 that reactions of orders other than the first do not have constant half-lives.

Most reactions that can be studied have been found to be mono- or bimolecular; trimolecular reactions are uncommon. The probability is small for the simultaneous collision of the three molecules that would partake in such a reaction. Most overall reactions with three or more starting molecules therefore do not show the true reaction mechanism, which should instead involve several mono- or bimolecular steps.

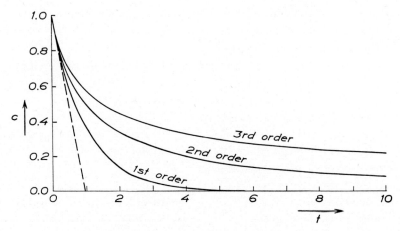

Fig. 8.2. Reactions of different order but the same initial rate.

c. Reaction rate and temperature. Most chemical reactions are hastened by increase in temperature. The formulas in 8a and b apply in those cases only at constant temperature. Nuclear reactions on the other hand are not at all affected by temperatures below about 10^6 °K.

In solutions at ordinary temperature, a temperature increase of 10°C usually leads roughly to a doubling of the reaction rate. Reactions in which very large molecules take part are often still more sensitive to temperature. Such is the case with most biological processes. Because of the great temperature dependency of the reaction rate this is measurable generally only within a quite small temperature range; outside this range it is too high or too low for measurement.

If the heat that is developed in a reaction is not carried away, the temperature of the system rises, causing an increase of the reaction rate. The ever faster development of heat facilitates the reaction further, which finally may reach a violent stage. In this way in general every combustion is initiated, even spontaneous combustion of easily oxidized substances. In especially violent cases one may speak of a "heat explosion".

The rate constant k in general changes with temperature according to the expression (Arrhenius, 1889)

$$k = be^{-E_A/RT} \tag{8.5}$$

The quantity E_A, which has the dimension of energy per mole, is constant as long as the reaction mechanism remains the same, and b may be assumed to be constant if the temperature region under study is not too great. The significance of these two quantities is discussed in the following section.

d. Collision number, activation energy, steric factors. In order for two molecules (or as a special case, atoms) A and B to be able to react they must collide with each other. The number of collisions per second of molecules A with molecules B within

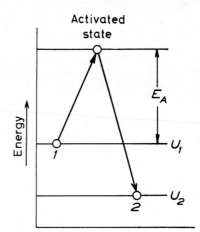

Fig. 8.3. Activation energy.

a volume of 1 cm³ is called the *collision number* of the system. In a gas at atmospheric pressure and room temperature the collision number is of the order of 10^{25}. If this collision number is compared with the number converted in a process that has a conveniently measurable rate, it is found that only a vanishingly small fraction of all collisions lead to reaction. The most important reason that the greater part of the collisions is ineffective is that only molecules with a certain minimum energy can react. Such molecules are said to be *activated*, and the amount of energy that must be added to a normal molecule in order that it shall reach the activated state is called the *activation energy*. This is generally expressed per mole and then is identical with the above-mentioned quantity E_A.

Fig. 8.3 shows an energy diagram for a system that goes from state 1 (total energy U_1) to state 2 (total energy U_2). However, the reaction can only take place if the system is first carried over to an activated state by the addition of the activation energy E_A. Regardless of this, the change of energy of the reaction is $\Delta U = U_2 - U_1$ (if the reaction takes place at constant pressure, U is replaced by H and the value of the activation energy is also changed somewhat).

The activation energy may be added in different ways, for example, by heating, by bombardment with particles or by electromagnetic radiation (8f). Activation may involve a contribution of one or more of the kinds of energy mentioned in 2-1b, with the exception of nuclear energy.

However, only in a certain fraction of the collisions do the molecules have a mutual position that allows reaction. The more complicated the structure of the molecules the more critical becomes the reaction orientation, and the greater role this *steric factor* plays.

Many reactions are assumed to go through an *activated complex* or *collision complex*. Let us assume the reaction AB + C → A + BC, in which A, B and C may be atoms or atomic groups. AB and C approach each other in the proper mutual orientation and the necessary activation energy is added. A bond then begins to form between AB and C at the same time that the bond A–B becomes weaker (A and B begin to move away from each other). We then have as a transition state an activated complex that might be indicated as A$\cdots$B$\cdots$C. This gives off energy, A is completely repelled and the bond B–C attains full strength.

The conditions for reaction mentioned above lead to the relation

number of reacting molecules =
(collision number)
× (fraction of molecules in proper orientation)
× (fraction of activated molecules) (8.6)

The fraction of activated molecules increases exponentially with temperature in the same way as the exponential function in (8.5). Therefore, it is this increase that accounts for the large temperature dependency of the reaction rate (the collision number is proportional to $\sqrt{T}$ and thus is only weakly dependent on temperature). The common doubling of the reaction rate with a temperature rise of about 10°C implies an activation energy of approximately 50 kJ mol^{-1} if the rise in temperature occurs in the vicinity of room temperature.

A comparison of (8.5) and (8.6) shows that the quantity b is proportional to the product (collision number) × (fraction of molecules in proper orientation).

The addition of extra energy of the activation energy type is required to bring about many different types of processes, and not only collision reactions. The need for activation energy implies an energy threshold (2-3 b; cf. fig. 2.1 and 8.3) and may be considered to be the most common cause of arrested reactions.

e. Ionic reactions. Certain homogeneous ionic reactions proceed extremely fast. To these belong especially transfer of protons, for example, in the process $H_3O^+ + OH^- \rightarrow 2H_2O$, which occurs in the neutralization of a strong acid by a strong base (chap. 12). Electron transfer (redox processes, chap. 16), on the other hand, may go slowly and often follows a more complicated path than the overall reaction indicates. For example, the reaction between iron(III) and tin(II) ions in hydrochloric acid solution does not take place according to the overall reaction $2Fe^{3+} + Sn^{2+} \rightarrow 2Fe^{2+} + Sn^{4+}$, but goes through several intermediate reactions in which complexes with chloride ions play a large role.

Arrested reactions are especially common when the reaction involves rearrangement of the atoms in molecules or complex ions.

f. Photoreactions, photochemistry. In reactions whose rates are increased by absorption of electromagnetic radiation (*photoreactions*), the reacting molecules are activated by the absorption of quanta. Sometimes the *quantum yield*, that is, the number of reacting molecules per absorbed quantum, $=1$ (*photochemical equivalence law*, Einstein, 1912). Most often deviations occur, however. An activated molecule may give off energy (be deactivated) before it has a chance to react. In complicated molecules it may happen that some portion of the molecule that does not have the ability to react absorbs a quantum. In such cases the quantum yield is less than 1. When hydrogen bromide decomposes under radiation with ultraviolet light, this takes place in the following steps:

$$HBr + h\nu \rightarrow H + Br$$
$$HBr + H \rightarrow H_2 + Br$$
$$Br + Br \rightarrow Br_2$$
$$\overline{2HBr + h\nu \rightarrow H_2 + Br_2}$$

For the overall reaction the quantum yield is thus 2. In the so-called chain reactions (8g) the quantum yield is much greater than 1.

Tab. 1.1 shows that the energy of the quanta in ultraviolet radiation is of the order of magnitude of 1000 kJ mol^{-1}. Thus, reactions that require very high activation energy can be initiated with such radiation.

A photoreaction often takes place in another manner than the reaction between the same substances without irradiation (*dark reaction*).

The science of photoreactions is often called *photochemistry*.

g. Chain reactions. In many cases there are reactions that do not follow the simple reaction types, as mentioned in 8a and b, and neither can they be divided up into steps of such types. These occur especially often in gas mixtures but sometimes also in liquid solutions and are often very sensitive to the presence of small amounts of foreign substances.

An example is the formation of hydrogen chloride from a gaseous mixture of chlorine and hydrogen. Under the influence of light (absorption of a light quantum) a chlorine molecule decomposes according to

$$Cl_2 + h\nu \rightarrow 2Cl$$

and thereafter occur the reactions

$$Cl + H_2 \rightarrow HCl + H$$
$$H + Cl_2 \rightarrow HCl + Cl$$

and so on.

Absorption of a single quantum thus initiates a *chain reaction*, which theoretically should be able to continue until the conversion is complete. In actual fact the reaction chain is broken off earlier because active hydrogen and chlorine atoms are sidetracked into other reactions. Thus, they may be combined with impurities and also with the walls of the vessel. If oxygen is present, it easily bonds with the hydrogen atoms. In oxygen-free mixtures the interruption occurs chiefly as a result of recombination of active chlorine atoms according to $2Cl \rightarrow Cl_2$. The reaction chain then has an average length of 10^4 HCl molecules formed per absorbed quantum, which means that the quantum yield of the reaction is of the order of 10^4.

Chain reactions take place in most flame combustions and in typical *gas explosions*. Inadequate removal of the heat of reaction increases the rate of reaction further (cf. 8c). However, in explosions the reaction chain is most often *branched*, that is, one active atom or molecule gives rise to two or more new ones. When this branching is repeated, extraordinarily high reaction rates are quickly reached.

A substance that stops a reaction chain is called an *inhibitor*. When tetramethyl or tetraethyl lead is added to motor fuel to prevent knocking it is probably the finely divided lead or lead oxide formed on burning that acts as inhibitor. The inhibitor is always consumed during the course of the reaction.

Very important chain reactions also occur in nuclear reactions (9-4a).

h. Homogeneous catalysis. As mentioned earlier in 2-3b a catalyst acts to hasten a reaction by making possible a new reaction route with a lower energy threshold (lower activation energy). Homogeneous catalysis, which as the name implies occurs in homogeneous systems, does not involve any special problems. It can be interpreted on the basis of the general laws of reaction rate and reaction mechanism described above.

A given starting substance may react in different ways with different catalysts. By choice of the proper catalyst a reaction may then be led in the desired direction.

Sometimes a reaction product catalyzes the reaction by which it was produced (*autocatalysis*). The reaction then begins very slowly (one speaks of an *induction period*), but when a sufficient quantity of the catalyzing product is formed, the rate sharply rises. It finally diminishes in the usual way because of the decrease in the amounts of the reacting substances.

A multitude of biological processes are made possible by homogeneous catalysis, in which various organic substances, *enzymes*, act as catalysts. Biological redox processes, for example, oxygen exchange in breathing, is catalyzed by respiratory enzymes containing transition metals (often iron), whose alternation between different oxidation states is a contributing factor.

CHAPTER 9

The Atomic Nucleus

9-1. The composition and stability of the atomic nucleus

a. General features of nuclear composition. As already stated in 1-3b an atomic nucleus consists of protons and neutrons, the so-called nucleons. These are held together by very strong forces. Typically, these forces fall off extremely rapidly as the distance between the nucleons grows; they are decidedly short-range forces.

The nucleons are arranged in a definite way in the nucleus and certain combinations are especially stable. As in the electron cloud one can distinguish definite energy levels in the nucleus and therefore speak of a *shell structure* for it. The nucleons in a particular nucleus can assume different energy levels under different conditions. However, all the energy states, except the ground state with the lowest energy, are often so short-lived that they cannot be observed. The properties of the nuclear species are then only dependent on the number of protons and neutrons. But sometimes states with higher energy, excited states, may have a measurable life. According to the definition in 1-3b nuclei with different energies are said to constitute different nuclides even if their nuclear compositions are the same. They are then said to be *nuclear isomers*. All isomeric nuclides have the same nuclear charge and their masses vary only by extremely small values depending on the differences in energy. Nuclear isomers therefore have the same chemical properties. Nuclear isomers with higher energies than the ground state are indicated by an "m" after the mass number, for example, $^{26m}_{13}$Al.

A very large number of nuclides are *unstable* even when they correspond to a ground state. The nucleus then decomposes with the emission of various kinds of particles and electromagnetic radiation (*radioactivity*). The probability for decomposition may be extraordinarily small, however. As an example, ^{204}Pb has a half-life of about 1.4×10^{17} years and therefore an extremely weak radioactivity. If no decomposition can be detected the nuclide is said to be *stable*. The boundary between the two classes naturally depends on the sensitivity of the method of measurement. In natural earthly material 266 stable nuclides have been found along with approximately 65 unstable ones (naturally radioactive nuclides). In addition, by nuclear reactions up to 1966, more than 1500 unstable nuclides (nearly 1300 in the ground state and over 250 isomers) have been synthesized. All told, therefore, more than 1800 nuclides are known.

The nuclear composition (number of protons = nuclear charge, Z; number of neutrons, N) of most of these nuclides are shown in fig. 9.1. All nuclides with the

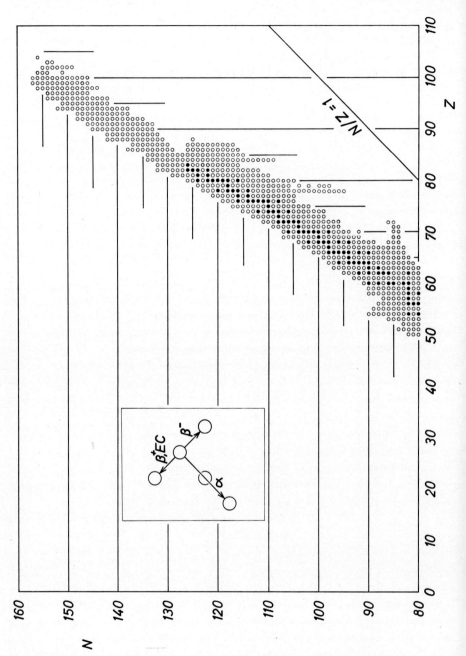

Fig. 9.1. Nuclear composition of known nuclides. ● = stable nuclide; ○ = unstable nuclide.

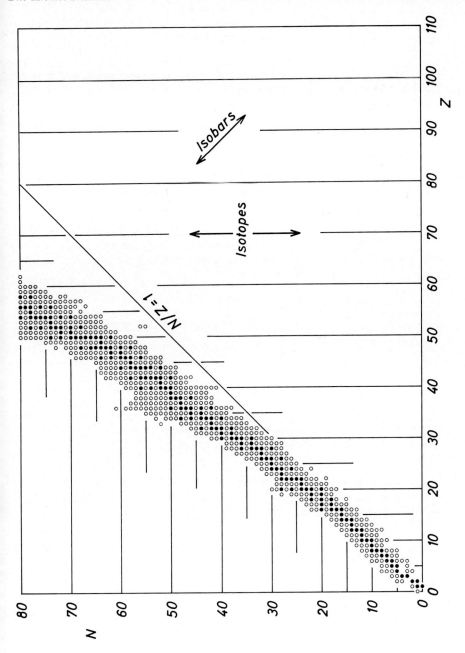

same Z value are isotopes of the same element. (For the most important nuclides up to $Z=10$ certain data are given in tab. 1.4; in addition, all stable nuclides are enumerated in tab. 1.3.)

In the NZ plane the stable nuclides follow fairly closely to a curve, which at low Z values nearly coincides with the line $N/Z=1$ and then slowly bends away from it toward the N axis. The quotient N/Z therefore is nearly $=1$ at the start but rises slowly to somewhat over 1.5. This leads to the result that the mass number $A=Z+N$ (1.4) is approximately twice as large as Z in the beginning but gradually overrides the doubled Z value. No stable nuclides heavier than $^{209}_{83}$Bi are known (even this nuclide may possibly be weakly radioactive). Among the lighter elements 43Tc and 61Pm lack stable isotopes. In 9-1b some general conditions for stability are stated.

The stable nuclides are surrounded on both sides by rather narrow zones of unstable nuclides. When an unstable nuclide decomposes spontaneously, the process involves a more or less direct transition to a neighboring stable nuclide through a change in N and Z. The most common processes of this kind are described in 9-2a.

b. Nuclear energy. If free nucleons are combined into an atomic nucleus, large amounts of energy are given off. This released energy implies that the system loses mass and because of the large energy changes the loss of mass is easily detected. Inversely, from the observed loss of mass the energy given off can be calculated according to Einstein's equation (1.1).

Each neutral atom is built up of Z protons $+Z$ electrons (in the electron cloud) $+N$ neutrons. The total mass of these elementary particles in the free state may be calculated, for which a simpler expression is obtained if each pair, proton $+$ electron, is taken together. The mass of such a pair is equal to the mass of the hydrogen atom, since this is formed from one proton $+$ one electron with so little evolution of energy that no appreciable decrease of mass results. The total mass of the free particles thus becomes $Zm_H + Nm_n$, where m_H and m_n are the masses of one hydrogen atom and one neutron, respectively. The values of m_H and m_n expressed in mass units u (1-3f; u $= 1.66043 \times 10^{-27}$ kg) may be taken from tab. 1.4. If M is the mass of the atom created expressed in the same units, then

$$\text{loss of mass} = (1.007825\,Z + 1.008665\,N - M)\,\text{u} \qquad (9.1)$$

A change of mass of one unit u corresponds, according to (1.1), to a change of energy of 1.4923×10^{-10} J $= 931.5$ MeV (1-2c). The loss of mass in (9.1) implies thus that the nuclear system gives off the energy

$$\begin{aligned}-\Delta E &= 1.4923 \times 10^{-10}\,(1.007825\,Z + 1.008665\,N - M)\,\text{J} \\ &= 931.5\,(1.007825\,Z + 1.008665\,N - M)\,\text{MeV}\end{aligned} \qquad (9.2)$$

With the formation, for example, of $^4_2$He ($M=4.002604$ u) from two hydrogen atoms ($Z=2$) and two neutrons ($N=2$) the loss of mass is 0.03038 u. The nuclear

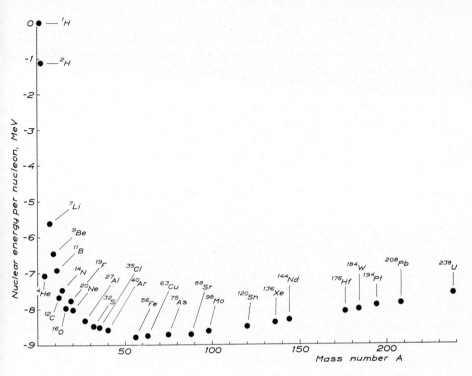

Fig. 9.2. Nuclear energy per nucleon for a number of natural nuclides.

system then gives off the energy 4.53×10^{-12} J $= 28.3$ MeV. The energy developed per mole of He created is obtained from this by multiplication by Avogadro's number (1-4d) and attains the considerable value 2.73×10^{12} J.

By dividing by the mass number A, that is, the number of participating nucleons, an average value per nucleon for the energy developed is obtained. In the example, $-\Delta E/A = 7.1$ MeV. The energy per nucleon of the nuclear system (*nuclear energy*) has thus been reduced by 7.1 MeV by this process. If the energy before the process is set equal to 0, after the process it is -7.1 MeV. This value and corresponding values for a number of other natural and reasonably stable nuclides have been plotted in fig. 9.2 as a function of A. The nuclear energy has a minimum approximately at $A = 56$, that is, in the vicinity of Fe. Here the nuclides are most stable. Clearly, energy may be gained either by combining nucleons (as in the example mentioned) or very light nuclides, or also by cleaving the heaviest nuclides so that fragments of medium weight are formed. Processes of the first kind give the largest amounts of energy per nucleon. The existence of all so-called stable nuclides with energies higher than the minimum shows, however, that their nuclear systems do not try to go toward the energy minimum by spontaneous transformation. The reason lies in arrested reactions, caused by extremely high activation energies (8d, 9-2b).

The energy developed when a nuclide is formed from free nucleons, that is the distance from a point in fig. 9.2 to the zero axis, is also called the *binding energy* in the nuclide in question. The binding energy is obviously greater, the more stable the nuclide is.

Often, the *packing fraction* $(M-A)/A$, of a nuclide is given, that is, the deviation of the atomic weight from the mass number per unit of mass number. The course of the packing fraction is very similar to that of the nuclear energy values $-\Delta E/A$ in fig. 9.2, as can be seen if one writes $(M-A) = -(Z+N-M)$. The latter expression according to (9.2) is very nearly proportional to $-\Delta E$. The packing fraction is less easily appreciated than the nuclear energy per nucleon, however.

The nuclear energies of all nuclides would be better visualized in a three-dimensional model in which the nuclear energy is expressed as a function of both Z and N. The nuclear energies of all the nuclides as given in fig. 9.2 would then be plotted perpendicular to the NZ plane in fig. 9.1. These values form an energy surface shaped like a narrow valley, whose projection on the NZ plane is quite well reflected by the positions of the known nuclides in fig. 9.1. The nuclear energies of the stable nuclides lie in the bottom of the valley, and this bottom line follows to a large extent the same height profile as the points in fig. 9.2. The walls of the valley rise steeply and on these lie the nuclear energies of the unstable nuclides. Nearly all of the natural, unstable nuclides lie on the neutron-rich sidewall (in fig. 9.1, the "northwest" wall).

It can be shown that if the protons were uncharged the nucleus would have maximum stability with $N=Z$. However, through their charges electrostatic, repulsive forces are introduced, which increase with rising Z, and these can only be compensated by greater and greater "dilution" of the nuclear contents with neutrons. In this way the main course of the energy valley in the NZ plane is explained. This course, and the increase in nuclear energy with mass number after the minimum is passed, make the nuclides more and more unstable, the higher the atomic numbers and mass numbers become. As already mentioned there are no stable nuclides with Z and A values higher than for $^{209}_{83}$Bi. $^{254}_{96}$Cf has been detected spectroscopically in supernovae, but in natural, earthly material, the highest values for unstable nuclides are represented by those for $^{239}_{94}$Pu (9-2c). Nuclides with still higher values are obtained only by artificial nuclear reactions (9-2b) and with increasing values they become more and more unstable. The highest Z value that has been attained with certainty up to 1968 is 104 (29a, see also 28-4).

The energy surface referred to above is not at all "smooth", but has a pronounced unevenness caused by the fact that the energy values of the nuclides show large individual variations. Even the energy values in fig. 9.2 show a rather irregular course, and if more unstable nuclides were included the irregularity would further increase because these would give higher-lying values. Three of the nuclides included, namely, $^4_2$He, $^{12}_6$C and $^{16}_8$O, give energy values that obviously lie lower than the neighboring values, and clearly possess especially stable nucleon combinations. The distribution of the stable nuclides in fig. 9.1 also shows the

The Atomic Nucleus

individual variations. Here, however, conformity to several laws can be traced, of which the following should be mentioned.

Even values of Z and N to a great extent increase the stability. From this viewpoint the known, stable nuclides show the following distribution:

$$Z = \text{even}, N = \text{even}: \quad 161 \text{ nuclides}$$
$$Z = \text{even}, N = \text{odd}: \quad 53 \text{ nuclides}$$
$$Z = \text{odd}, N = \text{even}: \quad 48 \text{ nuclides}$$
$$Z = \text{odd}, N = \text{odd}: \quad 4 \text{ nuclides}$$

The four nuclides of the last category are ^2_1H, ^6_3Li, $^{10}_5B$ and $^{14}_7N$, that is, the four lightest combinations with $Z = N = \text{odd}$. The distribution described is the reason that the stable nuclides follow a strikingly zigzag line in the NZ diagram (fig. 9.1). It shows that both protons and neutrons occur in the nucleus in pairs and that an unpaired proton or neutron makes the nucleus less stable. A combination of one proton pair and one neutron pair has particularly high stability. This group is represented by the very stable nucleus ^4_2He (cf. its energy in fig. 9.2), which also is often given off as an α particle (also indicated by $^4_2\alpha$) on decomposition of heavy nuclei.

From the above it follows that elements with $Z = \text{even}$ generally have more stable isotopes than elements with $Z = \text{odd}$ (see tab. 1.3). No element with $Z = \text{odd}$ has more than two stable isotopes and of the 20 elements with only one stable isotope, $Z = \text{odd}$ for 19 of them (the exception is 4 Be). On the other hand, all stable elements with $Z = \text{even}$, with the exception of 4 Be, have at least two stable isotopes. Often the number is greater; 50 Sn has 10 stable isotopes.

The dependency of the stability on Z and N is also shown by the fact that elements with $Z = 2n$ generally exist in nature in larger amounts than the neighbors with $Z = 2n \pm 1$ (*Harkins' rule*). The rule applies nearly universally for the relative, elemental abundances (cf. 17a) estimated for the known universe. A large and important exception is given by the most common element in the universe 1 H, and this must be related to the proton's function as nuclear building material. Of the stable isotopes of an element those with $N = \text{even}$ generally appear in the largest amounts. Therefore isotopes with even mass numbers most often predominate for elements with $Z = \text{even}$ and isotopes with odd mass numbers for elements with $Z = \text{odd}$.

Fig. 9.1 also shows that of all isobars with odd mass numbers, at most one is stable. Of isobars with even mass number, most often two are stable. Z (and thus also N) for stable isobars always changes in steps of two units. The pair $^{113}_{48}Cd-^{113}_{49}In$ appears to be an exception, but probably the first nuclide of the pair is unstable, although this has not yet been proved with certainty.

A nucleus is more stable than the average if its number of protons or neutrons is 2, 8, 20, 28, 50, 82, or 126. These *"magic numbers"* can be explained by the theory of the shell structure of the nucleus (9-1a), which also requires that the numbers 114, 164, and 184 ought to be included in the same series. The stability will be especially high if both the proton and neutron numbers are magic numbers. This may cause nuclides such as $^{278}_{114}X$, $^{298}_{114}X$ and $^{310}_{126}X$ to be so stable that they can be prepared.

9-2. Nuclear reactions

a. Spontaneous nuclear reactions. Spontaneous nuclear decomposition (decay) was first observed as natural radioactivity by Becquerel in 1896. Among the scientists who later carried out the basic studies on natural radioactivity, Pierre and Marie Curie and Rutherford should be mentioned. After new, unstable nuclides were successfully synthesized in nuclear reactions (the first time in 1932 by F. Joliot and Irène Joliot-Curie) our knowledge of the spontaneous nuclear reactions was rapidly increased.

A spontaneous nuclear reaction arises from a *mother nuclide* or *mother nucleus*, which more or less directly forms a *daughter nuclide* or *daughter nucleus*.

With a spontaneous nuclear decay the energy of the system decreases, and every nucleus of an unstable nuclide must have high enough energy in relation to the final state so that it can decompose. But this excess energy must be distributed in a certain way in the nucleus in order that decomposition can take place. The probability that this distribution will occur is proportional only to the number of nuclei. All nuclear reactions are therefore strictly monomolecular and thus of the first order (8a). To characterize the rate it is customary to give the half-life of the mother nucleus, which is inversely proportional to the probability of decomposition (8.3).

The most important spontaneous reactions are the following:

1. *Beta decay* (β^- *decay*). Involves the emission of electrons $_{-1}^{0}e$, β^- particles. These can be considered as being formed in the nucleus by the reaction $_0^1n \rightarrow _1^1p + _{-1}^{0}e$. For each β^- particle given off, Z thus increases by one unit and N decreases by one unit (see arrow in fig. 9.1). The mass number A is then unchanged, that is, mother and daughter nuclides are isobars.

2. *Positron decay* (β^+ *decay*). Emission of positrons $_1^0e$, β^+ particles. These can be considered as being formed in the nucleus by the reaction $_1^1p \rightarrow _0^1n + _1^0e$. The changes in Z and N thus go in the opposite direction from β^- decay (arrow in fig. 9.1). This decay is also "isobaric". The released positrons always very quickly encounter electrons in the surroundings and are consumed with the emission of annihilation radiation (1-3a). β^+ decay therefore is followed by electromagnetic radiation (γ radiation; however, γ radiation ordinarily arises in another way, as described below).

3. *Electron capture* ("EC"). Instead of emitting a positron, a nucleus can decrease its number of protons by capturing an electron from the electron cloud, following which the process $_1^1p + _{-1}^{0}e \rightarrow _0^1n$ takes place. Most often a K electron is captured since such electrons are most commonly nearest to the nucleus. The process is therefore also often called "K capture". The changes in Z and N are the same as in positron decay (arrow in fig. 9.1). The captured electron is immediately replaced by an electron from an outer portion of the cloud. In this way X rays are produced.

Since the electron cloud takes part in electron capture, alterations in it can affect the rate of the process slightly. In light atoms in which the cloud in the vicinity of the nucleus contributes to bonding, it has thus been found that changes in the bonding conditions can change the half-life up to approximately 0.1 per cent. However, the final product is always the same.

The Atomic Nucleus

4. *Alpha decay.* Here emission of α particles occurs, consisting of the especially stable nucleon group $^4_2$He (also written $^4_2\alpha$). α particles lack an electron cloud, and can thus be considered as helium nuclei with the net charge $+2$ (they should in complete form be written $^4_2$He^{2+}). Loss of an α particle leads to a decrease of Z and N by 2 units (arrow in fig. 9.1). The mass number A is thus decreased by 4 units. The emitted helium nucleus very quickly takes up two electrons from almost any atom whatever in the vicinity and becomes a neutral helium atom. Formation of helium can be detected with α decay.

Immediately after a decay process the remaining nucleons often still have not succeeded in arranging themselves in the most stable possible structure. The daughter nucleus is then excited, that is, its energy is higher than that corresponding to the ground level. In general, however, a rearrangement of the nucleons very rapidly takes place, which involves a lowering of the energy to the ground level of the daughter nucleus. The corresponding amount of energy is emitted as energy-rich (in the range 0.1–10 MeV) and therefore short-wave, electromagnetic radiation, γ *radiation* (cf. tab. 1.1). Sometimes the excited nucleus can persist long enough to be detected, so that nuclear isomerism (9-1 a) is said to arise. The γ radiation is then emitted only in connection with the transition to the lower energy level ("isomeric transition").

It may also happen that the γ radiation will knock out one of the inner electrons in the electron cloud of the atom. This electron leaves the atom with a kinetic energy equal to the γ energy minus the energy that was used in the removal of the electron. The ejected electron is replaced by an electron from an outer part of the cloud, whereby X rays are produced.

In all of the reactions mentioned above the total mass of the system is decreased. The corresponding quantity of energy appears partly in the form of kinetic energy of the particles produced as they are ejected from the nucleus, and partly in the form of electromagnetic radiation. In the reactions given in the following examples this energy quantity is most often omitted, however. But in each nuclear reaction the sum of the mass numbers as well as the sum of the nuclear charges will be the same before and after the reaction (if it is desired to show this clearly, complete symbols should be used; for example, $^0_{-1}$e). Since the electron cloud is without significance in the main course of a nuclear reaction, the net charge of the atom is irrelevant and is generally not given.

In spontaneous nuclear reactions the nuclear system in general seeks the shortest route to a stable state. Starting with a particular nucleus, only one of the nuclear reactions given above can occur at a time. If a stable nuclide is not reached after only one reaction, the nucleon system passes through one or several intermediate states, that is, unstable nuclides, by means of several subsequent reactions. In this way a *radioactive series* of unstable nuclides arises.

The conditions of stability and directions of reaction as shown in fig. 9.1 allow general conclusions concerning the occurrence of different reaction types. Nucleon

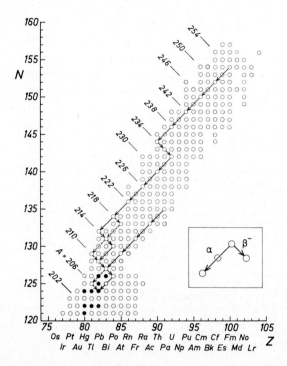

Fig. 9.3. Radioactive series with mass numbers $A = (4n+2)$. Solid arrows show the main reactions and the dashed arrows show less important reactions.

systems that lie on the neutron-rich ("northwest" in fig. 9.1) side of a stable isobar move toward it by one or several β^- decays. An example is the reaction $^{29}_{13}\text{Al} \rightarrow ^{29}_{14}\text{Si} + \beta^-$. Nucleon systems that lie on the proton-rich ("southeast" in fig. 9.1) side of a stable isobar move toward it either by β^+ decay or electron capture. Examples are the reactions $^{30}_{15}\text{P} \rightarrow ^{30}_{14}\text{Si} + \beta^+$ and $^{7}_{4}\text{Be} + ^{0}_{-1}\text{e} \rightarrow ^{7}_{3}\text{Li}$. Sometimes one portion of the nuclei of the same nuclide shows β^+ decay and the remainder electron capture. For nuclides with $Z > 83$ (Bi), however, β^+ decay is very rare. Here electron capture occurs mainly. If an unstable nuclide is surrounded on both sides by stable nuclides, one portion of the nuclei may go over to one isobar by β^- decay and the rest go over to the other isobar by β^+ decay or electron capture. This happens, for example, with the natural, unstable nuclide $^{40}_{19}\text{K}$ (see its location in fig. 9.1), which shows both β^- decay according to $^{40}_{19}\text{K} \rightarrow ^{40}_{20}\text{Ca} + \beta^-$ and electron capture according to $^{40}_{19}\text{K} + ^{0}_{-1}\text{e} \rightarrow ^{40}_{18}\text{Ar}$.

With sufficiently high mass numbers none of the isobaric reactions given lead to a stable nuclide (cf. fig. 9.1), so that here the mass number must therefore be reduced by α decay. An example is $^{210}_{84}\text{Po} \rightarrow ^{206}_{82}\text{Pb} + ^{4}_{2}\alpha$. Since no stable nuclide has higher mass number than 209, nearly all of the many unstable nuclides with $A > 209$ must enter into series with one or more α decays. These may alternate with β^- decay or electron capture, as neutron or proton surpluses must be corrected.

The great stability of the α particle causes some very light, unstable nuclides whose nuclear compositions nearly correspond to one, two or three α particles to decompose with the formation of α particles. As examples we may mention $^8_3\text{Li} \rightarrow 2{^4_2\alpha} + \beta^-$ and $^8_4\text{Be} \rightarrow 2{^4_2\alpha}$. Otherwise α decay occurs very rarely in nuclides with $Z < 83$ (Bi).

Often a spontaneous nuclear reaction is given with only the symbols of the mother and daughter nuclides connected with an arrow. Above the arrow the reaction type is written and under it the half-life may be given. Some of the reactions mentioned earlier may thus be written:

$$^{29}_{13}\text{Al} \xrightarrow[6.6 \text{ min}]{\beta^-} {^{29}_{14}\text{Si}}, \quad ^7_4\text{Be} \xrightarrow[54 \text{ d}]{\text{EC}} {^7_3\text{Li}}, \quad ^{210}_{84}\text{Po} \xrightarrow[138 \text{ d}]{\alpha} {^{206}_{82}\text{Pb}}.$$

With α decay the mass number can only be decreased in steps of four units. Thus, starting from a particular nuclide the mass numbers of all nuclides that follow each other in a radioactive series must differ from the mass number of the initial nuclide by multiples of four. Since all whole numbers can be expressed by one of the forms $4n$, $(4n+1)$, $(4n+2)$ and $(4n+3)$ where n is an integer, the mass numbers of all nuclides in the same series must thus be expressible by one of these forms. Since one can begin with different isobars, many series with the same form are of course possible. These series are often connected by β^- decay and electron capture, however.

In fig. 9.3, which is an enlargement of the upper part of fig. 9.1, two series with mass numbers $A = (4n+2)$ are shown. The longer series begins with ^{254}Fm and the shorter with ^{226}Pa; the series are united at ^{210}Bi and end in the stable nuclide ^{206}Pb. Since ^{206}Pb is the stable nuclide with the highest mass number of the form $A = (4n+2)$, it can be seen that it should be the final stage for all series of this type that begin with higher mass numbers.

One α decay and two β^- decays following each other (the sequence is arbitrary) brings the nuclide back to the same Z value while A is decreased by four units. In this way a series may include two or more isotopes of the same element. The long series in fig. 9.3, for example, contains three isotopes of Pb ($A = 214, 210, 206$).

The nuclides from ^{238}U to ^{206}Pb in the longer series are found in nature and have been known for the most part since the first decades of this century. They have been said to constitute a *natural radioactive series*, called the *uranium series*. The earlier members in the series are only known as artificially prepared nuclides.

Sometimes some nuclei of a nuclide show α decay and some β^- decay or electron capture. The series branches at such a nuclide (for example, at ^{218}Po in the uranium series) but a reunion generally occurs soon after. Often, but not always, one reaction type strongly dominates so that one can speak of a main branch and a side branch.

If one starts with the first nuclide of a series in pure form, the preparation in a very short time will contain more or less atoms of all the nuclides in the series.

Their relative amounts depend on the magnitudes of the respective half-lives and these may be very different.

Let us assume first the simple case in which the first nuclide A_1 of a series decays to A_2 and this in turn to the stable A_s. If A_1 decays much more slowly than A_2 (the half-life of A_1 much greater than for A_2) a stationary state is gradually reached in which the amount of A_2 is constant (equal amounts of A_2 are formed and decay per unit of time), if we consider times that are short in relation to the half-life of A_1. This stationary state is often but improperly called *radioactive equilibrium*. At radioactive equilibrium the numbers of atoms of the two nuclides are in the same relation as their half-lives. The number of remaining atoms of A_1 is thus much greater than the number of atoms of A_2. The number of atoms of A_s, of course, steadily increases.

If instead A_1 decays much faster than A_2, after a sufficiently long time, little remains of A_1 while there is still much of A_2.

In a longer series also, in which one nuclide has a much longer half-life than all the others, radioactive equilibrium is attained for all unstable nuclides that follow this nuclide. The number of atoms of each one of these nuclides at equilibrium is proportional to the half-life. The nuclides preceding the long-lived nuclide, on the other hand, are used up while there is still a large amount of the latter.

The main reaction sequence (the *main series*; side branches thus excluded) in the long series in fig. 9.3 is the following:

$$^{254}\text{Fm} \xrightarrow[3.2\,\text{h}]{\alpha} {}^{250}\text{Cf} \xrightarrow[13\,\text{y}]{\alpha} {}^{246}\text{Cm} \xrightarrow[5500\,\text{y}]{\alpha} {}^{242}\text{Pu} \xrightarrow[3.8 \times 10^5\,\text{y}]{\alpha} {}^{238}\text{U} \xrightarrow[4.5 \times 10^9\,\text{y}]{\alpha}$$

$$^{234}\text{Th} \xrightarrow[24\,\text{d}]{\beta^-} {}^{234}\text{Pa} \xrightarrow[6.7\,\text{h}]{\beta^-} {}^{234}\text{U} \xrightarrow[2.5 \times 10^5\,\text{y}]{\alpha} {}^{230}\text{Th} \xrightarrow[8.0 \times 10^4\,\text{y}]{\alpha} {}^{226}\text{Ra} \xrightarrow[1622\,\text{y}]{\alpha}$$

$$^{222}\text{Rn} \xrightarrow[3.8\,\text{d}]{\alpha} {}^{218}\text{Po} \xrightarrow[3.1\,\text{min}]{\alpha} {}^{214}\text{Pb} \xrightarrow[27\,\text{min}]{\beta^-} {}^{214}\text{Bi} \xrightarrow[20\,\text{min}]{\beta^-} {}^{214}\text{Po} \xrightarrow[1.6 \times 10^{-4}\,\text{s}]{\alpha}$$

$$^{210}\text{Pb} \xrightarrow[22\,\text{y}]{\beta^-} {}^{210}\text{Bi} \xrightarrow[5.0\,\text{d}]{\beta^-} {}^{210}\text{Po} \xrightarrow[138\,\text{d}]{\alpha} {}^{206}\text{Pb, stable.}$$

The most long-lived and therefore the most abundant nuclide in this series is ^{238}U, whose half-life is approximately 10^4 times greater than any other. Since the age of the earth is approximately 5×10^9 years (9-3d) all the nuclides before ^{238}U in the series have decomposed long ago, if they existed at all at the beginning (after 10^7 years already only one millionth of the original amount of these nuclides would be left as ^{242}Pu). But for ^{238}U hardly more than one half-life has gone by. The nuclides following after ^{238}U occur in nature in radioactive equilibrium.

In addition to the uranium series, two other natural radioactive series have long been known, the *thorium series* ($A = 4n$, ending with ^{208}Pb; the longest-lived preceding nuclide is ^{232}Th with half-life 1.4×10^{10} years) and the *actinium series* ($A = 4n + 3$, ending with ^{207}Pb; the longest-lived preceding nuclide is ^{235}U with half-life 7.1×10^8 years). Both these series are quite like the uranium series with some inserted β^- decays and some branching. A similar representative of a $(4n+1)$ series is the *neptunium series* (ending with ^{209}Bi; the longest-lived preceding nuclide is ^{237}Np with the half-life 2.2×10^6 years). The shortness of the half-life of the longest-lived nuclide in the neptunium series in relation to the age of the earth means that, even if it existed when the earth was formed, no detectable amounts remain. The extremely small amounts of some of the nuclides of the series (especially ^{237}Np) that can, in fact, be detected in nature are the result of a constant creation of these nuclides by the action of neutrons released in the spontaneous fission of ^{238}U (9-2c).

The above discussion shows that only the four stable nuclides having the highest mass numbers, ^{206}Pb, ^{207}Pb, ^{208}Pb and ^{209}Bi become the final stages in series with the four possible mass number forms.

Within the range between the heaviest known nuclides and the heaviest stable nuclides (fig. 9.3) the element $_{86}$Rn, radon, is found, which is gaseous under ordinary conditions. It may therefore be included as a gaseous member in several series. In nature radon is formed by α decay of radium isotopes and was formerly called "emanation". All radon isotopes are unstable and give polonium by α decay. These steps can be expressed generally by: $^{x}_{88}$Ra $\xrightarrow{\alpha}$ $^{(x-4)}_{86}$Rn $\xrightarrow{\alpha}$ $^{(x-8)}_{84}$Po. In the uranium series (cf. above) $x = 226$, in the thorium series $x = 224$ and in the actinium series $x = 223$ (Rn is not included in the neptunium series). As soon as a radioactive substance forms radon, it spreads into the surroundings. The polonium formed from it decays in turn. In this way the whole environment can become radioactive ("radioactive infection").

All unstable nuclides with mass number over 200 that are found in natural material on the earth belong to one of the four series described above. Only a few natural, unstable nuclides with mass numbers less than 200 are known. Two of these, ^{3}H and ^{14}C are relatively short-lived (β^- decays with half-lives of 12.26 years and 5730 years, respectively) and exist only because they are constantly created in the atmosphere by the action of cosmic rays (9-3c). The remainder have half-lives of at least 10^9 years and probably already existed in the original material from which the earth was formed. They are (decay type in parentheses): ^{40}K (β^-, EC, see p. 268), ^{50}V (β^-, EC), ^{87}Rb (β^-), ^{115}In (β^-), ^{123}Te (EC), ^{138}La (EC, β^-), ^{144}Nd (α), ^{147}Sm (α), ^{152}Gd (α), ^{174}Hf (α), ^{176}Lu (β^-), ^{180}Ta (β^-, EC), ^{187}Re (β^-), ^{190}Pt (α).

Unstable nuclides are characterized by the kind of radiation produced, its energy distribution (spectrum) and the half-life. For the determination of these data an unweighable amount of the nuclide suffices. If the number of particles (generally β^- or α particles) that a given quantity of a nuclide gives off per unit of time is known, the quantity of the nuclide in a sample can easily be determined by particle counting (most often with a Geiger counter or scintillation counter), even though it is far below the limit of weighability. This possibility for rapid quantitative analysis of extremely small amounts of substances is of very great practical significance (9-3e, 9-5c).

b. Induced simple nuclear reactions. In the year 1919 Rutherford showed that if nuclei of the most common isotope of ^{14}N are bombarded with α particles (obtained by decay of ^{214}Po) the oxygen isotope ^{17}O and protons are formed. Here the *induced* (to be distinguished from spontaneous) *nuclear reaction*

$$^{14}_{7}N + ^{4}_{2}\alpha \rightarrow ^{1}_{1}p + ^{17}_{8}O$$

takes place.

Since 1919 several thousands of similar nuclear reactions have been studied and the synthetic, unstable nuclides have in general been prepared by such reactions. The first time this was done was in 1932 when F. Joliot and Irène Joliot-Curie obtained ^{30}P by the reaction

$$^{27}_{13}Al + ^{4}_{2}\alpha \rightarrow ^{1}_{0}n + ^{30}_{15}P$$

Often a shorthand method of writing these reactions is used. The light bombarding particle ("projectile") and the light particle given off are written in this order in parentheses between the symbols of the initial nucleus ("target nucleus") and the final nucleus. The two reactions given above are thus written $^{14}N(\alpha,p)^{17}O$ and $^{27}Al(\alpha,n)^{30}P$ and are said to be (α,p) and (α,n) reactions, respectively.

It is believed that the projectile and the target nucleus generally are first combined into a *compound nucleus*, which, because of the kinetic energy of the projectile and the binding energy, becomes strongly excited and often very rapidly (after approximately 10^{-12} to 10^{-14} s) decomposes. The compound nucleus may also be preserved and its excess energy given off as γ radiation; for example, the reaction $^{27}Al(n,\gamma)^{28}Al$. It is customary then to say that the projectile has been *captured*.

The most common projectiles are positively charged particles (in the form of atomic nuclei) and neutrons. More rarely photons are used, most often in the form of γ radiation. The atomic nuclei are commonly protons, deuterons (nuclei of the hydrogen isotope deuterium, $^{2}_{1}H$, or $^{2}_{1}D$, usually indicated "d") or α particles, but also nuclei of heavier elements such as $^{12}_{6}C$, $^{14}_{7}N$ and $^{16}_{8}O$ have been used.

When a positively charged projectile comes very close to the target nucleus a strong electrostatic repulsion arises, which can be said to constitute a high energy barrier. If the projectile has high enough kinetic energy, however, it can pass through this barrier, that is, penetrate so deeply into the target nucleus that the short-range forces between the nucleons of the nucleus and the projectile begin to operate. Only then can the nuclear reaction set in. The electrostatic repulsion between a target nucleus with the charge Z_1 and a projectile with the charge Z_2 is among other things proportional to the product $Z_1 Z_2$. A projectile therefore more easily penetrates a target nucleus with low nuclear charge than one with high charge. Neutrons have a unique position because they have no charge ($Z_2 = 0$). They are not prevented by electrostatic repulsion from penetrating the nucleus and in fact do so even at very low velocity. Thus no energy barrier exists for neutrons.

In order to obtain sufficiently large amounts of a substance converted in a reasonable time the projectiles must be used in streams or fluxes of as great intensity (number of particles per unit area and unit of time) as possible.

As in the first two of the reactions mentioned above, α particles from natural, radioactive nuclides were used in the beginning as projectiles. By the reaction $^9\text{Be}(\alpha,n)^{12}\text{C}$ neutrons could also be obtained for bombardment. The kinetic energy of the α particles was low, however, and in all cases the intensity of the flux was very slight. After charged particles (ions) were successfully accelerated to very high velocities in electric fields in special *accelerators* (cyclotrons, linear accelerators, etc.), projectiles with very high energies and in fluxes significantly stronger than from radioactive decay could be obtained. In this way protons, deuterons, α particles and heavier ions are accelerated. In an "ion source" in the accelerator ions are formed, for example by strong heating of gas containing the desired nucleus: hydrogen for protons, deuterium for deuterons, helium for α particles, oxygen for oxygen nuclei, etc. The largest accelerator constructed up to 1968 can give proton energies of 7×10^{10} eV $= 7 \times 10^4$ MeV $= 70$ GeV.

Neutrons still can only be obtained by nuclear reactions and because of the absence of charge cannot be accelerated. Neutrons are customarily divided according to their kinetic energy into *slow* (energy <1 eV), *intermediate* (energy 1 eV $- 0.1$ MeV) and *fast* (energy >0.1 MeV). Neutrons ejected from a nucleus are generally fast. Nuclear reactions in nuclear reactors (9-4a) give very strong neutron fluxes, and reactors are incomparably the most important neutron sources. In these, however, the kinetic energy of the neutrons is often reduced by allowing them to collide elastically with certain kinds of atomic nuclei. The neutrons then finally come to temperature equilibrium (2-1 b) with these atoms. Their mean energy becomes $3kT/2 = 0.00013\ T$ eV, or 0.04 eV at room temperature, for example. These slow neutrons are also called *thermal neutrons*. They can be thought of as a neutron gas which permeates and fills the system.

In order to study the effect of particles with energies higher than can be obtained artificially, *cosmic radiation* is used. Cosmic radiation consists primarily to a great part of protons, but also of other nuclei. The energy of these particles is generally 10^4–10^6 MeV, and in isolated cases up to 10^{12} MeV. In the atmosphere they give rise to a multitude of new particles of very high energy.

Only nuclear reactions in reactors have any significance for the preparation of larger quantities of nuclear reaction products. Products synthesized with the help of accelerators are often obtained in such small amounts that they can only be studied if they are unstable. In such cases, however, their radiation can be used for both qualitative and quantitative analysis, and often the chemical character can be determined by *carrier techniques* (9-5a) and *ion exchange chromatography* (15-4). Of the two or three heaviest nuclides synthesized to date only a few atoms, perhaps 10–100, have been obtained. In reactions caused by cosmic radiation only the properties of individual atoms can be studied.

With bombardment by the same kind of projectile the result is often strongly dependent on what nuclei are bombarded. For the same nuclei the result also varies significantly with the energy of the projectiles. If the projectile can penetrate the nucleus with little energy (particularly neutrons and protons, where for the

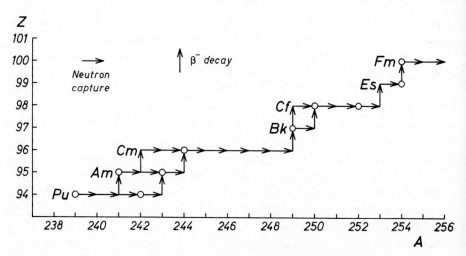

Fig. 9.4. Nuclear reactions for the synthesis of transuranium elements by neutron irradiation of ^{239}Pu. For each element those nuclides obtained in the largest amounts are indicated by circles.

latter the nuclear charge should not be too high) the excess energy of the compound nucleus often is so low that no particles can be ejected. Capture reactions are then obtained of the types (n,γ) and (p,γ). (n,γ) reactions are very important for the preparation of neutron-rich (subject to β^- decay) nuclides. An isotope is for this purpose exposed to irradiation by slow neutrons in an atomic reactor. Intensive neutron irradiation of ^{239}Pu in this way constitutes the most important method for synthesizing a series of *transuranium elements* (28-4). The nuclear reactions in this case are shown in fig. 9.4. Elements with atomic number higher than 100Fm have such short half-lives, however, that it is not possible to prepare them in this way in detectable amounts.

For different kinds of target nuclei the fractions of bombarding neutrons that give rise to (n,γ) reactions, and thus to a large degree also neutron absorption, are very individual. For example, there is no relation to the atomic number or mass number of the target nucleus. Also, different isotopes of the same element can vary considerably. When hydrogen nuclei are encountered by slow neutrons of the same energy, reaction with the nucleus ^{1}H takes place 575 times more easily than with the nucleus ^{2}H(=deuterium ^{2}D). Examples are known of even greater differences between isotopes of the same element.

If a certain target nucleus is bombarded with neutrons of varying energy, especially strong neutron capture in the range of intermediate neutrons is found when the energy of the neutrons is equal to the difference between particular energy levels in the nucleus. With such neutron energies very high neutron absorption (*resonance absorption*) occurs, which can be compared with the appearance of absorption lines in an optical spectrum.

If the energy of the projectiles is increased, particles may be ejected. With protons, for example, the reaction (p,n) first becomes dominant. At still higher proton energies the compound nucleus acquires such high energy that more than one particle may be ejected. Reactions such as (p,2n) or (p,pn) may then become predominant.

Of the reactions caused by deuterons, (d,p) reactions often go with notable ease. It has been assumed that the deuteron on approaching the nucleus is divided into a proton and a neutron. The neutron goes easily into the nucleus and the proton is repelled without ever having been incorporated into a compound nucleus.

α particles of low energy (for example, from radioactive decay without subsequent acceleration) can only penetrate light nuclei (cf. the earliest induced nuclear reactions mentioned earlier). In order to enter into heavier nuclei they must be accelerated; for example, for the reaction ^{238}U$(\alpha,n)^{241}$Pu. At high projectile energies a number of particles may be ejected; for example, ^{209}Bi$(\alpha,3n)^{210}$At. Reactions of these and similar types are important for the synthesis of transuranium elements in small amounts. They are also the only way by which the heaviest elements can be synthesized. Here, heavier nuclei than α particles have been used as projectiles. With carbon and boron ions the processes $^{246}_{96}$Cm $+ ^{12}_{6}$C $\to ^{254}_{102}$No $+ 4^1_0$n and $^{250}_{98}$Cf $+ ^{11}_{5}$B $\to ^{259}_{103}$Lr $+ 2^1_0$n have been produced.

With many of the reactions mentioned above, for example the pure capture reactions and reactions of types such as (d,n), (d,p) and (α,n), a heavier nuclide is built up of lighter constituents. With building-up reactions of these and other kinds among the lighter elements, especially large amounts of energy per participating nucleon are released (cf. fig. 9.2). Such reactions are therefore of great interest for the future production of energy, and it is mainly these that have been called *fusion reactions* (Lat. fundere, to melt, mix together). Among the fusion reactions that have been extensively studied in the laboratory may be mentioned the two reaction possibilities with bombardment of deuterium by deuterons (the so-called D–D processes). Here the reactions ^{2}H(d,n)^{3}He and ^{2}H(d,p)^{3}H take place with about equal probability. The possibility of using fusion reactions for the production of energy is discussed in 9-4 b.

Some more complicated types of induced nuclear reactions are treated in 9-2 c.

c. Fission and spallation. Under certain conditions a very heavy atomic nucleus can be split into two (in very rare cases three) parts of medium weight, which form new atomic nuclei. The phenomenon is called *cleavage* or *fission* (Lat. findere, to split) and occurs on the bombardment of some very heavy nuclei with neutrons. This happens, for example, with the nucleus ^{235}U, which first takes up one neutron with the formation of the compound nucleus ^{236}U. This can afterwards either lower its energy to the ground state with γ radiation—the result is then a (n,γ) reaction—or undergo fission. It is customary to say that it is the initial nucleus, that is, ^{235}U, that cleaves. Different ^{235}U nuclei cleave in different ways, and a sample in which fission occurs can therefore produce a large number of different nuclides.

In this case, however, the two fission products are found to have mass numbers especially around 95 and 140, corresponding to an unsymmetrical cleavage. Some examples are:

$$^{235}_{92}U + ^1_0n \rightarrow ^{139}_{56}Ba + ^{94}_{36}Kr + 3\,^1_0n + \text{energy}$$

$$^{235}_{92}U + ^1_0n \rightarrow ^{145}_{54}Xe + ^{89}_{38}Sr + 2\,^1_0n + \text{energy}$$

The fission of ^{235}U was first demonstrated in 1939 by Hahn and Strassman. On fission neutrons (fast) are always released, which is easy to understand from fig. 9.1. The quotient N/Z is noticeably greater for very heavy than for medium-heavy nuclides and the latter therefore cannot hold the original number of neutrons. For each cleaved ^{235}U atom an average of 2.5 neutrons are released. The primary products even so are most often neutron-rich and stability is generally reached only after one or more β^- decays following fission (decay chains). Among the primary products of the fission of ^{235}U over 80 nuclides have been found. Each of these undergoes an average of three β^- decays. Therefore, altogether about 300 nuclides appear as fission products, representing over 30 elements. (This is the reason for the remarkably large number of known, neutron-rich nuclides with mass numbers about 95 and 140; see fig. 9.1.)

Because the energy per nucleon is considerably less for the medium-weight than for the very heavy nuclides (cf. 9-1 b and fig. 9.2) large amounts of energy are released on fission. The energy values in fig. 9.2 give for the number of nucleons appropriate to a typical fission of ^{235}U an amount of released energy of $(140 \times 8.4 + 95 \times 8.6 - 235 \times 7.6) = 210$ MeV, that is, at least 20 times more than the largest amount of energy developed in simple nuclear reactions. Based on one mole of ^{235}U ($=235$ g) this corresponds to about 2×10^{13} J $= 230$ MWd (megawatt-days). The production of 1 MW thus requires the fission of approximately 1 g per day, a number that may be useful to remember. The fission energy consists for the most part of the kinetic energy of the fission products, and to a lesser extent of the energy of the γ radiation that is also emitted.

Fission energy is the only form of nuclear energy that can be practically produced up to now on a large scale and in a controllable way. An account of this follows in 9-4a.

Fission of this kind with slow neutrons has been demonstrated for altogether some twenty very heavy nuclides. However, only the fissions of ^{233}U, ^{235}U and ^{239}Pu have gained any practical significance.

Fast neutrons, like other projectiles of high energy, can cause fission of all the heavier nuclides. The probability that a collision with a fast neutron will lead to fission is somewhat greater for those nuclides that also are cleaved by slow neutrons. Therefore, also here ^{233}U, ^{235}U and ^{239}Pu play the greatest role and are often given as the truly fissionable nuclides.

In fission with fast neutrons, the lighter the nucleus the greater the projectile energy that is required. Fission has been observed down to 73 Ta, but for this α par-

The Atomic Nucleus

ticles with the energy of 400 MeV are required. With high projectile energies the distribution of the mass numbers of the products is different compared to fission with slow neutrons.

In many heavy nuclides there is also a tendency for *spontaneous fission*. This tendency depends on the nuclear structure and is therefore very individual, but becomes more evident the higher the mass number is. This is, of course, an expression of the fact that with increasing mass number, the energy of the heavy nuclides becomes increasingly greater than the energy of the fission products (cf. fig. 9.2). Sponteneous fission, however, requires a certain energy distribution in the nucleus, which with lower mass numbers occurs with very low probability. The spontaneous fission of ^{238}U thus corresponds to a half-life of about 10^{16} years (cf. the half-life of 4.5×10^9 years for the α decay of ^{238}U). As the mass number increases, this requirement is more easily fulfilled; the half-life for the spontaneous fission of ^{254}Cf is 56 days and for ^{256}Fm, 3 hours. Here spontaneous fission can take place faster than other spontaneous decays and this possibility may be the factor that limits the synthesis of new elements.

The neutrons formed on spontaneous fission of natural nuclides cause a constant, very slow creation of certain nuclides. Neutrons formed by spontaneous fission of ^{238}U thus initiate with other atoms of ^{238}U the reactions

$$^{238}_{92}\text{U (n,2n)} \; ^{237}_{92}\text{U} \xrightarrow[6.8\text{ d}]{\beta^-} \; ^{237}_{93}\text{Np} \quad \text{and} \quad ^{238}_{92}\text{U (n,}\gamma\text{)} \; ^{239}_{92}\text{U} \xrightarrow[23.5\text{ min}]{\beta^-} \; ^{239}_{93}\text{Np} \xrightarrow[2.3\text{ d}]{\beta^-} \; ^{239}_{94}\text{Pu}.$$

The last members have long half-lives and therefore can be detected in very small proportions in uranium minerals.

With projectile energies over about 100 MeV, so-called *spallation* also occurs. Here the target nuclei give off light nuclear constituents such as nucleons, deuterons and α particles in varying numbers. After spallation there therefore remains in the sample a multitude of different nuclei, representing nearly all possible NZ combinations within a large region below the target nucleus. A decrease of 20 units in the N value and of 15 units in the Z value of the target nucleus is not uncommon. As the extent of splitting-off is increased, however, it becomes harder to bring about, and therefore more and more rare.

9-3. Isotopes

If we exclude purely nuclear properties, which are expressed most strikingly as varying nuclear stability and varying behavior in nuclear reactions, in general the different isotopes of an element are extraordinarily similar, both from a physical and a chemical standpoint. Only the relative difference in nuclear mass can here have an effect. Since this is generally greater the lighter the atom is, the difference in properties of the isotopes increases as we go toward lighter elements.

The chemical differences between the isotopes are so small that they have no significance for ordinary chemical work. This, of course, is because the properties of the electron cloud are only extremely slightly affected by the nuclear mass. However they can be detected in many cases (*isotope effect*), especially in the light elements, partly because the relative nuclear mass differences are greater, and partly because the outermost portions of the electron cloud that are important for the chemical properties lie nearer the nucleus. For the light elements chemical differences can thus be used to separate isotope mixtures.

The similarity between the isotopes of an element, which causes difficulties in the analysis and separation of isotope mixtures, on the other hand, is the basis for the extremely important application of isotopes as *tracer substances*.

a. Isotope analysis. It a substance contains unstable, well-characterized isotopes, their amounts can often be determined by studying their radiation and its strength. This measurement can be carried out on quantities that are too small to detect by other methods.

The composition of an isotope mixture that only contains two isotopes can in principle be obtained by weighing (for example, of a certain volume, hence a density measurement). The accuracy is sufficient, however, only if the relative atomic weight difference is very large. The method has been used mainly for the analysis of mixtures of ^1H and ^2H, for which very accurate density measurements are carried out on water formed from the isotope mixture.

The *mass spectrograph* (1-3f) is used in the most generally applicable and most important method of isotope analysis.

Both the vibrational and rotational energies of a molecule are dependent on the masses of the constituent atoms (cf. the formulas (3.8) and (3.9)). These energy values therefore are somewhat different for molecules in which different isotopes are incorporated. The difference in vibrational energy can easily be observed in molecular spectra and this has therefore been used for isotope analysis. The isotopes ^{18}O, ^{15}N and ^{13}C were discovered spectroscopically.

b. Isotope separation. As will be shown later there is a need for pure isotopes both in research and technology. Methods for separating the isotopes of an element have thus been rapidly developed, and certain processes important for the production of nuclear energy are carried out on an industrial scale.

The most general method of separation in principle makes use of a powerful mass spectrograph in which the ion beam has a very large transport capacity and the isotopes are accumulated on the target surface. The method is customarily called *electromagnetic separation* and can deliver pure isotopes in kilogram amounts of a large number of different elements.

Several important methods for isotope separation make use of other differences in physical properties that are based directly on the mass difference. For example, use is made of the fact that the mean velocity of a molecule in a gas is inversely proportional to the square root of its molecular weight (cf. 2-1b). This means that the rate of diffusion and flow of a gas through fine canals (effusion) is inversely proportional to the square root of its molecular weight.

The enrichment of ^{235}U is accomplished in an industrially developed separation process. Natural uranium consists mainly of ^{238}U (99.274 atom %) and ^{235}U (0.720 atom %), which are thus to be separated. Here, in fact, effusion is used, although the process is customarily denoted as *gaseous diffusion*. It is carried out with uranium hexafluoride, UF_6, which is gaseous above the sublimation tem-

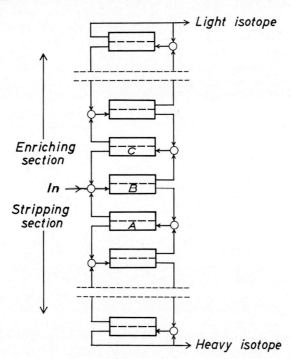

Fig. 9.5. Cascade for isotope separation by gaseous diffusion. Each rectangle indicates a diffusion cell and its barrier is shown by a dashed line. Each circle indicates a pump.

perature 56°C. The hexafluoride has the advantage that fluorine only exists as ^{19}F, so that the uranium isotopes are the only ones which are separated. Further, the atomic weight of fluorine is so low that it does not seriously decrease the relative molecular weight difference for the uranium isotopes. These advantages are counteracted by the great disadvantage of the high reactivity of the hexafluoride, which makes the choice of construction materials difficult.

In this case the relation between the diffusion rates for ^{235}UF$_6$ and ^{238}UF$_6$ is $\sqrt{352/349} = 1.0043$. The separation achieved by a single diffusion process therefore is extremely small and the process must be repeated many times to get any nearly complete separation. The best yield is obtained when the different diffusion stages are coupled in *cascade*, which also allows continuous operation.

The diffusion takes place at each stage in a diffusion cell, shown schematically (among others those marked A, B, C) in fig. 9.5. Each cell is divided into two parts by a porous wall ("barrier"). With the help of pumps the gas pressure is held higher below each barrier than above it (the designations "below", "above", etc. apply only to the figure).

Let us assume that the gas from the inlet is pumped into the cell B. Approximately half the amount of gas is allowed to pass through the barrier and the remainder leaves the high pressure section at the other end. The molecules containing the lighter isotope go through the barrier somewhat faster than those containing the heavier isotope. The former is therefore somewhat enriched in the gas that passes through the barrier.

The transmitted gas is pumped up to the cell C, where it undergoes the same procedure. Above the barrier in C the gas is thus somewhat more enriched in the lighter isotope. The enrichment continues in this way stage by stage. At each stage the processed gas volume becomes smaller and smaller, and the stages can therefore be constructed with gradually decreasing volume.

The gas that passes by the barrier in each cell is not discarded, however. It contains valuable material and is already enriched in earlier stages. It is therefore carried down to the next stage below (e.g. from C to B) and is there, together with the product from a still earlier stage, pumped in under the barrier, after which it takes part in a new diffusion process. This feedback takes place in all stages above the inlet and continues for as many stages below the inlet as seems necessary to recover a sufficient amount of the lighter isotope. Also in these stages an ascending gas stream must be arranged in order to carry the lighter isotope upward.

In the stages above the inlet a successive enrichment of the lighter isotope relative to its proportion in the inlet material takes place. In the stages below the inlet a successive depletion ("stripping") of the lighter isotope occurs, so that less and less of it remains as one moves away from the inlet.

The cascade is therefore said to consist of an *enriching section* and a *stripping section*. Throughout the whole cascade an ascending gas stream flows with ever-increasing proportion of the lighter isotope and a descending gas stream flows with an ever-increasing proportion of the heavier isotope. The final products are removed from the top and bottom of the cascade, and their compositions depend on the number of stages. If UF_6 prepared from natural uranium is fed in and it is desired that the products at the two ends of the cascade shall have 90% and 0.20% ^{235}U respectively, under ideal conditions 1660 enriching stages and 295 stripping stages are required. In this case clearly $(0.720-0.20)/0.720 = 72\%$ of the whole amount of ^{235}U is collected as a 90% product. The actual number of stages, however, must be considerably greater than the theoretical number.

In the large plants for separation by gaseous diffusion that have been built in the United States, the Soviet Union and England, the cascades contain several thousand stages with a total barrier area of several acres. Considering in addition the difficulties of building a cascade so that it is not attacked chemically by the hexafluoride, one can see the magnitude of the technical problem that has been solved.

In most other methods for isotope separation cascades with the same circuits for the transport of substances as in fig. 9.5 are used.

For the production of energy there also exists a great need for "heavy" water (9-4a). In natural occurrences of hydrogen (chiefly water), the element generally contains about 0.015 atomic percent ^2H (D) and the rest is ^1H (which can be written simply H when deuterium is expressly written D). This proportion corresponds approximately to 1 atom of D per 6700 atoms of H. In hydrogen gas there are the molecular species H_2, HD and D_2, and in water H_2O, HDO and D_2O ("heavy" water). Because of the very slight proportion of D in products from natural sources D is found much more often in the forms of HD and HDO than in the forms D_2 and D_2O, respectively.

For the preparation of pure D_2O on an industrial scale several methods have been used, often in combination with each other. The most important methods up to now have been *electrolysis* of water, *fractional distillation* of water and certain *exchange reactions*.

The electrolytic method is based on the fact that the overvoltage of hydrogen (16f), when it is formed by electrolysis of water, is somewhat higher with the splitting off of D_2 than of H_2. D_2O is therefore enriched in the remaining electrolyte.

To set up a material transport according to the cascade principle the $H_2 + D_2$ formed by the electrolysis is burned to $H_2O + D_2O$, which is returned to the preceding electrolytic cell. The method is extremely demanding of energy in the first enrichment stages where enormous volumes must be electrolyzed. If heavy water is to be prepared only in this way, it must therefore be a by-product of some other manufacturing process in which the electrolysis of large amounts of water takes part. In addition very inexpensive electrical energy is required. Such is the case at Norsk Hydro in Norway where hydrogen gas from the first large electrolytic stages is used in the manufacture of ammonia (Haber-Bosch process; see 22-6b) and thus is not burned.

The method of fractional distillation of water is based on the fact that the vapor pressure of H_2O at all temperatures is higher than for D_2O (the boiling point of D_2O is 101.42°C). It is carried out in principle as an ordinary fractional distillation (13-5c). The relative difference in vapor pressure increases as the temperature decreases and the most effective separation is therefore achieved with distillation at low temperature, that is, at reduced pressure. The enrichment by distillation is usually carried to the point where about 90 % of the total hydrogen content is D. The final enrichment up to 99.8 % D is accomplished electrolytically.

Exchange reactions give purely chemical separation methods. In the most modern plants for the production of heavy water the exchange of H and D between gaseous hydrogen sulfide and liquid water has been used. The reaction is applied at such a low degree of enrichment that D mainly occurs as HDS in hydrogen sulfide and as HDO in water. The hydrogen sulfide can therefore be considered as a mixture of H_2S and HDS and the water as a mixture of H_2O and HDO. The distribution of D between hydrogen sulfide and water is then determined by the exchange reaction

$$HDS(g) + H_2O(l) \rightleftharpoons H_2S(g) + HDO(l)$$

This equilibrium is shifted to the right (the equilibrium constant—see 10b—at 25°C is $k = 2.34$). When, for example, hydrogen sulfide and water with the same mole fraction of D come in contact with each other, an exchange HSD⇌HOH therefore takes place until the mole fraction of D is approximately 2.3 times as great in the water as in the hydrogen sulfide. D is thus enriched in the water. The process can be carried out (in cascade) at one temperature, but it is more economical to make use of the fact that the equilibrium is displaced less to the right at higher temperature (at 130°C $k = 1.86$). If the process is carried out alternately at low and high temperature D can therefore be transported back and forth between hydrogen sulfide and water so that every transfer to a new phase results in an enrichment of D. The enrichment by this method is generally carried to the point where about 15 % of the total hydrogen content consists of D. Further enrichment is accomplished by fractional distillation and finally by electrolysis.

It is possible that fractional distillation of liquid hydrogen may prove to be more economical for the initial enrichment of deuterium than distillation of water or the exchange reaction mentioned above.

The methods referred to above, with the exception of electrolysis, have also been applied to the separation on a smaller scale of other isotope mixtures. To these is also added *thermal diffusion*, which is based on the fact that when a temperature gradient is set up in a gas containing molecules of different masses, the heavier molecules move toward the colder and the lighter molecules toward the hotter region (the phenomenon was predicted theoretically in 1910 by Enskog and has been used and developed for isotope separation by Clusius; it is because of this phenomenon that dust in a room settles especially on places that are cold in relation to the surroundings, such as cold surfaces on outer walls, and walls and ceilings adjacent to hot pipes). The separation is carried out in a long vertical tube with cooled walls and an axial metal wire that is electrically heated. In the tube there arise partly thermal diffusion and partly convection currents which ascend next to the hot wire and descend next to the cold wall. The lighter molecules diffuse toward the wire and are carried upward and the heavier molecules diffuse toward the wall and are there carried downward. The former are thus accumulated at the top and the latter at the bottom of the tube.

Other procedures that have been tested on a small scale are *centrifugation*, and methods based on differences in the rates of *ionic migration*. Ions containing different isotopes of an element migrate more slowly toward an electrode, the greater the ionic mass is.

c. Natural isotope mixtures. The remarkably constant composition of terrestrial isotope mixtures shows that the isotopic composition must have been very constant before the formation of the present solid crust of the earth. In this crust, however, the possibilities for mixing the isotopes are very small. Therefore, local changes in the isotopic composition of an element in and on the crust in general cannot be evened out. Variations in the compositions of isotope mixtures that thus arise are common, although in most cases they are very small. The most common causes are the following.

Besides the amount of a stable nuclide that existed from the beginning, there may be a further amount that is "radiogenic", that is, formed by *radioactive decay of a mother nucleus*. Because of the addition of the radiogenic atoms, the amount of this nuclide in relation to the other isotopes becomes greater than normal in places where the mother nuclide occurs or has occurred. In the uranium, actinium and thorium series the nuclides ^{238}U, ^{235}U and ^{232}Th respectively have by far the longest half-lives (9-2a). The succeeding nuclides in the series are therefore found in such small amounts that they may be neglected in comparison with the stable terminal elements. The essential results of the decay in the three series are then well represented by the overall formulas:

$$\text{uranium series:} \quad ^{238}\text{U} \rightarrow {}^{206}\text{Pb} + 8\,{}^{4}\text{He} \tag{9.3}$$

$$\text{actinium series:} \quad ^{235}\text{U} \rightarrow {}^{207}\text{Pb} + 7\,{}^{4}\text{He} \tag{9.4}$$

$$\text{thorium series:} \quad ^{232}\text{Th} \rightarrow {}^{208}\text{Pb} + 6\,{}^{4}\text{He} \tag{9.5}$$

If a rock mass, for example, contains or has contained ^{238}U it will also thus contain radiogenic lead ^{206}Pb, formed by decay of ^{238}U. A rock mass having ^{232}Th will contain radiogenic lead in the form of ^{208}Pb. Lead from rock containing uranium or thorium may therefore show large variations in isotope composition. On the other hand, ^{204}Pb, which is not known as a terminal element in any radioactive series, probably is never radiogenic.

Certain nuclides are constantly formed in nature on the earth by *induced nuclear reactions*, especially as a result of neutron bombardment. Slow neutrons caused by cosmic radiation give ^{14}C from nitrogen in the upper atmosphere, according to the scheme ^{14}N(n,p)^{14}C. Fast neutrons from the same source give tritium according to the scheme ^{14}N(n,^{3}H)^{12}C. Both ^{14}C and ^{3}H are formed in such small amounts that they do not affect weighably the isotopic composition of carbon and hydrogen. However, they both decay with β^- radiation, which reveals their presence.

Finally, there occur on the earth's surface continual *isotope-separating processes* of the same type as several of the separation processes described in 9-3b. They may involve gaseous diffusion, thermal diffusion, distillation or exchange reactions. In nature these lead to very small variations in the isotopic composition, which, however, are often detectable by sensitive methods. They can take place both in non-living, natural systems and in organisms.

For example, it has been found that the D_2O content in rain and river water in general lies between 0.0133 and 0.0154 mole percent, while that in the oceans lies between 0.0153 and 0.0156 mole percent. D_2O is enriched in the oceans on vaporization because of its lower vapor pressure. The slight proportion of D, however, causes this variation to affect the atomic weight of hydrogen only in the fifth decimal place. In other cases the atomic weight may vary more noticeably. The atomic weight of sulfur, for example, varies within the range 32.064 ± 0.003.

Because of the rapidly increasing use of nuclear reactions and preparation of pure nuclides, more or less local variations in the isotopic composition of the elements may be expected to become noticeably more numerous.

d. Radioactive age determination. If an unstable nuclide at a certain moment of time is fixed within a volume that can be analyzed, in favorable cases it is possible, with the help of an expression similar to (8.2), to calculate how long ago the fixation occurred. This was pointed out in 1905 by Rutherford and the method has been used particularly for geologic and archeologic age determination.

A mother and a daughter nuclide generally have such different chemical properties that they soon become completely separated by the chemical processes that often take place in the course of geologic time, both on the surface and in the interior of the earth. The moment of separation then becomes the starting point of a decay process, which starts with pure mother nuclide. However, the half-life of the process must be proper for a reasonably certain time determination. On the one hand there must be a measurable amount of the mother nuclide remaining, and on the other hand there must be a measurable fraction of it that has decayed.

For example, the age of minerals that contain ^{238}U have been determined. Here, only the overall reaction (9.3) need be considered and in comparison with the half-life of 4.5×10^9 years for ^{238}U the much shorter half-lives of all the following members of the uranium series may be neglected (9-2a). By chemical determination of the proportions of uranium and lead and mass-spectrographic determination of the isotopic compositions of these elements, the concentrations [^{238}U] and [^{206}Pb] are obtained. Then in the expression (8.2) $[A]_0 = [^{238}U] + [^{206}Pb]$ and $[A] = [^{238}U]$. The age of the mineral t is thus obtained from the expression

$$kt = \ln \frac{[^{238}U] + [^{206}Pb]}{[^{238}U]}$$

There are several sources of error, however. The system has not always been completely isolated after the mineral was formed. Thus, sometimes uranium or lead may have been leached out by permeating fluids. The sample may also already have contained ^{206}Pb at the beginning. An attempt has been made to correct for this by making the assumption, although uncertain, that the nonradiogenic lead contains ^{206}Pb and ^{204}Pb in the proportion most often found in "ordinary" lead. From a measurement of the proportion of ^{204}Pb in the sample one should then be able to calculate the amount of radiogenic ^{206}Pb. The proportion of ^{4}He has also been used instead of the proportion of ^{206}Pb (cf. equation 9.3). This method is dangerous, however, because the helium may have diffused away from the sample.

The decays (9.4) and (9.5) can be used in an analogous way. The half-lives of ^{238}U, ^{235}U and ^{232}Th (9-2a) have values that make them usable for the determination of geologic ages greater than some millions of years.

Other natural nuclear reactions that are considered to by very useful for geologic age determination are:

$$^{40}K \xrightarrow[1.42 \times 10^9 \text{ y}]{EC} {}^{40}Ar \quad \text{and} \quad {}^{87}Rb \xrightarrow[5.2 \times 10^{10} \text{ y}]{\beta^-} {}^{87}Sr.$$

Since argon has a larger atomic radius than helium, the risk of losing argon by diffusion is less than the risk just mentioned of the loss of helium. If the rock is heated in connection with metamorphosis, all the argon formed is certainly given off, however, and the argon method therefore gives the "metamorphic age". ^{40}K decays in addition according to the scheme $^{40}K \xrightarrow{\beta^-} {}^{40}Ca$ (see 9-2a), but this can be corrected for since it is known that 11% of the ^{40}K nuclei decay according to the first-named reaction (the latter decay cannot be used for age determination because ^{40}Ca is so common that the newly-formed portion of it cannot be ascertained). The strontium method at the present time probably gives the best values.

The various methods for the most part give consistent age values. The oldest mineral studied in this way is about 3×10^9 years old. The age of the earth as a planet may be nearly the same as the age of the meteorites. By the same methods it is found that these probably are derived from one or several mother bodies in which a period of a general molten state ended approximately 4.5×10^9 years ago.

For the age determination of archeological samples and late Quaternary deposits the so-called C-14 *method*, proposed by Libby, is more and more used. It is based on the assumption that the proportion of ^{14}C in the carbon dioxide of the air remains constant throughout the time period under study. By the action of cosmic radiation on nitrogen in the upper atmosphere ^{14}C is constantly created (9-3c), which is very rapidly oxidized to $^{14}CO_2$ and mixed with the rest of the atmosphere. The nuclei of ^{14}C decay meanwhile by the reaction $^{14}C \xrightarrow[5730 \text{ y}]{\beta^-} {}^{14}N$ and it is assumed that a stationary state has been attained between creation and decay. The fraction of ^{14}C in the atmospheric carbon is, however, only 10^{-12}. In a living plant, which constantly assimilates carbon dioxide, the carbon also retains the proportion of ^{14}C of the atmospheric carbon. When the plant dies it no longer takes up carbon dioxide, and the ^{14}C content of the fixed carbon thereafter steadily decreases by nuclear decay. By measurement of the strength of the β^- radiation from a sample with a known total amount of carbon its content of ^{14}C can be calculated and thence the time that has elapsed since the plant died. Animal remains, whose carbon content ultimately came from living plants, can be dated in the same way. That the bases of the method are valid at least for ages up to 5000 years has been verified by determinations on archeologically datable finds. For objects older than approximately 40 000 years, however, the content of ^{14}C has decreased so much that the method is unusable. Fossil carbon in coal and oil no longer contains any ^{14}C.

Studies of wood from the last few hundred years up to the beginning of nuclear explosions on a large scale show that the proportion of ^{14}C in atmospheric carbon has decreased by 1–2 per cent in that time. This probably results from the use of fossil fuel, which has added large quantities CO_2 free of ^{14}C to the air. On the other hand, nuclear weapon explosions in recent times have caused the proportion of ^{14}C to again increase.

Tritium, ^3H, which according to 9-3c is constantly being created in the upper atmosphere, can also be used for age determination in an analogous way. The short half-life of tritium, 12.26 years, however, permits the measurement only of small age values. Mainly, attempts have been made to determine the age of water in precipitation samples. The problem is complicated by the fact that one can seldom assume that a stationary state has been attained during such short times. Nuclear weapons tests also have a very disturbing influence.

e. Tracer methods. In the true tracer methods the fact is utilized that the different isotopes of an element are practically speaking equal with respect to most chemical and physical properties but nevertheless can be identified and determined.

Let us assume that an element consists of one or more stable isotopes. If an unstable isotope of the element is added, the mixture will behave in nearly every way like the original element. In practically every process in which it takes part, therefore, the unstable isotope will always make up the same fraction of the element. (If the half-life of the unstable element is short, however, a correction must be made for the steady decrease in amount caused by the radioactive decay). From its radiation the unstable isotope can be detected and measured with far greater sensitivity than is possible with ordinary analytical methods. A portion of the atoms in a quantity of the element has been "labelled", so to speak, so that it can be traced wherever it goes in a system. Besides *labelled atoms* one also speaks of *trace atoms, trace elements*[1] and *tracers*. When the method was first proposed by Paneth and de Hevesy (1913), they referred to the trace element as a "radioactive indicator".

---

[1] The term trace element has also been used to denote an element that is found in very small amounts (as traces) in biological systems. In such cases microelement is a better expression.

The labelled atoms can be used in the same way, of course, if they occur bound in molecules or crystals. The labelling can also be done with a stable isotope so that the amount of element that is to be traced has an abnormal isotopic composition. However, it cannot then be traced with the same sensitivity. Further, mass spectrographic analysis, which most often becomes necessary, is much more difficult than radiation measurements. If the element does not have any unstable isotope with a proper half-life (this must not be so long that the radioactivity becomes too weak, nor so short that the isotope decays too rapidly) one may be forced to use stable isotopes. This is the case, for example, with the biologically so important elements oxygen and nitrogen, where ^{18}O and ^{15}N must be used. Other frequently used stable trace elements are ^{2}H (D) and ^{13}C, but the unstable isotopes ^{3}H (T) and ^{14}C have also been used. In using hydrogen isotopes it must be remembered that the chemical properties of ^{2}H, and still more of ^{3}H, may noticeably deviate from the properties of ^{1}H.

The great sensitivity of the detection of radioactive trace elements can be utilized for the *measurement of very small concentrations*, for example in the determination of low solubilities and gas pressures. The total radiation from a known amount of substance containing a radioactive isotope is then compared with the total radiation from a known volume of solution or gas standing in equilibrium with the substance. Coprecipitation, which is of great importance in analytical work (15-1 b), is also studied with radioactive trace elements. The trace element is mixed with an element that is subject to coprecipitation, which can then be accurately followed.

Radioactive trace elements have also been used to study *self-diffusion*, that is, the diffusion of one kind of atom or molecule in a phase consisting only of these atoms or molecules. For example, the radioactive isotope ^{212}Pb (a member of the thorium series) has been mixed with ordinary, inactive lead and an area of this mixture pressed against an area of inactive lead. The migration of the labelled lead atoms into the inactive lead can then easily be followed. In this way it has been found that at 200°C lead atoms migrate ten times more slowly in lead than tin atoms in lead.

Exchange reactions between atoms of the same element can only be studied with the help of trace elements. For example, if a bromide containing the radioactive isotope ^{82}Br is dissolved in water and inactive Br_2 is added to the solution, it is found that radioactive Br_2 molecules are formed extremely rapidly. The exchange may occur through a polybromide ion (20-5e) according to the reaction (the asterisk indicates a labelled atom):

$$Br^{*-} + Br_2 \rightleftharpoons Br_2Br^{*-} \rightleftharpoons Br^- + BrBr^*$$

but probably also via other reactions in which water takes part.

The isotope ^{35}S has been used for studies of exchange of sulfur atoms. If sulfur containing this isotope is boiled with inactive sulfite ion SO_3^{2-}, active thiosulfate ion $S_2O_3^{2-}$ is formed according to:

$$\left[\begin{array}{c} \text{O} \\ | \\ \text{O}-\text{S} \\ | \\ \text{O} \end{array}\right]^{2-} + \text{S}^* \rightleftharpoons \left[\begin{array}{c} \text{O} \\ | \\ \text{O}-\text{S}-\text{S}^* \\ | \\ \text{O} \end{array}\right]^{2-}$$

If the thiosulfate solution is acidified the equilibrium is shifted back to the left. Of the products then formed, the sulfite ion is inactive and the sulfur active. This shows that the two sulfur atoms in the thiosulfate ion are nonequivalent, and that no exchange takes place between them. Neither has it been possible to detect any exchange at 100°C between the sulfur atoms in SO_3^{2-} and SO_4^{2-} in the same solution.

The oxygen isotope ^{18}O has been used to study exchange of oxygen atoms. In neutral and alkaline solution, for example, no noticeable exchange takes place between the oxygen atoms in the ions ClO_4^-, ClO_3^-, NO_3^-, SO_4^{2-} or PO_4^{3-} and the oxygen atoms in the water. On the other hand the exchange becomes appreciable in acid solution, which may be related to the fact that these atoms are here in equilibrium with molecular species with higher or lower oxygen coordination; for example, free anhydride according to the reaction $SO_4^{2-} + 2H^+ \rightleftharpoons H_2SO_4 \rightleftharpoons H_2O + SO_3$. With the same oxygen isotope it has also been shown that the exchange between the water of hydration of an ion (5-2h, i) and the water molecules in the solution generally occurs very rapidly.

Tracer methods show that electron transfer often easily takes place between unequally charged atoms or between unequally charged molecules with the same composition. For example, this is the case between Fe^{2+} and Fe^{3+}, $Fe(CN)_6^{4-}$ and $Fe(CN)_6^{3-}$, MnO_4^{2-} and MnO_4^-, ClO_2^- and ClO_2.

All problems mentioned up to now require the use of trace elements. In many other cases these are used where they are not necessary in principle. This applies almost exclusively to radioactive trace elements, where use is made of the great sensitivity and convenience of detection. Thus, diffusion in general, wear of surface layers (a piston ring in an internal combustion engine can be made radioactive, following which extremely small traces of worn particles can be detected in the oil), gas leaks and flow measurements of various kinds (for example, transport of liquids in industrial processes) are now studied by means of radioactive tracers.

Perhaps the greatest importance of tracer methods, however, lies in biology and medicine, where, among other things, they have opened up whole new possibilities for following the path of an element in an organism.

9-4. Energy production by nuclear reactions

In 9-1 b it was shown that energy can be gained either by uniting nucleons or very light nuclides (fusion, 9-2b) or also by cleaving the heaviest nuclides so that fragments of medium weight are produced (fission, 9-2c). In the following, the possibilities for utilizing these processes for the production of energy on a large scale are discussed. We start with the fission process, the only one that until now can be applied for peaceful energy production.

Spontaneous nuclear decay other than fission is used for convenient energy production on a very small scale, for example in satellites. One has then to choose a nuclide having a half-life which permits the production of the necessary power but at the same time does not cause a too rapid consumption of the nuclide. Strong γ emitters, which cause shielding troubles, are avoided. Among nuclides used may be mentioned ^{90}Sr (β^- decay, half-life 28 years) and ^{238}Pu (α decay, half-life 90 years). The heat energy is converted thermoelectrically into electric energy.

a. Fission. The peaceful production of energy by fission is carried out in a plant called a *nuclear reactor* or simply *reactor*. The reactor, in fact, offers the only possibility for executing controlled nuclear reactions on a large scale. It is thus used both for the production of energy and for the production of certain nuclides. It is also an important research instrument, not the least because of the intense neutron flux it can deliver.

Since a reactor uses the fission process it must contain some of the fissionable nuclides ^{233}U, ^{235}U or ^{239}Pu (9-2c). The only one of these found in usable amounts in nature is ^{235}U, which must therefore constitute the original material. In natural uranium ^{235}U is present only to the extent of 0.7 %, while nearly all of the remainder consists of ^{238}U. In spite of this a reactor can be operated with natural uranium as "fuel".

Because of nuclear reactions caused by natural nuclear decay and cosmic radiation, small amounts of neutrons are constantly produced in all matter. If a fissionable nucleus encounters such a neutron, it may cleave. If the nucleus consists of ^{235}U, each fission releases an average of 2.5 fast neutrons (9-2c) according to the overall reaction (FP signifies fission products)

$$^{235}U + n \rightarrow FP + 2.5 n + energy \tag{9.6}$$

The released neutrons can cleave new ^{235}U atoms. If one of the released neutrons can always be used for a new fission, the fission process becomes a chain reaction (8g), which proceeds without further addition of neutrons. If more than one of the released neutrons causes a new fission, the chain reaction becomes branched and a rapidly growing number of atoms are cleaved per unit of time. The fission process can then spread explosively throughout the whole of the fissionable substance.

It is thus the supply of neutrons that becomes decisive for the course of the process. The number of neutrons formed per fission is so small that, in order to obtain a chain reaction, one must economize with them and be careful that as many as possible start new fissions. The conservation of neutrons is especially important when the fuel contains such a small proportion of fissionable material as natural uranium (0.7 % ^{235}U).

The most important processes that a fission neutron can undergo are:

1. Diffusion out of the system.
2. Capture by uranium isotopes without fission. This is especially important when the original, fast neutrons after collision become intermediate and thus come into the energy range (about 5–500 eV) where ^{238}U has a strong resonance absorption (9-2 b).
3. Capture by other atoms (impurities, construction materials, moderators—see below—cooling medium) without fission.
4. Capture by ^{235}U with fission.

From the above it is apparent that the conditions must be arranged so that the neutron-generating process 4 is favored and the other processes suppressed. If the ratio of the number of neutrons formed per unit of time by process 4 to the number of neutrons that are captured or diffuse away by the other processes is <1, no chain reaction is possible. The system is then said to be *subcritical*. If the ratio =1, a chain reaction of constant rate occurs, and the system is then *critical*. If the ratio is >1, the chain reaction becomes branched and the system is *supercritical*.

The fraction of neutrons that escape from the system (process 1) depends, among other things, on the size and shape of the system. For a system with a given composition the extent of processes 2–4 is proportional to the volume of the system. But neutron escape, which takes place at the surface of the system, is proportional to the surface area. The fraction of escaping neutrons is thus decreased the greater the ratio of volume to area is. For a certain shape this ratio increases as the volume increases (for example, for a sphere the ratio is proportional to the radius). A system of a certain composition and shape therefore has a *critical size* below which no chain reaction occurs, but at which a chain reaction can just maintain itself.

The experimentally determined probabilities for the capture of different kinds by ^{235}U and ^{238}U at different neutron energies show that, with the proportion of these isotopes in natural uranium, the system can be made critical only if the processes can take place in the range of slow (thermal) neutrons. The reactor is constructed as a *slow* or *thermal reactor*. The fast fission neutrons must therefore be slowed down, but this must be done in such a way that process 2 does not become too extensive. The slowing of the fission neutrons is accomplished by including a *moderator* in the system. This consists of atoms that do not absorb neutrons, but with which they can collide elastically until they are in temperature equilibrium with the moderator. The closer the mass of such an atom is to the mass of the neutron, the more kinetic energy the neutron loses in a collision. The moderator should thus consist of as light atoms as possible, but only those that have small neutron absorption. ^{1}H has the largest slowing capacity, but is so strongly neutron-absorbing that its use is possible (in the form of ordinary water) only if the system has a larger proportion of fissionable atoms than natural uranium. On the other hand ^{2}H can always be used (in the form of heavy water, D_2O), since it has significantly less neutron absorption than ^{1}H (cf. 9-2 b). Because of its lower cost, carbon is also used (in the form of very pure graphite), though it has less slowing capacity and greater neutron absorption than D_2O. The critical size of the system is therefore greater with a graphite moderator than with a D_2O moderator. Beryllium and beryllium oxide can also be used as moderators.

So that the resonance absorption shall not become too extensive when the neutrons on slowing pass through the dangerous intermediate range, the fuel is distributed as compact pieces, so-called fuel elements, in a definite way in the moderator. The reactor is then said to be *heterogeneous*. The fuel elements consist of uranium metal, uranium oxide UO_2 or uranium carbide UC, often in the form of rods or plates.

The heavy atoms in the fuel elements do not appreciably slow down the neutrons and most of these therefore still have a high velocity when they sooner or later leave the fuel element. They then enter the moderator where they are slowed down, but when they pass through the intermediate velocity range there are no resonance-absorbing atoms in the vicinity. When they again enter a fuel element they are generally slow and are not captured under resonance absorption.

The fuel and moderator form the *reactor core*. This is surrounded as a rule by a *reflector* of some moderator material, for example, graphite, heavy water or beryllium, which throws back a portion of the escaping neutrons and thereby decreases the critical size. Outside the reflector is placed finally a heavy radiation shield containing passages through which fuel elements can be exchanged, neutron beams passed out, and so on.

Process 3 must be made as insignificant as possible by excluding from the system other atoms with large absorption capacity for neutrons. Even very small amounts of such atoms in the reactor core are extremely detrimental. All construction materials must be chosen with this in mind, and certain impurities are completely forbidden.

During operation the reactor is held critical with rods or plates (control elements) of a strongly neutron-absorbing substance, for example, cadmium, which are introduced between the core and the reflector to the proper length. The reactor must therefore be constructed so that without the control elements it would be supercritical.

Whether or not the reactor is used for the production of energy, the heat energy developed by fission must be picked up and carried away by a cooling system. Later some questions regarding materials for the removal of heat will be considered.

During operation of the reactor the fissionable material is consumed and fission products are produced (9-2c and equation (9.6)), of which some are strongly neutron-absorbing. Both factors lead to a continual deterioration of the neutron balance. Gradually, therefore, the fuel elements become "burned out", whereupon they are successively replaced with new elements. Replacement may also be made necessary by a change in the shape and dimensions of the elements during operation.

When the fuel contains ^{238}U, another important process also takes place. The greater proportion of neutrons absorbed by ^{238}U induce the previously mentioned reaction series

$$^{238}_{92}\text{U}(n,\gamma)^{239}_{92}\text{U} \xrightarrow[23.5 \text{ min}]{\beta^-} {}^{239}_{93}\text{Np} \xrightarrow[2.3 \text{ d}]{\beta^-} {}^{239}_{94}\text{Pu}.$$

^{239}Pu has a quite small tendency to decay spontaneously (α decay, half-life 23 600 years) and its proportion in the fuel therefore increases, at least in the beginning. Under certain conditions a reactor can produce ^{239}Pu from ^{238}U in amounts equal or even greater than the amounts of ^{235}U cleaved. The reason for this is that each cleavage gives more than two fission neutrons.

^{239}Pu is one of the easily fissionable nuclides, however, and when its proportion increases in the fuel an ever-increasing amount of it is cleaved under the action of the neutron radiation. In this way the neutron balance is improved. If it is desired to recover ^{239}Pu, however, the fuel elements must be exchanged before so much is formed that an excessively large fraction undergoes fission. In addition, ^{239}Pu is consumed also by absorption of neutrons with the formation of heavier plutonium isotopes, of which, for example, ^{240}Pu is not fissionable with thermal neutrons. ^{240}Pu exhibits a very high probability for spontaneous fission, which means that its proportion must be kept very low in plutonium intended for use in weapons. However the plutonium is used, it is in any event clear that its creation from ^{238}U makes even this uranium isotope extremely valuable. It has been called a *fertile*, that is, fruitful material and makes it possible that the whole of the earth's supply of uranium in principle can be used for the production of energy.

The creation of plutonium can at most occur at approximately the same rate as the consumption of ^{235}U, that is, 1 g per MWd of developed energy (9-2c). A rate of production of 1 kg plutonium per day implies, therefore, a constant output of a least 1000 MW. As early as 1944 the reactors in Hanford, Washington, operating exclusively for the production of plutonium, developed a total output of 1500 MW. The energy was not recovered at that time but was discharged into the Columbia River! However, there is nothing to prevent a reactor from being used to produce energy and plutonium simultaneously. According to the above discussion, however, in this case the fuel elements must be removed earlier than would be necessary only for the production of energy.

The spent fuel elements can be processed chemically for the purpose of recovering the plutonium and the remaining uranium. Plutonium has different chemical properties than uranium and the separation of these two elements can therefore be carried out in principle with ordinary chemical methods. But the whole operation is made extremely complicated by the presence in addition of some 30 elements in the form of fission products. Some of these constitute valuable radioactive isotopes and are recovered, the others are discarded. The work is further made more difficult by the strong radioactivity, which makes necessary remote manipulations in all operations and creates great problems in the disposal of waste. A great inorganic chemical industry with a completely new aspect is here being developed.

The possibility of enriching or preparing in pure form the isotope ^{235}U (9-3b) and producing ^{239}Pu in reactors has lead to the fact that nowadays reactors are operated more and more with fuel containing larger amounts of fissionable material than natural uranium (*enriched fuel*). The probability of fission then becomes greater, which improves the neutron balance. The critical size of the reactor drops (enriched fuel is therefore desirable for movable reactors, for example, for driving ocean vessels) and the requirement of low neutron absorption in moderators, cooling system and other constructional parts can be reduced. Ordinary water can be used as moderator. With sufficiently high enrichment, the fuel can also be intimately mixed with the moderator material, whereby is obtained a *homogeneous reactor*. The mixture may be a solution or a melt, which can circulate and thus be cooled outside the reactor core. Only within the core is it given such a volume and shape that it becomes critical.

It is particularly in the utilization of the heat energy that great difficulties in choice of materials arise. This is especially true if the heat is to be developed at high temperature, which is necessary for economical production of power. The cooling system must be both physically and chemically resistant at high temperature and pressure, and must, especially if the fuel is natural uranium, have low neutron absorption. As a coolant in such cases heavy water and certain gases, for example, air and carbon dioxide, have been most used. Ordinary water is used mainly with enriched fuel. As construction materials with low neutron absorption (especially for the encapsulation of the fuel elements to protect them against chemical attack by air or coolants) aluminum, magnesium, zirconium and alloys of these metals have been extensively used up to now, but nowadays beryllium is also being tried. Zirconium must be completely free of hafnium, which has a high neutron absorption. With enriched fuel stainless steel has also been used. The difficulties of choice of materials is still one of the greatest problems in reactor technology, and much remains to be done here.

If the fuel is very highly enriched in fissionable material, it is most advantageous to work with fast neutrons instead of slow (thermal) neutrons. This is achieved by not using a moderator but allowing the fast neutrons to cause cleavage of the fissionable material. Elimination of the moderator makes the critical volume of the reactor core very small (of the order of 1-5 dm^3). A *fast reactor* is then obtained. Since fast neutrons are absorbed less easily than thermal neutrons the choice of material in the cooling system, etc., is not so difficult for a fast reactor. However, here velocity-slowing, that is, light atoms, are not tolerated. Water is therefore unsuitable as a coolant and instead, liquid metals have been used.

By surrounding the core of a fast reactor with a mantle of natural uranium a yield of ^{239}Pu can be obtained in the mantle that exceeds the consumption of fissionable material in the core. As a fertile material in the mantle natural thorium, which ordinarily contains only ^{232}Th, can also be used. Then the fissionable uranium isotope ^{233}U is formed by the processes

$$^{232}_{90}\text{Th}(n,\gamma)^{233}_{90}\text{Th} \xrightarrow[22.3 \text{ min}]{\beta^-} {}^{233}_{91}\text{Pa} \xrightarrow[27.0 \text{ d}]{\beta^-} {}^{233}_{92}\text{U}.$$

It is probable that supplies of ^{239}Pu and ^{233}U to be used for the enrichment of reactor fuel will come to be obtained with reactors of this type (breeder reactors).

The most common nuclear weapon is the *fission bomb*. Such a bomb in principle is a fast reactor in which the fission process is strongly branched. It contains fissionable material in the form of ^{235}U or ^{239}Pu. The total amount of this material in sufficiently concentrated form would be supercritical, but initially is distributed so that the escape of neutrons is high. This can be done either by division into two or more subcritical pieces or by forming into a sufficiently spread-out shape. By bringing the pieces together or changing to a more concentrated shape the system is made supercritical. The assembly of pieces or change of shape must take place extremely fast and is carried out with a properly shaped explosive charge. The total amount of fissionable material cannot exceed by very much the critical amount of favorable shape and is therefore fairly well predetermined. The size is about 20 kg, that is, approximately 1 dm^3.

The utilization of fission for the production of energy in reactors was made possible primarily by Fermi and his coworkers, who, among other things, introduced the moderator. The first reactor was built in Chicago and was made critical for the first time on December 2, 1942. Man had thus harnessed atomic energy to his service. Currently, power-producing reactors are built with a thermal output of about 2000 MW, which gives about 500 MW of electrical power.

b. Fusion. Fusion between two light nuclides is only possible if the nuclei collide with such velocity that the repulsive forces between them are overcome (the energy barrier is penetrated) and the nuclear forces can begin to operate. Fusion between individual nuclei can be brought about in accelerators (9-2b). On a larger scale fusion can only be achieved by raising the temperature of the system so high that the heat motion gives a sufficient number of nuclei the required energy. One speaks then of *thermonuclear reactions*. With pressures that can be obtained on the earth, thermonuclear reactions do not have any appreciable scope until temperatures above 10^7°K are reached, and for the practical production of energy temperatures of 10^8°K may be necessary. Until now such temperatures have been successfully achieved only by initial ignition with a fission bomb. Fusion then takes place explosively (*fusion bomb, thermonuclear bomb*).

Even at temperatures that have been successfully attained in this way, only those fusion reactions can be used that take place most easily. Thus reactions in which ^{1}H takes part, and which require still higher temperatures to take place with sufficiently high probablility, are excluded. On the other hand, fusions between the heavier hydrogen isotopes may be used (*hydrogen bomb*). The most important of these are two D-D processes (mentioned earlier in 9-2b), one T-D process and one T-T process, which here are written in the following form:

$$^2D + {}^2D \rightarrow {}^3He + {}^1n + 3.3 \text{ MeV}$$
$$^2D + {}^2D \rightarrow {}^3T + {}^1H + 4.0 \text{ MeV}$$
$$^3T + {}^2D \rightarrow {}^4He + {}^1n + 17.6 \text{ MeV}$$
$$^3T + {}^3T \rightarrow {}^4He + 2\,{}^1n + 11.4 \text{ MeV}$$

The T-D process goes most easily and may in any case be the reaction that is initiated in the first place by the fission bomb. Tritium has been manufactured by exposing the lithium isotope ^{6}Li (which occurs in natural lithium to 8%) to neutrons in a reactor, whereby the reaction ^{6}Li$(n,\alpha)^3$T takes place.

In order for the fusion to be effective the initial material must have quite high density. If the hydrogen isotopes are used in elementary form they therefore must be liquified, which introduces technical difficulties. It is better to incorporate them in compounds. One can go one step further and synthesize the tritium in the bomb at the instant of explosion. In this way the bomb may be made storable without limit (the half-life of tritium is only 12 years). For this case solid lithium deuteride ^{6}Li^{2}D is used. The neutrons from the igniting fission bomb first produce ^{3}T according to the reaction ^{6}Li$(n,\alpha)^3$T. Thereafter the T-D process mentioned above takes place.

In certain cases such a hydrogen bomb has been surrounded by a thick shell of natural uranium. The very fast neutrons that are formed in large numbers during fusion are able to cause a nonself-maintaining fission of ^{238}U. The outer shell has no critical size. It can therefore contain uranium by the ton and thus give off a very large amount of fission energy. In bombs of this type 80–90% of the released energy may be developed in the last fission stage. In contrast to purely fusion bombs these bombs produce very large amounts of fission products (they are "dirty"), which by being spread in the atmosphere can cause radiation damage over great distances. The extremely intensive neutron radiation also gives with ^{238}U very neutron-rich uranium isotopes (up to ^{255}U), which very rapidly transform to heavier transuranium elements by β^- decay. The first identification of 99 Es and 100 Fm was made in material from an explosion of a bomb of this type.

A search is in progress, of course, for a method for carrying out fusions in a controllable way for the peaceful production of energy. This may then be based on the earth's practically inexhaustable supply of deuterium. Also, the often unusable, strongly radioactive by-products that are obtained from every fission would be avoided. The chief difficulty is the maintainance of the controlled high temperatures within a confined system with sufficient density and for the requisite length of time.

Attempts have been made to set up a powerful magnetic field so that an ionized gas present in the field, when traversed by current of high density, is retained within a small volume even after a very sharp temperature rise. The gas then forms a plasma (5-2f). Providing the plasma has the required temperature, it is a necessary condition for the net production of energy that the product of the number n of atomic nuclei per unit of volume, and the time t that the high-temperature plasma can be held together, exceeds a certain value. It is thus necessary that for T-D processes $nt > 10^{20}$ sm^{-3} and for D-D processes $nt > 10^{21}$ sm^{-3}. In 1964 temperatures over 10^{7}°K and values of nt of about 10^{18} sm^{-3} had been attained.

Fusion processes certainly form the basis of the development of energy in the sun and the stars. The processes may vary, depending on the different properties of different stars. Probably also, different processes prevail at different depths of the same star. In the sun, for example, the temperature increases from about 6000°K on the surface to about 1.5×10^{7}°K at the center. At the extremely high pressures that occur in the deeper parts (at the center about 10^{11} bar), conditions exist for many types of fusion processes. At present it is believed that the most important process in the sun is the so-called proton–proton chain:

$$\begin{array}{r} 2 \cdot (^{1}\text{H} + ^{1}\text{H} \rightarrow {}^{2}\text{D} + \beta^{+}) \\ 2 \cdot (^{1}\text{H} + ^{2}\text{D} \rightarrow {}^{3}\text{He} + \gamma) \\ {}^{3}\text{He} + {}^{3}\text{He} \rightarrow {}^{4}\text{He} + 2\, {}^{1}\text{H} \\ \hline 4\, {}^{1}\text{H} \rightarrow {}^{4}\text{He} + 2\beta^{+} + 2\gamma \end{array}$$

The energy developed in the total process is 26.7 MeV, which corresponds to 2.6×10^{12} J = 30 MWd per mol (= 4 g) ^{4}He. A conversion of 1 g per day thus corresponds to an output of 7.5 MW.

9-5. Further chemical applications

a. Chemical work with radioactive nuclides. In all work with radioactive nuclides the dangerous effects of the radiation on the body must be kept in mind. Often operations must be done behind adequate radiation shields and the handling performed by means of manipulators. Work with short-lived substances must also be done rapidly. From a strictly chemical standpoint the processes are often the same as in ordinary chemical work but not uncommonly one is forced to use special methods because of the very low concentrations of the radioactive nuclides. If the radioactive nuclide formed by nuclear reaction is not an isotope of the starting material, it is obtained without the admixture of stable isotopes. If it then occurs, as is most often the case, in very small amounts, it very often evades the intended route of transfer. For example, it may be adsorbed (15-1 a) on the walls of the vessel, on filters or on solid phases in the system (for example, coprecipitation, 15-1 b).

A manageable amount of a *carrier* is then added. This may be an *isotopic carrier*, that is, consisting of a stable nuclide of the same element (of course, in the same state, and thus incorporated in the same compound as the radioactive nuclide). For example, if the amount of a radioactive nuclide is 10^{-11} g and 10^{-2} g of an isotopic carrier is added, an isotope mixture is obtained in which 1 atom in 10^{9} is radioactive. This mixture can be handled by ordinary chemical methods and the amounts that are unavoidably lost contain only an extremely small amount of the radioactive nuclide. Often the mixture of radioactive nuclide and isotopic carrier can then be used directly in tracer methods.

Not uncommonly a preparation of a radioactive nuclide already contains an isotopic carrier from the start. Such is the case, for example, when the radioactive nuclide is prepared by neutron irradiation of a stable nuclide of the same element, where a (n,γ) reaction takes place. A large amount of the stable nuclide always remains after the irradiation and serves as a carrier.

Often the radioactive nuclide is allowed to follow a *nonisotopic carrier*. This possibility depends as a rule on coprecipitation and is treated in more detail in 15-1b. If at the end of the process the radioactive nuclide must be separated from the nonisotopic carrier, this is done far more easily than with an isotopic carrier.

Many times a new radioactive nuclide can be identified chemically by determining which known element is the most effective carrier for the new nuclide. If the new nuclide represents a wholly new element, of course, no isotopic carrier can be found. Nevertheless, if an effective carrier is turned up, this means that the new nuclide has properties similar to those of the carrier. Several new elements could at first be characterized chemically only by such *carrier techniques*.

b. Chemical processes caused by high-energy radiation. If a system is exposed to sufficiently energy-rich radiation, whether it is typically electromagnetic or consists of particles, in general, changes take place in the system. If the electron clouds in the atoms of the system are struck by the radiation, excited states are produced, which most often are followed by chemical processes. The advance of science and technology has made possible types of radiation with higher and higher energy and intensity values, and their effect on the electron clouds and thus the bonds is nowadays under active study. The field has been called *radiation chemistry* and in fact implies a broadening of photochemistry (8f), although the latter has been taken to include only processes in which light quanta are absorbed.

Effects of other kinds arise if the radiation leads to nuclear reactions. If an element is thus transformed to another it is very probable that its bonds to neighboring atoms will be broken. In addition the daughter nucleus acquires a high kinetic energy both from the energy of the projectile and from the recoil energy it takes up when the compound nucleus decays. It therefore becomes reactive to a high degree.

In the (n,γ) reaction that commonly occurs with capture of thermal neutrons the daughter nucleus is only another isotope of the mother nucleus. In spite of this and even with the small kinetic energy of the thermal neutrons, the daughter nucleus, on the emission of a γ quantum, generally acquires such high recoil energy that it breaks its bonds. It may be bound again at a similar site but it may also serve another function. If a water solution of $NaClO_3$ is irradiated with neutrons the reaction $^{37}Cl(n,\gamma)^{38}Cl$, among others, occurs, whereupon the ^{38}Cl is split out of the ClO_3^- ion and forms Cl^- ion. In such cases (Szilard-Chalmers processes) the created isotope, because of its new function, is easily separated from the starting material by ordinary chemical means.

The high kinetic energy of the recoiling nuclei causes them in their effects to correspond to atoms at very high temperature (*hot atoms*). In (n, γ) reactions recoil energies often occur of the order of 100 eV, which corresponds to the mean energy at a temperature of approximately $10^{6} °K$ (2-1b). The resulting high reactivity may make possible reactions that could not be achieved in any other way (hot-atom chemistry).

c. Activation analysis. If a substance is exposed to neutrons or fast, charged particles, radioactive particles are frequently formed. These can be characterized by measurement of the energy of their radiation and their half-lives, and their amount can be determined by measurement of the intensity of the radiation (generally, by counting β^- or α particles). This *activation analysis* is often extremely sensitive. Europium, for example, can be determined in amounts of 3×10^{-11} g.

In general, the sample is exposed to thermal neutrons in a reactor together with a standard containing a known amount of the element sought. Thus, (n,γ) reactions are used and therefore neutron-rich nuclides are obtained, which as a rule are β^- radiators. Then the intensity of the radiation from this nuclide is measured in the sample and in the standard. If the sample and standard are reasonably similar and have been exposed to the same neutron radiation, an equal fraction of the element sought has been transformed to a radioactive nuclide in both. The original amount of the element in the sample can then be easily computed.

Since several elements in the sample may give radioactive nuclides, generally the nuclide in question must be separated after irradiation by chemical methods (see 9-5a) in both sample and standard. Only then is the radiation intensity measured.

CHAPTER 10

The Equilibrium Equation

a. Chemical potential and activity. In 2-3d the quantity μ_i, the chemical potential of a substance "i", was introduced. If this substance makes up the mole fraction x_i of an *ideal solution* (1-4i), which may be gaseous, liquid or solid, it is exactly true that

$$\mu_i = \mu_i^0 + RT \ln x_i \qquad (10.1)$$

If $x_i = 1$, that is, the substance is pure, then $RT \ln x_i = 0$ and thus $\mu_i = \mu_i^0$. The term μ_i^0 is thus equal to the chemical potential of the pure substance (at the same pressure and temperature as the solution in question). The expression shows that the absolute value of the chemical potential cannot be given, but only potential differences. It is stated only how much greater or less the potential is than the potential of a *standard state*, which thus, for the solutions in question, establishes the zero point on a potential scale. If the expression (10.1) is used, that is, if μ_i is expressed as a function of x_i, the pure substance "i" is clearly chosen as a standard state.

μ_i can also be expressed in terms of other independent variables besides x_i. For *ideal solutions* it is still true that (c_i = the molarity and m_i = the molality of the substance "i")

$$\mu_i = \mu_i^0 + RT \ln c_i \qquad (10.2)$$

$$\mu_i = \mu_i^0 + RT \ln m_i \qquad (10.3)$$

and in particular, for *ideal gas mixtures* (p_i = the partial pressure of "i").

$$\mu_i = \mu_i^0 + RT \ln p_i \qquad (10.4)$$

Here the standard states correspond respectively to $c_i = 1$, $m_i = 1$ and $p_i = 1$, that is, they are, like the values of μ_i^0, mutually different and different from the standard state in (10.1).

For example, (10.4) is obtained if $x_i = p_i/P$ (1.10) is substituted into (10.1), which gives $\mu_i = (\mu_i^0 - RT \ln P) + RT \ln p_i$. At constant pressure and temperature, $RT \ln P$ and thus the expression in parentheses is constant, and it is the latter that gives the μ^0 value in (10.4). The μ^0 values in (10.1) and (10.4) thus differ by the term $RT \ln P$.

If P and p_i are expressed in atm, for example, the standard state is the ideal gas mixture in which $p_i = 1$ atm. Only for $P = 1$ atm is this state equivalent to the pure substance "i".

The above expressions do not apply to nonideal solutions. However, it has been shown to be practical to introduce a_i, the *activity* of a substance "i" (G. N. Lewis,

1907; for a substance with the symbol A the activity is also indicated by {A}), which is defined by the expression

$$\mu_i = \mu_i^0 + RT \ln a_i \tag{10.5}$$

The formal analogy between (10.5) and the expressions (10.1)–(10.4) means that the laws of equilibrium which can be derived from the latter for ideal solutions, apply also to nonideal solutions, provided concentration and pressure terms are replaced by activities.

There are many ways to measure the difference in the chemical potential of a substance in two different solutions. From this difference, according to (10.5), if T = constant, the *ratio* between the activities of a substance in the two solutions is obtained. On the other hand, no absolute value for the activity can be determined. However, if a certain activity value for the substance in a certain solution can be agreed upon, a fixed point on the activity scale has been established by which the activity of the substance in any other solution can be given. In principle it is arbitrary what scale is chosen, but for each given case, generally one or two scales are more practical than any other that might be imagined. When the same substance occurs in different phases, different scales are often used for each phase. If the scale is changed for a substance, all the activities of the substance are multiplied by a constant factor, while at the same time a constant term is added to or subtracted from μ_i^0 in (10.5).

The activity scale for a substance is often defined by stating a *standard state* for the substance, that is, a state (real or imaginary) in which the activity is set equal to 1.

The activity scale for a substance is generally chosen according to the following rules:

1. *Pure, liquid or solid substances*: $a=1$, that is, the pure substance is chosen as the standard state.

2. *Liquid or solid substances that occur as solvents* (that is, are present in predominant amount in the solution): $a=1$ for the pure substance, which is thus chosen as the standard state. With increasing dilution of the solution, the activity of the solvent then approaches the limiting value of 1. In dilute water solutions, therefore, the activity of the water is often set as $a_{H_2O}=1$.

3. *Dissolved substances* (present in lesser amount in the solution). There is generally a great difference between the state of the building units of a substance when the substance exists pure and when it exists in a dilute solution. Therefore, it is often inappropriate to use the pure substance as a standard state for a dissolved substance. Instead we proceed from the fact that, however the activity scale is chosen, the quotients a_i/x_i, a_i/c_i or a_i/m_i with increasing dilution approach a limiting value (different for the different quotients). It is simplest to choose the scale so that the quotient containing the particular variable of concentration x_i, c_i or m_i

The Equilibrium Equation

to be used approaches the limiting value 1 as the variable approaches 0. The activity then approaches the value for the variable in question, which is convenient. In very dilute solutions a_i may be set equal to x_i, c_i or m_i. The standard state with $a_i=1$ in this case does not correspond to an easily described state.

The values of the activity of a dissolved substance referred to any of the concentration variables x_i, c_i or m_i are proportional to the values of these variables in very dilute solutions. If these activities are denoted by $a(x)_i$, $a(c)_i$, and $a(m)_i$, the relation $55.5\,a(x)_i = a(c)_i = a(m)_i$ therefore is nearly true in water solutions (1-4e).

4. *Gases.* For a gas the quotient a_i/p_i always approaches a limiting value as p_i approaches 0. The activity scale is therefore often defined so that the quotient a_i/p_i approaches 1 as p_i approaches 0. Even at rather high partial pressures the state is often nearly ideal, that is, the quotient a_i/p_i is nearly $=1$. In such cases the activity can thus be set $a_i = p_i$ and the standard state is the gas at partial pressure $p_i = 1$ atm.

5. In *special cases* it may be proper to choose the activity scale in another way. It must then be exactly defined. An example of this is given in 12-2a.

The departures from the ideal state can be expressed, according to the variable of concentration used, in terms of one of the quotients given above, which are all termed *activity factors* f_i. If the variable of concentration is shown, we may thus write

$$f(x)_i = a(x)_i/x_i \quad f(c)_i = a(c)_i/c_i \quad f(m)_i = a(m)_i/m_i \tag{10.6}$$

Since the type of the variable is generally clear from the context it is most often not shown. We write, for example,

$$f_i = a_i/c_i \quad \text{or} \quad f_i = \{i\}/[i] \tag{10.7}$$

According to the rules for the choice of the activity scale the activity factor of a dissolved substance always approaches a limiting value of 1 with increased dilution. For the solvent, where the pure substance is chosen as the standard state with the activity 1, this will only be true if a variable of concentration is used that also is 1 for the pure substance, that is, the mole fraction x_i. If c_i or m_i is used as the variable of concentration, for the pure substance it is nearly true that $c_i = m_i = 1000/M_i$, where M_i is the molecular weight of the solvent. The activity factor of the solvent then approaches the limiting value $f_i = 1/c_i = 1/m_i = M_i/1000$. In water solutions the limiting value is thus $f_{H_2O} = 18/1000 = 0.018$.

b. The equilibrium equation. Let us assume that a process occurs within a system according to the equation

$$a\mathrm{A} + b\mathrm{B} + \ldots \rightleftharpoons l\mathrm{L} + m\mathrm{M} + \ldots \tag{10.8}$$

According to (2.12), $G = a\mu_A + b\mu_B + \ldots$ for the system before the process, and $G = l\mu_L + m\mu_M + \ldots$ for the system after the process. Thus,

$$\Delta G = (l\mu_L + m\mu_M + \ldots) - (a\mu_A + b\mu_B + \ldots) \tag{10.9}$$

If the same process takes place so that all disappearing and appearing substances are in their standard states, the following analogous expression is obtained for ΔG, which is then indicated by ΔG^0:

$$\Delta G^0 = (l\mu_L^0 + m\mu_M^0 + \ldots) - (a\mu_A^0 + b\mu_B^0 + \ldots) \tag{10.10}$$

From (10.9) we may subtract (10.10), after which terms of the type $l(\mu_L - \mu_L^0)$ can be formed. Provided the activities are given in scales defined by the standard states in (10.10), according to (10.5) such a term becomes equal to $lRT \ln a_L = RT \ln a_L^l$. We have, therefore,

$$\Delta G = \Delta G^0 + RT \ln \frac{a_L^l a_M^m \ldots}{a_A^a a_B^b \ldots} \tag{10.11}$$

If ΔG^0 is known for the process, ΔG for the process occurring between the same substances in other states can be calculated from (10.11), provided that the activities of the substances are known in these states. One can often measure the electric work of a process, from which ΔG can be calculated (16b). If ΔG^0 is known, it is then possible to calculate the expression after the ln sign (the "activity quotient") which in simple cases can render activity values. Inversely, if ΔG and the activities are known, ΔG^0 may be calculated.

If ΔG^0 has been determined for a number of processes, ΔG^0 for completely different processes can be calculated from them (2-3b and below). A large number of ΔG^0 values are listed in the books referred to in 2-2a.

If the process (10.8) takes place at constant pressure and temperature, then $\Delta G = 0$ at *equilibrium* (2.10). From (10.11) we then have

$$\Delta G^0 = -RT \ln \frac{a_L^l a_M^m \ldots}{a_A^a a_B^b \ldots} \tag{10.12}$$

At constant temperature ΔG^0 is constant and consequently at equilibrium the "activity quotient" is constant. We then have

$$\left.\begin{array}{c} \Delta G^0 = -RT \ln K \\ \text{(for } 25°C, \quad \Delta G^0 = -5.7079 \log K \text{ kJ)} \end{array}\right\} \tag{10.13}$$

where

$$K = \frac{a_L^l a_M^m \ldots}{a_A^a a_B^b \ldots} \quad (P = \text{const.}, T = \text{const.}) \tag{10.14}$$

K is called the *equilibrium constant* of the process and the expression (10.14) the *equilibrium equation*. The constant is always chosen so that the substances on the right side of the reaction equation appear in the numerator of the equilibrium equation.

The equilibrium equation (10.14) applies to both ideal and nonideal states. The equilibrium constant K is thus always truly constant. In the ideal state the activi-

The Equilibrium Equation

ties a_i in (10.14) may be replaced by x_i, c_i, m_i or p_i, according to the standard state by which they are defined. The equilibrium equations then appear as

$$K_x = \frac{x_L^l x_M^m \cdots}{x_A^a x_B^b \cdots}; \quad K_c = \frac{c_L^l c_M^m \cdots}{c_A^a c_B^b \cdots}; \quad K_p = \frac{p_L^l p_M^m \cdots}{p_A^a p_B^b \cdots} \quad (10.15\text{--}17)$$

For the same equilibrium these different equilibrium constants generally have different values and dimensions. The values and dimensions also depend on the exponents, that is, on the number of moles in the reaction equation. Therefore, the type of constant used should always be stated, and in addition the reaction equation to which it corresponds should be given. Since the constant generally applies only at a definite temperature, this should also be given.

With increasing departure from the ideal state, K_x, K_c and K_p are no longer constants. Since equations of the type (10.16) are much used later in this book, we will continue here with it. If (10.14) is divided by (10.16) and according to (10.7) we let $f_i = a_i/c_i$, then we have

$$K_c = K \frac{f_A^a f_B^b \cdots}{f_L^l f_M^m \cdots} \quad (10.18)$$

As long as the properties of the system are such that all f values remain constant, K_c is then constant, so that (10.16) can be used for equilibrium calculations. In general $K_c \neq K$, but in the special case where all molecular species participating in the equilibrium are dissolved substances, $f=1$ at infinite dilution for all types of molecules and thus $K_c = K$. In the application of (10.16) there is the advantage of working with concentrations and not with the usually more difficultly measured activities. Care must be taken in applying the equation to cases where the K_c value cannot be assumed to be constant. In order to show that ratios of the types (10.15–17) are constant only under severely restricted conditions, k is often written instead of K in their application.

If a substance participating in the equilibrium is added to the system, that is, its concentration is increased, according to (10.15–17) the concentrations of other substances entering into the reaction equation must change. *The addition of a substance thus leads to a shift in the equilibrium such that the added amount is partly used up.*

For $\Delta G^0 = 0$, (10.13) gives $K = 1$. The equilibrium equation shows that the process then reaches equilibrium at what one might call a half-way stage. For $\Delta G° < 0$, $K > 1$ and for $\Delta G° > 0$, $K < 1$. In the first case the equilibrium swings to the right side of the reaction equation, in the latter case to the left side.

The equilibrium constant has a finite value, which shows that the state of the system does not correspond entirely to the left or to the right side of the reaction equation. This has led to the use of two oppositely directed arrows between the two sides. If a process goes in a certain direction or the direction has to be specified, a single arrow is used.

An equilibrium constant can be determined by measurement of the activities of the substances taking part in the equilibrium or, for ideal states, their concentrations or partial pressures. When the equilibrium constant is known, ΔG^0 may be calculated from (10.13), a quantity that makes possible the calculation of equilibrium constants for other equilibria.

For example, it is not possible to measure the equilibrium constant for the so-called water-gas equilibrium (for more detail, see 18d) $CO(g) + H_2O(g) \rightleftharpoons CO_2(g) + H_2(g)$ at room temperature. However, tabulated values of ΔG^0 at 25°C are available for all four gases, that is, ΔG for their formation (at 1 atm, 25°C) from the elements in the standard states. The values are: CO, -137.32 kJ; H_2O, -241.91 kJ; CO_2, -394.52 kJ; H_2, 0 kJ. In accordance with the discussion in 2-2a and 2-3b, we have for the above process $\Delta G^0 = (-394.52 + 0) - (-137.32 - 241.91) = -15.29$ kJ. From (10.13) we then obtain $\log K = -\Delta G^0/5.708 = 15.29/5.708 = 2.679$, from which $K = 478$. Thus, at 25°C $a_{CO_2} a_{H_2}/a_{CO} a_{H_2O} = 478$, in which equation the activities are defined by the standard states corresponding to ΔG^0, that is, the gases at partial pressures of 1 atm. As long as the gases are ideal, it is also true that $p_{CO_2} p_{H_2}/p_{CO} p_{H_2O} = 478$. The water-gas equilibrium as it is written above is therefore strongly shifted to the right at room temperature.

The equilibrium equation (10.14) and the more or less approximate forms derived from it are said to express the "law of mass action". In a form most similar to (10.16), this was proposed for the first time in 1867 by Guldberg and Waage and is therefore also called the Guldberg-Waage law. It is related, of course, to the dynamic nature of chemical equilibrium referred to in 1-5c. Its derivation can therefore also be obtained kinetically, and then becomes more easily visualized but less rigorous. The above thermodynamic derivation is preferable, although it uses expressions that here can only be postulated ((2.10), (2.12) and (10.1)).

c. Generalities concerning equilibria. An equilibrium in which the substances only participate in a single homogeneous system, is called a *homogeneous equilibrium*. An equilibrium between substances that are present in different phases of a heterogeneous system, is called a *heterogeneous equilibrium*. The equilibrium equation can be applied to both these cases.

Among the simplest of chemical reactions is *addition* (in special cases also called *complex formation*, *association*, etc.) $A + B \rightarrow AB$, and its inverse *dissociation* $AB \rightarrow A + B$. In both cases the same equilibrium occurs, which, however, depending on the initial state, is written either $A + B \rightleftharpoons AB$ or $AB \rightleftharpoons A + B$. According to the usual convention (10b) the equilibrium equation in the first case is $\{AB\}/\{A\}\{B\} = K$, where K is often called a *complex constant* or *stability constant*, and in the latter case $\{A\}\{B\}/\{AB\} = K'$, where K' is called a *dissociation constant*. Thus, for the same equilibrium the complex and dissociation constants are reciprocal.

In dissociation equilibria a *degree of dissociation* α is often given, which refers to the fraction of the original substance dissociated. For the calculation of α the equilibrium equation must be based on concentrations, and for the dissociation equilibrium $AB \rightleftharpoons A + B$ it is therefore written in the form $[A][B]/[AB] = k$. If the original concentration of AB was C ($=$ total concentration of AB) then $[A] = [B] = C\alpha$ and $[AB] = C(1 - \alpha)$. Substitution into the equilibrium equation gives

$$C\alpha^2/(1 - \alpha) = k \tag{10.19}$$

The Equilibrium Equation

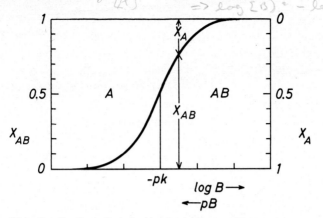

Fig. 10.1. The distribution of A in A plus AB at different B concentrations.

(10.19) shows that α increases as C decreases. If 1 mole of the original substance does not as in this case give 2 moles of dissociation products, other expressions are obtained. But α still increases with decreasing C.

Often a graphic picture is desirable of how an equilibrium is shifted by a change in concentration of a participating substance. For example, we wish to see how a homogeneous equilibrium $A + B \rightleftharpoons AB$ is changed by variation of the concentration of B. The equilibrium equation is written

$$[AB]/[A][B] = k \tag{10.20}$$

A larger range of concentration may be surveyed if (10.20) is written in logarithmic form:

$$\log [B] = -\log k + \log \frac{[AB]}{[A]} \tag{10.21}$$

According to (1.8), we may set $pB = -\log [B]$ and also $pk = -\log k$, from which we have

$$pB = -pk - \log \frac{[AB]}{[A]} \tag{10.22}$$

For graphical presentation, it is practical to introduce, instead of the concentrations of A and B, the fractions of all atoms A that are present partly as A and partly as AB. Denoting these fractions by x_A and x_{AB} we have

$$x_A = \frac{[A]}{[A]+[AB]}; \quad x_{AB} = \frac{[AB]}{[A]+[AB]}; \quad x_A + x_{AB} = 1 \tag{10.23}$$

Then, $[AB]/[A] = x_{AB}/x_A = x_{AB}/(1-x_{AB})$, that is,

$$pB = -pk - \log \frac{x_{AB}}{1-x_{AB}} \tag{10.24}$$

In fig. 10.1 x_{AB} is plotted as a function of pB. An S-shaped curve is obtained, whose midpoint lies at $x_{AB}=0.5$ and $pB = -pk$. The curve forms the boundary between the two areas labelled A and AB. At each pB value the length of the ordinate within the A area $= x_A$ and the length within the AB area $= x_{AB}$ (an example is indicated by arrows). The figure therefore shows at once the position of the equilibrium at a certain pB or log B value. It thus makes clear also how the dissociation equilibrium $AB \rightleftharpoons A+B$ depends on pB.

Particularly in the formation of metal complexes, a *stepwise addition* of ligands L to the metal ion M takes place. Each step gives an equilibrium equation:

$$M + L \rightleftharpoons ML \qquad [ML]/[M][L] = k_1$$
$$ML + L \rightleftharpoons ML_2 \qquad [ML_2]/[ML][L] = k_2$$
$$\cdots\cdots\cdots \qquad \cdots\cdots\cdots$$
$$ML_{n-1} + L \rightleftharpoons ML_n \qquad [ML_n]/[ML_{n-1}][L] = k_n$$

Here the atom M is present in M, ML, ML_2, ... ML_n, and the total concentration of M in all the substances belonging to the system M–L is thus $[M]_{tot} = [M] + [ML] + [ML_2] + ... + [ML_n]$.

The fraction of all atoms M present in each of these substances is denoted by x_n (n = the number of ligands in the complex ML_n) and becomes $x_0 = [M]/[M]_{tot}$, $x_1 = [ML]/[M]_{tot}$, ... $x_n = [ML_n]/[M]_{tot}$.

An example is the formation of ammonia complexes (ammines) of Cu^{2+}, which is shown in Fig. 10.2 (A = NH_3). The various x_n have been plotted as functions of pA, so that each complex has an area bounded by the area of the next complex along an S curve. pk_1 to pk_5 have been directly measured, but pk_6 is estimated from absorption spectra, so that the S curve between CuA_5^{2+} and CuA_6^{2+} is only approximately known (dashed). At any pA value, x_n = the length of the ordinate that falls within the area of the complex MA_n. For example, at pA = 4 (thus, $[NH_3] = 10^{-4}$ mol l^{-1}), $x_0 = 0.244$ (corresponding to Cu^{2+}), $x_1 = 0.497$ (CuA^{2+}), $x_2 = 0.234$ (CuA_2^{2+}), $x_3 = 0.024$ (CuA_3^{2+}), $x_4 = 0.001$ (CuA_4^{2+}), $x_5 = 1.5 \times 10^{-8}$ (CuA_5^{2+}). x_4 and x_5 are here too small to be read from the figure. The sum of these six x values is 1.

With CuA_4^{2+} the square coordination characteristic of Cu^{2+} (5-3i, j) is attained. For an appreciable addition of a fifth NH_3 molecule a significant increase in $[NH_3]$ is required, and this is still more true for the addition of a sixth NH_3 molecule, with which an octahedral NH_3 coordination would be completed. It must be remembered, however, that Cu^{2+} already in a NH_3-free water solution is certainly coordinated with at least four, perhaps six, H_2O (5-3i, 5-6c), and that the NH_3 "addition" thus in fact implies a replacement of H_2O by NH_3. The "addition" of the first NH_3 molecule thus does not refer to the equilibrium $Cu^{2+} + NH_3 \rightleftharpoons CuNH_3^{2+}$, but to something like $Cu(H_2O)_6^{2+} + NH_3 \rightleftharpoons Cu(H_2O)_5NH_3^{2+} + H_2O$. Since solvation is so common, similar processes may be nearly always assumed to occur when complexes are formed in solution. Extensive studies of ammine formation have been carried out by J. Bjerrum.

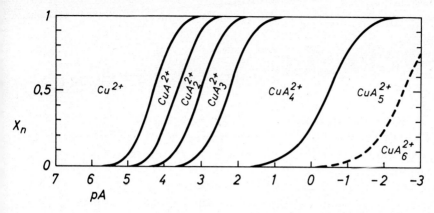

Fig. 10.2. Ranges of existence of ammines of Cu^{2+}. $A = NH_3$.

In stepwise addition or dissociation M is incorporated simultaneously in more than two kinds of molecule, whose concentrations are all dependent on each other. ($[M]_{tot}$ then consists of more than two terms.) Then the shapes of the S curves also become dependent on each other. In fig. 10.2, the S curve that forms the boundary between CuA_4^{2+} and CuA_5^{2+} is rather independent of the others, however, because of its large distance from them. In the range of this curve almost the entire amount of Cu is present only as CuA_4^{2+} and CuA_5^{2+} ($[Cu]_{tot} \approx [CuA_4^{2+}] + [CuA_5^{2+}]$) and the curve thus has practically the same shape as the S curve in Fig. 10.1. The four left-hand S curves, on the other hand, because of the close-lying pk values, are more dependent on each other and therefore deviate appreciably from the shape in fig. 10.1.

If a certain complex were to appear in a single addition process, it would be formed within a much smaller pA interval than is the case with stepwise formation according to fig. 10.2. If CuA_4^{2+} were formed directly, the transition from $x_{Cu^{2+}} = 0.99$ to $x_{CuA_4^{2+}} = 0.99$ would take place with a change in pA of one unit while the corresponding change in stepwise formation requires a change in pA of over 4.5 units.

When two different atomic or molecular species A and C compete for a third, B, the reaction can be treated as an interaction between two simpler equilibria according to

$$\begin{aligned} AB &\rightleftharpoons A+B \\ C+B &\rightleftharpoons CB \\ \hline AB+C &\rightleftharpoons A+CB \end{aligned}$$

The total reaction may be termed as a *substitution*, B *transfer* or B *exchange*. If the dissociation constants for AB and CB are defined by $\{A\}\{B\}/\{AB\} = K_{AB}$ and $\{C\}\{B\}/\{CB\} = K_{CB}$ we have

$$\{B\} = K_{AB} \frac{\{AB\}}{\{A\}} = K_{CB} \frac{\{CB\}}{\{C\}} \tag{10.25}$$

The activity of the transferred atom or molecule, $\{B\}$, evidently determines the position of the equilibrium in both the "AB system" and the "CB system", and therefore also the position of the total equilibrium.

Two examples are given here, of which I may occur in a melt and II in water solution:

I. $\quad 2MgO \rightleftharpoons 2Mg^{2+} + 2O^{2-}$ $\qquad$ II. $\qquad H_2O \rightleftharpoons OH^- + H^+$

$$\frac{SiO_2 + 2O^{2-} \rightleftharpoons SiO_4^{4-}}{2MgO + SiO_2 \rightleftharpoons 2Mg^{2+} + SiO_4^{4-}} \qquad \frac{NH_3 + H^+ \rightleftharpoons NH_4^+}{H_2O + NH_3 \rightleftharpoons OH^- + NH_4^+}$$

The equilibrium position is determined in I by the anion activity $\{O^{2-}\}$ and in II by the cation activity $\{H^+\}$. II is an example of a very important class of equilibria, the transfer of protons, hydrogen ions, between different molecular types. The significance of proton transfer rests on the fact that it is so prevalent in water, by far the most common solvent in natural inorganic systems on the earth's surface, in biological systems and in technology. The substances participating in proton transfer are called acids and bases, and equilibria between these are treated in detail in chap. 12.

d. The dependency of equilibrium on temperature and pressure. With *increase in the total pressure of a system* the equilibrium is generally shifted in such a direction that will lead in itself to a *decrease in volume*. From the ΔV^0 value (increase in volume when all participating substances are in their standard states) for the reaction, the dependency of the equilibrium on pressure may be derived. If only solid and liquid substances enter into the equilibrium the change in volume with reaction is small and the influence of pressure is noticeable only with very large changes in pressure. If gases participate, on the other hand, the volume change (measured at constant pressure) becomes large if the number of gas molecules is changed with reaction. The position of the equilibrium is then strongly dependent on the pressure.

For a given equilibrium between *ideal gases* we have

$$K_x P^{\Delta n} = K_c (RT)^{\Delta n} = K_p \qquad (10.26)$$

where Δn is the increase in the number of gas molecules in the process. If, for example, all the substances in the general process (10.8) are gaseous, then $\Delta n = (l + m + ...) - (a + b + ...)$. The constants K_c and K_p are independent of the total pressure P, but such is not the case with K_x. For $\Delta n = 0$, $K_x = K_c = K_p$.

Since K_c and K_p are independent of the total pressure at equilibrium between ideal gases, it is easy in such cases to show how a change in pressure shifts the equilibrium position. For example, for the dissociation equilibrium $PCl_5 \rightleftharpoons PCl_3 + Cl_2$ we have $p_{PCl_3} p_{Cl_2}/p_{PCl_5} = K_p$, where K_p is constant so long as the gases can be assumed to be ideal. If the degree of dissociation is α, the number of moles of undissociated PCl_5 is proportional to $(1-\alpha)$ and the number of moles of PCl_3 and Cl_2 are each proportional to α. The total number of moles is proportional to $(1-\alpha) + 2\alpha = (1+\alpha)$. The partial pressures are thus $p_{PCl_5} = (1-\alpha)P/(1+\alpha)$ and $p_{PCl_3} = p_{Cl_2} = \alpha P/(1+\alpha)$. If these are substituted into the equilibrium equation we get $\alpha^2 P/(1-\alpha^2) = K_p$. This expression shows that with increase in pressure the degree of dissociation α, that is, the number of gas molecules, decreases, and thereby also the volume.

A dissociation equilibrium in which the number of molecules does not change is $2HI \rightleftharpoons H_2 + I_2$. Here the equilibrium position is independent of pressure as long as the gases are ideal. If the degree of dissociation is α, we obtain analogously with the above case $\alpha^2/4(1-\alpha)^2 = K_p$, which shows that α is independent of the pressure.

The gas equilibrium $N_2 + 3H_2 \rightleftharpoons 2NH_3$ determines the yield when ammonia is prepared according to the Haber-Bosch process (22-6a, b). The number of molecules, and consequently the volume, is decreased if the equilibrium is shifted to the right. Increase in pressure thus causes a shift in this direction, that is, toward a larger proportion of NH_3 in the equilibrium mixture.

With an *increase in temperature*, the equilibrium of a system in general is shifted in such a direction that the reaction in itself will lead to *absorption of heat*. With most dissociations heat is taken up ($\Delta H^0 > 0$), which means that the degree of dissociation increases with a rise in temperature.

In the formation of ammonia according to $N_2 + 3H_2 \rightleftharpoons 2NH_3$ heat is given off ($\Delta H^0 < 0$) and the equilibrium is therefore shifted to the left with a rise in temperature. In order to obtain the greatest possible proportion of NH_3 in the equilibrium mixture the system should thus be at the lowest temperature that will permit a sufficiently rapid attainment of equilibrium. With a proper catalyst, it is possible to go to a quite low temperature before the attainment of equilibrium becomes too slow (22-6a, b).

The shift of equilibrium with change of pressure or temperature as well as concentration (10b, c) has commonly been said to be an example of a general law, *Le Chatelier's principle*, which may be stated in somewhat the following way: "If a variable of state of a system in equilibrium is changed, a reaction takes place that, if it occurred alone, would change the variable of state in the opposite direction." This statement, however, is not sufficiently precise and can lead to incorrect conclusions. It must therefore be broadened, but then the "principle" becomes too cumbersome to be of practical use. Instead it may simply be remembered that an increase in pressure or temperature shifts a chemical equilibrium so that the reaction in itself will lead to a decrease in volume or absorption of heat, respectively.

In the following, equilibria of many kinds will be discussed in various connections. Chemistry is much concerned with liquid solutions and the solution equilibria that can occur in them therefore play a large role. Very often ions exist in the solutions. General questions concerning ionic equilibria are discussed in chap. 11. Important types of homogeneous ionic equilibria are then considered in chap. 12. Heterogeneous equilibria are first treated in chap. 13 and then in several of the following chapters.

An extensive and very useful collection of constants for equilibria in which metal and hydrogen ions take part is to be found in Stability Constants of Metal-Ion Complexes (Chemical Society, London 1964). This book also contains values for solubility products and constants for redox equilibria (standard potentials).

CHAPTER 11

Ionic Equilibria

a. Electrolytes. If a phase contains ions that are sufficiently mobile, they possess the ability to carry electric current. As was mentioned in 6-3c the ionic mobility in solid phases is generally too small to carry current at room temperature. In such cases, therefore, the ionic conductivity is low. In liquid phases (solutions, melts) that contain ions, their mobility is sufficient, on the other hand, to give quite high conductivity, if the ionic content is not too low.

With the flow of current ions thus migrate between the electrodes, which implies transport of charge and matter simultaneously. At each electrode ions give electrons to, or take electrons from, the electrode. However, it is not necessarily the particular ion type that carries the main part of the current which exchanges electrons at the electrodes. At a particular electrode the exchange required by the direction of the current takes place mainly with the type of ion in the solution that has the greatest tendency for such an exchange, and is present in a sufficiently high concentration. No accumulation of charge occurs, since the quantity of charge exchanged at the electrodes always is equal to that transported between the electrodes. With the flow of current, however, the system is altered, partly by material transport between the electrodes, and partly by the exchange of electrons with them. Such processes were first observed at the beginning of the 19th century and were called *electrolysis* (Gk. λύσις, releasing, dissolution). Those substances that, pure or in solution, carried current under electrolysis were called *electrolytes*.

The ions in a phase can partake in a large number of equilibria. In the simplest case of two kinds of ions A^+ and B^- the equilibrium $AB \rightleftharpoons A^+ + B^-$ is very important. This may be said to express the *electrolytic dissociation* of the substance AB. The equilibrium equation becomes

$$\frac{\{A^+\}\{B^-\}}{\{AB\}} = K \tag{11.1}$$

and for an ideal state

$$\frac{[A^+][B^-]}{[AB]} = K_c \tag{11.2}$$

If the ions A^+ and B^- participate only in the equilibrium given above, then (10.19) applies to this electrolytic dissociation. When the total concentration of the electrolyte AB decreases, the degree of dissociation α thus increases. The same

Table 11.1. *Ionic mobilities* u_i *(cm² V⁻¹s⁻¹) at high dilution and 25°C.*

	H^+	Li^+	Na^+	K^+	OH^-	Cl^-	NO_3^-
$u_i \times 10^4$:	36.2	4.0	5.2	7.6	20.7	7.9	7.4

is true also for electrolytes that dissociate in more complicated ways than in this simple example.

b. Ionic migration during electrolysis. Let us assume that the electric current is carried only by ions. Electronic conduction is thus excluded. It must then be true that *the weight of a substance deposited at an electrode is proportional to the amount of electricity that has passed between the electrodes.*

From the definition of electrochemical equivalent (1-4d) it follows that *the weights of different substances deposited by the same amount of electricity are proportional to the electrochemical equivalent weights of the substances.*

These laws were shown to be true by Faraday's experiments in the early 1830's (*Faraday's electrolytic laws*).

In electrochemical context one also speaks of an equivalent of a formula unit that is composed of ions. An equivalent of $Al_2(SO_4)_3$ would contain one equivalent of each kind of ion, that is, $\frac{1}{3}$ mol $Al^{3+} + \frac{1}{2}$ mol SO_4^{2-}. It is made up thus of $\frac{1}{6}$ mol $Al_2(SO_4)_3$.

Here it is assumed that Ohm's law applies for ionic conduction. During electrolysis under given conditions (a definite phase; for example, a particular solution at a definite temperature) each kind of ion moves with a velocity that is proportional to the strength of the electric field. If in one case 10 V and in another case 20 V are applied across an electrolytic cell with a distance between electrodes of 5 cm, the field strength is 2 and 4 V cm⁻¹, respectively. Each ion in the latter field moves twice as fast as in the former. The velocity of an ion "i" in a field of strength 1 V cm⁻¹ is called the *ionic mobility* u_i of this ion in the given phase. The unit of u_i customarily is cm s⁻¹/V cm⁻¹ = cm² V⁻¹s⁻¹.

The ionic mobility generally can be easily measured. One method is to measure the shift of the boundary surface between two electrolytic solutions layered on each other during flow of current. Values for some common ions in very dilute water solutions at 25°C are listed in tab. 11.1.

For example, if a solution containing Na^+ ions is electrolyzed with 20 V across an electrode distance of 1 cm, the ions attain the velocity $20 \times 5.2 \times 10^{-4}$ cm s⁻¹ = 0.10 mm s⁻¹. The distance 1 cm is then traversed by a Na^+ ion in 100 s.

Now, in general, ions are solvated (5-2h, i). The water shell that surrounds the ion in a water solution is carried along during ionic migration and strongly affects the ionic mobility. This is the reason that in the series $Li^+-Na^+-K^+$ the u values increase with increasing ionic radius. Otherwise, for these ions one would expect that the smallest ion would most rapidly make its way among the water molecules. But as mentioned in 5-2h the degree of hydration of the alkali-

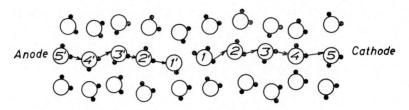

Fig. 11.1 Proton jumps toward the cathode, causing apparent migration of H_3O^+ toward the cathode and OH^- toward the anode.

metal ions decreases with increasing ionic radius. Of the three ions Li^+ is thus the most hydrated and moves most slowly, and K^+ is the least hydrated and moves fastest.

The high mobility of the ions H^+ and OH^- is remarkable. The true mobility of these ions, however, is surely of the same order of magnitude as for other ions. The reason the ions appear to move so fast depends on the jumping of protons from molecule to molecule.

Properly speaking, free H^+ ions (protons) are not found in water solutions, but these ions are always hydrated, in the first place to oxonium ions H_3O^+ (5-3d). Besides foreign molecules from dissolved substances, water solutions therefore always contain the molecular species H_2O, H_3O^+ and OH^- (more about this in 12-1a). The two last-named species are also formed in pure water by the transfer of a proton from one water molecule to a neighbor. In fig. 11.1, which is very schematic (among other things hydrogen bonds between the H_2O molecules are not shown), a proton has transferred from 1' to 1. In pure water equal numbers of H_3O^+ and OH^- ions are formed in this way, although their number is very small. If an electric field is applied across water or a water solution, proton jumps toward the cathode take place. These may begin with the H_3O^+ ion 1 to the H_2O molecule 2 and continue along the arrows in the sequence 2, 3, 4, 5. It is easy to see that this can apparently be interpreted as a migration of an H_3O^+ ion in the same direction. The proton jumps are extraordinarily fast and during the larger part of the time the protons are bound to oxygen atoms. A proton can also jump from the H_2O molecule 2' to the OH^- ion 1', whereupon follow proton jumps from 3', 4', 5'. This appears as a migration of an OH^- ion toward the anode. The reason that a greater apparent migration velocity is found for H_3O^+ than for OH^- depends on the fact that less energy is required to remove a proton from H_3O^+ than from H_2O.

c. Weak and strong electrolytes. When a substance dissolved in water is electrolytically dissociated (the substance is thus an electrolyte) the presence of undissociated molecules of the substance can often be established by various methods, at least if the concentration of the substance is not very low. The electrolyte then is clearly incompletely dissociated, that is, the degree of dissociation $\alpha < 1$. With decreasing concentration of the electrolyte α increases and may approach the value 1 as a limiting value for the concentration 0. The conductivity of the solution per equivalent of dissolved electrolyte (*equivalent conductivity*) therefore increases with decreasing concentration. A substance of this type is said to behave as a *weak electrolyte* in the solution in question.

For many electrolytes, however, the presence of undissociated molecules cannot be proved at any concentration. This generally happens when an electrolyte that in solid form occurs as ionic crystals (6-2a) is dissolved in water. Here the crystal is built up of ions, and it is quite natural that these would persist, although most often hydrated, also in solution. To be sure, in the solution there will be ion pairs and ion clusters (5-2a), but these have varying sizes and shapes and also have a short enough mean lifetime so that in a short space of time they are broken up and reformed. It is obvious that it is difficult to speak here of any limited electrolytic dissociation. These *strong electrolytes* are therefore considered as always completely dissociated. The electrostatic forces here play a very great role in addition to the part they play in pair and cluster formation. In general, they cause an ion to be surrounded by more ions of opposite charge than of the same charge. The distribution of ions in a solution from this viewpoint shows a certain similarity to the distribution in an ionic crystal, but, of course, is much more irregular. When an electric field is applied and ions of opposite charge are drawn in different directions, the electrostatic attraction has a strongly retarding effect. The retardation becomes greater, the more closely to each other the ions lie, that is, the greater the concentration of ions in the solution is. Here also, the equivalent conductivity is thus decreased as the concentration of the electrolyte is increased, and vice versa.

In solutions of weak electrolytes electrostatic forces between the ions are also present, but here these are much less prominent. The higher the total concentration of the electrolyte is, the smaller is its degree of dissociation. A high degree of dissociation can only be attained at low total concentration. The ionic concentration is therefore always low.

The majority of salts (for the concept of "salt" see 12-1 c) of alkali and alkaline-earth metals form typical ionic crystals and are strong electrolytes. The same is true, for example, of a large number of sulfates of divalent transition metals. Many other substances, which formerly were designated as salts, are weak electrolytes and also do not form typical ionic crystals. An example is $HgCl_2$, in which the crystal-building units are $HgCl_2$ molecules.

The "strong" acids are also counted as strong electrolytes in water solution, although in very concentrated solutions they exist to a greater or lesser degree in molecular form. The conditions that affect the strength of acids are treated in more detail in 12-2.

In comparing electrolytes that are weak in water solution with those that are strong, it is found that the bonds between the ions often have a more covalent character in the weak electrolytes than in the strong. However, there are cases in which undissociated molecules in solutions are held together mainly by electrostatic forces. These may then give typical molecules and not just accidental and poorly defined ion pairs and ion clusters. These molecules are usually discernible as building units in a crystal of the substance. If the strong acids are excepted, it may therefore be said that, in general, a strong electrolyte forms an ionic crystal, and a weak electrolyte a molecular crystal.

Because of the great possibility for variation both in bond type and bond strength, it is clear that no sharp boundary can be drawn between strong and weak electrolytes. Their character also depends on the solvent. The strength of the electrostatic forces depends on the nature of the solvent. The force between two charged bodies is inversely proportional to the dielectric constant (5-2g) of the medium that lies between them. The higher the dielectric constant of the solvent is, the smaller is the electrostatic attraction between oppositely charged ions and thus the greater is the solvent's "dissociating ability." Water has a very high dielectric constant (5-5) and its dissociating ability is therefore especially great. Weak molecule-forming forces may perhaps only be effective in solvents with low dissociating ability but cannot cause the formation of molecules in solvents with high dissociating ability such as water. The electrolyte in such cases may be "weak" in the former but "strong" in the latter.

In equilibrium calculations in which concentrations are used, the equilibrium equation may be applied for weak electrolytes, for example in the form (11.2), but for strong electrolytes complete dissociation is always assumed.

d. Activities in solutions of electrolytes. The electrostatic forces in a solution of an electrolyte cause the solution to depart from the ideal state. The activity factors for the various kinds of molecules in the solution therefore deviate from 1 even at low concentrations. Because of their charges the ions themselves are influenced much more strongly than uncharged molecules.

For a dissolved substance the activity scale is chosen so that the activity approaches the concentration with increasing dilution (10a). Thus, for an infinitely dilute solution, $f = 1$ for all dissolved substances. When the ionic concentration and thus the quantity of charged particles is increased, the f values of the ions drop off, at least at the beginning. For neutral molecules the departure from the value 1 is significantly smaller and often can be neglected in approximate calculations.

For a certain ion in dilute solution, f is very closely determined by the *ionic strength* I of the solution. If a solution contains the ionic species A, B, C, ... with the charges $z_A, z_B, z_C, ...$ and in the concentrations $c_A, c_B, c_C, ...$ its ionic strength is defined as

$$I = \tfrac{1}{2}(c_A z_A^2 + c_B z_B^2 + c_C z_C^2 + ...) = \tfrac{1}{2} \Sigma c_i z_i^2 \tag{11.3}$$

Thus we have for

$$0.01 \text{ M KCl}, \quad I = \tfrac{1}{2}(0.01 + 0.01) = 0.01$$
$$0.01 \text{ M CaCl}_2, \quad I = \tfrac{1}{2}(0.01 \cdot 2^2 + 0.02) = 0.03$$
$$0.01 \text{ M MgSO}_4, \quad I = \tfrac{1}{2}(0.01 \cdot 2^2 + 0.01 \cdot 2^2) = 0.04$$

Table 11.2. *Activity factors f_i calculated from (11.4).*

I	$\dfrac{\sqrt{I}}{1+\sqrt{I}}$	f_i		
		$z_i = 1$	$z_i = 2$	$z_i = 3$
0	0	1.00	1.00	1.00
0.001	0.03	0.97	0.87	0.73
0.002	0.04	0.95	0.82	0.64
0.005	0.07	0.93	0.74	0.51
0.01	0.09	0.90	0.66	0.40
0.02	0.12	0.87	0.57	0.28
0.05	0.18	0.81	0.43	0.15
0.1	0.24	0.76	0.33	0.1
0.2	0.31	0.70	—	—
0.5	0.41	0.62	—	—

In a solution that is 0.01 M with respect to both KCl and $MgSO_4$, $I = 0.01 + 0.04 = 0.05$.

For ionic strengths at least up to $I = 0.1$ the following expression, based on the analysis by P. Debye and E. Hückel of the way the activity is affected by the electrostatic forces, generally is approximately true:

$$-\log f_i = 0.5\, z_i^2 \sqrt{I}/(1+\sqrt{I}) \tag{11.4}$$

The factor 0.5 presupposes water as the solvent and a temperature of approximately 25°C.

At these low ionic strengths f_i is thus to a large degree a function only of z_i and I; $-\log f_i$ increases, that is, f_i decreases with increasing values of z_i and I. Tab. 11.2 lists a number of f_i values calculated from (11.4).

If the f values for the ions of a strong electrolyte decrease with increasing ionic strength, it is clear that they decrease with increasing concentration of the strong electrolyte itself. In this way, among other things, osmotic pressure, freezing-point depression and boiling-point elevation are affected in such a way that it was earlier believed that the degree of dissociation even of strong electrolytes decreased with increasing concentration. This was supposed to be confirmed by the fact that the equivalent conductivity of strong electrolytes also decreased with increased concentration (11 c).

At higher ionic strengths deviations from (11.4) set in. The deviations are characteristic of individual ions and as a rule increase rapidly with the ionic strength. This is particularly true of ions with high charge, whose activities even at constant ionic strength are often dependent on the nature of other ionic species in the solutions. Often f passes through a minimum between 0.5 and 2 in I, and then increases with increasing I. (11.4) must therefore be used with care. However, the expression is useful in approximate calculations and at moderate ionic strengths.

Since the activity of a substance according to the definition (10.5) depends on the difference in μ for the substance in two different states, the activity of an individual ion cannot be exactly measured. One ionic species cannot be moved from one solution to another without at the same time moving ions of opposite charge. It is only possible to measure exactly the *mean ionic activity* of an electrolyte, which is the geometric mean between the ionic activities of the different kinds of ions in the electrolyte. This does not prevent the measurement and application of approximate activity values of individual ions for the sake of convenience. A very important such case is the hydrogen ion activity (12-2b).

Let us assume an ionic equilibrium, for example, the electrolytic dissociation $AB \rightleftharpoons A^+ + B^-$. The equilibrium equation is

$$\{A^+\}\{B^-\}/\{AB\} = K \tag{11.5}$$

If activities are replaced by concentrations we may write

$$[A^+][B^-]/[AB] = k \tag{11.6}$$

By dividing (11.5) by (11.6) and introducing f values according to the definition (10.7) we have

$$k = K f_{AB}/f_{A^+} f_{B^-} \tag{11.7}$$

If all the molecular species taking part in the equilibrium are dissolved substances, then at the limiting value $I = 0$ their f values $= 1$ and thus $k = K$. As I increases f_{A^+} and f_{B^-} fall off while f_{AB} is changed relatively little because AB is uncharged. Since K is a true constant, k thus increases with I.

Inasmuch as k is dependent on the ionic strength of the solution, such is also the case with the degree of dissociation of the electrolyte, which expresses a concentration ratio. The degree of dissociation generally also increases with increasing ionic strength.

As long as the f values of the participating molecular species are constant, which often can be assumed at constant ionic strength, k is constant, however. If the k value is known for the prevailing ionic strength, equations of the type (11.6) can therefore be used for calculations of equilibria at this ionic strength. If the ionic strength changes, on the other hand, in more accurate calculations the fact that k does not remain constant must be taken into account. In the study of ionic equilibria the total ionic strength of the solution is often kept very high by the addition of a large amount of an electrolyte whose ions do not enter into the equilibrium in question. The changes in ionic strength caused by shifts in this equilibrium thus become proportionately small and can therefore be neglected.

e. Historical. When the electrolysis phenomena were first studied, it was believed that the electrolyte normally existed in molecular form and was split into ions only in the electric field. However, Svante Arrhenius showed in 1887 that the electrolytes in a solution are always split into ions to a greater or lesser degree ("theory of electrolytic dissociation"). Arrhenius thought it possible to divide electrolytes into strong electrolytes, which even in rather concentrated solutions are strongly dissociated, and weak electrolytes, which in such solutions are weakly dissociated. For the weak electrolytes it was possible to apply the equilibrium equation with advantage to the dissociation equilibrium. This was not successful, however, with the strong electrolytes. For these, different methods of measurement gave different values for the degree of dissociation. Niels Bjerrum found that the light absorption per unit of quantity of a colored electrolyte in water solution changed with changes of concentration for weak electrolytes, but was constant for strong electrolytes. This indicated that no dissociation equilibrium was involved in the latter case and Bjerrum assumed in 1909 that strong electrolytes, at least in water solution, are always completely dissociated. Through further studies carried out both by Bjerrum and other scientists, and theoretical work especially by Debye and Hückel, it gradually became quite clear that Bjerrum's assumption was correct.

CHAPTER 12

Acid-Base Equilibria

In 10c reactions were discussed that involve transfer or exchange of one kind of atom or molecule. As an especially important special case of this type of reaction, proton transfer was mentioned. It is the substances taking part in this proton transfer that are called acids and bases. Because of their great significance, the acid-base reactions are treated quite fully in this chapter. The material given here in principle can be applied also to other types of transfer reactions.

12-1. The acid-base concept

a. Brønsted's definition of acid and base. In every *proton transfer* (*proton exchange*, *protolysis*), there must be a *proton donor* and a *proton acceptor*. In 1923, Brønsted defined proton donors as *acids* and proton acceptors as *bases* (at approximately the same time the same definition was given by Lowry; however, since Brønsted treated the question more fully, the definition is often associated only with his name). The definition may be expressed by the scheme

$$\underset{\text{proton donor}}{\text{acid}} \rightleftharpoons \underset{\text{proton acceptor}}{\text{base}} + H^+ \tag{12.1}$$

Acids, like bases, may consist of uncharged as well as charged molecules (thus, in the latter case, ions). An acid and a base that are associated according to this scheme are called *conjugate* or *corresponding*, or said to form an *acid-base pair*. Examples are:
$HAc \rightleftharpoons Ac^- + H^+$, $NH_4^+ \rightleftharpoons NH_3 + H^+$, $H_2CO_3 \rightleftharpoons HCO_3^- + H^+$, $HCO_3^- \rightleftharpoons CO_3^{2-} + H^+$.
Acids and bases that can give up or take up more than one proton are called *polyvalent* (*divalent*, *trivalent*, etc.) or *polyprotic* (*diprotic*, *triprotic*, etc.). H_2CO_3 is a divalent acid, CO_3^{2-} a divalent base, PO_4^{3-} a trivalent base.

Acids and bases are collectively termed *protolytes*.

All protolytes, which like H_2CO_3, HCO_3^- and CO_3^{2-} may transform into one another by proton transfer, are said to belong to the same *acid-base system* or *protolytic system*. The system H_2CO_3–HCO_3^-–CO_3^{2-} contains two acid-base pairs. It is often practical to name a protolytic system after one of the member protolytes. The system just referred to, for example, may thus be called the carbonic acid system or the carbonate system.

Acid-Base Equilibria

In systems with several acid-base pairs each intermediate stage such as HCO_3^-, $H_2PO_4^-$, HPO_4^{2-} can function both as an acid and as a base. Such a protolyte is said to have *amphoteric* (Gk. ἀμφότερος, on both sides) or *amphiprotic* properties and is called an *ampholyte*.

The hydrogen atom is always attached to other atoms by a bond that has a strongly covalent character (the electronegativity of hydrogen lies in an intermediate range, 5-3f). Therefore, it never appears in any acid as a typical hydrogen ion (proton). Neither can a proton exist for any length of time in an acid-base system, but may at most quickly be carried over from an acid to a base, thereby forming a new acid. This implies that two acid-base pairs must function simultaneously. Schematically, we have (a = acid, b = base)

$$a_1 \rightleftharpoons b_1 + H^+$$
$$\underline{H^+ + b_2 \rightleftharpoons a_2}$$
$$a_1 + b_2 \rightleftharpoons b_1 + a_2 \tag{12.2}$$

The total reaction (12.2) shows proton transfer from the acid a_1 to the base b_2, the *acid-base process* or *protolysis*. The equilibrium is called an *acid-base equilibrium* or *protolysis equilibrium*.

If an acid is dissolved in a pure solvent it can be protolyzed ("dissociated") only if the solvent is a base. In the same way a base can be protolyzed only if the solvent is an acid. In a solvent that lacks basic properties, for example benzene, an acid can thus not form any ions but is dissolved in molecular form. If a base is also dissolved, however, protons are immediately transferred from the acid to the base. The "acidity" of the acid, that is, the tendency to donate protons, can thus only become effective in the presence of a base. Solvents that can neither take up protons (function as a base), or give up protons (function as an acid), can protolyze neither acids nor bases. Benzene is an example of such a *nonprotolyzing* or *aprotic* solvent.

H_2O enters as an amphoteric intermediate into the protolytic system $H_3O^+-H_2O-OH^-$. Therefore water can function as an acid according to

$$H_2O \rightleftharpoons OH^- + H^+ \tag{12.3}$$

and as a base according to

$$H_2O + H^+ \rightleftharpoons H_3O^+ \tag{12.4}$$

Both acids and bases are therefore protolyzed when they are dissolved in water. A considerable number of other solvents have an analogously amphoteric character.

The equations (12.3) and (12.4) are schematic. As mentioned earlier in 5-5 the H_2O molecules are bound together by hydrogen bonds. Single H_2O molecules are therefore present in quite small amounts and the major portion of the water consists of molecular groups of widely varying size and shape. For the same reason truly free OH^- and H_3O^+ ions are quite rare. In order to show water which has given up or taken up protons, however, we write most simply OH^- and H_3O^+. The large groups may then be considered as more or less hydrated OH^- or H_3O^+ ions. The ion H_3O^+, which is called *oxonium ion*[1], has been discussed in 5-3d.

The ion H_3O^+ has been found in the crystallized acid hydrates $HClO_4 \cdot H_2O$ (20-8e) and $H_2SO_4 \cdot H_2O$ (21-11b1). A more hydrated ion $H_5O_2^+$, the structure of which can be shown by the formula $[H_2OHOH_2]^+$, exists in the hydrates $HClO_4 \cdot 2H_2O$ (20-8e), $HCl \cdot 2H_2O$ and $HCl \cdot 3H_2O$ (20-5c). Several facts seem to indicate that the still more hydrated ion $H_9O_4^+$ may be of some importance in water solutions. In this each of the three hydrogen atoms of the H_3O^+ ion would take part in a hydrogen bond to the oxygen atom of a water molecule.

If acetic acid is dissolved in water, the proton exchange is thus shown by the equations

$$HAc \rightleftharpoons Ac^- + H^+ \qquad (12.5)$$
$$ a_1 \phantom{{}\rightleftharpoons{}} b_1$$

$$H^+ + H_2O \rightleftharpoons H_3O^+ \qquad (12.6)$$
$$\phantom{H^+ + {}}b_2 \phantom{{}\rightleftharpoons{}} a_2$$

which makes up an acid-base equilibrium according to the equation

$$HAc + H_2O \rightleftharpoons Ac^- + H_3O^+ \qquad (12.7)$$
$$a_1 \quad b_2 \quad\quad b_1 \quad a_2$$

What was earlier called the "dissociation" of acetic acid and written as in (12.5) is thus the result of a proton exchange according to (12.7). The free hydrogen ion or proton H^+ does not exist in the acetic acid solution, any more than in any other solution. For the sake of brevity, one still often speaks of the hydrogen ion and writes schematically H^+, in the same way that a protolysis is often written as in (12.5). If it is desired expressly to emphasize the proton exchange, however, the concept of the oxonium ion, H_3O^+, is used and equations written as in (12.7). Then the idea that the protons are bound to water is expressed in the simplest way.

If instead a base is dissolved in water, for example ammonia, it reacts according to (12.3) and the acid-base equilibrium becomes

$$NH_3 + H_2O \rightleftharpoons NH_4^+ + OH^- \qquad (12.8)$$
$$b_1 \quad a_2 \quad\quad a_1 \quad b_2$$

Water solutions of acids thus contain oxonium ions, and of bases, hydroxide ions. The appearance of these ions, however, is wholly dependent on the fact that they are the acid and base, respectively, which correspond to the solvent, water. In another solvent other ions appear. Ethanol functions as both acid and base according to the equations

$$C_2H_5OH \rightleftharpoons C_2H_5O^- + H^+$$
$$C_2H_5OH + H^+ \rightleftharpoons C_2H_5OH_2^+$$

---

[1] The nature of the ligands is stated expressly by the more complete name *hydroxonium ion*, which earlier was often shortened to *hydronium ion*. Since this name gives the impression that hydrogen is the central atom it has now been abandoned.

Every solution of an acid in ethanol is therefore characterized by the ion $C_2H_5OH_2^+$ and every solution of a base by the ion $C_2H_5O^-$.

A charged protolyte naturally cannot be handled as free but must be incorporated in a substance in which its charge is compensated by ions of opposite charge. The base Ac^- may occur in NaAc, for example, which consists of Na^+ and Ac^-. When NaAc is dissolved in a solvent, nothing aside from solvation happens to the Na^+ ion. On the other hand, the base Ac^- can be protolyzed in water according to

$$Ac^- + H_2O \rightleftharpoons HAc + OH^- \tag{12.9}$$

Protolysis of a charged acid takes place, for example, if NH_4Cl is dissolved in water. NH_4^+ is thereby protolyzed according to

$$NH_4^+ + H_2O \rightleftharpoons NH_3 + H_3O^+ \tag{12.10}$$

Reactions such as these have since early times been called *hydrolysis* but are not different in principle from protolysis of uncharged protolytes according to the examples (12.7) and (12.8).

Acid-base processes may be more complicated than those mentioned up to now, which all involved only a proton transfer. Sometimes a substance that does not contain protons, or cannot by itself donate them, functions as an acid together with a proton-containing substance, most often water. Equation (5.18) showed such a process, and an example completely analogous to this is the dichromate ion $Cr_2O_7^{2-}$, which together with water functions as an acid according to the scheme

$$Cr_2O_7^{2-} + H_2O \rightleftharpoons 2CrO_4^{2-} + 2H^+ \tag{12.11}$$

In order to apply Brønsted's acid-base definition the combination ($Cr_2O_7^{2-}$ + H_2O) must be considered as the acid and the chromate ion CrO_4^{2-} as the base. We have here a transition from the dinuclear ion $Cr_2O_7^{2-}$ to the mononuclear ion CrO_4^{2-}, and similar processes are common in connection with changes in the number of nuclei in complex ions.

An example of a process of another kind is the protolysis of boric acid, which can be written as $B(OH)_3 + H_2O \rightleftharpoons B(OH)_4^- + H^+$.

Acid-base processes, which only involve a proton transfer, reach equilibrium practically instantaneously in homogeneous systems. On the other hand if the process also involves a rearrangement of other atoms as in (12.11), the reaction rate may be significantly reduced and often becomes very slow.

b. Metal ions as acids. Water solutions of metal cations are often acid, which is often said to be a result of *hydrolysis*. In general, this behavior is caused by quite complicated processes. Sillén and his coworkers have contributed primarily to the elucidation of these processes.

The simplest case is that where only mononuclear complexes are formed. The acid function of a Hg^{2+} ion, for example, may be schematically represented by the equations

$$Hg^{2+} + H_2O \rightleftharpoons HgOH^+ + H^+ \tag{12.12}$$

$$HgOH^+ + H_2O \rightleftharpoons Hg(OH)_2 + H^+ \tag{12.13}$$

Analogously with the example of dichromate in 12-1a we may then refer to $(Hg^{2+} + H_2O)$ and $(HgOH^+ + H_2O)$ as the acids that cause the acid reaction. As a matter of fact we are dealing with acids in the original sense of Brønsted. The Hg^{2+} ion is certainly hydrated and behaves as an acid by donating protons from H_2O in the water shell. Since the coordination number of Hg^{2+} is probably always 2 the most probable reactions are

$$Hg(H_2O)_2^{2+} \rightleftharpoons Hg(H_2O)OH^+ + H^+ \tag{12.14}$$

$$Hg(H_2O)OH^+ \rightleftharpoons Hg(OH)_2 + H^+ \tag{12.15}$$

Polynuclear complexes are also frequently formed, which can grow into infinite complexes. In solutions of In^{3+}, in addition to the mononuclear complexes, also polynuclear complexes $In(OIn)_n^{(n+3)+}$ are thus formed according to the schematic equation (written without water of hydration since the degree of hydration is unknown)

$$(n+1)In^{3+} + nH_2O \rightleftharpoons In(OIn)_n^{(n+3)+} + 2nH^+ \tag{12.16}$$

If H^+ is bound up by the addition of base, n increases. If n goes to ∞, the formula of the polynuclear complex approaches $(OIn)_n^{n+} = (InO^+)_n$, where n is a very large number. Finally an oxide salt (formerly called a "basic" salt) precipitates containing exactly this infinite complex in the form of parallel layers between which the anions are packed. Fig. 12.1 shows the structure of the layer and the structure of indium oxide chloride InOCl. The polynuclear complexes that earlier occurred in the solution can probably be thought of as fragments of the layer in the oxide salt, although the complexes in the solution are surely hydrated.

In solutions of Fe^{3+} salts the hydrated ion $Fe(H_2O)_6^{3+}$ gives up first one and then one more proton according to the equations

$$Fe(H_2O)_6^{3+} \rightleftharpoons Fe(H_2O)_5OH^{2+} + H^+ \tag{12.17}$$

$$Fe(H_2O)_5OH^{2+} \rightleftharpoons Fe(H_2O)_4(OH)_2^+ + H^+ \tag{12.18}$$

Here the coordination number 6 (octahedral coordination) is maintained by the ion Fe^{3+}. At higher concentrations of iron, however, a dinuclear complex predominates, formed according to the equation

$$2Fe(H_2O)_6^{3+} \rightleftharpoons \left[(H_2O)_4Fe \begin{matrix} H \\ O \\ O \\ H \end{matrix} Fe(H_2O)_4 \right]^{4+} + 2H^+ + 2H_2O \tag{12.19}$$

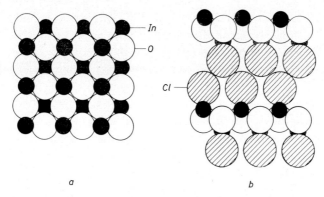

Fig. 12.1. (a) Fragment of $(InO^+)_n$ layer in InOCl. The layer has unlimited extent in the picture plane. (b) The structure of InOCl viewed along the plane of the layers. The geometric relationships are somewhat schematic.

The two OH^- ions form bridges between the two Fe^{3+} ions, which retain the coordination number 6. The ion may thus be considered to be composed of two coordination octahedra with a common edge. In more basic solutions, which, however, cannot be studied, possibly also complexes with larger nuclear multiplicity than two may occur.

The ion $Al(H_2O)_6^{3+}$ is believed to react analogously to (12.19), but as proton after proton is removed during the addition of base, more and more Al^{3+} ions are bound together by double OH^- bridges. The complexes thus grow and finally more or less irregular, giant molecules are obtained with the gross formula $Al(OH)_3$, in which the coordination is still sixfold. These may precipitate as a gel (15-4). If the gel is allowed to stand, it generally changes its properties, it "ages" and may finally crystallize. One modification of crystallized $Al(OH)_3$ is known, gibbsite, that is built up of parallel, infinite layers of gross composition $Al(OH)_3$ and having the structure shown in fig. 6.18b. The layer may be said to consist of $Al(OH)_6$ octahedra with common edges. Probably the polynuclear complexes in Al^{3+} solutions may in principle be thought of as fragments of such layers.

The above discussion shows how variable the course of acid-base processes of this kind can be. Moreover, it is impossible in a particular case to predict the reaction.

Examples of other important metal ions that function as acids are Be^{2+}, Ca^{2+}, Zn^{2+}, Cu^{2+} and Bi^{3+}.

c. Salts. The concept of salt lacks significance in the application of the acid-base definition of Brønsted. Therefore, it is of no particular consequence that it is difficult nowadays to define a salt exactly. It might be best to use the term to mean *a substance that forms ionic crystals*. This definition covers a large number of the classical salts, but does not entirely coincide with the old definition of a salt

(12-1 d). NaCl, Na_2CO_3 and NH_4Cl are still salts. This is not prevented by the fact that the ion CO_3^{2-} is a base and the ion NH_4^+ is an acid. But NaOH also becomes a salt, built up of the ions Na^+ and OH^-, of which the latter is a base. On the other hand $HgCl_2$ was formerly counted as a salt, but now cannot be included as such because the crystal consists of $HgCl_2$ molecules.

Since the transition between ionic bonding and molecular bonding is continuous this definition, of course, is not exact. It should be observed, however, that the concept of salt is associated entirely with *solid phases*.

A salt crystal may contain several different kinds of cations and anions. In a solid solution of KCl and KBr, for example, Cl^- and Br^- occur in the same role in anion sites. Another example is alum, $KAl(SO_4)_2 \cdot 12H_2O$, in which K^+ and Al^{3+} occur in lattice sites of different kinds and, of course, have different functions. Expecially salts of the latter type have been called *double salts*.

Salts whose anions cannot give up protons have commonly been called "normal" salts. Such salts are KF, Na_2CO_3 and Na_3PO_4. But salts such as KHF_2, $NaHCO_3$ and NaH_2PO_4 contain the ions HF_2^-, HCO_3^- and $H_2PO_4^-$ respectively, which can give up protons. Such salts have been called "acid" salts since the number of acid-anions per metal atom is greater than in the normal salt. If the number of acid-anions per metal atom is twice as great as in the normal salt the ion often was given the prefix "bi-". KHF_2 was thus called potassium bifluoride, $NaHCO_3$ was called sodium bicarbonate. Nowadays the above-mentioned acid salts are named potassium hydrogen fluoride, sodium hydrogen carbonate and sodium dihydrogen phosphate.

In all acid salts of known structure the protons are incorporated into hydrogen bonds. Accordingly, for example, the anions HF_2^- and HCO_3^- have the schematic structures

$$F-H\cdots F \qquad \begin{array}{c} -H\cdots O \\ | \\ C \\ / \ \ \backslash \\ O \quad O-H\cdots O \end{array} \qquad \begin{array}{c} O \ \ \ \ O-H\cdots O \\ \backslash / \\ C \\ | \\ O-H\cdots O \end{array} \qquad \begin{array}{c} O \ \ \ \ O- \\ \backslash / \\ C \\ | \\ O-H\cdots O \end{array}$$

Acid salts are primarily known for anions that contain oxygen or fluorine atoms, and this depends probably on the fact that hydrogen bonding between oxygen atoms and between fluorine atoms are among the strongest types of hydrogen bond (5-5). The cations in acid salts are almost exclusively ions of alkali or alkaline-earth metals. This can be explained in that other cations exert such a strong repulsive force on protons that these cannot be incorporated into the crystal (since both the repulsive and the polarizing action of an ion are electrostatic effects of the same kind, the discussion in 5-2e concerning the relative strength of the latter also can be applied to the former).

d. The development of the acid-base concept. The word *acid* is derived from the Latin *acidum*, from *acidus* meaning sour, and the alchemists classified substances with sour taste as acids. Boyle in 1663 defined acids as substances with acid taste and the ability to give a red color to blue, plant-dye substances such as litmus. As an opposite to acid, *alkali* was referred to (for the derivation of this word, see 25-2b). The distinguishing characteristics of alkalis were the soapy, slimy feel of the solutions, and especially their ability to destroy, "neutralize," the action of acids. They were able to restore the blue color of litmus colored red by acid, and in the reaction between acids and bases *salts* were formed, which lacked the typical properties of the initial substances.

The formation of salts led to the idea that a salt must consist of two constituents of opposite nature. The nonacid constituent was often thought to be what is now known as a metal oxide or hydroxide and was termed as the *base* (in the sense of "fundamental substance") of the salt. In 1774 Rouelle defined a base as any substance that reacted with an acid with salt formation, and this definition has persisted to the present day.

The first special theory of acids was presented in the 1770's by Lavoisier. He showed that a great many elements, for example carbon, phosphorus and sulfur, when they burned in oxygen, gave "acid oxides," which formed acids with water. Lavoisier therefore believed that oxygen was found in all acids and was the cause of their properties. He thus established the name oxygène (Lat. form oxygenium, acid former, from Gk. ὄξος, sour, and γεννάω, I create).

Berzelius for a long time defended Lavoisier's theory of acids but after Davy showed around 1810 that hydrogen chloride in spite of its acid properties does not contain oxygen, it was gradually given up. When it became clear that acids always contain hydrogen, which could be replaced by metals (whereby salts were formed), this property instead became accepted as a definition of acids. The bases were still considered to be substances that neutralized acids with salt formation, but no common feature in their makeup could be discovered.

The dissociation theory of Arrhenius made possible acid-base definitions which predominated up to the beginning of the 1920's. It was found that typical acid properties were associated with what was considered to be the hydrogen ion H^+. The higher the hydrogen ion concentration $[H^+]$ of a solution was, the more acid was the solution. An acid was therefore defined as an electrolyte that could split off hydrogen ions in solution. Since a water solution was more basic the greater its hydroxide ion concentration $[OH^-]$ was, bases were defined as electrolytes that could split off hydroxide ions in water solution. On reaction between an acid and a base, that is, neutralization, water was formed, and after evaporation of the water a salt appeared, which consisted of the cations of the base and the anions of the acid.

However, these definitions only consider the conditions in water solution and do not express the functional relationship between acid and base. It is apparent that if the definition of acid and base is to be based on the ions that appear in solution, a special definition must be given for each different solvent. Such is not the case, on the other hand, with Brønsted's more general acid-base definition.

Some further generalizations of the acid-base concept are considered in 12-1e.

e. New, more general, acid-base definitions. The transfer of protons may be thought of as a special case of still more general types of reaction and attempts have been made to characterize these in order to attain a common starting point and a unified treatment of all analogous reactions. This has also led to proposals of a generalization of the acid-base concept.

G. N. Lewis in 1923 started with the general definition that an acid is any substance that can neutralize a base such as sodium hydroxide, and a base is any substance that can neutralize an acid such as hydrochloric acid. He subsequently found that this stated property of an acid is related to the fact that it is an electron-pair acceptor (5-3c). The specified property of a base is related to the fact that it is an electron-pair donor. Lewis's ultimate definition therefore states that *an acid is an electron-pair acceptor* and *a base is an electron-pair donor*. Neutralization implies the formation of a covalent bond between the acid and the base (for example, according to equation (5.16)).

In a "Lewis acid" an electron pair is lacking in some valence-electron orbital. Examples are H^+, SO_3, BF_3. The Lewis acids clearly embrace another and much larger range of substances than the Brønsted acids. In a base it is the lone (unshared) electron pair that contributes to the bond. Examples are NH_3, OH^-. Since an electron-pair donor can always unite with the electron-pair acceptor H^+, it is always a base also according to Brønsted's definition. Brønsted and Lewis bases are therefore the same.

Gutmann and Lindqvist proposed in 1954 that an *ion transfer* (*ionolysis*) be regarded as the essential aspect of a general acid-base reaction. Depending on whether cations or anions are transferred we obtain either one of the definitions:

Cation transfer: acid $\rightleftharpoons$ base + cation   (12.20)
                 cation donor cation acceptor

Anion transfer: acid + anion $\rightleftharpoons$ base   (12.21)
              anion acceptor   anion donor

Brønsted's definition is a special case of cation transfer, applicable to protons.

Important cases of anion transfer are encountered in reactions between fused oxides of the type shown by example I in 10c. Since early times the oxides, because of their reactions with each other in melts, have been divided into "acid" oxides (for example, B_2O_3, SiO_2, P_2O_5, SO_3) and "basic" oxides (for example, Na_2O, MgO, CaO). Lux in 1939 took the view that oxide ion transfer was the essential feature of these reactions and proposed that they should be treated as acid-base reactions on the basis of the definition acid $+ O^{2-} \rightleftharpoons$ base. In the example, MgO and SiO_4^{4-} then become bases and SiO_2 and Mg^{2+} become acids. Expecially Flood has shown the usefulness of this approach. However, Lux's definition is included as a special case of the acid-base definition according to Gutmann and Lindqvist.

In anhydrous halogen systems anion transfer often occurs. Arsenic trichloride can react as an acid, for example according to $\underset{a_1}{AsCl_3} + \underset{b_2}{PCl_5} \rightleftharpoons \underset{b_1}{AsCl_4^-} + \underset{a_2}{PCl_4^+}$ (PCl_5 has predominantly basic character), or as a base, for example according to $\underset{b_1}{AsCl_3} + \underset{a_2}{TiCl_4} \rightleftharpoons \underset{a_1}{AsCl_2^+} + \underset{b_2}{TiCl_5^-}$. In pure liquid arsenic trichloride therefore there also exists the *autoionization equilibrium* $\underset{b_1}{AsCl_3} + \underset{a_2}{AsCl_3} \rightleftharpoons \underset{a_1}{AsCl_2^+} + \underset{b_2}{AsCl_4^-}$, which gives rise to appreciable conductivity. If both PCl_5 and $TiCl_4$ are dissolved in liquid arsenic trichloride, the equilibrium $\underset{b_1}{PCl_5} + \underset{a_2}{TiCl_4} \rightleftharpoons \underset{a_1}{PCl_4^+} + \underset{b_2}{TiCl_5^-}$ is naturally also

Acid-Base Equilibria

prevalent. Other examples are the autoionizations of PCl_5 and PBr_5 mentioned in 5-3i.

Both Lewis's and Gutmann and Lindqvist's generalizations have to a large degree increased our understanding of the associated types of reactions. In ordinary discussions, however, the terms acid and base still refer only to proton transfer.

12-2. Strength of protolytes

a. Strength of protolytes. The strength of an acid must be measured by its tendency to function as an acid, that is, give off protons. In the same way the strength of a base is measured by its tendency to take up protons. From definition (12.1) it then follows that the strength of a base is inversely proportional to the strength of the acid corresponding to the base. A strong acid thus corresponds to a weak base, and vice versa.

For the release of protons by an acid $a \rightleftharpoons b + H^+$, regardless of whether the proton H^+ exists free or not, the following equilibrium equation applies:

$$\{H^+\}\{b\}/\{a\} = K_a \tag{12.22}$$

If a solution contains several acid-base pairs, an equation of this type applies for each of them.

In all these equations, to establish $\{a\}$ and $\{b\}$ the standard state may be chosen in the usual way so that f_a and f_b approach 1 as $[a]$ and $[b]$ approach 0 (10a). On the other hand the hydrogen ion concentration $[H^+]$ cannot be measured directly in any solution, so that the standard state for the establishment of $\{H^+\}$ must be chosen in another way. One of the acid-base pairs a_0–b_0 in the solution is selected as a standard pair and the standard state is chosen so that the constant K_{a0} for this pair $= 1$. We thus set

$$\{H^+\}\{b_0\}/\{a_0\} = 1 \quad \text{or} \quad \{H^+\} = 1 \; \frac{\{a_0\}}{\{b_0\}} \tag{12.23}$$

In this way a scale is also established for all K_a and for the various acid-base pairs we may write

$$\{H^+\} = 1 \; \frac{\{a_0\}}{\{b_0\}} = K_{a1} \; \frac{\{a_1\}}{\{b_1\}} = K_{a2} \; \frac{\{a_2\}}{\{b_2\}} = \ldots = K_a \; \frac{\{a\}}{\{b\}} \tag{12.24}$$

If the solvent is a protolyte it is proper to choose a pair taking part in the solvent system as the standard pair because it occurs in all solutions.

The constant K_a for the acid-base pair a–b is called the *acidic constant* of the pair. It was formerly often called the *dissociation constant* of the acid a. Since the proton release by an acid does not take place as a simple, isolated dissociation, this term is less suitable.

The greater is an acid's tendency to give off protons, the greater K_a is, which therefore becomes a proper measure of the strength of the acid. K_a for an acid is the same in all solutions of this acid in a given solvent (whether or not it is a protolyte), as long as $\{H^+\}$ in every solution is defined by the same standard pair. In going over to another solvent, however, K_a changes even if the standard pair remains the same. Also, the *ratio* between the K_a values of different acids is different in different solvents. In changing from one solvent to another, the environment (for example dielectric constant, charge conditions) may be strongly altered, so that protolytes of different constitution may be differently affected.

Instead of the proton release by an acid, we may equally well start out with the absorption of protons by a base according to $b + H^+ \rightleftharpoons a$. We then have the equilibrium equation $\{a\}/\{H^+\}\{b\} = K_b$, where K_b is proportional to the strength of the base. If $\{H^+\}$ is defined as before we obtain in this case

$$\frac{1}{\{H^+\}} = 1 \cdot \frac{\{b_0\}}{\{a_0\}} = K_{b1} \frac{\{b_1\}}{\{a_1\}} = K_{b2} \frac{\{b_2\}}{\{a_2\}} = \ldots = K_b \frac{\{b\}}{\{a\}} \qquad (12.25)$$

The constant K_b for the acid-base pair a–b is called the *basic constant* of the pair. A comparison of (12.24) and (12.25) shows that K_a and K_b are inversely proportional. If $\{H^+\}$ is the same in both expressions, which means that the same standard pair was chosen in both cases to establish $\{H^+\}$, then for a given pair $K_a K_b = 1$. As a rule, however, two different standard pairs are chosen (see, for example, 12-2e).

Nowadays, usually only acidic constants are used in the treatment of acid-base equilibria.

b. Hydrogen ion activity and acidity. In spite of the fact that free hydrogen ions (protons) do not occur in measurable amounts in any solution, their activity $\{H^+\}$, as just shown, is a thermodynamically definable quantity. In addition it is subject to easy measurement.

The expression (12.24), which applies to proton exchange equilibria in all types of solvents, shows that the ratio $\{a\}/\{b\}$ for all acid-base pairs is determined by the *hydrogen ion activity* $\{H^+\}$ of the solution (this is a special case with $B = H^+$ of the general discussion on transfer reactions in 10c). The hydrogen ion activity of the solution is customarily also designated as its *acidity*.

The acidity of a solution is often given by its pH value, which decreases as the acidity increases. With the operator symbol "p" introduced in 1-4e, pH should signify $-\log [H^+]$. Since the hydrogen ion activity $\{H^+\}$ is easy to measure the definition in this case has been modified to

$$\text{pH} = -\log \{H^+\} \qquad (12.26)$$

The acidity is most accurately measured by potentiometric measurement of the electromotive force of an element in which one electrode potential is a function of $\{H^+\}$. The simplest in principle of such electrodes is the *hydrogen electrode* (16c). More convenient and often very useful is a so-called *glass electrode*. For rapid but less accurate determinations *indicators* (12-3g) are used.

Acid-Base Equilibria

c. Protolysis constant. For an acid-base equilibrium $a_1 + b_2 \rightleftharpoons b_1 + a_2$ the equilibrium equation $\{b_1\}\{a_2\}/\{a_1\}\{b_2\} = K$ applies. K is often called the *protolysis constant*. If in (12.24) the terms for the two pairs a_1–b_1 and a_2–b_2 are combined we get

$$K_{a1}\frac{\{a_1\}}{\{b_1\}} = K_{a2}\frac{\{a_2\}}{\{b_2\}} \quad \text{or} \quad \frac{\{b_1\}\{a_2\}}{\{a_1\}\{b_2\}} = \frac{K_{a1}}{K_{a2}} = K \qquad (12.27)$$

The protolysis constant thus is the quotient of the acidic constants of the two participating pairs.

For an equilibrium $a_1 + b_0 \rightleftharpoons b_1 + a_0$, where a_0–b_0 is a standard pair ($K_{a0} = 1$), we have $K = K_{a1}$. Thus, here the protolysis constant is equal to the acidic constant.

d. The autoprotolysis and acidic constant of water. The development in this section up to now has been entirely independent of the nature of the solvent. Next we consider applications to water solutions, which because of their great significance may be taken as a model for all solutions in protolyzing solvents. Acid-base equilibria in other solvents than water are considered in 12-5.

As an intermediate member of the protolyte system H_3O^+–H_2O–OH^-, H_2O is an ampholyte, which functions as an acid according to (12.3) and as a base according to (12.4). In any ampholyte proton exchange can take place between two ampholyte molecules (*autoprotolysis*, a special case of autoionization, 12-1e, 5-3i). For water we obtain by addition of (12.3) and (12.4) the autoprotolysis reaction

$$H_2O + H_2O \rightleftharpoons H_3O^+ + OH^- \qquad (12.28)$$

Pure water thus contains H_3O^+ ions and OH^- ions in the same concentration. If autoprotolysis takes place in a liquid the formation of ions always gives rise to electrical conductivity.

When (12.24) is applied to water solutions the pair H_3O^+–H_2O is chosen as standard pair and thus $K_{H_3O^+} = 1$. With this pair, the pair H_2O–OH^- and the dissolved pair a–b introduced in (12.24), we obtain for water solutions

$$\{H^+\} = 1 \cdot \frac{\{H_3O^+\}}{\{H_2O\}} = K_{H_2O}\frac{\{H_2O\}}{\{OH^-\}} = \ldots = K_a\frac{\{a\}}{\{b\}} \qquad (12.29)$$

In water solutions we thus have the definition

$$\{H^+\} = \{H_3O^+\}/\{H_2O\} \qquad (12.30)$$

In solutions so dilute that we may set $\{H_2O\} = 1$ (10a), we have

$$\{H^+\} = \{H_3O^+\} \qquad (12.31)$$

In very dilute solutions we may set $\{H_3O^+\} = [H_3O^+]$, whereby

$$\{H^+\} = \{H_3O^+\} = [H_3O^+] \qquad (12.32)$$

The oxonium ion concentration $[H_3O^+]$ then implies the collective concentration of all more or less hydrated hydrogen ions (12-1a).

It is easily seen that (12.32) and therefore also a definition such as pH = $-\log[H_3O^+]$ would not give a correct measure of the acidity in concentrated solutions. If, for example, an H_2SO_4 solution is made more and more concentrated, eventually $[H_3O^+]$ must decrease because of an insufficient supply of H_2O. In spite of this $\{H^+\}$ must increase and pH decrease as long as the solution is getting more concentrated. Therefore, here only the exact definitions (12.30) and (12.26) can be used. In order also in this case to gain a better appreciation of the situation, the activity quotient in (12.30) may be replaced by the concentration quotient $[H_3O^+]/[H_2O]$, but at high concentrations no calculations should ever be based on this expression.

The constant K_{H_2O} in (12.29) is the acidic constant for the pair H_2O–OH^- and is therefore called the *acidic constant of water* (formerly electrolytic dissociation constant). For this the internationally adopted symbol K_w is used in the following. If the three first terms in (12.29) are multiplied by $\{OH^-\}/\{H_2O\}$ we then get

$$\{H^+\}\{OH^-\}/\{H_2O\} = \{H_3O^+\}\{OH^-\}/\{H_2O\}^2 = K_w \qquad (12.33)$$

In solutions so dilute that $\{H_2O\} = 1$, (12.33) becomes

$$\{H^+\}\{OH^-\} = \{H_3O^+\}\{OH^-\} = K_w \qquad (12.34)$$

Because of the relationship (12.34), K_w has also been called the *ionic product* of water.

In *pure water* in which autoprotolysis is the only acid-base process, according to (12.28) $[H_3O^+] = [OH^-]$. Since these concentrations are very low in pure water (see below) they are equal to the corresponding activities. (12.34) then gives

$$\{H^+\} = \{H_3O^+\} = \{OH^-\} = \sqrt{K_w} \qquad (12.35)$$

Tab. 12.1 lists K_w and $pK_w = -\log K_w$ for water at different temperatures. It can be seen that autoprotolysis increases markedly with rising temperature. The ionic concentrations $[H_3O^+]$ and $[OH^-]$, which as stated above are equal to the corresponding activities, and thus $= \sqrt{K_w}$, however, are always extremely small. Because of this the conductivity of pure water is very low.

At room temperature K_w is often rounded to 10^{-14} and pK_w to 14 for rough calculations.

Since $\{OH^-\}$ in dilute water solutions as a rule is calculated by the expression (12.34) from directly measured $\{H^+\}$ values, pOH has nowadays been given the definition pOH = $-\log\{OH^-\}$. Taking logarithms of (12.34) then gives

$$pH + pOH = pK_w \qquad (12.36)$$

and rounded, at room temperature

$$pH + pOH = 14 \qquad (12.37)$$

For pure water, by taking logarithms of (12.35) we have

$$pH = pOH = \tfrac{1}{2}pK_w \qquad (12.38)$$

Table 12.1. *Acidic constants K_w of water*

Temp. °C	K_w	pK_w
0	0.12×10^{-14}	14.93
15	0.45×10^{-14}	14.35
20	0.68×10^{-14}	14.17
25	1.01×10^{-14}	14.00
30	1.47×10^{-14}	13.83
50	5.48×10^{-14}	13.26
100	37.5×10^{-14}	12.43

If in a water solution $\{H^+\} = \{OH^-\}$ or pH = pOH, the solution is said to be *neutral*. At room temperature, thus, in neutral solution, pH = pOH = 7. If pH < pOH, and thus at room temperature pH < 7, the solution is called *acid*. If pH > pOH, and thus at room temperature pH > 7, the solution is called *basic* or *alkaline*.

e. Acidic and basic constants in water solution. Any acid-base pair a–b dissolved in water takes part in the following equilibria with the water system:

$$a + H_2O \rightleftharpoons b + H_3O^+ \qquad (12.39)$$

$$b + H_2O \rightleftharpoons a + OH^- \qquad (12.40)$$

Since the pair H_3O^+–H_2O is chosen as the standard pair in establishing the K_a scale, according to 12-2c the constant (protolysis constant) for the equilibrium (12.39) is equal to the acidic constant K_a for the pair a–b. We may obtain directly from (12.29) the equilibrium equation

$$\{H_3O^+\}\{b\}/\{a\}\{H_2O\} = K_a \qquad (12.41)$$

Directly from (12.29) or by applying (12.30) to (12.41) we may further obtain the equivalent equation

$$\{H^+\}\{b\}/\{a\} = K_a \qquad (12.42)$$

In a water solution of the acid-base pair a–b the quantities $\{H^+\}$, [a] and [b] as a rule are easiest to measure. Therefore, another type of acidic constant is often used, defined by the expression

$$\{H^+\}[b]/[a] = k_a \qquad (12.43)$$

By dividing (12.42) and (12.43) and introducing f values, we obtain

$$k_a = K_a f_a/f_b \qquad (12.44)$$

As long as the quotient f_a/f_b is constant, which generally can be assumed to be the case for constant ionic strength, k_a is also constant. k_a values can therefore be used for equilibrium calculations at constant ionic strength (cf. 11d). The constant k_a defined according to (12.43) is called a *mixed* or *incomplete* acidic constant.

If a and b in (12.43) are both dissolved substances, then at infinite dilution $f_a = f_b = 1$ and thus $k_a = K_a$. On the other hand, if one of the two constitutes the solvent itself the corresponding f value is not $=1$ (10a). In water solutions the limiting value $f_{H_2O} = 0.018$ is reached at infinite dilution. In dilute water solutions, thus, according to (12.44) we have for the pair $H_3O^+-H_2O$, $k_{H_3O^+} = K_{H_3O^+} \times 1/0.018 = 55.5\,K_{H_3O^+} = 55.5$ ($pk_{H_3O^+} = -1.74$); and for the pair H_2O-OH^-, $k_w = K_w \times 0.018/1 = 0.018\,K_w = 1.82 \times 10^{-16}$ ($pk_w = 15.74$).

A protolyte can also participate in other equilibria than acid-base equilibria. In connection with heterogeneous equilibria it will be described how solubility equilibria can be related to acid-base equilibria (13-1e). In a homogeneous phase certain acids also take part in a hydration equilibrium.

In a water solution of carbonic acid there is thus set up as the first stage of protolysis the equilibrium $H_2CO_3 + H_2O \rightleftharpoons HCO_3^- + H_3O^+$, but in addition carbonic acid also takes part in the hydration equilibrium $CO_2 + H_2O \rightleftharpoons H_2CO_3$. Carbonic acid thus exists in the solution partly as directly dissolved CO_2 and partly as hydrated H_2CO_3. According to (12.43) for the first equilibrium we have (a): $\{H^+\}[HCO_3^-]/[H_2CO_3] = k_a$. For the second, if $[H_2O]$ is assumed to be constant we have (b): $[H_2CO_3]/[CO_2] = k_h$ (k_h is called the *hydration constant*). Since it is difficult to measure accurately $[H_2CO_3]$ and $[CO_2]$ separately, it is difficult to determine k_a and k_h. On the other hand it is easier to measure $([H_2CO_3] + [CO_2])$ and it is therefore practical instead of (a) to write (c): $\{H^+\}[HCO_3^-]/([H_2CO_3] + [CO_2]) = k_{ah} = k_a k_h/(1+k_h)$. If k_a and k_h are constants, k_{ah} is also a constant (*"apparent" acidic constant*). It is this (or the corresponding constant K_{ah} with activities only) that without further qualification is customarily given as the acidic constant for H_2CO_3 (for example in tab. 12.2). In general, for the sake of brevity $[H_2CO_3]$ is written as the denominator of the equation. The equation then has the same form as (a) and can be applied in the same way.

k_{ah} is always smaller than k_a. For H_2CO_3 k_{ah} is of the order of 10^{-7} while k_a is of the order of 10^{-4}. H_2CO_3 thus appears to be a much weaker acid than it really is, because a large part of the total amount is unhydrated CO_2.

Hydration equilibria adjust themselves in both directions much more slowly than acid-base equilibria and therefore often cause noticeable delayed-reaction phenomena. If strong base is added to a carbonic acid solution a portion of the existing H_2CO_3 is consumed immediately in the acid-base process, which leads to a slow displacement to the right of the hydration equilibrium. This becomes evident from the slow increase in the acidity from the value obtained immediately after the addition of base (easily shown by a color indicator).

A similar example is sulfurous acid with the hydration equilibrium $SO_2 + H_2O \rightleftharpoons H_2SO_3$.

Basic constants in water solution are commonly defined by starting with the reaction of the base with water, thus the equilibrium (12.40), which gives

$$\{OH^-\}\{a\}/\{b\}\{H_2O\} = K_b \qquad (12.45)$$

The basic constant (formerly the dissociation constant of the base b) K_b is a measure of the strength of the base. A comparison of (12.45) with (12.25) shows that the pair H_2O-OH^- has been chosen as the standard pair to establish the K_b scale.

In dilute solutions where $\{H_2O\} = 1$, (12.45) becomes

$$\{OH^-\}\{a\}/\{b\} = K_b \qquad (12.46)$$

Multiplying (12.42) by (12.46) and recalling (12.34) we have

$$K_a K_b = K_w; \quad pK_a + pK_b = pK_w \qquad (12.47)$$

K_a and K_b are, as always, inversely proportional, but because the K_a and K_b scales are determined by different standard pairs $K_a K_b \neq 1$. However, the reason that the relation is so simple is that both standard pairs belong to the water system.

A mixed basic constant is also used, defined by the expression

$$\{OH^-\}[a]/[b] = k_b \qquad (12.48)$$

Multiplying (12.43) by (12.48) we have

$$k_a k_b = K_w; \quad pk_a + pk_b = pK_w \qquad (12.49)$$

f. Values of the acidic constants. According to 12-1a an acid can give off a proton only if this can be taken up by some base. It can never be dissociated like a salt, and for any acid protolysis under certain conditions may be incomplete. Therefore, no acid is strong in the true sense but for a particular solvent those acids are usually called strong that in moderate concentration are practically completely protolyzed. Similarly a base is called strong in a particular solvent if in moderate concentration it is practically completely protolyzed.

Many anions are known that are believed not to bond to a proton under any conditions. To these belong a series of halo ions. For example, it has not been shown that the tetrafluoroborate ion BF_4^- is able to give a "fluoroboric acid" HBF_4. In spite of this, such an acid has been spoken of and said to be strong, although it cannot be detected even in very acid solutions of tetrafluoroborate. The same applies to such acids as H_2SiF_6, H_2SnCl_6, H_2TiCl_6, H_2PtCl_6 and $HAuCl_4$. None of these substances are known in anhydrous form. On the other hand hydrates can often be obtained, in which, however, probably the proton is bound to the water. It is likely, for example, that the hydrate which is usually written $HAuCl_4 \cdot 4H_2O$ is actually $(H_3O)AuCl_4 \cdot 3H_2O$. A completely different behavior is illustrated by perchloric acid $HClO_4$ whose monohydrate is certainly $(H_3O)ClO_4$, but which is also known in anhydrous form.

The solvent has an effect on the course of the protolysis, among other things because its dielectric constant determines the strength of the electrostatic forces between the ions (11 c). If the solvent is a protolyte itself and therefore participates in the proton-exchange equilibria, however, its protolytic properties take on decisive significance. Here water solutions will next be discussed, but in 12-5 a survey of the conditions in other solvents will also be given.

The typical effects of a solvent are associated with its high concentration in the solution. Its dielectric effect requires that it form a "medium" between the ions and many of its protolytic effects result from its presence in such amounts that it can give off or take up the major portion of the exchangeable protons in the solution. Very concentrated solutions of protolytes in water therefore do not possess the properties typical of water solutions.

In water solution the two acid-base pairs of the water system take part: H_3O^+–H_2O ($pk_{H_3O^+} = -1.74$; cf. 12-2e) and H_2O–OH^- ($pk_{H_2O} = pk_w = 15.74$; cf. 12-2e). Acids that in water solution are stronger than H_3O^+ ($pK_a < \sim -1.7$) are protolyzed, if the solution is not very concentrated, so completely that they are designated as strong. As long as the water concentration is sufficiently great, all of the protons are taken up by H_2O with the formation of H_3O^+ (schematically speaking). The acids are transformed to bases and the solutions behave in respect to proton exchange as though they all contained the same acid, namely H_3O^+, which in water solution is counted as a strong acid. The *levelling effect* of water on strong acids has therefore sometimes been spoken of. The most important of these acids are HI, HBr, HCl, $HClO_4$ and H_2SO_4 (but not HSO_4^- which is relatively weak). HNO_3 is also considered as a strong acid. However, these acids have among themselves very different strengths, which can be shown if they are dissolved in a solvent that has less tendency to take up protons, that is, is less basic than water (12-5). Here they have been listed in order of decreasing strength in water solution.

A similar levelling effect of H_2O on bases implies that all bases with greater base strength than OH^- ($pK_a > \sim 15.7$) are protolyzed practically completely if the solution is not very concentrated. Thus they are able to take up all the necessary protons from H_2O, whereby OH^- is formed.

No uncharged strong bases are known but on the other hand there are many strong anionic bases that are obviously anions of extremely weak acids. To these belong OH^-, but the amide ion NH_2^- and the oxide ion O^{2-} are stronger. The latter belong to the acid-base pairs NH_3–NH_2^- and OH^-–O^{2-}, which shows that NH_3 and OH^- also have extremely weak acid character, that is, are ampholytes. Evidently the complete ammonia and water systems are NH_4^+–NH_3–NH_2^-–NH^{2-}–N^{3-} and H_3O^+–H_2O–OH^-–O^{2-}, respectively. In water solutions, however, this is unimportant because of the complete protolysis of NH_2^- and O^{2-} with the formation of OH^-.

The anions of acids that are strong in water solution (for example, Cl^-, NO_3^-) are such weak bases in water solution that for practical purposes it is never necessary to consider their basic character.

It is seen thus how for a certain solvent the protolytes may be divided into classes according to strength, separated by the acid-base pairs of the solvent. In water, acids in all acid-base pairs with $pK_a < \sim -1.7$ appear as strong acids and bases in all pairs with $pK_a > \sim 15.7$ as strong bases. In the intervening range both the acid and the base in a pair are weak and it is the equilibria between such protolytes that become of interest in all fairly dilute water solutions.

In tab. 12.2 are listed pK_a values mainly for inorganic acid-base pairs. Also, a number of acids stronger than H_3O^+ and weaker than H_2O (the bases corresponding to the latter are thus strong bases) have been included. The values for these have necessarily been determined in other solvents than water and therefore are very approximate on the water scale.

Table 12.2. pK_a[1] values for common acids in water solution at 25°C (unless otherwise specified).

Acid	Formula	pK_a	Acid	Formula	pK_a
hydrogen iodide	HI	~ −9	hydrogen sulfide		
hydrogen bromide	HBr	~ −8	(20°C, $I=1$)	H_2S	6.88
hydrogen chloride	HCl	~ −7	hydrogen sulfite ion	HSO_3^-	7.20
perchloric acid	$HClO_4$	~ −7	dihydrogen phosphate		
sulfuric acid	H_2SO_4	~ −3	ion	$H_2PO_4^-$	7.21
oxonium ion ($pk_{H_3O^+}$)	H_3O^+	−1.74	hypochlorous acid	HClO	7.53
nitric acid	HNO_3	−1.35	boric acid	$B(OH)_3$	9.00
iodic acid	HIO_3	0.77	ammonium ion	NH_4^+	9.25
oxalic acid	$H_2C_2O_4$	1.19	hydrogen cyanide	HCN	9.40
sulfurous acid	H_2SO_3	1.89	hydrogen carbonate ion	HCO_3^-	10.33
hydrogen sulfate ion	HSO_4^-	1.99	hydrogen peroxide	H_2O_2	11.65
phosphoric acid (ortho)	H_3PO_4	2.15	hydrogen phosphate ion	HPO_4^{2-}	12.36
hydrogen fluoride	HF	3.17	hydrogen sulfide ion		
nitrous acid	HNO_2	3.35	(20°C, $I=1$)	HS^-	14.15
hydrogen oxalate ion	$HC_2O_4^-$	4.29	water (pk_w)	H_2O	15.74
hydrogen azide	HN_3	4.72	ammonia	NH_3	~ 23
acetic acid	CH_3COOH	4.76	hydroxide ion	OH^-	~ 24
carbonic acid	H_2CO_3	6.37			

[1] For H_3O^+ and H_2O, in comparing strengths of acidities the pk_a values given in 12–2e for dilute solutions must be used.

From tab. 12.2 it is apparent that for a polybasic acid pK_a increases, that is, K_a decreases, for each proton given off (for example, cf. H_3PO_4, $H_2PO_4^-$ and HPO_4^{2-}). This may be explained as an electrostatic effect resulting from the fact that the remaining ion becomes more and more negative or less and less positive with each proton given up. In this way the release of the next proton is made more difficult. The difference between subsequent pK_a values may be large if the protons in question lie near to each other in the acid molecule as in H_3PO_4. If they lie at greater distances from each other, as is often the case with organic acids, the difference may be rather small.

Many other regularities in the strengths of acids can be found and at least roughly explained. This is most easily done for the pure *hydrides*, especially the hydrogen halide acids with their simple constitution HX. In water solution HF is a weak acid, the other hydrogen halide acids strong. The strength increases with increasing atomic number of the halogen atom. The chief reason for this is that the bond strength in the hydrogen halide molecule is greatest for HF and then decreases as the atomic number of the halogen increases. Unlike what might be expected, the ionic bond contribution and the acid strength change in the opposite directions. The electronegativity of the halogen atom and consequently the ionic bond contribution in the hydrogen–halogen bond decrease with increasing atomic number of the halogen atom (5-3f).

The acid strength is, however, affected by several factors. This is best illustrated by dividing the process $HX(aq) \to H^+(aq) + X^-(aq)$ into different steps according to the scheme below (Y_{HX} is the hydration enthalpy and D the dissociation enthalpy for one mole $HX(g)$; other symbols have been used in 5-2).

$$HX(aq) \xrightarrow{+Y_{HX}} HX(g) \xrightarrow{+D} \begin{cases} H(g) \xrightarrow{+I} H^+(g) \xrightarrow{-Y_H} H^+(aq) \\ X(g) \xrightarrow{-E} X^-(g) \xrightarrow{-Y_X} X^-(aq) \end{cases}$$

$$\Delta H$$

$$\Delta H = Y_{HX} + D + I - E - Y_H - Y_X \qquad (12.50)$$

It may generally be assumed that for acids of similar constitution the ΔG values for the above process have the same sequence as the ΔH values (cf. 2-3b). The smaller ΔH is, the smaller will ΔG thus be, and the stronger will be the acid according to (10.13).

I and Y_H have the same values for all acids. Y_{HX} has such a low value compared with the other terms in (12.50) that it can be neglected for a rough comparison. Practically speaking, therefore, the quantity $(D - E - Y_X)$ is decisive for the sequence of ΔH values for different acids. This may be written $D - (E + Y_X)$ where $(E + Y_X)$ is a measure of the tendency of $X(g)$ to form the hydrated ion $X^-(aq)$. The dissociation enthalpies D for the gaseous hydrogen halides are: for HF, 570; HCl, 425; HBr, 365; HI, 295 kJ mol^{-1}. D thus decreases with increasing atomic number of the halogen atom. According to the values in tab. 5.2 and 5.4, $(E + Y_X)$ values become: for F, 858; Cl, 739; Br, 681; I, 612 kJ mol^{-1}. Consequently, $D - (E + Y_X)$ becomes: for HF, -288; HCl, -314; HBr, -316; HI, -317 kJ mol^{-1}. ΔH should thus be significantly greater for HF than for the other hydrogen halides, among which the differences are small.

A corresponding effect of bond strength is seen in the hydrides of the elements of the oxygen group, for which the bond strength falls and the acid strength rises in the sequence H_2O–H_2S–H_2Se–H_2Te.

For NH_3, H_2O and HF, that is, hydrides of elements in the same period, on the other hand, the tendency to form negative ions is decisive for the relative acid strength. Both bond strength and ionization tendency increase in the sequence NH_3–H_2O–HF. However, the latter increases so sharply, especially because of the large increase of the electron affinity in the element sequence N–O–F (table 5.2), that its effect takes precedence. The acid strength therefore increases in the sequence NH_3–H_2O–HF.

Many important inorganic acids are *oxo acids*. In the most common of these a central atom A attaches one or more OH groups with the bond A–O–H. The acid strength here depends on the electron distribution between A and O. The more the electron cloud can be displaced from A toward O, the more strongly H is bound to O. From a schematic electrostatic viewpoint it may be said that in this way O becomes more and more negative and therefore attracts the proton more strongly. Electron displacement toward O becomes stronger the lower the electronegativity and the positive charge of A are. When A has a sufficiently low electronegativity ($< \sim 1.7$) the A–O bond acquires so strong an ionic character that A–O–H becomes $A^+ + OH^-$. The substance then forms a salt containing the base OH^- (formerly the substance itself was termed a base).

An electron displacement in the opposite direction, that is from O toward A, decreases the ability of O to bind H. At the same time the ionic character of the A–O bond and thus the tendency of the substance to form $A^+ + OH^-$ decreases. Instead the substance can function as an acid, that is, give up H^+ to a base, whereby $A–O^-$ remains. The acid strength becomes greater, the higher the electronegativity and positive charge of A is. Generally the acid behavior becomes noticeable when the electronegativity is greater than about 1.7. The effect of electronegativity is apparent, for example, in the pK_a values for: ClOH, 7.5; BrOH, 8.6; and IOH, 11.

Changes in the charge of A may cause significantly greater changes in acid strength. For example, if one O atom is added to hypochlorous acid $|\overline{Cl}-\overline{O}-H$ with the formation of chlorous acid $|\overline{O}-\overline{Cl}-\overline{O}-H$, the new O atom will draw electrons away from Cl. In this way the positive charge of Cl is increased (its oxidation number is increased from I to III and its formal charge from 0 to $1+$). This leads to an electron displacement from the O atom bound to H, so that H is bound more weakly. The whole effect may be represented as an electron displacement along the arrows in the formula $O \leftarrow Cl \leftarrow O \leftarrow H$. OClOH is therefore a stronger acid than ClOH.

Addition of further O atoms, which gives chloric acid $|\overline{O}-\overset{|\overline{O}|}{Cl}-\overline{O}-H$ and perchloric acid $|\overline{O}-\overset{|\overline{O}|}{\underset{|\overline{O}|}{Cl}}-\overline{O}-H$, implies further increases in charge on the chlorine atom and still stronger acids.

These, like many other inorganic oxo acids, may be written with the general formula $AO_m(OH)_n$, which shows the constitution more correctly than the more common formulation $H_n AO_p$. The greater m is, the more O atoms draw electrons away from A, and the more easily a first proton is released by any of the OH groups. For *neutral oxo acids* of this type a rough estimate of the acid strength may be obtained from the expression $pK_a = 7 - 5m$. For the acids referred to above, which are generally formulated $HClO$, $HClO_2$, $HClO_3$ and $HClO_4$, the pK_a values 7, 2, -3 and -8, respectively, are thereby obtained. The corresponding measured values are 7.5, 2.0, -1 and ~ -7. This rough estimate takes no consideration of the electronegativity of A.

For (ortho)phosphoric acid H_3PO_4 with the structure shown on the next page, for example, $m = 1$, which gives $pK_a = 2$. In order to apply the rule to phosphorous acid and hypophosphorous acid it must be understood that they have the structures shown. In both acids one O atom thus draws electrons from the P atom

($m=1$). On the other hand its charge is not noticeably affected by the directly bound H atoms. In both cases, therefore, we have approximately $pK_a = 7-5 = 2$. The directly bound H atoms cannot be released as protons, which leads to the result that phosphorous acid is a divalent and hypophosphorous acid a monovalent acid.

phosphoric acid   phosphorous acid   hypoposphorous acid

The long-range effect that the addition of O atoms to A has on the O–H bond is an example of what is called *inductive effect*. The successive electron displacement easily provides a schematic explanation.

When a neutral oxo acid of the type $AO_m(OH)_n$ has given off one proton, pK_a for the second protolytic stage becomes approximately 5 units greater than for the first. For the third protolytic stage pK_a is another 5 units greater, and so on. This is shown by the values for the phosphoric acid system in tab. 12.2. This rule can be combined with the preceding one into a more general rule that applies to *oxo acids with the charge z* of the formula $AO_m(OH)_n^z$

$$pK_a = 7 - 5(m + 2z) \qquad (12.51)$$

For example, for $H_2PO_4^-$ ($m=2$, $z=-1$) and HPO_4^{2-} ($m=3$, $z=-2$), (12.51) gives pK_a values of 7 and 12, respectively.

It should be noted that (12.51) gives approximate pK_a values corresponding to *true* acidic constants. For H_2CO_3 we thus get the pK_a value of 2 while tab. 12.2 lists the value 6.37, corresponding to the apparent constant (12-2e).

Oxo acids with several central atoms do not follow these rules. It is also found that the difference between subsequent acidity constants becomes less, the further away from each other the corresponding protons lie. This can be seen, for example, in diphosphoric (pyrophosphoric) acid $H_4P_2O_7$. The ion $P_2O_7^{4-}$ consists of two tetrahedra with a common corner and can be written schematically

$$\begin{bmatrix} & O & & O & \\ & | & & | & \\ O- & P & -O- & P & -O \\ & | & & | & \\ & O & & O & \end{bmatrix}^{4-}.$$

In the acid two protons are bound to each tetrahedron. When the first proton is given off, a negative charge appears on the associated tetrahedron. The second proton is given off from the second tetrahedron and is therefore little attracted by this charge. The third proton, on the other hand, is given off from a tetrahedron that already has a negative charge, and the same applies to the fourth. This explains the sequence of pK_a values 1.5, 2.3, 6.6 and 9.3 with the differences 0.8, 4.3 and 2.7.

When *hydrated metal ions* function as acids (12-1b), Coulomb's law often serves as a guide to the estimation of acid strength. The water molecules are attached to the surface of the metal ion and the force with which the proton is repelled by the nucleus of the ion should therefore rapidly decrease as the radius of the ion increases. In addition it should be proportional to the charge of the ion. The monovalent aquo ions of the alkali metals have no acid properties. The divalent aquo ions of the alkaline-earth metals, with the exception of Be^{2+}, are very weak acids. If the radius of the central ion is reduced without changing the charge, the acid strength increases, as, for example, for the aquo ion of Zn^{2+} and still more for the very small ion Be^{2+}. The acid strength is still greater for aquo ions of trivalent metal ions with relatively small radii; for example, Al^{3+}. However, other factors contribute besides the electrostatic repulsion of the proton.

12-3. Equilibrium position and acidity

a. The equilibrium equation and total concentration. In calculations of acid-base equilibria one is mostly concerned with concentration values and therefore uses the equilibrium equation (12.43). In order to obtain reasonably exact results one should then as far as possible use k_a values that are valid for the solution under study. These can be calculated from (12.44) if the f values in the solution are known. If reliable f values are lacking, (11.4) may be used to estimate the f values of the dissolved substances. If we take logarithms of (12.44) and substitute the value of log f from (11.4), since always $z_b = z_a - 1$, we obtain

$$pk_a = pK_a + (z_a - 0.5)\sqrt{I}/(1+\sqrt{I}) \qquad (12.52)$$

The calculation of pk_a is easily accomplished with the help of the values of $\sqrt{I}/(1+\sqrt{I})$ in tab. 11.2. In a solution with $I = 0.1$, for example, we have for HAc ($z_a = 0$) $pk_a = 4.76 - 0.12 = 4.64$, and for NH_4^+ ($z_a = +1$) $pk_a = 9.25 + 0.12 = 9.37$.

For the water system, in dilute water solutions, according to 12-2e generally the values $k_{H_3O^+} = 55.5$, $pk_{H_3O^+} = -1.74$ and $k_w = 1.82 \times 10^{-16}$, $pk_w = 15.74$, may be used.

If a protolyte is dissolved with the total concentration C and the protolytes contained in the system participate only in acid-base equilibria, the sum of the concentrations of these protolytes $= C$. In this case, in a C molar solution of Na_2HPO_4, we have $C = [H_3PO_4] + [H_2PO_4^-] + [HPO_4^{2-}] + [PO_4^{3-}]$. The concentration of each individual protolyte is in addition dependent on the acidity and therefore cannot be immediately given. However, one or more of the protolytes in the system may also take part in other equilibria. For example, if Fe^{3+} ions are added

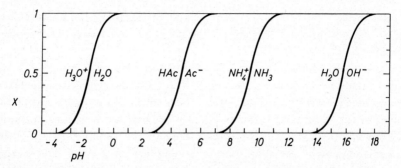

Fig. 12.2. The distribution between acid and base for some acid-base pairs at different acidities.

to the phosphate solution, a portion of the phosphate system is incorporated into complexes with Fe^{3+}. The sum of the four terms given above is then $<C$ and its magnitude depends on many conditions.

b. Acid fraction and base fraction. It is often practical to state how large a fraction of the total concentration of a system the concentration of a certain one of the protolytes of the system constitutes. This is often done especially when the system contains only a single acid-base pair. The portions of the system that are present as acid and as base are then called *acid fraction* x_a and *base fraction* x_b, respectively. Hence we have

$$x_a = \frac{[a]}{[a]+[b]}; \quad x_b = \frac{[b]}{[a]+[b]}; \quad x_a + x_b = 1 \tag{12.53}$$

If a and b do not participate in other equilibria, then $C = [a]+[b]$ and thus

$$x_a = [a]/C; \quad x_b = [b]/C \tag{12.54}$$

Starting with a certain protolyte, the *degree of protolysis* is often given. The degree of protolysis for an acid is the fraction of the acid-base pair that is transformed to a base, that is, equal to the base fraction. The degree of protolysis for a base is equal to the acid fraction.

Let us assume that the weak acid a is dissolved in water, whereupon the equilibrium $a + H_2O \rightleftharpoons b + H_3O^+$ is set up. If we neglect those H_3O^+ ions formed by the autoprotolysis of the water, then $[b] = [H_3O^+]$. According to (12.32), in sufficiently dilute solutions, $[H_3O^+] = \{H^+\}$. (12.43) may then be written as $[b]^2/[a] = k_a$. Introducing $[b] = Cx_b$ and $[a] = Cx_a = C(1-x_b)$, we obtain

$$Cx_b^2/(1-x_b) = k_a \tag{12.55}$$

Analogously, for the protolysis of a base on solution in water we have

$$Cx_a^2/(1-x_a) = k_b \tag{12.56}$$

These two expressions show that the degree of protolysis (according to older terminology, degree of dissociation or hydrolysis) increases with increasing dilution. They also show that the degree of protolysis at a given concentration is greater the stronger the acid or the base is.

c. Displacement of equilibrium with acidity. This problem is treated in a manner analogous to the corresponding general case in 10c. By taking logarithms of (12.43) we obtain

$$\mathrm{pH} = pk_a + \log \frac{[b]}{[a]} \tag{12.57}$$

If k_a is assumed to be constant, (12.57) immediately shows how the composition of an acid-base pair is shifted with change in pH. Often it is practical to introduce the base fraction. From (12.53) it follows that $[b]/[a] = x_b/x_a = x_b/(1-x_b)$. (12.57) may then be written

$$\mathrm{pH} = pk_a + \log \frac{x_b}{1-x_b} \tag{12.58}$$

If x_b in (12.58) is plotted as a function of pH, an S curve of the same form as in fig. 10.1 is obtained (midpoint at $x_b = x_a = 0.5$ and $\mathrm{pH} = pk_a$), which forms the boundary between an area for the acid a and an area for the base b. Such curves have been drawn in fig. 12.2 for the pairs H_3O^+–H_2O and H_2O–OH^- belonging to the water system, and for the pairs HAc–Ac$^-$ and NH_4^+–NH_3. The ionic strength has been assumed to be constant $= 0.1$, which with (12.52) gives the pk_a values -1.74, 15.74, 4.64 and 9.37, respectively. The figure thus is a superposition of a diagram for the two pairs of the water system and a diagram for each of the two other pairs. The interaction between the two pairs of the water system means that the corresponding S curves are dependent on each other (cf. 10c). However, the two pk_a values are here so different that this does not appreciably affect the shape of the curves. In similar cases pairs belonging to the same system can often be considered to be independent of each other.

From fig. 12.2 the equilibrium position at a certain pH value can be immediately read off, and thus also the displacements of equilibrium with pH changes, for example, with the addition of strong protolytes. It can be seen that the transition from acid to base for by far the greatest part takes place within a quite small pH range. From $\mathrm{pH} = pk_a - 1$ to $pk_a + 1$, x_b is changed approximately from 0.1 to 0.9. From $\mathrm{pH} = pk_a - 2$ to $pk_a + 2$, x_b is changed approximately from 0.01 to 0.99. Within a range of 4 pH units the equilibrium is thus shifted from 99 % acid to 99 % base.

Fig. 12.2 also shows how the two water pairs separate the region for weak protolytes from the regions for strong acids and strong bases (12-2f).

NH_3 is volatile. If pH and thus $[NH_3]$ is increased by the addition of base, the ammonia pressure of the solution increases, and at sufficiently high pH, the solution therefore gives off appreciable quantities of NH_3. If the solution contains a volatile acid, this is given off in an analogous manner when pH becomes sufficiently low.

d. Calculation of acid-base equilibria. If the pH value of a solution is known, it is generally an easy matter to calculate the concentrations of all acids and bases. For each pair (12.57) gives the ratio [b]/[a] and from the total concentration of the pair [b] and [a] are then obtained.

If the pH of the system is not known, the calculations become more difficult. Since all acid-base equilibria are determined by the pH value, the primary aim is to calculate it. However, the methods must be passed by, and only the simple case where the solution contains only one strong protolyte is mentioned here.

A *strong acid* dissolved in a sufficient amount of water is practically completely protolyzed according to the reaction $a + H_2O \rightarrow b + H_3O^+$. If the total concentration of the acid is C then $[H_3O^+] = C$ and according to (12.32) $\{H^+\} = C$, that is,

$$\text{pH} = -\log C \qquad (12.59)$$

For example, in 0.001 M HCl, pH = 3.0.

Analogously, a *strong base* is completely protolyzed according to the reaction $b + H_2O \rightarrow a + OH^-$, which gives $[OH^-] = C$ and in dilute solution $\{OH^-\} = C$. According to (12.34) then $\{H^+\} = K_w/\{OH^-\} = K_w/C$, that is

$$\text{pH} = pK_w + \log C \qquad (12.60)$$

For example, in 0.1 M NaOH, pH = 13.0.

e. Superacid and superbasic solutions. In 12-2f the levelling effect that the two water pairs exert on acids stronger than H_3O^+ and bases stronger than OH^- was discussed. When the proportion of water in very concentrated solutions becomes insufficient for complete protolysis of these strong acids or bases, the levelling effect of the water disappears. The pH value of the solution is then $< \sim 0$ (*superacid solution*) or $> \sim 14$ (*superbasic solution*). Thus, in sulfuric acid with 78% H_2SO_4 pH = -6, in pure H_2SO_4 pH = -10.6, and in saturated KOH solution pH = 19.

At least in strong superacid or superbasic solutions the water can no longer be considered to be the solvent, and as in solutions in other solvents the acid-base equilibria here may become very different from those in more dilute water solutions. The fact that the relative acid strengths may become altered because of the completely different properties of the medium contributes to this effect.

In superbasic solutions, for example of alkali oxides, the very strong base O_2^- mentioned in 12-2f may appear in equilibrium with OH^-. Here OH^- thus behaves as an acid. The proportion of water must be so small, however, that a liquid phase is obtained only at high temperature, and, in order that the water not be lost by vaporization, high pressure is also required.

Acid-Base Equilibria

f. Buffer action. Let us assume that a strong acid or base is added to a protolyte solution while the pH of the solution is being measured. If pH is plotted as a function of added acid or base a *titration curve* is obtained, whose appearance reflects the presence of weak protolytes in the solution. If it contains the weak acid-base pair a–b, according to 12-3c a rapid shift of equilibrium occurs when the pH passes the pk_a value of the pair. If at the start $pH < pk_a$, the pair is present mainly as a. If a strong base is added, which implies the addition of OH^- ions, as the pk_a value is passed the reaction $a + OH^- \rightarrow b + H_2O$ takes place. If at the start $pH > pk_a$, the pair is present mainly as b. Addition of strong acid, that is, H_3O^+ ions, on passing the pk_a value leads to the reaction $b + H_3O^+ \rightarrow a + H_2O$. This consumption of OH^- or H_3O^+ ions causes the pH of the solution to change much more slowly than if the pair a–b were not present in the solution. It is said that the pair serves as a *buffer* and increases the *buffer action* of the solution. If the titration curve is constructed as mentioned above, the buffer action of the solution at a particular pH value is inversely proportional to the slope of the curve, that is, its derivative, at that value.

The buffer action of the solution at a given pH value is equal to the sum of the contributions of all the acid-base pairs present at that value. The contribution of a particular pair is greatest at the pk_a value of the pair. It drops quite rapidly on both sides of this point. It is also clear that the contribution at a given pH value is proportional to the total concentration of the pair.

The presence of the pair H_3O^+–H_2O and H_2O–OH^- in all water solutions gives a high buffer action in the vicinity of $pH = 0$ and 14. If there are no weak protolytes dissolved in the water, on the other hand, the buffer action is low over a large intermediate pH range, at least from 2 to 12. For example, if a solution with $pH = 5.0$ is desired, theoretically 10^{-5} M HCl could be used. However, the traces of impurities that can never be avoided would cause a completely different pH value. In order to increase the pH of 1 l of this solution to 5.3, addition of only 5×10^{-6} mole of OH^- ions is required, for example in the form of 0.05 ml (approximately one drop) 0.1 M NaOH. A solution with the molarities 0.073 M HAc + 0.127 M NaAc has the same value of $pH = 5.0$ but with tremendously greater buffer action. To raise the pH to 5.3, 0.028 mol OH^- ions must be added to 1 l of this solution, that is, the amount present in 280 ml 0.1 M NaOH. A solution which is prepared in this way with the purpose of obtaining high buffer action, is called a *buffer solution*.

A buffer solution can often be prepared by mixing the acid and base of the pair a–b. In the above example, solutions of a = HAc and b = Ac^- (contained in NaAc) were mixed. Addition of strong base to a solution of a, or of acid to a solution of b gives the same result. Although high buffer action would be obtained with a high total concentration of the buffer system, in order to avoid complications caused by excessively high ionic concentration, concentrations higher than $C = 0.2$ are seldom used. For a certain acid-base pair then, relatively effective buffer action can be expected only within the interval $pH = pk_a \pm 1$. If a polybasic acid is used, a high buffer action can be obtained in several pH ranges or over a broader continuous region.

Most biological fluids are strongly buffered and therefore can maintain relatively independently of external influences the pH necessary for the processes, often very sensitive to pH, that occur in the organism.

g. pH indicators. A pH indicator is an acid-base pair in which the acid and the base have different colors, or only one is colored. The color (or for a one-color pair the strength of the color) of a solution containing the indicator system, therefore becomes dependent on the ratio [a]/[b], that is, also on pH.

It is best to refer to the expression (12.58). For $pH = pk_a$, $x_b = x_a = 0.5$. The solution here shows a mixed color consisting of equal parts of acid color and base color. If the pH of the solution is changed from the pk_a value of the indicator, on the acid side the contribution of the base color is decreased, and on the basic side the contribution of the acid color is decreased. It may be assumed that the contribution of a certain color cannot be seen after the amount of the corresponding protolyte falls below 10 % of the whole amount of the indicator system. The *transition range* within which the transition from one color to the other can be seen, is then symmetrically placed about the pk_a value, according to (12.58), and has approximately the limits $pH = pk_a \pm 1$. Its breadth thus is about 2 pH units (cf. the curves in fig. 12.2).

All pH indicators now in use are organic protolytes. The difference in color between the acid and the base is often caused by rearrangements in the indicator molecule which follow the proton exchange. However, these rearrangements do not affect the account of their method of action given here.

Since a given indicator can show the pH only within a limited interval, a series of indicators is required to cover a larger pH range. For quick tests a mixture of several indicators (*universal indicator*) is often used, as well as filter paper saturated with indicator solution and afterwards dried (*indicator paper*).

12-4. Polyvalent protolytes and ampholytes

a. Definitions. A molecule or ion that can function both as acid and base was defined in 12-1a as an *ampholyte*. It was shown also that any ion existing as an intermediate in the splitting off of protons by a polyvalent acid is an ampholyte. For example, in the phosphoric acid system H_3PO_4–$H_2PO_4^-$–HPO_4^{2-}–PO_4^{3-} there are two ampholytes, $H_2PO_4^-$ and HPO_4^{2-}.

An ampholyte, amf, would thus in the simplest case enter into the system a–amf–b (also a and b may then be ampholytes). On solution in water the ampholyte acts as an acid according to

$$\text{amf} + H_2O \rightleftharpoons b + H_2O^+ \tag{12.61}$$

and as a base according to

$$\text{amf} + H_2O \rightleftharpoons a + OH^- \tag{12.62}$$

Acid-Base Equilibria

In addition there occurs the *autoprotolysis* (12-2d)

$$\text{amf} + \text{amf} \rightleftharpoons \text{a} + \text{b} \tag{12.63}$$

Autoprotolysis can take place as well in the pure ampholyte (as with water) or in its solution in a nonprotolyzing solvent.

b. Equilibria in ampholyte solutions. The acid-base function of ampholytes is best characterized by the acidic constants of the two participating acid-base pairs a–amf and amf–b. These constants are often called primary and secondary acidic constants for the divalent acid a. If this is designated by H_2A we may thus consider mainly the equilibria

$$\underset{\text{a}}{H_2A} + H_2O \rightleftharpoons \underset{\text{amf}}{HA^-} + H_3O^+ \tag{12.64}$$

$$\underset{\text{amf}}{HA^-} + H_2O \rightleftharpoons \underset{\text{b}}{A^{2-}} + H_3O^+ \tag{12.65}$$

If we denote the primary and secondary acidic constants for H_2A by K_{a1} and K_{a2} we have for these equilibria

$$\{H_3O^+\}\{\text{amf}\}/\{\text{a}\} = K_{a1} \tag{12.66}$$

$$\{H_3O^+\}\{\text{b}\}/\{\text{amf}\} = K_{a2} \tag{12.67}$$

For a trivalent acid H_3A a further constant K_{a3} is obtained, and so on.

An ampholyte may also be characterized by an acidic constant K_a that determines the equilibrium (12.61) = (12.65), and a basic constant K_b that determines the equilibrium (12.62). We then have

$$\{H_3O^+\}\{\text{b}\}/\{\text{amf}\} = K_a \tag{12.68}$$

$$\{OH^-\}\{\text{a}\}/\{\text{amf}\} = K_b \tag{12.69}$$

For the two types of constant it is thus true that $K_a = K_{a2}$, $K_b = K_w/K_{a1}$.

When using concentrations [amf], [a] and [b] in equilibrium calculations, expressions analogous to (12.66–12.69), containing mixed constants k_{a1}, k_{a2}, k_a, k_b, are applied.

Fig. 12.3 shows the distribution of the various protolytes of the phosphoric acid system as a function of pH. Solutions containing this system may be obtained, for example by dissolving H_3PO_4, NaH_2PO_4, Na_2HPO_4 or Na_3PO_4. If these solutions are made 0.3 molar, their pH values are 1.3, 4.7, 9.8 and 12.8, respectively (note that the expression "acid" salt—cf. 12-1c—thus has nothing to do with the reaction of the solution). From the figure it is seen that the system in the first solution consists of 86% H_3PO_4 + 14% $H_2PO_4^-$. In the second and third solutions the system consists nearly entirely of $H_2PO_4^-$ and HPO_4^{2-} respectively. In the fourth solution it consists of 25% HPO_4^{2-} + 75% PO_4^{3-}.

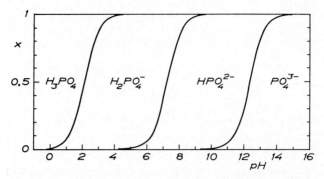

Fig. 12.3. The phosphoric acid system at different acidities.

Starting with the solution of H_3PO_4, if the pH is increased by addition of a strong base, the composition of the system at any pH value can be read off from the figure. Of course, all of the system's protolytes are present at any pH value whatever, although the quantity of certain ones may be too small to be seen on a linear concentration scale such as in fig. 12.3; these low proportions are read with more advantage on a logarithmic diagram.

If the pK_a values of a polyvalent acid lie sufficiently far apart, a solution may contain the intermediate ampholyte in high concentration over a rather large pH range. Such is the case with H_3PO_4, in which the system exists mainly as $H_2PO_4^-$ and HPO_4^{2-} over fairly large pH ranges (fig. 12.3). From solutions within these pH ranges salts of the hydrogen orthophosphate ions such as NaH_2PO_4 and Na_2HPO_4, respectively, often crystallize. If the pK_a values come closer together, the region of existence of the ampholyte is decreased; its proportion of the total concentration becomes less. An example of this is diphosphoric (pyrophosphoric) acid, $H_4P_2O_7$ (fig. 12.4; cf. the pK_a values in 12-2f). Here the ampholytes $H_3P_2O_7^-$ and $HP_2O_7^{3-}$ have narrow ranges of existence and the only hydrogen diphosphate that crystallizes from the solution contains the ion $H_2P_2O_7^{2-}$.

c. Isoelectric point. If the acidity of a solution of an ampholyte is increased, the equilibrium (12.61) is displaced to the left and the equilibrium (12.62) to the right. A decrease in the acidity causes the equilibria to be displaced in the opposite direction. At a certain acidity, therefore, the degree of protolysis according to (12.61) becomes equal to the degree of protolysis according to (12.62). The ampholyte is then equally strongly protolyzed as an acid and as a base. At this acidity, which is called the *isoelectric point* of the ampholyte, thus [a]=[b]. It can be shown that here $pH = \frac{1}{2} (pk_{a1} + pk_{a2})$.

For a given concentration at the isoelectric point, [a]+[b]=minimum and [amf]=maximum. Because of this, a number of the properties of ampholyte solutions show a maximum or a minimum at this point.

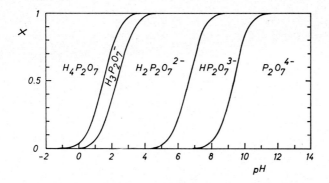

Fig. 12.4. The diphosphoric acid system at different acidities.

d. Amphoteric hydroxides. Many more or less difficultly soluble hydroxides are ampholytes. The slight solubility is probably related to the fact that they form complexes with a very large number of nuclei.

Reactions that occur during the formation and solution of amphoteric hydroxides are often very involved and therefore incompletely understood. A probably quite typical case is illustrated by the following highly schematic outline of the reactions of aluminum hydroxide.

Freshly precipitated $Al(OH)_3$ is easily dissolved in both acids and bases, thereby showing its amphoteric properties. The solubility has a minimum at about $pH = 7$, which is thus the most favorable pH value for the precipitation of $Al(OH)_3$. If the precipitate stands, its properties generally become altered; it "ages" and may finally crystallize. If it has aged, heating and strong reagents may be required to dissolve it.

$Al(OH)_3$ may be thought of as an infinite complex, in principle as in fig. 6.18b. The Al^{3+} ions are thus bound together by double OH^- bridges.

The amphoteric properties of $Al(OH)_3$ may be visualized by the schematic equations (12.70) and (12.71). The first shows only how an OH^- bridge (middle term) is dissolved on addition of acid (to the left) and on addition of base (to the right). The remaining parts of the molecule are denoted by R.

$$\left.\begin{array}{l} H_2O - R \\ H_2O - R \end{array}\right\} \underset{+H^+}{\overset{-H^+}{\rightleftarrows}} HO \overset{R}{\underset{R}{\diagup\!\!\!\diagdown}} + H_2O \underset{+H^+}{\overset{-H^+}{\rightleftarrows}} \left\{\begin{array}{l} HO - R \\ HO - R \end{array}\right. \qquad (12.70)$$

On addition of acid the OH^- bridges are thus broken and replaced on each fragment by attaching H_2O molecules. The complexes become smaller and smaller, and as a limit at high acidity they may be thought of as hydrated Al^{3+} ions $Al(H_2O)_6^{3+}$.

On addition of base the OH^- bridges are also broken, but replaced on each fragment by attaching OH^- groups. Experimental data suggest that we have here as a final stage the mononuclear (hydroxo)aluminate ion $Al(OH)_4^-$. The degree of hydration of this ion is unknown. It may possibly have the composition $Al(OH)_4(H_2O)_2^-$ so that the coordination number of aluminum is still six. However, we must not neglect the coordination number four, since crystals containing the anion $(OH)_3AlOAl(OH)_3^{2-}$ can be obtained from basic solutions. This ion consists of two AlO_4 tetrahedra with one O atom in common and the rest belonging to OH^- groups.

The overall reactions on the acid and the basic side of $Al(OH)_3$ may be summarized as in (12.71).

$$Al(H_2O)_6^{3+} \underset{+3H^+}{\overset{-3H^+}{\rightleftarrows}} \left\{ \begin{matrix} Al(OH)_3 \\ 3H_2O \end{matrix} \right\} \underset{+H^+}{\overset{-H^+}{\rightleftarrows}} Al(OH)_4^- + 2H_2O \qquad (12.71)$$

Between the middle term and the mononuclear final terms to the left and to the right lie many intermediate stages. Among these there occur to a large extent polynuclear complexes, whose nuclear multiplicity increases toward the middle term. The mechanism of hydrolysis in solutions of Al^{3+} salts mentioned in 12-1b involves the passage from the left term through such intermediate terms toward the middle term.

Other important amphoteric hydroxides are those of Be^{2+}, Zn^{2+}, Sn^{2+}, Pb^{2+}, Cr^{3+}, As^{3+} and Sb^{3+}. The acid and base functions of most of these hydroxides may probably be considered to be analogous to the one above, although the coordination numbers may vary.

12-5. Acid-base equilibria in other solvents than water

In 12-2f it was stated that the influence of the solvent on acid-base equilibria in solution among other things depends on the dielectric constant, but above all on its protolytic properties. On the basis of their protolytic properties, solvents may be divided into amphoteric solvents analogous to water, acid solvents, basic solvents and nonprotolyzing (aprotic) solvents.

Of the amphoteric solvents analogous to water, the alcohols are very important. Among the acid solvents, H_2SO_4, HF, HCN, formic acid and acetic acid have especially been studied. All of these acids can take up protons, however (with formation of $H_3SO_4^+$, H_2F^+ and H_2CN^+, for example), and therefore have amphoteric properties. Among the basic solvents may be mentioned NH_3, hydrazine N_2H_4, hydroxylamine NH_2OH and a considerable number of organic nitrogen compounds. Many of these may also be amphoteric, but there are some basic solvents that cannot function as acids (for example, diethyl ether). Among the nonprotolyzing solvents hydrocarbons and their halogen derivatives are important.

An acid is protolyzed more strongly the more basic the solvent is. If an acid is weak in water solution, it may become strong in a strongly basic solvent. Acetic acid is therefore a strong acid when it is dissolved in liquid NH_3. In a corresponding way the strength of a base is increased in more strongly acid solvents.

It was mentioned in 12-2a that also the relationships between the K_a values of different acids are changed in going from one solvent to another. The relative strengths of the acids are thus altered, and the changes may be so great that even the order of sequence of the different acids may be rearranged. However, if the strengths of acids having analogous constitution are compared in different solvents the relative changes in strength are generally small.

In water solution HF is a quite weak acid (12-2f) but if liquid HF is mixed with pure HNO_3, HF shows itself to be a stronger acid than HNO_3. An equilibrium is set up, $HF + HNO_3 \rightleftharpoons F^- + H_2O + NO_2^+$ (nitryl cation), which is strongly displaced to the right. HNO_3 thus functions here as a base.

Concentrated H_2SO_4 is a stronger acid than HCl. Thus, the basic properties of the Cl^- ion here become noticeable. If, for example, NaCl is added to concentrated H_2SO_4, HCl is therefore formed. The conversion is nearly complete because HCl is volatile and so removed from the system. H_2SO_4 gives up one proton to pure HNO_3 with the formation of water and nitryl cation. The equilibrium $H_2SO_4 + HNO_3 \rightleftharpoons HSO_4^- + H_2O + NO_2^+$ is thus strongly shifted to the right.

CHAPTER 13

Heterogeneous Equilibria

A general presentation of heterogeneous equilibria can only be done with the help of the phase rule, which is treated in 13-2b. Here we consider at the start an important type of a simple heterogeneous equilibrium, which involves the distribution of a substance between two different phases (*distribution equilibrium*).

13-1. Distribution equilibrium between two phases

If the substance "i" is contained in two phases (indicated by ′ and ″), and these are in equilibrium with each other, the chemical potential of the substance is equal in the two phases, that is, $\mu_i' = \mu_i''$ (2-3d). This equilibrium condition, analogously with the derivation of the equilibrium equation (10.14) also gives the condition

$$a_i''/a_i' = K \quad (P=\text{const.}, \ T=\text{const.}) \tag{13.1}$$

(13.1) is also obtained directly if (10.14) is applied to the equilibrium $i' \rightleftharpoons i''$. Some important cases of this equilibrium in practice are discussed in the following sections.

a. Vapor pressure relations. Let us assume that a liquid solution containing the component "i" is in equilibrium with a gas phase. If "i" is volatile, it will also be present in the gas phase. The equilibrium condition (13.1) is applied and we indicate the gas phase by ″ and the solution by ′. The gas can be assumed to be ideal in most cases and for now it is assumed that the solution is also ideal. Then, with proper choice of standard state according to 10a we have $a_i'' = p_i$ and $a_i' = x_i$, where p_i is the partial pressure of the substance "i" in the gas and x_i is its mole fraction in the solution. Substituting these values in (13.1) we have

$$p_i = Kx_i \tag{13.2}$$

If the solution is not ideal at all concentrations, it will be so at least at high dilution. In this ideal, dilute solution (1–4i) the component 1 is present in large excess and is termed the solvent. Other components, for example 2, have low concentration and are termed the dissolved substances. For the solvent it is thus true that

$$p_1 = Kx_1 \tag{13.3}$$

If the values for the vapor pressure of the *pure* solvent at the same temperature $p_1 = p_1^0$ and the mole fraction $x_1 = 1$ are substituted into (13.3), we obtain $K = p_1^0$. For the *solvent*, therefore, *Raoult's law* applies:

$$p_1 = x_1 p_1^0 \qquad (13.4)$$

Since x_1 in any solution is <1 and decreases with increasing concentration of dissolved substance, a *vapor-pressure depression* occurs, which increases with the concentration of dissolved substance. For example, Raoult's law is valid for the water-vapor pressure over cane-sugar solutions.

If the volatile component is a *dissolved substance* 2, Henry's law is obtained

$$p_2 = K x_2 \qquad (13.5)$$

which, for example, determines the solubility of oxygen in water as a function of the partial pressure of the oxygen.

Only if the solutions are ideal for all proportions from pure solvent to pure dissolved substance is K equal to the vapor pressure of the pure dissolved substance. In this rather rare case Raoult's law and Henry's law become identical. According to Henry's law, then, the partial pressure of the dissolved substance is proportional to its mole fraction. In dilute solutions, where the law is generally valid, the mole fraction of a substance is very nearly proportional to the molarity and molality of the substance, so that the partial pressure is also proportional to these quantities. Henry's law can also be expressed to state that the solubility of a gas in a liquid is proportional to the partial pressure of the gas over the liquid.

All these formulas apply only if a certain number of molecules of the substance in the gas give the same number of molecules in the liquid. Such is often the case with the solvent but is frequently not true of the dissolved substances. Henry's law does not apply, for example, for solutions of HCl in water, in which 1 molecule of HCl in the gas gives 2 molecules ($H_3O^+ + Cl^-$) in the solution.

b. Solubility. The equilibrium condition (13.1) may be applied to a pure solid or liquid substance "i" that is in equilibrium with a solution of "i". The two phases are indicated by ' and ", respectively. According to 10a, in this case $a_i' = 1$ and thus

$$a_i'' = K \quad (P = \text{const.}, \ T = \text{const.}) \qquad (13.6)$$

If the solution is ideal x_i'' and c_i'' are also constants. The concentration of "i" in the solution is thus constant at constant pressure and temperature. This constant concentration is the *solubility* s_i of "i" at the pressure and temperature in question. If the solution is not ideal, the solubility is no longer constant. One type of variation in solubility caused by departure from the ideal state is discussed in 13-1 d.

When as in this case both the phases taking part in the solubility equilibrium are condensed, K and thus the solubility are little affected by pressure (10d).

On the other hand, the solubility may be strongly dependent on temperature (see below).

A solution is said to be *saturated* with respect to a certain substance if it is in equilibrium with that substance as a pure phase. If the concentration under otherwise equal circumstances is less than the equilibrium concentration the solution is said to be *unsaturated*; if the concentration is greater, the solution is said to be *supersaturated*. A supersaturated solution is unstable. Solubility equilibria are often established very slowly. In the measurement of solubility it is therefore often necessary to start both with an unsaturated solution plus a sufficient amount of solid phase, and with a supersaturated solution.

The concentration at which a solution is in equilibrium with a pure, condensed phase, is dependent on the particle size of the pure phase. As described in 6-5b, the solubility increases when the radius of curvature of the particle surface decreases, that is, as the particle becomes smaller. A small particle will therefore be in equilibrium with a more concentrated solution than a large particle. When the particle size becomes sufficiently large (approximately 1 to 10 μm), however, the change in solubility with further increase in size becomes too small to be detectable. When a saturated solution of a substance is spoken of without further qualification, a solution is referred to that is in equilibrium with a condensed phase of particle size so great that its further increase does not measurably change the concentration. Unsaturation and supersaturation is then defined in relation to just this concentration of saturation.

The solubility values in the literature in general state the total concentration of the substance in question in a solution that according to the above is designated as saturated.

c. The magnitude of the solubility and its change with pressure and temperature. The solubility of different substances in a certain solvent can vary within extremely wide limits, and the same is true of the solubility of a single substance in different solvents. Only a few general rules concerning the magnitude of solubility in various cases are known.

Substances whose building units are held together by electrostatic forces (ion-ion, ion-dipole and dipole-dipole bonds) are generally best dissolved by solvents with high dielectric constant. In 11c it was stated that the force between two charged bodies is inversely proportional to the dielectric constant of the medium that lies between them. In the same way that this causes solvents with high dielectric constant (strongly polar solvents) to dissociate electrolytes easily, this fact also causes such solvents to have great ability to dissolve substances with electrostatically held building units. The very high dielectric constant of water (5-5) therefore explains its great solvent power for such substances.

For other substances (except metals) a rough survey of solubility behavior can be presented if a division is first made between substances in which the molecules are mainly held together by hydrogen bonds (for example, H_2O, NH_3, HF, sugars) and substances in which the molecules are mainly held together by van der Waals forces (for example, hydrocarbons, CCl_4, CS_2). The binding forces are greater in the first than in the second group (5-5, 5-4). If molecules belonging to both groups are present, those that can be bound together with hydrogen bonds therefore try to bond to each other and thus form a separate phase. It may be roughly stated, therefore, that substances with hydrogen bonds are easily soluble in each other and substances without hydrogen bonds are likewise, while the miscibility between substances that belong to different groups in general is limited. Cane sugar is easily soluble in water but does not dissolve in the hydrocarbon benzene. The hydrocarbon naphthalene is easily soluble in benzene and poorly soluble in water.

The alcohols may be taken as an example of substances that occupy an intermediate position. Between the hydroxy groups of the alcohol molecules hydrogen bonds occur (though fewer per molecule than in water) and between their hydrocarbon radicals van der Waals forces occur. When the hydrocarbon radicals are small the hydrogen bonds play the greatest role. If the hydrocarbon radicals increase in size, the van der Waals forces increase also and finally come to predominate. The solubilities of the aliphatic alcohols in water show this effect. Up to and including propanol they are completely miscible with water, butanol shows limited solubility in water, and from decanol onward they are practically insoluble. The solubility in hydrocarbons goes in the opposite direction. The result is that many substances in both the groups referred to above are at least moderately soluble in ethanol. Thus, both water and benzene are completely miscible with ethanol. If ethanol is added to a two-phase mixture of water and benzene, the two liquid phases, if enough is added, are converted to a single phase.

In cases where more or less covalent bonds hold the building units of a solid phase together in a one-, two- or three-dimensional giant molecule, the solubility usually becomes slight in most solvents.

It has been said above that the solubilities of solid and liquid substances are little dependent on the pressure, but the solubilities of gases, on the other hand, are strongly dependent on pressure. The influence of temperature on solubility can be great, and most often the solubility increases with rising temperature.

The direction in which the solubility is changed with rise in temperature depends on the sign of the partial molar enthalpy of solution in the saturated solution at the temperature in question (concerning partial molar quantities, see 1-5e). The sign is *positive* if heat is *taken up* on solution of the substance in a very large amount of nearly saturated solution, and in such a case the solubility *increases* with rise in temperature. With a *negative sign*, that is, with evolution of heat, the solubility *decreases* with rising temperature. The former case is far more common than the latter.

The partial molar enthalpy of solution most often has the same sign in dilute as in saturated solutions. If this sign is positive, heat is always absorbed when the substance is dissolved in the solvent and vice versa. There are exceptions, however. For NaOH the partial molar enthalpy of solution in saturated solution is positive, and in agree-

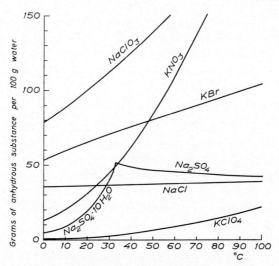

Fig. 13.1. Solubility curves for some salts.

ment with this the solubility of NaOH increases with rising temperature. In spite of this, heat is given off when NaOH is dissolved in water. This is because the partial molar enthalpy of solution in dilute solutions is negative.

Fig. 13.1 shows the temperature dependence of the solubility for a number of salts. We may note the slight temperature dependence for NaCl, and the different solubility curves for Na_2SO_4 and $Na_2SO_4 \cdot 10H_2O$. At lower temperatures the saturated solution is in equilibrium with the crystalline hydrate $Na_2SO_4 \cdot 10H_2O$, but if the temperature is increased to 32.38°C the crystal water is given up to the solution, and the solid phase above this temperature is Na_2SO_4. In stating the solubility of a substance, therefore, the composition of the solid phase must always be given. Only with rough, qualitative estimates of solubility may this perhaps be omitted.

The dependency of the solubility on pressure and temperature means that changes in pressure and temperature in the proper direction can lead to supersaturation. If the solubility increases with rising temperature a drop in temperature often gives a supersaturated solution (in this case one also speaks of *supercooling*).

The solubility is often given qualitatively by expressions such as "easily soluble", "difficultly soluble", and so on. In spite of the possibility of great variation in the expression and various interpretations of the boundary lines, an attempt is made in the following table to give an estimate of the significance of such expressions:
Solubility, g per 100 g solvent:

<0.01	0.01–0.1	0.1–1	1–10	>10
very difficultly soluble (insoluble)	difficultly soluble (poorly soluble)	moderately difficultly soluble (slightly soluble)	moderately easily soluble (soluble)	easily soluble (very soluble)

d. Solubility product. Let us assume that a solution of a substance A_aB_b is in equilibrium with the same substance in solid, pure form. For the solubility equilibrium $A_aB_b(s) \rightleftharpoons A_aB_b(soln)$ we have according to (13.6)

$$\{A_aB_b(soln)\} = K \tag{13.7}$$

If now the dissolved substance can be dissociated according to the equation $A_aB_b(soln) \rightleftharpoons aA + bB$, we have for the dissociation equilibrium

$$\{A\}^a\{B\}^b/\{A_aB_b(soln)\} = K' \tag{13.8}$$

from which by (13.7) we get

$$\{A\}^a\{B\}^b = KK' = K_s \tag{13.9}$$

K_s is called the *solubility product* of the substance A_aB_b. If it cannot be said that the dissociation takes place through dissolved molecules A_aB_b but directly from the solid substance, we may write $\{A\}^a\{B\}^b/\{A_aB_b(s)\} = K''$. By setting $\{A_aB_b(s)\} = 1$ we obtain from this $\{A\}^a\{B\}^b = K'' = K_s$, thus the same result.

If the solubility is small and in addition the concentration of other substances in the solution so slight that the solution may be assumed to be ideal, we may write

$$[A]^a[B]^b = k_s \tag{13.10}$$

At higher concentration k_s is no longer constant. If (13.10) is divided by (13.9) and $f_i = a_i/c_i$ is introduced according to (10.7) we obtain

$$k_s = K_s/f_A{}^a f_B{}^b \tag{13.11}$$

In general the solubility product is used when electrolytic dissociation occurs. In such cases f_A and f_B are ionic activity factors, which, as a rule, decrease with increasing ionic strength (11 d). Since K_s is a true constant, k_s then increases with increasing ionic strength.

Often $pK_s = -\log K_s$ or $pk_s = -\log k_s$ is given. In tab. 13.1 values for a number of difficultly soluble electrolytes are listed. The ionic strength in saturated solution is here throughout so small that it may be assumed that $k_s = K_s$. Many of the values are only known with very small accuracy, mainly because the solid phase cannot be defined with sufficient preciseness.

Sometimes reference is made somewhat improperly to the solubility product of a substance that is unknown in solid form. An example is AgOH, which is not known as a solid (36-7a), but when its solubility product is stated reference is made to the product $[Ag^+][OH^-]$ in a solution in equilibrium with solid Ag_2O.

Under certain conditions a simple relation is obtained between the solubility and the solubility product. The solubility of the electrolyte A_aB_b is equal to the total concentration of the electrolyte in the solution. If the ions A and B do not take part in any other equilibria than the dissociation equilibrium $A_aB_b \rightleftharpoons aA + bB$ (thus, for example, no formation of complexes of other kinds or reaction with the solvent such as protolysis), the solubility in pure water becomes $s_0 = [A_aB_b] + [A]/a = [A_aB_b] + [B]/b$.

23 – 689861 *Hägg*

Table 13.1. pK_s values at room temperature.

	pK_s		pK_s		pK_s
Fluorides		CuS	36	*Carbonates*	
CaF_2	10.5	PbS	28	$CdCO_3$	14
SrF_2	8.6	SnS	26	$PbCO_3$	13
MgF_2	8.2	CdS	26	$CoCO_3$	12
PbF_2	7.5	ZnS	25	$FeCO_3$	10.5
BaF_2	6.0	CoS	22	$ZnCO_3$	10
Chlorides		NiS	20	$MnCO_3$	10
AgCl	9.7	FeS	17	$SrCO_3$	10
CuCl	6.7	MnS green	13	$BaCO_3$	8.3
Hg_2Cl_2	17.2	MnS light red	10	$CaCO_3$	8.1
$PbCl_2$	4.7	Ag_2S	50	$NiCO_3$	8
Bromides		Cu_2S	48	$MgCO_3$	4
		Bi_2S_3	90	Ag_2CO_3	11.2
AgBr	12.2				
CuBr	8.3	*Hydroxides*		*Sulfates*	
TlBr	5	AgOH	7.7	$BaSO_4$	10.0
Hg_2Br_2	22.2	$Cu(OH)_2$	20	$PbSO_4$	7.8
$PbBr_2$	4.5	$Zn(OH)_2$	17	$SrSO_4$	6.6
Iodides		$Ni(OH)_2$	15	$CaSO_4 \cdot 2H_2O$	4.6
AgI	16.0	$Co(OH)_2$	15	Hg_2SO_4	6.1
CuI	12	$Fe(OH)_2$	15	Ag_2SO_4	4.8
TlI	7.2	$Mn(OH)_2$	13		
Hg_2I_2	28	$Cd(OH)_2$	14	*Chromates*	
PbI_2	8.2	$Mg(OH)_2$	11	$PbCrO_4$	13.8
		$Ca(OH)_2$	5.4	$BaCrO_4$	9.8
Sulfides		$Al(OH)_3$	32	$SrCrO_4$	4.4
HgS	52	$Cr(OH)_3$	30	Ag_2CrO_4	12

If the dissociation is complete $[A_a B_b]$ may be neglected. Then $[A] = as_0$ and $[B] = bs_0$. If these values are inserted into (13.10) the following relation between s_0 and k_s is obtained

$$(as_0)^a (bs_0)^b = a^a b^b s_0^{a+b} = k_s \qquad (13.12)$$

If $a = b = 1$, then $s_0^2 = k_s$.

If a solid electrolyte $A_a B_b$ is present in a solution in which the product $[A]^a[B]^b$ is less than k_s, the solid electrolyte goes into solution until this product reaches the value k_s. Thus the solution at the start is unsaturated with respect to $A_a B_b$. If instead the product $[A]^a[B]^b$ is greater than k_s, $A_a B_b$ precipitates until the product has fallen to the value k_s. When the solubility product is exceeded precipitation does not always take place, but the solution may remain supersaturated for some length of time.

Changes of ionic concentrations that lead to an exceeding of or a falling below a solubility product, and thus precipitation or solution of the corresponding substance, may come about in many different ways. An increase is easily effected by addition of the type of ion in question. Let us assume that the amount of SO_4^{2-} in a solution is determined by precipitation as $BaSO_4$. Ba^{2+} is added and

Heterogeneous Equilibria

when $[Ba^{2+}][SO_4^{2-}] = k_s = 10^{-10}$, $BaSO_4$ begins to precipitate. When an amount of Ba^{2+} equivalent to the SO_4^{2-} has been added, $[Ba^{2+}] = [SO_4^{2-}] = 10^{-5}$ mol l^{-1}. $[SO_4^{2-}]$ is then the same as in a saturated solution of $BaSO_4$ in water. If more Ba^{2+} is added, more $BaSO_4$ precipitates. Since $BaSO_4$ is completely dissociated in solution its solubility is $s = [SO_4^{2-}] = k_s/[Ba^{2+}]$, that is, inversely proportional to $[Ba^{2+}]$. If Ba^{2+} is added further so that $[Ba^{2+}] = 10^{-2}$ mol l^{-1}, then $s = [SO_4^{2-}] = 10^{-8}$ mol l^{-1}. An excess of the precipitating agent thus causes a significantly more complete precipitation than that occurring at the equivalence point.

Here k_s has been assumed to be constant. On the addition of ions, however, the ionic strength increases and consequently, in accordance with the earlier discussion, usually also k_s. In most cases this effect is outweighed by the equilibrium displacement caused by the addition of the ions, but the increase of k_s means that the solubility is not decreased as much as it would be if k_s were constant. There are examples, however, in which the activity factors decrease so rapidly that the solubility of an electrolyte increases even on the addition of one of its own ions. Also, addition of foreign ions that do not react with the substance in question or its ions will cause an increase in k_s. The solubility of the substance is thereby increased.

One type of ion can often be separated from others by the addition of a substance with which it forms a sparingly soluble compound, a *precipitate*. It should then be precipitated as completely as possible while the other types of ions remain in the solution. For this it is required that the solubility product of the precipitate be sufficiently less than the solubility products of the substances that the precipitating agent can form with the other kinds of ions. Furthermore, it must be possible to regulate the concentration of the precipitating agent so that it can attain a proper value.

At the extremely low concentrations in which radioactive nuclides often occur, in general no insoluble compound of the nuclide can be precipitated directly. As a rule the concentration is too small for the solubility product to be exceeded. But even if this were possible, the amount of the precipitate would be too small to collect, and at the same time, too great a fraction of the radioactive nuclide would remain in the solution. In this case a carrier (9-5a) must be added before the precipitation.

e. Solubility equilibrium in interaction with other equilibria. If one or more ions of a substance or the substance itself participates in other equilibria than the solubility and dissociation equilibria, the ionic concentrations and hence the solubility may be changed by displacement of the other equilibria.

Let us assume that a substance AB is dissociated according to $AB \rightleftharpoons A + B$. The solubility product for AB is defined by $[A][B] = k_s$. The solubility is $s = [AB] + [A]$. With complete dissociation, $[AB] = 0$ and $s = [A] = k_s/[B]$. If now a substance C is added that with B can form the compound BC, $[B]$ decreases. Then $[A]$ and thus s must increase until the product $[A][B]$ again reaches the value k_s. The more C is added, the more the solubility increases. The increase in solubility also becomes greater the more strongly B is bound to C, that is, the greater is the equilibrium constant for the equilibrium $B+C \rightleftharpoons BC$.

Salts of weak acids provide examples of this effect. The anion of the salt (corresponding to B above) is a base and by combining with protons (corresponding to C), its concentration is decreased. The solubility is determined by the solubility product of the salt, the hydrogen ion activity of the solution and the strength of the acid of the salt (corresponding to BC). The solubility increases as the solubility product and hydrogen ion activity increase and as the strength of the acid decreases. If a salt of a very weak acid is still difficultly soluble in a strongly acid solution, the solubility product of the salt must be extremely small.

The solubility of salts of strong acids according to this principle should be independent of the hydrogen ion activity. The anions in such cases have no tendency to take up protons. Here, however, addition of acid, as with any addition of ions, may increase the solubility somewhat by increasing the ionic strength.

For salts of polyvalent acids the solubility at different acidities is, of course, determined by the strength of the acid that is formed when the ion of the salt takes up the first proton. Because the acid HSO_4^- is weak, the solubility of the poorly soluble sulfates is dependent on the acidity of the solution. Ca, Sr, Ba and Pb sulfates therefore dissolve in significantly larger amounts even in dilute strong acids, except for sulfuric acid, than in pure water. In sulfuric acid of not too strong concentration the presence of SO_4^{2-} ions causes a decrease in the solubility.

If the weak acid itself is poorly soluble in water or volatile, it precipitates or escapes as a gas if its concentration on increase of the acidity of the solution exceeds the saturation concentration. The acid is removed from the equilibrium in this way, causing a continuing protolysis of the anion of the salt and thereby a strong increase in the solubility of the salt. However, this requires that the solubility of the salt be so high that the saturation concentration of the acid can be exceeded.

A poorly soluble substance can also itself be a protolyte. If it is an acid the solubility is, of course, increased if the protons are taken care of by addition of a base. If it is a base the solubility is increased by addition of an acid. If the substance is an ampholyte, according to 12-4c [amf] = maximum at the isoelectric point of the substance. With an increase in the total concentration of the ampholyte, its solubility will therefore be exceeded most easily when the pH of the solution corresponds to this point. The solubility of an ampholyte thus has a minimum at its isoelectric point. With increase or decrease of pH, the solubility increases.

Other equilibria besides acid-base equilibria can also strongly affect the solubility. Often, what are usually more particularly called complex equilibria are encountered. For example, Ag^+ forms with NH_3 the complex ions $AgNH_3^+$ and $Ag(NH_3)_2^+$. If NH_3 is added to a solution of AgCl that is in equilibrium with solid AgCl, $[Ag^+]$ is decreased by complex formation. In this way, $[Cl^-]$ and thus the solubility if AgCl must be increased.

Of great significance are the complexes that a considerable number of heavy metal ions (for example, Fe^{3+} and Cu^{2+}) form with organic compounds containing OH groups. Addition of such a compound, for example tartaric acid or citric acid, can prevent the precipitation of such a metal from its solution or bring about the solution of poorly soluble compounds of the metal. If tartrate ion is added to a Cu^{2+} solution, copper(II) hydroxide does not precipitate on the addition of alkali (cf. 36-6 b).

Complexes may also be formed by a further addition of ions that are included in the poorly soluble substance. If cyanide ions CN^- are added to a solution of Ag^+, AgCN precipitates when its solubility product is exceeded. However, if CN^- is added in excess, complex ions are formed, at first $Ag(CN)_2^-$. Since the complex constant (10c) for this ion is very high, the addition of CN^- depresses $[Ag^+]$ to a large degree and thereby causes a significant increase of the solubility of AgCN. On addition of a large excess of Cl^- to a Ag^+ solution complexes are also formed, at first the ion $AgCl_2^-$. The complex constant for $AgCl_2^-$, however, is significantly less than for $Ag(CN)_2^-$, so that the effect of the excess on the solubility is much less. Nevertheless, the complex formation leads to the result that, after the solubility of AgCl is at first depressed in the usual way by a small excess of Cl^-, it increases again with larger excess. The solubility minimum occurs at approximately $[Cl^-] = 5 \times 10^{-3}$ mol l^{-1}.

f. Two liquid solutions. Extraction. Let us assume that the substance "i" is dissolved in two different liquid phases ' and ", that are in equilibrium with each other. One condition for this is clearly that the two solvents have limited mutual solubility. (13.1) is the condition of equilibrium, and if the solutions are ideal, therefore, we also have

$$x_i''/x_i' = K_x \quad \text{and} \quad c_i''/c_i' = K_c \qquad (13.13-14)$$

The constant K_x or K_c is called the *distribution coefficient (factor, ratio)* for "i" between the two solvents.

If K departs very much from 1, a strong enrichment of "i" is obtained in one of the phases, and this can be used for analysis and preparation (*extraction*). Iodine, for example, can be extracted from a water solution by shaking the solution with tetrachloromethane CCl_4 ("carbon tetrachloride") until equilibrium is reached. At 25°C, $[I_2(\text{in } CCl_4)]/[I_2(\text{in } H_2O)] = 85.5$. After the CCl_4 solution has been removed, the aqueous solution can be shaken with a new amount of pure CCl_4, then this new solution removed, and the procedure repeated until practically all of the iodine has been extracted. Silver can be extracted from molten lead with molten zinc (Parkes process). Zinc (m.p. 419.5°C) and lead (m.p. 327.4°C) are very slightly soluble in each other at a temperature just above the melting point of zinc, and for the silver solutions in these melts $[Ag(\text{in } Zn)]/[Ag(\text{in } Pb)] = 270$, where the concentrations are expressed in molalities.

The above reasoning presupposes that the dissolved substance has the same molecular size in both solutions. Even if this is not the case, the equilibrium equation in the forms (10.15) or (10.16) can always be applied; however, the ratio of the total concentrations of the substance in the two phases will not be constant.

If a substance distributed between water and another solvent is strongly enriched in the latter, it can easily be extracted from the water solution. If this substance can form a complex with a metal ion in the water solution, this metal complex can also be extracted in most cases. This method of metal extraction has become widely used. The complex-forming substance is most often organic (as a rule chelate-forming; see 5-6 c). Because of the general rules of solubility given in 13-1 c, such a substance usually is considerably more soluble in an organic solvent than in water and is therefore enriched in the former.

Different metal ions form complexes of very varying strengths (different complex constants) with a given complex-forming substance. The different metal ions are therefore extracted with varying facility, a property that can be used for separation. Also of importance is the fact that the complex-forming substance is very often a weak acid. The acidity of the water phase therefore affects both the solubility of the substance in it (cf. 13-1 e) and the complex equilibrium, and thereby can be of great importance for the distribution. By varying the acidity it is possible conveniently to arrange extraction conditions suitable for a certain kind of metal ion.

Complex-forming substances frequently used in metal-ion extraction are diphenylthiocarbazone ("dithizone"), 8-hydroxyquinoline ("oxine") and the ammonium salt of N-nitrosophenyl hydroxylamine ("cupferron"). Extraction for all three substances can be carried out with chloroform, for example. Dithizone (abbreviated "Dz") in acid form ("HDz") dissolved in chloroform, $CHCl_3$ takes up Zn^{2+} from a water solution according to

$$Zn^{2+}(\text{in } H_2O) + 2HDz(\text{in } CHCl_3) \rightleftharpoons 2H^+(\text{in } H_2O) + ZnDz_2(\text{in } CHCl_3)$$

The complex $ZnDz_2$ is not very strong and the equilibrium is therefore sufficiently displaced to the right only if the water phase is made weakly basic so that H^+ is used up. The complex $CuDz_2$ is stronger and here the extraction can be done from weakly acid solution.

g. Osmotic pressure. If a substance is dissolved in a solvent, the chemical potential of the latter drops below the value μ_1^0 that it has in the pure solvent. The chemical potential of the solvent in the solution is thus $\mu_1 < \mu_1^0$. This can be seen from the fact that the vapor pressure of the solvent is lowered. It is evident also from the fact that a water solution of a substance is not in equilibrium with ice at 0°C. Before the substance was dissolved in the water the chemical potentials of the water and the ice were equal at the equilibrium temperature 0°C, but now it has been lowered for the water in the solution so that the ice goes over to the liquid phase—the ice melts.

However, if the partial molar volume of the solvent (1-5e) is positive, which is usually the case, the chemical potential of the solvent in the solution rises if the pressure on it is increased. By raising the pressure μ_1 can then be raised so

Heterogeneous Equilibria

that it reaches the value μ_1^0, which corresponds to that of the pure solvent at unchanged pressure. When $\mu_1 = \mu_1^0$ equilibrium is attained between the solution at elevated pressure and the pure solvent. The increase in pressure required is called the *osmotic pressure* of the solution (Gk. ὠσμός, impulse, push).

If the solution and the solvent are separated by a wall that is permeable to the solvent but not to the dissolved substance (*semipermeable* membrane), the solvent diffuses into the solution. If the solution is in an enclosed volume, the pressure in this way is increased in it, and the diffusion continues until the rise in pressure is equal to the osmotic pressure, at which point equilibrium is reached. If the solution chamber is open, the level of the solution rises in it and the osmotic pressure can be observed directly from the increase in hydrostatic pressure resulting from the rise. If one raises the pressure on the solution still more so that $\mu_1 > \mu_1^0$, the direction of diffusion will be reversed. The solvent then goes from the solution to the pure solvent (*reverse osmosis*).

If a dilute solution with volume V contains n_2 mol dissolved substance, for the osmotic pressure Π we have approximately (van't Hoff, 1885)

$$\Pi V = n_2 RT \tag{13.15}$$

The expression (13.15) thus has the same form as the gas law (1.9). However, since (13.15) always has only approximate validity (even if the solution is ideal), the osmotic pressure, unlike the chemical potential for example, is not a proper basis for theoretical derivations.

In the same way that the gas pressure is determined by all molecules of all kinds in the gas, all molecules of the dissolved substances present in the solution contribute to the osmotic pressure of the solution. If the solution contains several dissolved molecular types 2, 3, 4, ..., then n_2 in (13.15) is replaced by $(n_2 + n_3 + n_4 + ...)$.

The relation (13.15) permits a measurement of osmotic pressure to be used for the determination of the order of magnitude of a molecular weight. This is because (13.15) gives the number of mols a weighed amount of a dissolved substance forms in the solution.

13-2. The phase rule and the Clapeyron equation

a. Introduction. In the treatment of heterogeneous equilibria the concepts of degree of freedom (1-5d), phase (1-1c) and component play a large role. The number of independent *components* of a system may be defined as the minimum number of pure substances that are required to express the composition of all the phases of the system. This number is usually equal to, but can be greater than, the minimum number of substances required to make up the system (compare 1-4b). Systems with the number of independent components 1, 2, 3, 4, etc., are called *one-component systems, two-component systems*, etc., or *unary, binary, ternary, quaternary*, etc., systems. The system is customarily indicated by its components.

The equilibrium $H_2O(s)-H_2O(l)-H_2O(g)$ belongs to the system H_2O and the equilibrium $NaCl(s)-$(saturated water solution of $NaCl)-H_2O(g)$ belongs to the system $NaCl-H_2O$.

In the following, for a system we will indicate:

The number of degrees of freedom: f
The number of phases: p
The number of independent components: k

Processes in heterogeneous systems often proceed sluggishly (chap. 14), which leads to slow attainment of equilibrium. This is especially true when the process involves material transport in a solid phase, which can only take place by quite slow diffusion. Equilibrium is reached more rapidly the higher the temperature is.

b. Gibbs' phase rule. In 1-5d the number of degrees of freedom of a system that consisted of only *one* phase was discussed. In a heterogeneous system, which thus contains several phases, the variables of composition in the different phases at equilibrium become dependent on each other, so that the number of degrees of freedom decreases. In connection with his extensive thermodynamic treatment of chemical-equilibrium problems, Gibbs in 1876 derived the general relation among the number of degrees of freedom, phases and components in a system at equilibrium. The relationship is called the *phase rule*. With the symbols introduced in 13-2a the phase rule reads

$$f + p = k + 2 \tag{13.16}$$

The phase rule can be proved in the following way. Let us assume that the system consists of the components $1, 2, ..., k$ found in the phases $', '', ..., ^p$. If then μ_k^p represents the chemical potential (2-3d) of the component k in the phase p, according to (2.13) at equilibrium it is necessary that

$$\left. \begin{array}{l} \mu_1' = \mu_1'' = \cdots = \mu_1^p \\ \mu_2' = \mu_2'' = \cdots = \mu_2^p \\ \cdots\cdots\cdots\cdots\cdots\cdots \\ \mu_k' = \mu_k'' = \cdots = \mu_k^p \end{array} \right\}$$

Since the chemical potential μ is a partial molar quantity, according to 1-5d it is determined in each phase by pressure, temperature and $(k-1)$ variables of composition. For each new phase $(k-1)$ new variables of composition are added. The number of variables of composition in the p phases is thus $p(k-1)$. In addition we have the variables of pressure and temperature common to all phases, which implies that the total number of variables is $2 + p(k-1)$.

The equilibrium requirement above, however, involves $k(p-1)$ independent equations among the μ values. Each line represents $(p-1)$ equations and for all k lines $k(p-1)$ equations are thus obtained. Through these $k(p-1)$ of the $2 + p(k-1)$ variables are determined. The remainder are independent and give the number of degrees of freedom. Thus we have $f = 2 + p(k-1) - k(p-1) = k + 2 - p$.

c. The Clapeyron equation.

Often a phase is transformed to another phase of the same composition. A solid phase may melt or be vaporized, or a liquid may be vaporized. A solid crystalline phase may be transformed to another with another crystal structure. As long as the two phases have the same composition they may be assumed to belong to a one-component system ($k=1$).

When the two phases are in equilibrium with each other, $p=2$ and thus according to the phase rule, $f=1$. If one variable of state is fixed, for example the temperature, the state is therefore defined and thus the pressure determined. The *equilibrium pressure* is thus a function of the temperature, $P=\mathrm{f}(T)$. Correspondingly, the *equilibrium temperature* is a function of the pressure. If this is constant, the equilibrium temperature is constant. Phase transformations at a given pressure are, in fact, used as temperature calibration points.

In 1834 Clapeyron derived an expression for the temperature derivative of the equilibrium pressure (*Clapeyron equation*) which with more recent symbolization reads

$$\frac{\mathrm{d}P}{\mathrm{d}T} = \frac{\Delta H}{T \Delta V} \tag{13.17}$$

ΔH is the amount of heat that is added to the system at constant pressure P to transform a certain amount, and ΔV is the concomitant increase in volume.

Important phase transitions with rise in temperature are melting (s→l), sublimation (s→g) and vaporization of a liquid (l→g). In all these transitions $\Delta H > 0$. For the two last-named we also always have $\Delta V > 0$, which makes $\mathrm{d}P/\mathrm{d}T > 0$. The equilibrium pressure (in this case the "vapor presure") thus increases with rise in temperature. The same is most often true for s→l but in a few cases (among others, melting of ice, bismuth and gallium; see 1-4h), $\Delta V < 0$ so that $\mathrm{d}P/\mathrm{d}T < 0$. For ice $\mathrm{d}P/\mathrm{d}T = -132.5$ bar deg^{-1} and $\mathrm{d}T/\mathrm{d}P = -0.00755$ deg bar^{-1}. The melting point of ice thus is depressed by $0.00755°C$ with a rise in pressure of 1 bar.

For transitions from a condensed phase to a gaseous phase (s→g and l→g), $\Delta H/\Delta V$ is much smaller than for transitions between condensed phases (s→l, s→s'). For the former, therefore, $\mathrm{d}P/\mathrm{d}T$ is much smaller (thus $\mathrm{d}T/\mathrm{d}P$ much larger) than for the latter.

13-3. Lack of equilibrium and order–disorder transitions

a. Metastable equilibrium.

This concept has already been discussed in 1-5c and 2-3b. Let us assume that a system under definite conditions (pressure, temperature, composition) can exist in a series of states 1, 2, 3, ..., s, in which its G values (2-3b) are $G_1 > G_2 > G_3 > ... > G_s$. The state s with the smallest G value is thus the stable state under the given conditions. If the system starts out from state 1, for example, it often happens, however, that it does not pass directly to the stable state, but goes to and perhaps may remain for a long time in an intermediate state such as 2 or 3. If the transition to s is slow, the intermediate states are called *metastable states* and the prevailing, apparent equilibria *metastable equilibria*. Metastable equilibria, which very slowly move toward the stable state, can often be treated as true equilibria. Apparently, the phase rule can then also be applied, which is not the case with more rapidly changing intermediate states.

When a solid phase is formed from a liquid (pure substance or solution), the latter has become unstable in relation to the state in which the solid phase exists. But the solid phase is not always the most stable of the phases of the substance but may be a metastable phase, which with appreciable or inappreciable rate transforms to a more stable phase. If mercury(II) iodide is precipitated at room temperature from a solution of a Hg(II) salt to which an iodide solution has been added, a yellow modification is formed at first, which very rapidly transforms to the red form stable at room temperature (if this is heated, it goes over at 127°C to the yellow form, which is stable above this temperature).

Many minerals are metastable under the conditions under which they now exist. They may have either been formed under other conditions in which they were stable (for example, at high pressure and temperature), or been unstable already when they were formed. In either case they have not later been able to transform to the stable state. Such minerals are diamond (C, for which graphite is the stable modification at atmospheric pressure and ordinary temperature), aragonite ($CaCO_3$, for which calcite is stable) and marcasite (FeS_2, for which pyrite is stable).

b. Pseudosystems. A phase of a pure substance is often composed of different kinds of building units. In such cases the phase rule applies if the different kinds of building units are in equilibrium with each other and if this equilibrium is attained in a time that is significantly shorter than it takes to determine the properties of the phase. Liquid water contains a multitude of different building units such as H_2O, H_3O^+, OH^- and larger aggregates of varying size. However, these are in equilibrium with each other and this equilibrium is reached so rapidly that the water phase after a change of state (for example, a rise in temperature) promptly takes on the properties that are characteristic of the new state.

A considerable number of pure substances, however, show properties that are variable and are dependent on the history of the substance (for example, vapor pressure, melting point, mechanical properties). After a change in temperature the properties may be dependent on whether the new temperature was reached from a lower or a higher temperature, and how long after the change observations are made. Often, final, constant properties are attained only after a very long time, which, however, may sometimes be shortened by the addition of a catalyst. The reason, as a rule, is that the substance can exist in different forms in a single phase, for example in different molecular forms or in varying degrees of polymerization (1-4a), and that equilibrium among these forms is reached slowly. A phase of the substance A would thus possibly contain the molecular species A_2 and A_4, between which there is the sluggishly reached equilibrium $A_4 \rightleftharpoons 2A_2$. If this equilibrium on heating is shifted to the right, the phase after heating to a certain temperature at first will have a higher proportion of A_4 than that corresponding to equilibrium at that temperature. If instead this temperature is reached from a higher temperature, the phase will have too high a proportion of A_2 at first.

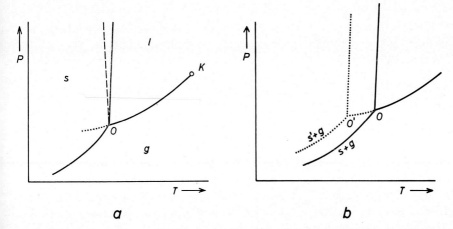

Fig. 13.2. (a) P,T diagram for a substance that occurs in a solid, a liquid and a gaseous phase. (b) The same as a with dotted curves added for metastable equilibria involving a metastable phase s'.

In such cases it is said that a *pseudosystem* prevails. For substances with widely varying degrees of polymerization the number of forms, of course, is practically unlimited. Pseudosystems are encountered, for example, in liquid sulfur (21-3b) and in phosphorus (22-3b).

c. Order–disorder transitions. Any normal phase transition is recognized by sudden changes in the functions of state of the system, for example volume, energy and enthalpy, at the transition point. In the order–disorder transitions, which were described in 6-3b, on the other hand, the functions of state often change continuously over the whole temperature interval within which the distribution order changes. In such cases from a thermodynamic standpoint the same phase prevails on either side of the temperature T_c. Aside from the varying degrees of order the structure is also the same on both sides of T_c. Order–disorder transitions do therefore not influence the application of the phase rule.

13-4. One-component systems

a. Phase diagrams. Heterogeneous equilibria are often best set forth by means of a *phase diagram* (equilibrium diagram, state diagram). In a phase diagram the whole of the contiguous region (volume, area, line or point) within which the same phase exists is called a *phase region*. For a one-component system the phase diagram is usually a P,T diagram, that is, is plotted on a coordinate system with P and T given along the two axes.

If a one-component system ($k=1$) consists of a single phase ($p=1$), according to the phase rule the number of degrees of freedom $f=2$. P and T can therefore vary independently of each other within a *one-phase region*, which occupies an area in the P,T diagram. As an example, we may take a system which besides the gas phase can exist in a solid and a liquid phase. (At moderate pressures and at temperatures above about $-100°C$, H_2O is such a system; however, cf. 21-5c.) Each of these phases occupies an area in the P,T diagram (fig. 13.2a).

For equilibrium between two phases in a one-component system the discussion in 13-2c is applicable. The number of degrees of freedom is thus $f=1$ and we may write, for example, $P = f(T)$. The two-phase equilibria $s+l$, $s+g$ and $l+g$ thus each prevails along its curve (*two-phase region*) in the P,T diagram. These curves must, of course, form the boundaries between the one-phase regions. The P,T curve for the equilibrium $s+l$ is called the *melting curve* for s, the curve for the equilibrium $s+g$ is called the *vapor-pressure curve* for s, and the curve for the equilibrium $l+g$ is called the *vapor-pressure curve* for l.

If three phases are in equilibrium with each other, the phase rule gives $f=0$. This equilibrium can thus occur only at a definite pressure and a definite temperature. In the example, the equilibrium $s+l+g$ occurs at a definite point on the P,T diagram (O in fig. 13.2a), which must be the intersection point of the three two-phase curves. This point, the *triple point*, thus constitutes the three-phase region. For water the triple point lies at $+0.01°C = 273.16°K$ and 6.105 mbar. Since the triple point is easier to determine than the melting point, it is used as a temperature calibration point in the International Practical Temperature Scale.

Even if the system should have more than three phases (often the substance may have several solid modifications), no more than three of these can occur together in equilibrium in a one-component system. However, the number of triple points then increases (see the P,T diagram for sulfur in fig. 21.2).

The slope dP/dT of a two-phase curve is determined by the Clapeyron equation (13.17) and as stated in 13-2c is much less for the vapor-pressure curves than for the melting curve. This is apparent also in fig. 13.2a. In this diagram two melting curves have been drawn in, one solid for the most common case $dP/dT > 0$ and one dashed for the less common case $dP/dT < 0$ (true for H_2O, for example).

The curve for the equilibrium $l+g$ ends at the *critical point* K (at the coordinates *critical pressure* and *critical temperature*) where the difference between liquid and gas ceases and the system becomes one-phase. For H_2O the critical point lies at $+374°C$, 221 bar.

A gas can be condensed to a liquid by compression or by cooling, or by both methods in combination. With compression the temperature must be below the critical temperature for a liquid to be formed. In the laboratory the cooling may be done by means of freezing mixtures, or by using the heat of vaporization on sublimation or vaporization of an already condensed phase (solid carbon dioxide, liquid air). On a larger scale, the gas is allowed to do work, whereby it is cooled. If it is allowed to increase its volume by ΔV at the constant pressure P, it does the work $P\Delta V$ (2-1c). However, liquid air is most often produced by another principle. Strongly compressed air is allowed to flow into a volume with significantly lower gas pressure. During such a "free expansion" an ideal gas would not perform any volume work. With any real gas, on the other hand, an exchange of work occurs because the ideal gas laws do not apply exactly, and this leads to a temperature change on free expansion (Joule-Thomson effect).

The phenomenon is complex but it can be seen that the work that the gas performs when its molecules are moved against the van der Waals forces must cause cooling. All gases except the two lightest, hydrogen and helium, in which the van der Waals forces are especially small, are then cooled on free expansion at ordinary temperatures or below. Hydrogen and helium must first be cooled in another way (hydrogen with liquid air, helium with liquid hydrogen) in order to reach a temperature region where free expansion is accompanied by cooling.

Two phases can often exist together within a P,T region in which they are not in equilibrium with each other. The system is then unstable. $H_2O(l)$ can easily be cooled to temperatures below the triple point (*supercooling*), and if the vapor pressure of the supercooled water is measured, it is found that it corresponds to the dotted extension of the vapor-pressure curve $l+g$ in fig. 13.2a. It is more difficult to realize the extension of the two other curves past the triple point, for example to superheat a solid phase. The extension of any of the equilibrium curves, however, must fall between the two other curves.

If a solid substance at higher pressure than the triple-point pressure is heated, first *melting* occurs, and then *boiling*. The *melting point* and the *boiling point* of a substance is usually given for $P=1$ atm ($=1.01325$ bar). They thus correspond to the T values on the curves $s+l$ and $l+g$ at this pressure. The much smaller slope of the latter curve means that the pressure must be defined much more precisely in giving a boiling point than a melting point.

If a solid substance is heated at a lower pressure than the triple-point pressure, it goes directly over to a gas phase—it *sublimes*. Sometimes a *sublimation point* at $P=1$ atm is given, which thus corresponds to the T value for this pressure on the curve $s+g$.

It is well known that on the addition of heat in free air at the air pressure 1 atm, a liquid can vaporize at temperatures below the boiling point, and in the same way a solid substance can sublime at temperatures below the sublimation point. In such cases the gas formed is so rapidly dissipated by convection and diffusion that its partial pressure in the boundary layer immediately adjacent to the condensed phase is less than 1 atm. If, for example, the atmosphere over water consisted only of water vapor or if the atmospheric pressure were transmitted to the water by a movable piston resting on the water surface, no gas formation would occur below the boiling point. However, on heating in free air the rate of gas formation is increased strongly at the boiling and sublimation points.

If heat is conducted to a solid phase, it can clearly sublime as long as the partial pressure of the gas formed in the boundary layer remains below the triple-point pressure. In order for sublimation to attain an appreciable rate, however, the gas pressure must be quite high, which means that the triple point pressure should not be too low. The latter can nevertheless be considerably less than normal air pressure. Iodine (triple point $+114°C$, 120 mbar) and camphor (triple point $+175°C$, 472 mbar) sublime noticeably in free air even at room temperature, and on heating the sublimation proceeds so rapidly that the solid phase often disappears before the triple point is reached and melting can occur. Since the triple-point pressure in both cases is less than ordinary air pressure, however, melting can take place in free air if the addition of heat is done rapidly, especially if at the same time the gas formed is not carried away too effectively by strong convection. A third example is ice, which, however, because of its rather low triple-point pressure (6.1 mbar) sublimes slowly and therefore can easily be melted. For a very large number of substances the triple-point pressure is so low that sublimation is not noticeable.

If the triple-point pressure is greater than the air pressure, on the other hand, only sublimation can occur on heating in free air. Examples of this are CO_2 (triple-point pressure 5.2 bar; it is well known that solid CO_2, "Dry Ice", does not melt but sublimes in free air), SO_3 (2.3 bar for the stable, α modification), As (36 bar), NH_4Cl, and many organic substances. To melt these the heating must take place in a closed vessel whose volume is so small in relation to the amount of the substance that the triple-point pressure is reached before vaporization is complete.

The triple-point pressure determines whether a substance that is kept under high pressure as a liquid can be drawn off as such in the free air. In a sufficiently full CO_2 cylinder there is liquid CO_2 (crit. temp. $+31°C$). The triple-point pressure is as previously stated 5.2 bar, and liquid CO_2 therefore cannot exist at 1 bar. If an inverted cylinder is opened, the cooling on expansion therefore leads to the formation of solid CO_2 ("carbon-dioxide snow"). NH_3 (crit. temp. $+133°C$), on the other hand, can be drawn off as a liquid, because its triple-point pressure is 0.06 bar. The same is true, for example, of Cl_2 and SO_2.

The dotted curve for the vapor pressure of supercooled water in fig. 13.2a shows that it is higher than the vapor pressure of the stable phase ice at the same temperature. It is generally true that *a metastable phase has a higher vapor pressure than the stable phase of the same substance at the same temperature.* In fig. 13.2b, in addition to the equilibrium curves for stable phases like those in fig. 13.2a, a dotted curve for the equilibrium $s'+g$ has also been drawn, along which a metastable phase s' is in equilibrium with the gaseous phase. This curve wherever s' is metastable must lie above the curve $s+g$. It intersects the curve for supercooled l in O', which thus becomes the triple point for the metastable equilibrium $s'+l+g$. O' lies at lower temperature than O and it is obvious that s' also has a lower melting point than s. It is generally true that *a metastable solid phase of a substance has a lower melting point than the stable solid phase of the same substance.* For example, there is a stable modification of iodine monochloride ICl with the melting point $27.2°C$, and a metastable modification with the melting point $13.9°C$. *Metastable solid phases also have greater solubility than the corresponding stable solid phases.*

b. Influence of bonding and structure on melting and vaporization. When a crystal *melts* there occurs a breaking up of certain bulding units (atoms or atomic groups),

Heterogeneous Equilibria

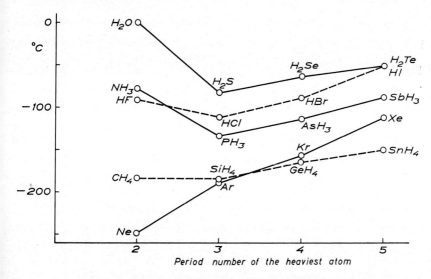

Fig. 13.3. Melting points of noble gases and hydrides.

which in the melt become much more irregularly distributed than in the crystal. This breaking up implies that the heat motion weakens or completely breaks down one or several of the bond types important in holding the crystal together, so that increased disorder becomes possible.

Any property that is related to a breaking-up process, that is, a process that acts against bonding forces in a substance (in addition to melting, for example, cleavage, vaporization, solution), is generally very dependent on the weakest bond type in the substance. A chain is as strong as its weakest link. However, the breaking up or division may act differently on different kinds of bonds, so that it is not only the strength of the bond that determines how easily it is affected. Bonds that are not covalent are inherently independent of, or only slightly dependent on direction (that such bonds nevertheless lead to ordering below the melting point depends only on the fact that the order gives the system the lowest energy). Therefore, these bonds persist, though weakened, even after melting. On the other hand, if the definite distribution of directions between atoms that were covalently bonded to each other is eliminated, this means that the bond is almost completely ruptured. In this a larger addition of energy per bond is often required than in the first case. Rupture of typical covalent bonds often signifies a decomposition of the substance, leading to wholly new substances.

The noncovalent bonds for the most part are less affected by heat motion, the stronger they are. As a rough guide, therefore, the following series in order of decreasing disintegration on melting may be used: van der Waals bond–hydrogen bond–ionic bond–covalent bond. The metallic bond is difficult to place in the series. It can be considered to be a kind of covalent bond, but is little dependent on direction because of strong orbital delocalization (6-4b), and it varies very much in strength (cf. the isomorphous metals cesium and tungsten with the melting points 28.7 and 3380°C, respectively).

A noble-gas crystal is held together only by van der Waals forces and should therefore have a low melting point. Since the van der Waals forces increase with increased molecular weight (5-4a), the melting point rises as one goes toward heavier noble-gas atoms (fig. 13.3). Similar conditions prevail in other crystals in which the building units are held together mainly by van der Waals forces. In such substances with mutually similar structures the melting point also rises with the molecular weight (the series CH_4–SnH_4 and th three heavier members of the other series in fig. 13.3). However, if other bond forces enter in, departures from this rule occur. In HF, H_2O and NH_3 hydrogen bonds are present between the molecules (5-5) and here the melting points are unexpectedly high (fig. 13.3).

Among the fluorides of the elements in period 3 SiF_4, PF_5 and SF_6 form individual molecules, with compositions like their formulas. Mainly van der Waals forces prevail between the molecules. The melting points are therefore low as shown in the table below:

	NaF	MgF_2	AlF_3	SiF_4	PF_5	SF_6
Melting point, °C: (For AlF_3, sublimation point)	980	1400	1260	-77	-83	-55

Al tries to bond 6 F octahedrally, but with the composition AlF_3 it can only do this if the AlF_6 octahedra are joined together with every F corner common to two octahedra. In this way a framework molecule arises held together by Al-F-Al bonds (of character between ionic and covalent). AlF_3 therefore remains solid all the way up to 1260°C, where its vapor pressure reaches the value of 1 atm. Also in MgF_2 and NaF (NaCl structure) the weakest bonds are M–F bonds so that the melting points are high. The large discontinuity between SiF_4 and AlF_3 thus does not signify any large discontinuity in bond character. This shifts only very slowly from predominantly covalent bonding to predominantly ionic bonding as one goes from right to left in the table. The reason for the discontinuity instead is that the weakest bond that must be ruptured to break down the crystal has changed from the van der Waals bond between the SiF_4 molecules to the significantly stronger bond between Al and F.

An extreme case, in which the entire crystal is held together by very strong, covalent bonds of a single kind, is diamond (fig. 5.16). The melting point here is very high (about 4000°C at a pressure of about 120 kbar).

The above observations, of course, are equivalent to what was said in 6-2d, that substances with high (low) lattice energies have high (low) melting points.

In *vaporization* of a liquid the same sequence of effect of the different types of bonds for the most part applies as in melting. In fig. 13.4 it can be seen how the boiling points of liquids consisting of individual molecules with van der Waals attraction behave in a rather similar manner to the melting points of the cor-

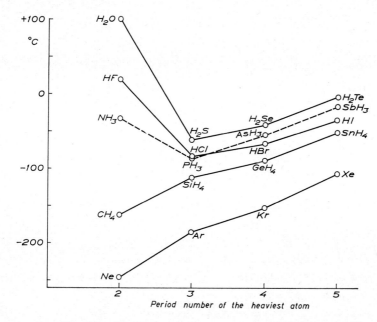

Fig. 13.4. Boiling points of noble gases and hydrides.

responding solid substances. Also here the boiling points of substances with analogous structures rise with the molecular weight. Similarly strong deviations are found for HF, H_2O and NH_3, which shows that a large proportion of the hydrogen bonds between these molecules remains in the liquids.

Liquid molecules with permanent electric dipole moments are bonded to each other with dipole-dipole bonds (5-2h). The consequent increased cohesion within the liquid causes a higher boiling point than would be expected on the basis of van der Waals forces alone. CH_4 and CCl_4, which have no permanent dipole moments, boil at -164 and $+76°C$. If the boiling point were to rise linearly with the proportion of chlorine, the intermediate chloromethane derivatives, CH_3Cl, CH_2Cl_2 and $CHCl_3$, would have the boiling points -102, -43 and $+17°C$, respectively. The dipole moments in these unsymmetrical molecules, however, account for the significantly higher values -24, $+40$, $+61°C$, respectively. The boiling points in fig. 13.4, except those for the noble gases and the hydrides XH_4, are, of course, also affected by dipole-dipole bonding.

13-5. Binary systems

a. Introduction. Binary systems occur in such different types that not even a general treatment can here be at all complete. Therefore, only some especially important cases will be covered.

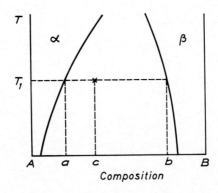

Fig. 13.5. Two-phase region in a binary system.

The greatest number of degrees of freedom in a binary system is 3 (for $p=1$). Usually, P, T and a variable of composition, for example mole fraction x or a percentage of proportion, are chosen. A complete phase diagram therefore becomes three-dimensional (for example, P,T,x diagram), but two-dimensional sections are often given, usually with $T=$const. (P,x diagram) or $P=$const. (T,x diagram). The latter is especially common for systems in which only condensed phases play any important role. The equilibria in that case are so little dependent on pressure (10d) that such a section is representative of a large pressure range. Most often the section is taken at $P=1$ atm.

For $p=1$, we have $f=3$. The system can thus exist as one phase while P, T and x are all varied (within certain limits). A one-phase region is therefore in general a space region in the P,T,x diagram. However, if the phase for some reason does not have a variable composition in the given system (for example, in the system $NaCl$–H_2O $NaCl(s)$ cannot dissolve H_2O and therefore has constant composition), only P and T can vary. A special case then arises in which the one-phase region degenerates from a P,T,x space to a P,T surface. In a T,x section the general T,x area degenerates to a line with $x=$const.

A one-phase region may be made up of a pure component or originate from one (solution in the pure component). It may also lie inside the P,T,x space, completely separated from the component phases. One speaks then of an *intermediate phase*.

The one-phase regions should always be considered as the basic regions of a phase diagram. In fact, the polyphase regions are only gaps between one-phase regions. Fig. 13.5 will show this for a general case of a two-phase region. The two components A and B are partially soluble in each other. These solutions are indicated by α and β and occur in three-dimensional, one-phase regions, which in the section of the figure become areas. Between these the two-phase region forms a gap (*miscibility gap*). The compositions of the phases that are in equilibrium with each other are determined by the boundaries of the one-phase regions. At the temperature T_1 the α phase of composition a is in equilibrium with the β phase of composition b. It can be shown that if the total composition of the system

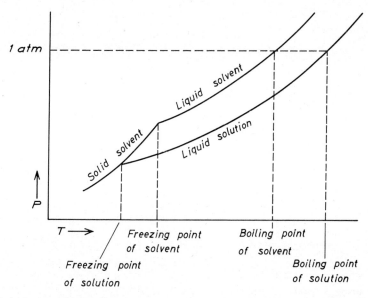

Fig. 13.6. Vapor-pressure lowering as the cause of freezing-point depression and boiling-point elevation.

is c the ratio of the quantity of α phase to the quantity of β phase (expressed in moles if the composition is expressed in mole percent and as weight amounts if the composition is expressed in weight percent) is inversely proportional to the lengths of the segments ca and cb ("*the lever rule*"). In the example, α phase thus prevails. If the total composition is changed within the two-phase region at constant temperature, only the relative amounts of the phases are altered, but not their compositions.

b. Dilute liquid solution in equilibrium with pure solvent. Often it happens that a liquid solution is in equilibrium with *pure* solvent. If a sufficiently dilute water solution is cooled, ice crystallizes out, which in general is pure. If the dissolved substance is not volatile, on boiling a water solution the gas phase consists only of water vapor. In such cases the solution has a lower vapor pressure than the pure solvent at the same temperature. This vapor-pressure lowering, which for an ideal solution is determined by Raoult's law (13.4), affects the freezing and boiling points of the solution.

Fig. 13.6 is a P,T diagram with vapor-pressure curves for pure solid and pure liquid solvent, analogous to those in fig. 13.2. The freezing point of the solvent (properly, the triple point) is the intersection of the vapor-pressure curves of the solvent. The boiling point of the solvent is the temperature at which its vapor pressure is 1 atm. For a given solution, at any temperature a lowering of the vapor pressure occurs. The vapor-pressure curve of the solution thus lies lower than that of the solvent. The freezing point of the solution is the intersection point between

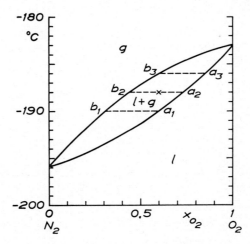

Fig. 13.7. Boiling-point diagram for liquid solutions of the components nitrogen and oxygen.

this curve and the vapor-pressure curve of the pure, solid solvent, and its boiling point is the temperature at which the vapor pressure of the solution is 1 atm. Thus the lowering of the vapor pressure causes a *freezing-point depression* and a *boiling-point elevation*.

If the solution is so dilute that it can be considered to be ideal, the following simple laws apply to the freezing-point depression and the boiling-point elevation.

The molality of the dissolved substance in the solution is assumed to be m, that is m moles of the substance are dissolved in 1 kg of solvent. If the heat of fusion of this amount of solvent (1 kg) is q_f, the freezing-point depression is

$$\Delta T = \frac{RT^2}{q_f} \, m = K_f m \qquad (13.18)$$

K_f is the *molal freezing-point depression* for the solvent in question. For water, $K_f = 1.867°C$ per mole of dissolved substance in 1 kg of solvent.

For the boiling-point elevation an analogous expression is obtained:

$$\Delta T = \frac{RT^2}{q_e} \, m = K_e m \qquad (13.19)$$

Here q_e is the heat of vaporization of 1 kg of solvent. K_e is the *molal boiling-point elevation* of the solvent. For water, $K_e = 0.513°C$ per mole of dissolved substance in 1 kg of solvent.

Most solvents have higher K_f and K_e values than water.

All molecules of the dissolved substance present in the solution contribute to the lowering of the vapor pressure and hence to the freezing-point depression and the boiling-point elevation. If the solution contains several kinds of dissolved molecules 2, 3, 4, ..., then m in (13.18) and (13.19) is replaced by $(m_2 + m_3 + m_4 + ...)$.

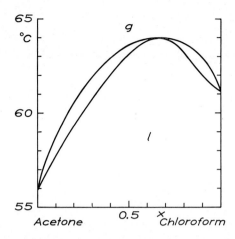

Fig. 13.8. Boiling-point diagram with a maximum.

Measurements of freezing-point depression and boiling-point elevation can be used to determine molecular weight. If the corresponding molal value for the solvent used is known, from such measurements it is possible to calculate with (13.18) and (13.19) how many moles a weighed quantity of dissolved substance constitutes in a solution that can be assumed to be ideal. Starting with a given molecular size for the dissolved substance, it is also possible then to determine how these molecules are associated or dissociated in the solution. Such calculations are not possible for electrolyte solutions where the electrostatic forces complicate the situation.

The formulas (13.18) and (13.19) were derived in 1885 by van't Hoff, but Raoult in 1882 had already found that the freezing-point depression for dilute solutions was proportional to the number of dissolved moles and that the constant of proportionality was characteristic of the solvent.

c. The two components are completely soluble in each other in the liquid phase. The equilibria between this liquid phase and the gas phase will be discussed here. Both components may be volatile and so appear in the gas phase.

A temperature-composition diagram for this case and $P=1$ atm is called a *boiling-point diagram*. Fig. 13.7 shows an ordinary type of such a diagram, applying to the system N_2–O_2. The diagram contains two one-phase regions indicated by l and g, separated by a two-phase region indicated by l+g. The boundary between the gas phase and the two-phase region is called the *gas curve* and the boundary of the liquid phase is called the *liquid curve*. Every point of the latter gives the boiling point of a solution of the corresponding composition, and the curve is therefore also called the *boiling-point curve*. Since the solution here is not in equilibrium with a pure gaseous component, the boiling point is not necessarily raised on addition of the other component. In the example the boiling point of N_2 is raised and the boiling point of O_2 is lowered.

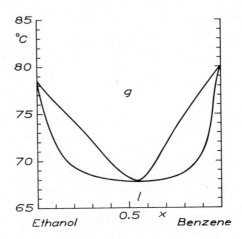

Fig. 13.9. Boiling-point diagram with a minimum.

For example, if a liquid solution with $x_{O_2}=0.6$ is warmed, it begins to boil at $-190°C$ (point a_1). It is then in equilibrium with, that is it gives, a gas mixture that contains more of the more volatile component N_2 than the liquid (point b_1 with $x_{O_2}=0.3$). When the vapor escapes, the liquid is enriched in the less volatile component O_2 and the boiling point rises. The boiling point thus moves up the liquid curve. If all of the liquid were vaporized, the whole, final quantity of gas would have the same composition as the starting liquid and thus correspond to point b_3. The last drop of liquid must be in equilibrium with this gas and thus have the composition a_3. To obtain a separation of the two components the distillation must be stopped when the liquid composition lies somewhere between a_1 and a_3, say, at a_2 (*fractional distillation*). The whole gas volume then has the composition b_2. The sample thus has been divided into an oxygen-rich liquid (a_2) and a nitrogen-rich gas (b_2), which is to be condensed. The relative amounts of these two fractions can be determined by the lever rule since the total composition is equal to the composition of the starting solution (the point x with the same composition as a_1). By repeating the procedure on the two fractions a further separation of the components is obtained, and by continued repetition, finally pure nitrogen and pure oxygen can be obtained.

Nowadays fractional distillation is nearly always carried out as a continuous process in a *column*. Every vaporization and condensation stage takes place on a so-called *plate* in the column. The plates are coupled over each other according to the cascade principle (9-3b) and the material transport takes place through the movement of vapor upward and the flow of liquid downward in the column. In this way a very effective separation can be achieved.

It is not always true that the gas phase contains more of the most volatile component than the liquid phase over the entire composition range. Figs. 13.8 and 13.9 show boiling-point diagrams with a maximum and a minimum, respectively. The curves are continuous and at the extremes the liquid and gas curves have a common tangent. The fact that an extreme occurs shows that the solutions in the vicinity of this point strongly deviate from the ideal state. The extreme does not give any indication, on the other hand, that the components form any compound at the composition in question (besides, this composition is generally dependent on pressure). At an extreme, liquid and gas of the same composition are in equilibrium with each other. At this point the liquid can thus be distilled unchanged and is said to be *azeotropic* (Gk. α not, ζέω I boil, τροπή change).

In an azeotropic system the two components cannot be separated from each other by fractional distillation. Depending on the starting composition, only one component can be separated from a solution with the composition of the extreme.

Boiling-point diagrams with a minimum are more common than those with a maximum. In addition to the examples in figs. 13.8 and 13.9 we may mention the system ethanol–water (at a pressure of 1 atm, a boiling-point minimum of 78.13°C at 95.6 wt % ethanol) and the system HCl–water (at a pressure of 1 atm, a boiling-point maximum of 108.58°C at 20.2 wt % HCl). Thus, a higher proportion of ethanol in the evolved gas phase than 95.6 wt % cannot be obtained by distillation. If hydrochloric acid is boiled at 1 atm either water or HCl goes off so that the remaining liquid finally contains 20.2 wt % HCl.

d. Liquids with limited solubility. In 13-5c only the case was considered where the two components were soluble in each other in all proportions so that only one liquid phase was present. Liquids may also be practically insoluble in each other, for example, water and mercury. In between there are many cases of limited solubility. In the liquid state there may then appear a two-phase region (miscibility gap) more or less of the type shown in fig. 13.5. Most often the mutual solubility increases with increase in temperature (as in fig. 13.5) and the miscibility gap may therefore disappear before the liquid curve is reached, so that the two one-phase regions coalesce. One such case would be shown by the diagram of fig. 13.11b if the liquid phase l were replaced by a gas phase g and the solid phases by liquid phases. If the miscibility gap persists, it is often joined to the liquid and gas curves in a way that would be shown by fig. 13.11c and d if also in these l were replaced by g and the solid phases by liquid phases.

e. Equilibria between only solid phases and a gas phase. Only certain important types can be covered here. One component is assumed to be a solid body and the other a gas. Only the latter is assumed to appear in the gas phase. Two cases can then be distinguished.

1. *The gas component forms one or more solid phases of constant composition with the solid component.* As the first example, let us take the dissociation of $CaCO_3$ according to the reaction $CaCO_3(s) \rightleftharpoons CaO(s) + CO_2(g)$. Here $p=3$ and $k=2$ (the simplest choice of components is CaO and CO_2). Thus, $f=1$, which implies that a

definite temperature corresponds to a definite carbon dioxide pressure (equilibrium pressure, dissociation pressure). That the carbon dioxide pressure is determined by the temperature is also apparent from the equilibrium equation (see 10b), which can be written $p_{CO_2} = K_p$, since CaO and $CaCO_3$ are pure, solid substances and thus have activities $=1$ (10a). The equilibrium pressure increases with rise in temperature, and reaches 1 atm at 898°C. At this temperature the dissociation thus commences on heating in an open vessel. If CO_2 is effectively removed, for example by a strong stream of air, the partial pressure of CO_2 over $CaCO_3$ becomes less than 1 atm and the dissociation can begin sooner.

The same case in principle can arise as one stage in a series of equilibria, for example in the decomposition or formation of a series of salt hydrates. As an example let us take the system $CuSO_4$–H_2O in which occur as solid phases the pentahydrate $CuSO_4 \cdot 5H_2O$, the trihydrate $CuSO_4 \cdot 3H_2O$, the monohydrate $CuSO_4 \cdot H_2O$ and the anhydrous $CuSO_4$ ("anhydride"). On dissociation of a hydrate, H_2O and a lower hydrate or anhydride are formed. For equilibrium between two such phases and water vapor, $p=3$ and thus $f=1$. For a certain pair of solid phases a definite water-vapor pressure is thus obtained at a definite temperature.

Fig. 13.10 shows a P,x diagram for the system $CuSO_4$–H_2O at 25°C. When $CuSO_4$ is dissolved in H_2O, the water-vapor pressure drops from 31.7 mbar along the short curve "1" and attains the value 30.8 mbar in the saturated solution ($x_{H_2O} = 0.976$). If water is removed from this solution the pentahydrate precipitates and the water-vapor pressure remains constant at 30.8 mbar until only pentahydrate is present. The pressure 30.8 mbar is thus the equilibrium pressure for the equilibrium (saturated solution) $\rightleftharpoons$ (pentahydrate + water vapor) at 25°C, and this equilibrium is maintained along the horizontal line "1+5." Over the pentahydrate alone the water-vapor pressure can drop to 10.5 mbar before anything further happens. However, if water is removed at this pressure the trihydrate is formed. The equilibrium (pentahydrate) $\rightleftharpoons$ (trihydrate + water vapor) with the equilibrium pressure 10.5 mbar (dissociation pressure of the pentahydrate) is thus maintained along the horizontal line "5+3." Further analogous stepwise stages occur along the lines "3+1" at 7.0 mbar and "1+0" at 0.03 mbar. Only when the system contains the anhydride as the only solid phase can the water-vapor pressure fall below 0.03 mbar.

From the figure one can decide whether a given hydrate is stable in air at a given water-vapor pressure. Thus, the pentahydrate is stable at 25°C when the partial pressure of water vapor is 10.5–30.8 mbar. If the partial water-vapor pressure is less than 10.5 mbar (corresponding to a relative humidity at 25°C of 32.8%) the pentahydrate gives off water, that is, it *effloresces*. If the water-vapor pressure is greater than 30.8 mbar (relative humidity 97.2%) the pentahydrate takes up water with the formation of saturated solution—it liquefies (*deliquesces*).

It should not be forgotten that the water-vapor pressure at a certain temperature has a definite value only if the vapor is in equilibrium with *two* condensed phases. In order that a given temperature correspond to a definite vapor pressure

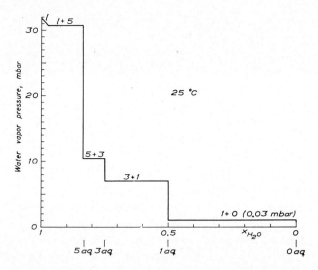

Fig. 13.10. Water-vapor pressure of the system $CuSO_4$–H_2O at 25°C. "l" signifies liquid solution, 5, 3, 1, and 0 signify the number of H_2O per $CuSO_4$ in the solid phases. The line "1 + 0" has been moved up to separate it clearly from the x axis.

f must $=1$, which requires that $p=3$, that is, a gas phase and two other phases. Mixtures of two salt hydrates or one hydrate and anhydride can be used to maintain a definite vapor pressure. For example, a mixture of monohydrate and anhydride of $CuSO_4$ will maintain a vapor pressure of 0.03 mbar at 25°C as long as enough of the two phases is present so that incidental variations in the proportion of water in the gas phase can be compensated.

Reactions in the solid phase are often strongly inhibited (chap. 14) and this is also often the case in a transition from one hydrate to another. Undamaged crystals of $Na_2SO_4 \cdot 10H_2O$ (Glauber's salt) thus persist for a long time in air with lower water-vapor pressure than the equilibrium pressure for the equilibrium $Na_2SO_4 \cdot 10H_2O \rightleftharpoons Na_2SO_4 + 10H_2O$. However, if a trace of Na_2SO_4 is added which can serve as "seed" for continued anhydride formation, efflorescence begins. The possibility of arrested reactions as well as the generally low rate of ractions in solid phases must be kept in mind when it is desired to regulate the water-vapor pressure with solid hydrates.

2. *The gas component forms a solid solution with the solid component.* Here no new phase is formed when gas is evolved or absorbed. The system therefore contains besides the gas phase only *one* solid phase ($k=2$, $p=2$, $f=2$). In order to establish the state at a definite temperature, one additional degree of freedom must thus be defined, either the pressure or the composition of the solid phase. There is thus no definite dissociation pressure, but rather the gas pressure is dependent on the proportion of the gaseous component in the solid phase and rises continuously with increase in this proportion. Instead of a steplike curve as

in fig. 13.10, a continuous curve is thus obtained within the composition range over which the solid solution can exist.

The classical example of this case is that of the zeolites (aluminosilicates of alkali and alkaline-earth metals; see 15-3 and 23-3h5), which contain water in solid solution. The H_2O molecules are here bound in such a way that a change in the proportion of water within certain limits does not destroy the crystal structure. They often lie between the layers of a layer structure or in open channels in the structure, and a change in their number at most leads to a change in the lattice dimensions. The zeolites often can also dissolve other molecules than H_2O (for example, NH_3) in a similar way. *Zeolitic water* is also encountered in hydroxides and hydroxide salts of divalent metals, which often have layer structures (6-2c), and in the clay minerals (23-3h4), which also crystallize in a similar way.

f. Equilibria between condensed phases only. Often the components and the phases formed from them are so nonvolatile that the gas phase is of little interest. Such is the case with most alloy systems and also in many other important systems as long as one does not go to too high a temperature. The same is true also of systems with water at atmospheric pressure and low temperature. As soon as a gas phase is present in the system (although this phase often for the most part consists of air), however, the gas phase must be counted in the number of phases in the application of the phase rule in its ordinary formulation (13.16).

In the most important systems in practice the components are completely soluble in each other in the liquid state, and this has been assumed to be true in all the following examples. At sufficiently high temperatures the liquid phase thus extends over the whole composition range. The phase diagrams thereby often have a formal similarity to the diagrams for equilibria between liquids and a gas phase, in which at sufficiently high temperatures the gas phase extends over the whole composition range.

In all diagrams for equilibria between liquid and solid phases the boundary surfaces of the homogeneity space region of a liquid phase, over which this is in equilibrium with solid phase, are called *liquidus surfaces*. Sections through these give *liquidus curves*. All the uppermost curves in figs. 13.11 and 13.13 are liquidus curves.

1. *Complete solubility also in the solid phase.* Here the building units by which the two components are distinguished must be so similar that they can replace each other completely in the crystals (6-3a). The phase diagrams become in principle the same as the diagrams in figs. 13.7–9. Diagrams are known without a maximum or minimum (the melting point of one component is lowered, that of the other raised) or with a minimum (the melting points of both components are lowered), while diagrams of this type with a maximum have not been established with certainty. Fig. 13.11a shows the diagram for the system Bi–Sb, which is of the first kind. The upper curve in the diagram, above which only liquid phase 1 exists, is a *liquidus curve* (see above); the lower curve, below which only solid

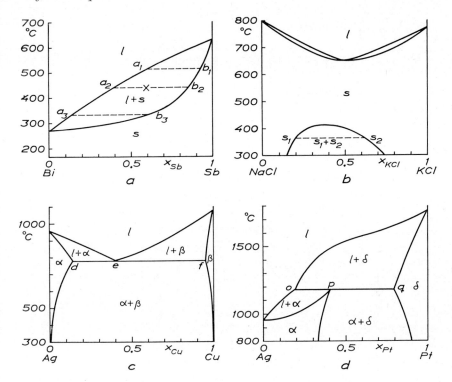

Fig. 13.11. Various types of T,x diagrams for binary systems.

phase s exists, is called a *solidus curve*. Between the two curves lies a two-phase region $l+s$.

If an alloy with $x_{Sb}=0.6$, for example, is melted and the temperature then lowered, crystals begin to form at 515°C. The melt here has the composition a_1, and the first crystals the composition b_1. Since the crystals are richer in Sb than the melt, its composition on further crystallization is moved downward along the liquidus curve. The composition of the crystals that form from the melt is shifted simultaneously downward along the solidus curve. For example, at 440°C the melt a_2 is in equilibrium with the crystals b_2. If the whole system is to consist of melt a_2 in equilibrium with crystals b_2, however, the composition of the crystals formed earlier, which are richer in Sb than b_2, must be altered in the direction of b_2. If the system is held at 440°C, such a change does occur by reaction with the melt. Since this involves diffusion in the solid phase, however, a long time may be required before equilibrium is reached. With too rapid cooling equilibrium cannot be established and on a previously formed crystal new, more and more Bi-rich, crystal layers are deposited. The crystal grains thus have variable composition (*coring, zoning*).

2. *Limited or no solubility in the solid phases.* This occurs if the building units of the two components are so different that they can replace each other in the

crystals only partly or not at all. In interstitial solutions (6-3a) the solubility, as a rule, is limited.

In the solid state there then arises a *miscibility gap* of the same type as in fig. 13.5. This may, as in fig. 13.11*b*, lie entirely separated from the solidus curve by a region of complete, mutual solubility. Within the miscibility gap at each temperature two solid phases with compositions corresponding to the limits of solubility at that temperature (for example s_1 and s_2) are in equilibrium with each other. Fig. 13.11*b* illustrates the description in 6-3a concerning the mutual solubility of NaCl and KCl.

More often the miscibility gap extends all the way up to the two-phase regions with a liquid phase. In the case where the melting points of both components are lowered a diagram such as in fig. 13.11*c* is obtained. In the example the miscibility gap is wide, that is, the mutual solubility in the solid phases is quite small. In the Ag structure, Cu atoms can replace Ag atoms up to a certain limit and these substitution solutions are designated as α phase. In an analogous way the β phase arises by replacement of Cu by Ag in the Cu structure. The two liquidus lines intersect each other in the *eutectic point e* with the coordinates *eutectic temperature* and *eutectic composition*. The eutectic temperature is the lowest temperature at which the liquid phase can exist under the pressure to which the T,x section corresponds (Gk. εὖ easy, well; τηκτός fusible). The phase then has the eutectic composition. The horizontal line *def* is called the *eutectic line* and is a *three-phase line* along which the liquid l with the composition e and the two solid phases α with the composition d and β with the composition f are in equilibrium with each other. Here, $p=3$ and thus $f=1$. If P is defined, T as well as the compositions of the three phases are therefore fixed.

If a melt with a Cu content less than e is cooled, α phase crystallizes as soon as the liquidus line is crossed, and thereafter on passage through the region $l+\alpha$ a course of crystallization is followed that is analogous to that described for fig. 13.11*a*. If the Cu content is less than d and the crystallization goes so slowly that equilibrium can be maintained, α phase is finally obtained with the original proportion of Cu. Then, when on further lowering of temperature the limit of solubility of the α phase is passed, β phase begins to separate out. If the cooling continues sufficiently slowly, more and more β phase separates. At 300°C the α phase cannot hold more Cu in stable solution than that corresponding to $x_{Cu}=0.01$.

If the Cu content of the melt lies between d and e the eutectic line is reached on lowering the temperature. Immediately before this point the α phase with practically the composition d is in equilibrium with the melt with practically the eutectic composition e. When the eutectic line is passed, the eutectic melt crystallizes all at once, whereupon most often a very fine-grained mixture (*eutectic*) of the two solid phases α (composition d) and β (composition f) is obtained. The behavior of melts with Cu contents between e and f is analogous, although the first crystals to separate then consist of β phase. When the temperature of any system with a

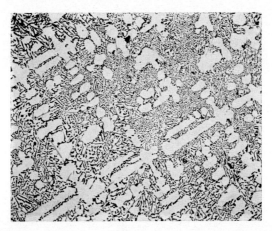

Fig. 13.12. Primary aluminum crystals (white) in a eutectic of aluminum and silicon (the latter black). The picture shows a flat-ground, polished and etched surface (cf. fig. 6.9a). Magnification 70 ×. (S. Modin.)

total composition between d and f passes the eutectic line there thus takes place a *eutectic reaction*:

$$1 \underset{\text{heating}}{\overset{\text{cooling}}{\rightleftarrows}} \alpha + \beta \tag{13.20}$$

Below the eutectic line only the two solid phases α and β exist, but the kind of crystal that was first formed from the melt occurs in larger, often more regularly formed crystals ("primary crystals"). These are embedded in a eutectic which consists of much smaller crystals of both crystal kinds (fig. 13.12, which shows the appearance of the similar system aluminum–silicon). The primary crystals are of different kinds on either side of the eutectic composition, but below the eutectic line the same two-phase region extends on both sides of this composition.

The phase regions in a T,x diagram for a two-component system are, in a general case, *one-phase areas*, *two-phase areas* and *three-phase lines* ($T=$const.). All three types are found in fig. 13.11c. If a phase cannot dissolve any component but has constant composition, its one-phase area degenerates to a *one-phase line* ($x=$const.). This is the case in fig. 13.13b–d. In fig. 13.13b and d one solid phase consists of ice, which dissolves most other substances to a negligible degree.

The junction between the miscibility gap and the two-phase regions with a liquid phase, for the case where the melting point of one component is raised and the other lowered, usually appears in principle as shown in fig. 13.11d. Also here there is a horizontal, three-phase line, but along this line a melt of composition o is in equilibrium with α phase of composition p and δ phase[1] of composition q. This equilibrium is said to be *peritectic* (Gk. περί around, beside), which suggests

---

[1] The Pt phase with dissolved Ag is designated δ because two other phases β and γ exist at temperatures below the range of the figure.

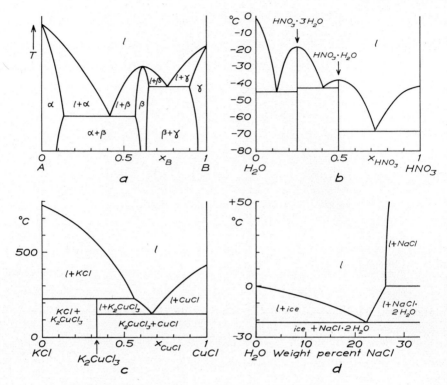

Fig. 13.13. Various types of T,x diagrams for binary systems.

that the liquid phase lies outside the region between the two solid phases. The three-phase line opq is called a *peritectic line* and its temperature *peritectic temperature*. When the temperature of any system with a total composition between o and q passes the peritectic line a *peritectic reaction* takes place:

$$l + \delta \underset{\text{heating}}{\overset{\text{cooling}}{\rightleftarrows}} \alpha \qquad (13.21)$$

If the total composition of the system lies between o and p, there is in addition an excess of liquid phase of composition o; if the total composition lies between p and q, there is an excess of solid phase δ of composition q.

Phase diagrams of the same type in principle as in fig. 13.11 c and d also occur frequently with equilibria between solid phases alone. The phases designated l in the figures then are solid phases. Since there are no melting phenomena here the terms "eutectic" and "peritectic" cannot be used, but the similarity to these equilibria is suggested by the use of the terms *eutectoid* and *peritectoid*. It can then be said in general that the decomposition of one phase into two phases lying on either side on cooling (see equation (13.20)) is a eutectic or eutectoid reaction. On the other hand, if a mid-lying phase decomposes on heating (see equation

(13.21)) this is a peritectic or peritectoid reaction. An important eutectoid reaction occurs in the Fe–C system (34-3a).

In the usual case where the solubility of a component in a solid phase increases with rise in temperature, the solid solution can often be supercooled from a certain temperature, so that at lower temperature it becomes supersaturated and unstable. In this case it is cooled very rapidly (*quenched*) to so low a temperature that the atomic mobility becomes small. The state that prevailed at the higher temperature thus becomes, so to speak, "frozen in." For example, by quenching from 800°C the homogeneous α phase in the Ag–Cu system with $x_{Cu}=0.1$, the homogeneity can be preserved down to room temperature. If this unstable phase is reheated, the atomic mobility increases and gradually β phase is separated out. By careful heating the degree of separation can be controlled very precisely. If an early stage is chosen where the precipitation is still very fine-grained (usually before it can yet be observed in a light microscope) a very large increase in the hardness (*precipitation hardening*) and strength is effected. The causes of the precipitation hardening are not wholly understood but in an elementary way it may be accepted that the incipient fine precipitation prevents gliding (6-5). Precipitation hardening is of great practical importance and makes possible, for example, the production of hard and strong, light-metal alloys.

If a solid component cannot dissolve impurities that are present, it can easily be obtained pure except for what may incidentally be bound by absorption (15-1). To separate components soluble in each other, *fractional crystallization* may be applied. To explain this we may continue the earlier discussion of crystallization in the system Bi–Sb associated with fig. 13.11a. If the whole melt with $x_{Sb}=0.6$ is crystallized, at equilibrium the final crystal mass has the same composition as the initial melt, that is, corresponding to point b_3. If it is desired to separate the two components the crystallization must be stopped earlier, for example at 440°C, at which point the two phases, liquid a_2 and crystals b_2, are separated mechanically. Melting and crystallization is then repeated with these two parts, and so on. The two-phase regions with liquid phase in diagrams such as fig. 13.11c and d are equivalent in principle to parts of the two-phase region in fig. 13.11a. In these cases also fractional crystallization can be carried out insofar as the extent of the respective two-phase regions allow.

For purification, nowadays (since 1952) repeated melting and crystallization is often used in a more continuous and particularly effective process, the so-called *zone melting*. A rod of the solid substance is used and a thin cross section (zone) of the rod is melted. The rod is appropriately supported and if the zone is sufficiently narrow the surface tension prevents the melt from running out. In fig. 13.14 is shown a metal rod that is assumed to contain dissolved impurity atoms. Heating of a narrow zone can here be done by high-frequency induction. In the upper picture the rod has been introduced into a high-frequency coil and a zone melted. When the rod is subsequently moved to the right (velocity 1–25 cm per hour), the zone migrates to the left in the rod and crystallization takes place at the right

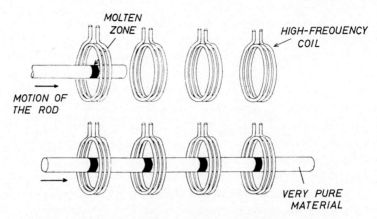

Fig. 13.14. Zone melting (adapted from T. Wallmark).

melt–crystal boundary. The figure shows the case where the impurity atoms lower the melting point of the metal and may therefore be associated with fig. 13.11a with Bi as an impurity in Sb. The newly formed crystals then become poorer in Bi, that is, purer, than the melt. As the melt zone moves to the left it therefore carries with it Bi atoms, which thus become concentrated in the trailing end of the rod. The growing crystals also reject incidental insoluble impurities, which thus are carried backward. To obtain a sufficient purity the zone must in general be allowed to traverse the rod several times. The number of passages required depends on the degree of purity desired, and on the relationship between the impurity contents of the crystal and melt in equilibrium, that is, on the character of the phase diagram. The number of passages is usually of the order of 5–50. Time is saved if, as shown in the figure, several independent zones are used, which are caused to move through the rod simultaneously.

If instead the impurity atoms raise the melting point of the substance (for example, if Bi is to be freed of an impurity of Sb), the crystals become less pure than the melt. As the zone traverses the rod the impurity in this case is concentrated in the leading end of the rod. If several different kinds of impurity atoms are present, these may be concentrated at both ends and the middle of the rod then becomes purest.

Zone melting has become indispensable for the preparation of the extremely pure substances required in the manufacture of semiconductors with a definite number of impurity atoms (6-4d). A considerable number of substances can be relatively easily prepared in this manner with less than 1 impurity atom in 10^8 atoms.

3. *Intermediate phases are present.* If there is only one intermediate phase and its melting point is lowered by addition of either component, in general a phase diagram is obtained (fig. 13.13a) that may be thought of as being formed by attaching together two diagrams of the type shown in fig. 13.11c. The intermediate phase (β) thus has a melting-point maximum.

It should be noted that the liquidus curve is horizontal (its derivative = 0) at the maximum. The reason for this is that the intermediate phase in the liquid state is always at least somewhat dissociated into the simple components. The curvature of the curve is less (the maximum flatter), the greater this dissociation is.

Fig. 13.13*b* shows a system with two intermediate phases, which both have melting point maxima and are separated by a eutectic.

If the intermediate phase has a melting point maximum, at this point liquid and solid phases of the same composition are in equilibrium with each other. The solid phase thus here fuses altogether to a melt of the same composition as the solid phase. However, the melting of an intermediate phase also often takes place as a peritectic reaction. For example the intermediate phase K_2CuCl_3 in the system KCl–CuCl melts peritectically (fig. 13.13*c*). When it reaches the peritectic temperature of 225°C it melts with the formation of a melt with $x_{CuCl}=0.56$ and solid KCl. Such a melting is also termed *incongruent*, in contrast to melting at a melting point maximum, which is called *congruent*. Starting with a melt that has a composition between the ends of the peritectic line, if this is cooled from a temperature above the liquidus line, KCl crystallizes first. If the temperature is further lowered, on passing through the peritectic line the remaining melt reacts with the KCl crystals with the formation of K_2CuCl_3. Such a reaction with a solid phase may proceed slowly, and therefore equilibrium is often not reached in peritectically melting phases.

Fig. 13.13*d* shows a portion of the system $NaCl–H_2O$, in which the hydrate $NaCl \cdot 2H_2O$ melts incongruently, as is very often the case with salt hydrates. The figure does not extend as far as the one-phase line for $NaCl \cdot 2H_2O$ which at 62 wt % NaCl goes from below up to the peritectic temperature of $+0.15°C$.

The freezing-point curves for NaCl and $NaCl \cdot 2H_2O$ are also the curves for the solubility of these phases in water. Correspondence with the solubility curve for NaCl in fig. 13.1 is obtained if the directions of increasing temperature and increasing NaCl content are made to coincide in the two figures. The unusual steepness of the freezing-point curve for NaCl expresses the minimal temperature dependency of the solubility of NaCl in water.

In a system salt–H_2O the eutectic temperature is the highest temperature at which ice and salt can exist in equilibrium with each other. Thus, if ice and salt are mixed at a higher temperature a solution is formed with the absorption of heat, which is caused primarily by the melting of the ice. If the mixture is made in a proportion corresponding to the eutectic composition, and the insulation is good enough so that the solid phases are not used up, the temperature drops to the eutectic temperature where ice, salt and solution can exist in equilibrium with each other (*freezing mixture*). The lowest temperature that can be reached by a certain freezing mixture is thus the corresponding eutectic temperature. With a mixture of NaCl and ice a temperature of $-21.2°C$ can thus be attained. Application of NaCl is used to melt ice, for example, in railroad track switches and on road surfaces.

13-6. Systems with more than two components

The maximum number of degrees of freedom rises by 1 with each new component, and already for a ternary system where this number is 4 ($k=3$; for $p=1$, $f=4$, namely P, T, x_1, x_2) a complete phase diagram cannot be represented by a three-dimensional model. Often, however, one is interested in equilibria with $P=$ const. and a three-dimensional model for this P value and the variables T, x_1 and x_2 is satisfactory. In this model two-dimensional sections are then cut for $T=$ const.

The large number of degrees of freedom has led to the result that, except for very simple types, only a few ternary systems are known reasonably completely. They have generally been investigated because of their great practical importance (alloy systems, systems of oxides that are especially important in metallurgy, the ceramic industry, mineralogy and other mineral systems). The number of fairly well-known quaternary systems is much smaller still. Space does not permit us to consider systems with more than two components here. It may only be pointed out that the distribution equilibria between two liquid solutions, such as those discussed in 13-1f, belong to ternary or quaternary systems. Also, we may note that a eutectic temperature is often lowered by the addition of a new component. In this way ternary or quaternary eutectics with very low eutectic temperatures may be obtained. In the system Bi–Pb–Sn there is thus a ternary eutectic at 96°C with 52 wt% Bi, 32 wt% Pb, and 16 wt% Sn, which constitutes so-called Rose's metal. If Cd is added to this system a quaternary eutectic is reached at 68°C with 50 wt% Bi, 25 wt% Pb, 14 wt% Sn and 11 wt% Cd (Wood's metal).

13-7. Determination and construction of phase diagrams

a. Methods for determination of phase diagrams. The surest and most general method is, after the system has attained equilibrium, to make a *phase analysis* at a sufficient number of points of the diagram defined by pressure, temperature and independent variables of composition. This involves the identification of the phases in equilibrium with each other and a determination of their composition. The identification is most often carried out microscopically or by X-ray diffraction methods (7-1). Recently, electron microprobe techniques together with X-ray spectroscopy have provided a powerful method in such studies. Sometimes equilibria that prevail at a higher temperature can be "frozen in" by quenching, and the phase analysis carried out at room temperature.

The temperature of a system can easily be changed while the other variables are held constant. Changes in the system with temperature variation generally are recognized by changes in many physical quantities. These change most rapidly on passage through boundaries of phase regions. The formation or disappearance of a phase (for example, crystallization or melting) can be directly observed, or measurements made of the evolution of heat of crystallization ("thermal analysis"), changes in the length or volume of the specimen, changes in conductivity or changes of magnetic properties. The work may be hampered because the system does not reach equilibrium sufficiently rapidly. The cause may be arrested or inhibited reactions in phase formation or excessively slow diffusion through a solid phase (chap. 14). This situation arises especially at lower temperatures where the atomic mobility is low.

b. Some general properties of phase diagrams. The following generally applicable rules are useful in the construction of new phase diagrams and the study of phase diagrams in the literature.

For any *three-dimensional diagram* (whether it is a P, T, x diagram for a binary system or a diagram for a system with $k > 2$ and one or several variables held constant), the following applies:

1. *Between two phase regions that bound each other over a surface, the number of phases always differs by* 1. A phase region that is itself a surface should be considered as a degenerate space bounded over surfaces by adjacent phase spaces. An upper limit for the number of phases is set by the phase rule.

2. *In any one-phase space all edges and corners are convex.*

For any *two-dimensional diagram* (whether it is a P, T diagram for a one-component system or a section through the above-mentioned three-dimensional diagrams) these rules become:

1. *Between two phase regions that bound each other along a line, the number of phases always differs by* 1. A phase region that is itself a line should be considered as a degenerate surface bounded along lines by adjacent phase surfaces.

2. *In any one-phase surface all corners are convex.*

The last two rules can be conveniently tested on the diagrams in this book.

CHAPTER 14

Reaction Kinetics in Heterogeneous Systems

Heterogeneous reactions take place at the boundary surface between two phases. To this category belong also the reactions that involve formation and growth of one phase at the expense of material from another phase.

14-1. The role of diffusion in heterogeneous systems

If the reaction at a phase boundary involves the transition of one phase to another of the same composition and practically the same mass per unit volume (that is, the same density), no obvious transport of material occurs. In the reaction atoms or atomic groups are moved only over distances of the order of atomic dimensions. Such is the case, for example, in the fusion and crystallization of a congruently melting substance (13-5f-3), and transformations in a solid phase without change of composition (transition of one substance from one structure—modification—to another).

In other cases material transport must take place. In a general case the total process can then be divided up into the partial processes: (1) transport of the reacting participants to the region of reaction; (2) the reaction itself; (3) removal of reaction products from the region of reaction. The rate of the total process is mainly determined by the slowest of these partial processes.

For example, if a solid phase is dissolved in a liquid solvent, the solution very quickly becomes saturated immediately outside the phase-boundary surface. From this surface outward the concentration within a surrounding layer falls off until practically the mean concentration of the whole solution is reached. Through this layer diffusion of the dissolved material takes place in a direction from the boundary surface outward, and of the solvent in a direction inward toward the boundary surface (*diffusion layer*). The diffusion layer can be made thin by vigorous stirring, but it never completely disappears. If the disintegration of the solid phase occurs rapidly, which is generally the case when the solubility is high, the diffusion is the slowest and thus the rate-determining process. The rate of solution then becomes greater, the thinner the diffusion layer is made by stirring, and, for different solid substances in the same solvent, the greater the diffusion rate of the building units is (that is, the smaller the units are). Other conditions being equal, the diffusion rate decreases as the drop in concentration through the diffusion layer decreases. The solution rate therefore decreases as more of the substance goes into solution.

If the disintegration of the solid phase proceeds sufficiently slowly, on the other hand, the rate of this process may become decisive for the total rate. The latter is then affected to a much smaller degree by stirring and by changes of concentration in the solution.

The conditions are entirely similar if the process at the phase boundary is a typical "chemical" reaction, for example the solution of a metal or a carbonate in acid.

If the material transport takes place through a solid phase, the diffusion rate is so small that it always becomes decisive for the total rate of the process.

As soon as the total reaction rate is determined by a diffusion rate it is increased by all factors that promote diffusion. First among these is increase in temperature. As mentioned earlier in 6-3c, the diffusion rate is also increased in a solid phase if the order is disturbed. If a solid phase has a transition point, in general the crystal structure undergoes a thorough rearrangement at this point. The strong disorder that is then introduced leads to increased reactivity (Hedvall). For example, if orthorhombic sulfur is oxidized by an acid solution of potassium permanganate at increasing temperatures the rate of oxidation shows a clear maximum when the orthorhombic sulfur is transformed to monoclinic at 95.5°C. The effect may be apparent even in reactions between solid phases alone.

14-2. Nucleation and phase growth

a. Nucleation. It is well known that the formation of a new phase in an originally homogeneous phase is often strongly inhibited. For example, a liquid phase can easily be carried into a temperature range where it is unstable. It can be supercooled below a crystallization temperature and become supersaturated or it can be superheated above a boiling temperature. Such arrested reactions can be released by introducing particles of the new phase, so-called *seeding*. These particles serve as starting points, *nuclei*, for further growth of the phase, which thereafter can proceed rapidly. Evidently, the phase formation in the first stage requires a nucleus and inhibition results from hindrance to the formation of nuclei.

The impediment to formation of nuclei is caused by the fact that a small particle is less stable than a larger one. As mentioned earlier in 6-5b, a smaller particle therefore has higher solubility (cf. also 13-1b) and vapor pressure than a larger particle.

For example, if a particle of a substance is introduced into a concentrated solution of the substance, and this particle has a size that gives it a solubility just corresponding to the concentration of the solution, nothing will happen. On the other hand, particles of other sizes are not in equilibrium with the solution; if they are smaller than the equilibrium size, they go into solution, and if they are larger, the dissolved substance will be deposited on them. These processes continue until equilibrium is attained. Thus, no particle growth would ever be able to take place in a system that does not already contain particles that are larger than the equilibrium size. A supersaturated solution that is not in contact with the dissolved substance therefore may remain homogeneous for a very long time under favorable circumstances.

However, a supersaturated solution, at least one with sufficiently high supersaturation, may spontaneously precipitate the dissolved substance as a separate phase. This may be considered to result from the fact that the building units that make up this phase here and there in the solution may acquire such energy values, and because of their heat motion come together in such a way that a sufficiently large particle can be formed directly. The probability for this nucleus formation increases, under otherwise equal circumstances, with the concentration of the solution. This is partly because the probability for a proper conjunction of building units increases with their number per unit of volume, and partly because the required minimum size of the nucleus becomes smaller, the more concentrated the solution is.

The above applies to precipitation *within* the solution. The conditions are different at boundary surfaces between the solution and other phases. On a surface that is flat, and still more on one that is concave toward the solution, the new phase can be deposited even at the start with a significantly larger radius of curvature than is possible within the solution. In fact, one often sees deposition begin on vessel walls, grains of dust, and such places. The familiar crystallization-initiating effect of a scratch on the vessel walls probably results partly from a fresh exposure of the surface, and partly from the formation of concave surfaces, which also may have gained an excess of energy through mechanical deformation.

We have compared above at the same temperature supersaturated solutions of different concentrations and thus different degrees of supersaturation. Similar results are obtained if one supercooled liquid phase is studied at different temperatures, thus at different degrees of supercooling. The lower the temperature, the smaller a crystal needs to be in order to grow further and thus act as a nucleus; the heat motion, which can detach building units from the crystal, decreases with falling temperature. On the other hand, nucleation is hampered by falling temperature because the phase then becomes less and less fluid (more viscous). A decrease in temperature therefore leads first to an increase in the number per unit of time of nuclei *formed*, which, after having passed through a maximum, falls off again. At the same time the increased viscosity causes the *growth* of incipient nuclei to be hindered. At sufficiently low temperature no appreciable crystallization occurs, and the liquid goes over to a glass. As mentioned in 6-3d the tendency to form a glass is especially great if in addition the building units that would make up the crystal have low mobility because of their size and shape or because they are kept together by strong forces even in the liquid phase; such liquids are characterized by a comparatively high viscosity.

Nowadays, crystallization of a glass is employed in the manufacture of a new type of construction material (trade name, Pyroceram). To a glass melt is added a small amount of a substance (for example, titanium dioxide) that is soluble in the glass at higher temperature, but at lower temperature precipitates in small crystals which can serve as nuclei. An object is molded from the melt in the ordinary way, which is then cooled to a temperature at which these nuclei are formed; thereafter the temperature is raised somewhat so that crystallization initiated by the nuclei proceeds sufficiently rapidly. The object then becomes completely polycrystalline and loses its transparency, but with appropriate melts many properties become superior to those of a glass (greater toughness and hardness, higher softening temperature, often low thermal expansion and therefore high resistance to rapid changes in temperature).

In the formation of a liquid in a gas phase the conditions are entirely analogous to those that prevail in crystallization. Inhibition of condensation in a supercooled gas phase is here explained by the fact that a small liquid drop has a higher vapor pressure than a larger drop. In the superheating and delayed boiling of a liquid, formation of the first gas bubbles is hindered by the fact that the molecules less readily go over from the liquid phase to the gas bubble because of the strongly concave shape of the phase-boundary surface toward the gas. Therefore, the molecules are more strongly held in the liquid phase than if the surface were flat, and the vapor pressure of the liquid in the small bubble is lower than it would be over a flat surface at the same temperature.

b. Phase growth. Once the nuclei have been formed, it remains for the new phase to grow. If this is crystalline, the building units must be arranged properly. This process may sometimes proceed very slowly and can then determine the total rate of the growth process. In other cases the material transport to and from the surface of the growing phase may instead be the slowest and thus the rate-determining process. If the phase growth requires heat exchange and the processes mentioned above go reasonably rapidly, the growth rate may be completely dependent on the rate at which heat is brought in or carried away.

The building up of a crystal has been partially treated in 6-1d. There it was shown among other things how difficult it is for a building unit to become attached permanently to an already completed, densely packed crystal surface and thereby begin a new lattice plane. This would be especially difficult when the building units of the crystal are in low concentration in the surrounding phase, that is, on crystallization from a gas phase or from a solution. One might expect in such cases to find extremely low growth rates normal to densely packed surfaces, but often much higher rates are observed.

The explanation is that the crystals in question contain a special kind of dislocation (6-5a), called *screw dislocations*, that make possible a completely new growth mechanism (Frank, 1949).

In fig. 14.1 a and b is shown schematically how a crystal is deformed so that a screw dislocation arises. From a fixed line (in the picture approximately the vertical mid-line of the crystal) along which no change occurs, a glide surface begins. The parts of the

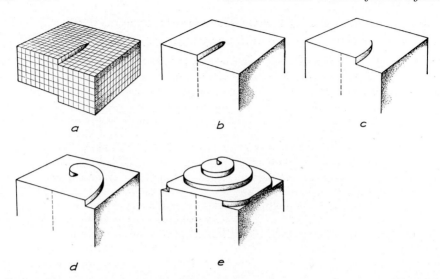

Fig. 14.1. Crystal growth by a growth spiral caused by a screw dislocation. (C. Kittel.)

crystal adjacent to this surface have glided against each other in the direction of the line a distance which becomes constant within a short distance from the line and amounts to a whole and often small number of identity periods. On the crystal surface there is thereby formed a step of corresponding height, and on addition of crystal-building units, these are easily deposited at the step. This causes the step to turn about the fixed line as an axis. If the supply of material per unit of length of the step is everywhere the same, the rate of turning falls off as the distance from the axis increases. An outer portion of the step therefore falls behind an inner portion and the step turns in a spiral (*c–e*). Continued addition of material leads to the result that this spiral seems to rotate about its axis so that step after step sweeps out from the axis over the whole surface of the crystal, which is thus raised. A *growth spiral* is obtained in which the locus of a favorable deposition of building material, the step, never disappears.

Growth spirals have been found to be very common, and the reason they have been observed only so recently is the generally small height of the spiral. Fig. 14.2 shows an electron micrograph of a beautiful growth spiral. The crystal is here built up of chain molecules 125 Å long, which are arranged in parallel nearly perpendicular to the plane of the picture. The height of the step can be calculated from the length of its shadow and corresponds to exactly one molecule. The growth rate in the picture plane is here different in different directions, which causes the spiral to depart from the Archimedean spiral of fig. 14.1*e*. Growth spirals can also be detected with the light microscope, especially with interference methods or by use of foreign substances that are preferentially bound to the steps and thus outline them. Interference methods have made possible the measurement of step heights down to about 10 Å.

With strong supersaturation or supercooling readily skeletal, often tree-shaped crystals, so-called *dendrites* (Gk. δένδρον, tree), are formed. The entire dendrite is built up of one continuous crystal. Snow crystals, which are formed from strongly supercooled water vapor, are typical dendrites (see also fig. 13.12). The addition of material through the surrounding phase proceeds faster at the parts around the edges

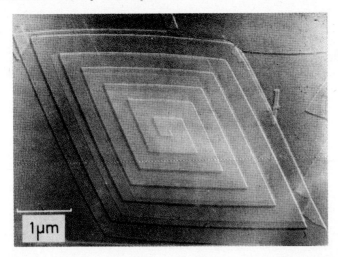

Fig. 14.2. Electron micrograph of a growth spiral on a crystal of the hydrocarbon normal hectane, $n\text{-}C_{100}H_{202}$. The object is "shadowed" with palladium as in fig. 1.2. (I. M. Dawson.)

of the crystal and especially the corners than at the more central portions of its surfaces. With sufficiently rapid growth of the former parts, the mid-portions of the surfaces therefore cannot keep up. On the edge and corner portions new edges and corners are formed, which in turn grow especially fast, and so on.

c. Number of crystals and crystal size. If every nucleus gives rise to a crystal, at the end of the crystallization the average mass of one crystal is equal to the quotient (total crystal mass)/(number of nuclei). On crystallization from a phase of definite composition and amount the total, final crystal mass is constant while the number of nuclei increases with increased supercooling. The mean size of the crystals thus decreases with increased supercooling. Solutions with different concentrations give different total crystal masses, but it has been shown that the number of nuclei at constant temperature increases so rapidly with increasing concentration that the value of the quotient, and thus the mean size of the crystals formed, decreases.

If no special precautions are taken, the number of nuclei is generally so large that the crystallized material becomes *polycrystalline*, that is, consists of a large number of crystals. At the start these grow freely and are then usually bounded by natural crystal faces. When the growing crystals contact each other *grain boundaries* arise, which are contact boundaries with no relation to natural crystal faces. If a liquid phase crystallizes completely to a single solid phase, the crystals everywhere contact each other in such grain boundaries (fig. 6.9). If the system contains substances that will not dissolve in the solid phase, at first these remain in the melt, dissolved or suspended in it. They are therefore rejected by the primary crystals and are incorporated, after the whole system has solidified, in the material between the crystals. If there are only small amounts of such substances (then often referred to as impurities), they form only thin layers in the grain boundaries, but nevertheless they can strongly influence the properties of the material. In unfavorable cases the grain boundaries may in this way have very low cohesion, which easily leads to intercrystalline fracture.

A solid phase is often formed by a chemical reaction in a solution. If the solid phase is poorly soluble in the medium, this will be strongly supersaturated with respect to the former just prior to its formation. The number of nuclei becomes large, which leads to a large number of crystals and small crystal size, that is a *precipitate*.

It is of importance to be able to prepare large *single crystals* of various substances. Single crystals of metals are used to determine exactly defined properties (for example, magnetic) and single crystals of semiconductors are required for electronics equipment. Single crystals of salts are used in the manufacture of optical parts for special wavelength regions, for example, infrared (polycrystalline material, because of the grain boundaries, is not transparent and glass can seldom be made of substances containing primarily ionic bonds). In general one starts with a melt of the substance in question (salt crystals, can, of course, be grown from solution but then they often are of poor quality) and arranges matters so that there is only a single nucleus or at least one that predominates. For example, the point of a small single crystal can be dipped into a melt of the same substance, which is held just above the melting point. By cooling the crystal holder heat is carried away so that the single crystal grows at the expense of the melt. The single crystal is drawn slowly up out of the melt in pace with its growth. Also, a melt may be contained in a crucible that is conically pointed downward, which is then slowly lowered into a colder region. The crystallization begins in the point and because of its small volume often only with one crystal, which then on further lowering grows throughout the whole contents of the crucible. If a zone melting (13-5f) is carried out on a polycrystalline rod with a single melt zone, a seed crystal in the leading end of the rod (often fused into it in predetermined orientation) can be made to grow throughout the entire rod.

d. Secondary processes in crystalline phases. When the crystal formation is terminated, the crystallized phase is often far from equilibrium. The phase may be extensively divided into small crystals, which also are usually of varying sizes. The latter may result from an uneven distribution of the original nuclei and an uneven supply of material during growth. The crystals may also be heterogeneous (cf. zoning, 13-5f), disordered or deformed. The system therefore tries to achieve equilibrium by the processes described in 6-5b. These are all favored by increased temperature and perhaps may only become noticeable at higher temperature. If material transport is possible between different crystals, growth of larger crystals takes place at the expense of smaller ones. If the crystals contact each other in grain boundaries, the larger crystals can grow by movement of the boundaries. However, this may be hampered or prevented if too much foreign material has separated out in the grain boundaries (14-2c). Heterogeneities are evened out by diffusion. Deformations generally can be eliminated only by gross structural rearrangement, *recrystallization*. The first-formed phase may also be unstable, in which case it may be transformed to a more stable phase. In precipitates all these secondary processes contribute to what is customarily called the *aging* of the precipitate.

The material transport between crystals can take place through a liquid or gaseous phase, but in addition the processes described occur in and between solid phases alone. In 14-3 such processes are treated in somewhat more detail.

14-3. Processes in solid systems

a. General features. Because of the low mobility of atoms and atomic groups in solid phases the reaction rate in solid systems is to a large degree dependent on properties such as grain size and consequent surface development, energy storage in the form of deformation or other disorder, and so on. The previous history of the system therefore often plays a very large role.

The low mobility is evident, whether a process involves material transport or only a rearrangement of building units. In the formation of a new phase, also in solid systems, a nucleus must appear before the phase can begin to grow.

The reaction rate in solid systems is always favored by increased temperature, which leads to greater mobility. Since the mobility increases very rapidly with rise in temperature the reaction rate often only becomes appreciable when a rather definite, but of course not sharply defined, temperature is reached (*reaction temperature*). Substances of similar structure often acquire appreciable reactivity at an absolute temperature that approximately corresponds to a certain fraction of the absolute melting temperature. For metals, for example, the absolute temperature at which recrystallization becomes clearly evident lies at approximately 35–55% of the absolute melting temperature. The recrystallization temperature for magnesium is thus about $+150°C$ ($\sim 420°K$, that is, 45% of the melting temperature $923°K$) and for tungsten about $+1200°C$ ($\sim 1470°K$, that is, 40% of the melting temperature $3653°K$).

The increase in surface energy per unit of area as the particle size decreases causes a smaller grain to react more easily than a larger one. Starting with a certain mass of a solid phase, its total contact area with other phases may also be significantly increased as the grain size diminishes. A deformed or otherwise distorted structure is more reactive than an undistorted structure because of its higher energy. The excess of energy also makes possible higher diffusion rates in the former, which is also easily understood in terms of the greater disorder.

The effect of small grain size and disordered structure is striking when oxidizable substances through fine subdivision become *pyrophoric*. Many metals do not react with dry air at room temperature if they are in "compact" form, that is, not very finely divided. In finely divided form, on the other hand, they often react so violently with air that spontaneous ignition occurs (for example, iron powder produced by reduction at low temperature, in which the grain size is very small, and the low temperature of formation has not allowed the atoms to order themselves as well as in a stable structure).

Atoms or atomic groups may have high mobility on the surface of a solid phase (6-1 d). Surface diffusion or migration can therefore play a large role, and this is true even in cases where the diffusion rate is low in the interior of the phase. It is favored, of course, by high surface development.

In reactions between solid phases where a gaseous phase (for example, air) is also present, this phase may enter into the process in various ways. This should not be forgotten in the study of these rections.

b. Powder reactions. Reactions between substances in powdered form, generally at temperatures above room temperature, illustrate the above discussion. Many such *powder reactions* have been studied by Hedvall.

Fe(III) salts whose anions can give a volatile oxide on heating leave a residue of Fe_2O_3. Many salts, for example the oxalate, give a brown- to purple-colored Fe_2O_3 powder with rather well-built crystallites. Other salts, especially the sulfate, if the heating does not exceed 700°C, give a beautiful, red Fe_2O_3 powder ("red ochre"). An X-ray diffraction study shows that the crystal structure is here much more distorted. If the red "sulfate oxide" is heated over 700°C it recrystallizes and gives a darker powder with well-built crystallites (cf. 7-3a).

The original red "sulfate oxide" has a much greater reactivity than the "oxalate oxide." For example, if the different oxide preparations are heated with CaO at 700°C, after the same time interval a significantly larger amount of "sulfate oxide" than "oxalate oxide" has been converted to $CaFe_2O_4$. The former oxide also dissolves appreciably more rapidly in sulfuric acid than the latter.

When a powder recrystallizes the crystal formation at the contact surfaces between the grains causes them to become more or less strongly bonded to each other. The powder *sinters*. If the powder is first compressed, at sufficiently high temperature sintering can lead to a very firm and strong material. The recrystallization of a powder takes place more easily, the more distorted and disordered the crystal structures of the grains are. At the same time sintering is thus facilitated. If compressed pellets of iron oxide are sintered for the same length of time and at the same temperature, pellets of "sulfate oxide" attain much greater strength than pellets of "oxalate oxide." Ceramic material is to a large extent sintered. Nowadays sintering is also much used in the manufacture of metal objects, which for one reason or another (for example, too high melting point of the material) cannot be obtained by melting (*powder metallurgy*).

Powder metallurgy is applied especially in the manufacture of metal objects that are to be used at high temperature or for their hardness (*hard metals*, for example edges of drilling and turning tools). The *hard substance* used for this purpose (particularly metallic carbides but also borides, silicides and nitrides) is practically always high-melting. Since as a rule it is also brittle, it is often sintered together with a metal that forms a tougher, uniform ground mass (*binder*).

Modern methods of ore enrichment (magnetic separation and flotation, 15-2) generally require finely ground ore. The enriched, likewise fine-grained ore concentrate packs densely so that contact with gases with which it is to react, for example in roasting and reduction, is hampered. The concentrate is then often sintered into pellets among which the gases can easily circulate.

c. Working and recrystallization of metals. Metals are often subjected to mechanical working at ordinary temperature, so-called *cold working*. This happens, for example, with cold rolling, hammering, pressing, drawing and bending, but also occurs with other processes. It often causes strong plastic deformation of all crystals within the worked region. To a great extent the deformation implies the occurrence of glides, but when these have become sufficiently extensive they lock each other so that further gliding becomes very difficult. The material therefore becomes significantly harder, which has great practical importance.

The cold-worked material has acquired an appreciable excess of energy and is therefore unstable. It tries to reduce its energy by rearrangement of the crystal-building units so that undistorted crystal regions are formed; that is, it recrystallizes. Recrystallization generally occurs at a noticeable rate only at higher temperatures. One can then observe how new crystals grow out from individual points, which evidently have especially high energy and therefore serve as nuclei. As would be expected, the larger the number of nuclei is at a certain point in a piece of metal, the larger the deformation was a that point. When the recrystallization has proceeded sufficiently far the material is about as soft as it was before the cold working; it is *soft annealed*. Often annealing must be done between different stages of a cold-working process, for example different cold-rolling steps. This is made necessary also because the cold working makes the material more brittle, so that further cold working could cause fracture.

Hot working (hot rolling, forging, and so on) is done at a high enough temperature so that recrystallization can take place immediately. This therefore leads to no increase in hardness.

14-4. Heterogeneous catalysis

Sometimes the rate of a reaction between one or more substances in a homogeneous phase is increased if these substances come in contact on the surface of another phase. The phenomenon is called *heterogeneous catalysis* (also *contact catalysis*). Commonly the active phase, the *catalyst*, is solid. Classical examples are the burning of hydrogen with Pt as the catalyst, the manufacture of sulfuric acid by the contact process, and the decomposition of hydrogen peroxide with Pt or MnO_2 as the catalyst. Heterogeneous catalysis plays a very great role in chemical technology. As examples of its use in inorganic chemistry we may mention, in addition to the manufacture of sulfuric acid by the contact process, the manufacture of ammonia according to the Haber-Bosch process, and of nitric acid (primarily NO) by the combustion of ammonia.

As in homogeneous catalysis (8h), heterogeneous catalysis makes possible a new reaction route with a lower energy threshold (lower activation energy; cf. also 2–3b). The principal reason for this is that one or more of the initial substances are *adsorbed*, that is, are bound to the surface of the catalyst (for adsorption, see 15-1) and at the same time are *activated*, that is, converted to a more reactive state. It was believed earlier that the activation was mainly caused by points and disorder sites on the surface of the catalyst, that is, the regions of higher energy ("active centers"). This interpretation has now been abandoned but it is clear that large contact area and thus a high state of subdivision of the catalyst is important to its capacity to convert large amounts of a substance per unit of time. Further, the reacting substances should not be too tightly bound to the catalyst surface. Metals such as Li, Mg and Al, which bond N_2 to form stable nitrides, cannot be used as catalysts in the synthesis of ammonia.

In typical, heterogeneous catalysis the catalyst is very often a metal or a semiconductor (6-4d), in the latter case usually an oxide, and it is the capacity of these substances to take up or give off electrons easily that causes the activation. The oxidation of CO according to the reaction $CO(g) + \frac{1}{2}O_2(g) \rightarrow CO_2(g)$ is catalyzed, for example, by NiO. NiO can acquire an excess of oxygen by the formation of Ni^{2+} vacancies, while charge compensation is maintained by the conversion of a corresponding number of Ni^{2+} ions to Ni^{3+} ions (in principle, according to fig. 6.20b). The Ni^{3+} ions become points of electron deficiency, electron holes. NiO with excess oxygen is thus a semiconductor of p type (band scheme of fig. 6.25f).

If now a CO molecule is adsorbed on the NiO surface it is forced to give up an electron to NiO, and an adsorbed, ionized molecule CO^+ is left behind: $CO(g) \rightarrow CO^+(ads) + e^-$. CO^+ is very reactive and reacts extremely fast with oxygen from the gas phase according to the reaction $CO^+(ads) + \frac{1}{2}O_2(g) + e^- \rightarrow CO_2(g)$. The electron required for this purpose is taken from NiO, which thus is returned to its original state. The CO_2 molecule formed goes out to the gas phase and leaves room for new adsorption.

The catalytic action of nickel oxide in the oxidation of carbon monoxide can be enhanced (that is, the activation energy further reduced) by the addition of Li_2O. Here a number of Ni^{2+} ions is replaced by Li^+ and for charge compensation the same number of the remaining Ni^{2+} ions is converted to Ni^{3+}. The electron deficiency thus increases and so also does the catalytic action. We have here a *mixed catalyst*, whose effect is enhanced by the *promoter* Li_2O. Addition of other substances can instead decrease the electron deficiency and thereby impair the catalytic action.

In other cases the catalyst may be a semiconductor of n type, for example ZnO, and some of the participating molecular types then are activated by electron absorption. With metallic catalysts both activation processes which involve giving up electrons to the metal and those which involve absorption of electrons from the metal are known. Here, alloying with another metal can markedly alter the effect of the catalyst.

The relevant states of oxidation of the metal in the oxide of a catalyst must have close-lying energies. The metal thus must easily be able to change its oxidation number, which explains why transition metal oxides are so often effective. But other oxides as well as substances of other types can also act as catalysts. Al_2O_3 can be mentioned as an example. In such cases the mechanism is often less obvious but also here the catalytic action can frequently be associated with shifts of electrons or electron pairs.

In many technologically important cases of heterogeneous catalysis the various reactions are not completely known. It is also often difficult to decide the exact state of the catalyst in which it is active. There is no doubt, however, that the reactions in principle are similar to those described above. Mixed catalysts are much used in industry. In the sulfuric acid contact process, besides Pt, V_2O_5 with K_2O added (in the form of K_2SO_4) is used; in the Haber-Bosch process, Fe with oxides added, especially Al_2O_3; and in the combustion of ammonia, besides Pt with added Rh, also Fe_2O_3 with added Bi_2O_3.

In order that a catalyst will have a large area and in addition not be so easily reduced in area by grain growth or recrystallization (sintering), it is generally spread on a porous, nonreacting material, a *catalyst support*. Porous SiO_2 (for example, kieselguhr or diatomaceous earth, and, with higher water content, "silica gel," 23-3h6) is a common catalyst support.

Not uncommonly a system can follow different reaction routes and give different reaction products with different catalysts (*selective catalysis*). The mixture $CO + H_2$, so-called water gas (18d), gives methanol with pure ZnO as the catalyst. With ZnO + large amounts of Cr_2O_3 higher alcohols are formed; with ZnO + Fe_2O_3 methane and higher aliphatic hydrocarbons are produced (the Fischer-Tropsch process).

A catalyst is often *poisoned* by foreign substances in the reaction mixture. The effect depends on the particular catalyst, but common catalyst poisons are CO, and compounds of S, P and As. The poisoning results from the fact that the substance in question is more easily adsorbed on the surface of the catalyst than that or those substances that must be adsorbed for the desired reaction to take place. The surface of the catalyst is thus blocked for the latter.

CHAPTER 15

Surface Chemistry and Colloids

15-1. Adsorption

a. General. In 6-5b we discussed the outward-directed, unsaturated bond forces on the free surface of a phase, and how these try to satisfy themselves by bonding to other atoms or atomic groups. This phenomenon is called *adsorption*. If a solution, that is, a phase with several components, is in contact with the above-mentioned phase, it is clear that the different components of the solution may be bound in this way with different strengths. If one component has been more strongly drawn to the boundary surface than the others and thus concentrated there, the adsorption of this component is said to be *positive*. If the component is more weakly drawn to the boundary surface than the others or perhaps even repelled (this occurs especially when electrostatic forces are active), its adsorption is said to be *negative*.

When a single layer of atoms or molecules is adsorbed on a phase-boundary surface (*monatomic* or *monomolecular layer*), further adsorption can take place, but then the bonding forces quickly become of another kind than in the first layer. For example, if water molecules are adsorbed from a gas phase onto a glass surface, the first layer of H_2O molecules is bonded by special bonds. Further out the adsorbed water layer is held together in approximately the same way as in liquid water.

Since phases with very large surface area are often encountered in chemistry, either because of a high degree of subdivision (6-5b), or great porosity, adsorption plays an important role. In the following discussion, positive adsorption of a dissolved substance at the boundary surface between a liquid solution and a solid phase (*adsorbent*) is of particular interest.

At constant temperature, the relation between the amount of a substance adsorbed and the concentration of this substance in the main body of the solution is often given by the empirical expression:

$$x = \alpha c^\beta \tag{15.1}$$

c is the concentration of the substance in the main body of the solution and x is the amount of the substance which, after equilibrium is attained, is adsorbed on the surface by a unit of weight of the solid phase. α and β are characteristic constants for the system. As the expression gives the adsorption per unit weight and not per unit area of the solid phase, α and β (especially the former) are depend-

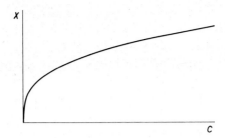

Fig. 15.1. An adsorption isotherm.

ent on the degree of dispersion of the solid phase and constant only if this is constant. An expression of this type is called an *adsorption isotherm*.

The essential relationship between c and x according to (15.1) is shown in fig. 15.1. Since β is always <1, the curve is always convex upward. This implies that the amount of a substance adsorbed changes more slowly than its concentration in the solution. For example, if $\beta = \frac{1}{2}$, and c is decreased to one fourth of its original value, x is decreased only to one half of the original value. This leads to the familiar difficulty in removing the last traces of an adsorbed substance by washing.

Adsorption of *ions* is of very great importance. If a certain kind of ion is adsorbed more strongly than the other ions in the solution, the adsorption layer acquires a net charge. This charge causes in turn an enrichment of ions of opposite charge in the solution immediately outside the adsorption layer. These latter ions are called *counter ions*. The situation is shown schematically in fig. 15.2.

If a solid phase is built up of ions from a solution, the bonds between these particular ions under the prevailing conditions are stronger than between other kinds of ions. The solid phase therefore adsorbs preferably its own kinds of ions from the solution, as soon as they are in sufficient concentration. Another reason for this is that the crystal structure is already, so to speak, adapted to these ions.

If a KI solution is added to an excess of $AgNO_3$ solution, crystalline AgI is formed, in which nearly all I^- ions are used up. The ions remaining in solution are therefore mainly K^+, Ag^+ and NO_3^- and of these Ag^+ is adsorbed preferentially on the AgI surface. Anions become counter ions, in this case mostly NO_3^-, the only anion present in any great concentration. Schematically one may thus depict a homogeneous region of the solid phase and its surface by $(n\text{AgI})Ag^+$, where n

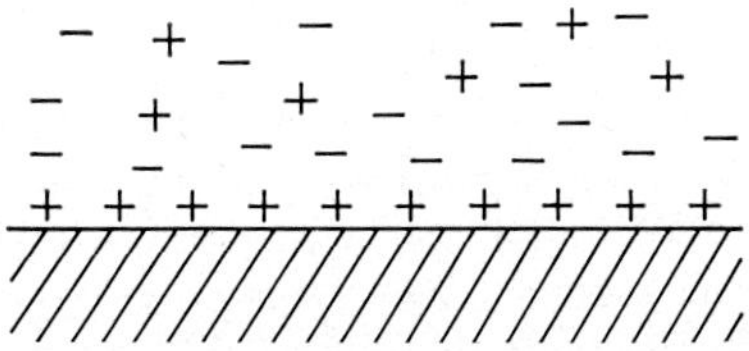

Fig. 15.2. Adsorption of a certain kind of ion gives a charged adsorption layer.

is a very large number. The expression within the parentheses corresponds to the interior of the solid phase and is followed by the adsorbed surface layer.

If instead AgI is precipitated by adding $AgNO_3$ solution to an excess of KI solution, Ag^+ is nearly completely used up and the remaining solution contains mainly K^+, I^- and NO_3^-. Here mostly I^- is adsorbed and the formulation becomes $(nAgI)I^-$. The adsorbed layer takes on a negative charge, and mostly K^+ acts as counter ion.

If solutions with equivalent amounts of $AgNO_3$ and KI are mixed, no particular kind of ion becomes favored in the adsorption layer, which thus acquires no net charge.

If an adsorbed ion is a protolyte, the charge of the surface layer becomes dependent on the acidity of the solution. In such cases it is convenient to consider the charged layer to be formed by the giving up to the solution of ions of a certain kind by the solid phase. If the solid phase consists of the difficultly soluble acid HA, this gives off protons to a sufficiently basic solution. The solid phase thus acquires the composition $(nHA)A^-$. The more basic the solution is, the more protons are given up and the more the negative charge is increased. However, the process can equally well be described as the solution and protolysis of molecules of HA followed by adsorption of A^- ions on the surface of the solid phase.

b. Coprecipitation. It was stated above that the shape of the adsorption isotherm makes more difficult the removal of an adsorbed substance by washing. If the adsorption layer is charged, electrical neutrality requires that also an amount of counter ions equivalent to this charge remains on washing. If AgI is precipitated by an excess of KI, the precipitate therefore retains KI.

Contamination of a precipitate by foreign substances is customarily called *coprecipitation*. As a crystal grows, not only the regular building units are attached to its surface, but also other atoms or atomic groups. In general the latter are successively exchanged with proper building units, but when these are used up in the surrounding phase and crystal growth is thereby halted, as stated above an adsorbed surface layer remains. During rapid growth it also may happen that the foreign substances are not able to be exchanged with proper building units but become built around and enclosed by the growing crystal (*occlusion*). The growing crystal may, of course, also take up foreign building units in *solid solution* if they fulfill the requirements for such (6-3a). A coprecipitation that results from adsorption is favored by a large area, that is, a high degree of dispersion of the solid phase.

If a precipitate is to be made use of, it must be freed of coprecipitated impurities. If the coprecipitation results from adsorption, the precipitate must be recrystallized in the mother liquor, so that its free surface area is reduced and the coprecipitated substances returned to the solution. After subsequent washing the precipitate may besides be redissolved in an appropriate solvent. In the new solution the concentration of the originally adsorbed substances usually is so reduced that a *reprecipitation* gives a significantly purer product. However, if the coprecipitation results from solid solution, reprecipitation has no effect.

A precipitate that takes up a radioactive nuclide by coprecipitation is often used as a *nonisotopic carrier* of such a nuclide (9-5a). A precipitate is then chosen that either especially strongly adsorbs the atoms or the atomic groups in which the nuclide is included, or whose crystals can take these up in solid solution. A number of finely divided precipitates, especially those containing more highly charged metal ions, cause very strong coprecipitation of various kinds of dissolved substances. If such a precipitate is formed, it can take up from the solution all radioactive nuclides that are present in small amounts and not accompanied by larger amounts of stable isotopes. An example of such a general, nonisotopic carrier ("scavenger") is iron(III) oxide hydroxide.

In work with a radioactive nuclide it also often happens that it is desired to remove other elements present by precipitation from a solution in which the radioactive nuclide is to remain behind. If this is present in small amount, however, it is also almost completely removed by coprecipitation. The remedy for this is to add to the solution a substantial amount of an isotopic carrier (9-5a), so that coprecipitation, which affects the total amount of the two isotopes, leads to the loss of only a small fraction of the radioactive nuclide.

15-2. Flotation

If there are no, or only very weak, bond forces present between a solid phase and a liquid phase, the solid phase is "wetted" only with difficulty by the liquid. If the solid phase consists of sufficiently small particles, these then will lie on the surface of the liquid. Such a phase is said to be *lyophobic* (Gk. λυειν, dissolve; φόβος, fear, flight); if it shuns water the special term *hydrophobic* is used. Carbon black and aluminum powder are hydrophobic and lie on the water surface without penetrating the water. With sufficiently strong bond forces, on the other hand, the solid phase is easily wetted, and even small particles penetrate the liquid. The phase is then said to be *lyophilic* (Gk. φίλος, friend) or, if the liquid phase is water, *hydrophilic*.

Certain atomic groups are typically hydrophobic, others, especially those that can form hydrogen bonds with water, typically hydrophilic. Thus a hydrocarbon chain is hydrophobic and a carboxyl group $-C\begin{subarray}{c}\diagup OH\\ \diagdown O\end{subarray}$ is hydrophilic. A fatty acid molecule, in which a carboxyl group is bound to a hydrocarbon chain, therefore has a hydrophilic end and a hydrophobic end. If a molecular species is adsorbed on a solid phase so that, for example, a hydrophobic group is facing outward, the whole surface becomes hydrophobic as soon as it is covered by a complete molecular layer.

Certain minerals are hydrophobic and others hydrophilic. This difference is exploited to separate minerals from each other in ore enrichment by *flotation*. In this method the material is crushed to a particle size less than 0.5 mm and stirred with water to which a small amount of a *frother* has been added. For this a moderately foam-forming substance is used, often "tall oil," or a higher alcohol, for example amyl alcohol. Air is blown through the mixture so that the foam bubbles give a liquid–air phase boundary of very large area, to which the hydrophobic particles are collected. The bubbles with their clinging particles rise to the surface and are there skimmed off. The hydrophilic particles, on the other hand, do not follow the bubbles.

Most silicate minerals are inherently hydrophilic, while many sulfide minerals are hydrophobic. The latter can thus be separated by flotation from the gangue, which most often consists of silicate minerals. The separating power of the method is increased by the addition of a *collector*, that is, a substance that is adsorbed on certain types of minerals thus making them hydrophobic. The amount of the collector may be very small, most often of the order of 100 g per ton of ore. A common collector is the xanthate ion $C_2H_5OCS_2^-$, which attaches itself to a sulfide phase by the sulfur end and faces the hydrophobic ethyl group outward. By proper choice of collector, regulation of the acidity and sometimes addition of other ions, a large number of separations, even of closely related minerals, can be processed with good yields. This has led to the result that flotation nowadays plays an extraordinarily large role in ore enrichment.

15-3. Ion exchangers

Many solid bodies contain framework ions (6-2a), which then form the actual skeletons of the bodies, while small, finite ions and often also uncharged molecules occupy the interstices in the skeleton. If these interstices form traversing channels that allow ions or molecules to pass to and from the surface of the body, often these may be exchanged between the solid body and a liquid or gaseous phase that surrounds it. The framework skeleton allows the solid phase to retain its homogeneity, but its composition can vary by the exchange within certain limits. We may therefore here speak of solid solutions. However, the equilibrium between the framework skeleton and the exchangeable atoms or atomic groups is analogous to an adsorption equilibrium, so that we may also consider the phenomenon as an adsorption on the internal walls of the channels mentioned above. Since the exchange of ions first took on practical importance, solid bodies of this type have been called *ion exchangers*.

Many different kinds of solid substances have ion-exchange properties, but only a rather small number have found practical application. Their importance, both for analytical and preparative work, is extremely great and continues to grow rapidly. For a quite long time *zeolites* have been used, which generally consist of crystalline aluminosilicates of alkali and alkaline-earth metals (23-3h5). Zeolites are, in fact, minerals, but since 1905 mainly synthetic, zeolitic materials (*Permutit*) have been used for ion-exchange purposes. An example is the zeolite natrolite, containing the framework ion $(Al_2Si_3O_{10}^{2-})_n$, where n is a very large number. This ion forms a stable skeleton traversed by open channels. In these in normal natrolite there are $2Na^+$ and $2H_2O$ per formula unit $Al_2Si_3O_{10}^{2-}$. However, the contents of the channels can be varied within wide limits without collapsing the skeleton. In 13-5e it was stated that the water content can vary, but in addition the Na^+ ions can be replaced by other ions, although the total cationic charge must remain constant. If a solution containing Ca^{2+} ions is passed through a column (15-4) of "sodium-ion-saturated" natrolite, $2Na^+$ are replaced by Ca^{2+} in the natrolite, which thus becomes "calcium-ion-saturated", starting in the end of the column into which the Ca^{2+} solution flows in. When the whole column has become calcium-ion-saturated, it can be regenerated by passing through NaCl solution. We thus have a reversible reaction which may be written

$$2NaR + Ca^{2+} \rightleftharpoons CaR_2 + 2Na^+ \qquad (15.2)$$

where R denotes one equivalent of the framework anion.

This process is used to remove Ca^{2+} from water and thus make hard water soft. If, instead of constantly supplying one kind of ion, a certain quantity of ion exchanger is shaken with a certain quantity of solution, the reaction (15.2) goes to an equilibrium.

Organic ion exchangers, which were introduced about 1935, have acquired greater importance than the inorganic ones. They contain stable and in most solvents insoluble skeletons of high-polymer, synthetic resins ("network polymers", which have irregular structure, that is, are practically amorphous), to which negative or positive groups are firmly bound. These groups in turn can bind cations or anions, respectively, which can be exchanged through the network. Thus, both *cation exchangers* and *anion exchangers* are obtained.

The negative groups in the *cation exchangers* are most often sulfonate groups $-SO_3^-$. These can bond hydrogen ions, and if this occurs completely a hydrogen-ion-saturated cation exchanger is obtained with the schematic formula HR. If this is treated with a solution containing Na^+, H^+ is replaced by Na^+ according to

$$HR + Na^+ \rightleftharpoons NaR + H^+ \qquad (15.3)$$

The reaction is reversible and the sodium-ion-saturated ion exchanger can thus be restored to hydrogen-ion-saturated form with acid. It can also react with other cations, for example, according to (15.2).

The positive groups in *anion exchangers* are most often quaternary-substituted ammonium ions $-N(CH_3)_3^+$. These groups can bond hydroxide ions. A hydroxide-ion-saturated anion exchanger can be written ROH. If this reacts with a solution containing Cl^- the following reaction takes place:

$$ROH + Cl^- \rightleftharpoons RCl + OH^- \qquad (15.4)$$

Since most metal hydroxides are poorly soluble, however, it is rarely possible to exchange the anions in a solution with hydroxide ions. On the other hand, other anions can often be exchanged with each other. For example, a sulfate can be converted to a chloride by allowing the sulfate solution to flow through a layer of chloride-ion-saturated anion exchanger.

If a salt solution is first passed through a column of hydrogen-ion-saturated cation exchanger, the cations of the solution are replaced by hydrogen ions according to (15.3). Then the solution is passed through a column of hydroxide-ion-saturated anion exchanger, whereupon the anions in the solution are replaced by hydroxide ions (15.4), which with the hydrogen ions form water. Since the cations and anions in the original solution were equivalent, this is also true of the hydrogen and hydroxide ions. The result is practically salt-free (deionized) water, which can often be used to replace distilled water, and is considerably cheaper. In this case the cation and anion exchangers can be used mixed together, but then they must be separated before regeneration.

An ion exchanger is also used to recover ions from dilute solutions. For cationic recovery, the solution is passed through a hydrogen-ion-saturated cation exchanger (15.3). When this is regenerated with a small volume of acid a solution is obtained with perhaps one thousandth the volume of the initial solution.

Many common constituents of soils and natural sediments are ion exchangers, among minerals especially the clay minerals. Ion-exchange processes therefore play a large role in the properties of soil.

The way in which *uncharged molecules* are taken up by and bound to an ion exchanger depends to a large degree on their dipole properties and dimensions. Thus, for example, a certain ion exchanger can absorb and bind the polar molecules water and ammonia but not the nonpolar molecules benzene and tetrachloromethane. Unsaturated hydrocarbons are bound more strongly than saturated hydrocarbons. Molecules that are too large to enter into the channels of the ion exchanger are not bound at all. We thus have a *molecular sieve*, which can separate molecules according to their sizes. Smaller molecules, which can enter into the channels, are retained while the larger molecules pass over (the sieve effect is thus reversed). Molecular sieves, mostly consisting of artificial zeolites, have found important application in the separation of branched from unbranched hydrocarbons, the efficient drying of gases and for chromatographic separation of gas mixtures (15-4).

15-4. Chromatography

Chromatography is a method for separation of different substances in a liquid or gaseous solution, which is used for both analysis and purification. It is characterized by the passage of the solution over a phase with large area that in one way or another binds in varying degrees the different solution components. The solution thus constitutes a moving phase while the binding phase is usually stationary. The large area of the latter is achieved, as a rule, by its being very finely divided or highly porous. It is often packed in a vertical tube which is then called a *column*.

Use is often made of the different adsorptivity of the components of the solution by an adsorbent, which then constitutes the stationary phase (*adsorption chromatography*). This may consist, for example, of Al_2O_3, MgO, silica gel, charcoal or cellulose. A not too large volume of the solution is introduced into the upper part of the adsorption column. The different substances are adsorbed with varying ease and therefore reach different distances down in the column. Then a pure solvent (the original solvent or another one) is allowed to flow through the column, whereupon the adsorbed substances are again released (*eluted*) and carried down through the column by the solvent. Those that are adsorbed most weakly move most rapidly and those that are adsorbed most strongly move most slowly. When the difference in adsorptivity is sufficiently great, the different substances finally appear in separate layers. This can easily be observed if the substances are colored (the method was originated for such cases and is therefore still called chromatography). Otherwise special methods may be used, for example, color reactions, which make the layers visible. Instead of eluting with a pure solvent a substance may also be added to it that is adsorbed more strongly than the previous substances, and therefore *displaces* them.

By dismantling and slicing up the column the different layers can be separated. However, it is more convenient to allow the elution to continue until one substance after another flows out of the tube.

For qualitative chromatographic analysis cellulose is often used in the form of filter paper. The liquids are allowed to flow into it and the different substances are collected at different places on the paper (*paper chromatography*).

A chromatographic column may also consist of an ion exchanger. One then speaks of *ion-exchange chromatography* but the technique is entirely analogous to that just described. The elution often is accomplished with a complex-forming reagent. Those ions with which this forms the strongest complexes are eluted most easily and migrate most rapidly downward. Zirconium and hafnium can be separated in this way by eluting with fluoride ions, which form fluoro complexes. The separation of rare earth metals with the same oxidation number, which is very difficult by other methods, can be carried out by elution with solutions of citrate or EDTA. Ion-exchange chromatography is also much used in nuclear-chemical work. Radioactive substances can be separated and identified conveniently and in very small amounts by constantly following the radioactivity of the solution that runs off during elution.

The different distribution coefficients (14h) of the solution components in distribution between two solvents can also be used (*partition chromatography*). The column is packed with a finely divided solid phase that will adsorb an appropriate solvent that is not appreciably miscible with the solvent of the solution whose components are to be separated. The adsorbed solvent thus forms a nearly stationary layer on the surface of the solid phase. When the solution then flows through the column a distribution of the solution components is set up between the stationary and the moving solvents. Paper chromatography can also be carried out as partititon chromatography.

In a very important form of partition chromatography the moving solution is gaseous (*gas chromatography*). The gaseous components that are to be separated are diluted with a carrier gas, usually hydrogen, helium or nitrogen. They are bound in varying strengths to the stationary solvent layer and are then eluted with pure carrier gas. When they leave the column they are identified in various ways, for example by the change they cause in the thermal conductivity of the gas, or the ionization they produce when they are passed over a flame. Gas chromatography, especially for preparative purposes, is carried out also with molecular sieves (15-3).

15-5. Colloids

If, in a system consisting of a continuous phase (dispersion medium) containing dispersed particles of another phase, the particle size is continually reduced, the system undergoes no sudden changes in properties. However, it is common to give such systems different designations according to the particle size. The boundaries between the different types are very diffuse. If the particle size is greater than approximately 10^{-7} m ($=0.1$ μm), the system is called *coarsely dispersed*; if it falls within the range 10^{-7}–10^{-9} m, the system is called *colloidally dispersed*; and if it is less than approximately 10^{-9} m ($=10$ Å) one speaks customarily of an *ordinary* or *true solution*. The colloidally dispersed phase is said to occur as a *colloid* (Gk. κόλλα, glue; the name was given by Graham, who in the 1860's made the first investigations of colloidally dispersed systems). Practically every substance can by proper methods be obtained in colloidal state, that is, as a dispersed phase with a particle size within the range referred to. The nature of the particles is unimportant; they may consist of single crystals, crystal aggregates, amorphous solid bodies, liquid drops or large molecules.

In order to illustrate the typical properties of colloidally dispersed systems we discuss here the continuous changes in properties that occur with decrease in particle size. The dispersion medium is assumed at the start to be a liquid or a gas.

By constant collisions between the particles of the system, these exchange kinetic energy so that at a uniform temperature all particles, independent of size, have the same kinetic energy (cf. 2-1 b). For sufficiently large particles gravity overrides the kinetic energy of the particles. The particles therefore sink downward in the system (*sedimentation*) if they have higher density than the dispersion medium, and rise upward if they have lower density. With decrease in the mass of the particles, gravity has less and less effect. Sedimentation finally takes place only slowly, and we then often speak of *suspensions* when the particles are solid and *emulsions* when they are liquid. As a rule the particles here are not visible to the naked eye.

When the particle size falls to the order of 10^{-7} m, gravity no longer prevails over the kinetic energy of the particles. The earth's gravity force field thus does not suffice for sedimentation, and this can then only be obtained in artificial gravity force fields (*ultracentrifugation*, developed primarily by The Svedberg, who, among other things, determined particle masses by measurements of sedimentation rate in an ultracentrifuge field).

When particle sizes are less than 10^{-7} m the resolution of the ordinary microscope is insufficient and the heterogeneity of the system (cf. 1-1 b) thus can no longer be demonstrated in this way. As long as the particles are not too small, however, the heterogeneity often causes the path of a strong light beam through the system to become visible, especially against a dark background (Tyndall effect). The Tyndall effect results from the fact that the incident light is diffracted by the particles and thus scattered diffusely to the sides. The light scattered by individual particles can often be observed (without being able to see the size and shape of the particle) in a microscope (*ultramicroscope*) in which the light beam is observed from the side. For direct observation of particles less than about 10^{-7} m the electron microscope is required.

The adsorption of molecules and ions from the dispersion medium on the surface of the particles becomes more and more important as the particle size decreases. For uniform particles the ratio (area)/(mass) is inversely proportional to the diameter of the particle. If it is assumed for the sake of simplicity that the amount adsorbed per unit of area is independent of the particle size, the ratio (amount adsorbed)/(particle mass) is also inversely proportional to the diameter of the particle. If the particles preferentially adsorb ions of the same kind or at least ions with the same charge sign, they acquire a charge in relation to the main body of the dispersion medium. The ratio (charge)/(particle mass) = the *specific charge* of the particle should thus also become inversely proportional to the diameter of the particle. For particle sizes of 10^{-7} m and below the specific charge can become appreciable. In an electric field the particles then move parallel to the lines of the field through the dispersion medium, and from the direction of motion the charge sign can be determined (*electrophoresis*).

Colloidal particles in liquid or gaseous dispersion media can pass through ordinary filters, which retain coarser particles. On the other hand many membranes (cellophane and plastic films, biological membranes) are too dense to permit passage of colloidal particles, but let through smaller molecules and ions. If a solution containing both these types of particles is separated from pure solvent by such a membrane, the small particles diffuse out through the membrane while the colloidal particles remain behind (*dialysis*). Dialysis is often used for clearing smaller particles from a colloidal solution.

Colloidal solutions are also called *sols*. Depending on the dispersion medium, there are, for example, hydrosols, alcosols and aerosols (smoke and mist are often aerosols). Even if the dispersion medium is a solid phase, one speaks of a colloidally dispersed system when the particle size of the dispersed phase lies between the limits mentioned earlier. However, it is clear that the typical, colloidal properties described above are lacking in solid systems. Neither does most of the following discussion apply to solid systems.

In 6-5b the strong cohesive forces that characterize small particles, and the resulting tendency to particle growth, was discussed. If this particle growth occurs in a colloidal solution, larger particles or particle aggregates settle out; the colloid is *precipitated* or *coagulated*. The fact that a colloidal solution can remain stable and not coagulate depends very often on the charge of the particles. All particles of the same kind through similar adsorption have acquired charges of the same sign and therefore repel each other. If the charge is neutralized, the colloid as a rule coagulates. The neutralization can be brought about by adsorbtion of ions of charge opposite to that of the particles, and this nearly always occurs if electrolytes in sufficient amounts are added to the colloidal solution. The added anions are effective with positively charged colloidal particles and cations are effective with negatively charged particles. This is apparent from the fact that the magnitude of the charge of the oppositely charged ion strongly influences the coagulation effect. For precipitation of an As_2S_3 sol in which the particles are negatively charged because of adsorption of HS^- and S^{2-} ions, for example, NaCl, $MgCl_2$ and $AlCl_3$ are required in concentrations of 51, 0.72 and 0.093 millimol l^{-1}, respectively.

By extensive dialysis also, the ionic layer on the particles can be so much reduced that precipitation occurs.

Colloids are often divided into *lyophilic* and *lyophobic* colloids, although no sharp boundary can be drawn between these two types. Lyophilic colloidal particles easily unite with the molecules of the dispersion medium (if the dispersion medium is water one speaks of *hydrophilic* colloids) and surround themselves with an envelope of these molecules. Solutions of lyophilic colloids can often be made very concentrated and with sufficiently high concentration frequently form a gelatinous mass, a *gel* (cf. a gelatine solution). They are also relatively insensitive to the addition of electrolytes. Lyophobic colloids unite only to a small degree with molecules of the dispersion medium (if the dispersion medium is water the term *hydrophobic* colloid is often used). The solutions can be obtained only relatively dilute (at most about 1 %) and are more fluid than lyophilic solutions of corresponding concentration. Lyophobic colloids are easily precipitated by addition of electrolytes.

The greater stability of lyophilic colloids depends on the envelope of solvent molecules that surrounds each particle and makes contact between the particles more difficult. If the envelope is removed, these colloids also become very sensitive to changes in charge. Protein molecules surround themselves with an envelope of water and therefore give stable solutions of colloidal nature. If the water shield is removed by the addition of alcohol, the stability is markedly reduced.

Certain substances, when added in sufficient amounts to a lyophobic colloid, can provide its particles with an envelope that interferes with precipitation. Such *protective colloids* are often themselves lyophilic colloids (for example, proteins).

Not infrequently a coagulated colloid can be brought into solution again, by *peptization*. This may occur on washing, which mainly removes the coagulating ionic species, so that the particles regain their charges. Addition of other ions that are adsorbed and thus give the particles new charges, or of substances of the protective colloid type, may also have a peptizing effect.

Colloids are common in biological systems and are then always lyophilic. Lyophobic colloids are too unstable to exist in such systems.

In order to obtain a substance in colloidally dispersed form, above all a dispersion medium must be chosen in which the substance is rather poorly soluble. In such a medium colloidal particles may be either built up (*condensed*) from more elementary constituents (atoms, molecules, ions), or split off (*dispersed*) from larger aggregates. The former often occurs on the precipitation of poorly soluble substances; the latter can often be effected by extremely fine grinding or by solution of a solid substance. Metals can often be obtained in colloidal form by striking an electric arc between electrodes of the metal immersed in the dispersion medium. In this case metal vapor formed in the arc is condensed.

In ordinary inorganic work colloids are often formed by precipitation of difficultly soluble substances. One condition for this process is that the particle growth should cease while the particles have colloidal dimensions. This occurs if the number of nuclei is so large in relation to the original concentration that the solution becomes impoverished in the dissolved substance before colloidal dimensions are exceeded. Supersaturation should thus be great.

Inorganic colloids are most often hydrophobic in water solution. The sign of the charge depends both on the nature of the colloidal particles and on the ionic contents of the dispersion medium. Colloidal particles of AgI are positively charged by Ag^+ ions in $AgNO_3$ solution, and negatively charged by I^- ions in KI solution (15-1a). If the solution contains equivalent amounts of Ag^+ and I^- ions, the particles have a zero net charge, which leads quickly to coagulation. Particles of acids give off protons and thus become negatively charged. In hydrosols of noble metals such as Ag, Au and platinum metals, the particles are commonly negative. Particles of less noble metals are for the most part oxidized in water and often become positive, as do particles of many metal oxides and metal hydroxides.

If the colloidal particles are protolytes, the magnitude of the charge becomes dependent on the acidity of the solution (15-1a). If the particles consist of an ampholyte (12-4), providing only protons are exchanged with the solution, they become uncharged when the pH value is equal to the isoelectric point of the ampholyte. In this solution the ampholyte molecules on the surface of the particles are protolyzed equally much as acid and as base, so that the total charge becomes zero. If the acidity is greater than at the isoelectric point, the base protolysis prevails and positive ions appear on the particle surfaces. If the acidity is less, the acid protolysis predominates and negative ions appear. Since other ions may also be adsorbed, however, the acidity corresponding to zero particle charge may deviate from the isoelectric point of the ampholyte. For lyophilic colloids, whose stability depends not only on the charge but also on the envelope of solvent molecules, the point of zero charge does not necessarily correspond to the stability minimum.

CHAPTER 16

Redox Equilibria and Electrochemistry

a. Oxidation and reduction. By *oxidation* was originally meant formation of an oxide, that is, absorption of oxygen, and by *reduction* the inverse process, the giving off of oxygen. An ordinary *combustion* in air or oxygen thus would involve oxidation.

When an oxygen atom is bound to any other atom except fluorine, the higher electronegativity of the oxygen leads to a displacement of electrons toward the oxygen atom. One or more atoms in the original substance therefore acquire an increased oxidation number (5-3g). Examples: the oxidation of metallic copper to copper(I) oxide, $2\,Cu^0 \to (Cu^I)_2O$; and the continued oxidation of copper(I) oxide to copper(II) oxide, $(Cu^I)_2O \to 2Cu^{II}O$. But other atoms with higher electronegativity than copper also cause analogous electronic displacements and therefore an increase in the oxidation number of the copper. If chlorine atoms are taken up, for example, we have $Cu^0 \to Cu^ICl \to Cu^{II}Cl_2$. This has led to a generalization of the concept of oxidation to: *oxidation of an atom involves an increase in its oxidation number*. Conversely: *reduction of an atom involves a decrease in its oxidation number*.

Since the basic process is an electron displacement, an oxidation and a reduction must always occur together. If magnesium is oxidized in oxygen gas, the reaction is $Mg^0 + \frac{1}{2}(O^0)_2 \to Mg^{+II}O^{-II}$. The oxidation number of oxygen has thus decreased, that is, the oxygen has been reduced.

The difference in electronegativity between Mg and O is so great that MgO can be assumed to be built up of Mg^{2+} and O^{2-} ions. On the other hand, between carbon and oxygen, for example, the electronegativity difference is so small that when carbon is oxidized according to the reaction $C^0 + (O^0)_2 \to C^{+IV}(O^{-II})_2$, the bonds are mainly covalent and thus we cannot speak of the formation of C^{4+} and O^{2-} ions. However, the electronegativity difference is sufficiently great so that there is no doubt concerning the oxidation numbers as they are written in the reaction equation.

When the electron transfer from one atom to another has become so nearly complete that it may be said that one atom gives off and the other takes up a certain number of electrons, the definitions just given may be expressed thus: *oxidation involves a giving off of electrons*; and *reduction involves a taking up of electrons*. The complete process, in fact, is often divided up into partial processes, involving respectively giving off of electrons and taking up of electrons. For the oxidation of magnesium we may write:

$$Mg \rightarrow Mg^{2+} + 2e^-$$
$$\tfrac{1}{2}O_2 + 2e^- \rightarrow O^{2-}$$
$$\overline{Mg + \tfrac{1}{2}O_2 \rightarrow Mg^{2+} + O^{2-}}$$

In the solution of zinc in an acid Zn is oxidized to Zn^{2+} and $2H^+$ are reduced to H_2:

$$Zn \rightarrow Zn^{2+} + 2e^-$$
$$2H^+ + 2e^- \rightarrow H_2$$
$$\overline{Zn + 2H^+ \rightarrow Zn^{2+} + H_2}$$

If such partial processes are written:

$$Na \rightleftharpoons Na^+ + e^-$$
$$Fe^{2+} \rightleftharpoons Fe^{3+} + e^-$$
$$2Cl^- \rightleftharpoons Cl_2 + 2e^-$$
$$Mn^{2+} + 4H_2O \rightleftharpoons MnO_4^- + 8H^+ + 5e^-$$

the direction to the right expresses an oxidation and the direction to the left a reduction. All molecules and ions on the left side are said to constitute the *reduced form* of the system and those on the right side the corresponding *oxidized form* of the system.

If a reduced form is indicated by "red" and an oxidized form by "ox", we have

$$red \rightleftharpoons ox + ne^- \tag{16.1}$$

The reduced form and the oxidized form constitute a *redox pair*. As redox pairs we have, for example, Fe^{2+}–Fe^{3+} and (Mn^{2+}, H_2O)–(MnO_4^-, H^+). All forms that can be converted into each other by taking up or giving off electrons can be considered together as a *redox system*. A redox system may contain several redox pairs; for example, the redox system Fe–Fe^{2+}–Fe^{3+}.

From the above discussion it follows that an oxidation process in a redox pair must always be accompanied by a reduction process in another redox pair. At least two redox pairs thus must always take part in the total process, the *redox process*. At the beginning of the process at least, the reaction direction in the first pair corresponds to an oxidation, $red_1 \rightarrow ox_1 + n_1 e^-$, and in the second pair to a reduction, $ox_2 + n_2 e^- \rightarrow red_2$. The equation for the redox process is obtained if these expressions are multiplied by n_2 and n_1 respectively, and then combined:

$$n_2 red_1 \rightarrow n_2 ox_1 + n_1 n_2 e^-$$
$$n_1 ox_2 + n_1 n_2 e^- \rightarrow n_1 red_2$$
$$\overline{n_2 red_1 + n_1 ox_2 \rightleftharpoons n_2 ox_1 + n_1 red_2} \tag{16.2}$$

The two processes will finally balance each other. (16.2) therefore expresses a *redox equilibrium*. This can be considered formally in the same way as other transfer or exchange equilibria (10c). The formal similarity to proton transfer in acid-base equilibria (12-1a) is very striking.

The tendency to give off electrons may be extremely different for different reduced forms. It is also clear that the greater the tendency for a reduced form to give up electrons, the less must be the tendency for the corresponding oxidized form to take up electrons.

In the derivation of the equation for a redox process, at least in more complicated cases it is proper first to derive the two partial processes and then combine these so that the numbers of electrons given off and taken up are equal.

We may assume that the principal initial and final substances are known. For example, the equation is desired for the oxidation of Fe^{2+} to Fe^{3+} by dichromate ion $Cr_2O_7^{2-}$ and it is known that the latter is reduced to Cr^{3+}. First, the complete equation is derived for the partial process $Cr_2O_7^{2-} \to Cr^{3+}$. In this equation all the kinds of atomic species must be equal in number and the sums of the charges equal on both sides. The numbers of atoms are made equal by choice of appropriate coefficients and, as usual for processes in water solution, by the proper addition of water and its ions (H_2O, H^+ or OH^-). Finally the charge sums are made equal by addition of electrons. The procedure is:

$$\begin{array}{ll} \text{Cr is balanced:} & Cr_2O_7^{2-} \to 2Cr^{3+} \\ \text{O is balanced:} & \phantom{Cr_2O_7^{2-} \to 2Cr^{3+}} + 7H_2O \\ \text{H is balanced:} & +14H^+ \\ \text{Charges are balanced:} & +6e^- \\ \hline & Cr_2O_7^{2-} + 14H^+ + 6e^- \to 2Cr^{3+} + 7H_2O \end{array}$$

The reaction for the process $Fe^{2+} \to Fe^{3+}$ may be written directly, $Fe^{2+} \to Fe^{3+} + e^-$. This must be multiplied by six to give the number of electrons that are consumed by the first partial process. Then, addition of the two partial processes gives the total process: $Cr_2O_7^{2-} + 6Fe^{2+} + 14H^+ \to 2Cr^{3+} + 6Fe^{3+} + 7H_2O$.

The final equation should be checked to see that the numbers of atoms and charges are equal on both sides. The equation obtained above shows that the process to the right uses up hydrogen ions and thus is favored if the solution is acid.

If H^+ appears as part of a reaction while OH^- is preferred (for example, if the medium must be basic), enough OH^- may be added to both sides of the equation to use up all H^+ according to $H^+ + OH^- \to H_2O$. If OH^- appears at first when H^+ is preferred, H^+ is added in a corresponding way.

b. Electrode and cell processes. Let us imagine an *electrochemical cell* consisting of a solution and two electrodes immersed in it, which are connected through an external circuit. Current can pass through the cell and the external circuit, in the process of which the current transport in the solution is accomplished by ions, and in the electrodes and external circuit by electrons. The electrodes are thus electronic conductors (usually metals but sometimes semiconductors), which carry electrons between the solution and the external circuit. The current may be generated within the cell itself (which is then a *galvanic cell*) or by a source in the external circuit (the cell is then an *electrolytic cell*). In both cases the exchange of free electrons at the electrodes is related to oxidation and reduction processes at each of them (*electrode processes*). At the electrode where electrons are supplied from the external circuit, a reduction of some oxidized form takes place. At the electrode where electrons are carried away to the external circuit, an oxidation of some reduced form takes place. During a given time an equal number of electrons must be supplied at one electrode as is carried away at the other, so that electrochemically equivalent amounts of substances are converted in both electrode processes.

In every cell a potential difference appears, and it is this which gives rise to the current when the cell itself is the current producer. If the cell contains only one solution, only the two potential differences between each of the electrode metals and the solution need be considered. These *electrode potentials* make up the electromotive force (emf) of the cell.

It also often happens that the two electrode metals are immersed each in its own solution, so that two "half cells" are obtained. The two solutions are connected to each other in some manner that prevents their becoming mixed. In the boundary surface between the liquid phases *liquid potentials* appear, and when the cell is used for emf measurements efforts are made to make these as small as possible (the two cells are then often connected by a salt solution, a "salt bridge"). In other cases the two solutions are in contact with each other through a porous wall. Incidental liquid potentials are neglected in the following discussion.

An electrode potential can be most simply understood as a consequence of the exchange of electrons in the electrode process. In one type of electrode process the electrode consists of a nonreacting (inert) material immersed in a solution containing a redox pair. If this pair is, for example, $Fe^{2+} \rightleftharpoons Fe^{3+} + e^-$, Fe^{2+} and Fe^{3+} ions will constantly collide with the electrode. The Fe^{3+} ions can thereby take up electrons and be converted to Fe^{2+} ions, and the Fe^{2+} ions can give up electrons and be converted to Fe^{3+} ions. When equilibrium is attained, the electrodes generally have a net charge—positive or negative—in relation to the solution; an electrode potential has been produced. This must be a function partly of the tendency of the oxidized form to take up electrons (or, what is equivalent, the corresponding tendency of the reduced form to give up electrons), and partly of the ratio between the number of collisions per unit of time of the ions of the two forms with the electrode surface. This ratio must be at least approximately equal to the ratio between the concentrations of the two forms. An electrode of this type is called a *redox electrode*. Since all electrode potentials result from electron transfers and thus redox processes, however, any electrode is, in fact, a redox electrode.

Sometimes one of the forms of the redox pair is a gas which is bubbled over the surface of the inert electrode metal. If the gas is hydrogen we have a *hydrogen electrode* characterized by the redox pair $H_2(g) \rightleftharpoons 2H^+ + 2e^-$ (platinum is used as electrode metal, which, in order to increase the surface area, is electrolytically platinized with very finely divided platinum, "platinum black"). If the gas is chlorine we have a chlorine electrode characterized by the redox pair $2Cl^- \rightleftharpoons Cl_2(g) + 2e^-$ (here an alloy of platinum and iridium resistant to chlorine is used as electrode metal).

If the electrode is not inert, it may itself be the reduced form in the redox pair that characterizes the electrode, and then take part in the redox process. For example, if a zinc rod is immersed in a solution containing Zn^{2+} ions, the redox pair is $Zn(s) \rightleftharpoons Zn^{2+} + 2e^-$. Then, on the one hand, Zn^{2+} ions can take up electrons from the rod and be converted to Zn atoms attached to the surface of the rod, and on the other hand, Zn atoms on the rod surface can give up electrons to the rod and thus become Zn^{2+} ions, which move out into the solution.

Let us now consider a cell consisting of a zinc electrode and a chlorine electrode in a water solution of $ZnCl_2$. The discussion will apply both to the case where this cell is a current source, that is, a galvanic cell, and the case where current from an external current source is passed in the opposite direction through the cell, which is then an electrolytic cell. The electrode processes must be as follows: at the zinc electrode, $Zn(s) \rightleftharpoons Zn^{2+} + 2e^-$; at the chlorine electrode, $2Cl^- \rightleftharpoons Cl_2(g) + 2e^-$. These are combined into the cell process, $Zn(s) + Cl_2(g) \rightleftharpoons Zn^{2+} + 2Cl^-$, in which one arrow applies to the case of the galvanic cell and the other arrow to the case of the electrolytic cell. However, without further information, it cannot be said which direction applies to a particular case. In a galvanic cell that process always takes place which would go spontaneously (toward equilibrium) if all the substances of the cell system were brought together; in the cell these substances are distributed among the two separate electrode systems, but they react in the same way as soon as electron exchange between the electrodes can occur through the external circuit. Now, it is known that the equilibrium position for the total process given above is displaced far to the right (zinc and chlorine react spontaneously with the formation of $ZnCl_2$). The cell system must seek the same equilibrium position, and the electrical energy that the cell gives off in the production of current corresponds to a greater or lesser degree (see further below) to the decrease in Gibbs free energy G of the system as it goes toward the equilibrium position. The cell process in the galvanic cell thus goes in the direction $\rightarrow$, so that the directions of the electrode processes for this case and the pole signs are also given. In the example we then have:

Galvanic cell

electrode process at the minus pole $\qquad$ $Zn(s) \rightarrow Zn^{2+} + 2e^-$
electrode process at the plus pole $\qquad$ $Cl_2(g) + 2e^- \rightarrow 2Cl^-$
cell process $\qquad\qquad\qquad\qquad$ $\overline{Zn(s) + Cl_2(g) \rightarrow Zn^{2+} + 2Cl^-}$

When the cell operates as a galvanic cell, clearly oxidation always occurs at the minus pole and reduction at the plus pole. If an actual galvanic cell is available, the current direction (pole sign) can be determined and from this the directions of the cell processes decided directly. If neither the direction of the spontaneous reaction nor the current direction of the galvanic cell are known, the two electrode potentials can be calculated, as will be shown later, which immediately gives the current direction.

If current from an external source is passed through the cell in the direction opposite to that when it functions as a galvanic cell, all reaction directions are reversed. The processes thus become:

Electrolytic cell

cathode process	$Zn^{2+} + 2e^- \rightarrow Zn(s)$
anode process	$2Cl^- \rightarrow Cl_2(g) + 2e^-$
cell process	$Zn^{2+} + 2Cl^- \rightarrow Zn(s) + Cl_2(g)$

The minus pole of the galvanic cell becomes the cathode of the electrolytic cell and the plus pole becomes the anode. The electrode process at a cathode is always a reduction, and at an anode always an oxidation. Often the electrode of a galvanic cell where reduction occurs, that is, the plus pole, is also called the cathode, and the electrode where oxidation occurs, that is, the minus pole, is called the anode. Thus, in both galvanic and electrolytic cells the anode is the electrode through which *positive* current flows to the electrolyte.

A cell is often described by a cell scheme that contains the formulas of the constituent substances in the order in which the current passes through them. Each phase boundary at a liquid solution is indicated by a vertical bar |. A salt bridge, whose composition is not given, is therefore indicated by a double bar ||. According to international agreement E for a cell is equal to the potential of the electrode metal shown at the right in the scheme, minus the potential of the electrode metal shown at the left. The cell scheme and E value

$$Zn \mid Zn^{2+}, aq, 1M \parallel Cu^{2+}, aq, 1M \mid Cu \qquad E = 1.10 \text{ V}$$

thus show that the Cu electrode is the plus pole. But the scheme can also be written

$$Cu \mid Cu^{2+}, aq, 1M \parallel Zn^{2+}, aq, 1M \mid Zn \qquad E = -1.10 \text{ V}$$

For a galvanic cell with known current direction it is nowadays customary to write the *plus pole at the right* in the scheme, so that E is positive. This implies that *inside* the cell the (positive) current moves from left to right. All negative charges (electrons and negative ions) thus move inside the cell from right to left. To avoid misunderstanding, the pole signs are sometimes indicated.

If the emf of a cell is balanced by an emf in the external circuit, no changes take place in the cell system. If this situation corresponds to a true equilibrium, the current direction and thus the direction of the cell process are reversed by an infinitely small change in external emf. The cell is then said to be *reversible*. The current through the cell is in this case infinitely small. With finite current strength no cell is strictly reversible, but many cells can be considered to operate reversibly with the passage of very small currents. In certain cells no equilibrium can ever be realized, and they can only operate *irreversibly*.

When a galvanic cell operates reversibly, the electrical energy obtained is equal to the decrease in Gibbs free energy G (2-3b) of the system in the cell process.

For each electrochemical equivalent (1-4d) of the amount of substances converted in the cell, an amount of electricity of 1 faraday ($1\,F$) passes through the cell. If n equivalents are converted, thus the amount of electricity nF is passed through. For a cell emf E, this amount of electricity corresponds to the energy nFE. If the cell operates reversibly, this electrical energy is equal to the decrease in Gibbs free energy, that is, $-\Delta G$. Thus we have

$$-\Delta G = nFE \qquad (16.3)$$

ΔG for the cell process can thus be calculated from the emf of the cell when it operates reversibly. This constitutes a very important method for obtaining this quantity.

If the cell process has the general form (10.8), and we set $-\Delta G^0 = nFE^0$, where E^0 is the emf of the cell if all the participating substances are in standard states, on insertion of the expressions for ΔG and ΔG^0 in (10.11) we obtain

$$E = E^0 - \frac{RT}{nF} \ln \frac{a_L^l \, a_M^m \cdots}{a_A^a \, a_B^b \cdots} \qquad (16.4)$$

In the application of this expression the rules for the activities of the different substances given in 10a should be observed. In particular, the activity of a pure, solid or liquid substance $= 1$. In dilute solutions we may often set $\{H_2O\} = 1$, and for a gas "i" with not too high partial pressure p_i we may set $a_i = p_i$.

For the cell with zinc and chlorine electrodes in $ZnCl_2$ solution, after inserting the constant values ($R = 8.3143$ J deg^{-1}mol^{-1}, $F = 96\,487$ C equiv^{-1}, with E expressed in V) and transforming to common logarithms, at $T = 298.15°$K $(=25°C)$ we have

$$E = E^0 - \frac{0.05916}{2} \log \frac{\{Zn^{2+}\}\{Cl^-\}^2}{p_{Cl_2}} \qquad (16.5)$$

If a cell operates irreversibly, the whole amount of energy ΔG cannot be obtained as electrical energy. It is clear that heat is developed as soon as the current strength is not infinitely small. Inhibited reactions also often lead to irreversibility. Finally, changes in the cell always occur with passage of current when the current density is sufficiently large. These cause an emf that opposes the applied external emf—a *counterelectromotive force*. It is said that the original cell is *polarized*, and it therefore operates irreversibly. Polarization may be caused by many different kinds of changes. The electrodes may be changed by the formation of completely new substances through the action of the current—*chemical polarization*. But even if this does not occur, with sufficiently high current density the normal constituents of the electrolyte may be formed or used up in the vicinity of the electrodes faster than they can be carried away or brought in by diffusion or convection. In this way changes in concentration occur at the electrodes, which lead to *concentration polarization*.

c. Electrode potential; redox potential.

The above discussion concerning the origin of an electrode potential shows that a redox pair as a whole has an oxidizing or reducing capacity that depends partly on the nature of the pair, and partly—at least approximately—on the ratio of the concentrations of the oxidized and reduced forms. The greater the oxidizing power of the pair is, the greater is the net number of electrons it has taken up from the electrode when equilibrium is reached, and the more positive the potential of the electrode then becomes in relation to the solution. *The electrode potential is defined as the potential of the electrode metal minus the potential of the solution.* The electrode potential thus increases with increase in the oxidizing power of the redox pair and can therefore be used as a measure of this power.

The electrode potential of an electrode consisting of an inert metal immersed in a solution containing one or more redox pairs is called the *redox potential* of the solution. The redox potential becomes a measure of the oxidizing power of the solution. Strongly oxidizing solutions have a high tendency to take up electrons from the electrode and have a high redox potential, while strongly reducing solutions have a high tendency to give up electrons to the electrode and have a low redox potential. Also, if the redox system contains several phases, one speaks of the redox potential of the system. If any of the phases is a metal, the redox potential is obviously equal to its electrode potential.

It is of great fundamental importance that the value of a single electrode potential cannot be measured, but only the emf of a galvanic element. After correction for possible liquid potentials, the emf of the element is equal to the difference in the potentials of the two electrodes. We are forced, therefore, to assign a certain electrode as a *standard electrode* with an agreed electrode potential. The *standard hydrogen electrode* as defined below has been chosen and its electrode potential is set $=0$, independently of temperature. If a cell contains a standard hydrogen electrode, then the other electrode potential is equal to its potential relative to that of the standard hydrogen electrode. The numerical value of the potential is equal to the emf of the cell (after the subtraction of liquid potentials) and its sign is $+$ or $-$ according to whether it is higher or lower than that of the standard hydrogen electrode. Any such relative potential is called without further qualification an electrode potential (sometimes redox potential).

In practice the potential of an electrode usually is not compared directly with a standard hydrogen electrode, but with some other more convenient electrode whose potential relative to the standard hydrogen electrode is already known.

A redox pair according to the general scheme

$$pP + qQ + \ldots \rightleftharpoons xX + yY + \ldots + ne^- \tag{16.6}$$

gives the electrode potential

$$e = e^0 + \frac{RT}{nF} \ln \frac{a_X^x a_Y^y \cdots}{a_P^p a_Q^q \cdots} \tag{16.7}$$

which at $T = 298°K$ may be written

$$e = e^0 + \frac{0.059}{n} \log \frac{a_X^x a_Y^y \cdots}{a_P^p a_Q^q \cdots} \qquad (16.8)$$

If the activities of all participating substances $= 1$, the last term in these expressions becomes $= 0$ and thus $e = e^0$. e^0 is called the *standard potential* of the pair and, as every electrode potential, is referred to the standard hydrogen electrode as a standard.

The validity of (16.7) can be seen by comparison with (16.4). Let us consider, for example, the simple cell process $A + D \rightleftharpoons B + C$, composed of the electrode processes $A \rightleftharpoons B + e^-$ and $D + e^- \rightleftharpoons C$. For the cell process, (16.4) gives

$$E = E^0 - \frac{RT}{F} \ln \frac{\{B\}\{C\}}{\{A\}\{D\}}$$

For the electrode processes, (16.7) gives $e_1 = e_1^0 + \dfrac{RT}{F} \ln \dfrac{\{B\}}{\{A\}}$ and $e_2 = e_2^0 + \dfrac{RT}{F} \ln \dfrac{\{D\}}{\{C\}}$, from which by subtraction we get

$$(e_2 - e_1) = (e_2^0 - e_1^0) - \frac{RT}{F} \ln \frac{\{B\}\{C\}}{\{A\}\{D\}}$$

The expression (16.4) presents a very important method for determining activities by measurement of the emf of a cell. Since at least two kinds of ions with opposite charges must contribute to a cell process, however, (16.4) can never be used for the determination of the activity of a single kind of ion. On the other hand, individual ionic types can participate in an electrode process and it would therefore be supposed that (16.7) could be used for the determination of ionic activities. However, any such measurement must be carried out in such a way that liquid potentials are involved. The values of these are always subject to some uncertainty, which therefore also jeopardizes all ionic activity values. This confirms the discussion in 11d.

If the transformation from reduced form to oxidized form involves only a change in charge of the same atom or atomic group, for example, $Tl^+ \rightleftharpoons Tl^{3+} + 2e^-$, (16.8) has the form

$$e = e^0 + \frac{0.059}{n} \log \frac{a_{ox}}{a_{red}} \qquad (16.9)$$

Here $e = e^0$ when $a_{ox} = a_{red}$.

The potential of a hydrogen electrode is determined by the redox pair $H_2(g) \rightleftharpoons 2H^+ + 2e^-$. The standard hydrogen electrode is defined as a hydrogen electrode with $\{H^+\} = 1$ and $p_{H_2} = 1$. It may be conceived as an inert metal (for example, Pt) immersed in a solution with hydrogen ion activity $= 1$ and swept over with hydrogen at a pressure of 1 atm.

It has been mentioned above that the definition of other electrode potentials is based on the choice of $e = 0$ for the standard hydrogen electrode. From the

Table 16.1. Standard potentials at 25°C.

The states of the substances are in general evident and are therefore not indicated. All ions are assumed to be in water solution.

red	ox	e^0, V
$Li \rightleftharpoons Li^+ + e^-$		-3.045
$K \rightleftharpoons K^+ + e^-$		-2.925
$Rb \rightleftharpoons Rb^+ + e^-$		-2.925
$Cs \rightleftharpoons Cs^+ + e^-$		-2.923
$Ba \rightleftharpoons Ba^{2+} + 2e^-$		-2.90
$Sr \rightleftharpoons Sr^{2+} + 2e^-$		-2.89
$Ca \rightleftharpoons Ca^{2+} + 2e^-$		-2.87
$Na \rightleftharpoons Na^+ + e^-$		-2.714
$La \rightleftharpoons La^{3+} + 3e^-$		-2.52
$Mg \rightleftharpoons Mg^{2+} + 2e^-$		-2.37
$Lu \rightleftharpoons Lu^{3+} + 3e^-$		-2.25
$Sc \rightleftharpoons Sc^{3+} + 3e^-$		-2.08
$Be \rightleftharpoons Be^{2+} + 2e^-$		-1.85
$Al \rightleftharpoons Al^{3+} + 3e^-$		-1.66
$Ti \rightleftharpoons Ti^{2+} + 2e^-$		-1.63
$Mn \rightleftharpoons Mn^{2+} + 2e^-$		-1.18
$H_2 + 2OH^- \rightleftharpoons 2H_2O + 2e^-$		-0.828
$Zn \rightleftharpoons Zn^{2+} + 2e^-$		-0.763
$Cr \rightleftharpoons Cr^{3+} + 3e^-$		-0.74
$S^{2-} \rightleftharpoons S + 2e^-$		-0.48
$Fe \rightleftharpoons Fe^{2+} + 2e^-$		-0.440
$Cr^{2+} \rightleftharpoons Cr^{3+} + e^-$		-0.41
$Cd \rightleftharpoons Cd^{2+} + 2e^-$		-0.403
$Co \rightleftharpoons Co^{2+} + 2e^-$		-0.277
$Ni \rightleftharpoons Ni^{2+} + 2e^-$		-0.250
$Sn \rightleftharpoons Sn^{2+} + 2e^-$		-0.136
$Pb \rightleftharpoons Pb^{2+} + 2e^-$		-0.126
$H_2 \rightleftharpoons 2H^+ + 2e^-$		0.0000
$Cu^+ \rightleftharpoons Cu^{2+} + e^-$		0.153
$H_2SO_3 + H_2O \rightleftharpoons SO_4^{2-} + 4H^+ + 2e^-$		0.17
$I^- + 6OH^- \rightleftharpoons IO_3^- + 3H_2O + 6e^-$		0.26
$ClO_2^- + 2OH^- \rightleftharpoons ClO_3^- + H_2O + 2e^-$		0.33
$Cu \rightleftharpoons Cu^{2+} + 2e^-$		0.337
$Fe(CN)_6^{4-} \rightleftharpoons Fe(CN)_6^{3-} + e^-$		0.36
$ClO_3^- + 2OH^- \rightleftharpoons ClO_4^- + H_2O + 2e^-$		0.36
$Cl_2 + 4OH^- \rightleftharpoons 2ClO^- + 2H_2O + 2e^-$		0.40
$4OH^- \rightleftharpoons O_2 + 2H_2O + 4e^-$		0.401
$S + 3H_2O \rightleftharpoons H_2SO_3 + 4H^+ + 4e^-$		0.45
$Cu \rightleftharpoons Cu^+ + e^-$		0.521
$2I^- \rightleftharpoons I_2 + 2e^-$		0.5355
$MnO_4^{2-} \rightleftharpoons MnO_4^- + e^-$		0.564
$MnO_2 + 4OH^- \rightleftharpoons MnO_4^- + 2H_2O + 3e^-$		0.588
$Br^- + 6OH^- \rightleftharpoons BrO_3^- + 3H_2O + 6e^-$		0.61
$H_2O_2 \rightleftharpoons O_2 + 2H^+ + 2e^-$		0.682

Table 16.1. (*Cont.*)

red	ox	e^0, V
Fe^{2+}	$\rightleftarrows Fe^{3+} + e^-$	0.771
$2Hg$	$\rightleftarrows Hg_2^{2+} + 2e^-$	0.793
Ag	$\rightleftarrows Ag^+ + e^-$	0.7991
$NO_2 + H_2O$	$\rightleftarrows NO_3^- + 2H^+ + e^-$	0.81
Hg	$\rightleftarrows Hg^{2+} + 2e^-$	0.850
$Cl^- + 2OH^-$	$\rightleftarrows ClO^- + H_2O + 2e^-$	0.89
Hg_2^{2+}	$\rightleftarrows 2Hg^{2+} + 2e^-$	0.920
$NO + 2H_2O$	$\rightleftarrows NO_3^- + 4H^+ + 3e^-$	0.96
$2Br^-$	$\rightleftarrows Br_2(l) + 2e^-$	1.0652
$Au + 2Cl^-$	$\rightleftarrows AuCl_2^- + e^-$	1.15
$ClO_3^- + H_2O$	$\rightleftarrows ClO_4^- + 2H^+ + 2e^-$	1.19
$I_2 + 6H_2O$	$\rightleftarrows 2IO_3^- + 12H^+ + 10e^-$	1.195
Pt	$\rightleftarrows Pt^{2+} + 2e^-$	~1.2
$HClO_2 + H_2O$	$\rightleftarrows ClO_3^- + 3H^+ + 2e^-$	1.21
$2H_2O$	$\rightleftarrows O_2 + 4H^+ + 4e^-$	1.229
$Mn^{2+} + 2H_2O$	$\rightleftarrows MnO_2 + 4H^+ + 2e^-$	1.23
Tl^+	$\rightleftarrows Tl^{3+} + 2e^-$	1.25
$2Cr^{3+} + 7H_2O$	$\rightleftarrows Cr_2O_7^{2-} + 14H^+ + 6e^-$	1.33
$2Cl^-$	$\rightleftarrows Cl_2 + 2e^-$	1.3595
$Br^- + 3H_2O$	$\rightleftarrows BrO_3^- + 6H^+ + 6e^-$	1.44
$Pb^{2+} + 2H_2O$	$\rightleftarrows PbO_2 + 4H^+ + 2e^-$	1.455
Mn^{2+}	$\rightleftarrows Mn^{3+} + e^-$	~1.5
$Mn^{2+} + 4H_2O$	$\rightleftarrows MnO_4^- + 8H^+ + 5e^-$	1.51
$Cl_2 + 2H_2O$	$\rightleftarrows 2HClO + 2H^+ + 2e^-$	1.63
$MnO_2 + 2H_2O$	$\rightleftarrows MnO_4^- + 4H^+ + 3e^-$	1.692
$2H_2O$	$\rightleftarrows H_2O_2 + 2H^+ + 2e^-$	1.77
Co^{2+}	$\rightleftarrows Co^{3+} + e^-$	1.842
$2SO_4^{2-}$	$\rightleftarrows S_2O_8^{2-} + 2e^-$	2.01
$2F^-$	$\rightleftarrows F_2 + 2e^-$	2.87

general expression for the potential of the hydrogen electrode it can be seen that this requires that $e^0 = 0$. Thus we have

$$e = e^0 + \frac{0.059}{2} \log \frac{\{H^+\}^2}{p_{H_2}} = \frac{0.059}{2} \log \frac{\{H^+\}^2}{p_{H_2}} \qquad (16.10)$$

In rough calculations with (16.7) and expressions derived from it, the accuracy is often sufficient if the activities of the dissolved substances are replaced by concentrations.

In tab. 16.1 the standard potentials for a number of important redox pairs are listed. The book by Latimer mentioned in 2-2a contains a large collection of standard potentials, but in using these it must be remembered that they are defined according to an older rule used mainly in the United States, which gives signs opposite to those given by the present, internationally accepted rule.

Since G is a function of state (2-3b) ΔG for a process is independent of the route followed by the system between the initial and final states. According to (16.3) the same must be true for the product nFE and, because F is a constant, also for the product nE. The same applies to electrode processes, for which thus the product ne for specified initial and final states is independent of the route. It is possible, therefore, to calculate from known electrode potentials (or ΔG values) unknown potentials for other electrode processes. For example, the standard potential of the electrode process $MnO_2(s) + 4OH^- \rightleftharpoons MnO_4^- + 2H_2O + 3e^-$ occurring in basic solution is known and it is desired to calculate the standard potential of the corresponding process in acid solution. In the equation of the process, therefore, H^+ will be involved instead of OH^-. This is accomplished by the addition and subtraction of equations and the corresponding ne^0 values, as set forth in the following scheme:

	e^0, V	ne^0, V
$+ (MnO_2(s) + 4OH^- \rightleftharpoons MnO_4^- + 2H_2O + 3e^-)$	$+0.588$	$+(+1.764)$
$- (2H_2(g) + 4OH^- \rightleftharpoons 4H_2O + 4e^-)$	-0.828	$-(-3.312)$
$+ (2H_2(g) \rightleftharpoons 4H^+ + 4e^-)$	0	$+(\ 0\ )$
$MnO_2(s) + 2H_2O \rightleftharpoons MnO_4^- + 4H^+ + 3e^-$	$+1.692$	$+5.076$

For the new electrode process, $ne^0 = +5.076$ V and since here $n = 3$, $e^0 = +5.076/3 = +1.692$ V.

In more extensive tables of standard potentials, values are often listed for processes in both acid and basic solution.

It is clear that the oxidizing or reducing power of a substance cannot be defined, but only that of a redox pair. The oxidizing power of a pair in a certain situation is given by the e value appropriate for that situation. A solution can neither contain only an oxidized form nor only a reduced form. Let us assume that a solution contains only Tl^{3+} and no Tl^+. Then we have $c_{ox}/c_{red} = \infty$, that is, according to (16.9) $e = \infty$. This is absurd, because with infinite redox potential any reduced form in the environment would be oxidized. For example, the OH^- ions in the solution would always be able to undergo oxidation with the evolution of O_2 according to the equation $4OH^- \rightleftharpoons O_2(g) + 2H_2O + 4e^-$. Then Tl^+ would be formed and e would very rapidly drop to a finite value. In the example, equilibrium would be reached after so little reaction that neither the Tl^+ ions formed nor the oxygen evolved could be detected. An analogous reasoning would show that the solution could not contain only Tl^+, either.

If all the substances participating in a redox equilibrium have activities (approximated by concentrations) $= 1$, the oxidized form in a certain pair can only oxidize ($=$ be reduced by) reduced forms in pairs with lower standard potentials. Since e is changed relatively slowly by changes in activities a or concentrations c, this is true even if the a and c values depart appreciably from 1, although then only on the condition that the standard potentials of the two pairs do not lie very close to each other. Without exception, it is true that a practically pure oxidized form of a pair with a higher standard potential will oxidize a practically pure reduced form of a pair with a lower standard potential. As a rule, the standard potentials therefore provide qualitative information about the relative oxidizing powers of redox pairs.

Thus reduced forms at the top of tab. 16.1 are strong reducing agents, and oxidized forms at the bottom of the table are strong oxidizing agents. All metals are reduced forms, and their tendency to form ions in water solution decreases with rising standard potential. It is customary to say in connection with this change that the metals go from *base metals* (at the top of the table) to the *noble metals* (at the bottom of the table). As mentioned in 5-2i, the tendency to form ions in water solution does not strictly follow the ionization energy of the metal atoms, because the hydration enthalpy plays so great a role in ion formation in water solution.

Since in a galvanic cell with electrodes of a noble and a base metal the former constitutes the positive and the latter the negative electrode, the noble metals have often been called electropositive and the base metals electronegative. In reference to the tendency to form positive ions, however, the reverse designations have also been used, which agrees in addition with the sense of the electronegativity scale. In order to avoid misunderstanding it is therefore advisable to suggest the position of a metal in the standard potential scale by its greater or lesser "nobility", and use the terms electropositive and electronegative only in connection with electronegativity as defined in 5-3f.

If no arrested reactions are involved, any metal for which $e^0 < 0$ V for any pair $M \rightleftharpoons M^{n+} + ne^-$ should dissolve in water or acids with the evolution of hydrogen. The metal then would be able to reduce H^+ according to $2H^+ + 2e^- \rightarrow H_2$. As mentioned in 16e, however, in such cases arrested reactions often are involved. On the other hand a metal for which $e^0 > 0$ V for all pairs will not dissolve in water or acids with the evolution of hydrogen. This is true for copper, for example, where for the conversion to Cu^+, $e^0 = 0.521$ V, and to Cu^{2+}, $e^0 = 0.337$ V. However, if other redox pairs are in a position to be effective, other reactions may become possible. If oxygen is added, it can be reduced according to $O_2 + 4H^+ + 4e^- \rightarrow 2H_2O$ with $e^0 = 1.229$ V. The presence of atmospheric oxygen, especially if it is dissolved in the liquid, therefore causes copper and even silver and gold to be attacked slowly by dilute hydrochloric acid. In solutions of a number of acids with oxidizing anions the redox potential may become high, especially at high acidity and temperature (*"oxidizing acids"*). Nitric acid and concentrated sulfuric acid thus dissolve copper and silver (to dissolve silver, the sulfuric acid must be hot) without evolution of hydrogen, but instead with reduction of the anion. Arrested reactions may arise also under the action of oxidizing acids. Very pronounced effects of this kind are caused by the formation by the acid of an insoluble oxide layer on the metal surface, which prevents solution of the metal (*passivity*, 16e).

If one kind of ion in a redox pair is bound by a substance into a complex or a poorly soluble compound, its activity (concentration) is altered. Complex formation can in this way to a high degree influence the effect of a solution on a metal. Thus, if the solution contains substances with which the metal ion forms complexes stronger than the always-present aquo complexes the activity of the metal ion in the solution will be lowered. More metal is then ionized, that is, goes into solution. The reason that aqua regia dissolves several very noble metals probably is in large part that its high chloride-ion activity leads to the formation of chloro complexes (22-10d1). Thus, in the solution of gold in aqua regia, the process $Au^{3+} + 4Cl^- \rightarrow AuCl_4^-$ is of great importance. Also, in concentrated hydrochloric acid the chloride-ion activity is sufficiently high to cause a number of metals with $e^0 > 0$ (for example, copper) to be attacked even in the absence of oxygen.

A wholly different example of the effect of complex formation is the following. For the pair $Co^{2+} \rightleftharpoons Co^{3+} + e^-$, $e^0 = 1.842$ V. The ion Co^{3+} is therefore a very strong oxidizing agent and because of this is extremely unstable in many environments (see 16e). On the other hand, for the pair $Co(NH_3)_6^{2+} \rightleftharpoons Co(NH_3)_6^{3+} + e^-$, in which the oxidation number of cobalt similarly changes from II to III but with both kinds of ions forming hexammine ions, $e^0 = 0.1$ V. Therefore, $Co(NH_3)_6^{3+}$, like a number of Co(III) complexes, is such a weak oxidizing agent that it often does not react appreciably with other substances (cf. 5-6c).

$e^0 = 0.1$ V is the e value for a solution in which $\{Co(NH_3)_6^{3+}\} = \{Co(NH_3)_6^{2+}\}$. If this e value is substituted in the expression $e = 1.842 + 0.059 \log \{Co^{3+}\}/\{Co^{2+}\}$, we find that in this solution $\{Co^{3+}\}/\{Co^{2+}\} = \sim 10^{-30}$. This shows how much stronger the complex $Co(NH_3)_6^{3+}$ is than $Co(NH_3)_6^{2+}$.

d. Equilibrium position in a redox equilibrium. If two redox pairs with different e values are brought together, the oxidized form in the pair with higher e will oxidize the reduced form in the other, providing no inhibitions occur. Meanwhile e for the first pair drops and e for the second rises. Equilibrium is reached when both pairs attain the same e value. Equilibrium among redox pairs must thus imply that the e values of all the pairs participating in the equilibrium are equal. If a solution contains redox pairs in equilibrium with each other, it thus shows a certain redox potential. This establishes the positions of all the redox equilibria in the solution in the same way that the pH of a solution establishes the position of all the acid-base equilibria.

In simpler cases the calculation of the equilibrium position is easy if the standard potentials of the participating pairs are known. For example, suppose it is desired to calculate how completely Fe^{2+} can be oxidized by Tl^{3+}. The equilibrium equation is $2Fe^{2+} + Tl^{3+} \rightleftharpoons 2Fe^{3+} + Tl^+$. Here the pair $Fe^{2+} \rightleftharpoons Fe^{3+} + e^-$ is thus involved, for which we have

$$e_{Fe} = 0.771 + 0.059 \log \frac{\{Fe^{3+}\}}{\{Fe^{2+}\}} = 0.771 + \frac{0.059}{2} \log \frac{\{Fe^{3+}\}^2}{\{Fe^{2+}\}^2}$$

and also the pair $Tl^+ \rightleftharpoons Tl^{3+} + 2e^-$, for which

$$e_{Tl} = 1.25 + \frac{0.059}{2} \log \frac{\{Tl^{3+}\}}{\{Tl^+\}}$$

At equilibrium, $e_{Fe} = e_{Tl}$ and thus

$$\log \frac{\{Fe^{3+}\}^2\{Tl^+\}}{\{Fe^{2+}\}^2\{Tl^{3+}\}} = \log K = \frac{2(1.25 - 0.77)}{0.059} = 16. \text{ Thus, } K = 10^{16}$$

This high value of the equilibrium constant shows that, if equilibrium can be attained, the oxidation of Fe^{2+} by Tl^{3+} will be practically complete.

As a second example, let us calculate the position of the equilibrium $Ag^+ + Fe^{2+} \rightleftharpoons Ag(s) + Fe^{3+}$. The participating redox pairs are $Ag(s) \rightleftharpoons Ag^+ + e^-$ and $Fe^{2+} \rightleftharpoons Fe^{3+} + e^-$ with e values $e_{Ag} = 0.7991 + 0.059 \log \{Ag^+\}$ and $e_{Fe} = 0.771 + 0.059 \log \frac{\{Fe^{3+}\}}{\{Fe^{2+}\}}$, respectively. The condition that at equilibrium $e_{Ag} = e_{Fe}$ gives $\frac{\{Fe^{3+}\}}{\{Ag^+\}\{Fe^{2+}\}} = K = 3.0$. The small difference between the standard potentials of the participating pairs causes K to be of the order of magnitude of 1. Therefore, as can be shown by a simple experiment, the direction of the reaction can easily be reversed by appropriate changes in the concentrations.

However, the difference between the normal potentials of two pairs participating in an equilibrium need not be great in order that the equilibrium be strongly displaced in one direction.

In reactions in water solution, when transitions between reduced and oxidized forms involve changes in oxygen content of the molecules or ions, very often H_2O and either H^+ or OH^- enter into the redox pair. In this case, H^+ is always associated with the oxidized form and OH^- with the reduced form (many examples in tab. 16.1). Therefore on increase of H^+ the system is displaced toward the reduced form and vice versa. We may thus write

$$\text{red} \underset{+H^+}{\overset{-H^+}{\rightleftharpoons}} \text{ox} + ne^- \qquad (16.11)$$

From this it follows that the oxidizing power of the redox pair increases as the acidity increases. For the pair $Mn^{2+} + 4H_2O \rightleftharpoons MnO_4^- + 8H^+ + 5e^-$ we thus have

$$e = 1.51 + \frac{0.059}{5} \log \frac{\{MnO_4^-\}\{H^+\}^8}{\{Mn^{2+}\}}$$

The redox potential and accordingly the oxidizing power here increases very rapidly with increase in the acidity of the solution.

In water solution consideration must often be paid to the redox pairs $H_2(g) \rightleftharpoons 2H^+ + 2e^-$ and $2H_2O \rightleftharpoons O_2(g) + 4H^+ + 4e^-$ (for basic solutions we may write $H_2(g) + 2OH^- \rightleftharpoons 2H_2O + 2e^-$ and $4OH^- \rightleftharpoons O_2(g) + 2H_2O + 4e^-$, respectively; cf. 16a). In these cases also the redox potential increases with increasing acidity.

When H^+ or OH^- enter into both of the redox pairs taking part in a redox equilibrium, the oxidizing power of both oxidized forms is affected by changes in pH. The manner in which the direction of reaction is influenced can then be decided from the equilibrium equation. However, not infrequently, completely different reactions may prevail at different acidities. When MnO_4^- is reduced in acid solution Mn^{2+} is formed as described earlier, but in neutral or weakly basic solution $MnO_2(s)$ is formed according to the reaction $MnO_2(s)+2H_2O \rightleftharpoons MnO_4^- + 4H^+ + 3e^-$, which requires only half as much H^+ per MnO_4^- as the former reaction. Again, if in the oxidation of Fe(II) to Fe(III) in water solution with O_2 the Fe system would take part only in the form of the pair $Fe^{2+} \rightleftharpoons Fe^{3+} + e^-$ ($e^0 = 0.771$ V), according to the above discussion, the oxidation should be favored by an increase in acidity. However, it is well known that this oxidation takes place more easily in basic than in acid solution. This can be explained by the formation of hydroxo complexes in basic solution. Since the hydroxoiron(III) complexes are much more difficultly soluble than the hydroxoiron(II) complexes, the equilibrium is more easily shifted toward Fe(III), the more basic the solution is made (cf. the discussion of the cobalt hexammines in 16c).

Often a substance is included in an oxidized form of one pair "1", and in a reduced form of another pair "2". If, then, $e_1^0 > e_2^0$, one molecule of the substance can oxidize another molecule. Such a reaction, in which the oxidation number of an element is converted partly to a lower and partly to a higher value, is called *disproportionation*.

The ion Cu^+ is the oxidized form in the pair $Cu(s) \rightleftharpoons Cu^+ + e^-$ with $e_1^0 = 0.521$ V, and the reduced form in the pair $Cu^+ \rightleftharpoons Cu^{2+} + e^-$ with $e_2^0 = 0.153$ V. For the disproportionation equilibrium $2Cu^+ \rightleftharpoons Cu(s) + Cu^{2+}$ we get from these e^0 values $\{Cu^{2+}\}/\{Cu^+\}^2 = K = 1.7 \times 10^6$ (25°C). Therefore, Cu^+ is nearly completely disproportionated. However, if a Cu(I) compound yields very small amounts of Cu^+ to a water solution, for example because it is very poorly soluble or is a very strong, that is, very slightly dissociated complex, $\{Cu^+\}$ may be so small that $\{Cu^{2+}\}$ also becomes small. The amount of Cu^{2+} is then insignificant, that is, the Cu(I) compound practically speaking does not disproportionate. Therefore, the oxidation state Cu(I) is found in equilibrium with water only in poorly soluble compounds (Cu_2O, Cu_2S, CuI) and in strong complexes ($CuCl_2^-$, $Cu(CN)_4^{3-}$).

If Cl_2 is dissolved in a basic solution, the two pairs $2Cl^- \rightleftharpoons Cl_2(g) + 2e^-$ with $e_1^0 = 1.3595$ V and $Cl_2(g) + 4OH^- \rightleftharpoons 2ClO^- + 2H_2O + 2e^-$ with $e_2^0 = 0.40$ V appear. Since $e_1^0 \gg e_2^0$ Cl_2 is disproportionated almost completely according to $Cl_2(g) + 2OH^- \rightleftharpoons ClO^- + Cl^- + H_2O$. Addition of OH^- increases the degree of disproportionation. In acid solution as the second pair there appears instead $Cl_2(g) + 2H_2O \rightleftharpoons 2HClO + 2H^+ + 2e^-$ with $e_2^0 = 1.63$ V. The corresponding equilibrium $Cl_2(g) + H_2O \rightleftharpoons HClO + Cl^- + H^+$ is therefore strongly displaced to the left, that is, disproportionation of Cl_2 is hardly noticeable. However, by adding a suspension of HgO, which ties up both Cl^- and H^+ according to $2HgO(s) + 2Cl^- + 2H^+ \rightarrow HgO \cdot HgCl_2(s) + H_2O$, even this equilibrium can be displaced so strongly to the right that HClO can be prepared in this way.

If in (16.9) activities are replaced by concentrations, an expression is obtained that is formally quite analogous to (12.57). The terms e and e^0 correspond to pH and pk_a. If the mole fraction $x_{ox} = c_{ox}/(c_{ox} + c_{red})$ is inserted we obtain

$$e = e^0 + \frac{0.059}{n} \log \frac{x_{ox}}{1 - x_{ox}} \quad (16.12)$$

which shows the corresponding analogy with (12.58). If x_{ox} is plotted as a function of e, curves analogous to the curves in fig. 12.2 are also obtained. The midpoint of the S-shaped curve for a redox pair lies at $x_{ox} = 0.5$ and $e = e^0$.

It is then evident that the rapid equilibrium displacement when e passes through e^0 causes a redox pair to act as a buffer when, for example, one attempts to change the redox potential of a solution by adding a strong oxidizing or reducing agent (*redox buffer*). The pair has its greatest buffer action at $e = e^0$. Therefore, titration curves analogous to those for acid-base equilibria are obtained, and their shape can be used for *redox titration*. If the oxidized and reduced forms of a redox pair have different colors the pair can be used as a color indicator of the e value of the solution (*redox indicator*). The transition range will lie in the vicinity of the e^0 value of the indicator pair.

e. Arrested reactions. In redox processes low reaction rates or completely arrested or inhibited reactions often occur. These are especially common in processes that involve molecular rearrangements. It is then often practically impossible to attain equilibrium, and equilibrium calculations according to 16 d lead in such cases to results that differ from experience. Nevertheless, the calculations are of value because they always give information concerning the one possible direction for a given reaction. The calculation thus always rules out the possibility of a reaction in the opposite direction. However, the theoretical treatment cannot be so generally applicable as with, for example, acid-base equilibria, where arrested reactions are far more rare.

Very important cases of arrested reactions occur in a number of cases where gases are formed. For example, if the equilibrium constant is calculated from the e^0 values in tab. 16.1 for the reaction $Zn(s) + 2H^+ \rightleftharpoons Zn^{2+} + H_2(g)$, we get $\{Zn^{2+}\}\{H_2\}/\{H^+\}^2 = 7 \times 10^{25}$, which may be written $[Zn^{2+}] = 7 \times 10^{25} \{H^+\}^2/p_{H_2}$. For pure water ($\{H^+\} = 10^{-7}$) and $p_{H_2} = 1$ atm, we have $[Zn^{2+}] = 7 \times 10^{11}$, which shows that zinc should dissolve in water with the evolution of hydrogen. However, in the separation of hydrogen an arrested reaction sets in, which acts as though the hydrogen has to overcome a very high pressure in order to be released. Thus p_{H_2} becomes very high, and $[Zn^{2+}]$ then has such a low value that it cannot be detected. Only after $\{H^+\}$ is appreciably raised by acidification of the solution does the zinc go into solution. The fact that zinc dissolves in alkaline solutions depends on the formation of hydroxozincate ions, which involves a completely different equilibrium.

For the alkali metals e^0 is so much less and the corresponding equilibrium constant therefore so much greater that the arrested reaction in the evolution of hydrogen cannot prevent them from dissolving even in pure water.

The same calculation for Pb ($e^0 = -0.126$ V) gives for the equilibrium constant 2×10^4. For $\{H^+\} = 10^{-7}$ and $p_{H_2} = 1$ atm we have $[Pb^{2+}] = 2 \times 10^{-10}$, which shows that even in the absence of arrested reactions, lead cannot at all dissolve in water in detectable amounts. Here $\{H^+\}$ must be increased much more than for zinc before any appreciable solution can take place.

Arrested reaction in the formation of hydrogen thus leads to the result that many metals dissolve with the evolution of hydrogen only at far higher acidity than would be calculated from the standard potential. Often solution may also be prevented by thin oxide layers. These are formed on the surface of most base metals when they are exposed to air, but in certain cases the layer is dense and resistant to attack and therefore protective, in other cases not. Aluminum ($e^0 = -1.66$ V) should dissolve more easily than zinc ($e^0 = -0.763$ V) but the opposite is true because aluminum acquires a protective oxide layer. Chromium should dissolve more easily than iron, but is protected by an oxide layer which gives chromium metal very great stability. Stainless steel is protected in a similar way. These coatings are generally very thin, often only a few tens of Å in thickness. A metal protected in this way is said to be *passive*. The passive-rendering oxide layer is often a good protection against oxidizing action, but may be destroyed in a reducing solution. Stainless steel is not attacked by strongly oxidizing acids, for example nitric acid, but is attacked on the other hand by hydrochloric acid and many organic acids.

Arrested reactions also appear in the formation of oxygen. If a salt of Co^{3+} is dissolved in water, oxygen is evolved. Here the redox pairs $Co^{2+} \rightleftharpoons Co^{3+} + e^-$ ($e^0 = 1.842$ V) and $2H_2O \rightleftharpoons O_2(g) + 4H^+ + 4e^-$ ($e^0 = 1.229$ V) participate in the equilibrium $4Co^{3+} + 2H_2O \rightleftharpoons 4Co^{2+} + O_2(g) + 4H^+$. The e^0 value of the Co pair is so high in comparison to that of the other pair that arrested reaction cannot prevent the reaction from going toward an equilibrium position strongly displaced to the right. For the same reason evolution of oxygen occurs when fluorine is passed into water. Chlorine also evolves oxygen slowly with water, although intermediate products appear. Many other oxidizing agents that from their e^0 values would be expected to evolve oxygen from water, on the contrary do not react because of an arrested reaction in the evolution of oxygen. In certain cases the reaction can be brought about if H^+ is effectively used up by the addition of large amounts of base, possibly in the presence of a catalyst.

In many other systems a redox process should lead to a more stable state, but arrested reactions prevent any noticeable changes from occurring. An important example is hydrogen peroxide H_2O_2. Hydrogen peroxide can act as a powerful oxidizing agent according to the scheme

$$2H_2O \rightleftharpoons H_2O_2 + 2H^+ + 2e^- \qquad e^0 = 1.77 \text{ V} \qquad (16.13)$$

and as a weak reducing agent according to the scheme

$$H_2O_2 \rightleftharpoons O_2(g) + 2H^+ + 2e^- \qquad e^0 = 0.682 \text{ V} \qquad (16.14)$$

In this way one H_2O_2 molecule can oxidize another H_2O_2 molecule according to the equation

$$2H_2O_2 \rightleftharpoons 2H_2O + O_2(g) \qquad (16.15)$$

The two e^0 values give $\{H_2O\}^2\{O_2\}/\{H_2O_2\}^2 = 8 \times 10^{36}$. Hydrogen peroxide is thus very unstable, but in pure form or in pure water solution it decomposes extremely slowly at ordinary temperature. On the addition of many substances that can function as catalysts (especially many transition metals and their compounds; see further 21-6b), a very rapid decomposition can take place.

f. Electrolysis. Let us assume that 1 M HCl is electrolyzed at 25°C with platinized Pt electrodes. It may be assumed that $\{H^+\} = \{Cl^-\} = 1$. On electrolysis H_2 is formed at the cathode and Cl_2 at the anode. The electrodes are thus polarized so that a hydrogen-chlorine cell is produced. In this cell we have

$$e_{\text{cath}} = 0 + \frac{0.059}{2} \log \frac{1}{p_{H_2}} \qquad e_{\text{an}} = 1.3595 + \frac{0.059}{2} \log p_{Cl_2}$$

and the emf of the cell (the counterelectromotive force, polarization voltage) is

$$E = e_{\text{an}} - e_{\text{cath}} = 1.3595 + \frac{0.059}{2} \log p_{H_2} p_{Cl_2} \tag{16.16}$$

During the passage of current, more and more H_2 and Cl_2 are formed at the electrode surfaces and at first dissolved in the liquid. The partial pressures p_{H_2} and p_{Cl_2}, and consequently the counterelectromotive force E, increase until E is as great as the externally applied emf. At this point the passage of current and electrolysis should cease if the gases dissolved in the liquid did not diffuse away from the electrodes. A stationary state sets in at which equal amounts of H_2 and Cl_2 are formed as are diffused away. The small current that is thus allowed is called *residual current*. If the applied emf is increased, p_{H_2} and p_{Cl_2} rise and thus the diffusion and the residual current. Because of the slowness of the diffusion, however, the residual current can never reach high values. Finally, p_{H_2} and p_{Cl_2} become equal to 1 atm, which according to (16.16) occurs when the applied emf is approximately 1.36 V. If the pressure on the cell is 1 atm the partial pressure cannot increase further, but rather the gases go off as bubbles. This makes possible rapid evolution of gas, the decomposition can go quickly and the current strength can be significantly increased with further increase in external emf. $E = 1.36$ V is called the *decomposition voltage* for this case. The decomposition voltage can thus here be set equal to the emf of the cell that is constituted by the formed substances at the prevailing concentrations and pressures.

However, this presupposes that the cell produced operates reversibly. Often this reversibility is upset by arrested reactions in the electrode processes. These then require an activation energy (8d), which leads to the result that a larger potential difference at the electrode is needed than that corresponding to the reversible process. This often occurs when gases are given off and is especially important in the formation of hydrogen and oxygen. The excess potential difference is called *overvoltage* and is caused by the type of arrested reaction described in 16e.

The magnitude of the overvoltage is strongly dependent on the substance on which the gas forms, and also on the earlier treatment of this substance. The value of the overvoltage therefore cannot be stated with any accuracy. Under otherwise constant conditions the overvoltage increases with increasing current density. The overvoltages of hydrogen and oxygen are also dependent on the acidity of the solution.

On platinum black the overvoltage of hydrogen may be assumed to be 0 at very low current density; in fact, a hydrogen electrode operates reversibly under such conditions. On the other hand the overvoltage of hydrogen on white, that is, compact platinum is always appreciable. However, the overvoltage of hydrogen on all platinum metals is small compared to other metals, a fact that is related to the ability of the platinum metals to dissolve hydrogen and catalyze the electrode process. On cadmium and mercury the overvoltage of hydrogen is especially high. The hydrogen isotopes with their large relative mass differences show mutually somewhat different overvoltages on the same metal. Thus $^2H_2(=D_2)$ is evolved at approximately 0.1 V higher overvoltage than $^1H_2$.

Oxygen shows overvoltage on all electrode materials at room temperature.

In any electrolysis only those electrode processes and that cell process occur that require the smallest supply of energy, that is, the smallest externally applied emf. The electrolysis of a solution of H_2SO_4 in which $\{H^+\}=1$ (pH=0) with platinized Pt electrodes occurs as follows:

cathode process	$4H^+ + 4e^- \rightarrow 2H_2(g)$	$e^0 = 0$ V
anode process	$2H_2O \rightarrow O_2(g) + 4H^+ + 4e^-$	$e^0 = 1.229$ V
cell process	$2H_2O \rightarrow 2H_2(g) + O_2(g)$	

In the H_2SO_4 solution the current is mainly transported by the ions H^+ (proton jumps, 11b), HSO_4^- and SO_4^{2-}. However, the oxidation at the anode does not involve either of the two anions that migrate to it, but instead the H_2O molecules, because this oxidation process requires the least energy. If the two stated electrode processes were reversible, the decomposition voltage for $\{H^+\}=1$ and at 1 atm would be 1.23 V. In addition, in the formation of oxygen an overvoltage always arises, which at low current densities is at least 0.4 V. The decomposition voltage is then at least 1.6 V. The oxidation of the ions HSO_4^- and SO_4^{2-} in this solution would require a far greater emf and therefore does not take place. The result is thus a *water decomposition*.

SO_4^{2-} can be oxidized according to $2SO_4^{2-} \rightleftharpoons S_2O_8^{2-} + 2e^-$ with $e^0 = 2.01$ V. Anodic oxidation according to this equation is actually used for the preparation of the peroxodisulfate ion $S_2O_8^{2-}$, but in this case the concentration of SO_4^{2-} is high, and the formation of O_2 is suppressed by the use of anodes of white Pt and very high current densities resulting in high oxygen overvoltage (on white Pt the oxygen overvoltage at 1 A cm^{-2} is about 1.4 V, which means a decomposition voltage of over 2.6 V).

If also the hydrogen is evolved with an overvoltage (some other cathode material than platinized Pt, high current density), H_2 will be evolved somewhat more easily than D_2 because of its lower overvoltage. Heavy water, D_2O, is therefore enriched in the remaining water (9-3 b).

Let us assume now that a 1 molar solution of a metal sulfate is electrolyzed with platinized Pt electrodes. If $\{H^+\}=1$ (H_2SO_4 added), H_2 is developed at the cathode before the precipitation of a metal with $e^0<0$ can begin. At the anode O_2 is developed. If the solution contains ions of metals with $e^0>0$, these are deposited on the cathode before any development of hydrogen can occur. If the solution contains Ag^+ with $[Ag^+]=1$ and we assume $e_{an}=1.60$ V, the precipitation of Ag begins at $E=e_{an}-e_{cat}=1.60-0.80=0.80$ V. If instead the solution contains Cu^{2+} with $[Cu^{2+}]=1$, precipitation of Cu begins at $E=1.60-0.34=1.26$ V. If both Ag^+ and Cu^{2+} are present in these concentrations, only Ag^+ begins to precipitate at 0.80 V. As $[Ag^+]$ decreases, E rises. When $[Ag^+]=10^{-7}$, then $e_{cat}=0.80+0.059$ log $10^{-7}=0.39$ V and thus $E=1.60-0.39=1.21$ V. Following further deposition of Ag an E value of 1.26 V is soon attained, when Cu also begins to be deposited. At $E=1.46$ V, $[Cu^{2+}]=10^{-7}$ and therefore the Cu is practically completely precipitated. At a current density as low as 10^{-4} A cm^{-2} the hydrogen overvoltage on Cu is 0.35 V, and therefore hardly any evolution of hydrogen begins before $E=1.60-(-0.35)=1.95$ V. Thus both metals can be completely precipitated and also easily separated. If the difference in e^0 for two metals is small enough it may become impossible to obtain a complete separation.

Electrolytic precipitation from water solution of certain metals with $e^0<0$ can occur, however, if the formation of hydrogen is inhibited by lowering $\{H^+\}$ in the solution and by choosing a cathode material on which the overvoltage of hydrogen is high. In this way even Zn can be precipitated. For metals with still lower e^0 values (alkali metals, alkaline-earth metals, Al) water must not be present, but rather an anhydrous melt is electrolyzed. Recently organic metal compounds that can be ionized in other solvents than water have been used. In all those cases where the ions are not in water solution, other e^0 values than those given in tab. 16.1 apply.

It is important to remember that complex ions containing heavy metals never are decomposed at the anode toward which they migrate. Instead, it is the ions of this kind that are found at the cathode that are there reduced. For example, if a solution of $K[Ag(CN)_2]$ is electrolyzed, K^+ migrates toward the cathode and $Ag(CN)_2^-$ toward the anode, but metallic deposition takes place at the cathode according to the equation $Ag(CN)_2^- + e^- \rightarrow Ag(s) + 2CN^-$. For continued electrolysis $Ag(CN)_2^-$ must be carried to the cathode by diffusion and convection (stirring) against the electrical forces.

In electrolytic precipitation, both for quantitative determination of a metal (*electroanalysis*) and for producing a protective layer (*electroplating*), it is important to obtain a dense and even layer of the metal. A first condition for this is that the evolution of hydrogen be avoided. Often crystals are formed that project out into the solution, and the continued deposition occurs preferably on the pointed parts of these crystals, so that dendritic forms arise. Frequently this can be prevented by means of rather special precipitation arrangements. In general a denser and more even layer of the metal is formed from a complex ion. This is the reason that solutions of cyano complexes (with ions of the type $Ag(CN)_2^-$) are so often used. Because of the toxicity of the cyanides, other media have also been sought recently, and it appears that very good metal layers can be obtained from solutions of, among others, fluoroborates, MBF_4 (24-2f).

Electrolysis is also used for *electrolytic refining* of metals, especially copper. In this process the impure copper is made into an anode, from which Cu and all the more base metals such as Zn and Fe go into solution, but only Cu is deposited on the cathode. Metals that are more noble than Cu, for example Ag, are dissolved only insignificantly and the major portion of these fall to the bottom of the vessel as "anode slime" while the anode is being attacked.

g. Primary cells as energy sources. A galvanic cell, in which the chemical energy of the substances introduced in the cell system is converted into electrical energy, is called a *primary cell* (battery). The energy-producing process thus occurs as a cell process. If this is reversible, theoretically the whole decrease in G in the process can be realized as electrical energy (cf. 16b). Thus, in this way the best possible utilization of the chemical energy is made. Large energy losses could be avoided if instead of openly burning coal and hydrocarbons, their oxidation could be allowed to take place in a reversibly operating primary cell (*fuel cell*). Much work has been expended on this problem, which is not yet solved. However, recently a special cell of this type has been successful, the *hydrogen cell* described below. Otherwise, such direct conversion of chemical to electrical energy has been carried out only for very special processes and on a very small scale. The primary cells produced for this purpose, however, have tremendously wide application as convenient energy sources for small amounts of energy.

One of the oldest primary cells ("Voltaic pile", 1798) was constructed according to the scheme $Zn(s)\,|\,H_2SO_4,\,aq\,|\,Cu(s)$. The processes are:

minus pole	$Zn(s) \rightarrow Zn^{2+} + 2e^-$
plus pole	$\underline{2H^+ + 2e^- \rightarrow H_2(g)}$
cell process	$Zn(s) + 2H^+ \rightarrow Zn^{2+} + H_2(g)$

The hydrogen evolved at the plus pole (Cu) polarizes it and makes the process to a large degree irreversible. Attempts were early made to eliminate the formation of hydrogen by surrounding the plus pole with some oxidizing agent, which was therefore called a "depolarizer". The depolarization action should not be considered, as it was in the beginning, as an oxidation of the hydrogen formed to water. The introduction of the depolarizer implies that another cell and another cell process is obtained that is more reversible than the original. The Voltaic pile was the basis for several types of primary cells, but none of these are presently in use.

In the following, some of the primary cells currently in use are described. The minus pole is generally Zn, but Mg is also used nowadays in certain cases where the weight must be kept low. In general, gelatin or starch is added to the electrolyte in sufficient amount so that it cannot flow (*dry cell*). It is very important that the cell have high stability in storage ("shelf life") and long life, which requires that no processes go on when the circuit is open ("self discharge"). For portable cells it is further desirable that the cell system have low weight and volume in relation to the amount of electricity delivered by the cell process. From this standpoint, the development of the battery has not yet come as far as would be desired.

Leclanché cell

$$\text{Zn(s)} \,|\, \text{NH}_4\text{Cl, aq} \,|\, \text{MnO}_2(\text{s}), \text{C(s)} \qquad E = \sim 1.5 \text{ V}$$

MnO_2 surrounds the carbon electrode and is mixed with graphite to obtain higher conductivity. The processes are involved and depend on the manner of discharge. At very low current density they probably are:

$$\text{Zn(s)} \rightarrow \text{Zn}^{2+} + 2\text{e}^-$$
$$\underline{\text{Zn}^{2+} + 2\text{MnO}_2(\text{s}) + 2\text{e}^- \rightarrow \text{ZnMn}_2\text{O}_4(\text{s})}$$
$$\text{Zn(s)} + 2\text{MnO}_2(\text{s}) \rightarrow \text{ZnMn}_2\text{O}_4(\text{s})$$
$$\text{zinc manganese(III) oxide}$$

At higher current density the following processes mainly occur:

$$\text{Zn(s)} \rightarrow \text{Zn}^{2+} + 2\text{e}^-$$
$$\underline{2\text{NH}_4^+ + 2\text{MnO}_2(\text{s}) + 2\text{e}^- \rightarrow 2\text{NH}_3 + 2\text{MnO(OH)(s)}}$$
$$\text{Zn(s)} + 2\text{NH}_4^+ + 2\text{MnO}_2(\text{s}) \rightarrow \underbrace{\text{Zn}^{2+} + 2\text{NH}_3}_{\text{Zn(NH}_3)_2^{2+}} + 2\text{MnO(OH)(s)}$$

In contrast to the first case the electrolyte is thus changed during the process. Very schematically the cell process for both cases can be written

$$\text{Zn}^0 + 2\text{Mn}^{\text{IV}} \rightarrow \text{Zn}^{\text{II}} + 2\text{Mn}^{\text{III}}.$$

The Leclanché cell was introduced as early as 1868 and is still the primary cell most used as a dry cell. Cells of this type have also been made with Mg instead of Zn.

Mercuric oxide cell ("RM cell")

$$\text{Zn(s)} \,|\, \text{KOH, aq} \,|\, \text{HgO(s)} \qquad E = \sim 1.35 \text{ V}$$

HgO is mixed with graphite to obtain higher conductivity. The electrolyte is saturated with hydroxozincate ions Zn(OH)_4^{2-} by solution of ZnO (ZnO + 2OH$^-$ +

$H_2O \rightarrow Zn(OH)_4{}^{2-}$). In this way no further hydroxozincate ions are formed at the zinc electrode, but rather $Zn(OH)_2(s)$. The processes therefore are:

$$Zn(s) + 2OH^- \rightarrow Zn(OH)_2(s) + 2e^-$$
$$HgO(s) + H_2O + 2e^- \rightarrow Hg(l) + 2OH^-$$
$$\overline{Zn(s) + HgO(s) + H_2O \rightarrow Zn(OH)_2(s) + Hg(l)}$$

Because of the precipitation of $Zn(OH)_2(s)$, the electrolyte remains practically unchanged. More schematically we may write $Zn^0 + Hg^{II} \rightarrow Zn^{II} + Hg^0$.

This type of cell has especially high capacity per unit of volume and is therefore useful for hearing aids, rocket instrumentation, and so on.

Metal-air cells

In these the minus pole is a metal and the plus pole an oxygen electrode. The latter is formed of air that in a porous body (mostly carbon or nickel) is in intimate contact with the electrolyte. The oldest metal-air cell is the *zinc-air cell* with the cell scheme

$$Zn(s)\,|\,NaOH,aq\,|\,O_2(g),\,C(s) \qquad E = 0.6\text{ V}$$

During storage the air supply should be cut off. If, as is often the case also with this cell, the electrolyte is saturated with hydroxozincate ions, the processes are:

$$2Zn(s) + 4OH^- \rightarrow 2Zn(OH)_2(s) + 4e^-$$
$$O_2(g) + 2H_2O + 4e^- \rightarrow 4OH^-$$
$$\overline{2Zn(s) + O_2(g) + 2H_2O \rightarrow 2Zn(OH)_2(s)}$$

In a later type, the *magnesium-air cell*, the cell scheme is $Mg(s)\,|\,NaCl,aq\,|\,O_2(g)$ and the cell process $2Mg(s) + O_2(g) + 2H_2O \rightarrow 2Mg(OH)_2(s)$.

Silver chloride cell

$$Mg(s)\,|\,H_2O\,|\,AgCl(s),\,Ag(s) \qquad E = \sim 1.5\text{ V}$$
$$Mg(s) \rightarrow Mg^{2+} + 2e^-$$
$$2AgCl(s) + 2e^- \rightarrow 2Ag(s) + 2Cl^-$$
$$\overline{Mg(s) + 2AgCl(s) \rightarrow 2Ag(s) + Mg^{2+} + 2Cl^-}$$

This cell is stored dry and activated by addition of water ("reserve-type cell"). In order that the cell may come into operation quickly the water should be conducting (for example, sea water).

Hydrogen cell

$$H_2(g)\,|\,KOH,aq\,|\,O_2(g) \qquad E = \sim 1\text{ V}$$

The gases are continuously passed in through porous electrodes, usually made of carbon or nickel, in whose pores they come in contact with the electrolyte. The cell process is the oxidation of hydrogen to water.

$$2H_2(g) + 4OH^- \rightarrow 4H_2O + 4e^-$$
$$O_2(g) + 2H_2O + 4e^- \rightarrow 4OH^-$$
$$\overline{2H_2(g) + O_2(g) \rightarrow 2H_2O}$$

Catalysts (nickel, silver, platinum, palladium), with which the electrodes are impregnated, are necessary if the electrode processes are to go sufficiently rapidly at room temperature. Higher operating temperatures are also under study. The hydrogen cell is distinguished from other primary cells in that the reacting substances are continuously brought in and the substances formed carried away during the course of the process. The electrodes serve only as conductors, catalysts and catalyst support, and provide a large area at the phase boundary gas | electrolyte. In principle, they should not be altered during the process. The efficiency has been said to be 65–80%.

h. Secondary cells, storage cells. Certain cells are suitable for providing a system in which, by electrolysis with strong chemical polarization, the electrical energy introduced is stored as chemical energy. The polarized cell is then used as a battery, during which the stored chemical energy is converted back to electrical energy. Such a *secondary cell* is called an electrical *storage cell*. For a storage cell to be of practical use it is necessary that after charge and discharge it be restored so nearly to the original state that even after a large number of such "cycles" it is not appreciably changed. Otherwise the requirements are approximately the same as for the primary cells. Up to the present only a few different types of storage cells have been found to be useful. Also in these the processes lag behind under the current densities that must be maintained in practice in charging and discharging. The amount of electricity put in therefore can never be totally regained.

Lead storage cell

The cell scheme is:
charged cell $\quad\quad\quad\quad$ Pb(s) | H_2SO_4, aq | PbO_2(s), Pb(s) $\quad\quad E = \sim 2.0$ V
discharged cell Pb(s), $PbSO_4$(s) | H_2SO_4, aq | $PbSO_4$(s), Pb(s)

The processes are (discharge to the right →, charge to the left ←):

$$Pb(s) + SO_4^{2-} \rightleftarrows PbSO_4(s) + 2e^-$$
$$\underline{PbO_2(s) + 4H^+ + SO_4^{2-} + 2e^- \rightleftarrows PbSO_4(s) + 2H_2O}$$
$$Pb(s) + PbO_2(s) + 4H^+ + 2SO_4^{2-} \rightleftarrows 2PbSO_4(s) + 2H_2O$$

H_2SO_4 is thus consumed on discharge and the course of the discharge can therefore be followed by measuring the specific gravity of the electrolyte (in a charged cell it is 1.2–1.3). In principle the cell process involves only changes in the oxidation number of Pb and can be most simply written as $Pb^0 + Pb^{IV} \rightleftarrows 2Pb^{II}$.

The electrode plates are made as lead grids whose interstices are filled mainly with PbO. During the first charge ("formation") the PbO is reduced to Pb at the cathode and oxidized to PbO_2 at the anode, after which the processes proceed according to the above scheme.

The operation of the lead storage battery is interfered with by very small amounts of certain substances, of which particularly Cl^-, Fe, Mn and Pt are of practical importance. Therefore only distilled water should be added to the battery. In a discharged storage battery the electrodes consist mainly of $PbSO_4$(s) and in this phase crystal growth gradually takes place ("sulfating"). It is then converted only with difficulty on charging. The storage battery therefore should not be allowed to stand while discharged. After all the $PbSO_4$ is used up on charging, water decomposition begins with the evolution of H_2 and O_2 ("gassing") and an increase in the counterelectromotive force.

The lead storage battery was introduced in very primitive form in 1859, and after many technical improvements, in spite of its high weight, its sensitivity to shock and its high self-discharge rate, it is still by far the most important type of storage battery. It may be said that it delivers about 80% of the amount of electricity put into it.

Alkaline nickel storage cells

Edison in the United States and Jungner in Sweden independantly of each other developed the most important forms of this type of storage cell.

The cell scheme for the Edison storage cell is:

charged cell $\qquad$ Fe(s) | KOH, aq | NiO(OH)(s), Ni(s) $\qquad E = \sim 1.35$ V
discharged cell Fe(s), Fe(OH)$_2$(s) | KOH, aq | Ni(OH)$_2$(s), Ni(s)

The processes are (discharge →, charge ←):

$$\frac{Fe(s) + 2OH^- \rightleftarrows Fe(OH)_2(s) + 2e^- \\ 2NiO(OH)(s) + 2H_2O + 2e^- \rightleftarrows 2Ni(OH)_2(s) + 2OH^-}{Fe(s) + 2NiO(OH)(s) + 2H_2O \rightleftarrows Fe(OH)_2(s) + 2Ni(OH)_2(s)}$$

The formula NiO(OH) is schematic. In the fully charged cell this oxide contains besides Ni(III) also Ni(IV), so that the formula lies between NiO(OH) and NiO_2. However, the cell process can most simply be written as $Fe^0 + 2Ni^{III} \rightleftarrows Fe^{II} + 2Ni^{II}$.

In order to avoid the development of hydrogen at the cathode (the Fe-containing electrode) during charging, in the Edison storage cell a little Hg is added to it, which increases the overvoltage of the hydrogen (16f). In the Jungner storage cell (also called Nife-storage cell after the original electrode metals) the same effect is obtained by replacing the major portion of the Fe by Cd (Fe is then replaced by Cd in the above schemes and formulas).

The alkaline nickel storage batteries have lower efficiency than the lead storage battery and are more expensive. Their advantages are that they are mechanically more durable, tolerate higher current densities and have smaller self-discharge rates. Under certain conditions (among others, overcharging is to be avoided) smaller cells can be hermetically sealed, thus eliminating the need of any maintenance.

Silver-zinc storage cell

charged cell $\qquad$ Zn(s) | KOH,aq | AgO(s), Ag(s) $\qquad E = \sim 1.5$ V
discharged cell $\qquad$ Zn(s), Zn(OH)$_2$(s) | KOH,aq | Ag(s)

The electrolyte is saturated with hydroxozincate ions. The processes (discharge →, charge ←) are:

$$Zn(s) + 2OH^- \rightleftarrows Zn(OH)_2(s) + 2e^-$$
$$AgO(s) + H_2O + 2e^- \rightleftarrows Ag(s) + 2OH^-$$
$$\overline{Zn(s) + H_2O + AgO(s) \rightleftarrows Zn(OH)_2(s) + Ag(s)}$$

A simplification has been introduced here in that no account has been taken of the fact that the cell process at the Ag electrode goes through the intermediate stage Ag_2O. This type of storage battery has lower weight and volume in relation to amount of stored electricity than the others and is mechanically much more durable. However, it does not tolerate as many charge-discharge cycles as the others. Nevertheless, it has been much used in robot weapons, aircraft, and so on.

i. Galvanic cells in metal corrosion. In the corrosion of metals galvanic cells formed in various ways play a very large role. Such cells often appear when the metal is not chemically or physically homogeneous.

Very pure Zn dissolves at a hardly detectable rate in dilute H_2SO_4 (16e). If a sheet of very pure Zn and a sheet of very pure Cu are immersed in such acid, therefore, nothing happens as long as they do not touch each other. However, the arrangement sets up an open galvanic cell (in this case, a "Voltaic pile", 16g), and if its external circuit is closed by contact between the sheets, the cell process given in 16g is immediately set in motion. Zn goes into solution and H_2 is developed on the Cu sheet. With a sheet of Pt instead of Cu the process goes still faster because the hydrogen overvoltage on Pt is less than on Cu (16f).

Less pure Zn by itself dissolves better in acid than the very pure metal, and this is because the impurities as a rule are unevenly distributed. For example, if Cu is present as an impurity, certain regions may contain more Cu than others. The regions with unequal Cu content are in metallic contact with each other and when the metal surface is covered by an electrolyte solution, the Cu-rich regions act in the same way as the Cu sheet in the experiment just described. Thus, Zn goes into solution at the regions poorer in Cu. It is said that the regions with different composition constitute *local elements*, which hasten the dissolution. The local-element action on Zn can easily be increased by adding Cu^{2+} ions to the acid. By the reaction $Zn(s) + Cu^{2+} \rightarrow Zn^{2+} + Cu(s)$, Cu is deposited on the Zn, which leads to significantly faster dissolution. Irregularities in the composition may result from the presence of several phases with different composition, and from variations of composition within the same phase (for example zoning, 13-5f1). Local elements are also often formed when the metal is unevenly covered by an oxide or hydroxide layer or by an adsorbed layer. In addition, a chemically homogeneous metal surface can, as a result of uneven cold working (14-3c), have different energy and thus different potential in relation to a solution at different places. In general, the most strongly worked parts dissolve most easily.

Iron plated with zinc ("galvanized") is protected against corrosion even after the zinc layer is broken. As soon as the iron is exposed, a galvanic cell is formed, in which only the zinc electrode goes into solution. The dissolution of the zinc often proceeds quite rapidly, but the iron is protected as long as any zinc remains in its vicinity. With a protective layer of tin the more base metal iron instead begins to dissolve where the layer is broken, and more rapidly than if the tin were not present.

Galvanic cells of larger size are often formed when different metals are in contact with each other (for example, iron with copper and copper alloys) in the presence of water. The result may be rapid attack. Salt water is then especially dangerous because of its higher conductivity. Iron that is covered by a continuous layer of water is sometimes protected against corrosion by placing magnesium plates in contact with the iron at appropriate places. Then only the magnesium plates are attacked, but they are easily replaced. Recently, as a protection against corrosion, electrodes (often platinum) isolated from the iron have been used, which with a special battery are given a positive potential relative to the iron just high enough to prevent the ionization of the latter.

CHAPTER 17

Distribution of the Elements

a. The elements in the universe. It is, of course, extremely difficult to estimate the relative amounts of the elements in the universe. However, tab. 17.1 shows the result of an attempt at such an estimate. Hydrogen is strongly dominant, a fact which is certainly connected with the role of the proton as a nuclear building unit. With the exception of hydrogen, *Harkins' rule* (9-1 b) applies nearly throughout, that is, that an element with even atomic number is more common than its nearest neighbors with odd atomic numbers. Both in the series with odd and the series with even atomic numbers, for the most part the relative amounts decrease rapidly with increasing atomic number up to about $Z = 30$, after which the decrease proceeds slowly. The elements 3Li, 4Be and 5B, however, appear in strikingly small amounts in spite of their very low atomic numbers. This results from their high nuclear energy (see fig. 9.2), which enables them to form heavier elements through thermonuclear reactions. In the series with even atomic numbers 26 Fe marks a maximum.

If the elements are arranged according to decreasing amounts, the sequence

Table 17.1. *Logarithms of the relative numbers of atoms of the elements 1–28 in the known universe. The number of atoms of Si is taken as* 1.0×10^6.
(*Suess and Urey*, 1956.)

Odd atomic numbers		Even atomic numbers	
1 H	10.6	2 He	9.5
3 Li	2.0	4 Be	1.3
5 B	1.4	6 C	6.5
7 N	6.8	8 O	7.3
9 F	3.2	10 Ne	6.9
11 Na	4.6	12 Mg	6.0
13 Al	5.0	14 Si	6.0
15 P	4.0	16 S	5.6
17 Cl	4.0	18 Ar	5.2
19 K	3.5	20 Ca	4.7
21 Sc	1.5	22 Ti	3.4
23 V	2.3	24 Cr	3.9
25 Mn	3.8	26 Fe	5.8
27 Co	3.3	28 Ni	4.4

The logarithm of the sum of the relative numbers of atoms of all the elements in the table is 10.63. The logarithm of the sum of the relative numbers of atoms of all other elements is 3.0, and these therefore constitute only about 1 part in 5×10^7 of the total number of atoms.

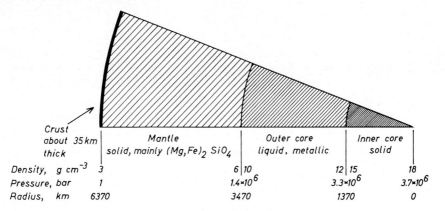

Fig. 17.1. Internal structure of the earth.

H, He, O, Ne, N, C, Si, Mg, Fe is obtained. The logarithm of the sum of the relative numbers of atoms (as in tab. 17.1 with the number for $Si = 1 \times 10^6$) of all the elements except these nine is 5.9, corresponding to 1 part in 5×10^4 of the total number of atoms and little more than the value for Fe.

b. Structure and composition of the earth. The chemical composition of the earth has only been directly studied within a very thin shell on its surface. The deepest drillhole goes down only a little over 10 km. Concerning the structure and composition of the major portion of the body of the earth, it has only been possible to propose hypotheses. These have been based primarily on the study of the propagation of earthquake waves (seismic waves) through the earth, and on deductions by analogy with meteoritic material.

Seismic investigations show that the structural material of the earth suddenly changes its properties at certain definite levels. The most important of these levels lie at the depths of 35 (under the continents), 2900 and 5000 kilometers below the earth's surface and are assumed to mark the boundaries between what are called the earth's crust, mantle, outer core and inner core (fig. 17.1).

The *earth's crust*, which is about 35 kilometers thick under the continents, consists of several layers. Below alluvium and sedimentary rocks the so-called granite layer lies nearest the surface, which consists of SiO_2-rich and therefore acid rocks (the terms "acid" and "basic" have long been applied to rocks, but can properly be thought of as an application to oxides of the acid-base definition given in 12-1e). The thickness of the granite layer is approximately 15 kilometers. Below this come layers of more basic rocks. Under the oceans the earth's crust is thinner than under the continents. The granite layer is missing and the basic layers are thinner so that the total thickness attains only about 5 kilometers.

The expression "earth's crust" comes from the time when it was believed that the underlying parts of the earth were fluid. The *mantle* however, has the properties of a solid body, although probably it is rather plastic. It probably consists of very basic ("ultrabasic") rocks, whose composition may most closely correspond to the formula $Mg_{2-x}Fe_xSiO_4$ (x probably $=0.2$–0.6). At the pressures that occur near the surface a similar phase crystallizes as olivine (23-3h1) but within a depth interval of about 400–800 kilometers it would transform into a denser structure, probably of the spinel type (24-3f). The liquid magma that is forced out by volcanic action seems to have a local origin. Volcanic action is always related to earthquake regions and it is probable that it results from the melting and degassing of underlying rocks caused by the release of pressure by earthquakes.

The *outer core* is assumed to be liquid but has metallic conductivity. It presumably consists of nickeliferous iron (Ni content up to about 9 wt %). According to another hypothesis it may, like the mantle, consist of an iron-containing magnesium silicate, which, however, because of the extreme pressure (see 6-4d and the pressure values in fig. 17.1) has acquired metallic properties. The earth's magnetic field has been explained by movements in the conducting mass (caused among other things by convection and "tide phenomena") which under the influence of a very weak external magnetic field produce strong electric currents and induced magnetic fields. Ferromagnetism cannot exist at such high temperatures as occur here.

The *inner core* probably consists for the most part of the same material as the outer core but is believed to be solid.

The temperature increases with depth, near the surface by approximately 30°C per kilometer, but more slowly at greater depth. At the center the temperature perhaps lies between 2000 and 5000°C. It is not known whether at the present time the temperature is falling or rising. Large amounts of heat are generated by radioactive decay, especially in the uranium and thorium series and the decay of ^{40}K (see 9-2a), but since the proportions of radioactive nuclides in the main body of the earth are not known, it is not possible to say whether the amount of heat produced is able to compensate the losses from heat radiation. In the crust the proportion of radioactive nuclides is greatest in the granite layer and this would be enough to keep the earth in a molten state if it were the same throughout.

The uncertainty in the estimates, especially with respect to the outer and inner cores, means that no accurate values can be advanced for the chemical composition of the earth as a whole. Whether or not the core consists of iron or iron-containing magnesium silicate, however, the elements O, Si, Mg and Fe make up about 90% of the mass of the earth. Of the other elements only S, Ni, Al and Ca are present in proportions over 1%.

Formerly it was believed that the planets were formed by the cooling of incandescent gas balls, but nowadays it is maintained that more probably they were formed by a sintering together of particles from cold dust clouds into larger and larger solid bodies. The rise in temperature can be explained by the heat developed, partly from radioactive decay (see above), and partly from compression resulting from the increased force of gravity. As the mass of the body grew and it became more compact, its heat radiation per unit of mass decreased. Finally the heat generation dominated so that the temperature began to rise.

If the earth was formed from a glowing gas ball, it is probable that it consisted of solar material from the outset, that, is, had almost the "universal composition" (tab. 17.1). In that case, it becomes difficult to account for many features of the present composition of the earth, particularly as to how the heavy noble gases Kr and Xe could have been reduced as much as is the case in relation to the elements with adjacent atomic numbers (this reduction seems to have occurred by a factor of about 10^{-6}). This could only have happened by such a strong rise in temperature that the velocity of the noble-gas molecules could have overcome the gravitational attraction of the earth (cf. 2-1b). But in that case many light elements would have also left the earth for the most part.

Formation from cold dust clouds can more easily account for the composition of the earth. The noble gases are so difficult to condense that at the temperatures that probably prevailed at the ultimate distance of the earth from the sun they would have remained in gaseous form. They could hardly have been chemically bound to any other element and therefore could not have been retained in the dust cloud other than in the small amounts that would be adsorbed or enclosed by the solid particles. Of the noble gases now present in and on the earth, Ne, Kr and Xe probably have this origin, while the major portion of He and Ar and all of the Rn were formed and are still being formed by nuclear reactions; see 19b.

At the distance of the earth from the sun considerable amounts of the elements H, C, N and O would probably have left the dust cloud as the uncondensed gases H_2, N_2, O_2, CH_4, CO and possibly CO_2 and NH_3. However, these elements could have been partly retained bound in solid phases. H and O would have been incorporated into mainly H_2O (ice particles are thought to have made up an important part of the dust cloud at the distance of the earth from the sun) and other oxides. C may have been present as the element and also bound in carbides, and N bound in nitrides. Starting with the nine most common elements H, He, O, Ne, N, C, Si, Mg and Fe (17a) in material of universal composition, it thus seems quite reasonable that O, Si, Mg and Fe would predominate as the structural material of the earth, and that the heavy noble gases would be so strongly reduced in amount. Meteorites also contain O, Si, Mg and Fe as the chief elements, as would be the case if they were fragments of one or several planets similar to the earth. However, it should be noted that analyses of meteorites have been used in figuring the values for the distribution of the elements in the universe, and also that the composition of the meteorites has had a bearing on the estimates of the composition of the interior of the earth.

However the different elements were combined in the increasingly dense aggregate of dust, the ultimate separation into core and mantle was to a large extent achieved when the aggregate became hotter and finally melted. To give a rough picture of the manner in which an iron core and a silicate mantle can be thought to have originated, we first list the following possible elemental proportions in the entire earth mass, expressed in atomic percent: O (48), Fe (19), Mg (16), Si (12), S (2), other elements (3). Let us assume that only the five elements named react with each other. Oxygen must combine in the first place with magnesium and silicon, and after that according to availability with iron. On the basis of 100 atoms, we then obtain 16MgO and 12SiO$_2$, whereby 40 atoms of O are combined. The remaining 8 O atoms are bound by Fe as 8FeO. Further, the 2 S atoms are bound by Fe as 2FeS. Thus 9Fe remain unbound. In the entire earth mass there arose in this way three liquid phases which were insoluble or only slightly soluble in each other: a magnesium-iron-silicate phase formed of the three oxides, an iron-sulfide phase and an iron phase. The relative amounts of these with proportions given above would be: 4 (4MgO, 2FeO, 3SiO$_2$), 2FeS and 9Fe, corresponding to about 75, 5 and 20 wt %. Large amounts of iron were thus left uncombined because of the lack of electronegative elements, and the liquid iron phase because of its density sank to the center and formed a core. The silicate phase formed a mantle outside the core and the sulfide phase was probably disseminated in both the core and mantle.

Which of these three phases will be chosen by another metal M depends on the position of equilibria of the types (M + Fe silicate) $\rightleftharpoons$ (M silicate + Fe) and (M + Fe sulfide) $\rightleftharpoons$ (M sulfide + Fe). If any of these equilibria is sufficiently strongly shifted to the right, M will clearly enter the silicate or sulfide phase. It may be said that M in the first case is bound to oxygen and in the second case to sulfur more strongly than iron is. The first type to a large extent includes metals that form noble-gas ions (5-2f), and the second type includes metals that form 18-shells, with the exception of gold.

On the other hand if the two equilibria mentioned are sufficiently strongly shifted to the left, M appears in metallic form and then enters the iron phase. In this case M is more weakly bound than iron to both oxygen and sulfur. To this type belong the metals of the latter parts of the transition-metal series (mainly groups 7–11, with the exception of copper and silver, which prefer the sulfide phase). The probably high proportion of nickel in the core is explained in this way.

It is clear that most substances will not enter only one phase. The above-mentioned equilibria are seldom shifted strongly enough to one or the other direction, and besides, the equilibrium positions are dependent on temperature, pressure and other chemical environment, factors that are very different in various parts of the earth's mass.

Even if the positions of the three main phases are determined by gravity, it is clear that the primary distribution of the remaining elements in these phases is not determined in this way. Thus, the distribution here does not depend on density or values of atomic mass, but on the affinities of these elements for the most important elements in the main phases.

Since probably all oxygen was combined when the earth was molten, it is believed that the original atmosphere was oxygen-free. Oxygen was later released, probably partly by the dissociation of water vapor in the upper atmosphere

Table 17.2. *The mean proportions of the most common elements in the accessible crust of the earth, the hydrosphere and the atmosphere.*

Element	Weight %	Atomic %	Element	Weight %	Atomic %
8 O	49.4	55.1	11 Na	2.6	2.0
14 Si	25.8	16.3	19 K	2.4	1.1
13 Al	7.5	5.0	12 Mg	1.9	1.4
26 Fe	4.7	1.5	1 H	0.9	15.4
20 Ca	3.4	1.5	22 Ti	0.6	0.2
			Totals	99.2	99.5

under the action of the strong radiation occurring there, and partly by the photosynthesis of green plants (21-2a).

c. Outermost layers of the earth. Of the directly accessible parts of the earth the outer layers of the solid crust of the earth have the largest mass. Outside the solid crust there are the water layers, which are often collectively termed the *hydrosphere* (this includes all liquid and solid water with dissolved substances, thus both fresh and salt water and ice), and the *atmosphere*. In addition it is customary to group all living matter on the earth together in the *biosphere*. Of these three regions the hydrosphere has the greatest mass, about 1.4×10^{23} kg, of which sea water constitutes over 98 %. The masses of the hydrosphere, atmosphere and biosphere are in the ratios of approximately 70 000:300:1. On the same scale the mass of the crust of the earth is approximately 1 400 000.

The distribution of the elements in the directly accessible portions of the earth is a result of a series of processes of which many stages are little known. Even at a very early stage of development a certain differentiation of elements at varying depths must have taken place, which surely was very important to the overall character of the outer layers. Thereafter there occurred in the earth's crust and the layers immediately underneath a constant mountain-building activity, involving folding, melting and crystallization. This activity is still going on and makes itself evident through earthquakes and volcanism. These deep-seated processes and the formations, rocks and minerals formed by them are usually called *endogenic*. The hydrosphere and atmosphere then work on the solid crust through extensive weathering, solution, erosion and transport processes. For the distribution of elements within regions where organic life has prevailed, the living chemical metabolism of the life processes plays a large role. All these externally operating processes and their products are called *exogenic*.

The differentiation causes difficulties in the estimation of the average composition even of the accessible portions of the earth, but values are now available that are probably fairly accurate. The mean proportions of the ten most common elements are listed in tab. 17.2. These proportions are little affected by the

Table 17.3. *The mean proportions of the elements in the accessible solid crust of the earth.*

The elements have been gathered in groups, each covering one decade of the unit p.p.m. ($=1$ part per million $=10^{-4}$ %). Within each group they are arranged according to decreasing proportion. The sequence is often somewhat uncertain. Elements with approximately equal proportion have been arranged according to atomic number inside parentheses.

p.p.m.		
> 100 000		O, Si
100 000	−10 000	Al, Fe, Ca, Na, K, Mg
10 000	− 1 000	Ti, H, P, Mn
1 000	− 100	F, S, Sr, Ba, C, (Cl, Cr), Zr, Rb, V
100	− 10	Ni, Zn, (N, Ce), Cu, Y, Li, (Nb, Nd), Co, La, (Ga, Pb), Th
10	− 1	{Sm, (Pr, Gd), (Sc, Dy, Hf), (B, Br, Sn, Er, Yb), (Be, Ge, As, Ta, U), (Mo, Cs, Eu, Ho, W, Tl)
1	− 0.1	Tb, Lu, Hg, I, (Cd, Sb, Tm, Bi), (Ag, In)
0.1 −	0.01	Se, Ar, Pd
0.01−	0.001	(Pt, Au), He, Te, (Ru, Rh, Re, Os, Ir)

composition of the hydrosphere and the atmosphere and are therefore in most cases practically equal to the proportions in the solid crust. The table shows that O and Si predominate, which is related to the abundance of water and the large predominance of quartz and silicates (especially feldspars) in the earth's crust. In the third place we find Al in terms of weight percent, and H in terms of atomic percent. In tab. 17.3 is shown schematically the mean proportions in the accessible, solid crust of all elements that are present over 0.001 p.p.m.

It is found that Harkins' rule often applies also to the proportions of elements in the earth's crust. However, exceptions are more common here than in the distribution of elements in the universe.

Many elements that one would expect to be classed as common, for example Cu and Sn, are obviously very rare in the earth's crust. The apparent "commonness" of these elements depends on the fact that they are concentrated in a quite small number of accessible localities, and the easy recovery from the minerals as well as their desirable properties make them very much used. On the other hand, an element such as Ti belongs to the most common elements (it has a larger proportion in weight percent than, for example, Cl, P, C, S, Cr, Zn, and Ni) but is rather evenly distributed and besides is difficult to recover and work. Other examples are As, Sb, and Cd, which are to be found in much smaller proportions than Y and Ce, which have been counted among the "rare-earth metals".

Minerals that contribute to the largest total amounts of an element in the earth's crust often cannot be exploited for economic recovery of that element. Instead recourse must be made to other minerals with a higher proportion of the element, which are sufficiently locally concentrated to make recovery profitable.

The differentiation of the elements often causes an element to occur in definite surroundings. Certain combinations of elements give especially stable systems and are, therefore, common. Elements that occur together with oxygen and are directly bound to oxygen atoms are, for example, nearly always H (H_2O and OH^-), B (borates), Al (oxides and silicates), C (CO_2 and carbonates), Si (SiO_2 and silicates), N (nitrates) and P (phosphates). Very often Cu, Ag, Zn, Cd, Hg, In, Tl and Pb occur as sulfides. More details are given under the different elements in part II.

On crystallization of the magma, elements that replace each other in a crystallizing phase, that is, form solid solutions (6-3a), often accompany each other. They must then have similar atomic or ionic radii. For direct substitution of one ion by another, the charges must be equal. Ions of this type that accompany each other are, for example (cf, the ionic radii in fig. 5.5): Mg^{2+}–Mn^{2+}–Fe^{2+}; Co^{2+}–Ni^{2+}; Al^{3+}–Cr^{3+}–Fe^{3+}; Zr^{4+}–Hf^{4+}; F^-–OH^-. In the pair Zr^{4+}–Hf^{4+} the ionic radii are so similar that Zr and Hf always faithfully follow each other. Hf is found only in Zr minerals, which, on the other hand, always contain Hf. The mutually nearly equal ionic size of the lanthanoids causes the lanthanoid minerals usually to contain simultaneously a number of different lanthanoids. Na^+–Ca^{2+} and Al^{3+}–Si^{4+} are examples of ions with different charges that, because of their closely similar sizes, often accompnay each other, although formation of solid solution here requires a charge-compensating process (6-3a).

In the layers outside the earth's crust, the hydrosphere and the atmosphere show a distribution of elements that is fairly easy to discern as far as the main features are concerned. The composition of the hydrosphere is represented quite well by the mean composition of sea water, that is, about 3.5 % salts, mainly NaCl. The total composition of the atmosphere is determined completely by the lower, denser layers of air, which consist chiefly of N_2, O_2, H_2O, CO_2 and noble gases. In the very thin layers at great height, however, the relative amounts of the components are altered. Within a layer between approximately 400 and 800 kilometers above the earth's surface oxygen predominates, which through the dissociating action of the strong ultraviolet radiation (21-3a) is present here mainly as O. Higher up the lightest elements begin to prevail. From approximately 800 to 1500 kilometers He predominates, after which hydrogen in the form of protons H^+ becomes more and more prevalent. Many of the chemical processes that take place in the hydrosphere and atmosphere, or in which these participate, are rather incompletely understood.

In the biosphere differentiation is naturally very great. Among the more important elements H, C and O take part in the overall processes in the photosynthesis of plants and in respiration and are therefore the energy-giving elements in life processes. Their proportions in organisms are also high. N, P, S, Ca, Mg, K, Na and Cl have lesser proportions. Of these Na and Cl do not appear to be necessary to plants, but on the other hand they play a large role in animal organisms. Here they are present in the form of Na^+ and Cl^- as important ionic constituents in the body fluids.

In addition there are present in organisms in smaller proportions a large number of elements that are essential, often in very special functions. The most important of these *microelements* are B, Si, F, I, Mn, Fe, Co, Cu, Zn, Mo.

Inorganic compounds formed in organisms by the above-named elements enter especially into *skeleton formations*.

Not uncommonly an element normally employed in a biological process is replaced by a similar element, analogous to the mineral replacements mentioned above. This can lead to severe disturbances if thereby an essential subsequent process is prevented or completely new processes are made possible (see, for example, under selenium, 21-2a).

Many of the elements existing in the biosphere are met with in large part in fossil remains of plants and animals (especially carbon in limestone, coal and oil). An appreciable enrichment of other elements also can take place in such deposits. This may result from the fact that the element in question has been taken up by the organism and is thereafter retained in the remains more strongly than other elements that form more easily soluble compounds. The reason may also be that the element during fossilization is especially strongly adsorbed from penetrating waters, this element being earlier dissolved from the surrounding rocks. An example is germanium, whose mean proportion in coal ash (coke ash) is about 100 times higher than in the earth's crust as a whole. Germanium contents of over 10 g kg^{-1} have several times been found in coal ash. A similar enrichment is also known for uranium.

PART II
Systematic Inorganic Chemistry

Arrangement of part II

In this part hydrogen, which has no generally obvious position in the periodic system (4-2c), is treated first. Thereafter the groups are taken up from the right in the periodic system as it is given in tab. 4.3, down to group 13. The electronegative elements are in this way encountered at an early stage, so that useful generalizations can be made in the discussion of anions in which these so often are incorporated. The greater portion of the metals belong to the remaining elements, which are then taken up group by group from the left beginning with group 1. The material already given under the anions is not in general repeated under the metals.

For the most part repetitions have been avoided as much as possible, but cross references are very often given. The reader should also look for the respective anion in the index.

Familiarity with the periodic table as it is shown in tab. 4.3 is assumed throughout. The isotopic composition of the elements is referred to only in certain cases, and for such data reference is made also to tab. 1.3 and 1.4, and fig. 9.1.

In general the properties and reactions of substances and systems have here been described as they occur under the conditions in which they are ordinarily encountered and used. Thus, temperatures and pressures are rarely considered to have any extreme values at which the properties may become completely different from those described here (cf., for example, the description of the effect of high temperature in 5-6d and of high pressure in 6-4d). However, it should not be forgotten that the development of technology makes possible operations under more and more extreme conditions and that the chemist may therefore soon have greater reason to consider even these conditions.

CHAPTER 18

Hydrogen

a. Occurrence. As stated in chap. 17 hydrogen is the most common element in the universe, and ranks in third place in the accessible regions of the earth in terms of number of atoms. Here it occurs in by far the largest part bound to oxygen in water and hydroxide ions, but also in organic compounds. Free hydrogen, on the other hand, is rare. It is formed in certain fermentation processes. In small amounts it may be found enclosed in rocks and contained in natural gases (well or mine gases and volcanic gases). At a height of some hundred kilometers in the atmosphere its proportion begins to rise noticeably. Here the hydrogen is primarily monatomic. Above a height of about 1500 kilometers hydrogen is the dominant element and here exists nearly completely ionized to protons H^+. In the upper air layers the hydrogen isotope $^3H(=T)$ is continually formed from nitrogen atoms under the action of cosmic radiation (9-3c).

b. History. Cavendish in 1766 found that a flammable gas was produced when metals were dissolved in acids. In 1783 he showed that water was formed on the combustion of this gas. Lavoisier in 1784 obtained the same gas by passing water vapor through a glowing iron tube, and interpreted Cavendish's result in accordance with modern concepts. Lavoisier gave the combustible gas the name *hydrogène* ("water former" from Gk. ὕδωρ, water; γεννάω, I create). The hydrogen isotope ^2H (deuterium) was discovered in 1932 by Urey.

c. Properties. Free hydrogen at the surface of the earth consists most often of molecules H_2, which at room temperature form a colorless and odorless gas. The bonding in the hydrogen molecule was discussed in 5-3a. Some physical data for molecular hydrogen are the following:

Melting point	Boiling point	Critical temp.	Critical pressure
$-259.2°C$	$-252.8°C$	$-239.9°C$	13.0 bar

Hydrogen gas has the least density of all gases. Liquid hydrogen is the lightest of all liquids (density at the boiling point 0.070 g cm^{-3}) and crystalline hydrogen the lightest of all solids (density at the melting point 0.088 g cm^{-3}). For the preparation of liquid hydrogen, see 13-4a.

Hydrogen dissolves only slightly in water; at 20°C and total pressure (that is, the partial pressure of hydrogen + the vapor pressure of water) 1 atm, 100 g water dissolves 0.16 mg H_2. Because of the small molecular weight hydrogen molecules have a high mean velocity (cf. 2-1b). This leads to a high diffusibility in the gas phase, and similarly to a high thermal conductivity. At red heat, hydrogen gas penetrates into transition metals in the form of atoms; palladium is permeated as low as 200°C.

The hydrogen isotopes ^{1}H (light hydrogen, protium), ^{2}H or D (heavy hydrogen, deuterium) and ^{3}H or T (tritium) and the special applications of the two heavier isotopes have already been considered several times (among others in 1-3e, 9-3, 9-4b, tab. 1.4). The relative mass differences among the different hydrogen isotopes is so large that the isotope effect (9-3) is more noticeable than for heavier elements.

Atomic nuclei of certain nuclides possess spin (*nuclear spin*). If two atoms with identical, rotating nuclei are combined into one molecule, the rotation axes become parallel. However, the direction of rotation of the two nuclei can either be the same (parallel spin) or opposite (antiparallel spin). The hydrogen atoms H, D and T all have nuclear spin and each of the molecules H_2, D_2 and T_2 therefore occur in two different forms. Molecules with parallel nuclear spin are called orthohydrogen and molecules with antiparallel nuclear spin are called parahydrogen. For hydrogen the two forms have noticeably different properties, but this is not the case for similar pairs that arise from heavier atoms (for example, Cl_2)

For H_2 the ortho form has higher energy than the para form. At 0°K, therefore, only parahydrogen exists at equilibrium, but with the addition of energy (rise in temperature) the equilibrium: (orthohydrogen)$\rightleftharpoons$(parahydrogen + energy), is shifted to the left. However, it approaches a statistically determined limiting value corresponding to 25% parahydrogen + 75% orthohydrogen, and this limit is practically reached at room temperature. Certain substances catalyze the approach to equilibrium (for example, Pt and Pd, which react with H_2 with rupture and reformation of the H–H bond; but also others, for example, active charcoal and iron(III) oxide hydroxide). Without catalysts this occurs very slowly at room temperature or below. If liquid hydrogen is prepared from hydrogen gas containing orthohydrogen (for example, corresponding to the room-temperature equilibrium), the conversion of the orthohydrogen leads to a slow development of heat. Such liquid hydrogen therefore cannot be kept without losses even if the thermal isolation were complete. To avoid such losses the liquid hydrogen is treated during its manufacture with an appropriate catalyst (iron(III) oxide hydroxide or other catalyst with secret composition) so that it is practically completely converted to parahydrogen.

The melting and boiling points of parahydrogen are approximately 0.1° lower than for ordinary hydrogen (25% parahydrogen + 75% orthohydrogen).

At ordinary temperature molecular hydrogen has little reactivity. It combines immediately only with fluorine, with chlorine on irradiation (8g) and with oxygen on ignition. Oxidation with oxygen occurs according to the reaction $H_2(g) + \frac{1}{2}O_2(g) \rightarrow H_2O(l)$; $\Delta H^0 = -286$ kJ. The high formation enthalpy of water is clearly caused by the high bond energy of the water molecule.

A hydrogen-oxygen gas mixture of the above composition explodes violently on ignition (*oxyhydrogen*). Mixtures of compositions that depart markedly from this are also explosive. As stated in 1-5b a hydrogen-oxygen gas mixture remains practically unchanged at ordinary temperature if no catalyst is present.

Hydrogen

Hydrogen gas burns in oxygen gas with an almost invisible, very hot flame (*oxyhydrogen flame*, 2700°C), which is used for melting and welding metals.

At higher temperatures hydrogen combines with numerous elements with the formation of *hydrides* (18f). The high bond energy of water causes hydrogen preferably to bond oxygen, and thus effectively reduce many oxides.

Hydrogen can also exist in monatomic form, *atomic hydrogen*. In 5-3a the bond energy in H_2 was given as 435 kJ per mole of bonds. This amount of energy must thus be supplied to dissociate 1 mol H_2. This can occur through a strong increase in temperature (at 3700°C the degree of dissociation of hydrogen is 62%) or by electric discharge in hydrogen gas at reduced pressure. As mentioned previously (8f, g), H atoms may also be formed as intermediates in chemical reactions.

In atomic hydrogen the atoms naturally recombine quite rapidly (at ordinary temperature practically complete recombination has taken place within one half second), but even so, with proper technique reactions that do not occur with molecular hydrogen can be carried out with atomic hydrogen. Nonmetals that are not attacked by molecular hydrogen at room temperature (for example, O_2, S, As) give hydrides, and a large number of substances are reduced. Recombination is catalyzed especially by many metals, in which the released bond energy can lead to strong heating. If hydrogen gas is blown through an electric arc between tungsten electrodes, it is dissociated, and when it then impinges on a metal surface it is so strongly heated (up to ~ 4000°C), that melting and welding can be carried out ("atomic arc welding"). One advantage of this method is the strongly reducing atmosphere, which prevents oxidation. Atomic arc welding is used especially with stainless and acid-resistant steel.

The ionized hydrogen atom ($^1H^+$ = proton, $^2H^+$ = deuteron, although both in ordinary chemical processes are most often written H^+ and called *proton*) plays an important role in *acid-base equilibria* (chap. 12). The participation of the proton in *hydrogen bonding* is treated in 5-5.

d. Preparation. Water and hydrocarbons are used as the main source materials for the preparation of hydrogen. *Electrolysis of water* is preferred when the cost of electric power is low. To obtain sufficient conductivity a solution of NaOH or KOH is used as an electrolyte. Fe cathodes and Ni anodes serve as chemically highly resistant electrodes with low overvoltage (16f). Hydrogen formed in the "chlor-alkali electrolysis" (20-4b) and electrolysis of waste hydrochloric acid is also recovered.

However, the largest amounts of hydrogen are produced by *reduction of water with carbon or hydrocarbons*. When water vapor is passed over incandescent carbon (coke in practice), H_2O is reduced in accord with the reaction

$$C(s) + H_2O(g) \rightarrow CO(g) + H_2(g) \tag{18.1}$$

Meanwhile, in the gas phase the *water-gas equilibrium* is set up

$$CO(g) + H_2O(g) \rightleftharpoons CO_2(g) + H_2(g) \tag{18.2}$$

for which $K_p = 1$ at 810°C. Below 810°C, $K_p > 1$ (cf. 10b where the value 478 was calculated for 25°C), and above 810°C, $K_p < 1$. With rising temperature the equilibrium is thus displaced more and more to the left, which means that the overall process corresponds more closely to (18.1). The gas mixture above 1000°C, consisting chiefly of CO and H_2, is called *water gas*. Heat is absorbed in the reaction and to maintain the high temperature, water vapor and air are blown in alternately so that the combustion of a certain amount of carbon raises the temperature again. Water gas is expensive as a fuel, but is used for certain heating purposes where a high heat of combustion is required. More often it is used, after modification of the composition, in various syntheses (*synthesis gas*). If only the hydrogen is desired it is best to convert the water vapor at a low enough temperature (400–500°C) so that the equilibrium (18.2) is greatly shifted to the right. CO then reacts with H_2O so that CO_2 and H_2 are formed. At this low temperature, however, a catalyst is required (iron or nickel oxide) in order to obtain a sufficiently rapid attainment of equilibrium. Addition of (18.1) and (18.2)—the latter with the reaction direction to the right—gives the overall reaction

$$C(s) + 2H_2O(g) \rightarrow CO_2(g) + 2H_2(g) \tag{18.3}$$

which shows that the theoretical amount of hydrogen obtained from a given amount of carbon is now doubled. The major portion of CO_2 in the gas mixture obtained is eliminated by solution in water under pressure and the remainder by absorption in NaOH solution. Traces of CO are removed with ammoniacal Cu(I) chloride solution (23-2f). A more recent method to obtain pure hydrogen from the gas mixture is to let the mixture at about 400°C pass along a thin wall of a palladium-silver alloy, which lets through hydrogen (35c) but not the other gas components.

Nowadays natural hydrocarbons are more and more being used as starting material. Hydrocarbon gases together with water vapor are passed over a catalyst (nickel) at about 900°C. For example, methane is thus converted according to $CH_4 + H_2O \rightarrow CO + 3H_2$ and $CH_4 + 2H_2O \rightarrow CO_2 + 4H_2$. The CO can then be converted with water vapor as above according to (18.2).

Since the 1950's, *partial oxidation of hydrocarbons* has been applied to a large extent to both natural gas and oil. Oxidation is carried out with oxygen at a temperature of over 1100°C and a total pressure of about 30 bar and a product obtained of the approximate composition $CO + H_2$.

To a more limited extent hydrogen is also produced industrially by *reduction of water vapor with iron* at about 600°C: $3Fe(s) + 4H_2O(g) \rightarrow Fe_3O_4(s) + 4H_2(g)$. Fe_3O_4 is then reduced to Fe with water gas and used over again. It is found that if the water-gas production is included the overall reaction is the same as (18.3).

In the manufacture of coke, particularly for use in the smelting of iron (23-2b), *coke-oven gas* is obtained, which consists in more than half its volume of hydrogen. Several of the large ammonia plants on the European continent manufacture their synthesis gas by cooling coke-oven gas so strongly that all of its components except hydrogen and nitrogen are condensed and can then be removed.

In commerce hydrogen gas is usually handled compressed at a pressure of 150 atm in steel cylinders. However, hydrogen has begun to be distributed also in liquid form.

On a laboratory scale, hydrogen is mostly prepared by dissolving metals in acids (most often zinc in hydrochloric or sulfuric acid). Light-weight sources of hydrogen for military uses among others are calcium hydride CaH_2 and lithium hydride LiH, which are made to react with water (for example, $CaH_2(s) + 2H_2O \rightarrow Ca(OH)_2 + 2H_2$).

e. Uses. Large quantities of hydrogen are consumed in the manufacture of ammonia according to the Haber-Bosch process (22-6b). In the synthesis of methanol and other alcohols as well as hydrocarbons according to the Fischer-Tropsch process (14-4), water gas is the starting material. Another important application of hydrogen is in the hydrogenation of unsaturated fatty acids contained in animal and vegetable oils, from which solid fats are obtained for the manufacture of margarine.

Hydrogen is used also in the production of hydrochloric acid. It is burned in this process with chlorine in special combustion chambers, after which the HCl formed is dissolved in water. The reduction of metal oxides with hydrogen has been used especially in the manufacture of Mo and W from MoO_3 and WO_3 in the incandescent-lamp industry and in the manufacture of hard metals (31-4a, 31-6c). Nowadays also iron ore is reduced with hydrogen–carbon monoxide gas mixtures (34-3b), and methods are being tested that use pure hydrogen gas.

The use of hydrogen gas as a fuel in welding has been mentioned earlier (18c). Liquid hydrogen has begun to be used as fuel in rockets. The heat of combustion is very high when figured per unit of weight of hydrogen, but less favorable when figured per unit of volume. One disadvantage, of course, is the low boiling point.

Because of its lightness hydrogen gas is much used in the inflation of balloons. However, the danger of inflammability of hydrogen has led to its replacement by helium where possible.

Tritium, incorporated in an appropriate compound mixed with a luminescent substance (6-4e), has been used in a self-luminous material. The advantage of tritium is that it emits only β rays with low penetrating power, which are thus easily shielded. The half-life of 12 years permits a useful life of about 20 years.

f. Simple hydrogen compounds. Here are considered only in the most general way simple hydrogen compounds, which for the simplest, binary cases have the general formula A_xH_y. These are usually collected under the term *hydrides*, but according to the rules in 5-2a and 5-3f this name should not be used when A is a halogen (belonging to the fluorine group) or a chalcogen (belonging to the oxygen group).

It may generally be said that those elements that in the periodic system (tab. 4.3) lie to the right of the transition elements form hydrogen compounds in which the bonds have strongly covalent character (*covalent "hydrides"*). For the most strongly electronegative elements (F, O, N, Cl, Br), however, the ionic contribution is pronounced, so that H, of course, becomes positive. The covalent hydrogen compounds most often consist of finite molecules and are then, as a rule, volatile. However, infinite molecules do occur, for example in high-polymer hydrocarbons. The stability of covalent hydrogen compounds of elements in the same group of the periodic system decreases generally as the atomic number of the element rises. In 5-6 b a number of conditions for the composition and stability of such hydrogen compounds was further discussed, where also charged molecules (ions) were considered. Cations in which protons are ligands are called *-onium ions*; for example, fluoronium ion H_2F^+, oxonium ion H_3O^+, ammonium ion NH_4^+, phosphonium ion PH_4^+. Anions are generally given names of the type tetrahydridoborate ion BH_4^-, but there are a few special names such as hydroxide ion OH^-, amide ion NH_2^-, imide ion NH^{2-}. It was also mentioned in 5-6 b that the hydrides of the elements in group 13 are often electron-deficient molecules. The hydrogen compounds of the most electronegative elements (especially the halogens but also the chalcogens) behave as acids in water solution.

The covalent hydrides are treated more extensively under the respective element. Many of them are very well-known and have special names, such as water H_2O, ammonia NH_3, methane CH_4. Usual methods of preparation are combination of the element (for example, halogens, chalcogens) directly with hydrogen, or decomposition of a metal compound of the element (for example, magnesium boride, carbide, silicide, phosphide) with water or acid. Other reduction methods (for example, cathodic reduction) of some compound of the element are also used.

The strongly electropositive metals in groups 1 and 2, on the other hand, form *hydrides with ionic bonding* (*saltlike hydrides*), containing the *hydride ion* H^-. The composition corresponds to the ordinary electrovalences, for examples, LiH, CaH_2. They are solid at ordinary temperature and can be prepared by heating the metal in hydrogen. At still higher temperatures they decompose again into the metal and hydrogen.

The transition elements form *metallic hydrides* (cf. 6-4 g), which can generally be prepared by heating the metal in hydrogen. The crystals are built up of metal atoms and hydrogen atoms and since these building units are uncharged the composition of the phase does not correspond to any electrovalence principle, and besides is easily varied. The metal-atom structure is unchanged with this variation and only the number of hydrogen atoms is changed (interstitial solution 6-3 a).

Intermediate states among the three bond types described above occur and are found especially in the neighborhood of the limits of the region of the transition elements.

CHAPTER 19

Group 0, Noble Gases

a. Introduction. The noble gases earned their name because of their great indifference to reaction, their "nobility". This results from their stable electron envelopes (4-1e) with filled orbitals, which discourage both ionization and stable covalent bonding (5-3a). However, ions such as He_2^+, Ne_2^+ and HNe^+ have long been known to be formed in electric discharges in gases (5-3a, 5-6b) but these ions very easily take up electrons and then quickly decompose. Beginning in 1962, however, a number of more stable compounds have been successfully prepared. In these the heavier, noble-gas atoms krypton, xenon or radon are bound to atoms with high electronegativity, namely, fluorine and oxygen.

b. Occurrence. All of the noble gases, helium, neon, argon, krypton, xenon and radon, are found in the earth's atmosphere. This is, in fact, the only source for the preparation of pure neon, argon, krypton and xenon. The proportions of the noble gases in the atmosphere at the surface of the earth are:

helium 0.0005 vol%	krypton 0,00011 vol%
neon 0.0015	xenon 0.000009
argon 0.94	radon varying traces

The proportion of argon is thus appreciable and more than 30 times greater than the average proportion of CO_2 (0.03 vol%). In 100 m³ of air (for example, a room 20 × 18 × 10 ft), there are 0.5 l helium, 1.5 l neon, 940 l argon, 110 ml krypton and 9 ml xenon at atmospheric pressure. At very high altitudes completely different relative amounts are found. Within a layer between approximately 800 and 1500 km helium is more common than all other elements.

In the universe helium is the most common element next to hydrogen (tab. 17.1), but neon and argon also show high proportions. As described in 17b, however, only small amounts of noble gases were incorporated into the earth when it was formed. The proportions of neon, krypton and xenon on the earth very likely have this origin, but the others have wholly or partly been formed later from constantly continuing nuclear reactions. The most common helium isotope ^4He (tab. 1.3) is thus formed in all α decays (9-2a), and the most common argon isotope ^{40}Ar is formed from the potassium isotope ^{40}K according to $^{40}_{19}K \xrightarrow{EC} {}^{40}_{18}Ar$ (half-life 1.42 × 10⁹ y). The origin of radon lies completely in nuclear reactions, since all radon isotopes are unstable with short half-lives. In nature they are produced by α

decay of radium isotopes (9-2a), and decay themselves to polonium. The longest-lived isotope is ^{222}Rn (half-life 3.8 d), which is formed in the uranium series from ^{226}Ra. The radon content of the air is less above the oceans than above the continents because of the low content of radium in sea water.

Noble gases produced by nuclear reactions under the earth's surface may become incorporated in natural gases. In these helium is found in richest supply. Helium-containing natural gases occur in sufficient amounts for the recovery of helium in several south-central states of the United States, in Canada and in the Soviet Union. These areas are also rather rich in uranium minerals. In most of the gas wells used the gas runs around 1 vol% helium but contents up to 7 vol% have been found. Helium may also be retained in the minerals containing the radiating element from which it is formed, and it can be released from the mineral by fusing or dissolving it.

c. History. In 1785 Cavendish made some experiments, which he did not completely understand, but they can now be said to have shown that air contains a small amount of a nonreactive gas. The next development came in 1892 when Lord Rayleigh found that air, freed from all oxygen and assumed to be nitrogen, had higher density than nitrogen prepared from ammonia. This discrepancy led Lord Rayleigh and Ramsay to investigate carefully atmospheric "nitrogen", on the basis of which they were able to establish in 1894 the existence of a new gaseous element in air. Because of its nonreactive (inert) properties it was given the name *argon* (Gk. ἀργός, lazy, sluggish).

In the year 1868 some lines were observed in the solar spectrum which could not be interpreted as being caused by any known element. Lockyer and Frankland therefore ascribed them to a hypothetical element, to which they gave the name *helium* (Gk. ἥλιος, sun). In 1895 Ramsay and Crookes found that a gas that was obtained when the uranium-bearing mineral cleveite was dissolved in sulfuric acid gave the same spectrum, and was thus helium. In the same year Kayser found helium lines in a spectrum of an argon preparation obtained from air. In "argon" obtained from air, condensed to a liquid and then subjected to fractional distillation, Ramsay and Travers in 1898 established the presence besides helium of three further gaseous elements, which were given the names *neon* (Gk. νέος, new), *krypton* (Gk. κρυπτός, hidden) and *xenon* (Gk. ξένος, stranger). Ramsay and Soddy showed in 1903 that α particles consisted of helium (in particular, helium nuclei $^4_2$He^{2+}, see 9-2a).

In 1900 Rutherford concluded that the gaseous "emanation" (9-2a), formed by decay of ^{224}Ra in the thorium series, was a noble gas. It was possible later to interpret the corresponding emanations in the uranium and actinium series as other isotopes of the same element. This now carries the name *radon* (earlier, radon was called niton and the three isotopes mentioned were called thoron, radon and actinon).

d. Properties. The noble gases commonly occur as monatomic gases, which are colorless and odorless. Their solubilities in water increase with the atomic weight and are quite high for the heavier gases. At 0°C the solubilties in 100 volume units of water are for He, 0.9; Ne, 1.1; Ar, 5.8; Kr, 11; Xe, 24; and Rn, 53 volume units at atmospheric pressure.

Both in the crystal and liquid the noble-gas atoms are almost entirely held

Table 19.1. *Physical properties of the noble gases.*

	Melting point °C	Boiling point °C	Crit. temp. °C	Crit. press. bar
2 He	−272.1[a]	−269.0	−267.9	2.29
10 Ne	−248.6	−246.0	−228.7	27.2
18 Ar	−189.4	−185.9	−122.4	50.6
36 Kr	−157.2	−152.9	− 62.5	55.0
54 Xe	−111.8	−107.1	+ 16.6	60.0
86 Rn	− 71	− 61.8	+104.5	63.2

[a] At a pressure of 25.3 bar.

together by van der Waals forces, which according to 13-4b accounts for the magnitudes and behavior of the melting and boiling points (cf. tab. 19.1).

Concerning the preparation of liquid helium, see 13-4a. Strangely enough, liquid helium does not solidify under its own vapor pressure, no matter how low the temperature is. Solidification takes place only at a minimum pressure of 25.3 bar, at which the solidification temperature is −272.1°C = 1.1°K. No other liquid can be cooled to so low a temperature without going over to a solid form (the melting point of hydrogen is −259.2°C = 14.0°K). At very low temperatures liquid helium has properties that have never been observed in other liquids, and which can almost be said to characterize a new kind of aggregation state.

When helium is condensed to a liquid, so-called helium I is first formed, which behaves as a normal liquid. However, with a lowering of the temperature this liquid is transformed at 2.17°K to another liquid modification, helium II. Among other things, this form is extraordinarily fluid (viscosity values have been measured that are about 10^{-3} times that of gaseous hydrogen, and peculiarly enough the viscosity decreases as the temperature is lowered), and it has a thermal conductivity that is 800 times that of copper at room temperature. These properties can only be explained through quantum mechanics.

e. Recovery. In the recovery of the rare gases from air, both condensation of the greater proportion to "liquid air" (13-4a) followed by fractional distillation (13-5c), and selective adsorption (15-1a) on active carbon (23-2b) at low temperature are used.

The processes of the first kind are based on the following boiling points of the components in dry, CO_2-free air:

He	Ne	N_2	Ar	O_2	Kr	Xe
−269	−246	−196	−186	−183	−153	−107°C

Selective adsorption is based on the fact that the adsorptivity of the noble gases on active carbon increases with increasing atomic weight of the gas, thus in the sequence He < Ne < Ar < Kr < Xe. Thus at −190°C all noble gases except He are adsorbed, but at −100°C only Ar, Kr and Xe. With rising temperature a lighter, adsorbed gas is liberated before a heavier one, thus making separation possible.

Liquid air contains mainly the five gases to the right in the boiling-point table above. A large part of N_2 remains gaseous as well as He and Ne. Separation of these is achieved by condensation of N_2 and adsorption of Ne on active carbon. Fractional distillation gives finally the three main fractions N_2, $Ar + O_2$ and $O_2 + Kr + Xe$. O_2 can be removed by burning H_2 (added in excess; the remaining H_2 is caused to reduce CuO, after which the gas is dried). In this way Ar and a mixture of $Kr + Xe$ containing about 10 vol% Xe are obtained.

The largest quantities of helium are separated from natural gas by similar methods. In this way about 18 million m³ at atmosphere pressure were produced in the United States in 1960, and the rate of production is rapidly increasing.

f. Uses. All noble gases except radon are used in illumination technology. Electric discharge through a glass tube containing neon at low pressure gives the well-known red "neon light", whose color can be changed by the addition of other noble gases or mercury, or by the use of colored glass. Argon is used in incandescent lamps, in which it reduces the rate of evaporation of the tungsten filament, thus making higher filament temperatures possible. Nowadays a mixture of 93% Ar and 7% nitrogen at a pressure of about 700 mbar is most commonly used. Furthermore, for this purpose the heavier noble gases are still more effective and also permit smaller lamp dimensions because of their lower heat conductivity. However, because of their high cost these gases are used mostly for special lamps. In such lamps the mixture of krypton and xenon mentioned in 19e is most often used. The noble gases are also used in a considerable number of electronic instruments (for example, Geiger counters).

Helium, and still more the cheaper argon, are much used to protect substances that are attacked by oxygen and nitrogen, especially under strong heating (shield gas). Such is the case in modern metallurgy, for example in the production of titanium and zirconium, and in the welding of light metals and highly alloyed steels (argon arc welding).

In the United States, where the supply of helium is incomparably greater than anywhere else, large quantities of helium are used for inflating balloons. Also, a great deal of helium is used to dilute oxygen in diving gas. Since helium only insignificantly dissolves in the blood and body fluids, diving sickness (bends) is avoided in this way.

The low molecular weight of helium enables it to diffuse and flow through fine orifices. Since it also is easily detected spectroscopically it is used as a leak detector. Because of the possibility it offers for reaching very low temperatures, helium has also become very important in low-temperature research.

In 1960 it was decided by international agreement that 1 m shall consist of 1 650 763.73 wavelengths in vacuum of the light of a certain (orange-colored) line in the spectrum of the krypton isotope ^{86}Kr.

g. Noble-gas compounds. In 1962 Bartlett and coworkers found that gaseous platinum hexafluoride PtF_6 combines with molecular oxygen to give a quite stable product $O_2(PtF_6)$, which they considered to be made up of the ions O_2^+ and PtF_6^-. The ion O_2^+ had not been previously encountered in any stable compound, but PtF_6 clearly had such a high electron affinity that it was able to ionize O_2 to O_2^+. Bartlett then noticed that the ionization energies for the processes $O_2 \rightarrow O_2^+ + e^-$ and $Xe \rightarrow Xe^+ + e^-$ are practically the same. PtF_6 should therefore also be able to ionize Xe to Xe^+ and thus be able to combine with this noble gas. An experiment carried out by Bartlett confirmed this hypothesis. A yellow-red, solid substance $Xe(PtF_6)_x$ with x varying between 1 and 2 was formed. This discovery led to intensive research activity in which numerous noble-gas compounds were prepared and studied.

The noble-gas compounds characterized chemically up to now contain krypton, xenon or radon. The stability toward decomposition increases with increasing atomic number, a behavior connected with the fact that the noble-gas atoms are more easily ionized the larger their electron clouds are (5-2b, fig. 5.4). Because of the rarity and strong radioactivity of radon, however, its compounds have been little studied. The best known compounds, therefore, are those of xenon, which are taken as a model in the following. The noble-gas atoms are always bound to strongly electronegative atoms; apparently only compounds with fluorine and oxygen have so far been established. The compounds show many similarities with those of the elements one or two places to the left in the periodic system; thus compounds of Xe are similar to those of Te and I.

The *xenon fluorides* XeF_2, XeF_4 and XeF_6 are known, which are all formed on the addition of activation energy to gaseous mixtures of xenon and fluorine. As a flask material nickel has here been mostly used. XeF_4 is most easily obtained by heating to about 400°C and then quenching. All three of the fluorides at room temperature are colorless, solid and strongly oxidizing phases. The vapor pressures increase with increasing fluorine content and are so high that the fluorides can be sublimed at room temperature. At this temperature they decompose only slowly and can be preserved for a long time at lower temperatures.

All three xenon fluorides react with water, more rapidly the higher the fluorine content is. XeF_2 decomposes completely according to the reaction $XeF_2(s) + H_2O \rightarrow Xe(g) + \frac{1}{2}O_2(g) + 2HF$. A yellow color appears transiently on the surface of the solid phase during the reaction.

The overall reaction for the hydrolysis of XeF_4 can be written: $3XeF_4 + 6H_2O \rightarrow 2Xe(g) + XeO_3 + \frac{3}{2}O_2(g) + 12HF$. Thus, disproportionation of Xe(IV) takes place into Xe(0) and Xe(VI) concurrently with the oxidation of H_2O. Here also a yellow color appears on the surface of the fluoride at first, and also a slowly dissipating yellow color in the solution. *Xenon(VI) oxide (xenon trioxide)* XeO_3 is obtained on evaporation of the solution in the form of colorless, highly explosive crystals. In the solution Xe(VI) occurs as a polyvalent weak acid, "xenic acid", which has been written $Xe(OH)_6$ or H_2XeO_4. The solution is stable, but strongly oxidizing.

When the solution is made basic, Xe(VI) disproportionates to Xe(0) and

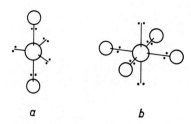

Fig. 19.1. Schematic structure and electron-pair distribution in compounds of the types (a) XeF_2 and (b) XeF_4.

Xe(VIII), usually with simultaneous oxidation of H_2O so that oxygen is produced. The same disproportionation of Xe(VI) occurs when XeF_6 is hydrolyzed in basic solution. In both cases Xe(VIII) forms a *xenate*(VIII) ("perxenate") ion XeO_6^{4-}. In order to convert practically all Xe(VI) into xenate(VIII) ozone may be passed through a solution of XeO_3 in 1 M NaOH, whereupon very stable hydrates of *sodium xenate*(VIII) Na_4XeO_6 precipitate. Many heavier metal ions give poorly soluble xenates(VIII). Water solutions of xenate(VIII) ion are strongly oxidizing. They also decompose according to the reaction: $Xe(VIII) + H_2O \rightarrow Xe(VI) + \frac{1}{2}O_2(g) + 2H^+$. This decomposition proceeds rather slowly in strongly basic solution but the rate increases rapidly as the acidity is increased.

With slow and complete hydrolysis of XeF_6 with water, XeO_3 is obtained: $XeF_6 + 3H_2O \rightarrow XeO_3 + 6HF$. If the amount of water is adjusted to $1H_2O$ per $1XeF_6$, the reaction $XeF_6 + H_2O \rightarrow XeOF_4 + 2HF$ takes place. *Xenon*(VI) *oxide tetrafluoride* $XeOF_4$ is obtained as a colorless liquid (m.p. $-46°C$).

Xenon(VIII) *oxide* (*xenon tetroxide*) XeO_4 is formed by reaction between sodium xenate(VIII) and concentrated sulfuric acid: $Na_4XeO_6 + 2H_2SO_4 \rightarrow XeO_4(g) + 2Na_2SO_4 + 2H_2O$. The process should be carried out carefully and in the cold (best about $-5°C$). The XeO_4 gas is condensed as a yellow substance, which at $0°C$ has a vapor pressure of 25 mm. The oxide decomposes rapidly, sometimes explosively, even below $0°C$.

XeF_2 has a linear, and XeF_4 a square structure. If the valence electrons are distributed as uniformly as possible around the central xenon atom according to the rules in 5-6e, the distribution in fig. 19.1a (trigonal bipyramid, cf. fig. 5.6g; probably sp^3d hybridization), and fig. 19.1b (octahedron, probably sp^3d^2 hybridization) is obtained. Two and four of these electron pairs, respectively, form bonds with fluorine atoms, which in turn are distributed as uniformly as possible over the electron pairs. In $XeOF_4$ an additional electron pair in fig. 19.1b has been used for bonding, now with an oxygen atom. XeF_6 has a total of seven electron pairs about the xenon atom, so that the six ligands coordinate with lower than octahedral symmetry.

XeF_2, XeF_4, XeF_6 and $XeOF_4$ with their lone electron pairs are electron-pair donors, and therefore can attach electron-pair acceptors. For example, the addition compounds $XeF_2 \cdot 2SbF_5$, $XeF_6 \cdot BF_3$ and $XeF_6 \cdot AsF_5$ are known, whose structures, however, are still undetermined. With solution of alkali-metal fluorides (except LiF) in liquid XeF_6, compounds of the type $MXeF_7$ are first formed. These easily give off XeF_6: $2MXeF_7(s) \rightarrow M_2XeF_8(s) + XeF_6(g)$. Octofluoroxenates(VI) of Rb and Cs withstand heating to 400°C and are the most stable xenon compounds known.

The noble-gas halides show many similarities to polyhalide ions (20-6). IF_4^- and IF_6^- have the same structures as XeF_4 and XeF_6, respectively, with which they are isoelectronic. IBr_2^-, for example has a structure analogous to XeF_2 (IF_2^- is not known).

XeO_3 is isoelectronic with the iodate ion IO_3^- (20-8d) and has the same pyramidal structure (cf. the structural formulation for the analogously formed chlorate ion ClO_3^- in 20-8a and its geometry according to the rules in 5-6e). In XeO_4 the xenon atom is surrounded by four electron pairs, which leads to a tetrahedral structure (cf. SO_4^{2-}). XeO_6^{4-} is isoelectronic with the orthoperiodate ion IO_6^{5-} (20-8e), and the six electron pairs around the central atom make the coordination octahedral.

Radon fluorides have been prepared but are still not completely characterized. Among *krypton fluorides*, KrF_2 and KrF_4 are known. These are more volatile and decompose more easily into the elements than the corresponding xenon fluorides.

Noble-gas atoms can be enclosed in growing crystals of certain substances that provide suitable cavities for them (clathrate compounds, 6-3a). Gas hydrates of noble gases are formed in this way, but only van der Waals forces are involved between the noble-gas atoms and the atoms of the enclosing structure.

CHAPTER 20

Group 17, Halogens

20-1. Introduction

Among the halogens, astatine has only unstable isotopes. The half-life of the longest-lived of these, ^{210}At, is only 8.3 hours, and consequently the chemistry of astatine is quite incompletely known. Its chemical characterization has been carried out chiefly with carrier techniques (9-5a). Therefore, data for this element often cannot be given, and it is omitted in many of the comparisons among the different halogens made here and in the following sections.

Fluorine is the most electronegative of all the elements (5-3f). As in all groups of the periodic system which do not include transition elements, the halogens also become less and less electronegative as the atomic number increases. However, only in the case of the heaviest halogen, astatine, has this decrease proceeded so far that the element may be considered to be a semimetal. The other halogens are nonmetals, although iodine without doubt lies on the boundary of the semimetals (note, for example, its metallic luster).

The high electronegativity of the halogens, especially the lighter ones, give them a great tendency to form anions. They therefore readily form typical salts. The word *halogen*, in fact, means salt former (4-2c).

Fluorine often deviates particularly strongly in its properties from the other halogens. This depends mainly on the factors described in the introduction to 5-6. It is found also that the abnormally low dissociation energy of the F_2 molecule (20-3a) leads to deviations, but this in turn can be considered to be a consequence of the same factors.

The fact that a noble-gas shell is achieved by the addition of one electron causes the oxidation number $-I$ to be particularly common for all halogens. As previously mentioned in 5-3g, fluorine takes this oxidation number in the presence of all other elements.

From tab. 20.3 it is seen that the ions XO^-, XO_2^-, XO_3^- and XO_4^- are represented among the oxohalate ions of chlorine, bromine and iodine. On the other hand, no oxofluorate ions or corresponding acids are known, which in 5-6b is explained for FO_2^-, FO_3^- and FO_4^- by the fact that fluorine is more electronegative than oxygen. Since only fluorine and oxygen are more electronegative than chlorine, chlorine has positive oxidation numbers only in compounds with these elements. Tab. 20.3 shows that in the oxygen compounds of chlorine all oxidation numbers from I to VII except II exist. The maximum oxidation number is thus equal to the unit's digit of the group number (5-3g). The odd numbers are the most important. With few exceptions, bromine and iodine also have mainly positive oxidation numbers in oxygen compounds, although here fewer oxidation numbers are represented than for chlorine. Other cases of positive oxidation numbers for bromine and iodine are shown by compounds between halogens (20-5e), and the existence of the ion I^{3+}, for example in IPO_4, I acetate and several other organic compounds. The existence of I^{3+} demonstrates the more electropositive character of iodine.

20-2. Occurrence and history

a. Occurrence. The free halogens are all very reactive and therefore do not occur in nature in this state.

Fluorine is the most common halogen (tab. 17.3). Large amounts are found in *apatite* $Ca_5(OH,F)(PO_4)_3$ (the ions OH^- and F^- can here substitute for each other, so that the fluorine content varies; cf. 5-2c and 22-2a) and *fluorite* (*fluorspar*) CaF_2 (6-2c), both very widespread minerals. Fluorite is the main source for the production of fluorine. A third fluorine mineral is *cryolite* Na_3AlF_6, which has been used primarily in the manufacture of aluminum (24-3c). Cryolite has been mined only in Greenland (Ivigtut), but this source is now exhausted. Fluoride ions are found in small amounts in both fresh water and sea water.

Vertebrate-animal skeletons contain some fluorine, but it is not clear in what way it is incorporated. It has been shown that a certain proportion of fluorine in drinking water increases the resistance of teeth against caries (the optimum concentration is about 1 part in 1 million parts of water, larger amounts causing mottled teeth; NaF is used as the additive).

Chlorine and *bromine* have to a large extent been eliminated from older magmas either in the form of volatile compounds or by solution in water. They therefore now occur for the most part as ions in the oceans (mean molarities: 0.53 M Cl; 0.0008 M Br) and in natural salt solutions and salt deposits, which arose from concentration of ocean waters or water in inland seas. Chlorine as Cl^- is essential to animal organisms, but generally occurs in very small amounts in plants and may not be essential to them.

Iodine has also been carried into the oceans where it is to a large extent organically bound in plants and animals. The largest local enrichment of iodine is found in the saltpeter deposits of Chile, where iodine occurs mostly as sodium iodate $NaIO_3$. In the United States large amounts of iodine are nowadays produced from brines associated with oil deposits, in which the iodine occurs mainly as iodide.

In mammals iodine is taken up especially in the thyroid gland in the form of amino acid derivatives (thyroxine and di-iodotyrosine). A deficiency of iodine can cause goiter. Since the body takes up iodine and concentrates it in the thyroid gland, unstable iodine isotopes are dangerous. Several such isotopes are formed by nuclear fission, but only ^{131}I has a long enough half-life (8 days) to be important.

Various unstable isotopes of *astatine* occur in nature in extremely small amounts as members of side branches in each of the natural radiaoactive decay series (cf. fig. 9.3). It has been calculated that the outermost 1 km layer of the earth's crust altogether contains only 43 mg astatine.

b. History. The first time a free halogen was prepared was when Scheele in 1774 obtained *chlorine* by oxidation of hydrochloric acid with pyrolusite. According to the phlogiston theory (21-2b) Scheele assumed logically that the new greenish-yellow gas was "dephlogisticated" hydrochloric acid. After the phlogiston theory was abandoned and the gas was proved to be undecomposable, Davy in 1810 designated it as an element and gave it the name *chlorine* (Gk. χλωρός, light green).

Next *iodine* was discovered by Courtois in 1811 in seaweed ash and was named after the color of the vapor (Gk. ἰωειδής, violet-colored). *Bromine* was prepared by Balard in 1826 from the mother liquor from ocean-salt pans and was named for its strong, unpleasant odor (Gk. βρῶμος, stench).

Fluorite was used very early as a "flux", that is, as a means of making oxides more easily fusible. This use is known from the middle ages, when the mineral was given the name, among others, of "fluores" (Lat. fluere, to flow). Scheele prepared "fluorspar acid" or "flux acid", that is, hydrofluoric acid. After Davy pronounced chlorine to be an element, Ampère in 1810 pointed out that fluorspar acid should be a hydrogen compound of an analogous element. This was given the name *fluorine*. The high reactivity of fluorine prevented its preparation, however, until 1886 when Moissan successfully isolated it.

Astatine was first prepared in 1940 by the nuclear reaction $^{209}_{83}\text{Bi}(\alpha,2n)^{211}_{85}\text{At}$. Later its occurrence in branches of the natural radiactive series was demonstrated. The name alludes to the fact that only unstable isotopes are known (Gk. ἄστατος, changeable).

20-3. Properties of the Halogens

In free form and at moderate temperatures, fluorine, chlorine, bromine and iodine in all aggregation states occur mainly as *diatomic molecules*. The bonds in these have been discussed in 5-3a. In tab. 20.1 it can be seen that the dissociation energies for these molecules rises at first as one goes upward from the bottom of the table. The iodine molecule is thus the most easily dissociated (in iodine vapor the degree of dissociation is 18% at 460°C and 1 atm). The value for F_2 interrupts this trend, however. The bond strength in the F_2 molecule is thus significantly less than would be expected judging from the bond strengths in the other halogen molecules. This results, among other things, in the fact that F_2 has greater oxidizing ability than Cl_2 even though F has a smaller electron affinity than Cl (see 5-2b).

As would be expected, the melting and boiling points rise with increasing molecular weight.

All the halogens have higher molecular weight than O_2 and N_2 and thus have

Table 20.1. *Physical properties of free halogens.*

Halogen	Color and State at normal temp. and pressure	Melting point °C	Boiling point °C	Dissociation energy D, kJ mol^{-1}
F_2	light-yellow gas	-223	-187	158
Cl_2	yellow-green gas	-101.6	-34.6	242
Br_2	red-brown liquid	-7.3	58.8	193
I_2	gray-black, lustrous crystals	113.5	184	152

greater gas density than air. However, the difference for F_2 is insignificant. Liquid Br_2 is a heavy fluid (density at 20°C = 3.12 g cm^{-3}).

Bromine has such a high vapor pressure at room temperature that the liquid readily gives off red-brown vapors. Even the vapor pressure of iodine at room temperature is high enough so that iodine sublimes (strong odor; concerning sublimation of iodine see 13-4a). On heating, violet iodine gas is formed. The colors of the gaseous halogens are caused by an absorption band, which for F_2 lies in the violet and ultraviolet (so that the color appears yellow) and then with increasing atomic number is shifted toward longer wavelengths. For I_2 the band lies in the middle of the visible spectrum so that only red and violet light is transmitted.

Their strong tendency for ion formation makes the halogens very reactive. The reactivity, like the electronegativity, decreases with increasing atomic number. The relative tendency for ion formation of the halogens is also evident from the e^0 value for the redox pair $2X^- \rightleftharpoons X_2 + 2e^-$, which decreases as the atomic number increases (tab. 16.1). A lighter halogen in free form therefore oxidizes the anion of a heavier halogen; for example, $Cl_2 + 2I^- \rightarrow 2Cl^- + I_2$.

The extremely high reactivity of fluorine causes great difficulties both in the preparation and handling of the element. Fluorine can form compounds with all elements except possibly the lighter noble gases (cf. 19g), and reaction will always occur at sufficiently high temperature. A large number of substances react at ordinary temperature and often with strong evolution of heat. However, with several metals, for example copper, nickel, aluminum and magnesium, a dense fluoride layer is formed, which at moderate temperatures protects against further attack. The layer of FeF_3 on pure iron is porous and is not protective, but on certain types of steel, on the other hand, is dense and protective. Dry chlorine at room temperature does not attack certain metals such as chromium, iron, nickel copper, silver and lead. In the presence of water the chlorine is hydrolyzed (see below), after which the hydrolysis products also attack the metals mentioned.

In the vigorous reaction between fluorine and water atomic oxygen is initially formed: $F_2 + H_2O \rightarrow 2HF + O$. The O atoms for the most part combine with each other to give O_2, so that the main overall process is $F_2 + H_2O \rightarrow 2HF + \frac{1}{2}O_2$ (cf. 16e).

But in addition, some O atoms combine with O_2 molecules to give ozone (21-3a) according to $O + O_2 \rightarrow O_3$. Chlorine and bromine dissolve in water (at 20°C and a total pressure of 1 atm 100 g water dissolves 0.72 g Cl_2 and 3.41 g Br_2). The solutions (*chlorine water* and *bromine water*) slowly decompose. In this process the equilibrium (disproportionation, 16d) $X_2 + H_2O \rightleftharpoons HX + HXO$ (hypochlorous and hypobromous acid) is first set up, which, however, is strongly displaced to the left. Decomposition of HXO according to $HXO \rightarrow HX + \frac{1}{2}O_2$, which is markedly hastened by sunlight, leads to further decomposition of the halogen. Iodine is poorly soluble in pure water, but on the other hand is easily soluble in many organic solvents. The solutions in strongly polar (5-2g, h, i) solvents (for example, water, alcohols, diethyl ether and acetone) are brown, while solutions in weakly polar solvents (such as carbon disulfide, chloroform and tetrachloromethane) are violet. In the former, iodine forms compounds with the molecules of the solvent (solvates), while in the latter there are relatively free iodine molecules, so that the color is approximately the same as that of iodine vapor.

The intense blue color that it produces with starch paste is characteristic of iodine. The color, which disappears on heating, results from a clathrate compound (6-3a).

All free halogens have a penetrating odor and in gaseous form severely attack the mucous membranes. Bromine produces difficultly healing sores when it comes in contact with the skin.

20-4. Production and uses of the halogens

Free halogens are most often prepared by oxidation of halides, that is, halogen in the oxidation state $-I$. The oxidation may take place either in a purely chemical way, for example with oxidizing salts ($KMnO_4$), certain oxides (MnO_2), free oxygen, or a free, more electronegative halogen; or by anodic oxidation on electrolysis. Fluorine can be prepared only by electrolysis, and for the industrial production of chlorine this method is the most important.

a. Fluorine. Fluorine is produced by electrolysis of a liquid fluoride phase, in which water must be absent. Pure fluorides are too high-melting, and therefore a melt in the system KF–HF is used. The most common has a composition in the vicinity of $KF \cdot 2HF$ with a melting point of about 70°C and a working temperature of 100°C. The problem of finding a resistant material for the electrolytic cell and electrodes has been considerable. Nowadays, the cell as well as the cathode are usually made of low-carbon steel, while the anode is made of carbon. The hydrogen evolved at the cathode is prevented from coming in contact with the fluorine at the anode by a steel barrier immersed in the melt between the electrodes. The fluorine is freed of the accompanying HF by being passed over solid NaF:

NaF+HF→NaHF$_2$. Fluorine under pressure can be stored in steel cylinders, and is now marketed in this manner.

In inorganic chemistry fluorine has been especially used in the preparation of fluorides, among others sulfur hexafluoride SF$_6$, which is used in high-voltage technology because of its high insulating ability, and uranium hexafluoride UF$_6$, which is used in the separation of uranium isotopes (9-3b). Theoretically, liquid fluorine should be very favorable as an oxidizing agent for rocket fuels, but practical difficulties have so far prevented its use.

Fluorine also has been of great importance in the manufacture of organic fluoro compounds (23-2e).

b. Chlorine. On a laboratory scale chlorine is often prepared by oxidation of Cl$^-$ by pyrolusite MnO$_2$ (by gentle heating of conc. hydrochloric acid or NaCl + conc. sulfuric acid with MnO$_2$): $4H^+ + 2Cl^- + MnO_2(s) \rightarrow Mn^{2+} + Cl_2(g) + 2H_2O$. A more convenient oxidizing agent is potassium permanganate KMnO$_4$ (conc. hydrochloric acid is dropped on KMnO$_4$ crystals): $16H^+ + 10Cl^- + 2MnO_4^- \rightarrow 2Mn^{2+} + 5Cl_2(g) + 8H_2O$. Calcium hypochlorite or bleaching powder (20-8b) can also be used. For the laboratory preparation of very pure chlorine in small amounts, a chloride of a heavy metal may be decomposed thermally: $2AuCl_3 \rightarrow 2Au + 3Cl_2(g)$.

At the end of the 19th century the most important method for the industrial production of chlorine was the oxidation of HCl by atmospheric oxygen at about 450°C in the presence of a catalyst (*Deacon process*): $4HCl + O_2 \rightarrow 2H_2O + 2Cl_2$. It was replaced at the turn of the century by the electrolytic method, but again became important during World War II. As a catalyst CuCl$_2$ was used earlier, and later also oxides and chlorides of other transition metals.

Chlorine is produced mainly by electrolysis of nearly saturated NaCl solution. The electrode and cell processes (cf. 16f) are:

cathode process (cathode of iron)	$2H_2O + 2e^- \rightarrow 2OH^- + H_2(g)$
anode process (anode of graphite)	$2Cl^- \rightarrow Cl_2(g) + 2e^-$
cell process	$2H_2O + 2Cl^- \rightarrow 2OH^- + H_2(g) + Cl_2(g)$

Chlorine is thus obtained at the anode and hydrogen and NaOH at the cathode, which makes the method very economical (*chlor-alkali electrolysis*). However, the electrode products Cl$_2$ and OH$^-$ must not come in contact with each other because then the disproportionation mentioned in 16d would take place: $Cl_2 + 2OH^- \rightarrow ClO^- + Cl^- + H_2O$. To avoid this a porous wall can be inserted between the electrodes (*diaphragm process*). Also, the iron cathode can be replaced by a mercury cathode, consisting of a pool of mercury on the bottom of the electrolytic cell (*mercury cathode process*). The hydrogen overvoltage over mercury is so high (16f) that the cathode process instead becomes $2Na^+ + 2e^- \rightarrow 2Na$(in amalgam). Sodium forms an amalgam with the mercury in the cathode and this amalgam is then allowed to react with water outside the cell: 2Na (in amalgam) $+2H_2O \rightarrow 2Na^+ + 2OH^- + H_2(g)$. The total process thus is equivalent to the cell process given above. For the preparation of KOH chloralkali electrolysis is also carried out with KCl solution.

Chlorine is also obtained in the production of metals (especially Mg and also Na) by electrolysis of molten metal chlorides. In addition, chlorine is produced by the electrolysis of hydrochloric acid obtained as a by-product in the chlorination of organic substances. Chlorine is marketed in liquid form in steel pressure cylinders.

When chlor-alkali electrolysis was introduced in the 1890's, alkali was the main product and chlorine the by-product. Beginning with World War II, however, the demand for elementary chlorine has risen so sharply that a troublesome surplus of alkali would arise if all of the needed chlorine were produced by chlor-alkali electrolysis. For this reason other methods of production have again begun to be important.

Over two thirds of the world's production of chlorine is used by the organic chemical industry. In addition large amounts go into the bleaching of paper pulp and the disinfection of water by chlorination.

c. Bromine. In the laboratory bromine can be prepared by heating an alkali bromide + MnO_2 with conc. sulfuric acid.

In the industrial production of bromine, mother liquors obtained in the processing of salt beds or natural brines are used as a source, and also more recently, unconcentrated sea water. In all cases the initial substance is Br^-. Oxidation with chlorine gives Br_2: $2Br^- + Cl_2(g) \rightarrow 2Cl^- + Br_2$. The bromine is blown out of the solution with air, but the proportion of bromine in the air mixture is so low that it must be concentrated. Several methods have been used. In one of these, the bromine is absorbed in water together with SO_2, which reduces it to Br^-: $Br_2 + SO_2 + 2H_2O \rightarrow 2Br^- + 4H^+ + SO_4^{2-}$. Bromine is again liberated from this more concentrated solution by introducing chlorine.

Bromine is used mostly for making organic bromine compounds. By far the largest part goes into the production of ethylene bromide $C_2H_4Br_2$, which is added to gasoline to prevent lead or lead oxide deposits caused by the antiknock agents tetramethyl- and tetraethyllead (23-6d). In the presence of ethylene bromide $PbBr_2$ is formed instead, which is vaporized. Some bromine is also used in photographic products (36-8).

d. Iodine. In the laboratory iodine can be prepared in a manner analogous to bromine. From the most important source of iodine, $NaIO_3$, iodine is produced industrially by reduction with sulfite ion. The process may be considered to occur in two stages: $IO_3^- + 3SO_3^{2-} \rightarrow I^- + 3SO_4^{2-}$, after which the remaining IO_3^- reacts with I^- according to $IO_3^- + 5I^- + 6H^+ \rightarrow 3I_2 + 3H_2O$.

Iodine is used in the organic chemical industry as a constituent of many drugs, and in photographic products (36-8).

20-5. Halides

a. General properties. Under the heading of halides there should be included halogen compounds in which the halogen has the oxidation number $-I$. According to the rules in 5-2a and 5-3f, however, the compounds of Cl, Br and I with N, and of I with C and S, should be called halides, although the halogens in these compounds are not more electronegative than the other elements and therefore cannot be assigned the oxidation number $-I$. The halometallate ions such as FeF_6^{3-} or $ZnCl_4^{2-}$ are also included here as halides.

Because of the high electronegativity of the halogens they have the oxidation number $-I$ in bonds with most elements, fluorine in bonds with all other elements (cf. fig. 5.23). However, because of the great range over which the electronegativity of the other elements can vary, the bond character may be very different from one case to another. It can also change significantly among halides of the same element in different oxidation states; in 5-2e, f it was described how the covalent contribution to a bond with a particular kind of atom increases with oxidation number of the atom. In comparing analogous compounds of one element with different halogens, it is found, of course, that for all elements that are not more electronegative than iodine, the covalent contribution increases as the halogen becomes heavier and thus less electronegative. The fluorides of most metals can be considered as ionic compounds, while the other halogens form typical ionic compounds only with the most electropositive elements, alkali metals and alkaline-earth metals except Be. However, the elements at the beginning of the transition-element series, especially the $4f$ and $5f$ elements and certain later transition elements in the lowest oxidation state (for example, Cr(II), Mn(II), Fe(II)), with the other halogens and especially with chlorine give compounds with quite strong ionic character. Other elements and most especially the nonmetals are bound to halogens by predominantly covalent bonds.

Neutral halides with *strong ionic bond contribution* are considered generally as *typical salts*. As a rule they form decidedly ionic crystals (NaCl, CsCl, CaF_2 types; see 6-2c), have comparatively high melting and boiling points (13-4b) and for the most part are soluble in polar solvents (13-1c).

When the electronegativity increases to the values for Be and the metals to the right of group 3 in the periodic system, the covalent contribution becomes very pronounced. Many di- and trihalides in this group crystallize in layer structures (as in fig. 6.18), in which the layers may be considered as two-dimensional, infinite molecules.

When the halogen atoms are bound with strong covalent contribution, finite molecules are generally formed, which in solid and liquid phases are held together by van der Waals forces. In such cases the halides have rather low melting and boiling points, which increase with the molecular weight (13-4 b). The same molecules most often also exist in the gas of the substance. The solubility in nonpolar solvents may be quite high (13-1 c).

The halide ions are colorless when they are more or less unpolarized, that is, enter into bonds with little covalent contribution (7-3 b). Such is the case, for example, in the halides of the alkali metals and the alkaline-earth metals. With higher polarization they may absorb light in the visible spectrum and thus show color. This absorption increases with increasing polarizability of the halogen, that is, in the sequence $F < Cl < Br < I$. Iodides therefore have the strongest color.

The behavior of halides with water is very variable. Among the metal fluorides many are easily soluble in water, exceptions being fluorides with especially high lattice energies like LiF, and fluorides of the alkaline-earth metals except BeF_2 (6-2 e). Other metal halides are in general easily soluble in water as long as the covalent-bond contribution is not large. With appreciable covalent contribution they become difficultly soluble, and for a given metal the solubility then decreases as the covalent contribution increases, that is, it changes in the sequence chloride > bromide > iodide. The most important of these difficultly soluble halides are those of Cu(I), Ag(I), Au(I), Hg(I), Tl(I) and Pb(II).

Ions of alkali metals and alkaline-earth metals except Be show a very slight tendency to form complexes with halide ions in water solution. On the other hand, halo complexes may always be expected to occur in water solutions of halides of Be and the metals to the right of group 3. The mononuclear complexes are most often octahedral. However, important tetrahedral complexes are formed by Be (Be has only four orbitals in the valence-electron shell), Zn, Cd and Hg(II). The existence of plane, square complexes has been discussed in 5-3 i, j. Linear halo complexes are formed by Cu(I), Ag(I) and especially by Au(I) and Hg(II). Many polynuclear complexes are also known. The discussion in 5-6 b concerning chloro complexes of Cu(II) shows that complexes may be completely different in solution and in the crystal, even if the coordination of the central atom is often the same in both cases.

Most halides with strong covalent bonding are decomposed by water (*hydrolysis*). Let us assume the simple case in which halides AX are decomposed by one molecule of water. In the latter the oxygen atom through its lone electron pairs becomes an electron-pair donor. It is bound thereby to that atom of A and X away from which the electrons have been most strongly drawn, that is, to the atom with the lowest electronegativity. However, this atom must have at least one free, stable orbital that can accept an electron pair from oxygen (5-3 c). The oxygen atom is accompanied by one hydrogen atom, that is, it is an OH^- ion that is added, while the remaining H^+ ion in the H_2O molecule is bonded to the other of the atoms A or X. Since the halogens have high electronegativity, OH^- is frequently added to A and H$^+$ to X. In this case the hydrolysis process can be written

$$\begin{array}{c} \text{H—O} \\ | \\ \text{H} \end{array} + \begin{array}{c} \text{A} \\ | \\ \text{X} \end{array} \rightarrow \begin{array}{c} \text{H—O—A} \\ + \\ \text{H—X} \end{array}$$

A simple example in which A is also a halogen (and the halide is thus an interhalogen compound, 20-5e), is ICl. Since I has lower electronegativity than Cl, the hydrolysis takes place according to $ICl + H_2O \rightarrow HOI + HCl$.

Often several H_2O molecules may be attached, and sometimes oxide halides are formed, possibly as intermediate products:

$$SCl_4 + 3H_2O \rightarrow H_2SO_3 + 4HCl$$
$$PCl_5 + H_2O \rightarrow POCl_3 + 2HCl$$
$$POCl_3 + 3H_2O \rightarrow H_3PO_4 + 3HCl$$

PCl_5 thus gives first phosphoryl(V) chloride, which with additional water gives phosphoric acid.

If A is the least electronegative atom but lacks free, stable orbitals, the halide is not decomposed at all by water, or only very slowly. This is the case with several halides of elements in period 2 such as CF_4, CCl_4 and NF_3, in which all four orbitals in the valence-electron shell of A are occupied by bonding or lone electron pairs. In BF_3 and BCl_3, on the other hand, the boron atom has a vacant, stable valence-electron orbital so that hydrolysis takes places easily. If the A atom belongs to period 3, for example SiF_4, $SiCl_4$, PF_3, PCl_3 and PCl_5, there are sites for the lone electron pair of the H_2O-oxygen atom in some of the 3d orbitals of the A atom, and here also hydrolysis easily takes place. The same applies if the A atom belongs to a period higher than 3. But also NCl_3 is hydrolyzed appreciably faster than NF_3 and in this case the reason is that N and Cl have the same electronegativity, so that Cl with its free, stable $3d$ orbitals can become the electron-pair acceptor. This is evident from the fact that the hydrolysis products are ammonia and hypochlorous acid according to the reaction $NCl_3 + 3H_2O \rightarrow NH_3 + 3HOCl$, and not nitrous acid and hydrogen chloride according to $NCl_3 + 2H_2O \rightarrow HNO_2 + 3HCl$.

The fluorides SF_6 and SeF_6 are not attacked by water, but this is probably because the chalcogen atoms are very effectively shielded by the octahedrally coordinated fluorine atoms.

Quite saltlike halides are also hydrolyzed in a similar manner, but in general for this higher and higher temperatures are required as the ionic contribution becomes larger. From a water solution of $MgCl_2$ the hexahydrate $MgCl_2 \cdot 6H_2O$ crystallizes and by careful drying it can be dehydrated to the monohydrate $MgCl_2 \cdot H_2O$. But if it is more strongly heated, hydrogen chloride is given off. The equilibrium $MgCl_2 \cdot H_2O(s) \rightleftharpoons Mg(OH)Cl(s) + HCl(g)$ is thus shifted to the right with the formation of magnesium hydroxide chloride. With very strong heating an additional molecule of hydrogen chloride is eliminated and magnesium oxide remains. The only possible way to prepare the anhydrous chloride from the hydrate is to heat it under a high enough pressure of hydrogen chloride so that the equilibrium is strongly displaced to the left. Similar behavior is found in heating, for example, $SnCl_2$, $AlCl_3$ and $FeCl_3$. With calcium chloride, in which the ionic-bond contribution is larger than in magnesium chloride, the hydrolysis is less pronounced and the anhydrous chloride can be obtained from the hydrate as long as the temperature is not raised too high.

The transfer of halogen ions between different halide molecules in solutions in nonaqueous solvents has been discussed in 12-1e.

Melts of saltlike halides often dissolve metals. In many cases, such as solutions of alkali metals in melts of alkali halides, the metal atoms are probably ionized with the formation of relatively free electrons: $Na \rightarrow Na^+ + e^-$. The electrons are assumed to occupy interstices among the cations and form analogs of the F centers found in alkali halide crystals with excess metal (6-3a). With increase in the amount of dissolved metal the increased proportion of free electrons leads to an increasingly dark color in the melt (7-3) and increased electronic conductivity. At high metal contents metallic conductivity is reached. When salt melts containing cations with higher oxidation number than I dissolve the corresponding metal, this probably often involves the formation of cations with lower oxidation number. Thus, a process takes place in which equilibria such as $Pb + Pb^{2+} \rightleftharpoons 2Pb^+$ or $Cd + Cd^{2+} \rightleftharpoons Cd_2^{2+}$ are displaced to the right. Here no metallic conductivity appears. Processes similar to these can, of course, occur in melts of other salts than halides, but they are best known in halide melts. When a salt melt with dissolved metal crystallizes, at least the greater part of the metal precipitates in finely divided form.

b. Formation of halides. Halides can be formed by a multitude of different processes. The element in question may react with a halogen or a halogen compound: $H_2 + Cl_2 \rightarrow 2HCl$ (cf. 8g); $2Al + 3Cl_2 \rightarrow 2AlCl_3$; $Mg + 2HF \rightarrow MgF_2 + H_2$. But often a compound of the element may also be used as a starting point (if this is an oxide, as a rule a reducing agent, for example carbon, must be added): $H_2S + I_2 \rightarrow 2HI + S$; $CS_2 + 3Cl_2 \rightarrow CCl_4 + S_2Cl_2$; $TiO_2 + C + 2Cl_2 \rightarrow TiCl_4 + CO_2$. Water solutions of halides are often prepared by the action of hydrogen halide on oxides, hydroxides or carbonates.

Higher halides or other halogen compounds may also be decomposed (often by heating): $2VCl_4 \rightarrow 2VCl_3 + Cl_2$; $KClO_4 \rightarrow KCl + 2O_2$. Poorly soluble halides may be precipitated from solution: $Ag^+ + Cl^+ \rightarrow AgCl(s)$.

The special methods for the preparation of hydrogen halides are described in 20-5d.

c. Hydrogen halides. All hydrogen halides at room temperature and atmospheric pressure are colorless gases. They have a strongly acrid odor and fume in air, since they take up water and form mist droplets.

If the properties of the hydrogen halides in pure form are compared, large

Table 20.2. *Properties of the hydrogen halides.*

	Melting point °C	Boiling point °C	ΔH^0 kJ	ΔG^0 kJ
HF	-83	19.5	-269	-271
HCl	-114	-85	-92.34	-95.30
HBr	-87	-67	-36.2	-53.24
HI	-51	-35.5	26	1.3

deviations are often found for HF in all states of aggregation. This is related to the great tendency for the formation of the hydrogen bond F–H····F, which results in strong association in the form of chains and rings (5-5). The behavior of the melting and boiling points, and especially the unexpectedly high values for HF caused by the hydrogen bond, were discussed earlier in 13-4b but are also shown in tab. 20.2.

From the values of ΔG^0 in tab. 20.2 (both ΔH^0 and ΔG^0 refer to the formation of 1 mole from the elements in the standard state, thus the processes $\frac{1}{2}H_2(g) + \frac{1}{2}X_2(g) \rightarrow HX(g)$; see 2-2a and 2-3b) it can be seen that the stability toward decomposition to the elements is especially high for HF and then falls off with increasing molecular weight. This change in stability is usually explained on a purely electrostatic basis. The Coulomb attraction at the surface of the halogen atom decreases as the radius of the atom increases. It should not be forgotten, however, that at the same time the ionic-bond contribution in the hydrogen halide molecule decreases (5-3f). This can also be stated by saying that the halogen atom is polarized more strongly by the hydrogen atom, the larger it becomes.

All hydrogen halides dissolve very easily in water (at 20°C and a total pressure of 1 atm 100 g water dissolves 72 g HCl, corresponding to 442 times its volume) and appear in water solution as acids, called hydrofluoric acid, hydrochloric acid, etc. In water solution HF is a weak acid, the others are strong acids (tab. 12.2). The low acid strength of HF results from the great bond strength in the HF molecule (12-2f). Crystals of the acid hydrates $HCl \cdot 2H_2O$ (m.p. $-17.7°C$) and $HCl \cdot 3H_2O$ (m.p. $-24.9°C$) are composed of $H_5O_2^+ + Cl^-$ and $H_5O_2^+ + H_2O + Cl^-$ respectively (12-1a).

When the hydrogen halides dissolve in the mucous membranes these are strongly attacked, but in addition, HF produces specific, toxic effects. In water solution HF is also marked by especially strong attack on the tissues.

The formation of chains gives liquid HF a high dielectric constant, so that it has a good dissolving power for electrolytes (13-1c) and also a quite high dissociating capacity (11c). HF also shows analogy with H_2O in its autoprotolysis to fluoronium and fluoride ions: $2HF \rightleftharpoons HFH^+ + F^-$.

Characteristic for HF is its ability, as long as it is not completely anhydrous, to attack SiO_2 and silicates, for example glass, with the formation of gaseous SiF_4: $SiO_2 + 4HF \rightarrow SiF_4 + 2H_2O$. This ability is utilized in the solution of silicates for analyses, for which a water solution of HF is used, and also in the etching of glass. The etched surface becomes smooth if a water solution is used, and dull with HF gas.

d. Production and uses of hydrogen halides. *Hydrogen fluoride* HF is produced industrially almost entirely by heating fluorite with conc. sulfuric acid: $CaF_2 + H_2SO_4 \rightarrow 2HF + CaSO_4$. Impurities in the gas are removed by washing and cooling, and after fractional distillation a product is finally obtained that is 99.95 wt % HF. This nearly anhydrous acid can be transported and stored in cylinders of carbon steel. For dilute acid it is best to use vessels of certain kinds of plastic (for example, polyethylene).

The production of HF is sharply rising. The largest amount is used in the manufacture of AlF_3 and synthetic cryolite for the electrolytic production of aluminum (24-3c). After this the largest amounts go into the production of organic fluoro compounds (23-2e) and the manufacture of UF_6 for uranium isotope separation (9-3b). Here F_2 is often an intermediate product, obtained by electrolysis of KF–HF melts (20-4a).

Hydrogen chloride HCl was formerly produced exclusively by heating sodium chloride with conc. sulfuric acid: $NaCl + H_2SO_4 \rightarrow HCl(g) + NaHSO_4$ (at about 150°C); $2NaCl + H_2SO_4 \rightarrow 2HCl(g) + Na_2SO_4$ (at about 650°C). The rise of chloralkali electrolysis at the turn of the century produced meanwhile a surplus of chlorine, which led to an increasing use of direct combination of chlorine and hydrogen: $H_2 + Cl_2 \rightarrow 2HCl$. The gas mixture $H_2 + Cl_2$ is explosive (8g), but the reaction goes quietly if H_2 and Cl_2 are passed into a combustion chamber where they are combined in a hydrogen-chlorine flame. This direct synthesis is still used, but the rising demand for elementary chlorine in recent decades has caused the first-mentioned method again to be the most important. HCl is also obtained as a by-product of the chlorination of hydrocarbons (for example, $CH_4 + Cl_2 \rightarrow CH_3Cl + HCl$).

In general, the water solution is marketed, called *hydrochloric acid*. An HCl solution saturated at 20°C and total pressure of 1 atm contains 42 wt % HCl The "concentrated" acid of the trade is about 38 wt % (density 1.19 g cm^{-3}). The most common acid contains about 30 wt % HCl. Pure hydrochloric acid is colorless, but the ordinary technical grade (muriatic acid; Lat. muriaticus, pickled) is yellow-colored, mainly by Fe(III) complexes. The system HCl–H_2O is azeotropic (13-5c).

The largest amounts of hydrochloric acid are used for the surface treatment of metals (oxide layers are removed by "pickling" with acid), manufacture of metal chlorides and in the organic chemical industry, among others the sugar industry and the manufacture of dye substances and synthetic rubber.

Hydrogen bromide HBr and *hydrogen iodide* HI cannot be prepared in gaseous form with satisfactory yield by the action of conc. sulfuric acid on a bromide or an iodide respectively. Both these hydrogen halides are, in fact, too easily oxidized by conc. sulfuric acid to the halogen. A nonoxidizing acid, for example phosphoric acid, can be used, but the corresponding trihalide can also be hydrolyzed with water: $PX_3 + 3H_2O \rightarrow H_2PHO_3 + 3HX$. In this process red phosphorus, the halogen in free form and water can be used directly as starting materials. However, HBr and HI are best prepared by direct combination of the elements. Because of the low stability of these hydrogen halides (see above) the temperature must be held low, and a sufficient reaction rate can be achieved only in the presence of a catalyst (finely divided platinum).

e. Interhalogen compounds and polyhalides. The halogens form the following *interhalogen compounds* with each other, all of the types XX', XX'_3, XX'_5 and XX'_7, in which X' is more electronegative and at the same time a smaller halogen atom than X (state of aggregation at 25°C and atmospheric pressure is indicated):

X'\X	Cl			Br			I			
F	ClF (g)	ClF$_3$ (g)	ClF$_5$ (g)	BrF (g)	BrF$_3$ (l)	BrF$_5$ (l)	—	IF$_3$ (s)	IF$_5$ (l)	IF$_7$ (g)
Cl				BrCl (g)	—	—	ICl (s)	ICl$_3$ (s)	—	—
Br							IBr (s)	—	—	—

All of the interhalogen compounds can be prepared by direct reaction between the corresponding free halogens. No interhalogen compounds with more than two kinds of halogen are known (this is not true, on the other hand, of the closely related polyhalide ions; see below). In the formula units given above there is always one atom of the halogen X that has a larger atomic radius and lower electronegativity, and an odd number of the other halogen X' (however, crystalline ICl$_3$ is made up of dimeric molecules I$_2$Cl$_6$). Since X' is coordinated around X, the number of the former may become larger the greater the ratio $r_X/r_{X'}$ is.

Compounds of the type XX' have structures like halogen molecules, which can be considered as a special case (with $X = X'$). Compounds of this type, however, become more and more unstable, and therefore more and more reactive, the further apart X and X' are separated in the halogen group. Thus, ClF is more reactive than F$_2$, and IF is too unstable to exist. As the number of atoms X' in the molecule increases the stability increases. IF$_3$, IF$_5$ and IF$_7$ are known, but the first decomposes even at room temperature very rapidly into I$_2$ and IF$_5$.

In the gas phase the interhalogen compounds exist in molecular form, but in the liquid phase several of them are more or less autoionized (cf. 5–3i and 12–1e). They then have quite high conductivity. Thus, for example, in liquid bromine trifluoride

there exists the equilibrium $BrF_3 + BrF_3 \rightleftharpoons BrF_2^+ + BrF_4^-$. The symbols a and b refer
 b₁ a₂ a₁ b₂
to the acid-base definition of Gutman-Lindqvist (12-1e). The autoionized interhalogen compounds will also dissolve substances that according to this definition can function as acid or base with reference to the transferable ion: $KF + BrF_3 \rightleftharpoons K^+ + BrF_4^-$, or
$AuF_3 + BrF_3 \rightleftharpoons AuF_4^- + BrF_2^+$. Many reactions of the last type make bromine trifluoride a very effective fluorinating agent, often equally as effective as free fluorine.

In molecules XX'_5 and XX'_7 higher orbitals than s and p must participate in the bonding of 5 or 7 X' around X (5-3i).

The ion BrF_2^+ mentioned above is a *polyhalide ion*. However, the best known polyhalide ions are anions. With few exceptions, these have unit charge and consist of an odd number (3 to 11) of halogen atoms. It is well known that the solubility of iodine in a water solution containing I^- is considerably greater than in pure water, and that this depends on the formation of triiodide ions according to the equilibrium $I^- + I_2 \rightleftharpoons I_3^-$. Other such polyhalide ions are IBr_2^-, $IBrCl^-$, BrF_4^-, I_7^-, Br_9^- and I_{11}^-. But I_8^{2-}, for example, is also known.

All these polyhalide ions have an even number of electrons. If one of the halogen atoms is larger than the others, it usually functions as the central atom. The structures of ions of the types IBr_2^- and BrF_4^- can be visualized by reference to fig. 19.1a and b. The shared and lone electron pairs associated with the central atom are thus placed as uniformly as possible around it, and further, the ligands are placed as uniformly as possible on the electron pairs (5-6e).

Polyhalide ions are generally prepared by addition of free halogen (as in the formation of I_3^-) or an interhalogen compound to a halide. The larger a polyhalide ion is, the larger must be the cation in order to build up a crystal. If a small monatomic cation takes part, it is always found that it is made larger by solvation (the crystal contains water of crystallization, benzene of crystallization, or similar solvent molecules). The following examples of crystalline polyiodides illustrate this: CsI_3, RbI_3, $NaI_3 \cdot 2H_2O$, $RbI_7 \cdot H_2O$.

The polyhalides are easily dissociated into the simple halide plus halogen or interhalogen compound.

In interhalogen compounds as well as polyhalide ions the halogens according to the ordinary oxidation-number rules (5-3g) often have oxidation numbers that are not found, for example, in oxides or oxo acids (for example, $Br^{III}F_3$, $I^{III}Cl_4^-$; cf. tab. 20.3). However, the most electronegative halogen always takes the oxidation number $-I$, which means that both types of compound may be considered as halides. Interhalogen compounds were therefore also mentioned in 20-5a.

20-6. Pseudohalogens

A number of atomic groups are known, consisting of two or more atoms of which at least one is a nitrogen atom, that show a strong analogy to halogens, and therefore are called *pseudohalogens* (also *haloids*). Like the halogens they can be ionized to monovalent anions, *pseudohalide ions*, whose salts often resemble halide salts, for example in their solubility relationships. The silver salts are thus poorly soluble in water. The pseudohalogens can also be bound by more strongly covalent bonds, for example to hydrogen. The hydrogen compounds are acids, like the hydrogen halides. Some pseudohalogens are known in free form, combined as dimers, analogous to the diatomic, free halogens.

Group 17, Halogens

The most important ions, acids and dimers are:

Anion		Acid		Dimer	
CN^-	cyanide ion	HCN^a	hydrogen cyanide	$(CN)_2$	cyanogen
OCN^-	cyanate ion	$HOCN^a$	cyanic acid		
SCN^-	thiocyanate ion (rhodanide ion)	$HSCN^a$	thiocyanic acid	$(SCN)_2$	thiocyanogen (rhodane)
$SeCN^-$	selenocyanate ion			$(SeCN)_2$	selenocyanogen
ONC^-	fulminate ion	$HONC$	fulminic acid		
NCN^{2-}	cyanamide ion	H_2NCN	cyanamide		
N_3^-	azide ion	HN_3	hydrogen azide		

[a] The esters RNC, RNCO and RNCS exist, corresponding to the "iso acids" HNC, HNCO and HNCS. However, in the free acids it is not possible to separate the iso acids from the above-mentioned normal acids.

Several of these ions and molecules are discussed in more detail in 22-6f and 23-2k. The cyanide ion is isoelectronic with carbon monoxide and may be considered to have the same electron distribution (5-3a). The triatomic pseudohalide ions have the same electron distribution as carbon dioxide (5-3e) and are thus at least nearly linear.

20-7. Halogen oxides

The halogen oxides are grouped here according to the oxidation number of the halogen atom (tab. 20.3). However, the oxygen compounds of fluorine are completely unique. Of these the strongly oxidizing OF_2 and O_2F_2 with the structures FOF and FOOF have been isolated. Also the existence of the odd molecules OF and O_2F has been proved. Since fluorine is more electronegative than oxygen these compounds must be called oxygen fluorides.

The best known of the oxygen fluorides is *oxygen difluoride* OF_2, a colorless, comparatively stable gas, which can be condensed to a liquid with a boiling point of $-145°C$. It is best prepared by passing fluorine into a sodium hydroxide solution: $2F_2 + 2OH^- \rightarrow 2F^- + OF_2 + H_2O$. OF_2 is somewhat soluble in water, but the solution is not acid, which shows that the compound, in contrast to Cl_2O and Br_2O, is not an acid anhydride. However, the structure is analogous to that of these oxides (20-7a).

a. Halogen(I) oxides. To this class belong the dihalogen oxides, of which *dichlorine oxide* Cl_2O and *dibromine oxide* Br_2O with the structure $|\underline{O}-\underline{\overline{X}}|$ are known. Both can be prepared by passing the halogen at low temperature over dry HgO: $2X_2 + 2HgO \rightarrow X_2O + HgO \cdot HgX_2$.

Cl_2O is a yellow-red gas, which is easily condensed (b.p. 4°C). It is unstable ($\Delta G° = 93.8$ kJ) and explodes easily on heating or in contact with oxidizable substances. Br_2O is still more unstable. Both are acid anhydrides. The formation of acid with H_2O can be considered as a hydrolysis according to 20-5a, in which OH⁻ is bound to the least electronegative atom X:

$$H-\overline{\underline{O}}| \quad |\underline{X}| \qquad H-\overline{\underline{O}}-\underline{X}|$$
$$| \quad + \quad | \quad \rightarrow \quad +$$
$$H \quad |\underline{O}-\underline{X}| \qquad H-\overline{\underline{O}}-\underline{X}|$$

b. Halogen(IV) oxides. These are *halogen dioxides*, of which *chlorine dioxide* ClO_2 and *bromine dioxide* BrO_2 are known. ClO_2 is the most important chlorine oxide, while BrO_2 has no practical significance.

ClO_2 is an odd molecule (5-3a), and has the properties of high reactivity (cf. also 5-6d), color and paramagnetism typical of such molecules. The unpaired electron is delocalized over the whole molecule (5-3e), so that the structure can be written

$$|\overline{Cl}-\overline{O}| \qquad \cdot \overline{Cl}-\overline{O}| \qquad |\overline{Cl}-\dot{O}|$$
$$| \quad \leftrightarrow \quad | \quad \leftrightarrow \quad |$$
$$\cdot \underline{O}| \qquad |\underline{O}| \qquad |\underline{O}|$$

Chlorine dioxide at ordinary temperature is a red-yellow gas with a strongly pungent odor, and can be easily condensed (b.p. 9.9°C). It is very unstable ($\Delta G^0 = 124$ kJ). With gentle heating or on contact with oxidizing substances it decomposes explosively into chlorine and oxygen. It dissolves easily in water and the solution keeps a long time without reaction. In alkaline solution disproportionation (16d) takes place to chlorite and chlorate ions: $2Cl^{IV}O_2 + 2OH^- \rightarrow Cl^{III}O_2^- + Cl^VO_3^- + H_2O$.

Chlorine dioxide is most often prepared by reduction of chloric acid or chlorate ion: $ClO_3^- + 2H^+ + e^- \rightleftharpoons ClO_2 + H_2O$ ($e^0 = 1.15$ V). If conc. sulfuric acid is added to a chlorate, one chlorate ion can act as a reducing agent ($ClO_3^- + H_2O \rightleftharpoons ClO_4^- + 2H^+ + 2e^-$; $e^0 = 1.19$ V) on another chlorate ion, so that the overall reaction becomes $3ClO_3^- + 2H^+ \rightarrow ClO_4^- + 2ClO_2 + H_2O$. The reaction goes in this direction because of the high acidity and also because the water formed becomes bound to the concentrated sulfuric acid. In the laboratory oxalate ion $C_2O_4^{2-}$ is best used as a reducing agent: $2ClO_3^- + C_2O_4^{2-} + 4H^+ \rightarrow 2ClO_2 + 2CO_2 + 2H_2O$. Thus, sulfuric acid is caused to react with a mixture of chlorate and oxalic acid. The chlorine dioxide becomes mixed with carbon dioxide and is therefore less dangerous. In industry, SO_2 was formerly used as a reducing agent. Nowadays the reduction is often done with chloride ion in acid solution (HCl): $2ClO_3^- + 2Cl^- + 4H^+ \rightarrow 2ClO_2 + Cl_2 + 2H_2O$. The method has been made more economical by allowing a $NaClO_3$ solution produced by electrolysis of NaCl solution (20-8d) to run directly into a reaction vessel where it is reduced with HCl.

With the development of cheaper methods of production ClO_2 has recently become important for bleaching paper pulp and for disinfecting water, where it has become supplementary to chlorination. Because of the instability of chlorine dioxide it is manufactured at the site where it is to be used.

Group 17, Halogens

Table 20.3. *The most important oxygen compounds of chlorine, bromine and iodine.*

Oxidation Number	Cl		Br		I	
VII	$\begin{cases} Cl_2O_7 \\ HClO_4 \\ ClO_4^- \end{cases}$	dichlorine heptoxide perchloric acid perchlorate ion	$\begin{cases} HBrO_4 \\ BrO_4^- \end{cases}$	perbromic acid perbromate ion	$\begin{cases} H_5IO_6 \\ IO_6^{5-} \end{cases}$	orthoperiodic acid orthoperiodate ion
VI	Cl_2O_6	dichlorine hexoxide				
V	$\begin{cases} HClO_3 \\ ClO_3^- \end{cases}$	chloric acid chlorate ion	$\begin{cases} HBrO_3 \\ BrO_3^- \end{cases}$	bromic acid bromate ion	$\begin{cases} I_2O_5 \\ HIO_3 \\ IO_3^- \end{cases}$	diiodine pentoxide iodic acid iodate ion
IV	ClO_2	chlorine dioxide	BrO_2	bromine dioxide		
III	$\begin{cases} Cl_2O_3{}^a \\ HClO_2 \\ ClO_2^- \end{cases}$	dichlorine trioxide chlorous acid chlorite ion				
I	$\begin{cases} Cl_2O \\ HClO \\ ClO^- \end{cases}$	dichlorine oxide hypochlorous acid hypochlorite ion	$\begin{cases} Br_2O \\ HBrO \\ BrO^- \end{cases}$	dibromine oxide hypobromous acid hypobromite ion	$\begin{cases} HIO \\ IO^- \end{cases}$	hypoiodous acid hypoiodite ion

^{a} Existence not definitely confirmed.

c. Halogen(V) oxide. The only representative is *diiodine pentoxide* I_2O_5, which is obtained as a strongly oxidizing, white, crystalline powder when iodic acid is heated to about 180°C: $2HIO_3 \rightarrow I_2O_5 + H_2O$. Diiodine pentoxide is thus the anhydride of iodic acid. It is the most stable of all the halogen oxides. ΔG^0 is unknown for the formation process $I_2(s) + \tfrac{5}{2}O_2(g) \rightarrow I_2O_5(s)$, but $\Delta H^0 = -201$ kJ. For the formation of all the other halogen oxides, $\Delta H^0 > 0$. However, on heating to about 300°C I_2O_5 decomposes into iodine and oxygen.

d. Halogen(VI) oxide. *Dichlorine hexoxide* Cl_2O_6 is known, and can be prepared by oxidation of chlorine dioxide with ozone: $2ClO_2 + 2O_3 \rightarrow Cl_2O_6 + 2O_2$. At room temperature it is a brown-red liquid, which boils at about 200°C with decomposition. In the liquid a small portion is dissociated to ClO_3 (weak paramagnetism) but in the gas phase the dissociation is greater. The structure is unknown.

e. Halogen(VII) oxide. *Dichlorine heptoxide* Cl_2O_7 is the anhydride of perchloric acid and can be obtained by removing water from this acid with phosphorus pentoxide: $2HClO_4 \rightarrow Cl_2O_7 + H_2O$. It is a colorless, quite volatile (b.p. 90°C), oily liquid. It is less reactive than other halogen oxides but even so easily explodes with heating or shock. The molecule consists of two ClO_4 tetrahedra with a common O corner.

20-8. Oxohalic acids and oxohalates

a. General. Tab. 20.3 lists the formulas and common (not systematic) names of the most important oxohalic acids and oxohalate ions. A number of acids and ions are known for chlorine, bromine and iodine, but not for fluorine on the other hand, which may be attributed to the high electronegativity of fluorine (5-6b, 20-1). Of the known acids and ions the oxohalogen(III) acids and oxohalate(III) ions (halous acids and halite ions) are the least important.

The structural formulas for the oxohalate ions can be typically written:

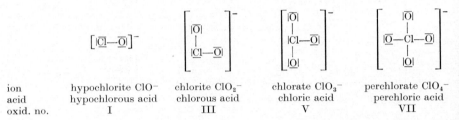

ion	hypochlorite ClO⁻	chlorite ClO$_2^-$	chlorate ClO$_3^-$	perchlorate ClO$_4^-$
acid	hypochlorous acid	chlorous acid	chloric acid	perchloric acid
oxid. no.	I	III	V	VII

In these formulas the octet rule (5-3h) is fulfilled for all atoms (cf., however, the discussion in 5-3h concerning double-bond contribution in oxo anions). More highly negatively charged ions are not known. The number of electrons would then exceed that required by the octet rule (the charge 2 − would besides give ions with an odd number of electrons).

As described in 5-3d, in all these ions tetrahedrally directed sp^3 orbitals extend outward from the chlorine atom, and the resulting geometrical forms were shown schematically in fig. 5.19. The chlorine and bromine atoms are too small to coordinate more than four oxygen atoms, but the iodine atom is large enough to allow higher coordination numbers (5-6b). With the oxidation number VII, then, octahedral coordination also appears in the orthoperiodate ion IO_6^{5-}. Polynuclear oxohalate ions and corresponding acids are rare.

If a proton is added to one oxygen atom in the oxohalate ion, in principle the structure of the oxohalic acid is obtained. In spite of the H–O bond, schematic formulas such as HXO, HXO$_2$, etc. are often written, analogous to the usual method of writing most oxo acids.

The acid strengths of the oxochloric acids were discussed in 12-2f. For all oxohalic acids of the general formula HAO$_n$ it is true that the acid strength increases as n and thus the oxidation number of the halogen increases. In 12-2f it was also shown that for analogous oxo acids with different central atom A, the strength increases as the electronegativity of A increases. Analogous oxo acids of different halogens therefore become weaker, the heavier the halogen is.

The oxohalate ions are generally colorless (7-3b). However, periodates with strongly polarizing cations (for example, Ag⁺) are colored.

The pure oxohalic acids in general decompose easily into oxygen and other halogen compounds. For the oxo acids of the same halogen, the stability generally increases with increasing oxidation number. It has been possible to prepare anhydrous perchloric acid, periodic acid (the most important is H$_5$IO$_6$; see below) and iodic acid, but no other anhydrous acids. Corresponding variation in stability is found for the oxohalates, and for both the acids and ions this is probably related

Table 20.4. *Standard potentials for redox pairs containing oxohalic acids and oxohalates.*

Redox pair	e^0, V		
	Cl	Br	I
Acid solution			
$\frac{1}{2}X_2 + 4H_2O \rightleftharpoons 8H^+ + XO_4^- + 7e^-$	1.42	—	1.34[b]
$\frac{1}{2}X_2 + 3H_2O \rightleftharpoons 6H^+ + XO_3^- + 5e^-$	1.47	1.52	1.195
$\frac{1}{2}X_2 + 2H_2O \rightleftharpoons 3H^+ + HXO_2 + 3e^-$	1.64	—	—
$\frac{1}{2}X_2 + H_2O \rightleftharpoons H^+ + HXO + e^-$	1.63	1.59	1.45
Basic solution			
$X^- + 8OH^- \rightleftharpoons XO_4^- + 4H_2O + 8e^-$	0.56	—	0.39[c]
$X^- + 6OH^- \rightleftharpoons XO_3^- + 3H_2O + 6e^-$	0.63	0.61	0.29
$X^- + 4OH^- \rightleftharpoons XO_2^- + 2H_2O + 4e^-$	0.78	—	—
$X^- + 2OH^- \rightleftharpoons XO^- + H_2O + 2e^-$	0.89	0.76	0.49

[a] The acids $HClO_4$, $HClO_3$, $HBrO_3$ and HIO_3 are so strong that they may be assumed to be completely protolyzed, even in acid solution.
[b] H_5IO_6 in the oxidized form.
[c] $H_3IO_6^{2-}$ in the oxidized form.

to the increasing coordination number around the central atom, which both leads to more symmetrical coordination and decreases the number of their reactive, lone electron pairs.

Typical of the oxohalic acids and oxohalates is their oxidizing power. Standard potentials for a number of important redox pairs of this group may be found in tab. 20.4. When the halogen X is changed, the oxidizing power as a rule decreases in the order Cl > Br > I, but for the ion XO_3^- in acid solution the order is Br > Cl > I.

Of the chlorine compounds in water solution HClO and $HClO_2$ are the most strongly oxidizing. For the chlorine compounds in acid solution the following standard potentials apply for each step in oxidation number:

Oxid. no.	−I	0	I	III	IV	V	VII
mol. spec.	Cl^-	Cl_2	HClO	$HClO_2$	ClO_2	ClO_3^-	ClO_4^-
e^0, V		1.36	1.63	1.65	1.27	1.15	1.19

The comparatively low potentials for the first reducing steps from ClO_4^- and ClO_3^- are probably the reason that these ions are sluggish oxidizing agents in water solution. Once the reaction has started, however, the evolution of heat in concentrated solutions may increase the reaction rate violently.

In acid solution all the chlorine and bromine compounds can decompose water with the evolution of oxygen ($2H_2O \rightleftharpoons O_2 + 4H^+ + 4e^-$; $e^0 = 1.229$ V). Arrested reaction (16e) often causes the decomposition to proceed slowly, but it can be catalyzed, for example, by many transition metal ions. Disproportionation reactions of various kinds are also very common.

Their high oxidizing power especially makes the oxohalic acids and oxohalates highly important in chemistry and chemical industry. On the other hand their oxidizing power together with their instability causes these compounds in high concentrations to be very explosive, especially in the presence of easily oxidizable substances.

b. Halogen oxidation number = I. The acids HXO are called *hypohalous acids* and correspondingly the salts (containing the ion XO$^-$) *hypohalites*.

Hypohalite ion is formed by reaction in the cold (at higher temperatures XO$_3^-$ ion is formed; see below) between the free halogen and hydroxide ion: $X_2 + 2OH^- \rightarrow XO^- + X^- + H_2O$ (see the discussion of this disproportionation in 16d). For the direct preparation of a water solution of HXO alone, the halogen is allowed to react with a suspension of HgO in water: $2X_2 + 2HgO(s) + H_2O \rightleftharpoons 2HXO + HgO \cdot HgX_2(s)$. The formation of the poorly soluble mercury(II) oxide halide causes the equilibrium to be strongly displaced to the right (16d).

For both acids and salts the stability decreases in the order Cl > Br > I. Therefore it is only the chlorine compounds that have any great significance.

A moderately strong HClO solution (highest attainable proportion about 5 wt %) is yellow and has a characteristically penetrating odor. It decomposes slowly in the dark and cold, more rapidly in the light and with heating, mainly according to $2HClO \rightarrow 2HCl + O_2$. A hypochlorite solution decomposes analogously according to: $2ClO^- \rightarrow 2Cl^- + O_2$.

Hypochlorite solutions are much used as a bleaching agent for paper pulp and textile goods. They are produced by the above-mentioned reaction in the cold between chlorine and corresponding hydroxide. The fact that the bleach solution also contains chloride ion does not interfere with its use. Nowadays the most important bleaching agent is *sodium hypochlorite* NaClO, which can be obtained by chlor-alkali electrolysis (20-4b), in which chlorine and sodium hydroxide are formed in just the required proportions.

Since the close of the 18th century, *bleaching powder* has been used as a bleaching agent, which is obtained as a white powder when chlorine is passed over solid but moist calcium hydroxide ("slaked lime"). X-ray diffraction studies show that bleaching powder consists of several phases, of which the most effective is a solid solution with a composition close to $Ca(ClO)_2$. The overall composition of bleaching powder is often given for the sake of simplicity as CaCl(ClO). Bleaching powder destroys mustard gas and therefore has been used for gas decontamination by being spread over the areas exposed to mustard gas. As a bleaching agent an aqueous solution of calcium hypochlorite is also used, which is obtained either by leaching bleaching powder with water, or by chlorination of a water suspension of calcium hydroxide ("milk of lime").

Any solution containing ClO$^-$ and Cl$^-$ gives chlorine on acidification: $ClO^- + Cl^- + 2H^+ \rightarrow Cl_2 + H_2O$. If bleaching powder is treated with a weak acid, hypochlorous acid (HClO) is released (under the action of carbon dioxide in the air bleaching powder therefore smells of HClO). If a strong acid is added, a rapid reaction takes place between HClO and the Cl⁻ ions present in the bleaching powder: $HClO + Cl^- + H^+ \rightarrow Cl_2 + H_2O$. Chlorine is therefore often prepared in the laboratory from bleaching powder + strong acid.

HXO is more strongly oxidizing than XO⁻, and in a solution containing comparable amounts of both these molecules they react for the most part according to: $2HXO + XO^- \rightarrow XO_3^- + 2X^- + 2H^+$. In a hypohalite solution the proportions of HXO and XO⁻ are more or less equal when the pH of the solution lies in the vicinity of the pK_a value for HXO. For HClO, $pK_a = 7.53$ and it is thus apparent that the reaction in question in this case occurs most easily in nearly neutral solution. A strongly basic hypochlorite solution, in which the proportion of HClO is small, is relatively stable on the other hand. The reaction is also favored by heating.

c. Halogen oxidation number = III. Only chlorine compounds are known with certainty. The acid $HClO_2$, usually called *chlorous acid*, is only known in water solution. Salts containing the ion ClO_2^- are called *chlorites*.

A water solution of chlorous acid can be obtained by reduction of chlorine dioxide with hydrogen peroxide: $2ClO_2 + H_2O_2 \rightarrow 2HClO_2 + O_2$. If the solution is alkaline, the chlorite ion is obtained directly, and this is the most important process for the production of *sodium chlorite* $NaClO_2$, which is a very effective bleaching agent and to a certain extent is used for bleaching paper pulp and textiles. If chlorine dioxide is passed into an alkaline solution without hydrogen peroxide, chlorite and chlorate ions are obtained as mentioned earlier.

d. Halogen oxidation number = V. Since early times the names have been *halic acids* HXO_3 and *halates* (containing the ions XO_3^-). The only known anhydrous acid is *iodic acid* HIO_3, which can be prepared by oxidation of iodine. It forms colorless water-soluble crystals (on the other hand it is difficultly soluble in conc. nitric acid, so that it can easily be obtained pure if nitric acid is chosen as the oxidizing agent). When iodic acid is heated the anhydride, diiodine pentoxide I_2O_5, is formed (20-7c).

By reacting barium chlorate with sulfuric acid ($Ba(ClO_3)_2 + H_2SO_4 \rightarrow 2HClO_3 + BaSO_4(s)$), filtering and evaporating at low pressure, a solution of up to 40% *chloric acid* $HClO_3$ can be obtained. An analogous preparation of *bromic acid* $HBrO_3$ can also be made.

It was mentioned earlier that in a hypohalite solution at proper pH the reaction $2HXO + XO^- \rightarrow XO_3^- + 2X^- + 2H^+$ takes place. Here halate ion is thus formed. This reaction is utilized in the production of chlorates. If chlorine is passed into an alkali hydroxide solution, hypochlorite is first formed according to $Cl_2 + 2OH^- \rightarrow ClO^- + Cl^- + H_2O$, but when the greater part of the OH⁻ ions is used up and the solution becomes nearly neutral, the chlorate reaction sets in. Therefore, in the manufacture of chlorate the hydroxide solution is saturated with chlorine and is also kept warm (~70°C). The overall reaction is $3Cl_2 + 6OH^- \rightarrow ClO_3^- + 5Cl^- +$

$3H_2O$. Most often a sodium chloride solution is electrolyzed with the electrodes set close to each other (cathode, iron; anode, graphite or magnetite) so that the electrolysis products are mixed. The electrolyte is held at about pH = 7 and the electrolytic cell cooled moderately so that the temperature is about 70°C.

Both *sodium chlorate* $NaClO_3$ and *potassium chlorate* $KClO_3$ are produced electrolytically. The latter is more easily obtained pure because it is much less soluble at low temperature than at high (the solubility of $KClO_3$ in 100 g water at 20°C is 7.3 g, at 100°C 56 g). On cooling the electrolyte $KClO_3$ therefore crystallizes, and can be easily made very pure by recrystallization. Sodium chlorate is cheaper but is more difficult to obtain pure (easily soluble even at low temperature) and easily liquifies in moist air (deliquescence, 13-5e).

All halates decompose quite easily on heating. If pure potassium chlorate is heated just to the melting point (368°C), it disproportionates to perchlorate and chloride: $4KClO_3 \rightarrow 3KClO_4 + KCl$. If the temperature is raised further, the perchlorate decomposes into chloride and oxygen: $KClO_4 \rightarrow KCl + 2O_2$. This decomposition, and probably also a direct decomposition according to $2KClO_3 \rightarrow 2KCl + 3O_2$, is catalyzed by certain metal oxides (for example MnO_2, Fe_2O_3). The formation of oxygen then becomes appreciable already at about 150°C (classical method for the preparation of oxygen gas; when catalysts are used, however, the gas contains a small proportion of chlorine). On rapid heating of a halate, an explosion may occur. *In the presence of oxidizable substances the strong oxidizing power of the halates makes them very dangerous even at room temperature.* Explosion can then take place even with very light mechanical disturbance.

Chlorate is used both for the preparation of chlorine dioxide (20-7b) and perchlorate (see below). Sodium chlorate is used to a large extent as a herbicide, potassium chlorate mostly as an oxidizing agent, in match heads (22-14) and in fireworks (for oxidation of solid rocket fuels, however, it has been replaced by various perchlorates, which are more reliable).

Iodates are more stable and less soluble than the other halates. They often crystallize together with iodic acid, for example $KIO_3 \cdot HIO_3$ (formerly called potassium biiodate). The important occurrence of *sodium iodate* $NaIO_3$ in the saltpeter deposits of Chile was mentioned in 20-2a.

The electropositive properties of iodine (20-1) are evident from the fact that two oxygen compounds, earlier assumed to be the oxides I_2O_4 and I_4O_9, on the basis of many of their properties are considered probably to be the iodates $I^{III}O(I^VO_3)$ and $I^{III}(I^VO_3)_3$, respectively.

e. Halogen oxidation number = VII. The acids since early times have been called *perhalic acids* and their salts *perhalates*. Corresponding compounds of different halogens are rather different from each other and are treated here separately.

As mentioned above, perchlorates can be obtained by moderate heating of chlorates. However, they are produced industrially by anodic oxidation of chlorate, often in direct association with the manufacture of chlorates: $ClO_3^- + H_2O \rightarrow$

$ClO_4^- + 2H^+ + 2e^-$. Perchlorates are generally easily soluble in water, but the perchlorates of potassium, rubidium and cesium are poorly soluble in cold water. In the manufacture of perchlorate a pure product is therefore easily obtained by precipitation of *potassium perchlorate* $KClO_4$.

Solid perchlorates give off oxygen on heating (cf. above under potassium chlorate) and are therefore used as oxidizing agents for solid rocket fuels and in fireworks. Most used are potassium perchlorate and *ammonium perchlorate* NH_4ClO_4, of which the latter has the advantage of giving only gaseous products (larger final volume and smokeless). Perchlorates are significantly more dependable than chlorates, but must still be handled very carefully (for example, when perchlorates are used as drying agents—see 26-1—the presence of appreciable amounts of organic substances should be avoided).

As mentioned in 5-6c the perchlorate ion ClO_4^- has an especially small tendency to act as ligand to metal ions in water solution.

Perchloric acid $HClO_4$ can be prepared anhydrous by distillation at low pressure of a solution in which the acid is formed without interfering by-products (for example, by the reaction $KClO_4 + H_2SO_4 \rightarrow KHSO_4 + HClO_4$). At ordinary temperature it is a colorless liquid, strongly corrosive to the skin. On heating it becomes brown-red and finally explodes. It also decomposes slowly at ordinary temperature, and explosion may occur. Combustible substances are oxidized explosively.

Perchloric acid is a very strong acid (12-2f) and therefore forms salts even with very weak bases. Its monohydrate $HClO_4 \cdot H_2O$ is known in two forms, both of which are composed of the ions H_3O^+ and ClO_4^-. They may thus be regarded as oxonium perchlorates. The dihydrate $HClO_4 \cdot 2H_2O$ (m.p. $-20.7°C$) contains the cation $H_5O_2^+$ (12-1a).

The system $HClO_4$–H_2O is azeotropic (13-5c) with boiling point maximum at $203°C$ and 72 wt% $HClO_4$. If a water solution of $HClO_4$ is boiled, the remaining liquid will finally have this composition. Such a solution ("conc. perchloric acid") is quite stable and is much used as an oxidizing agent in analytical chemistry and for dissolving metals. The dissolving power of perchloric acid is aided by its oxidizing power, high acid strength and the high solubility of most perchlorates. However, it should never be used without experience and the necessary precautions.

Perbromic acid $HBrO_4$ and *perbromates* were first synthetized in 1968 and are as yet little known.

There are several *periodic acids* whose compositions may be expressed by the general formula $I_2O_7 \cdot nH_2O$. For n the values 1, 2, 3 and 5 are known, of which 5 corresponds to the most common acid *hexaoxoiodic*(VII) *acid* or *orthoperiodic acid* H_5IO_6. The reason that the most common acid is not *tetraoxoiodic*(VII) *acid* HIO_4 (corresponding to $n=1$; analogous to perchloric acid $HClO_4$) is that, as mentioned in 20-8a, the iodine atom is so large that it permits octahedral coordination by six oxygen atoms. This leads to an equilibrium $HIO_4 + 2H_2O \rightleftharpoons H_5IO_6$, displaced far to the right.

H_5IO_6 can be prepared by anodic oxidation of iodic acid ($HIO_3 + 3H_2O \rightarrow H_5IO_6 + 2H^+ + 2e^-$) and at room temperature forms colorless crystals. If H_5IO_6 is heated at low pressure lower-hydrated periodic acids are formed, following which oxygen is given off leaving iodic acid. This then gives off water with the formation of diiodine pentoxide I_2O_5:

$$2H_5IO_6 \xrightarrow[80°C]{-3H_2O} H_4I_2O_9 \xrightarrow[100°C]{-H_2O} 2HIO_4 \xrightarrow[140°C]{-O_2} 2HIO_3 \xrightarrow[180°C]{-H_2O} I_2O_5$$

Thus, no iodine(VII) oxide I_2O_7 is obtained.

H_5IO_6 is a quite weak acid even in its first protolytic step, which is an accord with the rules for acid strength given in 12-2f. Even at about pH = 14 the protolysis does not proceed beyond $H_2IO_6^{3-}$. *Orthoperiodates* of the type $M_2^IH_3IO_6$ often crystallize out of periodate solutions. Addition of Ag^+, however, leads to precipitation of Ag_5IO_6. That the ion IO_4^- is also present in a water solution of H_5IO_6 is evident from the fact that salts of this ion can precipitate if these have sufficiently low solubility. $NaIO_4$ is found in small amounts in the Chilean saltpeter deposits.

In contrast to the behavior of the chlorine compounds, in general no periodate is formed on heating an iodate. On the other hand, they can be prepared by oxidation of iodate with strong oxidizing agents, for example chlorine in alkaline solution ($IO_3^- + Cl_2 + 4OH^- \rightarrow H_2IO_6^{3-} + 2Cl^- + H_2O$), or anodic oxidation. Most periodates are poorly soluble.

CHAPTER 21

Group 16, Chalcogens

21-1. Introductory survey

Since the heavy metals in the earth's crust occur mainly combined with oxygen and sulfur, and since the heavier elements in group 16 are similar to sulfur in their behavior, the elements of the group have been collectively referred to as *chalcogens*, that is, ore formers (derivation in 4-2c).

As in the case of the halogens, the electronegativity of the chalcogens decreases very markedly as the atomic number increases. Parallel with this, the metallic character increases. Tellurium may be considered as a semimetal and polonium as a metal. In addition, it will be noted that, as usual, the first element of the group, oxygen, has more distinct properties of its own (5-6, where also a survey is given of the structures of these elements).

As would be expected (5-3g), the highest oxidation number is equal to the group number, that is, VI, and the lowest oxidation number is 8 units lower, that is, $-$II. Furthermore, the most important types of compounds represent even oxidation numbers.

It was stated earlier in 5-3g that oxygen, because of its high electronegativity has the oxidation number $-$II in compounds with all other elements except fluorine. In OF_2, for example, the valences should be considered to be $O^{+II}(F^{-I})_2$. In compounds in which oxygen atoms are bonded to each other the rules in 5-3g lead also to other oxidation numbers. The most important of these cases is the oxygen in peroxide bonding, for example in the peroxide ion $[|\overline{O}\text{-}\overline{O}|]^{2-}$ where the oxidation number is $-$I (see also the hyperoxide ion O_2^- in 21-6c).

The most important oxidation numbers of sulfur are $-$II (sulfides), 0, IV (SO_2, sulfurous acid, sulfite ion) and VI (SO_3, sulfuric acid, sulfate ion). In oxidizing medium in contact with the oxygen in the air, VI is the most prevalent. However, especially in sulfur-sulfur bonding, other oxidation numbers often appear. Such is true, for example, in the disulfide ion S_2^{2-}, analogous to the peroxide ion, with the oxidation number $-$I.

The oxidation numbers $-$II, 0, IV and VI are also the most important for selenium, tellurium and polonium. However, a decrease in the tendency toward the oxidation number VI is noticeable already for selenium (compounds in this valence state try to go to the state IV and are thus strong oxidizing agents). With polonium the oxidation number II also exists in the ion Po^{2+}.

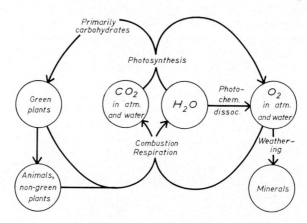

Fig. 21.1. The oxygen cycle in nature.

21-2. Occurrence and history

a. Occurrence. As stated in chap. 17 *oxygen* is believed to be the third most abundant element in the universe in terms of number of atoms. In the accessible earth's crust as well as the oceans and the atmosphere, oxygen is the most common element with a proportion both in weight and atom percent of about 50 %. In the earth's crust it occurs for the most part bound in various silicates, in the ocean bound to hydrogen, and in the atmosphere as free oxygen, mainly O_2. The proportion of oxygen in dry air at sea level is 20.9 vol % or 23.2 wt %. Oxygen is also an important constituent of living organisms, in which besides water it is bound primarily in carbohydrates. In the photosynthesis of carbohydrates by green plants oxygen is released, which becomes combined again when the carbohydrates or the carbon compounds formed secondarily from them are oxidized by respiration or other oxidizing processes (by decay or directly by burning of fuels). In fig. 21.1 this *oxygen cycle* is shown very schematically.

The figure also shows the formation of free oxygen by photochemical dissociation of water vapor in the upper atmosphere (the hydrogen thus formed probably to a great extent escapes from the earth). It is possible that the first free oxygen in the atmosphere originated by this process (cf. 17 b). At the same time oxygen is bound up by the weathering of rocks (for example, in the oxidations $Fe(II) \rightarrow Fe(III)$, $Mn(II) \rightarrow Mn(IV)$, $S(0) \rightarrow S(VI)$). However, it has not been possible to detect any change in the oxygen content of the atmosphere during the time that accurate analysis has been possible.

Sulfur is found in the earth's crust both in free form (native) and combined, for the most part in sulfides or sulfates. Free sulfur occurs endogenically (17 c) in volcanic regions, but the occurrence of free sulfur in the coastal regions around the Gulf of Mexico, which at present produces nearly half of the world's sulfur, like the classical occurrences in Sicily, is exogenic in origin.

In the occurrences around the Gulf of Mexico the sulfur is disseminated in limestones that overlie salts, of which $CaSO_4$ (*anhydrite*) lies immediately beneath the sulfur-bearing rock. This implies ocean deposits, and the nearby occurrence of oil shows that organic material was at hand during the formation of the sulfur. It is probable that there existed here an inland sea with high salt content and limited convection and consequently a lack of oxygen on the sea bottom. Anaerobic bacteria living on the plant and animal remains reduced the sulfate ions in the water to hydrogen sulfide. Thereafter the hydrogen sulfide was in some way oxidized to sulfur. The remaining sulfate ions may have been the oxidizing agent, for example according to $SO_4^{2-} + 3S^{2-} + 8H^+ \rightarrow 4S + 4H_2O$. The Sicilian deposits probably were formed in a similar way.

Even in the absence of sulfate, anaerobic bacteria in a similar oxygen-poor medium can in the process of protein degradation reduce the protein sulfur to hydrogen sulfide. If at the same time cations of heavy metals are brought in by water, sulfides can be precipitated. The final product is very often *pyrite* FeS_2, which is the most common sulfide mineral in the earth's crust and is an important contributor to the sulfur content of coal and oil shales. In these deposits, as in crude oil, however, there is still organically bound sulfur. Crude oil also often contains hydrogen sulfide, and natural gas is sometimes hydrogen-sulfide-bearing, for example in the large new gas fields in Lacq (Basses-Pyrénées) and in western Canada.

For the industrial production of sulfur and sulfur dioxide endogenic pyrite and sometimes anhydrite $CaSO_4$ are often used. The minable occurrences of pyrite are geographically much more evenly distributed than occurrences of free sulfur.

Because sulfur is an important constituent of protein, plants must have access to it (generally as sulfate ions), although only small amounts are required.

Selenium often accompanies sulfur in small amounts in sulfides. Usually sulfur atoms in the sulfides are replaced by selenium atoms so that a solid solution is formed. In this way selenium is quite widespread, although pure selenium minerals are rare. Where the bedrock contains larger amounts of selenium, as in parts of North and South America (especially in Wyoming) certain plants may take up selenium instead of sulfur and thus become harmful as animal feed. However, the animal organism requires a certain supply of selenium and a lack of it can cause damage to the liver and muscular atrophy.

Tellurium has too large an atomic radius to be able to substitute for sulfur and occurs in specific minerals or native, although always very sparingly.

Polonium has only unstable isotopes, but these are included in all three of the natural radioactive series (9-2a). Polonium therefore is found in nature, but only in extremely small amounts.

b. History. The nature of combustion was studied during the 17th century by several scientists, and especially Boyle, Hooke and Mayow did important and even quantitative experiments on this problem. Unfortunately, this development was interrupted by the advent of the *phlogiston theory*, whose particular pioneer was Stahl. According to this theory combustible substances contain *phlogiston* (gk. φλογιστός, burn, from φλόξ, flame), which was given off on burning and then became evident in the form of heat and light phenomena. On heating zinc in air the metal burns, and this was written: zinc → zinc lime + phlogiston. On heating zinc lime, that is, zinc oxide, with carbon, which because of its great combustability was assumed to contain much phlogiston, this phlogiston would be given to the zinc lime according to: zinc lime + phlogiston → zinc.

The phlogiston theory was soon accepted by most chemists and prevailed for the greater part of the 18th century. The fact that the combustion products of a substance weighed more than the substance itself caused little concern in the beginning. When, gradually, scientists were forced to account for this observation, the phlogiston theory was so firmly entrenched that it was extended by the assumption that phlogiston had negative weight.

Oxygen. At least by 1772, but probably as early as 1769, Carl Wilhelm Scheele had found that air consists of two gases, of which only one, "fire air", could support combustion. The other gas he called "foul air". He also succeeded in preparing fire air, among other ways by heating mercuric oxide HgO and silver oxide Ag_2O. However, Scheele's paper describing these discoveries was first published in 1777. When Priestley prepared oxygen from mercuric oxide in 1774, he was therefore ignorant of Scheele's work. Both Scheele and Priestley were followers of the phlogiston theory, and Priestley in accordance with the theory designated oxygen as "dephlogisticated air". Because of its smaller content of phlogiston in comparison with natural air, this would be able to draw phlogiston from combustible substances and thus cause them to burn violently.

Oxygen was further studied by Lavoisier, who may be said to have finally explained the true nature of combustion. The reason for his choice of the name *oxygène* was mentioned in 12-1d.

Sulfur has been known since prehistoric times (Gk. θεῖον; Lat. sulfur) but was first clearly established as an element by Lavoisier.

Tellurium was discovered in 1783 by F. J. Müller and was named somewhat later after the earth (Lat. tellus). When Berzelius in 1817 found that a reddish sludge that appeared in the lead chambers in the sulfuric acid plant at Gripsholm contained a similar element, he called it *selenium* (Gk. σελήνη, moon).

Polonium was isolated in 1898 from the uranium mineral pitchblende by Pierre and Marie Curie and was the first radioactive element to be discovered. The preparation contained mainly the most long-lived isotope of the uranium series, ^{210}Po (9-2a). The name was given after Marie Curie's homeland, Poland.

21-3. Properties of the chalcogens

a. Oxygen. The stable and unstable isotopes are listed in tab. 1.4.

Ordinary, elementary oxygen in all states of aggregation consists of diatomic molecules O_2. As described in 5-3a the two oxygen atoms in this molecule are joined by a single bond and two **three-electron** bonds, so that it can be written $|O\vdots\vdots O|$. The two unpaired electrons make oxygen paramagnetic in all states in which O_2 molecules appear.

Some important properties of ordinary oxygen are the following:

Melting point	Boiling point	Crit. temp.	Crit. press.
$-218.8°C$	$-182.97°C^a$	$-118.8°C$	50.4 bar

[a] Calibration point on the International Practical Temperature Scale.

The gas is colorless, odorless and tasteless. At 20°C and 1 atm total pressure, 100 g of water dissolves 4.35 mg O_2. Oxygen is enough more soluble in water than nitrogen that water in equilibrium with air contains an oxygen-nitrogen mixture in which the proportion of oxygen is approximately 35 %, thus significantly higher than in the air. Both liquid and crystalline O_2 (there are several crystalline modifications with different arrangements of the O_2 molecules) have a light-blue color.

Ordinary oxygen combines directly with all other elements except the noble gases, the halogens and some of the noblest metals. In this oxide formation or *oxidation* in the original sense (cf. 16a), often so much heat is released that the element is heated to incandescence, that is, it burns. In many cases, however, the reaction rate is low at ordinary temperature and becomes appreciable only when the temperature is raised.

With the addition of sufficient energy the oxygen molecule is dissociated: $O_2(g) \rightleftharpoons 2O(g)$; $\Delta H = 495$ kJ. The added energy may come from heating, radiation with sufficiently energy-rich quanta (495 kJ per mole corresponds according to 1-2b to the energy in radiation of wavelength 2420 Å, that is, ultraviolet radiation) or chemical reactions. In the upper atmosphere, where the short-wavelength, solar radiation is not yet absorbed, atomic oxygen is continuously formed. The number of free oxygen atoms per unit of volume therefore increases with the altitude above the earth's surface up to about 100 km. At greater heights it decreases again because of the general diminution of the density, but here the major portion of the oxygen is atomic (17c).

Atomic oxygen reacts with, among other things, O_2 molecules with the formation of a triatomic oxygen molecule O_3, *ozone* (trioxygen, in contrast to dioxygen): $O(g) + O_2(g) \rightleftharpoons O_3(g)$. Ozone is unstable with respect to O_2; for the process $\frac{3}{2}O_2(g) \rightleftharpoons O_3(g)$ at 25°C, $\Delta G = 163.5$ kJ. At low temperature and in the absence of catalysts, however, ozone decomposes very slowly. At high temperature, on the other hand, decomposition takes place rapidly, and if ozone is produced by strong heating of oxygen gas, it will for the most part decompose before the gas is cooled. For this reason, and even more because the equilibrium mixture also at very high temperature has a low proportion of ozone, for the preparation of ozone atomic oxygen should instead be made at low temperature by the addition of electrical, radiant or chemical energy. The most common method up to now has been to supply electrical energy by means of a dark electric discharge between glass-covered, cooled electrodes with large surface area, over which a stream of oxygen gas is passed. In this way a gas mixture with at most up to 15 % ozone can be obtained.

Recently, attempts have been made for the preparation of ozone to use anodic oxidation of water at high anode potential and low temperature, in which initially atomic oxygen is formed: $H_2O \rightarrow O(g) + 2H^+ + 2e^-$ ($e^0 = 2.42$ V). As an electrolyte, among others, perchloric acid solution at -50°C has been used, which is electrolyzed with high current density.

The fact that ultraviolet radiation forms atomic oxygen and thus also ozone is evident from the formation of ozone in the air around a burning "quartz lamp". In the upper atmosphere ozone is formed continuously in this way, and the ozone present nearer the earth's surface has mainly this origin. Most of the ozone is found at heights of 10–30 km where the air density is low enough so that the ultraviolet radiation is still strong, but high enough to give appreciable amounts of ozone. At greater heights not only the absolute quantity of ozone per unit of volume decreases, but also its proportion in relation to atomic oxygen (cf. above). The ozone formed is itself strongly absorbing both in the ultraviolet and in the infrared. It is believed, therefore, to play an important role in stopping ultraviolet radiation that is destructive to many organic molecules essential to life processes. The absorption of infrared radiation serves to reduce heat radiation from the earth.

Under the action of sunlight hydrocarbons and nitrogen oxides in motor exhaust fumes form organic peroxides (cf. 21-6 a), which decompose into ozone plus aldehydes and acids. In some cities like Los Angeles, where weather conditions inhibit movement of air and millions of automobiles are operating, these products have been found occasionally to accumulate in the atmosphere to the point where they become a severe health hazard.

In many chemical processes, especially those involving strong oxidizing agents, appreciable formation of ozone may occur. Initially, atomic oxygen has then been formed. An example is the reaction between fluorine and water mentioned in 20-3.

Pure ozone can be obtained by condensation and fractional distillation of an ozone-containing mixture. Liquid ozone is dark blue (ozone gas at atmospheric pressure is decidedly blue in color) and boils at $-112°C$. At $-250°C$ it freezes to black crystals. Because of its instability pure ozone is very explosive. On the other hand, dilute ozone can be handled without danger, although one should avoid breathing it. One hour spent in an ozone-containing atmosphere will cause lung damage even at a concentration of 0.0005 %. Ozone has a very characteristic odor (the name is derived from Gk. ὄζω, I smell), which is recognizable even at extremely low concentrations.

As stated in 5-3e the electronic structure of ozone corresponds to a mesomeric state:

$$\begin{array}{c} \diagup \overline{O} \diagdown \\ |\underline{O}| \quad |\underline{O}| \end{array} \quad \leftrightarrow \quad \begin{array}{c} \diagup \overline{O} \diagdown \\ |\underline{O}| \quad |\underline{O}| \end{array}$$

Chemically ozone is characterized especially by its high oxidizing power, which in acid solution is evident from the high standard potential for the redox pair $O_2(g) + H_2O \rightleftharpoons O_3(g) + 2H^+ + 2e^-$ ($e^0 = 2.07$ V). Metallic silver is thus blackened by the formation of silver peroxide Ag_2O_2, and black lead sulfide PbS is oxidized to white lead sulfate $PbSO_4$. Organic substances are generally vigorously oxidized, and also, ozone is often added to double bonds. Rubber is thus destroyed extremely rapidly. The oxidizing power of ozone is used in organic syntheses and for disinfection, for example of water.

The equilibrium $2O_2 \rightleftharpoons O_4$ should also be mentioned, which at very low temperature (in liquid and solid oxygen) is shifted far enough to the right so that O_4 can be detected. In this molecule two O_2 molecules are bound to each other by a weak bond, which, however, is stronger than a van der Waals bond.

b. Sulfur. Sulfur forms a series of different kinds of molecules in all the solid, liquid and gaseous states. Transitions among these often occur very sluggishly, so that their behavior becomes complicated. Therefore, only a very schematic presentation can be given here.

The most important sulfur molecules and the reasons for their occurrence have been described earlier in 5-6a. These are the S_8 ring (fig. 5.27b), the open, twisted chains of varying lengths (fig. 5.27a) and the S_2 molecule analogous to O_2.

Crystalline sulfur occurs in two different structures, *orthorhombic sulfur* (S_α) and *monoclinic sulfur* (S_β). Both consist of S_8 rings but these are packed in the two structures in different ways. S_α as well as S_β is soluble in carbon disulfide, and similar solutions are produced in both cases. The lowering of the freezing point (13-5b) shows that in both the sulfur is present as S_8 molecules. From these solutions S_α crystallizes at room temperature. Both S_α and S_β form brittle crystals of light-yellow color and a density of about 2 g cm^{-3}. The S_β crystals have a more needle-like habit than S_α crystals. At atmospheric pressure S_α is stable up to 95.5°C, at which point it is in equilibrium with S_β, which is then stable up to the melting point at 119.0°C. Freshly melted sulfur immediately above the melting point is a light-yellow, mobile liquid consisting of S_8 rings and commonly designated S_λ.

A schematic P,T diagram (13-4a) of the equilibria among S_α, S_β and S_λ is reproduced in fig. 21.2. The diagram shows the triple points for the equilibrium $S_\alpha + S_\beta + g$, which corresponds very closely to the above-mentioned transition point at a pressure of 1 bar and 95.5°C, and for the equilibrium $S_\beta + S_\lambda + g$, which corresponds very closely to the melting point of S_β at a pressure of 1 bar and 119.0°C. In addition the curves for the two-phase equilibria $S_\alpha + S_\beta$ and $S_\beta + S_\lambda$ intersect each other in a triple point for the equilibrium $S_\alpha + S_\beta + S_\lambda$ at 151°C and 1305 bar. Because of the sluggishness of the reactions, a triple point for the metastable (2-3b) equilibrium $S_\alpha + S_\lambda + g$ can also be realized at 112.8°C and approximately 1 bar. This corresponds to a metastable melting point for S_α and it can be attained if S_α is heated so rapidly that the transition to S_β does not have a chance to occur. With sufficiently slow cooling of a S_λ melt S_β first crystallizes, which can then easily be supercooled. For example, as soon as a crust has formed on the surface of the melt a hole can be punched in it and the melt poured off so that the cooling of the remaining S_β can take place more rapidly. This supercooling is indicated in fig. 21.2 by the short, dashed line below 95.5°C. After a few days at room temperature the S_β structure transforms to the S_α structure stable at this temperature.

Liquid sulfur is a pseudosystem of the kind described in 13-3b. In a stable sulfur melt an equilibrium exists between S_8 rings and open chains of varying lengths, which are formed by the breaking of the rings and joining of the fragments. Sulfur with such open chains is commonly designated S_μ and the equilibrium in the melt can thus be roughly indicated as $S_\lambda \rightleftharpoons S_\mu$. A pure S_λ melt is therefore not stable as suggested in fig. 21.2. Even if a quite pure S_λ melt is obtained just at the melting point it gradually approaches an equilibrium in which immediately above the melting point approximately 3.6% of the sulfur is S_μ (in equilibrium with the equilibrium melt $S_\lambda + S_\mu$, S_β also shows a lower melting point than that mentioned above, namely, 114.5°C). With further heating of the melt, more and more of the S_8 rings are broken, that is, the equilibrium $S_\lambda \rightleftharpoons S_\mu$ is shifted more and more to the right. The melt becomes thicker and thicker (viscous), because the open chains become tangled with each other. If the S_8 rings had persisted on the other hand, the viscosity would have decreased because of the increased thermal motion. With increasing S_μ content the color also darkens, because of unpaired electrons at the ends of the chains, for example as in $\cdot\underline{S}{-}\underline{S}{-}\underline{S}{-}\underline{S}{-}\underline{S}\cdot$. The viscosity reaches a maximum at about 200°C and here the melt can no longer be poured. With further rise in temperature the viscosity falls again, probably because the thermal motion increases and the chains become shorter; at 400°C the melt is again quite fluid. If molten sulfur is rapidly cooled, for example by being poured into water, the S_8 rings but not the chains have time to arrange themselves in crystals (6-3d). S_λ thus crystallizes, while S_μ forms an amorphous mass which is insoluble in carbon disulfide and may be thought of as a sulfur glass. If the experiment is carried out with the fluid melt at 400°C, which contains mainly S_μ, the sulfur forms a rubbery, brown-yellow mass (*plastic sulfur*). Gradually, the plastic sulfur hardens because of crystallization (at room temperature, as to be expected, to S_α).

Sulfur boils at 444.6°C (calibration point in the International Practical Temperature Scale). The gas formed at this temperature is orange-yellow, and with rising temperature becomes first red and then more and more light-colored. At 650°C the gas is straw-yellow. At the boiling point it consists mainly of S_8 rings. As the temperature is raised the rings are opened, after which the chains formed are successively broken down. At 650°C the largest portion consists of S_2 molecules and at 2000°C approximately half of these are dissociated into S atoms. On condensation of sulfur vapor on a cold surface solid sulfur is obtained in finely divided form ("flowers of sulfur"). This contains also S_μ.

Solid sulfur in all forms is an insulator. It is always insoluble in water. Many organic solvents dissolve S_α and S_β appreciably. However, the solubility is generally low, and is high only in carbon disulfide CS_2 and disulfur dichloride S_2Cl_2. Under the same conditions the rate of dissolution is smaller for S_μ than for S_α and S_β.

Under special circumstances it is possible to obtain more or less unstable sulfur rings other than the S_8 ring. If concentrated hydrochloric acid is added with cooling to a saturated solution of $Na_2S_2O_3$ (cf. 21-11f), sulfur is formed as S_6 rings in addition to S_8 rings and open chains. If the sulfur is extracted with *e.g.* toluene, the solution gives, among other modifications, crystals of *trigonal (rhombohedral) sulfur*, containing the S_6 rings. These rings, which are not plane, break up in a rather short time. In 1966 an S_{12} ring was synthetized, which can build up crystals of a new orthorhombic sulfur modification, more stable than the trigonal one.

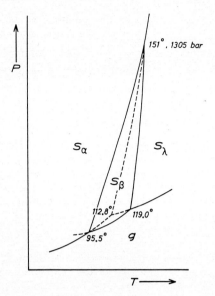

Fig. 21.2. P,T diagram for sulfur (not to scale).

Sulfur combines even at moderate temperature directly with many nonmetals and metals. It can be ignited when heated in air and it burns with a blue flame to sulfur dioxide ($S+O_2 \rightarrow SO_2$). It combines with hydrogen ($S+H_2 \rightarrow H_2S$) and with halogens (for example, $2S+Cl_2 \rightarrow S_2Cl_2$). On heating with iron filings, $\sim FeS$ is formed with strong evolution of heat. Copper burns in sulfur vapor with the formation of $\sim Cu_2S$, and if mercury is mixed with flowers of sulfur, black mercury(II) sulfide HgS is formed.

c. Selenium, tellurium and polonium. *Selenium* in many ways is similar to sulfur. Se_8 rings, open and zigzag chains of varying lengths, and Se_2 molecules are all known. The only stable, solid modification is dark gray with reddish cast and has a trigonal crystal structure (*gray* or *trigonal selenium*), which is built up of parallel, infinite, helical chains (fig. 5.27a). This form of selenium is a photoconductor (6-4d) and is therefore sometimes called, although improperly, metallic selenium (another, also improper name is hexagonal selenium). The melting point is 220°C and the boiling point is 688°C. The melt contains mainly Se chains and if it is cooled rapidly they are unable to arrange themselves in the trigonal structure but rather disordered, *glassy selenium* is produced (6-3d).

Selenium can be precipitated in elementary form from solution (for example by reduction of selenious acid with sulfur dioxide: $H_2SeO_3 + 2SO_2 + H_2O \rightarrow Se + 4H^+ + 2SO_4^{2-}$). A colloidal, red solution is first formed and then a red precipitate. This has a disordered structure but certainly contains Se_8 rings. A similar product is obtained if selenium vapor is chilled rapidly. Both these solid phases dissolve in carbon disulfide and from the solution one or the other of two types of *red (monoclinic) selenium* is obtained depending on the conditions. These are insulators and are both composed of Se_8 rings, although the rings are arranged in a somewhat different way in the two types. All forms of red selenium at ordinary pressure are unstable at all temperatures. The transition to stable trigonal selenium takes place very slowly at room temperature but can go quite rapidly at 100°C.

Selenium compounds are very poisonous.

The only modification of *tellurium* stable at ordinary pressure has the same trigonal structure as gray selenium and is thus made up of infinite Te chains. It has a silver-white metallic luster but is only a semiconductor. The melting point is 450°C and the boiling point 1390°C. On reduction from solution a brown precipitate of amorphous tellurium is obtained, which crystallizes at higher temperature. Tellurium compounds are less poisonous than selenium compounds. When even very small amounts of tellurium are absorbed by the body, the breath for a long time has a strong garlic odor (from methyl telluride).

Polonium has metallic conductivity. Its crystal structure is simple cubic and is related to the trigonal structures of selenium and polonium in that the bonds between the infinite chains have increased in strength to the point where they are equal to those along the chains. In its physical properties polonium resembles the preceding elements in the period, thallium, lead and bismuth, much more than it does tellurium. On the other hand, its chemical properties show a great similarity to those of tellurium.

21-4. Production and uses of the chalcogens

a. Oxygen. Formerly, when oxygen was prepared on a laboratory scale, this was most often done by heating potassium chlorate with manganese dioxide added as a catalyst (20-8d). Very pure oxygen can be obtained by heating certain easily decomposable metal oxides, especially oxides of noble metals (silver, gold, platinum metals).

Industrially, some oxygen is obtained by electrolysis of water (16f), but then only as a by-product when hydrogen is the main product. By far the largest amount of oxygen is obtained by the production of liquid air (13-4a) and its fractional distillation (13-5c). Oxygen is shipped more and more as liquid oxygen in tank vehicles of the Dewar-flask type. Production is very great and is increasing rapidly. At least half goes to the iron and steel industry, where oxygen nowadays plays a large role in the manufacture of steel. Large amounts are also used in the chemical industry and in welding. For rockets with liquid fuels liquid oxygen ("lox") up to now is the most used oxidizing agent.

It should never be forgotten that liquid oxygen, because of its high oxygen concentration, can cause very violent explosions in contact with oxidizable substances. Also liquid air for the same reason must be handled carefully, and with time it gradually becomes richer in oxygen because the more volatile nitrogen vaporizes more rapidly than the oxygen. As a cooling medium, therefore, liquid nitrogen is more preferable than liquid air.

b. Sulfur. Of the world's total production of sulfur approximately 55% comes directly from free sulfur. Free sulfur as a rule occurs disseminated in sulfur-bearing rocks and is obtained from them by sweating or distillation. In some places (Sicily, Japan, the Andes) the sulfur can be mined directly, but around the Gulf of Mexico the sulfur deposits are covered by layers of clay and quicksand several hundred feet thick, which makes direct mining uneconomical. Since the sulfur here is disseminated in limestone that lies between dense rocks (anhydrite below, sulfur-free limestone above), it is possible to melt the sulfur out of the sulfur-bearing limestone bed itself with superheated water and, after it has collected, force it up to the surface with compressed air. This process, invented by Frasch, is continuous, and the introduction of hot water (about 160°C) and compressed air and removal of the sulfur take place in the different passages of an aggregate of concentric tubes inserted into a drill hole.

Next in importance after free sulfur in the production of sulfur is pyrite. In addition sulfur is recovered from anhydrite and as a by-product of oil refining, gas manufacture and processing of oil shale. Recently, hydrogen-sulfide-bearing natural gas has also become an important source.

Of the total production of sulfur approximately 80% is used in the manufacture of sulfuric acid. In this and in the equally important production of sulfites the raw material is sulfur dioxide SO_2, and it is therefore clear that the processing of sulfur compounds often is carried directly to sulfur dioxide. Pyrite is roasted, that is, heated in air ($2FeS_2(s) + 5\frac{1}{2}O_2(g) \rightarrow Fe_2O_3(s) + 4SO_2(g)$), and anhydrite is reduced by heating with coke (overall reaction $2CaSO_4(s) + C(s) \rightarrow 2CaO(s) + CO_2(g) + 2SO_2(g)$).

In gases from oil refining, gas works and oil shale processing and also in natural gas the sulfur occurs as hydrogen sulfide. An important method for recovering the sulfur from these is the following. After purification the hydrogen sulfide is partially burned ($3H_2S + \frac{3}{2}O_2 \rightarrow 2H_2S + SO_2 + H_2O$), following which the H_2S–SO_2 mixture is passed over a heated catalyst (bauxite, 24-3a), whereby sulfur vapor is obtained ($2H_2S + SO_2 \rightarrow 3S + 2H_2O$).

Sulfur is one of the great strategic chemical substances, and the world's yearly consumption of sulfur is rapidly increasing. The largest producers are the United States and Mexico. In the future it is probable that the world's consumption can be supplied only by an ever-increasing recovery from pyrite.

As mentioned above about 80% of the sulfur production goes to the manufacture of sulfuric acid. A large part of the remainder is used in the production of sulfite for sulfite-cellulose processing, making carbon disulfide (mainly for rayon manufacture; 23-2j), in the synthetic organic chemical industry and the vulcanization of rubber.

c. Selenium and tellurium. Since *selenium* so often accompanies sulfur (21-2a), it is generally recovered in the processing of sulfur products, especially in the manufacture of sulfuric acid (cf. the discovery of selenium, 21-2b), and from anode slime

(16f) formed in the electrolytic refinement of copper obtained from selenium-bearing sulfide ores.

Selenium is used primarily in the manufacture of semiconductor rectifiers (6-4d) and photocells, and to a lesser extent in the coloring of glass and enamel (red color) and as an alloy substance. Selenium and selenium compounds are also used as catalysts in organic chemical processes.

Tellurium is used mainly as a glass-coloring agent and as an alloy substance.

21-5. Oxides

a. General. Since, next to fluorine, oxygen is the most electronegative element it has the oxidation number $-\mathrm{II}$ in bonds with all other elements except fluorine (5-3g). Oxygen may be said to have an oxide function in all such cases, but usually it is only the relatively simple oxygen compounds that are called oxides. Charged oxide molecules, that is, *oxo ions*, should also be considered as oxides. The particular descriptions of these, which constitute by far the most important of the complex anions, and the corresponding oxo acids, are given under the respective central-atom element.

A glance at fig. 5.23 makes it immediately clear that the nonmetals are bound to oxygen by predominantly covalent bonds, and, on the other hand, that the covalent contribution must be quite small in oxides of the more electropositve metals. The covalent oxides often occur as discrete molecules, and the weak van der Waals forces between them give the condensed phases low melting and boiling points (for example, Cl_2O, SO_2, NO, CO_2 and H_2O; the last, however, more easily condensed because of hydrogen bonding). However, in many cases the element bound to oxygen has a strong tendency toward a higher oxygen coordination number than the composition allows for a finite molecule. These molecules are then often formed with infinite extension in one or more dimensions (5-6b). The tendency of the silicon atom to coordinate tetrahedrally four oxygen atoms can be satisfied in the dioxide SiO_2 by having each oxygen atom join two silicon atoms, giving rise to a framework molecule. In such cases the melting point of the solid phase may be high and the volatility slight.

The oxides of the most electropositive metals at ordinary temperature give quite typical ionic crystals, in which the oxide ion O^{2-} is the anion. As a rule the melting points here are high, often very high (for example, for MgO, 2800°C).

For different oxides of the same element the covalent contribution increases as the oxidation number of the element increases. Thus, $Cr_2^{III}O_3$ has a weakly covalent, but $Cr^{VI}O_3$ a strongly covalent, character. This can be explained by the fact that the polarizing ability of the chromium atom increases as the charge becomes greater and its radius smaller (5-2e). Consequently the covalent contribution increases.

Most of the volatile covalent oxides (for example, Cl_2O, Cl_2O_7, SO_2, SO_3, N_2O_4, N_2O_5, CO_2) dissolve in water, producing an acid solution: for example, $SO_2 + H_2O \rightarrow$

$2H^+ + SO_3^{2-}$. Nonvolatile covalent oxides, for example SiO_2, are generally poorly soluble in water but dissolve easily in a strongly basic solution. Consequently, all these covalent oxides can be considered as *acid anhydrides*. Since early times they have also been called *"acid" oxides*. However, some covalent oxides are inert. This is true especially for N_2O, NO and CO.

Of the oxides with predominantly ionic bonding, the oxides of the alkali and alkaline-earth metals except beryllium and magnesium dissolve easily or fairly easily in water to give a basic solution: $O^{2-} + H_2O \rightarrow 2OH^-$. The O^{2-} ion is a very strong base, while the OH^- ion is a very weak acid (tab. 12.2). Most other metal oxides are difficultly soluble in water, but many of them dissolve in strong acids, because OH^- ions become bound up, or because of the reaction $O^{2-} + 2H^+ \rightarrow H_2O$. Like the more soluble metal oxides, they have long been called *"basic" oxides*.

The discussion in 12-2f showed that a hydroxo compound AOH functions as an acid according to $AOH \rightarrow AO^- + H^+$ when the A–O bond has covalent character, so that the oxide of A should also have covalent A–O bonding. On the other hand the dissociation $AOH \rightarrow A^+ + OH^-$ takes place when the A–O bond has sufficiently strong ionic character, which should then also be true of the oxides of A. With intermediate character in the A–O bond, AOH can react in one way with respect to a base and in the other way with respect to an acid, that is, be amphoteric.

Long ago it was known that a "basic" and an "acid" oxide could react with each other *on heating*, for example according to $MgO + SiO_2 \rightarrow Mg^{2+} + SiO_3^{2-}$, and recognition of such reactions served to divide oxides into "acid" and "basic" classes. Especially after Lux and Flood interpreted all these reactions in terms of oxide-ion transfer (12-1e), the principles inherent in these concepts became clear, and through the acid-base definition of Gutmann-Lindqvist (12-1e) they could be incorporated into a more general relationship (also, in the reactions of the oxides with protolytes mentioned above, the overall process becomes an oxide-ion transfer to or from the oxide). By this definition an oxide functions as a *base*, for example according to

$$\underset{\text{acid}}{A^{2+}} + \underset{\text{base}}{O^{2-}} \rightleftharpoons AO \qquad (21.1)$$

This applies to oxides with high ionic-bond contribution. The oxides of the alkali and alkaline-earth metals are most strongly basic. When the electronegativity of A increases (mainly on going to the right in the periodic system), the basic strength of the oxides decreases. When A has high electronegativity, and consequently the covalent contribution is strong, the oxide behaves preferably as an *acid*, for example according to

$$\underset{\text{acid}}{AO} + \underset{\text{base}}{O^{2-}} \rightleftharpoons AO_2^{2-} \qquad (21.2)$$

For example, the following are strongly acid oxides: B_2O_3, SiO_2, P_2O_5, SO_3 (the last two, however, are apparently less strong because of their volatility).

When A is moderately electronegative (electronegativity in the vicinity of 1.5) the oxides are *intermediate*. They can then function as a base with an acid oxide according to (21.1), and as an acid with a basic oxide according to (21.2). The oxide thus has an amphoteric function. Examples are BeO, Al_2O_3, Cr_2O_3, Fe_2O_3, ZnO.

Since the bonding in different oxides of the same element becomes more covalent, as the oxidation number of the element increases (cf. 5-2e) the oxides at the same time become more acid. While Cr_2O_3 may be said to be intermediate, CrO_3 is acid.

An oxo ion, formed as in (21.2), can often take up further oxide ions, for example according to

$$\underset{\text{acid}}{AO_2{}^{2-}} + O^{2-} \rightleftharpoons \underset{\text{base}}{AO_3{}^{4-}} \qquad (21.3)$$

It then functions as a base according to (21.2) and as an acid according to (21.3), that is, is amphoteric.

From what has been said above and in 5-6b it follows that the reactions of the types (21.2) and (21.3) most often involve solution or formation of polynuclear or infinite complexes. The two equations are therefore entirely schematic. When SiO_2, which forms a framework molecule and therefore can be designated $(SiO_2)_n$, takes up O^{2-} from MgO the Si can attain tetrahedral oxygen coordination by formation of chains, which can be written schematically in a plane as

$$\begin{array}{ccccccc} & & O & & O & & O \\ & & | & & | & & | \\ -Si & -O- & Si & -O- & Si & -O- \\ & & | & & | & & | \\ & & O & & O & & O \end{array}$$

Such chains of varying lengths are certainly present in the melt, and when this crystallizes to $MgSiO_3$ the crystals contain similar "infinite" chains in parallel position, and Mg^{2+} ions. The building units are thus Mg^{2+} and $(SiO_3{}^{2-})_n$, and the process can be written: $(SiO_2)_n + nO^{2-} \rightarrow (SiO_3{}^{2-})_n$.

If more MgO is added to the melt the chains are broken up. At the composition $2MgO + SiO_2$ the number of O^{2-} ions per Si is sufficient for the formation of single tetrahedral $SiO_4{}^{4-}$ ions: $(SiO_2)_n + 2nO^{2-} \rightarrow nSiO_4{}^{4-}$. These occur now as anions both in the melt and in the crystals of composition Mg_2SiO_4 (olivine) formed from it. If the amount of MgO is further increased, the oxygen coordination of silicon is not increased beyond four, but the ions in the melt now become Mg^{2+}, O^{2-} and $SiO_4{}^{4-}$. In this case, on crystallization the two phases MgO and Mg_2SiO_4 are formed, but other silicates are known, for example the mineral andalusite $Al_2O(SiO_4)$, in which both kinds of anions O^{2-} and $SiO_4{}^{4-}$ are included in the structure.

It can thus be seen how the competition between the oxide ions may be reflected in the crystal structure. Another example of this is the formation of different borate ions, which was also discussed in 5-6b. For example, we have $(BO_2{}^-)_n + nO^{2-} \rightarrow nBO_3{}^{3-}$.

Other salts of oxo acids in principle may be considered as being formed in an analogous way. Thus K_2SO_4 can be made according to $K_2O + SO_3 \rightarrow 2K^+ + SO_4{}^{2-}$.

Sulfur(VI) tries to achieve the coordination number four. In SO_3 this is accomplished by the formation of open or closed chains (21-10b) and the formula would therefore be better written as $(SO_3)_n$. By addition of oxide ions from the basic oxide K_2O the chains are broken down and free SO_4^{2-} ions formed.

The acidity of an oxide melt is determined by the oxide ion activity $\{O^{2-}\}$ (see 10c). In an acid melt this activity is low, in a basic melt it is high.

The acid-base nature of an oxide depends on many factors often difficult to appraise, for example conditions in the melt depending on its other components. Also, the relative acid strengths of different oxides can thus vary.

Oxide reactions of the kind mentioned above have great practical importance. Many pure oxides have high melting points, but if they are allowed to react as above, substances ("salts") are formed that are often more easily melted. This is important in many metallurgical processes in which extraneous oxides, for example from the gangue, can be melted in this way and then more easily removed (*slag formation*). For example, if the ore is contaminated by quartz (the acid oxide SiO_2) limestone is added (in which the basic oxide CaO is active) as a slag-forming agent. On a metal surface covered by an oxide layer molten metal will not adhere in soldering or welding. However the metal oxide is basic and therefore reacts with an added acid oxide (soldering flux) so that an easily melted salt is formed that can be displaced by the molten metal. On the other hand, care must be taken that a crucible or construction material made of an acid (basic) oxide does not come in contact with a basic (acid) oxide at high temperature. Since many salts are more easily soluble in water than pure oxides, the latter can often be converted to soluble form by melting them with another appropriate oxide ("fusion").

If two oxides of relatively electropositive elements are melted together, the oxide-ion transfer will certainly lead to changes in the oxide ion activity, but from a purely geometrical standpoint this will be hardly noticeable. Both CaO and MgO are made up of metal cations and oxide ions, and their melts will consist largely of Ca^{2+}, Mg^{2+} and O^{2-}. In melts of $MgO + Al_2O_3$ the oxide-ion transfer will be greater and probably lead to the formation of oxoaluminate complexes (schematically according to the equation (21.2)) in the melt. But in the phase (the mineral *spinel* $MgAl_2O_4$) that crystallizes out of the melt the bonds Mg–O and Al–O are so like one another that the structure gives no reason to speak of a magnesium aluminate. The building units are thus more exactly $Mg^{2+} + 2Al^{3+} + 4O^{2-}$ and not $Mg^{2+} + 2AlO_2^{-}$. Spinel should therefore be named magnesium aluminum oxide, and generally an oxide of this kind is called a *double oxide* (oxides with three metals all with similar bonds to oxygen are called *triple oxides*, and so on).

Spinel $MgAl_2O_4$ has given its name to the *spinel type* of structure (cubic symmetry), in which a large number of other double oxides also crystallize. If its general formula is written $MM_2'O_4$, electrical neutrality requires that $z_M + 2z_{M'} = 8$, and this is satisfied most often when $z_M = 2$ and $z_{M'} = 3$ (for example, $MgAl_2O_4$,

$ZnAl_2O_4$, $Mg(Fe^{III})_2O_4$, $Mg(Cr^{III})_2O_4$, $Fe^{II}(Fe^{III})_2O_4 = Fe_3O_4$) or when $z_M = 4$ and $z_{M'} = 2$ (for example, $Ti^{IV}Mg_2O_4$, $Sn^{IV}Mg_2O_4$, $Ti^{IV}(Co^{II})_2O_4$). Solid solutions in which more than two metals are present simultaneously are also common.

Many double oxides with the general formula $MM'O_3$ ($z_M + z_{M'} = 6$) and metal-atom radii of proper size belong to the *perovskite type* (cubic or slightly deformed cubic symmetry). Examples are $CaTi^{IV}O_3$ (the mineral *perovskite*), $NaNb^VO_3$, $Y^{III}AlO_3$.

An interesting double oxide is $AlPO_4$, which thus is not an aluminum phosphate. This is especially evident here in that the compound is completely analogous to SiO_2 and shows many of the different structural modifications that characterize SiO_2. Al and P in fact stand on either side of Si in the periodic system and may be said to constitute an "average atom" $\frac{1}{2}(Al^{III} + P^V)$, which is clearly closely similar to Si^{IV}.

The reactions between oxides described above generally take place only on heating, but they may often begin before melting occurs (14-3a, b).

The more or less free, little polarized oxide ion does not absorb visible light; it is "colorless". With bonding to strongly polarizing cations (large covalent-bond contribution), however, color may appear even if the cation is colorless (thus, silver oxide Ag_2O is black).

b. Formation and reduction of oxides. Many oxides are easily formed by direct reaction of the element in question with oxygen (possibly air), although often only at higher temperature: $Mg + \frac{1}{2}O_2 \rightarrow MgO$; $4Cu + O_2 \rightarrow 2Cu_2O$; $C + \frac{1}{2}O_2 \rightarrow CO$. The element may also react with an oxygen-containing compound, also here often only at higher temperature: $C + H_2O \rightarrow CO + H_2$. Oxygen-containing compounds of the element in question may be decomposed, often by heating, so that oxide remains: $CaCO_3(s) \rightarrow CaO(s) + CO_2(g)$. Such a process can sometimes be used when it is difficult to oxidize the element to the desired oxide, which, as a rule, implies that this oxide is unstable: $2HClO_4 \rightarrow Cl_2O_7 + H_2O$ (here the decomposition is carried out by means of a dehydrating agent, 20-7e).

Since the elements in the earth's crust are incorporated in oxide compounds to so large an extent, the *reduction of oxides* to the corresponding element is an extraordinarily important technical process, which is used to a large extent in the production of metals. Even when the metal occurs as a sulfide, the final step is often an oxide reduction. The sulfide is first converted to an oxide, usually by heating in air (*oxidizing roasting*), for example, $ZnS(s) + \frac{3}{2}O_2(g) \rightarrow ZnO(s) + SO_2(g)$. Oxygen dissolved in molten metal is often very harmful to the final material and is then usually removed (*deoxidation*) by the addition of an element that forms a stable oxide, which is collected as slag on the surface of the melt (for example, Si, Al, Na, Mg, Ca, Mn). A steel melt is deoxidized particularly with Mn and Si, in certain cases also with Al.

The processes occurring in these oxide reductions can generally best be considered as a transfer of oxygen atoms (not oxide ions as in the oxide equilibria discussed in 21-5a). Here the oxygen activity (not the oxide ion activity) becomes

Table 21.1. *Values of log K_p for gas equilibria with oxides.*

The values are based on p expressed in atm (mainly after Sillén, et al.)

Reaction	log K_p		
	1 000°K	1 500°K	2 000°K
2 CaO(s) $\rightleftharpoons$ 2 Ca(s) + O_2(g)	−55.80		
2 CaO(s) $\rightleftharpoons$ 2 Ca(l) + O_2(g)		−33.54	
2 CaO(s) $\rightleftharpoons$ 2 Ca(g) + O_2(g)			−21.17
ThO_2(s) $\rightleftharpoons$ Th(s) + O_2(g)	−54.64	−33.29	
2 BeO(s) $\rightleftharpoons$ 2 Be(s) + O_2(g)	−54.46	−33.07	
2 BeO(s) $\rightleftharpoons$ 2 Be(l) + O_2(g)			−22.25
2 MgO(s) $\rightleftharpoons$ 2 Mg(l) + O_2(g)	−52.07		
2 MgO(s) $\rightleftharpoons$ 2 Mg(g) + O_2(g)		−30.06	−17.23
2 TiO(s) $\rightleftharpoons$ 2 Ti(s) + O_2(g)	−49.03	−29.55	−19.86
ZrO_2(s) $\rightleftharpoons$ Zr(s) + O_2(g)	−46.69	−28.12	−18.92
$\frac{2}{3} Al_2O_3$(s) $\rightleftharpoons 1\frac{1}{3}$ Al(l) + O_2(g)	−46.59	−27.84	−18.46
2 BaO(s) $\rightleftharpoons$ 2 Ba(s) + O_2(g)	−44.89		
$\frac{2}{3} Ti_2O_3$(s) $\rightleftharpoons 1\frac{1}{3}$ Ti(s) + O_2(g)	−44.09	−26.15	−17.26
TiO_2(s) $\rightleftharpoons$ Ti(s) + O_2(g)	−38.38	−22.53	−14.64
2 VO(s) $\rightleftharpoons$ 2 V(s) + O_2(g)	−37.96	−22.71	
SiO_2(s) $\rightleftharpoons$ Si(s) + O_2(g)	−36.06	−20.89	
SiO_2(l) $\rightleftharpoons$ Si(l) + O_2(g)			−13.11
$\frac{2}{5} Ta_2O_5$(s) $\rightleftharpoons \frac{4}{5}$ Ta(s) + O_2(g)	−35.33	−20.97	
NbO_2(s) $\rightleftharpoons$ Nb(s) + O_2(g)	−32.87	−18.81	−11.78
2 MnO(s) $\rightleftharpoons$ 2 Mn(s) + O_2(g)	−32.62	−19.23	
2 MnO(s) $\rightleftharpoons$ 2 Mn(l) + O_2(g)			−12.27
$\frac{2}{3} Cr_2O_3$(s) $\rightleftharpoons 1\frac{1}{3}$ Cr(s) + O_2(g)	−29.96	−16.96	−10.46
2 Na_2O(s) $\rightleftharpoons$ 4 Na(l) + O_2(g)	−29.50		
2 Na_2O(s) $\rightleftharpoons$ 4 Na(g) + O_2(g)		−11.60	
2 V_2O_3(s) $\rightleftharpoons$ 4 VO(s) + O_2(g)	−29.46	−17.02	
$\frac{2}{5} P_2O_5$(g) $\rightleftharpoons \frac{2}{5} P_2$(g) + O_2(g)	−21.57		
2 CO(g) $\rightleftharpoons$ 2 C(s) + O_2(g)	−20.83	−16.94	−14.99
CO_2(g) $\rightleftharpoons$ C(s) + O_2(g)	−20.63	−13.77	−10.34
2 FeO(s) $\rightleftharpoons$ 2 Fe(s) + O_2(g)	−20.59	−11.55	
2 FeO(l) $\rightleftharpoons$ 2 Fe(l) + O_2(g)			−7.42
$\frac{2}{3} WO_3$(s) $\rightleftharpoons \frac{2}{3}$ W(s) + O_2(g)	−20.46		
2 CO_2(g) $\rightleftharpoons$ 2 CO(g) + O_2(g)	−20.43	−10.60	−5.68
WO_2(s) $\rightleftharpoons$ W(s) + O_2(g)	−20.37		
2 H_2O(g) $\rightleftharpoons$ 2 H_2(g) + O_2(g)	−20.02	−11.44	−7.15
MoO_2(s) $\rightleftharpoons$ Mo(s) + O_2(g)	−19.78		
2 Fe_3O_4(s) $\rightleftharpoons$ 6 FeO(s) + O_2(g)	−19.55	−8.67	
2 SiO(g) $\rightleftharpoons$ 2 Si(s) + O_2(g)	−18.62	−14.73	
$\frac{2}{3} MoO_3$(s) $\rightleftharpoons \frac{2}{3}$ Mo(s) + O_2(g)	−16.92		
2 CoO(s) $\rightleftharpoons$ 2 Co(s) + O_2(g)	−16.84	−8.69	
2 CoO(s) $\rightleftharpoons$ 2 Co(l) + O_2(g)			−4.51
2 NiO(s) $\rightleftharpoons$ 2 Ni(s) + O_2(g)	−15.25	−6.74	
2 NiO(s) $\rightleftharpoons$ 2 Ni(l) + O_2(g)			−2.34
SO_2(g) $\rightleftharpoons \frac{1}{2} S_2$(g) + O_2(g)	−15.14	−8.84	−5.68
4 VO_2(s) $\rightleftharpoons$ 2 V_2O_3(s) + O_2(g)	−15.11	−7.59	

Table 21.1. *Cont.*

Reaction	log K_p		
	1000°K	1500°K	2000°K
2 PbO(s) $\rightleftharpoons$ 2 Pb(l) + O$_2$(g)	−12.68		
6 Fe$_2$O$_3$(s) $\rightleftharpoons$ 4 Fe$_3$O$_4$(s) + O$_2$(g)	−11.36		
2 Cu$_2$O(s) $\rightleftharpoons$ 4 Cu(s) + O$_2$(g)	−10.84		
2 Cu$_2$O(s) $\rightleftharpoons$ 4 Cu(l) + O$_2$(g)		−4.84	
2 Cu$_2$O(l) $\rightleftharpoons$ 4 Cu(l) + O$_2$(g)			−2.45
2 Ag$_2$O(s) $\rightleftharpoons$ 4 Ag(s) + O$_2$(g)	+3.46		
2 Ag$_2$O(s) $\rightleftharpoons$ 4 Ag(l) + O$_2$(g)		+4.44	+4.83

a proper main variable, which as a rule can be represented by the oxygen-gas pressure.

In tab. 21.1 is listed log K_p (see 10b; p in atm) at three given temperatures for oxide equilibria, which are all written so that 1 mol O$_2$ appears on the right-hand side. The equilibria have been arranged in order of increasing equilibrium constant at 1000°K. Thus, the most stable oxide at this temperature is at the top of the table, the least stable at the bottom. It can be seen that the alkaline-earth oxides are very stable. After these for the most part come the oxides of the alkali metals and the earlier metals in the transition-metal series. The stability decreases further as one goes to the right in these series, and the noble-metal oxides come last.

The oxygen pressure p_{O_2} determines all of these equilibria in the same way as the hydrogen ion activity does in ordinary acid-base equilibria, as the redox potential does in redox equilibria of the types discussed in chap. 16, or as the oxide ion activity does in the oxide equilibria described in 21-5a. At each temperature, for example, a metal and its oxide are in equilibrium with each other only at a particular value of p_{O_2}; if the actual oxygen gas pressure is lower, the metal is stable, if it is higher the oxide is stable. If all the substances except O$_2$ are pure, condensed substances so that their activities = 1, the equilibrium equation is given immediately by $p_{O_2} = K_p$ (cf. the analogous case in the dissociation of CaCO$_3$ in 13-5e). In general it is practical to use log p_{O_2} and in the case mentioned, thus, log p_{O_2} = log K_p. The condition that the condensed substances be pure is not satisfied in many of the equilibria in tab. 21.1. Solid solutions with variable oxygen content occur often both in metal and oxide phases (6-3a). Here these variations are not considered.

Assuming pure, condensed substances, the substances on the right side of a system will reduce the substances on the left side for any system which has a higher K_p value at the same temperature. Ca is thus a very effective reducing agent. At the three given temperatures a large number of oxides is reduced also by Al, for example according to 2Al + Cr$_2$O$_3$ → 2Cr + Al$_2$O$_3$ ("aluminothermic" reduction according to H. Goldschmidt). Ag_2O is dissociated (reduced) even with gentle heating. At 1000°K, log p_{O_2}=3.46, that is, p_{O_2}=2880 atm; the oxygen pressure reaches 1 atm already at 463°K (190°C).

Tab. 21.1 also lists log K_p values for some very important equilibria containing other gases than just O_2, in which the conditions of participation of O_2 are less simple. For the equilibria $2CO_2(g) \rightleftharpoons 2CO(g) + O_2(g)$ and $2H_2O(g) \rightleftharpoons 2H_2(g) + O_2(g)$, by taking logarithms of the equilibrium equations we obtain

$$\log p_{O_2} = \log K_p + 2 \log (p_{CO_2}/p_{CO}) \tag{21.4}$$

$$\log p_{O_2} = \log K_p + 2 \log (p_{H_2O}/p_{H_2}) \tag{21.5}$$

Thus, in each equilibrium the oxygen pressure determines the relationship between the partial pressures of the other gases.

For example, suppose we wish to find the conditions for the reduction of FeO(s) with CO at 1500°K according to the reaction $FeO(s) + CO(g) \rightleftharpoons Fe(s) + CO_2(g)$. Equilibrium between FeO(s) and Fe(s) is attained at this temperature if log p_{O_2} = -11.55. If this value is inserted in (21.4), for log $K_p = -10.60$ we find that at equilibrium, $p_{CO_2}/p_{CO} = 0.34$. Thus, if the ratio p_{CO_2}/p_{CO} at 1500°K is >0.34, Fe will be oxidized, and if it is <0.34, FeO will be reduced.

For the reduction of $SiO_2(s)$ with CO at 1500°K, we find in an analogous way that at equilibrium, $p_{CO_2}/p_{CO} = 7.2 \times 10^{-6}$. Inasmuch as it is hardly possible to hold p_{CO_2} at such a low value that this ratio is reached, this reduction cannot be carried out in practice.

Similar reasons make it impractical to reduce very stable oxides with hydrogen gas. As an example we may take the reduction of $Cr_2O_3(s)$. If log p_{O_2} for the equilibrium between $Cr_2O_3(s)$ and Cr(s) is inserted in (21.5), for 1000°K, 1500°K and 2000°K we obtain the values $p_{H_2O}/p_{H_2} = 1.07 \times 10^{-5}$, 1.74×10^{-3} and 2.21×10^{-2}, respectively. Hydrogen gas from pressure cylinders always contains some oxygen, which on heating forms water vapor so that the ratios given can never be reached. Reduction will take place, however, if hydrogen gas is used that has been sufficiently cleaned of oxygen gas and then carefully dried. Clearly, reduction will go more easily the higher the temperature is.

Sometimes on reduction with CO or C carbides, and with H_2 hydrides are formed, so that the conditions are complicated. Therefore, these reducing agents cannot be used in the preparation of the pure element.

In the reduction of oxides of nonmetals of high electronegativity the element is frequently not produced, but the reduction goes further to negative ions.

c. Water. Dihydrogen oxide, *water*, is one of the most important of all substances. Water is one of the main constituents of living organisms and their environment. A very large proportion of chemical reactions on the earth's surface takes place in water solution, and this is especially true of life processes. Water is also the most frequently used solvent.

In 1871 Cavendish found that water is formed when hydrogen burns in oxygen.

Although he interpreted the phenomenon in terms of the phlogiston theory, it may be said that he drew from it the correct conclusion concerning the composition of water. After Lavoisier established hydrogen and oxygen as elements in the modern sense, he concluded that water consisted of hydrogen and oxygen.

The structure of the water molecule was discussed in 5-3d. This molecule is the building unit in all modifications of water whose structures are known. The most important of these are ordinary (hexagonal) ice, liquid water and water vapor. The phase diagram for these three phases was discussed in 13-4a. The temperatures of the triple point (ordinary ice) + (liquid water) + (water vapor) and the equilibrium (liquid water) + (water vapor) at 1 atm are calibration points for the International Practical Temperature Scale, and are given the values $+0.01°C$ and $100°C$, respectively.

Ice and liquid water in thick layers show a blue-green transmission color.

In the different modifications of water the water molecules are held together by hydrogen bonds (5-5) and the significance of these bonds in the physical properties has been considered in 13-4b. Ordinary ice has a hexagonal crystal structure and in it the oxygen atom in each water molecule is surrounded tetrahedrally by four neighboring molecules. The structure was discussed in 5-5 (see also 6-2c, last part) and a schematic picture of it is given in fig. 5.26.

In addition to hexagonal ice several other ice modifications are known. At pressures over ~ 2000 bar at least six stable modifications have been established. These can be preserved at atmospheric pressure for a long time provided that before the pressure is removed the temperature is dropped so low that the motion in the structure becomes very small (liquid air, about $-180°C$). In this way the study of these modifications is facilitated, and it has been possible to determine the oxygen atom positions in five of them by X-ray diffraction. The coordination in all cases is tetrahedral as in hexagonal ice, but the relative positions of the tetrahedra are different so that more dense packings are achieved.

It is possible that hexagonal ice at atmospheric pressure is stable down to a very low temperature. However, if water vapor is condensed on a cold surface at temperatures below about $-160°C$, a well-ordered structure has not had time to form. "Amorphous" ice is obtained (probably no long-range order, that is, liquid structure, glass; cf. 6-3d). At temperatures above about $-160°C$ the heat motion is sufficient for creation of long-range order. The resulting crystalline ice modification on condensation in the range of approximately -160 to $-120°C$ is cubic, although the coordination here is also tetrahedral. Cubic ice is also formed when amorphous ice is warmed to temperatures in this range. On the other hand, it is never obtained by cooling hexagonal ice. This may be because cubic ice is only a metastable member (13-3a) of the process amorphous → cubic → hexagonal, in which the transition to the last, stable phase is very slow when the temperature does not exceed $-120°C$. Above this temperature, on the other hand, the transition cubic → hexagonal takes place, more rapidly the higher the temperature is. Thus, hexagonal ice is here certainly stable.

The very special properties of the ice structure (now and in the following we refer to ordinary ice) causes it to take up hardly any other substance in solid solution. One exception is ammonium fluoride NH_4F, whose structure is completely analogous with that of ice (6-2c).

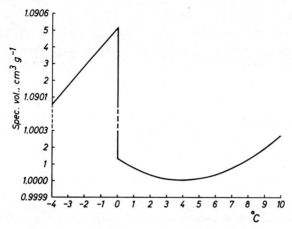

Fig. 21.3. Specific volume of ice and water.

As mentioned in 5-5, many hydrogen bonds are ruptured when ice melts, so that the structure of liquid water is much more irregular and changing. However, hydrogen bonds still play a large role and much of the short-range order (1-4g) of ice remains. Because the structure of ice is very open and the molecules can be packed more densely in the liquid phase with its less rigid geometrical restrictions, a considerable decrease in volume occurs on melting (cf. 1-4h). When the temperature is raised above the melting point, further hydrogen bonds are broken and the packing can become still more dense. The volume decreases still more, but at the same time this decrease is counteracted by the increased heat motion. The result is a volume minimum and consequently a density maximum at $+4°C$. This behavior, which is shown in fig. 21.3, is of exceedingly great importance in nature. Since melting is followed by a volume decrease ($\Delta V < 0$), according to 13-2c the melting point of ice drops with increase of pressure. This is the reason that the curve for the equilibrium s+l in the general phase diagram in fig. 13.2a has a slope as shown by the dashed curve for the water system.

Liquid water as a solvent was treated in 13-1c, and its ability to dissociate electrolytes in 11c. Water as a member of the protolytic system H_3O^+–H_2O–OH^-–O^{2-} was discussed in chap. 12. The conductivity of absolutely pure water ("conductivity water") is very small, because of the very low concentration of the current-carrying ionic species H_3O^+ and OH^-. Concerning their occurrence and the conductivity mechanisms themselves, see 12-2d and 11b.

In 5-2h, i the extremely common *hydration* of ions, especially cations, existing in water solution was described. The resulting *aquo complex ions* were discussed also in 5-3j and 5-6c. It was also mentioned that a greater or lesser portion of the water shield often accompanies the ion when it enters into a crystal. The hexaquo-magnesium ion $[Mg(H_2O)_6]^{2+}$ thus becomes a building unit, for example, in crystals grown from a solution of magnesium chloride and these can therefore be logically given the formula $[Mg(H_2O)_6]Cl_2$. Sometimes a crystal contains water molecules that are not bound directly to an ion, but nevertheless are necessary for the formation of a certain crystal structure. In the crystallization of an iron(II) sulfate solution crystals are formed from the building unit $[Fe(H_2O)_6]^{2+}$ and SO_4^{2-}, but in addition $1H_2O$ for each Fe is included. A logical formula for the crystals formed is therefore $[Fe(H_2O)_6]SO_4 \cdot H_2O$. In many cases, the impression is gained that these latter water molecules fill cavities that are not compatible with a stable structure. Both the former and latter types of water molecules have been termed *water of crystallization*, a term that therefore does not uniquely define the function of the water in the crystal. Water of crystallization is often given collectively at the end of the formula, for example, $MgCl_2 \cdot 6H_2O$ and $FeSO_4 \cdot 7H_2O$. A crystalline phase with water of crystallization is often called a *hydrate*. A crystal hydrate has a definite proportion of water and a definite dissociation pressure (13-5e1), but water in a solid phase may also be *zeolitically* bound, as described in 13-5e2, so that the proportion of water can vary without breaking down the crystal structure. Solid phases consisting of very small particles, because of their large area, may have extremely high adsorptivity (15-1a). This is further increased if the structure is disordered. Here large and continuously variable amounts of *adsorbed water* may be found. Many precipitates thus have a high and indefinite water content, at least before their particles become larger through aging (14-2d) and have achieved a more ordered structure.

In nature water always contains greater or lesser amounts of dissolved substances, especially salts and gases (ice and rain water are most free of salts). The preparation of more or less salt-free water is therefore of great importance.

The purest water is obtained by *distillation*. If completely pure water is desired for physical or chemical work the distillation must be repeated several times, and the whole apparatus must be made of quartz or noble metal. Glass is somewhat soluble in water and contaminates it. A convenient test of the purity can be made by measurement of the conductivity. As mentioned in 15-3 quite salt-free ("deionized") water can be obtained more cheaply with an *ion exchanger*.

A third method, which has recently become of practical importance, is *electrodialysis*. Crude water is allowed to flow slowly through an electrolytic cell, which in its simplest form is divided by two membranes into three compartments, one around each electrode and one between the two membranes. The membrane around the anode consists of a material with anion-exchange properties (15-3), so that it passes anions and blocks cations. The membrane around the cathode has cation-exchange properties and thus passes cations and blocks anions. Therefore, when voltage is applied, ions are preferentially drawn out of the middle chamber while only a few migrate into it. The water issuing from the middle chamber thus becomes deionized.

Sea water has a high salt content. The production of commercial water by the desalting of sea water (desalination) therefore requires large amounts of energy and can only be considered when the shortage of fresh water is severe. Among other methods being tested are distillation, electrodialysis, reverse osmosis (13-2c), and freezing out of ice (on freezing not-too-concentrated salt solutions mainly pure ice crystallizes; cf. fig. 13.13d). An acceptable economy may be attained through the use of nuclear energy in connection with power production.

Ground water often contains calcium, magnesium and iron(II) salts, which make it *hard* and unsuitable for many purposes. It can be made *soft* ("*softened*") by removing the corresponding ions, but often does not need to be completely deionized (more about this in 26-5c).

Concerning *heavy water* and its enrichment, see 9-3b.

d. Hydroxides, hydroxide and oxide salts. A hydroxide is characterized by the hydroxide ion OH^-, which is the base corresponding to the acid H_2O (12-1a). For this ion to exist in a compound containing the atomic group OH, it must be bound by a rather pronounced ionic bond. It was shown in 12-2f that a substance AOH functions as an acid according to $AOH \rightarrow AO^- + H^+$ when the A–O bond is sufficiently covalent but that the dissociation $AOH \rightarrow A^+ + OH^-$ occurs when the A–O bond has strong enough ionic character. Only in the latter case should AOH be called a hydroxide. If A is a metal with rather high electronegativity, however, the first of these reactions can take place on addition of base, and the second on addition of acid. In such cases AOH has since early times been designated as an amphoteric hydroxide. It is an ampholyte also in Brønsted's nomenclature if the A^+ ion is written as hydrated, in the simplest case according to

$$A(H_2O)^+ \underset{+H^+}{\overset{-H^+}{\rightleftarrows}} AOH \underset{+H^+}{\overset{-H^+}{\rightleftarrows}} AO^-$$

Amphoteric hydroxides are discussed in more detail in 12-4d.

As the electronegativity of the metal increases, and thus the covalent bonding, the solubility of the uncharged hydroxide in water decreases. For the hydroxides of the alkali metals and the alkaline-earth metals the solubility thus decreases in the order Cs, Rb, K, Na, Li, Ba, Sr, Ca, Mg, Be. In cold water, the hydroxides of the first four metals are very soluble, those of Li and Ba fairly easily soluble, and those of the last four poorly soluble. In $Be(OH)_2$ the amphoteric character begins to appear.

Many poorly soluble di- and trihydroxides crystallize, as mentioned in 6-2c, in layer structures with infinite molecules (fig. 6.18). The fact that these hydroxides often give colloidal solutions or precipitate in the form of gels (see 12-1b) is certainly related to the tendency to form giant molecules. The gelatinous precipitates are usually very poorly defined. It is possible that the layer structures have here already formed but the crystals are very small (of "colloidal" size) and the order low. Thus the adsorptivity is great and especially water is retained in indefinite amount (21-5c). On aging (14-2d) of the precipitates the order increases and the water content decreases.

Poorly soluble hydroxides are precipitated by making a solution containing ions of the metal in question sufficiently high in hydroxide concentration. The large differences in the solubility products of the different hydroxides make it possible to separate dissolved metal ions by precipitating certain of them as hydroxides while the others remain in solution (cf. 13-1d, e). Since the concentration of the precipitating ion, the hydroxide ion, is a function of the acidity (12-2d), it can easily be given the proper value for separation by controlling the acidity.

The most important method of preparation of the easily soluble alkali-metal hydroxides is electrolysis of a salt solution (chloralkali electrolysis 20-4b). Hydroxides of alkaline-earth metals can be prepared, besides by precipitation with hydroxide ion, by reaction between the oxide and water: $CaO(s) + H_2O \rightarrow Ca^{2+} + 2OH^-$.

By dehydration (for example, on heating) a hydroxide can be converted to an oxide: $Ca(OH)_2 \rightarrow CaO + H_2O$. Sometimes an *oxide hydroxide* appears as an intermediate stage in dehydration, but may also be formed directly on precipitation. For Al, Cr, Mn, Fe, Co, Ni in the trivalent state the poorly soluble oxide hydroxides MO(OH) are known, which, like the poorly soluble di- and trihydroxides, often crystallize in layer structures. Especially in freshly precipitated form the oxide hydroxides can also take up by adsorption large and indefinite amounts of water.

Through the transitions $H_2O \underset{+H^+}{\overset{-H^+}{\rightleftarrows}} OH^- \underset{+H^+}{\overset{-H^+}{\rightleftarrows}} O^{2-}$ which also can be applied to H_2O, OH^- and O^{2-} as ligands, the *aquo, hydroxo* and *oxo complexes* are closely related to each other. For examples of this, reference should be made to 12-1b and 12-4d. Hydroxo and oxo complexes occur in more basic solution than the corresponding aquo complexes and often crystallize from the solution with appropriate anions as *hydroxide* and *oxide salts*, earlier called basic salts.

Like the hydroxides, the hydroxo and oxo complexes in such salts very often form infinite sheets, of which fragments certainly exist in the solutions. An example of the formation of an oxo complex and the structure of the corresponding oxide salt was given by indium oxide chloride InOCl in 12-1b and fig. 12.1. A series of hydroxide salts with the compositions M(OH)X and M(OH)$_{1.5}$X$_{0.5}$, in which M is a divalent metal (for example, Mg, Mn, Fe, Ni, Zn) and X a halogen, have structures that are obtained from the MX$_2$ structure in fig. 6.18a if half or three fourths of X are replaced by OH.

Many difficultly volatile oxides become more or less volatile in the presence of water vapor, and in general the volatility increases as the temperature and the vapor pressure rises. Usually molecules containing OH groups are formed. It has long been known that B_2O_3 is made volatile by water vapor. Here mainly BO(OH) is formed according to $B_2O_3(s) + H_2O(g) \rightleftarrows 2BO(OH)(g)$, but $B(OH)_3$ is also present. The equilibrium $SiO_2(s) + 2H_2O(g) \rightleftarrows Si(OH)_4(g)$ is shifted appreciably to the right when the temperature and pressure of the water vapor lie above the critical values. When heavy-metal oxides are volatilized by water vapor at sufficiently high temperatures volatile oxide hydroxides are believed mainly to be formed; for example, according to $MoO_3(s) + H_2O(g) \rightleftarrows MoO_2(OH)_2(g)$. On cooling, the equilibrium is again displaced to the left.

21-6. Peroxides and hyperoxides

In peroxides and hyperoxides the oxygen atoms are bonded to each other, so that the oxidation number of oxygen deviates from $-\text{II}$.

a. Peroxides. Peroxides contain an O_2 group, which may function as a peroxide ion $[|\overline{\underline{O}}-\overline{\underline{O}}|]^{2-}$ (for structure see 5-3e) or as a covalently, singly ($-\overline{\underline{O}}-\overline{\underline{O}}|$) or doubly ($-\overline{\underline{O}}-\overline{\underline{O}}-$) bound group. The O_2 group as a ligand is designated by the name *peroxo*. In all cases the oxygen atoms have the oxidation number $-\text{I}$.

Peroxides with peroxide ions are known only for alkali metals, alkaline-earth metals (except Be) and the metals in groups 11 (except Au) and 12. These are all solid substances. They all have high oxidizing power and give off oxygen on heating. Important peroxides of this type are Na_2O_2 and BaO_2, which can be prepared respectively by combustion of sodium in air ($2Na + O_2 \rightarrow Na_2O_2$) and by heating barium oxide in a stream of air at about 500°C ($2BaO + O_2 \rightarrow 2BaO_2$).

The peroxide ion is a very strong base and when the above-mentioned peroxides are dissolved in cold water, the acid-base equilibrium $O_2^{2-} + H_2O \rightleftharpoons HO_2^- + OH^-$ sets in, strongly displaced to the right. The ion HO_2^- is called *hydrogen peroxide ion*. This is easily decomposed, especially on heating, according to $2HO_2^- \rightarrow O_2 + 2OH^-$. On addition of acid, hydrogen peroxide is formed: $HO_2^- + H^+ \rightarrow H_2O_2$.

Among the peroxides with strongly covalently bound peroxo groups hydrogen peroxide is the most important. This is described more closely in 21-6b. Often when the group $-\overline{\underline{O}}-\overline{\underline{O}}|$ or $-\overline{\underline{O}}-\overline{\underline{O}}-$ occurs as a covalently bound ligand, it can be formally considered as replacing an atom $-\overline{\underline{O}}|$ or $-\overline{\underline{O}}-$ in a known molecule. In many cases the new molecule is then named by giving the trivial name of the original molecule the prefix peroxo-. Thus, the ions

$$\left[\begin{array}{c}|\overline{O}|\\|\\|\overline{\underline{O}}-S-\overline{\underline{O}}-\overline{\underline{O}}|\\|\\|\overline{\underline{O}}|\end{array}\right]^{2-} \text{ and } \left[\begin{array}{ccccc}|\overline{O}| & & & & |\overline{O}|\\| & & & & |\\|\overline{\underline{O}}-S-\overline{\underline{O}}-\overline{\underline{O}}-S-\overline{\underline{O}}|\\| & & & & |\\|\overline{\underline{O}}| & & & & |\overline{\underline{O}}|\end{array}\right]^{2-}$$

are called peroxosulfate and peroxodisulfate ion, corresponding to the acids peroxosulfuric acid and peroxodisulfuric acid, respectively. Such *peroxo ions* and *peroxo acids* can be prepared by anodic oxidation (peroxidosulfate ion according to 16f) or by reaction of hydrogen peroxide or sodium peroxide with anions, acids and especially acid anhydrides: for example, $P_2O_5 + 2H_2O_2 + H_2O \rightarrow 2H_3PO_5$. They are all strong oxidizing agents. Peroxo acids are hydrolyzed by water to the corresponding normal acids and hydrogen peroxide: $H_3PO_5 + H_2O \rightarrow H_3PO_4 + H_2O_2$, which implies a reversal of the previous reaction.

The prefixes peroxo- and per- must not be confused! This warning is even more urgent because peroxo ions and peroxo acids were formerly called per ions and per acids (for example, persulfate and persulfuric acid).

It should also be noted that peroxides were earlier called superoxides, and then also certain oxides of metals in higher oxidation states but with no O_2 groups were erroneously called superoxides (for example, MnO_2 and PbO_2, which should be called dioxides, or -(IV) oxides). When superoxides began to be called peroxides, many authors transferred the name superoxide to what is now called hyperoxide (21-6 c). Caution is thus required in interpreting these names in the literature.

b. Hydrogen peroxide. The structure of hydrogen peroxide can be written schematically as
$$\begin{array}{cc} H & H \\ | & | \\ |O & O| \end{array}$$
. In crystalline hydrogen peroxide each bond angle
$$\begin{array}{c} H \\ \backslash \\ O{-}O \end{array}$$
is 97°, and in addition, the two H–O–O planes make a dihedral angle of 94° with each other. If one looks along the O–O direction, so that the two oxygen atoms are superimposed on each other, the molecule therefore has the appearance
$$\begin{array}{c} H \quad H \\ \backslash \quad / \\ O \end{array}$$
.

Pure hydrogen peroxide at room temperature is a viscous liquid. In thick layers it shows a blue transmission color, but appears colorless in thin layers. It freezes at $-0.5°C$. At 80°C it has a vapor pressure of 63 mbar but on heating above this temperature vigorous decomposition sets in (see below). Therefore, it can be distilled without decomposition only at reduced pressure. In the condensed phases the molecules are strongly bound together by hydrogen bonds. The liquid has a higher dielectric constant than water, and consequently is an excellent solvent for electrolytes (13-1 c). Hydrogen peroxide dissolves in water in all proportions and behaves in the solutions as a very weak acid, which, however, is stronger than water (tab. 12.2). The splitting off of the first proton gives the hydrogen peroxide ion HO_2^-.

The function of hydrogen peroxide as a strong oxidizing agent and a relatively weak reducing agent has been discussed in 16e. There its great instability was mentioned, which is further apparent from the strongly negative ΔG value for the process (25°C):

$$H_2O_2(g) \to H_2O(g) + \tfrac{1}{2}O_2(g); \quad \Delta G = -125 \text{ kJ}; \quad \Delta H = -96 \text{ kJ} \qquad (21.6)$$

Decomposition is catalyzed by many substances and may then become extraordinarily violent. Concentrated H_2O_2 solutions may then become so strongly heated (cf. ΔH value above) that they are completely vaporized. Many transition metals and their compounds as well as some enzymes are especially effective catalysts. Bases have the same effect, probably because the decomposition passes through the hydrogen peroxide ion HO_2^- (since glass gives off hydroxide ions glass vessels are not suitable for storage). Large contact surface hastens the process, so that rough surfaces, dust and dirt may cause rapid decomposition. However, if catalysts can be excluded decomposition takes place very slowly at ordinary temperature. Appropriate materials for storage are plastic and, for larger vessels, aluminum. Against the possible presence of decomposition catalysts, small amounts of stabilizing substances are added, which act by tying up the catalysts, often by complex formation (as, for example, with diphosphate ion $P_2O_7^{4-}$, added as $Na_4P_2O_7$, a common stabilizer).

Hydrogen peroxide is formed, although in small amounts, in many oxidations with oxygen, which may be because peroxide ion probably is often the first stage in the reduction of O_2: $O_2(g) + 2e^- \rightarrow O_2^{2-}$. However, there are only a few reactions that are suitable for industrial production of hydrogen peroxide. Before 1910 it was produced practically exclusively by treating barium peroxide with dilute sulfuric acid: $BaO_2 + H_2SO_4 \rightarrow H_2O_2 + BaSO_4(s)$. Nowadays nearly all manufacture of hydrogen peroxide is carried out by hydrolysis of peroxodisulfate ion, which is obtained by anodic oxidation of sulfate ion in the manner described in 16f. The electrolyte is here a solution of a hydrogen sulfate, and thus contains the ion HSO_4^-. The peroxodisulfate solution is evaporated at reduced pressure and when it becomes more concentrated the peroxodisulfate ions are hydrolyzed according to: $S_2O_8^{2-} + H_2O \rightarrow HSO_4^- + HSO_5^-$; $HSO_5^- + H_2O \rightarrow HSO_4^- + H_2O_2$ (for the intermediate product HSO_5^- see 21-11 b5). The hydrogen peroxide formed evaporates together with water, and can be obtained pure by fractional distillation at about 60 mbar. The hydrogen sulfate ions are returned to the electrolytic cell. On a large scale hydrogen peroxide is usually marketed as a 30% water solution, but for many purposes it is diluted to 3%. Rising demands for higher concentrations have led nowadays more and more to the manufacture, with careful avoidance of decomposition-catalytic substances, of solutions with up to 90% hydrogen peroxide. Because of the violence with which these strong solutions can decompose, they must be handled very carefully.

Hydrogen peroxide has great practical importance, especially because of its oxidizing power. It is thus used to a large extent for bleaching textile fibers, waxes, fats, feathers, hair (3%), wood (sometimes 30%) and so on, and as an oxidizing agent in the laboratory. It is used also as an antiseptic (3%), where the oxygen gas evolved in the tissues mechanically facilitates the cleaning of wounds. Hydrogen peroxide is also used to counteract chlorine, but then functions as a reducing agent: $H_2O_2 + Cl_2 \rightarrow 2HCl + O_2$. It is the starting material in the preparation of organic peroxy compounds.

In recent years hydrogen peroxide has been extensively used for the oxidation of liquid fuels, both in rockets and in submarine engines intended for submerged operation (Walter turbines). In addition the decomposition reaction (21.6) has been used directly to give, after addition of a catalyst, a gas mixture at high pressure that can be used in jet engines or turbines. The military importance of these methods has led to an enormous increase in the production of hydrogen peroxide after World War II.

c. Hyperoxides. These contain the hyperoxide ion O_2^-, whose structure was stated in 5-3a to be $[\overline{O} \cdots \overline{O}]^-$, that is, the oxygen atoms are bonded with one three-electron and one single bond. The oxygen here has the oxidation number -0.5.

Hyperoxides are always solid and are known only for the heavier alkali and alkaline-earth metals Na–Cs, Ca–Ba. They are more easily formed the larger the metal-atom radius is. Thus, if alkali metals are oxidized with oxygen in excess, hyperoxides MO_2 are obtained of K, Rb and Cs. On the other hand Na gives mainly the peroxide Na_2O_2 and only very little hyperoxide NaO_2.

The unpaired electron in the hyperoxide ion makes the hyperoxides paramagnetic and yellow- to orange-colored (5-3a, 7-3, 7-4b). The pure peroxides of the alkali and alkaline-earth metals, on the other hand, are colorless (the pale yellow color of commercial sodium peroxide is caused by the above-mentioned impurity of hyperoxide).

21-7. Sulfides, selenides and tellurides

Sulfides, selenides and tellurides are analogous to oxides. Thus, in these compounds the chalcogen has the oxidation number $-II$.

a. Sulfides. Much of the discussion in 21-5a concerning the oxides can be applied to the sulfides. In the sulfides of the nonmetals the sulfur atoms are predominantly covalently bound, while the sulfides of the most electropositive metals can be considered as ionic compounds. The fact that sulfur has significantly lower electronegativity than oxygen, however, causes the covalent bond type to extend farther to the left in the periodic system. This may also be thought of as resulting from the fact that the sulfur atom is more easily polarized than the oxygen atom (5-2e). Most sulfides also thus become more colored than the corresponding oxides. Further, it is notable that the sulfides have a greater tendency than the oxides toward semiconducting properties (for example, galena PbS, in contrast to PbO).

In sulfide melts different elements compete for the sulfide ions in the same way that they compete for oxide ions in oxide melts. Analogously, we may speak here of *basic* and *acid sulfides*. When an acid sulfide takes up sulfide ions a *thio ion* is formed: for example, $As_2S_5 + 3S^{2-} \rightarrow 2AsS_4^{3-}$ (tetrathioarsenate ion). Thio ions are also formed by reaction between acid sulfides and sulfide ions present in solution. Thio ions are especially easily formed by As(III), As(V), Sb(III), Sb(V), Sn(IV), Au(III) and Pt(IV). However, the corresponding thio acids (for example, H_3AsS_4) are unknown. When acid is added to a solution containing these thio ions the sulfide ions are tied up by the protons so that the sulfide precipitates out again: $2AsS_4^{3-} + 6H^+ \rightarrow As_2S_5(s) + 3H_2S$. However, from trithiocarbonate (obtained according to $CS_2 + S^{2-} \rightarrow CS_3^{2-}$) trithiocarbonic acid can be prepared by adding acid.

Many oxothio ions are known, for example, all the intermediate stages between AsO_4^{3-} and AsS_4^{3-} (22-14). An important oxothio ion is thiosulfate $S_2O_3^{2-}$, which is discussed in more detail in 21-11f1.

The sulfides of the alkali metals are easily soluble in water, the alkaline-earth metal sulfides are moderately soluble and thereafter the solubility of the sulfides decreases to a great extent as the ionic-bond contribution drops off. Most of the other metal sulfides are thus poorly soluble in water and can, therefore, be obtained by precipitation from a metal salt solution with sulfide ion.

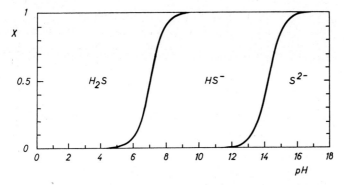

Fig. 21.4. The hydrogen sulfide system at different acidities.

The sulfide ion S^{2-} is a part of the hydrogen sulfide system H_2S–HS^-–S^{2-}. Hydrogen sulfide H_2S is a weak acid ($pk_a = 6.88$ at $I = 1$), the hydrogen sulfide ion HS^- a very weak acid ($pk_a = 14.15$ at $I = 1$; the sulfide ion is thus a very strong base). The distribution of the hydrogen sulfide system among the different protolytes of the system as a function of pH is shown by fig. 21.4 (cf. 12-3c, 12-4b). In sulfide solutions in pure water the major portion of the system consists of $HS^- + S^{2-}$. If acid is added the concentration of S^{2-} is decreased, and $[S^{2-}]$ may be made still lower because the concentration of hydrogen sulfide can increase so much that it exceeds the solubility value and is given off as a gas. By lowering $[S^{2-}]$ very insoluble sulfides can be dissolved. Their "decomposition" takes place if H_2S is given off (13-1e). On the other hand the easily soluble sulfides of the alkali metals can be prepared by the action of hydrogen sulfide on an alkali hydroxide solution. In fact, here the pH and consequently $[S^{2-}]$ can be given such high values that the alkali sulfides crystallize out on evaporation.

If hydrogen sulfide is passed into an ammoniacal solution, the pH value after saturation with hydrogen sulfide lies in a region where according to fig. 21.4 $[HS^-]$ is wholly predominant. Therefore the product is usually considered formally as a solution of *ammonium hydrogen sulfide* NH_4HS. Crystalline NH_4HS can be obtained by passing hydrogen sulfide into liquid ammonia.

The great variation in the solubility products of the metal sulfides makes possible separations by sulfide precipitation, which formerly had great analytical importance. For this purpose the sulfide-ion concentration can be given the proper value by controlling the acidity of the solution.

Sulfides can also be obtained by allowing sulfur to react directly with the element, usually at elevated temperature. Sulfides of alkali and alkaline-earth metals are prepared industrially by reduction of sulfate with carbon under heat: $M_2SO_4 + 2C \rightarrow M_2S + 2CO_2(g)$.

In the "cooking liquor" of the kraft cellulose process OH^-, HS^- and S^{2-} are primarily active in the solution of lignin. Na_2S is therefore an important constituent of the liquor. In the "black liquor" formed by the cooking, S(–II) has been converted to

S(VI) but after the liquor is evaporated S(VI) is reduced again to S(–II) by the carbon present when the residue is melted. Losses are compensated by simultaneous addition of Na_2SO_4.

In combination with OH^-, HS^- and S^{2-} dissolve keratin, and Na_2S is therefore used in tanning as a hair remover.

b. Hydrogen sulfide. Dihydrogen sulfide H_2S (in the following shortened to hydrogen sulfide) has a structure analogous to that of H_2O and may thus be written
$$\begin{array}{c} H \\ | \\ |\underline{S}\!-\!H \end{array}$$
. At ordinary temperature and atmospheric pressure hydrogen sulfide is a colorless, combustible and extremely poisonous gas, which even when highly diluted smells like rotten eggs. It can be condensed to a liquid with boiling point $-60.7°C$ (freezing point $-85.6°C$, critical temp. $100.4°C$, vapor pressure at $20°C$ 18.1 bar). It is thus considerably more volatile than water in spite of its higher molecular weight, as a result of the presence of hydrogen bonding in water (13-4 b).

Hydrogen sulfide dissolves easily in water (*hydrogen sulfide water*; the molarity of a saturated solution at $20°C$ and total pressure of 1 atm is 0.11) and functions in the solution as an acid in the system H_2S–HS^-–S^{2-}. The acid strength of H_2S and HS^- is greater than that of H_2O and OH^- respectively (tab. 12.2), which is a result mainly of the fact that the H–S bond is weaker than the H–O bond (12-2f).

Hydrogen sulfide is a reducing agent, in solution as well as in gaseous form. The most common oxidation product is sulfur. Note the position of the standard potential for the system $S^{2-} \rightarrow S + 2e^-$ in tab. 16.1.

Hydrogen sulfide reacts with most metals with the formation of sulfides, especially in the presence of water or heat. On heating, sulfides are also formed from many metal oxides.

For the process $H_2(g) + \frac{1}{2}S_2(g) \rightleftharpoons H_2S(g)$ at $25°C$, $\Delta G = -71.1$ kJ (the ΔG^0 value referred to formation from orthorhombic sulfur $= -33.0$ kJ). The molecule is thus stable at this temperature. However, the equilibrium is shifted to the left as the temperature is raised and the equilibrium constant reaches the value 1 at approximately $1500°C$. If this equilibrium is utilized for the preparation of hydrogen sulfide, the temperature should not be raised too high, and in order to obtain a sufficient reaction rate a catalyst is used, often bauxite (24-3a).

As stated in 21-7a, hydrogen sulfide is formed by the reaction of acids with sulfides, and a classical laboratory method of preparation is by the action of hydrochloric acid on iron(II) sulfide: $FeS + 2HCl \rightarrow Fe^{2+} + 2Cl^- + H_2S(g)$. For technical purposes, hydrogen sulfide from natural gas and from the raw gases produced by gas and coke works, oil refineries and oil shale processing is most used. Hydrogen sulfide is absorbed by a weakly basic solution, from which it later can be recovered by heating (several different, usually organic bases are used). The major portion of commercial hydrogen sulfide is used in the production of sulfur and sulfuric acid (by combustion to SO_2), but certain amounts are condensed and shipped in steel cylinders under pressure. Most of this is used in the preparation of sulfides and organic sulfur compounds.

c. Selenides and tellurides. The change in properties from oxides to sulfides continues as one goes to selenides and tellurides. Covalent and metallic properties thus become more and more pronounced.

H_2Se can be prepared from the elements, and H_2Te as well as H_2Se by reaction of tellurides and selenides respectively with acids. Both are ill-smelling and poisonous gases; H_2Se is extremely irritating to the mucous membranes. In contrast to H_2S they are unstable at ordinary temperature. The following $\Delta G°$ values (kJ, 25°C) illustrate the decrease in stability of the hydrogen chalcogenides as the chalcogen atom becomes heavier: H_2O, -228.7; H_2S, -33.0; H_2Se, 71.2; H_2Te, 138.5. This decrease in stability is caused by the decrease in strength in the hydrogen-chalcogen bond. The decrease is apparent also in that the acid strength increases simultaneously (12-2f).

21-8. Polysulfides

The structures of elementary sulfur, selenium and tellurium (5-6a, 21-3b, c) show the tendency of these atoms to form chains. If these chains are open, the unpaired electrons are found at the ends (for example, .S̄—S̄—S̄·) and therefore the chains try either to close themselves into rings, or connect with each other so that longer chains are formed (which in crystals may be "infinite"). Electron pairs can also be formed when the chains take up electrons from other kinds of atoms, either by ionization or by acting as an electron-pair acceptor in covalent bonding. They may then either become ions, for example, $[|\bar{S}-\bar{S}-\bar{S}|]^{2-}$, or be incorporated into molecules bound at one end, \S̄—S̄—S̄|, or at both ends, \S̄—S̄—S̄\.

The chemistry of sulfur, selenium and tellurium is characterized to a large degree by the appearance of such chainlike poly ions or poly groups. Here we will only mention briefly the *polysulfide ions* and *polysulfide groups*. Polysulfide groups occurring in oxosulfuric acids are discussed in connection with these acids (21-11).

If sulfides of alkali metals or alkaline-earth metals are fused with sulfur, the latter is added with the formation of polysulfide ions S_n^{2-} ($n=2$ to at least 6). The *disulfide ion* S_2^{2-} is an analog of the peroxide ion. Water solutions containing polysulfide ions are also obtained by allowing sulfide solutions to react with sulfur.

In these solutions, as in solutions of polysulfides made from melts, among the polysulfide ions only S_4^{2-} and S_5^{2-} can be definitely established. Equilibria are probably present of the type $nS_n^{2-} + H_2O \rightleftharpoons (n-1)S_{n+1}^{2-} + HS^- + OH^-$. For $n=2$ and 3 these equilibria are shifted wholly to the right so that S_4^{2-} is formed. For $n=4$ the equilibrium position is intermediate so that S_4^{2-} and S_5^{2-} are present in comparable amounts.

When a polysulfide solution is run into a large volume of conc. hydrochloric acid with cooling, a yellow, oily liquid separates, containing *hydrogen polysulfides* H_2S_n, which can be obtained pure by fractional distillation. The hydrogen polysulfides are very unstable and decompose especially easily in a basic medium.

If ammonium hydrogen sulfide solution (21-7a) absorbs sulfur, polysulfide ions are also formed and the solution takes on a more and more yellow color, the higher the sulfur content is.

Disulfide groups occur in certain sulfide minerals, particularly *pyrite* FeS_2.

21-9. Halogen compounds of the chalcogens

The most important halogen compounds of oxygen, that is, halogen oxides and oxohalogen acids, have been discussed earlier in chap. 20. Here we take up the halogen compounds of the other chalcogens.

a. Sulfur halides. These are commonly made by direct combination of the elements. The most important are:

S_2X_2 $\quad$ SX_2 $\quad$ SX_4 $\quad$ SX_6
X = F, Cl, Br $\quad$ X = Cl $\quad$ X = F, Cl $\quad$ X = F

In SX_4 and SX_6 the octet rule is not fulfilled and probably here the d orbitals participate (5-3i). In SF_4 fluorine lies at four of the five corners of a trigonal bipyramid (fig. 5.6g). The lone electron pair may be considered to lie at the fifth corner. SF_6 has octahedral fluorine-atom coordination.

The most important of the sulfur chlorides is *disulfur dichloride* S_2Cl_2, which is formed when chlorine is passed over molten sulfur. It is condensed to an orange-yellow liquid (b.p. 138°C), which has an obnoxious, pungent odor and fumes in moist air (hydrolysis with the formation of HCl). Disulfur dichloride easily dissolves sulfur and is therefore used in cold vulcanization of rubber.

Sulfur dichloride SCl_2 is formed by passing chlorine into S_2Cl_2, during which process the liquid gradually becomes dark red. In this liquid the equilibrium $2SCl_2 \rightleftharpoons S_2Cl_2 + Cl_2$ is set up, which is displaced to the right when the temperature is raised and above about 40°C leads to rapid decomposition.

Sulfur tetrachloride SCl_4 exists only at low temperature.

Sulfur hexafluoride SF_6 is formed as the chief product when sulfur is burned in fluorine and at room temperature consists of a colorless gas. Sulfur hexafluoride is notable for its stability and very low reactivity. It is odorless, insoluble in water, and can be heated to about 400°C in hydrogen gas, oxygen gas and free halogens without reaction. Molten sodium hydroxide and even molten sodium have no effect. This inertness probably results from the very complete protection of the sulfur atom by the fluorine-atom envelope. The electrical-insulation effect is also very high, which, together with its stability, makes sulfur hexafluoride useful for high-voltage insulation.

b. Halides of the heavier chalcogens. These are generally more stable than the corresponding sulfur halides. The best known are the tetrahalides. SeF_6 and TeF_6 have also been prepared. However, the former is noticeably and the latter considerably less stable than SF_6.

c. Oxide halides. In these the chalcogen coordinates both oxygen and halogen. The most important sulfur oxide chlorides are:

S(IV)

|Cl̄|
|
|O—S|
|
|C̄l|

sulfinyl chloride
(thionyl chloride)

S(VI)

|Cl̄|
|
|O—S—O|
|
|C̄l|

sulfonyl chloride
(sulfuryl chloride)

S(VI)

|Cl̄|
|
|O—S—O|
|
|O—H

chlorosulfuric
acid

Sulfinyl chloride (the older name *thionyl chloride* is also accepted) $SOCl_2$ is obtained by reaction between phosphorus pentachloride and sulfur dioxide: $PCl_5 + SO_2 \rightarrow SOCl_2 + POCl_3$. Commercially the preparation is carried out mainly according to the reaction $SCl_2 + SO_3 \rightarrow SOCl_2 + SO_2$. $SOCl_2$ is a colorless liquid (b.p. 79°C) with a pungent odor. It is hydrolyzed by water ($SOCl_2 + H_2O \rightarrow 2HCl + SO_2$), and therefore fumes in moist air. Sulfinyl chloride is frequently used in preparative organic chemistry as a chlorinating agent.

Sulfonyl chloride (the older name *sulfuryl chloride* is also accepted) SO_2Cl_2 is prepared by the reaction $SO_2 + Cl_2 \rightarrow SO_2Cl_2$ at room temperature (with cooling) and in the presence of a catalyst (often active carbon). It is a colorless liquid (b.p. 69°C) with a pungent odor. Sulfonyl chloride is hydrolyzed by water to chlorosulfuric acid (see below), which with one additional molecule of water gives sulfuric acid:

Cl
|
O—S—O   →
|
Cl + H OH

Cl + H OH
|
O—S—O   →
|
OH

OH
|
O—S—O
|
OH

Sulfonyl chloride fumes in moist air, as a result of the formation of both HCl and H_2SO_4. Like sulfinyl chloride it is used as a chlorinating agent in organic chemistry.

Chlorosulfuric acid $ClSO_3H$ (the name can be considered as an abbreviation of "chlorotrioxosulfuric acid"; the earlier name chlorosulfonic acid should be avoided), is usually prepared by combination of sulfur trioxide and hydrogen chloride ($SO_3 + HCl \rightarrow ClSO_3H$), which illustrates the tendency sulfur trioxide has to be an electron-pair acceptor (21-10b). Chlorosulfuric acid is a colorless liquid with a penetrating odor, and decomposes quite easily on heating. The hydrolysis to HCl and H_2SO_4 as described above makes chlorosulfuric acid an effective and widely used smoke generator (most often chlorosulfuric acid mixed with sulfur trioxide is used—"smoke acid"). Chlorosulfuric acid is also used in preparative organic chemistry.

21-10. Oxides of sulfur, selenium and tellurium

For sulfur the oxides S_2O, S_2O_3, SO_2, SO_3, S_2O_7 and SO_4 are known with certainty. Of these, however, all except sulfur dioxide SO_2 and sulfur trioxide SO_3 are of minor importance and are not treated further here. For selenium and tellurium the dioxides and trioxides are well known, of which only the dioxides will be considered. The following treatment will therefore be confined to the dioxides of chalcogens(IV), and sulfur trioxide.

a. Dioxides. The molecular structure of *sulfur dioxide* SO_2 has been described in 5-3e. Sulfur dioxide is a colorless gas with a sharply penetrating odor. It is strongly corrosive to the mucous membranes, and is highly injurious to plants. It can easily be condensed to a colorless liquid (b.p. $-10.0°C$, f.p. $-75.5°C$, crit. temp. $157.2°C$, vapor pressure at $20°C$, 3.4 bar). The heat of vaporization of the liquid is high, 402 J g^{-1} at the boiling point. Sulfur dioxide dissolves easily in water, and in the solution the hydration equilibrium (12-2e) $SO_2 + H_2O \rightleftharpoons H_2SO_3$ is set up. Sulfur dioxide is thus the anhydride of sulfurous acid H_2SO_3 (21-11d).

Liquid sulfur dioxide is a good proton-free solvent for many inorganic as well as organic substances. The conductivity of pure liquid sulfur dioxide is somewhat greater than that of water and probably results from the autoionization equilibrium (12-1e) $2SO_2 \rightleftharpoons SO^{2+} + SO_3^{2-}$ (the first ion is sulfinyl ion—older name thionyl ion—the second sulfite ion). Many salts show about equally high equivalent conductivity (11c) in sulfur dioxide as in water solution and are therefore about equally dissociated in both these solvents.

Sulfur dioxide often functions as a reducing agent ($S(IV) \rightarrow S(VI)$), more rarely as an oxidizing agent ($S(IV) \rightarrow S(0)$). The latter is the case with strong reducing agents such as when sulfur dioxide oxidizes hydrogen sulfide to sulfur ($2H_2S + SO_2 \rightarrow 3S + 2H_2O$). In most cases in which sulfur dioxide takes part in redox equilibria, however, the reactions occur in water solution where sulfurous acid is involved (21-11d).

Sulfur dioxide is prepared commercially as a rule by burning sulfur in air or by roasting a sulfide mineral, most often pyrite ($2FeS_2(s) + 5\frac{1}{2}O_2(g) \rightarrow Fe_2O_3(s) + 4SO_2(g)$. It is shipped in liquid form in steel cylinders. In the laboratory it is most simply prepared by dropping conc. sulfuric acid into a concentrated solution of sodium hydrogen sulfite: $HSO_3^- + H^+ \rightarrow H_2O + SO_2$.

The largest portion of commercially produced sulfur dioxide is used in the manufacture of sulfuric acid (21-11b2) and for the preparation of sulfite for the sulfite-cellulose process (21-11 d1). The reducing power of sulfur dioxide also makes it an effective bleaching agent. Sulfur dioxide formed directly by burning sulfur has been frequently used for sterilization purposes, although many things are damaged by it. This method is still important in the fumigation ("smoking") of malt or wine vats. Sulfur dioxide is also used in the preserving of foodstuffs. In beet-sugar manufacture it is used both as a bleaching agent and to prevent fermentation processes. Injurious insects are combated with sulfur dioxide. Because of its high heat of vaporization and the conveniently spaced boiling and freezing points (see above) sulfur dioxide is also used as a cooling medium in refrigerators.

Selenium dioxide SeO_2 and *tellurium dioxide* TeO_2 are obtained by burning the element in air. The former forms colorless, lustrous crystals built up of "infinte" chains with the structure

On heating, the crystals sublime (vapor pressure 1 bar at 317°C) to a foul-smelling gas consisting of SeO_2 molecules analogous to SO_2. Selenium dioxide dissolves easily in water and forms selenious acid H_2SeO_3 (21-12a).

Tellurium dioxide forms colorless crystals poorly soluble in water. The covalent-bond contribution is here much less than in SeO_2, as is evident from the fact that the crystal structure is the same as that of, for example, SnO_2, PbO_2 and MnO_2. The acid character is also very much less pronounced, as shown by the fact that tellurium dioxide dissolves both in basic and acid solutions with the formation of tellurites (21-12b) and tellurium(IV) salts, respectively.

b. Sulfur trioxide. The structure of the molecules in gaseous *sulfur trioxide* SO_3 has been described in 5-3e. The S(VI) atom has a strong tendency for fourfold, tetrahedral coordination. This is noticeable in the properties of sulfur trioxide, both in the pure state and in its reactions with other substances. It was stated in 5-3a that the tendency of the sulfur atom to take part in double bonding is relatively small. The preference for four-coordination can therefore be explained by noting that a state in which the sulfur atom fulfills the octet rule by bonding four ligands with single bonds is more stable than the state with double bonding in the free molecule. The trioxide is thus a typical electron-pair acceptor (strong acid according to Lewis, 12-1e). Formally, the structure $\overline{|O|} - S$ may be thought of as the electron-pair acceptor and, for example, the addition of the oxide ion O^{2-} with the formation of sulfate ion may be written according to

$$\text{|\overline{O}|}\text{—}\underset{\underset{\text{|\underline{O}|}}{|}}{\overset{\overset{\text{|\overline{O}|}}{|}}{S}} + [\,|\overline{\underline{O}}|\,]^{2-} \rightarrow \left[\text{|\overline{O}|}\text{—}\underset{\underset{\text{|\underline{O}|}}{|}}{\overset{\overset{\text{|\overline{O}|}}{|}}{S}}\text{—}|\overline{\underline{O}}|\right]^{2-}$$

When absolutely dry sulfur trioxide gas is cooled a colorless, glass-clear, solid phase (γ SO$_3$, "icy" SO$_3$) is formed which melts at 16.8°C. This is built up of S$_3$O$_9$ molecules with a ring structure as in *a* below.

a. γ SO$_3$ *b.* α and β SO$_3$

Here an oxygen atom is thus an electron donor to another SO$_3$ molecule. If SO$_3$ is kept for a long time at room temperature, white silky needles are formed in parallel bundles ("asbestoslike" SO$_3$), in which the tetrahedral coordination is achieved by forming more or less open chains as, for example, in *b* (the chains are zigzag). There are two asbestoslike modifications, one low-melting (β SO$_3$, m.p. 32.5°C) and one high-melting (α SO$_3$, triple-point temperature 62.2°C). α SO$_3$ is the only stable modification; β and γ SO$_3$ are mestastable (2-3b).

The behavior of sulfur trioxide according to the equation SO$_3$ + O^{2-} → SO$_4^{2-}$ makes it a strongly acid oxide also according to Gutmann-Lindqvist's definition (21-5a). With metal oxides it therefore reacts with the formation of sulfates. It combines with water to give sulfuric acid with strong evolution of heat: SO$_3$ + H$_2$O → H$_2$SO$_4$. Since the solid trioxide is rather volatile at room temperature it fumes strongly in moist air because of the sulfuric acid mist generated above it.

The formation of sulfuric acid from sulfur trioxide and water is used in one of the two methods for the industrial production of sulfuric acid. Sulfur trioxide is produced for this purpose by oxidation of sulfur dioxide with oxygen: SO$_2$(g) + $\frac{1}{2}$O$_2$(g) ⇌ SO$_3$(g); $\Delta H = -99.1$ kJ. The heat evolved by this process causes the equilibrium to be displaced to the left with rise in temperature. Without catalysts the process goes at a measurable rate only at so high a temperature that the SO$_3$ yield is very small. With a proper catalyst, however, a good rate of reaction can be obtained at a low enough temperature so that the equilibrium position is shifted sufficiently to the right (*contact process*). As a catalyst only finely divided platinum was used originally, but now to a very large extent mixed catalysts are used with vanadium(V) oxide V$_2$O$_5$ as the chief component together with a certain amount of potassium sulfate. The support is porous SiO$_2$ (kieselguhr). The contact temperature is held at about 450°C. The vanadium catalysts probably work mainly by changing the valence state between V(IV) and V(V).

Group 16, Chalcogens

The problem of converting sulfur trioxide to sulfuric acid with water is discussed in 21-11 b2.

In the laboratory sulfur trioxide can be prepared either by heating fuming sulfuric acid (21-11 b1) or a disulfate: $Na_2S_2O_7 \rightarrow Na_2SO_4 + SO_3$.

21-11. Oxosulfuric acids and oxosulfates

a. General. A large number of oxosulfate ions are known but the corresponding oxosulfuric acids are often so unstable that only a few can be obtained in pure form.

When oxygen atoms are bonded in increasing number to a central sulfur atom the following *mononuclear oxosulfate ions* are the most obvious species obtained (double-bond contributions, which are certainly often present, are not indicated, cf. 5-3 h):

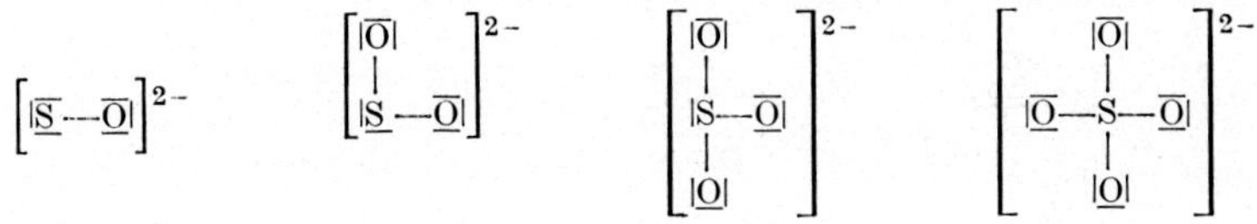

ion	oxosulfate(0) SO^{2-}	sulfoxylate SO_2^{2-}	sulfite SO_3^{2-}	sulfate SO_4^{2-}
acid	oxosulfuric(0) acid	sulfoxylic acid	sulfurous acid	sulfuric acid
oxid. no. of S	0	II	IV	VI

(The ion SO^{2-} as well as the acid H_2SO is unknown, but the corresponding group SO exists in organic derivatives; the names of this ion and acid are the only ones given above that are rational.)

The ions mentioned above are isoelectronic with the oxochlorate ions (20-8 a). Here also there is presumably sp^3 hybridization, and the tetrahedrally directed orbitals are occupied by shared or lone electron pairs. For example, the sulfite ion is thus pyramidal like the chlorate ion (cf. fig. 5.19).

The acids probably have the same sulfur–oxygen arrangement as the ions. The sulfuric acid molecule is formed by two hydrogen atoms being bonded each to one oxygen atom in the sulfate ion, but in the other ions the lone electron pair of the sulfur atom can be involved in bonding the hydrogen atoms. Thus, two formulations are possible for sulfurous acid:

In water solution these two forms are probably in equilibrium with each other. Equilibrium is attained very rapidly so that isolation of one form or the other is not possible. Analogously, sulfoxylic acid may be considered to exist in three forms:

$$
\begin{array}{ccc}
\text{H--O|} & \text{|O|} & \text{|O|} \\
| & | & | \\
\text{|S--O|} & \text{H--S--O|} & \text{H--S--O|} \\
| & | & | \\
\text{H} & \text{H} & \text{H}
\end{array}
$$

If the highly mobile protons are completely or partly replaced by alkyl groups, which are less mobile, derivatives corresponding to these different acid structures can be isolated. For example the following *derivatives of sulfurous acid* can be obtained:

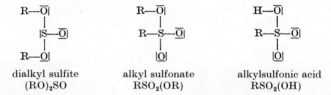

dialkyl sulfite   alkyl sulfonate   alkylsulfonic acid
$(RO)_2SO$ $RSO_2(OR)$ $RSO_2(OH)$

Also, the following *derivatives of sulfoxylic acid* can be prepared:

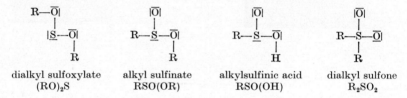

dialkyl sulfoxylate   alkyl sulfinate   alkylsulfinic acid   dialkyl sulfone
$(RO)_2S$ $RSO(OR)$ $RSO(OH)$ R_2SO_2

In polynuclear *oxodi-* and *oxopolysulfate ions* chains of SO_4 tetrahedra are present. In these one oxygen atom is common to two tetrahedra so that an endless zigzag chain $-O-S-O-S-O-$ is formed. The chains may also be branched. On the other hand, two tetrahedra never have more than *one* common oxygen atom, which implies that they have only common corners but never common edges or faces. By the linking of tetrahedra the sulfur atoms can achieve tetrahedral coordination even though the ratio $O/S < 4$ (cf. 5-6b). The simplest members of this series are (crimping of chains not indicated):

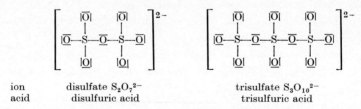

ion   disulfate $S_2O_7^{2-}$ trisulfate $S_3O_{10}^{2-}$
acid   disulfuric acid   trisulfuric acid

In all these ions the sulfur atom, as in the sulfate ion, has the oxidation number VI. The general formula of the polysulfate ions is $(S_nO_{3n+1})^{2-}$, and of the poly sulfuric acids is $H_2S_nO_{3n+1}$. The proportion of water in the acid decreases as n increases, and the formula approaches $(SO_3)_n$ as a limit. They thus constitute the transition between sulfuric acid ($n=1$) and sulfur trioxide ($n=\infty$; cf. the structure in 21-10b).

Group 16, Chalcogens

In certain of the oxosulfate ions mentioned above one oxygen atom can be replaced by a peroxo group (21-6a), so that *peroxosulfate ions* are formed. Such ions are the following:

$$\left[\begin{array}{c} |\overline{O}| \\ | \\ |\overline{O}{-}S{-}\overline{O}{-}\overline{O}| \\ | \\ |\overline{O}| \end{array}\right]^{2-} \qquad \left[\begin{array}{c} |\overline{O}| \quad\quad |\overline{O}| \\ | \quad\quad\quad | \\ |\overline{O}{-}S{-}\overline{O}{-}\overline{O}{-}S{-}\overline{O}| \\ | \quad\quad\quad | \\ |\overline{O}| \quad\quad |\overline{O}| \end{array}\right]^{2-}$$

ion   peroxosulfate SO_5^{2-} peroxodisulfate $S_2O_8^{2-}$
acid   peroxosulfuric acid   peroxodisulfuric acid

Here also sulfur has the oxidation number VI (in the peroxo group each oxygen atom has the oxidation number $-I$).

The tendency of sulfur to form S_2 groups and longer sulfur chains appears in a large number of oxosulfate ions. The root "thio" (Gk. ϑεῖον, sulfur) is often incorporated into their names. The most important of such ions, in which one S_2 group coordinates three to six O atoms, are the following:

$$\left[\begin{array}{c} |\overline{O}| \\ | \\ |\overline{O}{-}S{-}\overline{S}| \\ | \\ |\overline{O}| \end{array}\right]^{2-} \quad \left[\begin{array}{c} |\overline{O}| \\ | \\ |\overline{O}{-}\overline{S}{-}S{-}\overline{O}| \\ | \\ |\overline{O}| \end{array}\right]^{2-} \quad \left[\begin{array}{c} |\overline{O}| \;\; |\overline{O}| \\ |\quad\;\; | \\ |\overline{O}{-}S{-}S{-}\overline{O}| \\ | \\ |\overline{O}| \end{array}\right]^{2-} \quad \left[\begin{array}{c} |\overline{O}| \;\; |\overline{O}| \\ |\quad\;\; | \\ |\overline{O}{-}S{-}S{-}\overline{O}| \\ |\quad\;\; | \\ |\overline{O}| \;\; |\overline{O}| \end{array}\right]^{2-}$$

ion   thiosulfate $S_2O_3^{2-}$ dithionite $S_2O_4^{2-}$ disulfite $S_2O_5^{2-}$ dithionate $S_2O_6^{2-}$
acid   thiosulfuric acid   dithionous acid   disulfurous acid   dithionic acid
oxid. no. of S   } II (ave.)   III   IV (ave.)   V

Ions with chains of three or more sulfur atoms are called *polythionate ions* and the corresponding acids *polythionic acids*. The best known representatives are the following (the formulas are simplified with respect to those given above to save space; the sulfur chains are always zigzag):

$$\left[\begin{array}{c} O \quad\;\; O \\ O\;S\;S\;S\;O \\ O \quad\;\; O \end{array}\right]^{2-} \quad \left[\begin{array}{c} O \quad\quad\;\; O \\ O\;S\;S\;S\;S\;O \\ O \quad\quad\;\; O \end{array}\right]^{2-} \quad \left[\begin{array}{c} O \quad\quad\quad\;\; O \\ O\;S\;S\;S\;S\;S\;O \\ O \quad\quad\quad\;\; O \end{array}\right]^{2-} \quad \left[\begin{array}{c} O \quad\quad\quad\quad\;\; O \\ O\;S\;S\;S\;S\;S\;S\;O \\ O \quad\quad\quad\quad\;\; O \end{array}\right]^{2-}$$

ion   trithionate $S_3O_6^{2-}$ tetrathionate $S_4O_6^{2-}$ pentathionate $S_5O_6^{2-}$ hexathionate $S_6O_6^{2-}$
acid   trithionic acid   tetrathionic acid   pentathionic acid   hexathionic acid
oxid. no. of S (ave.)   } $3\frac{1}{3}$ $2\frac{1}{2}$ 2 $1\frac{2}{3}$

The polythionic acids thus have the general formula $H_2S_nO_6$ ($n \geqslant 3$). The acids with $n > 6$ are little known.

It will be noted that all of the above-mentioned ions have the charge $2-$[1]. The oxosulfuric acids, in which every sulfur atom is surrounded by a complete

---

[1] In water solution of peroxosulphates, however, the ion SO_5^{2-} never appears, but only the ion HSO_5^-. The peroxosulfuric acid molecule thus never gives up one of its two hydrogen atoms as a proton. This hydrogen atom is believed to be bonded to the peroxo group.

oxygen-atom tetrahedron, are most stable against decomposition and easiest to obtain in anhydrous form.

In the following the most important oxosulfuric acids and oxosulfates are described in more detail, in the order of decreasing oxidation number of sulfur (average oxidation number when the molecule contains nonequivalent sulfur atoms).

b. Sulfur oxidation number = VI. 1. *Sulfuric acid.* Sulfuric acid H_2SO_4 in pure form is a colorless, heavy (density at $15°C = 1.836$ g cm^{-3}), viscous liquid with the freezing point 10.4°C. It constitutes one of a series of definite sulfur trioxide hydrates existing in the system SO_3-H_2O, which have the ratios $SO_3:H_2O =$ 1:7, 1:5, 1:4, 1:3, 1:2, 1:1 (sulfuric acid H_2SO_4), 2:1 (disulfuric acid $H_2S_2O_7$, m.p. 35°C; see below under 4). Crystals of the hydrate 1:2, which can also be considered as the monohydrate of sulfuric acid $H_2SO_4 \cdot H_2O$ (m.p. 8.5°C), are composed of the ions H_3O^+ and HSO_4^-.

The liquid solutions of sulfur trioxide and water have a boiling-point maximum (13-5c) that at atmospheric pressure lies at 338°C and the composition 98.3 wt% H_2SO_4. If these solutions are boiled, they give off sulfur trioxide and water until a solution of these proportions remains. This constitutes the ordinary "concentrated" sulfuric acid of commerce. Solutions with a higher proportion of sulfur trioxide than corresponds to the composition H_2SO_4 give off primarily sulfur trioxide even at room temperature and therefore fume in air, the so-called *fuming sulfuric acid* (oleum).

Sulfuric acid is a very strong acid ($pK_a \sim -3$; the second proton is given up less freely, on the other hand, and thus for HSO_4^-, $pK_a = 1.99$). Because the equilibrium $H_2SO_4 + H_2O \rightleftharpoons H_3O^+ + HSO_4^-$ is thus so strongly shifted to the right, water is very effectively absorbed by concentrated sulfuric acid as well as the more SO_3-rich solutions. They are therefore effective drying agents for gases. Water is also extracted from many compounds, from organic compounds often so strongly that for the most part only carbon remains. Textiles and paper are thus completely destroyed. The skin is also very rapidly attacked.

On absorption of water a large amount of heat is released. When concentrated sulfuric acid is to be diluted with water, the acid should therefore be carefully poured into the water with stirring. If water is poured into the acid the heat evolved leads to boiling and splashing.

Metals with $e^0 < 0$ (16c, last part, and tab. 16.1) dissolve in dilute sulfuric acid with evolution of hydrogen ($2M(s) + 2H^+ \rightarrow 2M^+ + H_2(g)$) provided arrested reaction or protective coating does not appear (16e; of practical importance, for example, is the layer of lead sulfate $PbSO_4$ that protects lead against attack by up to 78% sulfuric acid). The redox potential of dilute sulfuric acid is quite low. In concentrated sulfuric acid, on the other hand, it becomes so high, especially with heat, that even, for example, copper, silver and mercury (but not gold and platinum) dissolve with the evolution of sulfur dioxide. Carbon and sulfur are also oxidized. Steel can resist sulfuric acid if it can be made passive. For more dilute acid alloy steel is required but in concentrated sulfuric acid even carbon steel becomes sufficiently passive.

2. *Manufacture and uses of sulfuric acid.* The methods for the industrial preparation of sulfuric acid in principle involve oxidation of sulfur dioxide to sulfur trioxide and the reaction of the latter with water. As mentioned in 21-10b a catalyst is required for the oxidation, and one of these methods, the *contact process*, was described. This was developed about 1900 into an economical process, which now accounts for 70–80 % of the world's production of sulfuric acid.

If sulfur trioxide gas comes in contact with water or water vapor a large part of it is converted to a sulfuric acid mist, which is very difficult to condense or absorb in water. The gas obtained from the contact process is therefore instead reacted with about 98 % sulfuric acid, in which it is quickly and completely absorbed. Simultaneously, water is added so that the proportion of sulfuric acid is held approximately constant. Fuming sulfuric acid (oleum) is obtained by using an excess of sulfur trioxide.

During the 19th century sulfuric acid was manufactured mainly by the *lead-chamber process*, which still, although often in strongly modified technical form, is of considerable importance. The catalysis is here effected with the help of nitrogen oxides. The process takes place in a series of complicated reactions, but these can be written schematically thus:

$$\begin{array}{c} NO + \tfrac{1}{2}O_2 \rightarrow NO_2 \\ \underline{NO_2 + SO_2 + H_2O \rightarrow NO + H_2SO_4} \\ SO_2 + \tfrac{1}{2}O_2 + H_2O \rightarrow H_2SO_4 \end{array}$$

The reaction takes place in reaction chambers, formerly lead chambers, that is, large, lead-lined rooms, but now most often of other types, for example reaction towers filled with resistant material with large free surface.

The major portion of the process is carried out at about 80°C, and approximately 65 % sulfuric acid (chamber acid) is thus formed. The outgoing gases, which contain NO and NO_2, are led from the bottom up through the niter tower (Guy-Lussac tower, often two or more connected in series), in which it comes in contact with 78 % sulfuric acid flowing down from above. Here *nitrosyl hydrogen sulfate (nitrosyl sulfuric acid)* $NOHSO_4$ (22-9b) is formed, which dissolves in excess sulfuric acid to give a solution called nitrose.

Nitrosyl hydrogen sulfate takes part in the equilibrium $NO + NO_2 + 2H_2SO_4 \rightleftharpoons 2NOHSO_4 + H_2O$, which in the above-mentioned reaction is displaced to the right. When the supply of water is too low in the reaction chamber, nitrosyl hydrogen sulfate may be formed in the same way here and precipitate as so-called *chamber crystals*. If the supply of water is increased, the equilibrium is shifted to the left and the crystals redissolve.

The nitrose is pumped up to a denitrating and concentrating tower (Glover tower) where it is combined with the chamber acid pumped up from the reaction chamber. The mixture flows down through the tower and there comes in contact with hot sulfur dioxide gas (brought in from sulfur combustion furnaces or pyrite roasting ovens; cf. 21-10a) moving upward in the tower. The nitrosyl hydrogen sulfate thus reacts with the water in the relatively dilute chamber acid, and with the sulfur dioxide according to $2NOHSO_4 + 2H_2O + SO_2 \rightarrow 3H_2SO_4 + 2NO$. Thus, sulfuric acid is also formed in the Glover tower. The acid is concentrated by this process and by the hot gases, and when it reaches the bottom of the tower is approximately 78 % (Glover acid). At the top of the tower the released nitrogen oxide goes out along with unused sulfur dioxide, air and water vapor. This gas mixture is led to the reaction chamber where the nitrogen oxide can once more take part in the main process. Additional water is also sprayed in. Unavoidable losses of nitrogen oxide are compensated by addition of nitric acid to the Glover tower.

A part of the Glover acid is pumped to the Gay-Lussac tower (see above), but the major portion is drawn off and becomes the primary product. Formerly the Glover acid was concentrated by evaporation, but nowadays more concentrated acid is generally produced by the contact process. The Glover acid is used directly in applications that do not require higher concentration.

Sulfuric acid is one of the most important substances in the chemical industry. Its properties of strong acidity and high boiling point make it very useful for driving more volatile acids, for example hydrochloric acid, hydrofluoric acid and nitric acid, out of their salts. As a strong acid it is also used to convert calcium phosphate to calcium hydrogen phosphate, which is more useful for plants (superphosphate, see 22-12e4; phosphate refinement for fertilizer consumes about 40 % of the world's production of sulfuric acid) and in the metal industry for removing oxides from metal surfaces (*pickling*; see 23-5c, 37-4b). Further uses are found in the preparation of sulfates, for example sodium sulfate from sodium chloride from which hydrochloric acid is obtained at the same time (20-5d), ammonium sulfate and the very insoluble sulfates constituting color pigments (for example, lead(II) sulfate and barium sulfate). In organic chemical syntheses conc. sulfuric acid is used as a water-extraction agent. This is the case, for example, in the explosives industry in the production of nitric acid esters or nitro compounds, but here nitryl cation appears formed from nitric acid by the sulfuric acid (12-5), and this ion is probably the effective agent. Large quantities of sulfuric acid are used in the petroleum industry as a catalyst in the production of high-octane fuels by alkylation of unsaturated hydrocarbons. Its electrochemical applications (for example, in lead storage batteries, 16h) should also be mentioned.

3. *Sulfates.* Salts of sulfuric acid containing the tetrahedral ion SO_4^{2-} have since early times been called sulfates, and this name has been kept as an accepted abbreviation of the rational name *tetraoxosulfate*(VI).

"Normal" sulfates, for example Na_2SO_4, have long been distinguished from "acid" sulfates, for example $NaHSO_4$, which later came properly to be called *hydrogen sulfates* (12-1 c). Hydrogen sulfates are known mainly for the alkali metals.

The sulfate ion in itself is colorless. Most sulfates are rather easily soluble in water, and alkali and ammonium sulfates are very soluble. The solubility of the sulfates of the alkaline-earth metals decreases with increasing atomic number; calcium sulfate is rather poorly soluble, strontium and barium sulfates are insoluble. Silver sulfate is relatively difficultly soluble, mercury(I) sulfate poorly soluble, and lead(II) sulfate insoluble.

The sulfates often crystallize with water of crystallization, the same compound sometimes forming several hydrates with varying proportions of water of crystallization (cf. the copper sulfate hydrates, 13-5e). A number of types of hydrates with a given crystal structure exist for several different sulfates.

Many sulfate hydrates were formerly called vitriols (from Lat. vitrum, glass, because they often form large, glasslike crystals), for example, zinc vitriol or white vitriol $ZnSO_4 \cdot 7H_2O$ (orthorhombic structure, which is also found in, for example, $NiSO_4 \cdot 7H_2O$ and $MgSO_4 \cdot 7H_2O$), and iron vitriol or green vitriol $FeSO_4 \cdot 7H_2O$ (monoclinic structure, also found, for example, in $CoSO_4 \cdot 7H_2O$). Copper vitriol or blue vitriol was $CuSO_4 \cdot 5H_2O$. In the heptahydrates mentioned above, six of the water molecules are incorporated into octahedral hexaquo ions, for example, $Fe(H_2O)_6^{2+}$, while the seventh is not considered to be associated with any ion.

Sulfates form some important types of double salts (12-1 c). The best known is the *alum type*, named after *alum* $KAl(SO_4)_2 \cdot 12H_2O$. The general formula can be written $M^+M^{3+}(SO_4)_2 \cdot 12H_2O$. M^+ is a rather large monovalent ion, especially K^+, Rb^+, Cs^+, NH_4^+, Tl^+, while M^{3+} is a rather small trivalent ion, especially Al^{3+}, Ga^{3+}, In^{3+}, V^{3+}, Cr^{3+}, Mn^{3+}, Fe^{3+}, Co^{3+}. The alums crystallize in the cubic system (usually as octahedra, which can easily be made quite large), but depending on the size of the ion M^+, three structures exist which are slightly different from each other. However, in all of them the M^{3+} ions form hexaquo ions $M(H_2O)_6^{3+}$ with six of the water molecules.

Similar to the alum type in many respects is the *schönite type*, which gets its name from the mineral schönite $K_2Mg(SO_4)_2 \cdot 6H_2O$. The general formula is $(M^+)_2M^{2+}(SO_4)_2 \cdot 6H_2O$ where M^+ can be the same ions as in the alums and M^{2+} can be ions such as Mg^{2+}, Mn^{2+}, Fe^{2+}, Co^{2+}, Ni^{2+}, Cu^{2+}, Zn^{2+}. All schönites crystallize with the same monoclinic crystal structure. The six water molecules here are incorporated in a hexaquo ion $M(H_2O)_6^{2+}$.

Alkali sulfates can be fused without decomposition (for example, for Na_2SO_4 the melting point is 884°C). At red heat the alkaline-earth sulfates and lead sulfate still remain undecomposed, but many other sulfates break down. Metal oxides are then formed (for the noble-metal sulfates, the metal) and initially sulfur trioxide, which decomposes into sulfur dioxide and oxygen at sufficiently high temperature. The sulfates of trivalent metals, for example $Fe_2(SO_4)_3$, are especially easily decomposed. Sulfur trioxide was formerly prepared from the latter by ignition, for making sulfuric acid.

4. *Di- and polysulfuric acids and their salts.* As mentioned earlier in 21-11b1, in the system SO_3–H_2O the definite hydrate $H_2S_2O_7$ occurs, which is called *disulfuric acid* (formerly pyrosulfuric acid). This is an important constituent of fuming sulfuric acid. Disulfuric acid has been used as a smoke generator, since with the moisture in the air it forms a very stable mist of sulfuric acid droplets. Salts of disulfuric acid, *disulfates*, can be obtained by heating hydrogen sulfates, whence the old name pyrosulfate: $2KHSO_4 \rightarrow K_2S_2O_7 + H_2O$. Potassium disulfate is used to give an acid melt (m.p. 414°C), which dissolves many basic oxides ("fusion") with the formation of water soluble sulfates: $Al_2O_3(s) + 3S_2O_7^{2-} \rightarrow 2Al^{3+} + 6SO_4^{2-}$.

In fuming sulfuric acid there are also found *polysulfuric acids*, among others *trisulfuric acid* $H_2S_3O_{10}$. Solutions of sulfates and disulfates in fuming sulfuric acid contain *polysulfate ions*. On dilution of such solutions, as with the solution of disulfates in water, the bonds between the SO_4 tetrahedra are broken. Equilibria of the type $S_2O_7^{2-} + H_2O \rightleftharpoons 2HSO_4^-$ are then shifted more and more to the right.

5. *Peroxosulfuric acids and their salts.* Formation of *peroxodisulfate ion* (older name "persulfate ion") $S_2O_8^{2-}$ by anodic oxidation of sulfate ion was discussed in 16f. *Peroxodisulfuric acid* $H_2S_2O_8$ occurs in acid peroxodisulfate solutions. As described in 21-6b the peroxodisulfate ion can be hydrolyzed with the formation of hydrogen peroxide, which forms the basis for the most important method for the preparation of that substance. The peroxodisulfates are strong oxidizing agents and have therefore found many technical applications. The oxidizing power is greater than that of hydrogen peroxide; thus, for example, Mn(II) is oxidized to Mn(VII) in the presence of Ag^+ as a catalyst.

In the hydrolysis of peroxodisulfate ion, the ion HSO_5^- was stated in 21-6b to be an intermediate product. The corresponding acid is *peroxosulfuric acid* H_2SO_5 ("Caro's acid"). The hydrolysis reaction can be reversed with sufficiently high concentration of H_2O_2 so that H_2SO_5 is formed according to $H_2SO_4 + H_2O_2 \rightarrow H_2SO_5 + H_2O$ (cf. 21-6a).

Both $H_2S_2O_8$ and H_2SO_5 are known in anhydrous form and at room temperature form colorless crystals.

c. Sulfur oxidation number = V. The *dithionate ion* $S_2O_6^{2-}$ can be formed by moderately strong oxidation of S(IV), for example by passing sulfur dioxide into a suspension of manganese dioxide. Dithionic acid, which is a strong acid, cannot be obtained anhydrous.

d. Average sulfur oxidation number = IV. 1. *Sulfurous acid and sulfites.* Water solutions of sulfur dioxide contain *sulfurous acid* H_2SO_3. The hydration equilibrium SO_2 (soln.) + $H_2O \rightleftharpoons H_2SO_3$ is strongly displaced to the left. If one attempts to concentrate the solution, sulfur dioxide gas is given off, which leads to reaction to the left. Anhydrous H_2SO_3 therefore cannot be prepared. The two possible structures for sulfurous acid have been discussed in 21-11a.

In water solutions of sulfur dioxide the acid-base system H_2SO_3–HSO_3^-–SO_3^{2-} appears. Besides the *sulfite ion* SO_3^{2-} and the *hydrogen sulfite ion* HSO_3^- there is also in these solutions the *disulfite ion* $S_2O_5^{2-}$, which participates in the equilibrium $2HSO_3^- \rightleftharpoons S_2O_5^{2-} + H_2O$. Both this equilibrium and the above-mentioned hydration equilibrium are therefore coupled with the acid-base equilibrium. As is usual in such cases apparent acidic constants (12-2c) are given, and sulfurous acid is therefore a stronger acid in water than these constants (tab. 12.2) would imply.

For the manufacture of sulfite solutions, generally sulfur dioxide is led into solutions or suspensions of the corresponding hydroxide or carbonate. At the begining SO_3^{2-} predominates among the ions, but with excess of sulfur dioxide HSO_3^- and $S_2O_5^{2-}$ predominate. Solid *sulfites* can be obtained from the solutions. Solid *hydrogen sulfites* have possibly been obtained for rubidium and cesium, but not for any other metals. From the more acid solutions, on the other hand, *disulfites* are often obtained.

Alkali and ammonium sulfites are easily soluble in water, other sulfites are poorly soluble. The latter can be obtained by precipitation from a metal salt solution with a soluble sulfite.

On heating wood with solutions containing sulfite, hydrogen sulfite or disulfite ions, the lignin is dissolved, leaving the cellulose (*sulfite cellulose*). For this purpose, solutions with excess of sulfur dioxide are used, in which the cation is calcium (oldest method), magnesium, sodium or ammonium ion.

Sulfurous acid and the ions formed from it can be both oxidized and reduced (cf. sulfur dioxide, 21-10a). Often oxidation to sulfate ion occurs. Sulfite is therefore used as a reducing agent in many processes. Sodium sulfite is added to photographic developing solutions to prevent the oxidation of the developer by atmospheric oxygen (36-8).

2. *Disulfurous acid and disulfites.* As mentioned above, solid *disulfites* can be obtained from acid sulfite solutions. The *disulfite ion* $S_2O_5^{2-}$ is composed of one SO_3 and one SO_2 group with a bond between the sulfur atoms (cf. 21-11a). The two sulfur atoms thus are nonequivalent and have the formal oxidation numbers V and III. The mean oxidation number of sulfur is thus IV. The disulfite ion is not analogous to the disulfate ion, which consists of two tetrahedra with a common O atom, according to which one might have expected the disulfite ion to consist of two SO_3 pyramids with a common O atom. $K_2S_2O_5$ is used, among other things, in photography for acidification of fixing baths (often under the old name potassium metabisulfite). Anhydrous disulfurous acid is unknown.

e. Sulfur oxidation number = III. On reduction of S(IV), for example of acid hydrogen sulfite solution with zinc or cathodic reduction, the *dithionite ion* $S_2O_4^{2-}$ is formed. In this ion, whose structure has been determined in crystalline $Na_2S_2O_4$, the S–S distance is unusually large. This indicates a weak bond between the two SO_2 groups in the ion, and in water solution a weak dissociation according to $S_2O_4^{2-} \rightleftharpoons 2SO_2^-$ is noticeable. The solution thus shows paramagnetism caused by the unpaired electron in SO_2^- (7-4b).

$Na_2S_2O_4$ is used as a reducing agent in dying and as a reducing bleaching agent.

f. Average sulfur oxidation number = II. 1. *Thiosulfuric acid and thiosulfates.* In the thiosulfate ion $S_2O_3^{2-}$ one of the oxygen atoms in the sulfate ion is replaced by a sulfur atom. The two sulfur atoms in the ion are nonequivalent, but the average oxidation number of sulfur is II. Thiosulfate ion can be obtained by boiling a sulfite solution with finely divided sulfur ($SO_3^{2-} + S(s) \rightarrow S_2O_3^{2-}$) or by oxidation of disulfides with atmospheric oxygen ($CaS_2 + \frac{3}{2}O_2 \rightarrow CaS_2O_3$). When a thiosulfate solution is acidified $HS_2O_3^-$ is probably formed at first, but this ion decomposes rather quickly: $H^+ + S_2O_3^{2-} \rightarrow HS_2O_3^- \rightarrow HSO_3^- + S(s) \rightarrow SO_2 + OH^- + S(s)$. Therefore, thiosulfuric acid and hydrogen thiosulfates have not been isolated.

The thiosulfate ion easily forms complexes with ions of heavy metals such as Cu^+ and Ag^+. This is the basis for the use of thiosulfate for photographic fixing, in which the silver halide is dissolved mainly with the formation of the complex ions $Ag(S_2O_3)_2^{3-}$ and $Ag(S_2O_3)_3^{5-}$ (cf. 36-7a, 36-8). The fixing bath is usually made up from $Na_2S_2O_3 \cdot 5H_2O$, often called "hypo" after the old name sodium hyposulfite.

Strong oxidizing agents oxidize thiosulfate ion to sulfate ion: $S_2O_3^{2-} + 5H_2O \rightarrow 2SO_4^{2-} + 10H^+ + 8e^-$. Chlorine, for example, acts in this way, so that sodium thiosulfate is used after chlorine bleach to destroy the remaining chlorine ("antichlor"). Weaker oxidizing agents oxidize thiosulfate to tetrathionate ion (21-11g): $2S_2O_3^{2-} \rightarrow S_4O_6^{2-} + 2e^-$. Iodine acts in this way and can therefore be determined quantitatively (*iodimetry*) by titration with a thiosulfate solution of known strength: $2S_2O_3^{2-} + I_2 \rightarrow S_4O_6^{2-} + 2I^-$. The point at which the amount of thiosulfate equivalent to the amount of iodine has been added is indicated by a change in color when the iodine disappears (brown → colorless, or blue → colorless if starch is added, 20-3). Since many oxidizing agents oxidize iodide ion to iodine, such substances can also be determined iodimetrically.

2. *Sulfoxylic acid and sulfoxylates.* In solutions of dithionite the *sulfoxylate ion* SO_2^{2-} participates in the equilibrium $S_2O_4^{2-} \rightleftharpoons SO_2^{2-} + SO_2$. If the products on the right side are taken care of, this equilibrium is shifted strongly to the right. This can be done, for example, by adding Co^{2+} and OH^-: $S_2O_4^{2-} + Co^{2+} + 2OH^- \rightarrow CoSO_2(s) + SO_3^{2-} + H_2O$. Only a few inorganic sulfoxylates are known. As mentioned in 21-11a several structures are possible for sulfoxylic acid, and organic derivatives of these are known. Certain alkyl sulfinates are important as reducing agents in dying.

g. Polythionic acid and polythionates. Acids $H_2S_nO_6$ ($n \geqslant 3$) form the class of *polythionic acids*. For the best-known polythionic acids, $n = 3$ to 6, but the series continues with n values greater than 6. Dithionic acid $H_2S_2O_6$ (21-11c) is not included in the polythionic acids. These acids are closely related to each other in manner of formation and properties, but are quite different from dithionic acid. In fact, in dithionic acid sulfur has a significantly higher oxidation number than in the polythionic acids (21-11a). The polythionic acids are strong acids.

In water solution the polythionic acids can be obtained, among other ways, by reduction of S(IV). On the other hand, dithionate ion, which contains S(V), is obtained by *oxidation* of S(IV). As a reducing agent S(–II) in the form of H_2S can be used, which is passed into a water solution of SO_2. The solution thus formed ("Wackenroder's solution") contains a mixture of polythionic acids and colloidal sulfur, which can be removed. From this solution polythionates can be prepared. The formation of tetrathionate ion by oxidation of thiosulfate ion with iodine has been mentioned earlier (21-11 f 1). The polythionic acids and polythionate ions decompose quite rapidly in water solution into sulfate ion, sulfite ion and sulfur.

Anhydrous polythionic acids have been prepared by reaction between chlorosulfuric acid and hydrogen polysulfides in ether solution at $-78°C$; for example, $HO_3SCl + HS_8H + ClSO_3H \rightarrow H_2S_{10}O_6 + 2HCl$.

21-12. Oxoselenic and oxotelluric acids

Since for analogous oxo acids with different central atom the strength increases with increasing electronegativity of the central atom (12-2f), the strength of analogous acids decreases in the central-atom sequence $S > Se > Te$.

a. Oxoselenic acids. *Selenious acid* H_2SeO_3 is formed if selenium dioxide is dissolved in water and can also be obtained by oxidation of selenium with dilute nitric acid. It is much more stable than sulfurous acid and at room temperature forms colorless crystals. Selenious acid can be oxidized to selenic acid only with the strongest oxidizing agents (chlorine, anodic oxidation); on the other hand, it is easily reduced to selenium. As a consequence, selenious acid most often shows oxidizing properties, in contrast to sulfurous acid.

Selenic acid H_2SeO_4, which in water solution is a strong acid, can be obtained as colorless crystals (m.p. 58°C). It is a considerably stronger oxidizing agent than sulfuric acid, and, for example, oxidizes hydrochloric acid to chlorine: $H_2SeO_4 + 2HCl \rightarrow H_2SeO_3 + Cl_2 + H_2O$. The solubilities of the *selenates* are similar to those of the sulfates, but the selenates may be distinguished from them by their greater oxidizing power.

b. Oxotelluric acids. *Tellurites* with the ion TeO_3^{2-} are formed on solution of tellurium dioxide in alkali hydroxide: $TeO_2(s) + 2OH^- \rightarrow TeO_3^{2-} + H_2O$. On acidification *tellurous acid* H_2TeO_3 is formed, a very weak acid, which easily decomposes to $TeO_2(s) + H_2O$ and has not been obtained in anhydrous form.

Telluric acid is obtained by oxidation of tellurous acid with strong oxidizing agents. From the solution *orthotelluric acid* H_6TeO_6 crystallizes. Telluric acid of the formula H_2TeO_4 thus has $2H_2O$ more added, which is explained by the fact that the space around the central atom has become great enough so that it can coordinate six oxygen atoms. Orthotelluric acid is a weak acid. As salts of this acid both *orthotellurates*, for example Ag_6TeO_6 and Hg_3TeO_6, and *hydrogen orthotellurates*, for example $Na_2H_4TeO_6$, are known (the latter were formerly thought to be salts of the telluric acid H_2TeO_4 with water of crystallization; for example, $Na_2TeO_4 \cdot 2H_2O$).

CHAPTER 22

Group 15, The Nitrogen Group

22-1. Preliminary survey

As would be expected, the electronegativity of the elements of the nitrogen group decreases as the atomic number increases. At the same time the metallic properties increase. Black phosphorus is a semiconductor. The stable modifications of arsenic, antimony and bismuth all have metallic, although low, conductivity. However, arsenic and antimony also exist in metastable, nonconducting modifications, so that they are customarily considered as semimetals. On the other hand, bismuth is counted as a metal. A survey of the structures was given in 5-6a. The increase in metallic properties is accompanied by corrresponding changes in the chemical properties.

The first element of the group, nitrogen, shows properties very divergent from those of the others (5-6, introd.). This is striking for the oxygen compounds, where the small radius of the nitrogen atom precludes the coordination number four with respect to oxygen, which is usual for the following elements phosphorus and arsenic. The nitrate(V) ion is therefore NO_3^- while the simplest phosphate(V) and arsenate(V) ions are PO_4^{3-} and AsO_4^{3-} respectively. Beginning with antimony, six-coordination with oxygen is also permissible (for example, $Sb(OH)_6^-$).

The highest and lowest oxidation numbers are V and $-$III. Among the more common compounds odd numbers are most represented (5-3g). For nitrogen all oxidation numbers over the given range are found, but, as for the whole group, the numbers V, III and $-$III are the most important. The last occurs for all the elements in the group in the hydrides of the type NH_3, which, however, rapidly become less stable the higher the atomic number of the central atom is (for nitrogen and phosphorus, ions of the type NH_4^+ also occur). Of the positive oxidation numbers V is the most prominent for nitrogen and phosphorus, but thereafter decreases in importance. Sb(V) in acid solution tends so strongly to go over to Sb(III) that the oxidizing effect becomes quite pronounced (iodide ion is oxidized to iodine.) For bismuth the number III has become wholly predominant. This behavior may be compared with the analogous shift in valence in neighboring groups, for example the chalcogens (21-1).

22-2. Occurrence and history

a. Occurrence. Estimates of the abundance of the elements in the universe indicate that *nitrogen* is number five in terms of the number of atoms (tab. 17.1).

On the earth the relative amount is significantly less (tab. 17.3). It has been estimated that approximately 1/50 of the total amount occurs as elementary nitrogen in the atmosphere, of which at sea level it constitutes the largest portion. Dry air contains 78.1 vol% or 75.5 wt% nitrogen. The rest of the nitrogen is for the most part bound in organic compounds. However, it has recently been found that a quite large amount of nitrogen is probably bound in silicates as ammonium ion. In certain desert regions the lack of rainfall has made possible the accumulation and retention of nitrates, but their origin and manner of deposition is not wholly understood. The sodium nitrate deposits of northern Chile (Chile saltpeter) constitutes the largest known localized occurrence of combined nitrogen.

As an essential element in proteins, nitrogen is indispensible to life on the earth. In connection with the formation and decomposition of organisms nitrogen takes part in a cycle of the greatest importance, which is described in 22-16.

Phosphorus occurs mostly in *apatite* $Ca_5(OH,F)(PO_4)_3$, which is widespread in eruptive and metamorphic rocks, but in these is quite rarely concentrated in recoverable amounts. The largest amounts are obtained from marine deposits of impure and poorly crystallized apatite, so-called *phosphorite* (phosphate rock). The ions OH^- and F^- substitute for each other completely (5-2c), and depending on which ion predominates one speaks of *hydroxyapatite* or *fluorapatite*. Larger proportions of F^- in the structure are found only in deep-lying rocks. Apatite crystals easily adsorb other ions on their surfaces and this effect increases for poorly formed crystals with disordered structure and large surface area, for example as in phosphorite. Phosphorite therefore often contains fluorine, although it consists mainly of hydroxyapatite.

Phosphorite occurs in many places, and especially in North Africa (Algeria and Tunisia) and the United States (especially Florida and Tennesee). Larger amounts of apatite have begun to be mined from eruptive rocks in Canada and the Kola Peninsula. Phosphorus in iron ores is made use of in the form of Thomas phosphate (22-12e4, 34-3c).

In plants and animals different kinds of phosphoric acids are incorporated in many ways: in nucleic acids and consequently in nucleoproteins, in phosphoproteins and phosphatides. Carbohydrate degradation, the most important of the energy-supplying processes of organisms, is carried out by phosphoric acids bound to a large number of the reacting substances. Hydroxyapatite occurs in shells of crustaceans and molluscs, and most especially in the bone of vertebrate animals, which is composed to a large proportion of this compound. Animal waste is very rich in phosphorus; a grown man takes in and gives up about 2 g of phosphorus per day. The accumulation of phosphoric acid from wastes left by man and domestic animals has been used by archeologists to localize prehistoric living places. The great importance of phosphoric acid to plant life makes necessary an ample supply of phosphate fertilizer in farming.

Arsenic occurs native in nature, but usually combined with heavy metals in minerals with metallic or semimetallic properties. In these it very often accompanies sulfur. Most of these minerals can be roughly assigned one of the formulas MAs, MAs$_2$ or MAsS, in which most often M = Fe, Co or Ni. Of these *arsenopyrite* FeAsS is the most important source of arsenic. As a sulfide arsenic occurs in *realgar* As$_4$S$_4$ and *orpiment* As$_2$S$_3$. Some arsenic minerals contain thio ions, for example, *proustite* Ag$_3$AsS$_3$ (silver trithioarsenite). Arsenic is so widespread in nature, especially in connection with sulfide minerals, that it is very common as an impurity both in metals and in sulfur obtained from such minerals.

Antimony occurs mainly as *stibnite* Sb$_2$S$_3$, but also often occurs combined with heavy metals in minerals analogous to the corresponding arsenic minerals.

Bismuth is a quite rare element. It occurs native, but a more important occurrence is *bismuthinite* Bi$_2$S$_3$, with the same structure as stibnite. Different isotopes of bismuth are members of each of the natural radioactive series (9-2a).

One of the world's most important sources of arsenic is the sulfide ores of northern Sweden, where it occurs mainly as arsenopyrite. Stibnite is mined primarily in China. The largest production of bismuth comes from Bolivia and Australia.

b. History. *Nitrogen.* As mentioned in 21-2b, at least by 1772 Scheele had shown that air consists of "fire air" and "foul air". The latter could not support combustion. Lavoisier gave Scheele's "foul air" the name "azote" (Gk. ἀ not, ζωτῖκός for life). When the relationship of the gas to nitric acid and saltpeter became known Chaptal in 1790 introduced the name *nitrogène* ("saltpeter former"; Gk. νίτρον, Lat. nitrum, originally meaning sodium and potassium carbonate but from the end of the 16th century used to mean potassium nitrate, saltpeter; γεννάω, I make).

Phosphorus was prepared first in free form in 1669 by the German alchemist Brand, who in his search for the philosopher's stone ignited in a closed vessel the residue obtained from evaporating urine, and condensed the gases. The name (Gk. φως-φόρος, light bearer) was given because the substance glows in the dark. It was first stated to be an element by Lavoisier, who illustrated his theory of combustion, among other ways, by the oxidation of phosphorus.

Arsenic. The mineral orpiment As$_2$S$_3$ was mentioned in the 4th century B.C. by Theophrastus as ἀρσενικόν. The alchemists termed the roasting product of arsenic sulfides, that is, As$_2$O$_3$, as "white arsenic" and also prepared free arsenic by reduction of As$_2$O$_3$ with carbonaceous substances. The Swedish chemist Georg Brandt in 1733 designated it as a semimetal (by semimetal Brandt meant a substance that looks like metal but is not malleable). Lavoisier regarded arsenic as a metallic element.

Antimony. Stibnite, which in ancient times was used as an eye cosmetic, was called *stibium* by the Romans. Later the name antimonium, of disputed origin, was used for this mineral and was also carried over to the element. It was doubtless prepared very early, although at least in the Middle Ages it was considered to be lead.

Metallic *bismuth* was probably first prepared at the end of the 15th century. The name is of German origin but its derivation is unclear.

22-3. Properties of the elements of the nitrogen group

a. Nitrogen. The element nitrogen ordinarily consists in all states of aggregation of diatomic molecules N$_2$. The triple bond in this molecule was discussed in 5-3a. Some important properties are:

Melting point	Boiling point	Crit. temp.	Crit. pressure
−209.9°C	−195.8°C	−147.1°C	33.9 bar

The gas is colorless, incombustible, odorless and tasteless. At 20°C and 1 atm total pressure, 100 g water dissolves 1.89 mg N_2. Crystalline N_2 occurs in at least two modifications with different packings of the N_2 molecules.

The bond energy of the triple bond in the nitrogen nolecule is high (5-3k), and the dissociation of the molecule therefore requires a considerable supply of energy: $N_2(g) \rightleftharpoons 2N(g)$; $\Delta H = 942$ kJ. The molecule is consequently extremely stable and inert. On the other hand, atomic nitrogen is highly reactive, but in nitrogen gas the high dissociation energy causes the proportion of atomic nitrogen to be extremely small as long as the temperature is not very high. However, nitrogen gas frequently reacts even at moderate temperature, for example with many metals (lithium, alkaline-earth metals, aluminum, many transition metals), with the formation of *nitrides*. An appreciable proportion of atomic nitrogen is most simply obtained by dark electric discharge in nitrogen gas. In the atmosphere the proportion of atomic nitrogen begins to exceed that of molecular nitrogen only at altitudes above about 300 km.

The high bond energy of nitrogen causes it to be easily formed from nitrogen compounds, which thus are unstable with respect to molecular nitrogen. This instability and the large amounts of heat developed in the process cause many nitrogen compounds to be explosive.

b. Phosphorus. Phosphorus exists in several solid modifications. In the preparation of phosphorus, when phosphorus vapor is condensed, in general liquid phosphorus is formed first, and from this solid, colorless, so-called *white phosphorus*, which is unstable under all conditions. The gaseous, liquid, and solid white phosphorus all contain tetrahedral P_4 molecules (5-6a, fig. 5.28a). As mentioned in 5-6a the bonds in the P_4 tetrahedra are probably highly strained, which may account for the instability of the molecule. White phosphorus next transforms to *red phosphorus* (see below). The change is favored by rise in temperature and illumination. White phosphorus, when exposed to light, thus becomes first yellow ("yellow phosphorus") and finally red, at least on the surface. However, at reasonably low temperature and in the dark (oxygen must not be present, so that phosphorus is often kept under water) white phosphorus remains unchanged for a long time. It also has a well-defined melting point, 44.1°C, and the melt boils at 280°C.

White phosphorus exists in two modifications, both of which are metastable. Below −76.9°C it has a hexagonal structure, and from this temperature to the melting point the structure is cubic. Both are insulators.

At low temperatures white phosphorus is brittle, at room temperature rather soft. At 0°C the density is 1.84 g cm^{-3}. It is usually marketed cast in stick form, but from solution or gas phase it can easily be obtained as colorless, lustrous crystals. The solubility is slight in water, greater in, for example, ether and benzene, and very high in carbon disulfide. White phosphorus is extremely reactive. In finely divided form it ignites itself in air at room temperature, and in larger pieces at slightly higher temperature. Phosphorus should therefore be cut or worked only under water. Burning phosphorus causes deep, difficultly healed burns. The combustion takes place with a yellow-white flame and strong evolution of heat to give phosphorus pentoxide: $2P(s) + \frac{5}{2}O_2(g) \rightarrow P_2O_5(s)$; $\Delta H^0 = -1550$ kJ.

White phosphorus is quite volatile at room temperature and the gas is oxidized in air, giving off a blue-green light. The glow of white phosphorus in the dark by this slow oxidation gives it its name (22-2b). The development of light ceases when the partial pressure of oxygen exceeds 800 mbar at 15°C.

White phosphorus also reacts easily with halogens, sulfur and many metals (in the latter case, phosphides are formed). It is also a powerful reducing agent. On being heated with white phosphorus sulfuric acid is reduced to sulfur dioxide, nitric acid to nitrogen oxides, while the phosphorus is oxidized to orthophosphoric acid H_3PO_4. From solutions of salts of the nobler metals (for example, gold, silver, copper, lead) the metal or a phosphide (for example, copper phosphide) is precipitated.

White phosphorus is very poisonous (lethal dose, about 0.1 g). As an antidote a 0.2 % copper(II) sulfate solution is used, in which the above-mentioned precipitation of copper phosphide occurs, and which also causes vomiting. Phosphorus that sticks to the skin on burning is made harmless by a more concentrated solution of copper(II) sulfate. Prolonged breathing of phosphorus vapor causes severe chronic injury, particularly decay of the gums and jawbone (phosphorus necrosis).

The *more stable modifications of phosphorus* contain phosphorus-atom groups that are unlimited in one or more dimensions, and they are therefore more difficult to melt and vaporize than white phosphorus. Their nature has been more difficult to establish, but probably a pseudosystem exists here (13-3b). With changes in the external conditions transitions among these different modifications take place sluggishly, because the transformations require the breaking of many covalent bonds in the large atomic groups.

At room temperature *black phosphorus* is the stable modification up to about 400°C. The structure, which has orthorhombic symmetry, was discussed in 5-6a (fig. 5.28c). Black phosphorus is a semiconductor. In black phosphorus the atoms are more densely packed (density 2.7 g cm^{-3}) than in any other phosphorus modification, and its formation is favored by pressure. However, the rate of formation is generally very low, and when white phosphorus is brought into the region of stability for black phosphorus, usually only an intermediate form is reached, *red phosphorus*, which is still metastable. Red phosphorus is a highly undefined product, and has a very disordered crystal structure. The formation of red phosphorus by irradiation of white phosphorus has been mentioned earlier. Also, red phosphorus is formed when white phosphorus is heated in a closed vessel to a maximum of about 400°C. At 250°C the process still goes rather slowly, but the rate is increased by raising the temperature, and adding certain substances, among others iodine, in small amounts. Even at relatively low pressure, however, the stable, black phosphorus can be obtained by heating white or red phosphorus with bismuth or mercury in a closed vessel at about 350°C. This is probably a result of the solution of the phosphorus in the metal in atomic form, so that the availability of single phosphorus atoms in the solution facilitates the building up of the black phosphorus. In order to obtain black phosphorus without the presence of other substances, very high pressure is required: at 200°C about 12 kbar, and at 100°C about 45 kbar.

Above about 400°C and at atmospheric pressure a new modification is stable, which is usually called *violet phosphorus*. Near to 400°C this is formed with quite low order and is then red, but with increasing temperature of formation the order increases simultaneously as the color deepens (cf. 7-3a). Well-ordered violet phosphorus has a complicated crystal structure, which is built up of phosphorus sheets. Its region of stability extends up to the melting point at about 600°C. Since violet phosphorus is stable at temperatures where the atomic motion is high, it is quite easily formed by sufficiently strong heating of the other forms.

Red, and still more black and violet phosphorus react more sluggishly than white phosphorus. All three forms are nonpoisonous if they are completely freed from white phosphorus (for example, by treatment with carbon disulfide, which dissolves only white phosphorus). Neither red, black nor violet phosphorus is attacked by dry air, but all are oxidized slowly in moist air. Especially red phosphorus is thus easily contaminated with handling and storage. The heat of oxidation can, in fact, cause spontaneous ignition particularly if a larger amount is collected in one place. All three forms should therefore be handled carefully, especially in the presence of oxidizing agents.

Red, black and violet phosphorus all are quite easily vaporized if they are heated at a not too high pressure. On vaporization the atomic aggregate is broken down and the gas consists of P_4 molecules. If this is condensed at low enough temperature white phosphorus is formed.

c. Arsenic. Arsenic also exists in several solid modifications. *Gray* or *metallic arsenic* is stable. It forms steel-gray, metallic crystals with metallic conductivity and a density of 5.78 g cm^{-3}. It has a layer structure with trigonal symmetry, which was described in 5-6a (fig. 5.28b). If metallic arsenic is heated at atmospheric pressure in the absence of air, it sublimes without melting and forms a lemon-yellow gas, which up to 800°C consists of As_4 molecules, and at still higher temperatures of As_2 molecules.

When arsenic gas is condensed at very low temperature, for example by cooling with liquid air, metastable *yellow arsenic* is obtained, which, as far as properties and structure are concerned, is completely analogous to cubic white phosphorus. It is thus an insulator, composed of As_4 molecules. It also dissolves easily in carbon disulfide. At room temperature yellow arsenic transforms to the metallic form.

If the condensation takes place on surfaces that have a temperature of 100–200°C, a lustrous black coating is obtained (*black arsenic*), which is often mirror-like ("arsenic mirror"). The coating contains layers of the same kind as in metallic arsenic, but the individual layers have not yet attained complete order with respect to each other. Black arsenic is an insulator. A metastable orthorhombic modification of arsenic analogous with black phosphorus is also believed to exist.

On heating in air arsenic burns with a bluish flame with the formation of a white smoke of arsenic trioxide As_2O_3. At the same time a garliclike odor is noticeable, which can probably be ascribed to As_4 molecules in the air. Arsenic reacts with chlorine with strong development of heat. Strongly oxidizing acids (conc. nitric acid, aqua regia) oxidize arsenic to arsenic acid; less strongly oxidizing acids (dilute nitric acid, conc. sulfuric acid) to arsenious acid. Arsenic combines readily with metals to give arsenides.

All soluble arsenic compounds are very poisonous (see also 22-4 c).

d. Antimony. The stable solid modification is *metallic antimony*, which is silver-white, highly lustrous, brittle and coarsely crystalline. It has metallic, although low, conductivity and density 6.69 g cm^{-3}. The structure is analogous to that of metallic arsenic (5-6a). Metallic antimony melts at 630.5°C. When the melt is heated in the absence of oxygen it boils at 1625°C. The gas contains Sb_4 and Sb_2 molecules. When it is condensed on a strongly cooled surface *black antimony* is obtained with a disordered structure analogous to that of black arsenic. The structure becomes ordered at about 0°C with the formation of metallic antimony.

If a strong solution in hydrochloric acid of an antimony trihalide, for example $SbCl_3$, is electrolyzed, so-called *explosive antimony* is formed at the cathode, which is not a pure antimony phase, but contains halogen. On abrasion, pulverizing or rapid heating, the product begins to glow and gives off halide until metallic antimony remains.

At ordinary temperature metallic antimony remains unchanged in the air. When heated above the melting point it burns to give the trioxide Sb_2O_3. In a finely powdered state it combines with halogens, for example to give a pentachloride $SbCl_5$. Metallic antimony will not dissolve in hydrochloric acid or dilute sulfuric acid. Depending on the concentration, nitric acid oxidizes it to trioxide or pentoxide Sb_2O_5.

e. Bismuth. At ordinary pressure bismuth occurs in only one solid modification, which has the same crystal structure as the metallic forms of arsenic and antimony. Its conductivity is metallic, although very low (6-4a). The color is lustrous white with a tinge of red. It is brittle and usually coarsely crystalline. The density is 9.78 g cm^{-3}. The melting point is 271°C and the boiling point 1490°C. The gas contains Bi and Bi_2.

At the melting point the solid phase has a greater specific volume than the liquid and thus on solidifying an expansion occurs (1-4h). This is greater than the decrease in volume of the solid phase on cooling to room temperature, as can be seen from the specific volumes: 0.099 (l, 271°C), 0.103 (s, 271°C) and 0.102 (s, 20°C). If bismuth is cast in a mold, the mold will therefore be filled out with excess pressure even at room temperature.

At high pressures (18–100 kbar) several other solid bismuth modifications have been found.

Bismuth is not attacked in air at room temperature but burns at red heat to give the trioxide Bi_2O_3. Oxidizing acids (nitric acid and hot conc. sulfuric acid) dissolve bismuth with the formation of bismuth salts. The best solvent is nitric acid.

22-4. Production and uses of elements of the nitrogen group

a. Nitrogen. On a laboratory scale pure nitrogen is often prepared by oxidation of ammonia or ammonium ion. This is most simply done by heating to about 70°C a concentrated solution of either ammonium nitrite or ammonium chloride and sodium nitrite. The ammonium ion is then oxidized by the nitrite ion, in which process the latter also gives nitrogen: $NH_4^+ + NO_2^- \rightarrow N_2(g) + 2H_2O$.

Nitrogen that is not to be used to make nitrogen compounds is nowadays recovered commercially from the air by condensation to liquid air (13-4a) and fractional distillation (13-5c). However, atmospheric nitrogen is used to a far greater extent in the production of nitrogen compounds, above all ammonia (22-6b).

Nitrogen in elementary form is used especially as a protective gas to prevent oxidation, particularly in the heat treatment of metals. It is also used in incandescent lamps to decrease the evaporation rate of the filament. In commerce nitrogen gas is commonly compressed to about 150 atm in steel cylinders. Liquid nitrogen is used as a cooling medium, for which it has the great advantage over liquid air in that it is not oxidizing and avoids the risks this leads to (cf. 21-4a).

b. Phosphorus. Phosphorus is produced nowadays exclusively from "phosphate rock" (mostly apatite) by reduction with carbon at high temperature in the presence of SiO_2 (quartz). The heating is usually carried out electrically, more rarely by the combustion of carbon, and the temperature is held at 1300–1400°C. SiO_2 drives out phosphorus pentoxide P_2O_5 and forms calcium silicate in the form of a fluid slag. The phosphorus pentoxide is then reduced by the carbon: $2P_2O_5 + 10C(s) \rightarrow P_4(g) + 10CO(g)$. The phosphorus distills off and is condensed as white phosphorus under water.

White phosphorus is used for the production of red phosphorus and phosphorus compounds and alloys (phosphor bronze 36-5a). During World War II it was used to a great extent as a smoke generator (burning to phosphorus pentoxide, which with the moisture in the air gives a white smoke of phosphoric acid droplets), and for phosphorus incendiary bombs. In the first matches introduced in the 1830's the match head contained white phosphorus, but because of their toxicity and flammability such matches are now forbidden in most countries.

Red phosphorus is produced by heating white phosphorus in the absence of air, first up to 260°C and finally to 350°C. The remaining white phosphorus is removed, usually by washing with sodium hydroxide solution. Red phosphorus found its widest application with the introduction of the so-called safety matches (22-14), for which it is incorporated in the striking surface.

c. Arsenic, antimony and bismuth. The most important source minerals for these elements are sulfides. These can all be processed in the usual way (21-5b) by roasting in air to oxide followed by reduction of the oxide, in these cases with carbon. Occasionally the oxide is used directly as starting material. The common ores arsenopyrite FeAsS and stibnite Sb_2S_3 are also processed by special methods. Arsenopyrite is heated in absence of air, so that arsenic sublimes out: $4FeAsS \rightarrow 4FeS + As_4(g)$. Stibnite is fused with iron: $Sb_2S_3 + 3Fe \rightarrow 2Sb + 3FeS$. The latter process, however, is used only for rich ores.

Arsenic in elementary form has very little application. It is used mainly for alloy fabrication: among other things, for the manufacture of lead shot (23-6c). Arsenic compounds are produced mainly from diarsenic trioxide, which is generally obtained directly as a troublesome by-product in the processing of arsenic-containing ores. The applications of arsenic compounds are based primarily on its poisonous properties (extermination of injurious animals, prevention of rot, weed killing, and in medicine for combating parasitic diseases).

Antimony is most valuable as an alloy metal, especially in tin and lead alloys (23-5c and 23-6c respectively), in which it increases the hardness.

Bismuth in elementary form is used in low-melting alloys (13-6) and as starting material for bismuth compounds. The majority of these are used in medicine (see, for example, 22-13d).

22-5. Nitrides, phosphides, arsenides and antimonides

According to the rules in 5-2a and 5-3f compounds between metals, B, Si and C on the one hand and N, P, As and Sb on the other should be called nitrides, phosphides, arsenides and antimonides, respectively (compounds also exist mutually among N, P, As and Sb, of which only the phosphorus nitrides are of any great importance). In these compounds generally N, P, As or Sb is the most electronegative element. However, with the exception of the nitrides and phosphides of alkali and alkaline-earth metals, for example Li_3N, Mg_3N_2, AlN, Ca_3P_2 in which we may consider the existence of the nitride ion N^{3-} and phosphide ion P^{3-} (5-2e), the ionic contribution is very small.

In 6-4g it was mentioned that phases containing transition elements and nonmetals are often metallic. This is found to be the case with nitrides, phosphides and arsenides of the transition elements as long as the proportion of nonmetal is not high (the phase then becomes a semiconductor). The antimonides of most metals are metallic. Bismuth always behaves as a metal and if it combines with a metal the phase formed is always metallic. We therefore never speak of bismuthides. The composition of the metallic phases is determined by other factors than ionic charges and usually cannot be predicted (Fe_4N, Fe_2P, $FeSb_2$). Solid solutions with wide regions of homogeneity are common.

A large number of nitrides can be made by heating the other element in a stream of nitrogen, and in many cases more effectively if the heating is done in a stream of ammonia gas. Sometimes in the latter case an amide is formed first (22-6a), which on further heating gives off ammonia and converts first to an imide and then to a nitride. Phosphides, arsenides and antimonides are most simply prepared by heating the elements together.

Many nitrides, phosphides, arsenides and antimonides are decomposed by acids with the formation of the corresponding hydride (22-7). If the ionic bond contribution is sufficiently great decomposition may even occur with water:

$$Mg_3N_2(s) + 6H_2O \rightarrow 2NH_3 + 3Mg(OH)_2.$$

22-6. Compounds with nitrogen-hydrogen bonds

The most important compounds with nitrogen bonded to hydrogen are collected in tab. 22.1. According to the rules in 5-2a and 5-3f the compounds of nitrogen and hydrogen alone should generally be called hydrides in spite of the fact that nitrogen is more electronegative than hydrogen. As can be seen, however, all also have special names.

In all cases where all the ligands of the nitrogen atom are bound by single bonds sp^3 hybridization (5-3d) may be assumed, and thus tetrahedral bond directions.

a. Ammonia. Ammonia (the name is derived from "sal ammioniac"; see ammonium chloride, 22-6c) in its various aggreagation states consists of the molecule NH_3, whose structure has been described in 5-3d. Some important properties are:

Melting point	Boiling point	Crit. temp.	Crit. press.	Vap. press. 20°C
−77.7°C	−33.4°C	132.8°C	113.8 bar	8.57 bar

At ordinary temperature ammonia is a colorless gas, strongly irritating to the mucous membranes and with very characteristic odor. Even a few percent of ammonia in the air rapidly causes cramps in the throat, which can lead to suffocation. The above data show that ammonia can be easily condensed, giving a colorless liquid (density at 25°C: 0.60 g cm^{-3}). The heat of vaporization of the liquid is high: 1369 J g^{-1} at the boiling point.

Table 22.1. *Compounds with nitrogen-hydrogen bonds.*

Oxid. no. of nitrogen				
$-\frac{1}{3}$ (ave.)		$\underset{\text{H}}{\overset{\text{H}}{\mid}}$ N=N=$\underline{\text{N}}$ ↔ $|\underline{\text{N}}$—N≡N$|$ with H on first N hydrogen azide HN$_3$		
$-\text{I}$	$\left[\begin{array}{c}\text{H}\\ \mid\\ \text{H—N—O}|\\ \mid\ \ \ \mid\\ \text{H}\ \ \text{H}\end{array}\right]^{+}$ hydroxylammonium ion NH$_3$OH$^+$	$\begin{array}{c}\text{H}\\ \mid\\ |\text{N—O}|\\ \mid\ \ \ \mid\\ \text{H}\ \ \text{H}\end{array}$ hydroxylamine NH$_2$OH		
$-\text{II}$	$\left[\begin{array}{c}\text{H}\ \ \text{H}\\ \mid\ \ \ \mid\\ \text{H—N—N—H}\\ \mid\ \ \ \mid\\ \text{H}\ \ \text{H}\end{array}\right]^{2+}$ hydrazinium(2+) ion N$_2$H$_6^{2+}$	$\left[\begin{array}{c}\text{H}\\ \mid\\ \text{H—N—}\underline{\text{N}}\text{—H}\\ \mid\ \ \ \mid\\ \text{H}\ \ \text{H}\end{array}\right]^{+}$ hydrazinium(+) ion N$_2$H$_5^+$	$\begin{array}{c}\text{H}\\ \mid\\ \text{H—}\underline{\text{N}}\text{—}\underline{\text{N}}\text{—H}\\ \mid\\ \text{H}\end{array}$ hydrazine N$_2$H$_4$	
$-\text{III}$	$\left[\begin{array}{c}\text{H}\\ \mid\\ \text{H—N—H}\\ \mid\\ \text{H}\end{array}\right]^{+}$ ammonium ion NH$_4^+$	$\begin{array}{c}\text{H}\\ \mid\\ |\text{N—H}\\ \mid\\ \text{H}\end{array}$ ammonia NH$_3$	$\left[\begin{array}{c}\text{H}\\ \mid\\ |\underline{\text{N}}\text{—H}\end{array}\right]^{-}$ amide ion[a] NH$_2^-$	$\left[\underline{\text{N}}\text{—H}\right]^{2-}$ imide ion[b] NH^{2-}

[a] NH$_2$ as substituent in organic compounds is called an amino group.
[b] NH as substituent in organic compounds is called an imino group.

Liquid ammonia shows many analogies with water. As mentioned in 5-5 its molecules, like those of water, are linked by hydrogen bonds and this leads here also to higher melting and boiling points than would otherwise be expected (13-4b). Since the hydrogen bonds in ammonia are weaker than in water, those properties that depend on the association are less pronounced than for water. The dielectric constant for liquid ammonia is significantly less than for water (at room temperature: ammonia, 17; water, 80) but nevertheless is large enough so that liquid ammonia is a quite good solvent for electrolytes (13-1c). However, the relative solubility of different substances is often quite different in liquid ammonia and in water. For example, AgI is very soluble in liquid ammonia but CaI$_2$ is difficultly soluble. When ammonia solutions of AgI and Ca(NO$_3$)$_2$ are mixed, CaI$_2$ therefore precipitates out.

As described in 12-2f ammonia is part of the acid-base system:

$$MH_4^+ \quad - \quad NH_3 \quad - \quad NH_2^- \quad - \quad NH^{2-} \quad - \quad N^{3-}$$

ammonium ion — ammonia — amide ion — imide ion — nitride ion

analogous to the water system:

$$H_3O^+ \quad - \quad H_2O \quad - \quad OH^- \quad - \quad O^{2-}$$

oxonium ion — water — hydroxide ion — oxide ion

Molecules with the same charge have been lined up vertically. In many respects, however, the hydroxide ion corresponds to both the amide and imide ions, just as the oxide ion corresponds to the nitride ion. These properties of the ammonia system lead to the result that a whole series of acid-base processes are analogous in liquid ammonia and in water.

Thus, in liquid ammonia the autoprotolysis equilibrium $2NH_3 \rightleftharpoons NH_4^+ + NH_2^-$ exists. Just as the oxonium ion is a strong acid in water solution, the ammonium ion is a strong acid in liquid ammonia. Solutions of ammonium salts in liquid ammonia therefore dissolve many metals with the evolution of hydrogen. Ammonium ion can be neutralized by amide ion, which means that the above equilibrium is shifted to the left. In the same way that metal hydroxides can be precipitated by hydroxide ion from solutions of metal salts in water, metal amides can be precipitated by amide ion from solutions of metal salts in liquid ammonia: $Ca^{2+} + 2NH_2^- \rightarrow Ca(NH_2)_2(s)$. Sometimes imides or nitrides precipitate (Pb^{2+} + $2NH_2^- \rightarrow PbNH(s) + NH_3$), corresponding to the occasional cases where oxides are precipitated by addition of hydroxide ($2Ag^+ + 2OH^- \rightarrow Ag_2O(s) + H_2O$). Amides of alkali and alkaline-earth metals are formed by the action of ammonia gas with heat on the metals: $Na(l) + NH_3(g) \rightarrow NaNH_2(s) + \frac{1}{2}H_2(g)$.

A large number of metal cations form *ammine complexes* (5-2h, 5-3j, 5-6c; note the spelling with two *m*'s) in liquid ammonia, just as they form aquo complexes in water. Thus, when AgI dissolves in liquid ammonia the ion $Ag(NH_3)_2^+$ is formed. Ammine complexes of metal cations are also formed in ammoniacal water solutions. The equilibria which then arise were discussed in 10c.

Liquid ammonia is very remarkable in its ability to dissolve alkali metals, calcium, strontium and barium. When dilute such solutions are blue, and concentrated solutions have a more metallic, coppery color resulting from high reflectivity. The absorption spectrum is independent of the kind of metal dissolved. In all proportions the equivalent conductivity (11c) is higher than for any salt solutions and the concentrated solutions have metallic conductivity. The latter therefore behave almost like metals.

It is assumed that the metal atoms in these solutions are ionized with the formation of, for example, Na^+, which forms the ammine ion $Na(NH_3)_6^+ + e^-$. The electrons are assumed to be located in cavities in the liquid structure of the ammonia ("solvated electrons"; cf. F centers in alkali halides with excess of metal, 6-3a, 20-5a), but are easily displaced so that the conductivity is high. The electrons with their environment are responsible for the blue color (here also cf. F centers).

If the solutions are pure, they last some days or weeks. Decomposition takes place slowly with the formation of amides and the evolution of hydrogen ($Na^+ + e^- + NH_3 \rightarrow NaNH_2 + \frac{1}{2}H_2$) and is strongly catalyzed by certain substances, for example Fe, Pt and Fe_2O_3, and by ultraviolet radiation. The solutions are powerful reducing agents and are used in the reduction of organic substances.

Ammonia dissolves very easily in water with the evolution of heat. At 20°C and total pressure of 1 atm, 100 g water dissolves 53.1 g NH_3, corresponding to 702 times the volume of the water. In the solution NH_3 functions as a base, so that the equilibrium $NH_3 + H_2O \rightleftharpoons NH_4^+ + OH^-$ is set up. This has been previously considered in several connections in 12-1, 12-2 and 12-3c. The existence of ammonium hydroxide NH_4OH cannot be assumed. In the system NH_3–H_2O the hydrates $NH_3 \cdot \frac{1}{2}H_2O$ and $NH_3 \cdot H_2O$ crystallize on cooling to $-80°C$. In both these compounds the water molecules appear as such, so that they must be considered as true hydrates and not as ammonium oxide $(NH_4)_2O$ and ammonium hydroxide NH_4OH, respectively.

As is apparent from the ΔG^0 value for the process

$$\tfrac{1}{2}N_2(g) + \tfrac{3}{2}H_2(g) \rightleftharpoons NH_3(g); \quad \Delta H^0 = -46.2 \text{ kJ}; \quad \Delta G^0 = -16.6 \text{ kJ} \quad (22.1)$$

at ordinary temperature the equilibrium is strongly displaced to the right. According to (10.13), from this ΔG^0 value at 25°C the equilibrium constant is $K_p = 822$ atm^{-1}. Since $\Delta H^0 < 0$, the equilibrium is shifted to the left with increase in temperature (10d). On heating to moderate temperatures, however, the new equilibrium is established very slowly, but the rate can be strongly increased by the presence of catalysts.

Ammonia burns in oxygen with a pale yellow flame, to give nitrogen: $2NH_3 + \tfrac{3}{2}O_2 \rightarrow N_2 + 3H_2O$. In air the heat of combustion generally is not sufficient to keep the temperature high enough so that the combustion will continue; heat must be supplied, and ammonia will burn, for example, when it is fed into a gas flame. However, not only ammonia-oxygen mixtures but also ammonia-air mixtures are explosive when their compositions lie within certain limits.

Certain substances catalyze the formation of nitrogen oxide on the oxidation of ammonia by air, so that the process becomes $2NH_3 + \tfrac{5}{2}O_2 \rightarrow 2NO + 3H_2O$. Since nitrogen oxide can be converted to nitric acid (22-9d, 22-10d1) this process has become very important commercially (*Ostwald process*). Either Pt with added Rh, or Fe_2O_3 with added Bi_2O_3, are used as a catalyst industrially. The temperature is held at 700–800°C and since the equilibrium (22.1) at this temperature is strongly displaced to the left, the heating of the ammonia-air mixture and the combustion must be carried out so rapidly that no appreciable decomposition of ammonia is able to occur. The catalyst must thus be very effective.

Ammonia is also easily oxidized by other oxidizing agents than oxygen. If chlorine is led into ammonia gas, spontaneous ignition occurs: $2NH_3 + 3Cl_2 \rightarrow$

$N_2 + 6HCl$; $NH_3 + HCl \rightarrow NH_4Cl(s)$. Oxidation also occurs in water solution, giving various oxidation products (for example, hydrazine, nitrogen, nitrite ion, nitrate ion).

b. Production and uses of ammonia. On a laboratory scale ammonia is most simply prepared by decomposition of an ammonium salt with a strong base: $NH_4^+ + OH^- \rightarrow NH_3 + H_2O$. Often a mixture of ammonium chloride and calcium hydroxide is heated. When larger quantities are required, however, it is more appropriate to draw it from liquid ammonia, which is nowadays marketed in steel cylinders.

The world's demand for nitrogen is accounted for at the present time almost entirely by ammonia, and the major portion of this nowadays is synthetic. Nitrogen products are used most especially for two purposes: nitrogenous fertilizers and explosives. Up to World War I Chile saltpeter (22-2a) was the greatest source of nitrogen, and next to that came ammonia, which was obtained as a by-product at gas and coke plants and which was converted to ammonium sulfate. The danger that these sources of nitrogen would become exhausted led in the beginning of the 20th century to attempts to fix the nitrogen in the air, and the result was three substantially different methods. In the *Birkeland-Eyde process* atmospheric nitrogen and oxygen were combined into nitrogen oxides in an electric arc (22-9b); in the *calcium-cyanamide process* atmospheric nitrogen was combined with calcium carbide to give calcium cyanamide (23-2k6). The Birkeland-Eyde process is nowadays no longer economically competitive and has been abandoned. The calcium-cyanamide process, on the other hand, is still used, although far less extensively than the third method, ammonia synthesis by the *Haber-Bosch process*. The world production of combined nitrogen in the form of Haber-Bosch ammonia, by-product ammonia, calcium cyanamide and Chile saltpeter in 1967 reached 22 Mt in terms of elementary nitrogen. Of this, Haber-Bosch ammonia accounted for more than 90%.

In the Haber-Bosch process ammonia is formed from nitrogen and hydrogen according to the equilibrium (22.1). In order to obtain a high proportion of ammonia in the gas mixture, according to 10d the pressure should be high and the temperature low. A reasonable proportion of ammonia can only be attained at so low a temperature that the establishment of equilibrium must be hastened with catalysts. As mentioned earlier in 14-4, Fe with added oxides, especially Al_2O_3, is used as the catalyst. In most versions of the method now in use the pressure is 200–300 bar and the temperature about 500°C (at 300 bar and 500°C the equilibrium mixture contains 26% ammonia). In the technical development of the process great difficulty was encountered in finding a material for the synthesis apparatus that was resistant to hydrogen under these conditions. Nowadays steel alloyed with nickel and chromium is mostly used.

The cost of elementary hydrogen is of great importance for the economy of the Haber-Bosch process. Depending on the price of power and the value of the by-products one or another of the different methods of hydrogen production mentioned in 18d is chosen. A mixture of nitrogen and hydrogen can be had directly by starting with either water gas (18d) together with producer gas (23-2f) or with coke-oven gas (18d). Nitrogen can also be produced separately from liquid air (29-4a), at the same time obtaining liquid oxygen, which has a rapidly growing industrial use.

The ammonia formed in the Haber-Bosch process is condensed by cooling (still at high pressure) and stored and transported in liquid form. In gas and coke plants the ammonia is first absorbed in water (gas water), whereby the gas is washed. The ammonia is present in this solution mostly as ammonium carbonate and is liberated by distillation after adding calcium hydroxide (milk of lime).

Ammonia is converted mainly to ammonium salts and urea (23-2g), both of great importance as fertilizers, and to nitric acid and nitrates (22-10d). Nitric acid is used in the explosives industry as well as in other large areas of both inorganic and organic chemical industry; nitrates are mainly used in fertilizers. In recent years ammonia gas obtained directly from liquid ammonia has begun more and more to be added to the earth as fertilizer.

Because of its high proportion of hydrogen per unit of volume, liquid ammonia is often a significantly more economical medium for the transport of hydrogen than compressed hydrogen gas. At 25°C 1 l of liquid ammonia contains 106 g hydrogen while 1 l of hydrogen gas at 150 atm. contains only 12.2 g hydrogen. Before use the ammonia is dissociated by heating over a catalyst (finely divided iron or nickel), and the mixture of nitrogen and hydrogen used in cases where the nitrogen content does not interfere, for example in reduction of oxides and protection against oxidation, especially in the heat treatment of metals.

For larger refrigeration installations ammonia is the most-used cooling medium. For this purpose its very high heat of vaporization is exploited, besides which the boiling and freezing points have convenient values.

c. Ammonium compounds. The ammonium ion NH_4^+ in many respects is similar to the alkali-metal ions. It is tetrahedral but behaves as though it has a nearly spherical effective surface with a radius of approximately the same size as that of the rubidium ion. Ammonium ions therefore easily replace rubidium and also potassium ions in crystals so that solid solutions are formed. Like the alkali-metal ions, the ammonium ion is colorless.

Ammonium salts thus resemble the alkali salts, for example in solubility (most are easily soluble in water) and in their crystal structures (6-2c). However, they readily decompose on heating. If the anion is not particularly oxidizing (for example, halide ion, SO_4^{2-}, CO_3^{2-}), ammonia is then formed ($NH_4Cl \rightarrow NH_3 + HCl$). If the anion is more strongly oxidizing (for example, NO_2^-, NO_3^-, ClO_3^-, $Cr_2O_7^{2-}$), it oxidizes the ammonium ion to nitrogen or some nitrogen oxide ($NH_4NO_2 \rightarrow N_2 + 2H_2O$; $NH_4NO_3 \rightarrow N_2O + 2H_2O$). All ammonium salts of the latter category should be considered as possible detonators and handled carefully. Ammonium salts are mostly prepared by passing ammonia gas into a solution of the corresponding acid.

The acid properties of the ammonium ion also distinguish it from the alkali ions. However, the ammonium ion in water solution is so weak an acid (tab. 12.2) that the reaction becomes only slightly acid. For example, a 0.1 M solution of NH_4Cl has $pH=5.2$, and does not taste acid. However, if a strong base is added to an ammonium salt solution, ammonia is formed ($NH_4^+ + OH^- \rightarrow NH_3 + H_2O$), and with sufficiently high ammonia concentration ammonia gas is given off.

If the hydrogen atoms in the ion NH_4^+ are replaced by organic radicals R, the ions RNH_3^+, $R_2NH_2^+$ and R_3NH^+ are obtained also as weak acids, while the ion R_4N^+, in which all four hydrogen atoms have been replaced, is no longer an acid at all. Ions of this type can therefore exist in strongly basic solution and, for example, tetramethylammonium hydroxide $(CH_3)_4NOH$ is a strong electrolyte, which can be isolated in solid form.

The similarity of the ammonium ion to the alkali-metal ions has led to attempts to prepare the radical *ammonium* NH_4 in metallic form, corresponding to the alkali metals, but so far without success. However, ammonium can exist in *ammonium amalgam*, which can be prepared either by treating alkali metal amalgam with conc. ammonium chloride solution ($Na(\text{in Hg}) + NH_4^+ \rightarrow NH_4(\text{in Hg}) + Na^+$), or by electrolysis of an ammonium salt dissolved in liquid ammonia with a mercury cathode ($NH_4^+ + e^- \rightarrow NH_4$ (in Hg)). At room temperature ammonium amalgam is a spongy, soft mass, which quite rapidly decomposes ($NH_4(\text{in Hg}) \rightarrow NH_3 + \tfrac{1}{2}H_2$), but at $-85°C$, for example, it is hard and stable.

The following ammonium salts should be mentioned here in particular:

Ammonium chloride NH_4Cl has been called since early times sal ammoniac (Lat. sal ammoniacum, salt from Ammon, that is, from the Ammon Temple in the Siva Oasis). Ammonium chloride is easily sublimed and on condensation forms a fibrous crystalline mass, which is the usual commercial form. In the gas the dissociation equilibrium $NH_4Cl \rightleftharpoons NH_3 + HCl$ prevails. The dissociation is considered to be practically complete when the temperature rises to $250°C$. When gaseous NH_3 and HCl are mixed at room temperature, a smoke of sal ammoniac is formed. Ammonium chloride is produced in this way, or also extracted as a by-product in the ammonia-soda process (Solvay process, 25-5 b).

Many metal oxides on being heated with sal ammoniac give volatile chlorides. In this way an oxidized metal surface can be freed of oxide coating after which it can be wetted by tin in soldering or tinning. Sal ammoniac is therefore used as a soldering flux. However, the greater use of sal ammoniac is in Leclanché cells (16g). Taken internally, sal ammoniac facilitates loosening of phlegm in the lung passages.

Ammonium perchlorate NH_4ClO_4 is used as an oxidizing agent in solid rocket propellants (20-8e). *Ammonium sulfate* $(NH_4)_2SO_4$ has wide applications as nitrogenous fertilizer and is, therefore, produced in larger amounts than any other ammonium salt. *Ammonium nitrate* NH_4NO_3 decomposes on heating to about

250°C in a controllable way according to $NH_4NO_3 \rightarrow N_2O + 2H_2O$. With rapid heating to temperatures above about 300°C or under the action of a primary detonator, the decomposition $NH_4NO_3 \rightarrow N_2 + \frac{1}{2}O_2 + 2H_2O$ ($\Delta H = -117$ kJ) takes place with explosion. Decomposition with explosion causing very great damage (Oppau, 1921; Texas City, 1947) has occurred under circumstances which show that storage of larger amounts of ammonium nitrate should be avoided. For this reason ammonium nitrate is no longer used in pure form as a fertilizer, but rather mixed with, for example, ammonium sulfate or calcium carbonate. The hygroscopic properties of pure ammonium nitrate, which make it difficult to store and to spread, are also eliminated in this way. Large amounts of ammonium nitrate are used in the explosives industry, both as an additive to organic explosives, and as an explosive itself in mixture with easily oxidizable substances (nowadays especially about 6 wt% diesel oil or aluminum powder). *Diammonium hydrogen phosphate* $(NH_4)_2HPO_4$ and *ammonium dihydrogen phosphate* $NH_4H_2PO_4$ are used as fertilizers (see also 22-12e4). These phosphates are also used to fireproof wooden structures and to make textiles and paper fire-resistant. On heating they give off ammonia and leave a phosphoric-acid-rich protective coating. The last five ammonium salts mentioned are produced by passing ammonia into a solution of the corresponding acid.

Ammonium carbonate $(NH_4)_2CO_3$, *ammonium hydrogen carbonate* NH_4HCO_3 and *ammonium amidocarbonate* NH_2COONH_4. The last is the ammonium salt of amidocarbonic acid NH_2COOH (23-2g). If carbon dioxide is reacted with gaseous or liquid ammonia, ammonium amidocarbonate is formed according to $2NH_3 + CO_2 \rightarrow NH_2COONH_4(s)$. Also, if carbon dioxide is passed into an ammonia solution this reaction predominates since it takes place much faster than the hydration of CO_2 to H_2CO_3, which must precede the formation of carbonate and hydrogen carbonate. However, there exists in the solution the equilibrium

$$NH_2COO^- + H_2O \underset{+NH_3}{\overset{-NH_3}{\rightleftarrows}} HCO_3^- \underset{+H^+}{\overset{-H^+}{\rightleftarrows}} CO_3^{2-}$$

from which it is apparent that the formation of HCO_3^- and CO_3^{2-} is favored by removal of NH_3 (for example, by boiling) and addition of base. Solid ammonium carbonate gives off ammonia even at room temperature with the formation of the hydrogen carbonate, which at about 60°C decomposes into ammonia, carbon dioxide and water according to

$$(NH_4)_2CO_3 \xrightarrow{-NH_3} NH_4HCO_3 \rightarrow NH_3 + CO_2 + H_2O$$

The commercial product *hartshorn salt* (formerly prepared by dry distillation of horn) contains mainly amidocarbonate and hydrogen carbonate, possibly as a double salt. It smells of ammonia (it has been used as smelling salts) and its decomposition into gases on heating makes it useful for baking powder.

d. Hydrazine. The structure of hydrazine N_2H_4 (tab. 22.1) also suggests the name diamide. At room temperature hydrazine is a colorless liquid, which fumes

Group 15, The Nitrogen Group

However, the difference between the electronegativities of these elements on the one hand, especially phosphorus, and of hydrogen on the other, is very small (5-3f). Some trivial names such as phosphine, arsine and stibine are still accepted.

The hydrogen compounds of the elements in the nitrogen group become less and less stable to decomposition into the elements, the farther down in the group the element lies. The hydrogen compounds of phosphorus are already significantly less stable than those of nitrogen, and for bismuth they are extremely unstable. At the same time their reducing power increases.

a. Phosphorus hydrides. The most common is *phosphorus trihydride* (*phosphine*) PH_3 with the same structure as ammonia. At room temperature it is a colorless, poisonous gas with an odor that is commonly said to be carbidelike, because calcium carbide most often contains traces of calcium phosphide which with water gives phosphine (the simultaneously formed acetylene is odorless). Condensed phosphine boils at $-87.8°C$ and solidifies at $-133.8°C$. These temperatures are lower than the corresponding ones for ammonia (22-6a; cf. also fig. 13.3 and 13.4) although the molecular weight is higher for PH_3 than for NH_3. As mentioned in 13-4b this is because of the molecular association due to hydrogen bonds in NH_3, which is much weaker in PH_3.

Phosphine is a much weaker base than ammonia. This implies that the *phosphonium ion* PH_4^+ is so strong an acid that the equilibrium $PH_4^+ \rightleftharpoons PH_3 + H^+$ is strongly displaced to the right. Solid phosphonium halides are known, but they dissociate even at low temperature and on solution in water give off practically all their PH_3. Phosphine is a stronger reducing agent than ammonia. This is shown by the fact that it easily ignites itself in air, burning to give phosphoric acid: $PH_3 + 2O_2 \rightarrow H_3PO_4$. Noble-metal ions in water solution are reduced to metal (mixed with phosphide) if phosphine is passed into the solution.

Phosphine can be prepared by hydrolysis of phosphides, for example of calcium phosphide with water, or better, aluminum phosphide with dilute acid: $P^{3-} + 3H^+ \rightarrow PH_3$. Another method is by heating white phosphorus with a solution of a strong base, on which the phosphorus disproportionates: $4P^0 + 3OH^- + 3H_2O \rightarrow P^{-III}H_3 + 3P^IH_2O_2^-$ (hypophosphite ion, 22-12b).

In the preparation of phosphine by these methods there is also formed a small amount of *diphosphorus tetrahydride* or *diphosphine* P_2H_4, which is analogous to hydrazine. Diphosphine ignites spontaneously in air and therefore makes the mixture self-igniting. However, the diphosphine can easily be removed by condensation (b.p. 52°C; diphosphine has therefore been called "liquid phosphorus hydride").

Diphosphine decomposes easily, especially on exposure to light, to phosphine and so-called "solid phosphorus hydride", which can be considered as a high polymer (1-4a) of phosphorus and hydrogen. Its composition often lies close to P_2H.

If the hydrogen atoms in phosphines or phosphonium ions are replaced by organic radicals, appreciably more stable derivatives are obtained.

b. Hydrides of arsenic, antimony and bismuth. For arsenic, antimony and bismuth the *trihydrides* are known, the two first ones being called also *arsine* and *stibine*, respectively. At room temperature they are all poisonous, foul-smelling gases. They easily decompose into the corresponding element and hydrogen, the arsenic and antimony hydrides on heating, bismuth hydride even at room temperature. With increasing molecular weight the reducing power also increases. Here also, increased stability is obtained by replacing hydrogen by organic radicals.

The trihydrides are formed when arsenides, antimonides or bismuth alloys with hydrogen-displacing metals are dissolved in acids. They are also obtained when soluble compounds of the element in question are reduced with zinc in acid solution. Because of the great instability of bismuth trihydride, however, its yield is very small.

The last-named method of preparation has been used for detecting the elements, especially arsenic ("*Marsh test*"). Hydrochloric acid and zinc is added to the sample to be tested, and the gas developed passed through a glass tube that is heated at one point. If the sample contains arsenic, antimony or bismuth, the gas contains the corresponding trihydride, which decomposes on heating. Where the tube following the hot region is again cool enough, a black, lustrous film of the element is formed (arsenic, antimony or bismuth mirror; of these the arsenic mirror is appreciably more volatile than the others, but they can be more certainly distinguished by the fact that only the arsenic mirror is soluble in sodium hypochlorite solution, while only the arsenic and antimony mirrors dissolve in yellow ammonium polysulfide solution).

Diarsine As_2H_4 is also known, which above $-100°C$ decomposes into AsH_3 and a solid high polymer with a composition near the formula As_2H.

22-8. Halogen compounds of the elements of the nitrogen group

a. Compounds with nitrogen-halogen bonds. If hydrogen in ammonia is replaced by halogen we obtain

$$\begin{array}{ccc} H & |\underline{X}| & |\underline{X}| \\ | & | & | \\ |N-\underline{X}| & |N-\underline{X}| & |N-\underline{X}| \\ | & | & | \\ H & H & |\underline{X}| \end{array}$$

for which several representatives are found although not all have been isolated. According to the rules in 5-2a and 5-3f the compounds NX_3 should be called nitrogen halides although the relative electronegativity values of NBr_3 and NI_3 would suggest their designation as nitrides. The above formulations indicate nearly tetrahedral bond angles (sp^3 hybridization) and thus pyramidal structures (fig. 5.31j).

The general method of formation is by the action of halogen on ammonia or ammonium ion. *Nitrogen trichloride* NCl_3 and *chloramine* NH_2Cl (for preparation see

22-6d; not to be confused with organic chloramines which are used as antiseptics) at room temperature are oily liquids, which are extremely explosive. *Nitrogen trifluoride* NF_3 (under normal conditions a gas) on the other hand is amazingly stable, a fact which is probably related to the greater ionic contribution. Pure *nitrogen triiodide* NI_3 has probably not been synthesized. When iodine is treated with ammonia solution solid and extremely explosive products containing less iodine than what would correspond to NI_3 are obtained, among other things, $NI_3 \cdot NH_3$. The hydrolysis of NF_3 and NCl_3 is discussed in 20-5a.

Nitrogen pentahalides NX_5 cannot be synthesized since the nitrogen atom cannot offer more than the four $2s2p^3$ orbitals for bonding (5-3i).

Concerning *nitrosyl halides*, see 22-9b.

b. Phosphorus halides. These are usually formed by direct combination of the elements. The most important are

$$\begin{array}{c} |\underline{X}| \\ | \\ |P-\underline{X}| \\ | \\ |\underline{X}| \end{array} \quad \text{and} \quad \begin{array}{c} |\underline{X}| \quad |\underline{X}| \\ \diagdown \, \diagup \\ P \\ \diagup \,|\, \diagdown \\ |\underline{X} \quad | \quad \underline{X}| \\ |\underline{X}| \end{array}$$

X in these formulas can represent any halogen, except that PI_5 is unknown. Several different halogens may also enter into the same molecule. Concerning the bonding conditions reference should be made to 5-3i.

If chlorine is passed over heated white phosphorus mainly *phosphorus trichloride* PCl_3 is formed, but also in smaller amounts *phosphorus pentachloride* PCl_5. They go off as gases and can easily be condensed in an ice-cooled trap, the trichloride as a colorless liquid (b.p. 76°C, m.p. −93.6°C), the pentachloride as a solid mass. To obtain pure trichloride some white phosphorus is added to the distillate ($3PCl_5 + 2P \rightarrow 5PCl_3$) and a new distillation carried out.

The trichloride is easily hydrolyzed by water ($PCl_3 + 3H_2O \rightarrow H_2PHO_3 + 3HCl$; for the formulation H_2PHO_3 see 22-12a) and therefore fumes in moist air. With chlorine it sets up the equilibrium $PCl_3 + Cl_2 \rightleftharpoons PCl_5$ (the lone electron pair of the phosphorus atom of the trichloride makes it an electron donor), which at room temperature is displaced rather far to the right.

The pentachloride forms greenish-white, easily sublimed crystals. The above-mentioned equilibrium is shifted to the left on raising the temperature, and at 300°C and atmospheric pressure the pentachloride vapor is nearly completely dissociated. The pentachloride is hydrolyzed by water (cf. 20-5a), on which the first step is

$$\begin{array}{c} Cl \\ \diagdown \\ Cl-P \\ \diagup \\ Cl \end{array} \begin{array}{c} Cl \quad H \\ \diagdown \\ + O \\ \diagup \\ Cl \quad H \end{array} \longrightarrow \begin{array}{c} Cl \\ | \\ Cl-P-O + 2HCl \\ | \\ Cl \end{array}$$

The phosphoryl(V) chloride $POCl_3$ (see below) that is formed reacts with additional water: $POCl_3 + 3H_2O \rightarrow H_3PO_4 + 3HCl$. Analogous processes occur with many other, including organic, substances that contain OH groups:

$$\begin{array}{c} Cl \\ \diagdown \\ Cl-P \\ \diagup \\ Cl \end{array} \quad \begin{array}{c} Cl \quad H \\ \diagup \\ +O \\ \diagdown \\ Cl \quad R \end{array} \quad \longrightarrow \quad POCl_3 + HCl + RCl$$

Phosphorus pentachloride is therefore much used for replacing OH groups in organic compounds with Cl.

Phosphoryl(V) *chloride* $POCl_3$ (see above) can also be prepared according to the reaction $P_2O_5 + 3PCl_5 \rightarrow 5POCl_3$. At room temperature and atmospheric pressure it is a colorless liquid (b.p. 105.8°C, m.p. 1.1°C). Phosphoryl(V) chloride is also used analogously with the pentachloride in preparative organic work.

c. Halides of arsenic, antimony and bismuth. Arsenic halides are prepared by combination of the elements. The best known is *arsenic trichloride* $AsCl_3$ (colorless liquid with b.p. 130°C). However it is more simply prepared by passing hydrogen chloride over heated diarsenic trioxide: $As_2O_3 + 6HCl \rightleftharpoons 2AsCl_3 + 3H_2O$. In water solution this equilibrium is displaced far to the left, but all of the arsenic can be converted to arsenic trichloride with conc. hydrochloric acid if the trichloride is distilled off.

Antimony trichloride $SbCl_3$ is often prepared by dissolving powdered stibnite in hot, conc. hydrochloric acid: $Sb_2S_3 + 6HCl \rightarrow 2SbCl_3 + 3H_2S$. It forms a colorless, soft but crystalline mass ("antimony butter") with melting point 73.5°C and boiling point 219°C. $SbCl_3$ with a small amount of water gives a clear solution, but on dilution this precipitates oxide chlorides (21-5d; cf. also 22-13c), for example SbOCl and $Sb_4O_5Cl_2$ ("Algaroth powder", formerly used in medicine). If the solution is boiled, all of the antimony is converted to Sb_2O_3.

Antimony pentachloride $SbCl_5$ (at ordinary temperature a lemon-yellow liquid) is obtained by treating the trichloride with chlorine: $SbCl_3 + Cl_2 \rightleftharpoons SbCl_5$. It easily adds many atoms or atomic groups so that the antimony atom attains octahedral coordination; for example, $SbCl_5 + Cl^- \rightarrow SbCl_6^-$ (*hexachloroantimonate*(V) *ion*).

Bismuth halides BiX_3 are formed by solution of bismuth oxide in hydrogen halide acids: $Bi_2O_3 + 6HX \rightleftharpoons 2BiX_3 + 3H_2O$. They are decomposed by water to oxide halides (22-13d): $BiX_3 + H_2O \rightleftharpoons BiOX + 2HX$.

22-9. Nitrogen oxides

Nitrogen forms probably at least eight oxides, of which the most important are listed in tab. 22.2. Concerning the structure of N_2O_3 some doubt still prevails. The structure shown for N_2O_4 is that most stable under ordinary conditions.

Table 22.2. *The most important oxides of nitrogen.*
A number of easily derived, possible mesomers have not been included.

Oxid. no. of nitrogen	
V	dinitrogen pentoxide N_2O_5
IV	nitrogen dioxide NO_2 dinitrogen tetroxide N_2O_4
III	dinitrogen trioxide N_2O_3
II	$\|N{\cdots}O\|$ nitrogen oxide NO
I average	$\underline{N}{=}N{=}\overline{O} \leftrightarrow \|N{\equiv}N{-}\overline{O}\|$ dinitrogen oxide N_2O

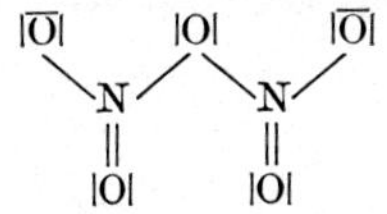

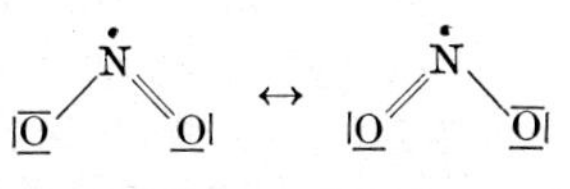

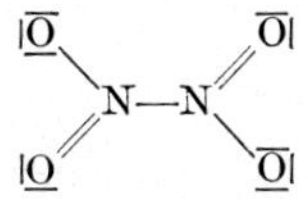

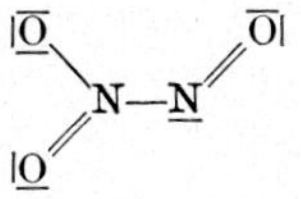

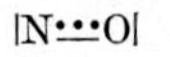

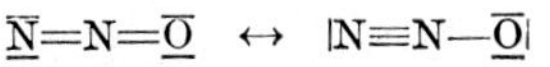

a. Dinitrogen oxide. Dinitrogen oxide was formerly called nitrous oxide. The molecule N_2O is linear and its structure is often described as a mesomerism between the two limiting structures shown in tab. 22.2. A third conceivable limiting structure, $|\overline{N}-N\equiv O|$, is usually neglected because the two neighbors around the triple bond would have the formal charge $1+(5\text{-}3g)$.

Dinitrogen oxide at room temperature consists of a colorless gas with a faint, sweetish odor. The critical point lies at 36.5°C and 72.6 bar, so that the gas is easily condensed by compression. It is marketed in liquid form in steel cylinders. The boiling point of the liquid is $-89.5°C$. If dinitrogen oxide is inhaled in small amounts it causes a kind of intoxication with a tendency for convulsive laughter (*laughing gas*). In larger amounts it acts as a narcotic and is used very much as an anesthetic.

For the formation of dinitrogen oxide from the elements according to $N_2(g)+\frac{1}{2}O_2(g)\rightleftharpoons N_2O(g)$, $\Delta H^0=81.6$ kJ and $\Delta G^0=103.6$ kJ. It is thus unstable with respect to decomposition into its elements. In fact, it readily gives up oxygen to easily oxidizable substances. Sulfur, phosphorus and carbon burn in it almost as vigorously as in oxygen gas. With hydrogen or ammonia, explosive mixtures are produced. However, dinitrogen oxide cannot support respiration, so that oxygen must always be added when it is used as an anesthetic.

Dinitrogen oxide is most easily prepared by careful heating of ammonium nitrate (reaction equation in 22-6e, where also the reaction is given for explosive decomposition, which can occur on strong heating of the nitrate).

b. Nitrogen oxide. Nitrogen oxide NO at room temperature is a colorless gas. NO is an odd molecule (5-3a), and the two atoms can be considered to be bonded to each other by a three-electron bond and two single bonds (5-3a; formula also shown in tab. 22.2). The unpaired electron makes the molecule paramagnetic. In several other respects, among others in the lack of color and relatively low reactivity, however, nitrogen oxide departs from the usual behavior of odd molecules. Liquid nitrogen oxide (which is more intensely blue than liquid oxygen) consists for the most part of dimeric molecules. This tendency for association causes nitrogen oxide to be more easily condensed than nitrogen or oxygen (crit. temp. $-93°C$, b.p. $-151.7°C$, m.p. $-163.7°C$).

For the sytem $\frac{1}{2}N_2(g) + \frac{1}{2}O_2(g) \rightleftharpoons NO(g)$, $\Delta H^0 = 90.4$ kJ and $\Delta G^0 = 86.7$ kJ, which shows that at ordinary temperature nitrogen oxide is very unstable with respect to decomposition into the elements. According to (10.13), with this ΔG^0 value at 25°C the equilibrium constant is $K_p = 6.48 \times 10^{-16}$. At room temperature, however, the decomposition rate is very low. Since $\Delta H^0 > 0$, the equilibrium is shifted to the right on increase in temperature (10d). Thus, air heated to 4000°C contains 14 vol% NO at equilibrium, and here equilibrium is established very rapidly. Therefore, if air is passed into a region of extremely high temperature, for example an electric arc, and then quenched so that its temperature reaches the value corresponding to low reaction rate before the equilibrium has a chance to shift back to the left, a large part of the nitrogen oxide formed in the arc can be recovered. This is the basis for the *Birkeland-Eyde process* for the fixation of atmospheric nitrogen, although nowadays this method is no longer of practical importance (22-6b).

When nitrogen oxide at ordinary temperature comes in contact with oxygen, nitrogen dioxide NO_2, which is brown in color, is immediately formed:

$$NO(g) + \tfrac{1}{2}O_2(g) \rightleftharpoons NO_2(g); \quad \Delta H^0 = -56{,}5 \text{ kJ}; \quad \Delta G^0 = -34.9 \text{ kJ} \qquad (22.2)$$

For 25°C, (10.13) gives the equilibrium constant $K_p = 1.3 \times 10^6$ atm$^{-\frac{1}{2}}$, but since $\Delta H^0 < 0$ the equilibrium is shifted to the left on heating (10d) and at approximately 500°C, $K_p = 1$ atm$^{-\frac{1}{2}}$.

Strong reducing agents can remove the oxygen from nitrogen oxide. Thus, phosphorus and magnesium will burn in nitrogen oxide.

The technical production of nitrogen oxide is extraordinarily important since it forms the basis for the production of nitric acid and nitrates (22-10d). This is carried out nowadays by the oxidation of ammonia (*Ostwald process*, 22-6a) In the laboratory nitrogen oxide is usually prepared by reduction of nitric acid:

$NO_3^- + 4H^+ + 3e^- \rightarrow NO + 2H_2O$; $e^0 = 0.96$ V. The reducing agent must belong to a redox pair with $e^0 < 0.96$ V, and to avoid the evolution of hydrogen gas, it should also have $e^0 > 0$ V. Copper is commonly used ($e^0 = 0.337$ V). The concentration of nitric acid is always important for the course that the reduction takes. Dilute acid gives rather pure nitrogen oxide and concentrated acid gives predominantly nitrogen dioxide. Intermediate strengths give mixtures of the two gases. However, in addition, other reduction products are formed (22-10d 1). Nitrogen oxide is only slightly water soluble and therefore can be collected over water, while at the same time most of the other reduction products are dissolved. Purer nitrogen oxide is obtained by reduction of nitric acid with Fe^{2+} ion ($e^0 = 0.771$ V). This reduction process is utilized to detect nitric acid and nitrate ion (after acidification with conc. sulfuric acid). In this test a solution of iron(II) sulfate is introduced as a layer over the acid solution, and if nitrogen oxide is formed this gives in the layer boundary a brown color from the ion $Fe(NO)^{2+}$ (see below; the same reaction is also given by nitrite ion, which is also reduced to nitrogen oxide by Fe^{2+} ion).

The *nitrosyl group* NO is often bound to other atoms or atomic groups, in which its function may be very variable. This results partly from the unpaired electron in the NO molecule. This unpaired electron can be removed with the formation of *nitrosyl cation* $[|N \equiv O|]^+$, which is isoelectronic with N_2, CO and CN^- and has a structure analogous to that of these molecules (5-3a). But one electron may also be taken up with the formation of an *oxonitrate*(I) *ion* $[\overline{N} = \overline{O}]^-$. The nitrosyl group may also be bound with a covalent bond, which is believed always to be associated with the nitrogen atom. Most often there is both covalent- and ionic-bond contribution, in which either of the ions mentioned can appear. Such versatility is uncommon, and may be most nearly compared to that of the hydrogen atom, which can exist as H^+, H^- and as H in covalent bonding. The nitrosyl group can also probably be assigned an electronegativity value of approximately the same intermediate value as hydrogen.

A fairly typical NO^- ion is only known in compounds with very electropositive metals, for example in NaNO. The NO^+ ion, on the other hand, is common in bonds with more or less covalent contribution. With fluorine, chlorine and bromine, nitrogen oxide gives *nitrosyl halides* $\overline{O} = N\begin{subarray}{c}|\overline{X}|\\|\end{subarray}$, usually written NOX, in which the covalent contribution, as in the hydrogen halides, is quite high. More typically, NO^+ appears in *nitrosyl perchlorate* $NOClO_4$, which can be written $NO^+ClO_4^-$, and *nitrosyl hydrogen sulfate* $NOHSO_4$ ($NO^+HSO_4^-$, 21-11b 2).

In many cases where the NO group is a ligand to atoms of transition metals, it probably functions as NO^+. The odd electron is then absorbed into the unfilled d shell of the metal atom so that the NO group in itself no longer causes paramagnetism. An example of this is $[Fe(CN)_5NO]^{2-}$, earlier called *nitroprusside ion*. The nitrosyl group in this ion was earlier assumed to be neutral and the iron therefore to be Fe(III). With NO^+ the iron instead becomes Fe(II). The bonding situation in the above-mentioned ion $Fe(NO)^{2+}$, as in many other nitrosyl compounds, is not wholly understood. Further examples of nitrosyl as a ligand with transition-metal atoms are given in 23-2f.

c. Dinitrogen trioxide. Dinitrogen trioxide N_2O_3 enters into the equilibrium $NO(g) + NO_2(g) \rightleftharpoons N_2O_3(g)$, which at room temperature and atmospheric pressure is displaced quite far to the left. At 25°C and 1 atm approximately 90 % of N_2O_3 is dissociated into $NO+NO_2$. Lowering the temperature and raising the pressure shifts the equilibrium to the right. On reduction of moderately strong nitric acid with copper, both NO and NO_2 are formed (see 22-9b) and on cooling this gas mixture to $-20°C$ a blue liquid is obtained, which for the most part consists of N_2O_3, but which also contains NO and NO_2 (N_2O_4) in solution. Pure N_2O_3 probably occurs only in the crystals (m.p. $-103°C$) obtained on cooling this liquid.

Since the equilibrium mentioned above is rapidly established, a mixture of the composition $NO+NO_2$ often acts chemically like N_2O_3. If this mixture is passed into water, nitrous acid is thus formed at first ($NO+NO_2+H_2O \rightarrow 2HNO_2$), of which N_2O_3 may be considered to be the anhydride. However, nitrous acid quickly disproportionates: $3HNO_2 \rightarrow H^+ + NO_3^- + 2NO + H_2O$. If the gas mixture is passed into a solution of a strong base, nitrite ion is formed: $NO+NO_2+2OH^- \rightarrow 2NO_2^- + H_2O$. With conc. sulfuric acid nitrosyl hydrogen sulfate is obtained: $NO+NO_2+2H_2SO_4 \rightleftharpoons 2NOHSO_4+H_2O$ (cf. 21-11 b2).

d. Nitrogen dioxide and dinitrogen tetroxide. *Nitrogen dioxide* NO_2 consists of an odd molecule (5-3a) in which the unpaired electron is placed at the nitrogen atom (tab. 22.2) but is appreciably delocalized. The oxide shows the properties typical of an odd molecule, namely, strong color (red-brown), paramagnetism and high reactivity. At room temperature it is usually encountered as a gas. This has a characteristic odor and is very poisonous. The molecule at room temperature is appreciably dimerized to *dinitrogen tetroxide* N_2O_4 (structure in tab. 22.2), in which the two odd electrons are paired together. For the dimerization equilibrium $2NO_2(g) \rightleftharpoons N_2O_4(g)$ at 25°C, $\Delta H = -58.1$ kJ and $\Delta G = -5.4$ kJ. The ΔG value according to (10.13) gives $K_p = 8.8$ atm^{-1} at 25°C. This implies a degree of dissociation of N_2O_4 at 1 atm of 17 %. At the same time the quite strongly negative ΔH value shows that the equilibrium is shifted rather rapidly to the left with rise in temperature (10d). Thus, the degree of dissociation of N_2O_4 at 1 atm is 40 % at 50°C, 89 % at 100°C, and 98.7 % at 135°C. The increasing proportion of nitrogen dioxide with rising temperature becomes evident in the stronger and stronger red-brown color. However, if the temperature is raised above approximately 150°C, the dioxide begins to dissociate (22-9b), so that the color again becomes lighter. If the gas mixture is cooled below room temperature it is easily condensed to a liquid that boils at 22.4°C and at this point is red-brown. As the liquid is cooled, its color becomes lighter and lighter, and when it crystallizes at $-10.2°C$ it is nearly colorless. At this temperature the dinitrogen tetroxide predominates almost completely.

Because of the ease with which it gives off oxygen, nitrogen dioxide, as well as dinitrogen tetroxide, is a powerful oxidizing agent, which supports combustion significantly more easily than N_2O and NO. For this reason these oxides have been used as oxidizing agents in rockets. The use of the dioxide in the lead-chamber process has been discussed in 21-11 b2.

With water both nitrogen dioxide and dinitrogen tetroxide give nitric acid and nitrous acid: $2NO_2 + H_2O \rightarrow HNO_2 + H^+ + NO_3^-$. As mentioned in 22-9c, however, the nitrous acid decomposes. If the nitrogen oxide thus formed is allowed to be oxidized by atmospheric oxygen to nitrogen dioxide, this can react again with water so that finally all of the original nitrogen dioxide becomes converted to nitric acid according to the overall reaction $2NO_2 + H_2O + \frac{1}{2}O_2 \rightarrow 2HNO_3$. This process is used in the manufacture of nitric acid (22-10d1).

With a water solution of a strong base, nitrogen dioxide gives nitrite and nitrate ion: $2NO_2 + 2OH^- \rightarrow NO_2^- + NO_3^- + H_2O$.

Nitrogen dioxide is an intermediate product in the industrial production of nitric acid. In the combustion of ammonia according to the Ostwald process (22-6a) nitrogen oxide is first formed, which with additional oxygen gives the dioxide. This is then converted with water as described above. In the laboratory nitrogen dioxide can be prepared by reduction of nitric acid (22-9b), but it is obtained more conveniently and in purer form by heating nitrates of heavy metals, especially lead nitrate: $Pb(NO_3)_2 \rightarrow PbO + 2NO_2 + \frac{1}{2}O_2$.

The *nitryl cation* (formerly called nitronium ion) NO_2^+ and *nitrite ion* NO_2^- (22-10a) may be considered to be derived from nitrogen dioxide by the loss of its unpaired electron, or by completing the pair by absorption of an electron, respectively. The nitryl cation is isoelectronic with carbon dioxide (5-3e) and is believed to have the structure $[\overline{O}=N=\overline{O}]^+$. It appears in a number of solid salts, for example nitryl perchlorate NO_2ClO_4, and in solid dinitrogen pentoxide (22-9e). It is formed, among other ways, by the action of conc. sulfuric acid on pure nitric acid (12-5).

e. Dinitrogen pentoxide. Dinitrogen pentoxide N_2O_5 forms colorless, hygroscopic crystals that melt at 30°C. The liquid boils with decomposition at 47°C and the compound has generally low stability. In fact, it may easily explode. With water it gives nitric acid: $N_2O_5 + H_2O \rightarrow 2HNO_3$. The gas contains molecules with the structure given in tab. 22.2. The crystals are made up of nitryl cations NO_2^+ (22-9d) and nitrate ions NO_3^-, and are thus nitryl nitrate NO_2NO_3.

Dinitrogen pentoxide is prepared by dehydration of nitric acid with phosphorus pentoxide ($2HNO_3 + P_2O_5 \rightarrow 2HPO_3 + N_2O_5$), or by the action of ozone on nitrogen dioxide ($2NO_2 + O_3 \rightarrow N_2O_5 + O_2$).

22-10. Oxonitric acids and oxonitrates

a. General. The following *mononuclear oxonitrate ions* with even numbers of electrons are known:

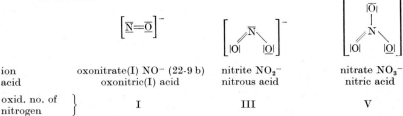

ion	oxonitrate(I) NO^- (22-9b)	nitrite NO_2^-	nitrate NO_3^-
acid	oxonitric(I) acid	nitrous acid	nitric acid
oxid. no. of nitrogen	I	III	V

In the nitrite and nitrate ions three of the electron pairs around the nitrogen atom can enter into sp^2 orbitals (5-3 d, e, h) directed toward the corners of an equilateral triangle. In the nitrite ion two of these electron pairs are shared with oxygen, in the nitrate ion all three. The remaining p orbital is used for π bonding (5-3 a, d, e) and thus gives rise to a delocalized double bond.

Oxonitrate ions with tetrahedral bond directions, for example HNO_3^{2-} or NO_4^{3-}, have not been prepared. Such ions are known for phosphorus, and the difference is usually considered to be the result of the fact that the nitrogen atom is smaller than the phosphorus atom (5-6 b). However, it should be noted that, for example in NCl_3 (22-8 a), nitrogen can have tetrahedral bond directions.

The lone electron pair of nitrogen in the nitrite ion cause nitrous acid to exist in two forms (22-10 c).

A mononuclear oxonitrate ion with an odd number of electrons is also known, namely, the dioxonitrate(II) ion NO_2^{2-}. The corresponding acid has been called *nitroxylic acid*.

There are also *oxodinitrate ions* with an N_2 group coordinating two, three or four oxygen ligands. The first two of these are:

ion	hyponitrite $N_2O_2^{2-}$	trioxodinitrate(II) $N_2O_3^{2-}$
acid	hyponitrous acid	trioxodinitric(II) acid (nitrohydroxylaminic acid)
oxid. no. of nitrogen	I	II (ave.)

In the following the hyponitrite, nitrite and nitrate ions and their corresponding acids are described in more detail.

b. Hyponitrous acid and hyponitrites. *Hyponitrous acid* $H_2N_2O_2$ at room temperature forms colorless, explosive crystals. In water solution it appears as a very weak, dibasic acid. The water solution decomposes (slowly when cold, more rapidly when hot) according to $H_2N_2O_2 \rightleftharpoons H_2O + N_2O$. This equilibrium is very strongly displaced to the right, so that hyponitrous acid cannot be obtained in detectable amounts by reaction between dinitrogen oxide and water.

The salts of hyponitrous acid, *hyponitrites*, are also unstable, although less so than the free acid.

Hyponitrous acid can be obtained in water solution, for example by oxidation of hydroxylamine with certain heavy metal oxides such as HgO, or by reduction of nitrous acid, either with sodium amalgam or electrolytically with a mercury cathode. Both these processes can be written

$$\begin{array}{c} O + H_2 NOH \\ O + H_2 NOH \end{array} \rightarrow \begin{array}{c} NOH \\ \parallel \\ NOH \end{array} \quad \text{and} \quad \begin{array}{c} 2H + O NOH \\ 2H + O NOH \end{array} \rightarrow \begin{array}{c} NOH \\ \parallel \\ NOH \end{array}$$

The poorly soluble, yellow silver hyponitrite $Ag_2N_2O_2$ can be precipitated from water solution.

c. **Nitrous acid and nitrites.** In 22-9c it was stated that the gas mixture $NO + NO_2$ (equivalent to N_2O_3) gives *nitrous acid* HNO_2 with water but that the latter rapidly decomposes. It was also mentioned that the same gas mixture, when passed into a solution of a strong base, gives *nitrite ion* NO_2^-. Solid nitrites can be obtained by heating nitrates, usually in the presence of a weak reducing agent such as lead: $NaNO_3 + Pb \rightarrow NaNO_2 + PbO$.

Nitrous acid is unstable and it has not been possible to prepare it in anhydrous form. Nitrous acid and nitrites are poisonous.

The intermediate oxidation state of nitrogen in the nitrite ion means that it can act both as an oxidizing and a reducing agent. The oxidizing effect is the most pronounced. In this case the nitrite ion is often reduced to nitrogen oxide but with certain reducing agents, for example Sn^{2+} ion, the oxidation number of nitrogen can be reduced to negative values. In its reducing action, which gives nitrate ion, a quite strong oxidizing agent is required because the standard potential of the system is rather high: $HNO_2 + H_2O \rightleftharpoons 3H^+ + NO_3^- + 2e^-$; $e^0 = 0.94$ V.

Nitrous acid is a relatively weak acid (tab. 12.2). Therefore, in a nitrite solution the proportion of nitrous acid generally becomes so high that, because of the instability of the acid, the solution cannot be kept for long. Addition of strong base decreases the acid content and therefore increases the preservability of the solution.

Most nitrites are easily soluble, with the rather difficultly soluble silver nitrite $AgNO_2$ as an important exception. Alkali nitrites have a pale yellow color.

The nitrite ion often occurs as a ligand to metal ions, especially transition metals such as Co^{3+}, Fe^{2+}, Cr^{3+}, Cu^{2+} and Pt^{2+}. The electron-pair donor may be either one of the oxygen atoms with the formation of a *nitrito complex* M-ONO or the nitrogen atom (which in the nitrite ion has one lone electron-pair) with the formation of a *nitro complex* M-NO$_2$. The nitro complex is noticeably more stable than the nitrito complex. The possibility for bond formation by either an oxygen atom or a nitrogen atom is apparent also from the fact that nitrous acid exists in two forms, either as HONO (*a*) or HNO_2 (*b*):

$$\begin{array}{cc} H-\overline{O}| & |\overline{O}| \\ | & | \\ |N=\overline{O} & H-N=\overline{O} \\ a & b \end{array}$$

These two forms are in a very rapidly established equilibrium with each other, in which the form *a* predominates. The rapidity with which the equilibrium is reached makes it impossible to isolate them from each other.

By replacing the easily mobile protons with organic groups that are more difficultly displaced, separate organic *nitrites* R–ONO and *nitro compounds* R–NO$_2$, respectively, can be obtained.

Nitrites, especially *sodium nitrite* $NaNO_2$, have great industrial application, primarily in the manufacture of azo dyes. Nitrite ions are used to protect iron and steel against corrosive attack by an aqueous phase. In this case the environment must be basic, and a protective oxide layer is probably formed by the oxidizing action of the nitrite ions.

d. Nitric acid and nitrates. 1. *Nitric acid*. Nitric acid HNO_3 in pure form at room temperature is a colorless liquid (density 1.522 g cm^{-3} at 25°C). It freezes at $-41.6°C$ to colorless crystals, and boils at 84°C with decomposition: $2HNO_3 \rightarrow H_2O + 2NO_2 + \frac{1}{2}O_2$. Even at ordinary temperature this decomposition takes place under the action of light. Nitrogen dioxide then remains partly dissolved in the acid and colors it yellow, and in higher proportions red.

The system HNO_3–H_2O has a boiling-point maximum (13-5c) at 68.2 wt% HNO_3. Thus, this is the most concentrated nitric acid solution that can be obtained by fractional distillation alone, and it is this that is often called "concentrated" nitric acid (density 1.410 g cm^{-3}). From this, solutions with smaller proportion of water can be prepared by vacuum distillation with water-absorbing agents added (conc. sulfuric acid, magnesium nitrate). In these solutions the HNO_3 pressure is so high that they fume in moist air. By the decomposition process mentioned above they also become yellow- or red-colored ("red fuming nitric acid" is an old commercial name).

The nitric acid molecule probably has the structure

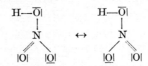

in which the nitrogen atom lies in the plane of the oxygen atoms. A third limiting structure with a double bond to the oxygen atom bonded to the hydrogen atom is less probable, since then both this oxygen atom and the nitrogen atom have the formal charges $1+(5-3g)$.

At low temperature two hydrates of HNO_3 crystallize, namely $HNO_3 \cdot H_2O$ and $HNO_3 \cdot 2H_2O$. It was formerly thought that the former consisted of orthonitric acid H_3NO_4 but this is not the case. In the crystals of these hydrates as well as of HNO_3 there are only NO_3 triangles, and these are held to each other by hydrogen bonds.

Nitric acid in water solution is a strong acid (tab. 12.2). As mentioned in 12-5, however, in anhydrous systems it can extract a proton from acids, which under such conditions are still stronger, for example HF, $HClO_4$ and H_2SO_4. It then acts as a base according to $HNO_3 + H^+ \rightarrow H_2O + NO_2^+$ (nitryl cation, 22-9d). The amphoteric behavior of nitric acid in liquid, anhydrous HNO_3 gives rise to weak autoprotolysis according to $2HNO_3 \rightleftharpoons H_2O + NO_2^+ + NO_3^-$.

Nitric acid constitutes one of the most common and most useful oxidizing agents. It is reduced usually either to nitrogen oxide: $NO_3^- + 4H^+ + 3e^- \rightleftharpoons NO + 2H_2O$, $e^0 = 0.96$ V; or to nitrogen dioxide (dinitrogen tetroxide): $NO_3^- +$

$2H^+ + e^- \rightleftharpoons NO_2 + H_2O$, $e^0 = 0.81$ V. From these equations for the redox pairs we get the equilibrium equation $NO + 2H^+ + 2NO_3^- \rightleftharpoons 3NO_2 + H_2O$, which shows that the formation of nitrogen oxide is favored by low acid concentration, and the formation of nitrogen dioxide by high acid concentration (cf. 22-9b).

The rate of reaction of oxidation with nitric acid is significantly increased if nitrogen dioxide is present. Therefore red (fuming) nitric acid reacts very vigorously, but if nitrogen dioxide is not present at the start the reaction rate increases rapidly when the oxide is formed in sufficient amount by the reduction of the acid.

Strong reducing agents, for example zinc, can reduce nitric acid to ammonium ion NH_4^+, and in 22-6e the cathodic reduction to hydroxylammonium ion NH_3OH^+ was mentioned. Other oxidation states of nitrogen can also be obtained. In most of these processes strong reaction inhibition occurs, which often becomes decisive for the result. For example, the reduction to free nitrogen seldom occurs, although it probably corresponds to the highest standard potential ($e^0 = 1.24$ V).

The standard potentials given above, when compared with the standard potentials for different metal–metal-ion pairs in tab. 16.1, show that most metals, even copper, silver and mercury, are dissolved by nitric acid, but that gold and platinum are not attacked (cf. also end of 16c). Since nitric acid can thus separate gold and silver, it was once called in German "Scheidewasser" (cf. 36-4b). However, a number of base metals under certain conditions are not attacked by concentrated nitric acid because they form protective oxide layers ("passivation"; see 16e). Thus nitric acid at room temperature does not attack aluminum if the proportion of acid is above about 80 wt % HNO_3 and does not attack iron if the proportion is above about 70 wt %. Chromium resists attack by nitric acid in all proportions. Stainless steel at room temperature resists attack in all proportions except the highest, over 97 wt % HNO_3. Therefore acid of over 95 wt % HNO_3 is kept in larger amounts in aluminum vessels and weaker acid in vessels of stainless steel.

To dissolve the noblest metals a mixture of conc. nitric acid and conc. hydrochloric acid (often in the volume ratio 1:3) is used. Since the mixture dissolves the king of metals, gold, it has since early times been called *aqua regia*. Aqua regia has the reddish-yellow color of nitrosyl chloride (22-9b): $4H^+ + NO_3^- + 3Cl^- \rightleftharpoons NOCl + Cl_2 + 2H_2O$. The action of aqua regia probably depends to a large degree on its high chloride ion activity, which causes the formation of chloro complexes with the metal ions (16c). Also, the metal certainly reacts with the nitrosyl chloride and chlorine.

Nitric acid also oxidizes many nonmetals, for example sulfur to sulfuric acid, phosphorus to phosphoric acid, carbon to carbon dioxide. Also, many organic substances are oxidized (organic packing material can sometimes be ignited by conc. nitric acid). With alcohols nitric acid esters, and with aromatic hydrocarbons nitro compounds (22-10c) are formed. These reactions are much used in synthetic organic chemistry (among others, in the explosives industry). Here a mixture of conc. nitric acid and conc. sulfuric acid is commonly used. The sulfuric acid causes the formation of the nitryl cation (see above), which probably is the main attacking agent. The sulfuric acid also absorbs the water formed. Nitric acid is highly destructive to biological tissues.

The manufacture of nitric acid was formerly always accomplished by heating alkali nitrates (in earliest time potassium nitrate, saltpeter; later sodium nitrate when this could be obtained from Chile) with fairly concentrated sulfuric acid ($H^+ + NO_3^- \rightleftharpoons HNO_3$) and distilling off the nitric acid formed. Nowadays atmospheric nitrogen is the greatest source of nitrogen, and nitric acid is obtained mainly through ammonia synthesis by the *Haber-Bosch process* (22-6b), combustion of the ammonia to nitrogen oxide by the *Ostwald process* (22-6a), followed by oxidation of the nitrogen oxide to nitrogen dioxide and conversion of the dioxide with water to nitric acid as described in 22-9d.

Nitric acid is one of the basic products of the chemical industry. The major portion is used in the manufacture of fertilizers, mainly ammonium nitrate (22-6c); the next greatest use is in the organic chemical industry, especially the explosives industry (in the United States in 1957, of the year's total production 75% was used for fertilizer production). Anhydrous HNO_3 has also begun to be used as an oxidizing agent for rocket fuels.

2. *Nitrates*. The salts of nitric acid containing the ion NO_3^- have since early times been called *nitrates*, and this name is accepted as a shortening of the rational name *trioxonitrate*(V). As stated in 22-10a, the nitrogen atom lies at the center of a triangle of oxygen atoms.

The nitrate ion in itself is colorless. All pure nitrates of simple metal ions are easily soluble in water. Certain oxide and hydroxide nitrates and nitrates of a number of organic bases are poorly soluble. Therefore, no purely inorganic precipitation reaction is available for the detection of nitrate ion, but rather the reaction with Fe^{2+} and H_2SO_4 mentioned in 22-9b has often been used. However, it must be remembered that nitrite ion gives the same reaction.

The nitrates are prepared generally by reaction of nitric acid with the corresponding metal, oxide, hydroxide or carbonate. Of course, nitrates of easily oxidized cations cannot be obtained in this way.

On heating, the nitrates decompose in various ways. The alkali nitrates in the first place give nitrites and oxygen ($NaNO_3 \rightarrow NaNO_2 + \frac{1}{2}O_2$); when the metal is more noble, the formation of metal oxide, nitrogen dioxide and oxygen predominates ($Cu(NO_3)_2 \rightarrow CuO + 2NO_2 + \frac{1}{2}O_2$). The nitrates of the most noble metals give metal, nitrogen dioxide and oxygen ($AgNO_3 \rightarrow Ag + NO_2 + \frac{1}{2}O_2$). Ammonium nitrate, which on careful heating gives water and dinitrogen oxide (22-6c), is a special case. Because of the oxygen evolved, nitrate melts are strongly oxidizing.

22-11. Phosphorus oxides

For phosphorus the P(III) and P(V) oxides are known, for which the formulas P_2O_3 and P_2O_5 are generally given, although they often occur as dimers or poly-

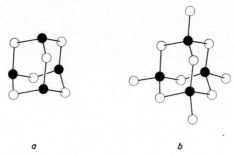

Fig. 22.1. Molecules of (a) P_4O_6, and (b) P_4O_{10}.

mers of these molecular units. In addition there are oxides whose compositions lie between these two and possibly with extended homogeneity ranges. These have as a rule been summarily given the formula P_2O_4, thus implying the existance of a P(IV) oxide, but they are actually P(III, V) oxides.

All phosphorus oxides have acidic character (21-5a).

a. Diphosphorus trioxide. Diphosphorus trioxide P_2O_3 is formed by oxidation of phosphorus when the oxygen supply is limited and the temperature low. The simultaneous formation of diphosphorus pentoxide P_2O_5 cannot be avoided, but the former can be separated from the latter because of its greater volatility. Diphosphorus trioxide forms colorless, very poisonous crystals, which melt at 23.8°C. The boiling point of the liquid is 175.4°C. The liquid and gas are composed of P_4O_6 molecules, whose structure is shown in fig. 22.1a. Thus, the phosphorus atoms are here arranged tetrahedrally as in the P_4 molecule (fig. 5.28a). By single bonding in this structure, the valence electrons can clearly be distributed so that the octet rule is fulfilled, but the observed angles and interatomic distances indicate the presence of double bonding.

On heating in air P_2O_3 ignites at about 70°C and burns to P_2O_5. With stronger heating in the absence of air the oxide disproportionates according to $4(P^{III})_2O_3 \rightarrow 3P^{III}P^{V}O_4 + 2P^0(\text{red})$. With cold water P_2O_3 slowly gives phosphorous acid ($P_2O_3 + 3H_2O \rightarrow 2H_2PHO_3$); warm water gives with vigorous reaction a number of products such as phosphorus hydrides, red phosphorus and phosphoric acid.

b. Diphosphorus pentoxide. Diphosphorus pentoxide P_2O_5 is formed by oxidation of phosphorus with a generous supply of oxygen. The heat of formation is very high: $2P(s) + \frac{5}{2}O_2(g) \rightarrow P_2O_5(s)$, $\Delta H^0 = -1550$ kJ. In this way P_2O_5 is obtained as a white, odorless powder, which sublimes at atmospheric pressure at 350°C. Both the solid phase and the gas consist of P_4O_{10} molecules, with the structure shown in fig. 22.1b. One additional oxygen atom has thus been attached to each phosphorus atom in P_4O_6 so that they all are surrounded tetrahedrally by four oxygen atoms. Double bonding is also present in the P_4O_{10} molecule.

If this volatile P_2O_5 phase is heated at about 500°C in a closed vessel so that it cannot sublime, polymerization takes place. P–O bonds are broken and giant molecules of infinite extent in two or three dimensions are formed. In these the phosphorus atoms are still tetrahedrally surrounded by four oxygen atoms, but three of them are common to two PO_4 tetrahedra, as in the following schematic figure, for example:

```
      \   /           \   /
       O   O           O   O
        \ /             \ /
     O—P—O—P—O       O—P—O—P—O
        / \             / \
       O   O           O   O
      /     \         /     \
 —O—P—O    O—P—O—P—O      O—P—O—
      \     /         \     /
       O   O           O   O
        \ /             \ /
     O—P—O—P—O       O—P—O—P—O
        / \             / \
       O   O           O   O
      /     \         /     \
```

The polymerization leads to greater stability. The volatility is reduced, because covalent bonds must now be broken on vaporization. Polymerized oxide melts at about 570°C. Large molecular aggregates remain even above the melting point, causing the melt to be easily supercooled to a glass (6-3d).

Volatile diphosphorus pentoxide takes up water extremely vigorously. With excess of water a solution of phosphoric acid is formed ($P_2O_5 + 3H_2O \rightarrow 2H_3PO_4$), while with smaller amounts of water condensed phosphate ions, mostly tetrametaphosphate ions $(PO_3^-)_4$, are produced (22-12a). Water is absorbed even from concentrated sulfuric acid so that sulfur trioxide is produced. When the oxide is used as a drying agent it becomes covered by a layer of phosphoric acid which hinders further access to water. It should therefore be spread over a large area.

22-12. Oxophosphoric acids and oxophosphates

a. General. A large number of definite oxophosphoric acids and oxophosphates are known, and many new such compounds have been prepared in recent years. In addition high-polymer acids and phosphates of varying molecular dimensions exist. Common to all is the central phosphorus atom, which by sp^3 hybridization and thus with tetrahedral bond directions binds its ligands. The structures and formulas therefore are completely different from those of the oxonitric acids and oxonitrates, where the small size of the nitrogen atom prevents tetrahedral coordination and sp^2 hybridization results instead.

For the *mononuclear oxophosphates* one would at first expect the ions PO^{3-}, PO_2^{3-}, PO_3^{3-} and PO_4^{3-}, isoelectronic with the corresponding oxochlorates and oxosulfates (cf. 20-8a and 21-11a). In the first three of these ions, in which the phosphorus atom has lone electron pairs, however, protons are very strongly bound to these pairs. Therefore, in water solution only the following ions are known (possible double bonding in these and following oxophosphate ions is not indicated; see 5-3h):

ion	hypophosphite $PH_2O_2^-$	phosphite PHO_3^{2-}	(ortho)phosphate PO_4^{3-}
acid	hypophosphorous acid	phosphorous acid	(ortho)phosphoric acid
oxid. no. of phosphorus	I	III	V

These structures show that hypophosphorous acid is a monovalent acid and phosphorous acid a divalent acid. The different functions of the hydrogen atoms in the acid molecules is apparent if the formulas are written HPH_2O_2 and H_2PHO_3. In 12-2f it was shown that the pK_a values for these acids and phosphoric acid H_3PO_4 are practically the same and of approximate magnitude 2, which can be explained if in all the molecules there is one oxygen atom that does not bond a hydrogen atom.

Analogously with the case of the corresponding sulfur compounds (21-11a), there exist polynuclear *oxodi-* and *oxopolyphosphate ions*, made up of PO_4 tetrahedra which form chains with one O atom common to two tetrahedra. They thus contain continuous, zigzag –O–P–O–P–O– chains. The chains may be closed into rings, or branched. The simplest examples are:

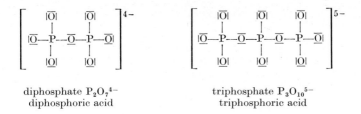

ion	diphosphate $P_2O_7^{4-}$	triphosphate $P_3O_{10}^{5-}$
acid	diphosphoric acid	triphosphoric acid

In all ions and acids of this type the oxidation number of phosphorus is V. Provided no ring formation occurs, the general formula of the polyphosphate ions is $P_nO_{3n+1}^{(n+2)-}$, and of the polyphosphoric acids, $H_{n+2}P_nO_{3n+1}$. As limiting formulas with $n = \infty$ we obtain $(PO_3^-)_n$ and $(HPO_3)_n$ respectively. As the limiting formula is approached, infinite molecules appear. When unbranched chains close into rings the composition becomes the same as that of the limiting formula, that is, independent of n. For example, the following types are known (unshared electron pairs omitted):

trimetaphosphate ion
$(PO_3^-)_3$

tetrametaphosphate ion
$(PO_3^-)_4$

Formerly all phosphates with ions corresponding to the formula $(PO_3^-)_n$ were called *metaphosphates* and the corresponding acids $(HPO_3)_n$ *metaphosphoric acids*. However it is now more common to refer only to substances with rings of the above type as metaphosphates. As shown by the names below the formulas the number of phosphorus atoms is indicated by a prefix. Nevertheless it is often convenient to say that a phosphate(V) with the ratio $O/P = 3$ has *metaphosphate composition*.

It may be noted that ions with metaphosphate composition $(PO_3^-)_n$ are isoelectronic with $(SO_3)_n$ and that the structures of these ions are very similar to the SO_3 structures (21-10b). The central atoms in fact try in both cases to attain tetrahedral coordination, which with this composition can be achieved by the chain formation described.

The polyphosphates may contain a great variety of ions. The rings may be of various types and they may be combined with each other in various ways and with tetrahedral chains, which also may be branched. The ratio O/P may also be <3. However, for any oxophosphate(V) ion $O/P > 2.5$, and the composition P_2O_5 can therefore be taken as a limit (concerning P_4O_{10} and polymeric P_2O_5, see 22-11b). All phosphates with ions that consist of several PO_4 tetrahedra, bound together in varying ways by common O atoms, are also called *condensed phosphates*. The corresponding *condensed phosphoric acids* all have lower water content than phosphoric acid H_3PO_4.

Oxophosphate ions and oxophosphoric acids are also known with P_2 groups and with longer chains of phosphorus atoms. Of these we may mention here the hypophosphate ion $P_2O_6^{4-}$ with the structure

and the associated hypophosphoric acid $H_4P_2O_6$, in which phosphorus has the oxidation number IV.

b. Hypophosphorous acid and hypophosphites. Hypophosphites are usually obtained by the disproportionation of white phosphorus as described in 22-7a for the preparation of phosphine. If base is added in the form of $Ba(OH)_2$, barium hypophosphite is formed on evaporation, which on reaction with an equivalent amount of sulfuric acid gives a solution of hypophosphorous acid alone: $Ba(PH_2O_2)_2 + H_2SO_4 \rightarrow BaSO_4(s) + 2HPH_2O_2$. The acid can be obtained as colorless crystals (m.p. 26.5°C).

Hypophosphorous acid and hypophosphites are strong reducing agents (for example, according to $HPH_2O_2 + H_2O \rightleftharpoons H_2PHO_3 + 2H^+ + 2e^-$, $e^0 = -0.50$ V), although they often react slowly. The metals are precipitated in free form from solutions of silver, gold and mercury salts. From a Cu(II) solution copper hydride CuH is precipitated.

c. Phosphorous acid and phosphites. Phosphorous acid can be prepared by hydrolysis of phosphorus trichloride: $PCl_3 + 3H_2O \rightarrow H_2PHO_3 + 3HCl$, and can be obtained as colorless crystals (m.p. 70.1°C).

Phosphorous acid and phosphites are also strongly reducing. However, the standard potentials are less negative than for the corresponding redox equilibria with hypophosphorous acid or hypophosphites as the reduced form (for example, $H_2PHO_3 + H_2O \rightleftharpoons H_3PO_4 + 2H^+ + 2e^-$, $e^0 = -0.276$ V). Here also the reducing process is generally slow.

d. Hypophosphoric acid and hypophosphates. Hypophosphoric acid $H_4P_2O_6$ is formed together with phosphorous and phosphoric acids when white phosphorus oxidizes slowly in moist air. However, it is usually prepared by oxidation of phosphorus with certain oxidizing agents in water solution. For example, hypophosphate is obtained by oxidation of red phosphorus with chlorite ion or hydrogen peroxide in alkaline solution. The poorly soluble barium hypophosphate can be precipitated with barium ion and from this the free acid obtained by addition of sulfuric acid. The anhydrous acid melts at 70°C.

e. Phosphoric(V) acids and phosphates(V). To this class belong *(ortho)phosphoric acid* H_3PO_4 and the *condensed phosphoric acids*, that is *diphosphoric acid* $H_4P_2O_7$ and all *polyphosphoric acids* with P–O–P bonding including the *metaphosphoric acids* $(HPO_3)_n$. All these acids can be formally written $P_2O_5 \cdot nH_2O$. H_3PO_4 has the highest proportion of water, and it therefore has the special name "orthophosphoric acid". The corresponding anions are thus the *(ortho)phosphate ion* PO_4^{3-}, and the *condensed phosphate ions*, that is, the *diphosphate ion* $P_2O_7^{4-}$ and all the *polyphosphate ions* with P–O–P bonding including the *metaphosphate ions* $(PO_3^-)_n$. All contain PO_4 tetrahedra, which in the polynuclear molecules are linked together by sharing an oxygen atom (for structure and nomenclature, see 22-12a).

The tendency of the phosphorus atom to bond four oxygen atoms tetrahedrally explains the existence of the different phosphate ions and their relation to each other (cf. 5-6b and 21-5a). The supply of oxide ion in a phosphate melt becomes decisive for the structure, and a change in the oxide-ion concentration changes the structure. As an example, consider the equilibrium between orthophosphate ion and diphosphate ion, $2PO_4^{3-} \rightleftharpoons P_2O_7^{4-} + O^{2-}$. If the concentration of oxide ion is increased, for example by heating with a basic oxide such as MgO, a greater number of independent PO_4^{3-} ions can be formed, which leads to the breaking down of the $P_2O_7^{4-}$ ions.

On heating phosphoric acid water is given off, and at first diphosphoric acid is formed (a process that gave rise to the old name "pyrophosphoric acid"): $2H_3PO_4 \rightarrow H_4P_2O_7 + H_2O(g)$.

1. *Phosphoric acid* H_3PO_4. If diphosphorus pentoxide is allowed to react with water in excess, a solution of phosphoric acid is formed: $P_2O_5 + 3H_2O \rightarrow 2H_3PO_4$. A water solution containing mainly phosphoric acid is also obtained when calcium phosphate is treated with dilute sulfuric acid and the poorly soluble calcium sulfate formed filtered off: $Ca_3(PO_4)_2 + 3H_2SO_4 \rightarrow 3CaSO_4(s) + 2H_3PO_4$. By evaporation at atmospheric pressure there is obtained from this solution a colorless, syrupy solution containing about 85 wt% H_3PO_4 (in commerce called "concentrated phosphoric acid"). From this, pure H_3PO_4 can be prepared by evaporation at low pressure. Pure H_3PO_4 forms colorless crystals (m.p. 42.4°C) that are deliquescent in air. Its acidic properties in water solution have been discussed in 12-2f and 12-4b.

In the industrial manufacture of phosphoric acid, phosphate rock, that is chiefly apatites of different kinds, is generally the starting product. From this, either phosphorus is first produced by the reducing process described in 22-4b (hence the designation "furnace process") following which the phosphorus is oxidized and the diphosphorus pentoxide allowed to react with water as described above; or the above-mentioned treatment with dilute sulfuric acid ("wet process") is used. The wet process gives a more impure product, which, however, is suitable for the manufacture of fertilizer. The major portion of the phosphoric acid production is immediately converted to phosphate, especially phosphate fertilizer (22–12e4).

2. *Phosphates.* The salts of phosphoric acid H_3PO_4, which contain the ions PO_4^{3-}, HPO_4^{2-} and $H_2PO_4^-$ have formerly been called orthophosphates, but more and more the simple term phosphate is used, which may be considered as a shortening of the rational name *tetraoxophosphate*(V).

The phosphate ions named are in themselves colorless. *Silver phosphate* Ag_3PO_4 is yellow because of the strong polarizing effect of the Ag^+ ion (7-3b).

All alkali and ammonium phosphates $(M^I)_3PO_4$, $(M^I)_2HPO_4$ and $M^IH_2PO_4$ (except Li_3PO_4) as well as the dihydrogen phosphates of alkaline-earth metals $M^{II}(H_2PO_4)_2$, are easily soluble or fairly soluble in water. Other phosphates are difficultly soluble in water. The poor solubility of magnesium ammonium phosphate $MgNH_4PO_4 \cdot 6H_2O$ (the mineral struvite) has been used in analysis. For example, this phosphate precipitates if an ammoniacal solution containing Mg^{2+} and NH_4^+ ("magnesia mixture") is added to a phosphate solution.

Phosphate ions easily form complexes with polyvalent cations. Phosphoric acid nearly completely decolorizes a Fe(III) solution because of the formation of phosphatoiron(III) ions.

The soluble phosphates are usually prepared by mixing phosphoric acid with a solution of the hydroxide or carbonate of the corresponding metal. Poorly soluble phosphates are best obtained by precipitation, for example, $3Ag^+ + PO_4^{3-} \rightarrow Ag_3PO_4(s)$ (the precipitation of yellow Ag_3PO_4 is used as a test for phosphate ion).

Of the natural phosphates, above all *apatite* should be mentioned, which was described in 22-2a. The semiprecious stone *turquoise* is a copper aluminum hydroxide phosphate with a blue to green color.

3. *Condensed phosphoric acids and phosphates.* Condensed phosphates have not been found as minerals in nature, but on the other hand they do play a very large role in biological systems, in which carbohydrate esters of, among others, diphosphoric and triphosphoric acid take part in energy-transfer processes (cf. 22-2a).

Inorganic condensed phosphates can be prepared by heating hydrogen phosphates, ammonium phosphates or ammonium hydrogen phosphates of the metal in question:

$$2Na_2HPO_4 \rightarrow Na_4P_2O_7 + H_2O(g)$$
$$2MgNH_4PO_4 \rightarrow Mg_2P_2O_7 + 2NH_3(g) + H_2O(g)$$
$$nNaH_2PO_4 \rightarrow (NaPO_3)_n + nH_2O(g)$$
$$nNaNH_4HPO_4 \rightarrow (NaPO_3)_n + nNH_3(g) + nH_2O(g)$$

Sodium ammonium hydrogen phosphate (last equation) has been used in qualitative analysis under the name *phosphor salt*. A few grains when melted give a "phosphor-salt bead", which dissolves basic oxides, often with characteristic color.

However, with the above reactions, no lower O/P value than 3 is reached, that is, the metaphosphate composition. Lower values are obtained by heating P_2O_5, either with phosphates of metaphosphate composition, or with basic oxides, for example $Ca(PO_3)_2 + P_2O_5 \rightarrow CaP_4O_{11}$ or $CaO + 2P_2O_5 \rightarrow CaP_4O_{11}$ (gross formulas only, molecular size not indicated).

The diphosphates are the only ones of all these products that are well defined. In other cases mixtures of various structures are believed to appear, and their constitution depends on the conditions of the experiment, especially on the manner in which the heating is carried out (temperature, time). If the ions are not larger than those of triphosphate, trimeta- and tetrametaphosphate, however, the salts can be obtained quite pure.

Condensed alkali phosphates with small ions are easily soluble in water. The solubility decreases with increasing ionic size, and the phosphates having giant ions and $O/P \leq 3$ are poorly soluble in water at room temperature. Phosphates with $O/P > 3$ are rapidly hydrolyzed (see below) and thus go rapidly into solution. The larger the ions are, the greater is the tendency of a phosphate melt to supercool, and thus the greater its ability to form a glass (6-3d). The more acid phosphates very easily form glasses.

Mixtures of condensed phosphoric acids are produced by the action of moderate amounts of water on diphosphorus pentoxide. The ring-shaped metaphosphoric acids (tri- and tetrametaphosphoric acids) are strong acids. The acidity constants of diphosphoric acid were discussed in 12-2f.

In water, hydrolysis of the condensed phosphates takes place, during which the P–O–P bonds are broken down according to the schematic equation

$$-\underset{|}{\overset{|}{P}}-O-\underset{|}{\overset{|}{P}}- + H_2O \rightleftarrows -\underset{|}{\overset{|}{P}}-O^- + {}^-O-\underset{|}{\overset{|}{P}}- + 2H^+$$

The equilibria of this type are strongly displaced to the right, but for $O/P \leqslant 3$ they are very slowly established at room temperature. However, the rate increases rapidly with rise in temperature, and already above 50°C all condensed phosphates under the action of water are rather rapidly decomposed into orthophosphate ions.

Condensed phosphate ions often form strong complexes with polyvalent cations. The complexes can therefore prevent these from being precipitated by other anions, or bring about the solution of solid phases containing such cations. Polyphosphate ions ($n \geqslant 3$) form the strongest complexes while diphosphate and ring-shaped metaphosphate ions are weaker complex formers.

4. *Uses of phosphoric acids and phosphates.* By far the largest portion of the world's consumption of phosphate is in the form of *fertilizer*. However, phosphate rock is so poorly soluble in water that it cannot be added to the soil as such, but must be converted to more easily soluble phosphates. The oldest and still the most important method involves treatment with about 70% sulfuric acid, resulting in the formation of a mixture of the soluble phosphate $Ca(H_2PO_4)_2 \cdot H_2O$ and $CaSO_4$. This mixture is called *superphosphate*. If phosphoric acid is used instead of sulfuric acid only soluble phosphate is formed. This product is often called *triple phosphate*.

Diammonium hydrogen phosphate $(NH_4)_2HPO_4$ and *ammonium dihydrogen phosphate* $NH_4H_2PO_4$ have high plant nutrient content and are being used to an increasing extent as fertilizer. They are made by passing ammonia into phosphoric acid (see also 22-6c).

Phosphate fertilizers made by heating processes, "*calcined phosphate*", are also important. In particular *basic slag (Thomas phosphate)* should be mentioned, which is obtained as a by-product of the manufacture of steel from phosphorus-containing pig iron (34-3c), and whose chief constituent is a calcium phosphate silicate. A quite new calcined phosphate is "*calcium metaphosphate*" which is prepared by allowing P_2O_5 to react with phosphate rock powder in a kiln. The calcined phosphates are less soluble than the previously mentioned fertilizers, and are therefore more slowly taken up by plants.

In the laboratory the phosphoric acid systems are used as buffer systems (12-3f) to obtain a desired pH value. Many alkali phosphates are used as water softeners (26-5c), especially Na_3PO_4, Na_2HPO_4, $Na_4P_2O_7$ (for these, cf. 12-4b, fig. 12.3 and 12.4) and $Na_5P_3O_{10}$. The softening effect is often favored by the solution becoming basic, but depends to a large degree on the formation of complexes between the phosphate ions and the polyvalent cations in the water (see above). This effect is utilized when water is softened with *sodium triphosphate* $Na_5P_3O_{10}$ and with a preparation often improperly called sodium hexametaphosphate (this substance, also termed "Graham's salt", contains tri- and tetrametaphosphate ions and high-polymer polyphosphate ions; it gives a weakly acid solution and is therefore mixed with $Na_4P_2O_7$). For these reasons the sodium phosphates are much used in cleansing, in which the removal of dirt is also favored by the dispersing effect of the highly charged phosphate ions (15-4).

Many sodium and calcium phosphates are used as fodder phosphate and for the addition of phosphoric acid to foodstuffs. In the food industry also phosphoric acid is used in acid light drinks, and hydrogen phosphates are used as a constituent of baking powder (mixed with sodium hydrogen carbonate so that carbon dioxide is given off).

Acid solutions of phosphates of heavy metals are much used to produce, by reaction with metal surfaces, especially iron and zinc, strongly adhering metal phosphate layers (*phosphating*). Thick layers, which in themselves give some protection against corrosion, are obtained by dipping the object in hot solution. Thin layers, which give a good base for lacquer, enamel, and so on, can be obtained by a hot spray.

22-13. Oxygen compounds of arsenic, antimony and bismuth

a. General. In these compounds the predominant oxidation numbers are III and V. A formal oxidation number of IV is found in certain compounds but in these probably equal numbers of atoms with the numbers III and V are present. Even for arsenic the acidity of the oxides is less pronounced than for phosphorus. However, they are predominantly acid and are acid anhydrides. Antimony oxides are acid and bismuth oxides are basic.

b. Oxygen compounds of arsenic. The form of *diarsenic trioxide* As_2O_3 stable at ordinary temperature forms colorless crystals with cubic symmetry, which are composed of As_4O_6 molecules analogous to the P_4O_6 molecule (fig. 22.1a). It sublimes on heating in an open vessel. The gas consists of As_4O_6 molecules, which above 700°C begin to dissociate appreciably into As_2O_3 molecules. On heating the cubic form in a closed vessel it is transformed to a monoclinic form, which is composed of infinite layer molecules. Fragments of these molecules are present in a melt of As_2O_3 and cause it to be easily supercooled to a glass (6-3d).

Diarsenic trioxide is the most important arsenic compound and is formed by heating arsenic and many arsenic compounds in air. Formerly it was often called "arsenic" or "white arsenic". It is very poisonous. The lethal dose is given as 0.1 g, but the body, by gradual exposure to increased amounts of the poison, can tolerate doses up to about 1 g.

Diarsenic trioxide is rather poorly soluble in cold water but more easily soluble in hot water (the rate of solution is very slow, however). In the solution the weak acid *arsenious acid* H_3AsO_3 appears, which is a tribasic acid unlike phosphorous acid (22-12a). It cannot be obtained free; on evaporation of the solution As_2O_3 crystallizes out. With alkali hydroxide solutions *arsenite* salts are obtained.

Diarsenic trioxide is easily reduced to arsenic, for example by heating with carbon or potassium cyanide: $2As_2O_3 + 3C \rightarrow 4As + 3CO_2$; $As_2O_3 + 3KCN \rightarrow 2As + 3KOCN$. Most other arsenic compounds are reduced in a similar way. The arsenic released sublimes and deposits on colder surfaces, often as an arsenic mirror (22-3c). On the other hand the oxide can be easily oxidized (see below). Arsenious acid can also be easily reduced (to arsenic) or oxidized (to arsenic acid).

Diarsenic trioxide is produced for the world market in more than sufficient amounts as a by-product of the processing (roasting) of arsenic-containing sulfide ores. It is the most important raw material for the production of other arsenic compounds. It is used directly for "fining" and decolorizing glass melts (23-3i), and in medicine as a stimulant. A number of Cu(II) arsenites are used to combat insects injurious to plants, and were earlier also used as color pigments. These include, for example, *Scheele's green* $CuHAsO_3$ and *Paris green*, which is a Cu(II) arsenite acetate.

On oxidation of diarsenic trioxide with conc. nitric acid, *arsenic acid* is produced. From the solution colorless crystals of the hydrate $H_3AsO_4 \cdot \frac{1}{2}H_2O$ can be obtained. These give off water on heating with the formation of diarsenic pentoxide As_2O_5, which easily redissolves in water to give arsenic acid. It has not been possible to obtain the pentoxide by oxidation of arsenic or diarsenic trioxide in air.

Arsenic acid is a tribasic acid, about as strong as phosphoric acid. Both the acid and its salts, *arsenates*, are in many ways analogous to phosphoric acid and phosphates (condensed arsenates are also known). Arsenic acid is significantly easier to reduce to compounds of lower oxidation numbers than V than phosphoric acid is. *Silver arsenate* Ag_3AsO_4 is chocolate brown.

Sodium, calcium and *lead arsenates* are extensively used as insecticides. Arsenates are also used in a very effective method of impregnating lumber against decay and injurious animals ("Boliden impregnation"). The wood is allowed to soak up an acid solution containing sodium arsenate, zinc sulfate and sodium dichromate. Substances present in the wood reduce the dichromate ions to chromium(III) ions, following which the poorly soluble zinc and chromium(III) arsenates precipitate out inside the wood fiber. The overall equation is usually written: $5AsO_4^{3-} + 3Zn^{2+} + Cr_2O_7^{2-} + 17H^+ + 6e^- \rightarrow 3ZnHAsO_4(s) + 2CrAsO_4(s) + 7H_2O$. The insoluble arsenates then cannot be leached out of the wood by water.

c. Oxygen compounds of antimony. *Diantimony trioxide* Sb_2O_3 is colorless and crystallizes in a cubic form composed of Sb_4O_6 molecules of the same type as P_4O_6 (fig. 22.1a) and As_4O_6. An orthorhombic form contains infinite chains of the type

```
     \   /O\   /O\   /O\
      Sb     Sb      Sb
      |      |       |
      O      O       O
      |      |       |
      Sb     Sb      Sb
     /   \O/   \O/   \O/
```

Diantimony trioxide is formed on heating antimony in air. It volatilizes quite easily and the gas even at 1500°C consists mainly of Sb_4O_6 molecules. The trioxide is poorly soluble in water but dissolves in alkali hydroxide solution, with the formation, among other things, of *tetrahydroxoantimonate*(III) *ion* $Sb(OH)_4^-$. Salts containing this ion, for example $KSb(OH)_4$, have formerly been erroneously called "antimonites" and written $KSbO_2 \cdot 2H_2O$. The trioxide dissolves in concentrated acids, and these solutions are assumed to contain the positive ion SbO^+, "antimonyl". From such solutions generally *antimony*(III) *oxide* or *hydroxide salts* (12-1 b, 21-5d) crystallize, such as $SbONO_3$ and $Sb_2O_2SO_4$. These have earlier been called antimonyl salts, but it is neither proven nor probable that they contain independent antimonyl ions. A tartrate belonging to this group is the double salt *potassium antimony*(III) *oxide tartrate*, which has been used as an emetic under the name "tartar emetic".

If antimony is oxidized with conc. nitric acid, an undefined, colloidal hydrate of *diantimony pentoxide* Sb_2O_5 is formed, which has been called "antimonic acid". By careful heating the pentoxide can be obtained from this product. When this oxide is dissolved in potassium hydroxide solution, *hexahydroxoantimonate*(V) *ion* $Sb(OH)_6^-$ is formed. $LiSb(OH)_6$ and $NaSb(OH)_6$ are poorly soluble. $NaSb(OH)_6$ has been erroneously designated as acid sodium pyroantimonate $Na_2H_2Sb_2O_7 \cdot 5H_2O$.

By heating metal oxides with Sb_2O_3 and Sb_2O_5 compounds with formulas such as $MgSb_2O_4$, $ZnSb_2O_4$ and $NaSbO_3$, $ZnSb_2O_6$, respectively, can be obtained. They can also be obtained by sufficiently strong heating of the above-mentioned hydroxoantimonates, for example $NaSb(OH)_6 \rightarrow NaSbO_3 + 3H_2O$. In spite of their formulas, these compounds are not antimonites or antimonates but double oxides (21-5a).

On heating Sb_2O_3 in air an oxide is formed of the composition Sb_2O_4 ("antimony tetroxide"), which is a double oxide $Sb^{III}Sb^VO_4$.

d. Oxygen compounds of bismuth. The common bismuth oxide is *dibismuth trioxide* Bi_2O_3. It is obtained as a yellow powder on combustion of the metal or heating bismuth hydroxide, nitrate or carbonate, and can easily be reduced to the metal. The oxide is somewhat soluble in strongly basic solutions, probably forming $Bi(OH)_4^-$. It easily dissolves in acids with the formation of the corresponding bismuth salts. Thus it is primarily a basic oxide. Crystalline salts, for example $BiCl_3$ and $Bi(NO_3)_3$, can be obtained if sufficiently acid solutions are evaporated. At lower acidity (for example, after adding water), the Bi^{3+} ions form oxo or hydroxo complexes (12-1 b) and from such solutions poorly soluble oxide and hydroxide salts precipitate, for example *bismuth oxide chloride* BiOCl and *bismuth hydroxide nitrate* Bi(OH)$_2$NO$_3$. The latter salt is used as a mild antiseptic as well as analogous salts of carbonic acid and some organic acids.

If a bismuth salt solution is made strongly basic, an undefined bismuth oxide hydrate forms as a white, flocculent precipitate. The dried precipitate is often formulated BiO(OH).

On oxidation of bismuth oxide hydrates by strong oxidizing agents in alkaline solution, brown products are obtained, which have been considered to be bismuthates. However, they are actually double oxides (21-5a). Such a compound is *sodium bismuth*(V) *oxide* NaBiO$_3$.

22-14. Sulfur compounds of the nitrogen-group elements

By the action of liquid ammonia on sulfur there is obtained, among other things, N$_4$S$_4$, which according to the nomenclature instructions (5-2a, 5-3f) is called *tetranitrogen tetrasulfide*, but chemically acts more like a sulfur nitride: 16NH$_3$ + 10S → N$_4$S$_4$ + 6(NH$_4$)$_2$S. The compound forms orange-colored crystals, which sublime in vacuum at 100°C, but easily explode at higher temperature or with shock. It consists of molecules with the structure shown in fig. 22.2, which is indicated by the schematic formula

$$\begin{array}{ccc} |\underline{N}-\underline{S}-\underline{N}| \\ |\ \ \ \ \ \ \ \ \ \ \ \ \ \ | \\ |S\ \ \ \ \ \ \ \ \ \ \ \ \ S| \\ |\ \ \ \ \ \ \ \ \ \ \ \ \ \ | \\ |\underline{N}-\underline{S}-\underline{N}| \end{array}$$

Six *phosphorus sulfides* are known, the formulas of which can be written P$_4$S$_n$ with n = 2, 3, 5, 6 ± 0.5, 7, and 10. The sulfides with n = 2, 3, 7, and 10 can be obtained as yellow crystals if red phosphorus is melted with sulphur. The two other sulfides melt incongruently (13-5f), which makes it difficult to obtain them pure from a melt. P$_4$S$_{10}$ has the same structure as P$_4$O$_{10}$ (fig. 22.1b). Especially P$_4$S$_{10}$ (older name, "phosphorus pentasulfide") is used to exchange sulfur for oxygen in organic compounds. P$_4$S$_3$ is used in the igniting tips of "strike-anywhere" matches.

The following *arsenic sulfides* are known and can be prepared by fusion of the elements: As$_4$S$_4$, As$_2$S$_3$ and As$_2$S$_5$. As$_4$S$_4$ is the red mineral *realgar*. The crystals (m.p. 320°C, b.p. 565°C) are made up of As$_4$S$_4$ molecules, analogous to N$_4$S$_4$ but with the atoms interchanged (fig. 22.2). At the boiling point the gas also consists of these molecules. As$_2$S$_3$ forms the golden-yellow mineral *orpiment*.

As$_2$S$_3$ and As$_2$S$_5$ can also be obtained by precipitation with hydrogen sulfide from acid solutions of As(III) and As(V), respectively. Both sulfides are then citron yellow. As$_2$S$_3$ precipitates from a strongly acid arsenite solution: 2H$_3$AsO$_3$ + 3H$_2$S → As$_2$S$_3$(s) + 6H$_2$O. If the solution is not acid enough an As$_2$S$_3$ sol is formed, and only on sufficient acidification or addition of other precipitating ions does the sulfide come down (15-4). As$_2$S$_3$ readily takes up sulfide ions with the formation of *thioarsenite ions* (21-7a), possibly mainly as AsS$_2^-$. It therefore dissolves in

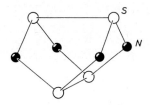

Fig. 22.2. The N_4S_4 molecule. An As_4S_4 molecule has the same form but N is replaced by S and S by As.

alkali or ammonium sulfide solution: $As_2S_3(s) + S^{2-} \rightarrow 2AsS_2^-$. In basic solutions As_2S_3 is converted to thioarsenite and arsenite: $2As_2S_3(s) + 4OH^- \rightarrow 3AsS_2^- + H_2AsO_3^- + H_2O$. Therefore, hydrogen sulfide will not precipitate As_2S_3 from a basic solution. Solutions that also contain polysulfide ions (21-8) dissolve As_2S_3 with oxidation to tetrathioarsenate ion: $As_2S_3(s) + 2S_2^{2-} + S^{2-} \rightarrow 2AsS_4^{3-}$. Thioarsenites can also be obtained by fusing As_2S_3 with the corresponding metal sulfide.

Thioarsenious acid is not stable, and on acidification of a thioarsenite solution As_2S_3 precipitates: $2AsS_2^- + 2H^+ \rightarrow As_2S_3(s) + H_2S$.

As(V) reacts very similarly, with As_2S_5 and *tetrathioarsenate ion* AsS_4^{3-} taking part in the processes. In weakly acid arsenate solutions, however, As(V) is reduced by H_2S to As(III) with the slow formation of $As_2S_3 + 2S$. In strongly acid solutions, on the other hand, As_2S_5 is obtained. By treating As_2S_5 with alkali hydroxide solution, and also by passing hydrogen sulfide into basic arsenate solution, in addition to thioarsenate ion, *oxothioarsenate ions* $AsOS_3^{3-}$, $AsO_2S_2^{3-}$ and AsO_3S^{3-} are also formed. In all these the arsenic atom has tetrahedral coordination.

From moderately acid solutions containing Sb(III) and Sb(V) hydrogen sulfide precipitates the orange-red antimony sulfides Sb_2S_3 and Sb_2S_5, respectively, which behave rather like the arsenic compounds. Thus, they are dissolved by sulfide-ion-containing solutions with the formation of thio ions, which are easily decomposed by acids. Unlike the arsenic sulfides, however, they dissolve in strong, conc. acids. Both the precipitated sulfides have a very disordered (amorphous) structure. On heating precipitated Sb_2S_3 it is transformed to a gray-black, crystalline phase, which is also known as *stibnite*, the most important antimony mineral. Stibnite is used in safety matches.

Safety matches were invented by the Swede G. E. Berggren-Pasch (patented 1844). The head usually consists mainly of stibnite or sulfur as the combustible substance and potassium chlorate as oxidizing agent. The head is ignited by rubbing on the striking surface, which contains red phosphorus, stibnite and glass powder. The flame is carried easily to the wood because the latter is dipped in paraffin near the head. Afterglow is prevented by impregnating the match with $NH_4H_2PO_4$, for example (cf. 22-6c), which also keeps the burned-out match and match head from crumbling.

Sb_2S_5 is used as a pigment, especially for the red coloring of rubber, in which it also takes part in the vulcanization process.

The common *dibismuth trisulfide* Bi_2S_3 constitutes the mineral *bismuthinite*, which is very reminiscent of stibnite. It can be obtained by heating together bismuth and sulfur, or by precipitation with hydrogen sulfide from solutions containing Bi(III). The precipitate is black-brown, and unlike the arsenic and antimony sulfides, does not dissolve in solutions of alkali hydroxides or sulfides.

22-15. Nitrogen-phosphorus compounds

The most important of these are the *phosphorus nitride dichlorides* $(PNCl_2)_n$. These are prepared by reaction between phosphorus pentachloride and ammonium chloride: $PCl_5 + NH_4Cl \rightarrow PNCl_2 + 4HCl$ (heated in a sealed tube or in boiling tetrachloroethane solution). The product contains ring molecules in which n is mainly 3 (formula a below) and 4, but may also be 5, 6 or 7.

At 250–350°C the phosphorus nitride dichlorides are further polymerized, which involves the breaking of the rings (ring rupture is believed to require the participation of oxygen atoms) and formation of chains of the type b (although zigzag). These may be very long and may also be branched. $(PNCl_2)_n$ is isoelectronic with $(SO_3)_n$ and $(PO_3^-)_n$ and analogous structure types clearly appear in these three classes of compounds (21-10 b, 22-12 a).

Phosphorus nitride dichlorides with high-polymer chain molecules resemble rubber to a high degree in their physical properties ("inorganic rubber"). They may find uses because of their ability to tolerate heat up to about 500°C, but their tendency to hydrolyze is a disadvantage.

22-16. The nitrogen cycle in nature

Though many features of the nitrogen cycle are still unclear, the current understanding of it is shown schematically in fig. 22.3. This will require only a few short comments.

Atmospheric nitrogen is fixed by several kinds of lower plant organisms, especially bacteria. These can live in symbiosis with a host plant (for example, bacteria of the genus *Rhizobium*, which forms small nodules on the roots of leguminous plants), or independently in the soil (especially species of the genus *Azotobacter*). Nitrogen fixation by bacteria requires the presence in the earth of

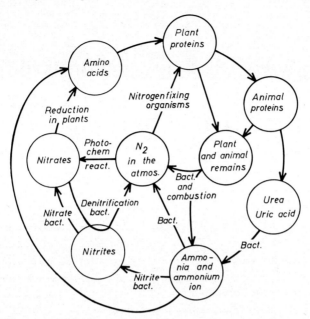

Fig. 22.3. The nitrogen cycle in nature.

traces of certain transition elements such as molybdenum, iron, cobalt and copper, and also boron. Blue-green algae can also fix free nitrogen, which is important, for example, in the growing of rice.

Rainfall always contains traces of nitrates, which undoubtedly come from nitrogen oxide formed in the atmosphere. This probably originates in the stratosphere, where photochemical reactions caused by the strong ultraviolet radiation play a fundamental role (contrary to earlier belief, on the other hand, electrical discharges in the lower levels of the atmosphere give only small amounts of nitrogen oxide). The nitrogen oxide is oxidized to nitrogen dioxide, which with the moisture in the air gives nitric acid (22-9d).

The rainfall also carries ammonia, which is formed in the combustion and decay of plant and animal remains.

Denitrification involves the loss of combined nitrogen. This process probably occurs in many ways, but an important one follows, among others, the path nitrate→nitrite→dinitrogen oxide→nitrogen.

Nitrogen soil enrichment, which is intended to compensate for losses caused by harvests, enters into the nitrogen cycle at various points. As nitrogen fertilizers, urea, ammonia or ammonium ion, and nitrate ion, for example, are used.

CHAPTER 23

Group 14, The Carbon Group

23-1. Survey

Many properties of the elements in this group are related to the fact that the valence electron shell is half-filled (ns^2np^2). The electronegativity is intermediate and bonds with strong covalent character are consequently very common. The oxidation number therefore often has only a very formal significance. However, with the valence-electron configuration assumed, especially the oxidation numbers $-$IV, II and IV would be expected. The oxidation number $-$IV may possibly be considered in carbides of strongly electropositive metals (5-2f), but IV is the most common. For germanium, tin and lead II also occurs, where the "inert" electron pair ns^2 is not given up on ionization (5-2f). The oxidation number II becomes more important in relation to IV as one goes down in the group.

In the carbon group as in other groups the electronegativity decreases with increasing atomic number. Along with this decrease the metallic properties increase. Carbon is a nonmetal. However, graphite because of its peculiar structure has very special properties as an electrical conductor (23-2b). Silicon and germanium at ordinary pressures are intrinsic semiconductors (6-4d), but at pressures above about 120 kbar are metallic conductors. Tin at ordinary pressure occurs both as an intrinsic semiconductor (gray tin) and as a metallic conductor (white tin). Lead is always metallic.

The bonding and chemical properties follow a generally similar pattern. The covalent contribution even in bonds with strongly electronegative elements is wholly dominant with carbon and is still strong with silicon and germanium. Ionic character begins to be more noticeable with Sn(II) where it is stronger than with Sn(IV). With Pb(II) it is very prominent and one can speak of rather typical Pb^{2+} ions. The ionic contribution in bonds with Pb(IV) is less. Thus the rules given in 5-2e are followed, according to which the bonds from a certain atomic species are more covalent the higher the oxidation number of the atom is. The basicity of the oxides and hydroxides increases with the ionic bond contribution (21-5a, d). Thus SnO is intermediate, SnO_2 acid, PbO predominantly basic, and PbO_2 intermediate although mainly acid.

As usual (5-6, beginning) the first element of the group, carbon, is quite distinct from the others. Carbon has unusually special properties, notable among which are its tendency to form carbon chains and its ability to form covalent bonds with many different types of atoms. In this behavior the tetrahedral sp^3 hybridizaton is especially prominent.

The changes in orbital supply and atomic radii (5-6b) as one goes through the carbon group show their effect beautifully. The number of orbitals available for bonding in carbon is at most four, so that, for example, halides with higher coordination number than four cannot be formed. Beginning with silicon the possibility of sixfold coordination arises through the advent of d orbitals usable for bonding. Because of space restrictions such coordination occurs for silicon only with the smallest halogen atom (SiF_6^{2-}), but for subsequent elements larger and larger halogen atoms can be bonded in sixfold coordination (SnI_6^{2-}). With oxygen the coordination number four is not permissable for the small carbon atom, so that the carbonate ion becomes CO_3^{2-} and not CO_4^{4-}. Beginning with silicon there is room for fourfold coordination with oxygen (SiO_4^{4-}, other silicate ions and SiO_2 in which fourfold coordination is achieved by linking tetrahedra; cf. 23-3g, h); beginning with tin sixfold coordination also occurs (SnO_2 and PbO_2 with linked octahedra).

The hydrogen compounds of the elements of the carbon group (hydrides) have covalent structures. As is generally true for such compounds, the stability against decomposition decreases with increasing atomic number. Tin, and still more, lead hydrides are thus quite unstable.

Because of the great variation in properties within the carbon group, the different elements will be treated separately in the following sections.

23-2. Carbon

The greater part of the chemistry of the compounds of carbon falls in the province of so-called organic chemistry. In this book the presentation must be limited to a few very simple carbon compounds, mainly those that do not contain carbon–carbon bonds.

a. Occurrence and history. All living organisms contain carbon compounds as essential substances, and through these carbon has been concentrated in the biosphere and as plant and animal remains (to a large extent used as fossil fuels) in the solid earth's crust. In addition it occurs in the earth's crust as carbonates, especially calcium carbonate, in the hydrosphere as dissolved carbonic acid and in the atmosphere as carbon dioxide. Through life processes and human influence carbon participates in a cycle described in 23-2i.

In nature carbon is found in elementary form as *diamond* and *graphite*. Coal has a very high proportion of carbon but never consists of pure carbon. Like petroleum, coal constitutes a major source of carbon in industry.

Diamonds are mined especially in South Africa, but a large amount, especially of "industrial diamonds", is produced by Brazil. More recently eastern Siberian and

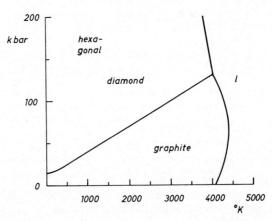

Fig. 23.1. P,T diagram for carbon. The triple point for the equilibrium graphite + l + g lies at about 105 bar and 4100°K, that is, so close to the abcissa that it cannot be plotted.

central African fields have become important. In South Africa and Siberia diamonds occur partly as a "primary" deposit in olivine rocks, in which they were probably formed, but also in "secondary" form, in sedimentary formations. In other locations they occur usually as secondary deposits.

Graphite is a widely distributed mineral, which, however, does not so often occur in minable concentrations. Long-known occurrences of rather well-crystallized graphite are found in Ceylon. Other important occurrences are found in Canada, Mexico, the Soviet Union, South Korea and Madagascar.

Carbon (Lat. carbo) has been known since prehistory, but it may be said to have first been recognized as an element in the modern sense by Lavoisier. After the Latin, it was somewhat later given the French name *carbone*.

b. Properties and uses. Among the isotopes of carbon (tab. 1.4), both the radioactive ^{14}C and the stable ^{13}C are used in tracer techniques (9-3e), and ^{14}C is used for age determination by the C-14 method (9-3d). The most common stable isotope ^{12}C is presently the atomic weight standard (1-3f).

As can be seen from fig. 23.1, graphite is the stable, solid modification of carbon at moderate pressures. Diamond is thus unstable at all ordinary pressures, and only a very low rate for the process diamond→graphite prevents diamond from being transformed at lower temperatures. On heating to about 1500°K (absence of oxygen), however, the transition begins to proceed at an appreciable rate. Since the triple point for the equilibrium graphite +l+g lies at about 105 bar graphite sublimes on heating to sufficiently high temperature at pressures lower than 105 bar (13-4a). Only at pressures above 105 bar can graphite be melted.

At room temperature the stability region of diamond is reached at a pressure of about 20 kbar, but the *synthesis of diamond* from graphite cannot be successfully accomplished by raising the pressure alone because the transition graphite → diamond also proceeds too slowly at room temperature. The rate is increased by heating, but then the pressure must be further raised, leading to great technical difficulties. However, it was found in 1953 that the presence of certain transition metals increases the rate of transformation so that synthesis can be brought about at about 2000°K and 70 kbar (ASEA in Sweden and General Electric Co. in USA). For this purpose Ni is usually used, but other metals in groups 8, 9 and 10 as well as Cr and Mn have a similar effect. By this method synthetic diamonds up to about 5 mm in size have been made so far (weight somewhat over 1 carat = 0.2 g). However, they are generally colored. During 1962 pure graphite was successfully converted to very small diamond crystals at pressures above 130 kbar and temperatures above 3000°K. In these experiments the graphite was heated by strong electric pulses of such short duration that its environment did not reach too high a temperature. Earlier claims for the synthesis of diamond have proved to be erroneous.

Diamond usually forms crystals of cubic symmetry (density 3.51 g cm^{-3}). In the pure, undisturbed state diamond is colorless, but traces of impurities and color centers produced by radiation (6-3a) can cause color. The structure of cubic diamond has been described in 5-3d (fig. 5.16) and 6-2c, and its insulator properties in 6-4d. The very great hardness of diamond (6-5c), its general durability as well as its high refractive index which permits total reflection over a wide range of angles, all make it highly prized as a gem stone. Because of its hardness it is also much used for drilling, cutting, polishing and for drawing very fine metal filaments (industrial diamonds). For this purpose diamonds that are not suitable for jewelry are used, among others often the black forms found in Brazil, and nowadays also synthetic diamonds.

In 1966 it was found that carbon can crystallize with tetrahedral coordination also in a hexagonal structure (*hexagonal diamond, lonsdaleite*). This can be considered as a wurtzite structure (6-2c), in which all atoms are carbon atoms. As a consequence the tetrahedra are arranged in another way than in cubic diamond. The hexagonal diamond can be obtained from graphite in about the area labelled "hexagonal" in fig. 23.1. A condition is that the pressure be applied perpendicular to the six-ring planes of the graphite. This requires a single crystal of graphite or well-crystallized polycrystalline graphite in which the crystals are parallel. The hexagonal diamond material that has been prepared is polycrystalline, with crystallite size of the order of 0.1 μm. Polycrystalline hexagonal diamond has also been found in meteorites.

It is uncertain whether hexagonal diamond is stable within a certain area of the P,T diagram (in which case new equilibrium lines must be added) or is a metastable phase.

Graphite is soft, black and has a submetallic luster. The two forms *hexagonal* and *trigonal (rhombohedral) graphite* were shown in fig. 5.21 and the bonding situation was discussed in 5-3e. The weak bonds between the six-ring sheets are often not strong enough to order the sheets well enough mutually in the formation of a graphite crystal so that well-ordered hexagonal or trigonal graphite is attained. The sheets extend to a large extent parallel to each other, but otherwise their mutual orientation is very irregular. The softness of the crystals also makes them easily deformed. The density of an ideal graphite crystal is 2.26 g cm^{-3}, and thus significantly less than that of diamond. However, much lower values are usually measured.

The structure is very highly disordered when graphite is formed at low temperature ("amorphous carbon"). The first crystallites are then also irregularly oriented with respect to each other. Ordering and crystal growth requires that the strong bonds within the sheet be broken before rearrangement (6-3 d). Only above 2000°C does this process, which involves so-called *graphitization*, proceed reasonably rapidly. If the carbon is deposited directly from the gas phase on a solid substrate at sufficiently high temperature, quite highly ordered graphite crystals are obtained immediately, with the hexagonal sheets parallel to the surface of the substrate. Such *pyrolytic graphite* is prepared by allowing a hydrocarbon to decompose in contact with a graphite surface heated to over 2000°C. The ordering is further increased by a subsequent heat treatment at temperatures over 3000°C.

The structure of graphite leads to a very strong anisotropy (6-1, introduction) in a graphite crystal with respect to many of its properties. In 5-3e the conductivity of graphite was discussed. Parallel to the hexagonal sheets the conductivity in good graphite crystals at 0°C may reach 2.8×10^4 ohm^{-1} cm^{-1}, while at right angles to the sheet it probably is not higher than 1–10 ohm^{-1} cm^{-1}. Thus graphite is a metallic conductor parallel to the sheet but a semiconductor (intrinsic semiconductor) perpendicular to them (6-4a). The heat conductivity shows a similar behavior. The weak bonding forces between the sheets lead to a strong cleavage parallel to them. Atoms and molecules of the most disparate types (K, F, H_2SO_4, $FeCl_3$) can also be absorbed between the sheets, causing an increase in the interlayer distance. If the interlayered substances are removed, the graphite structure is restored if their proportion was not so high that the sheets were broken down. When graphite is oxidized with strong oxidizing agents, for example conc. nitric acid, oxygen atoms are bound between the sheets to form so-called graphite oxide. The oxygen in graphite oxide also causes it to take up water easily.

The lubricating properties of graphite and its "greasy" character are generally explained by the ease with which the sheets glide over each other. However, it has recently been found that the lubricating ability falls off with decreasing air pressure, which indicates that the surfaces gliding over each other consist of layers of impurities, probably mainly water. The ability of graphite to leave a mark is used in pencil lead, which consists of graphite and kaolin mixed in various proportions to produce varying hardnesses (Gk. γράφω, I write; black lead was originally a name for the mineral graphite itself, whose marking ability was thought to depend on the presence of lead).

If a hydrocarbon, for example, is burned with a deficient oxygen supply, *soot* is produced, which for the most part consists of carbon that, however, still contains some combined hydrogen. It has wide industrial application under the name *lamp black* or *carbon black*, primarily in the manufacture of rubber (especially for automobile tires) since the admixed carbon black increases the toughness and durability of the rubber. Large amounts are also used for printing ink and other pigments.

Amorphous carbon has a large surface area and therefore high adsorptive power (15-1a). Amorphous carbon of various kinds (originally mainly charcoal and bone black) is, in fact, much used as an adsorption medium (purification and adsorption analysis of gases and liquids, gas masks). The free surface and thus the adsorptive power can be increased by gentle oxidation of the carbon, for example by heating in water vapor or carbon dioxide (*activation* to *active carbon*).

Fossil carbon generally has a larger proportion of carbon the older it is. *Bituminous coal* contains 75–90 %, *anthracite* 90–95 % C. All coals can largely be considered as a hydrocarbon with a low hydrogen content and a large proportion of aromatic ring systems. In addition, oxygen, nitrogen and sulfur are present. However, Precambrian carbon with up to 99 % C (schungite) is known. It exhibits hardly any long-range order and may therefore be considered as a carbon glass (6-3d).

If coal is heated in the absence of air (dry distillation) volatile constituents are given off and a product richer in carbon, *coke*, remains.

In a *gasworks* the volatile constituents of coal, and among them especially a highly volatile gas mixture with high heat content, *artificial gas* (formerly "illuminating gas"), is the main product. This consists mainly of hydrogen, methane, nitrogen, carbon monoxide and carbon dioxide. With the dry distillation heavier hydrocarbons, nitrogen and sulfur compounds (ammonia, hydrogen cyanide, hydrogen sulfide, carbon disulfide) are also produced. The heavier hydrocarbons are condensed as *coal tar* and the other substances are removed in various ways. Artificial gas nowadays contains admixed water gas (18d) or cracking gas obtained in the cracking of petroleum. The resulting excessively dangerous content of carbon monoxide is decreased by "conversion" with water vapor according to the water gas equilibrium (18.2), for which $K_p > 1$ at temperatures below 810°C. To shift this equilibrium as far as possible to the right low temperature and catalysts (usually iron(III) oxide with 10% chromium(III) oxide added) are used. In the United States artificial gas has largely been replaced by natural gas (23-2d).

Dry distillation of coal is also carried out in *coke plants* in iron works, but here the coke is the main product (however, see 18d for the application of coke-oven gases). The coke is used in metallurgical processes both as fuel and reducing agent.

Also, in oil refining, residues with high carbon content (*petroleum coke*) are obtained.

Finely divided carbon (often ground coke), mixed with a binder (coal tar) and heated to about 1000°C, provides material for electrodes, arc carbons, etc. To obtain a softer product, it is often heated higher (by Joule heating from current passing through the material itself) so that graphitization sets in. This is favored by the presence of SiO_2, which first gives silicon carbide SiC, which at about 2200°C decomposes into carbon and silicon (which vaporizes). The carbon formed in the decomposition can more easily build up graphite crystals (Acheson graphite).

Graphite is widely used for electrodes, crucibles and other high-temperature applications; for pencil leads and lubrication. Graphite is also used as a moderator in nuclear reactors (9-4a) and must then be completely free of neutron-absorbing substances. Pyrolytic graphite, which can be obtained in large pieces in practically parallel-oriented crystals of high order, has found special applications where use is made of its strongly anisotropic properties (for example, as a thermal insulator in the direction normal to the sheets).

Carbon is chemically resistant in all its forms, and is attacked mainly by oxidation, either with oxygen at high temperature or with strong oxidizing agents. Diamond is most resistant, graphite less so, and for the more disordered forms attack occurs more easily the greater the disorder and the free surface area are.

c. Carbides. According to the rules in 5-2a and 5-3f, compounds of metals or boron and silicon with carbon are called carbides. The carbides of the alkali and alkaline-earth metals have appreciable ionic character, and one usually considers the ion C^{4-} as a building unit (5-2f). An important type of alkaline-earth metal carbide is represented by *calcium carbide* CaC_2 ("carbide" in popular terminology 26-5b) in which C_2 groups, which can be thought of as acetylide ions $[|C \equiv C|]^{2-}$, are present as building units. Carbides of the transition elements have metallic conductivity if the carbon content is moderate, and are semiconductors if the content is high. The composition of the metallic carbides is entirely independent of ordinary valence rules and is often variable. A typical representative is *cementite* Fe_3C (34-3a), which, like many other carbides, occurs in steel. In boron carbide $\sim B_4C$ (24-2h) and silicon carbide SiC (23-3k) the bonding is strongly covalent. Especially $\sim B_4C$ and SiC, but also many metallic carbides, are very hard and high-melting and are thus technically important.

Carbides generally can be obtained by reaction between the element in question with carbon at high temperature. In commercial production the starting material is often an oxide so that the first step of the process is a reduction of the oxide with carbon (cf. 23-3k, 26-5b).

Many carbides react with hydrogen ions (most with acids, some even with water) with the formation of hydrocarbons, for example $Al_4C_3 + 12H_2O \rightarrow 3CH_4$ (methane) $+ 4Al(OH)_3$ or $CaC_2 + 2H_2O \rightarrow C_2H_2$ (acetylene) $+ Ca(OH)_2$. The latter reaction forms the basis for the wide use of calcium carbide (26-5b). Often several different hydrocarbons are formed simultaneously.

d. Hydrogen compounds. Carbon forms a very large number of binary hydrogen compounds, *hydrocarbons*, which are usually considered under organic chemistry. These include *methane* CH_4 ("marsh gas"; natural gas consists mostly of methane), *ethylene* C_2H_4, *acetylene* C_2H_2 and *benzene* C_6H_6, whose structures were discussed in 5-3d, e.

e. Halogen compounds. The hydrogen atoms in hydrocarbons can be replaced by halogen atoms, thus making possible an enormous number of halogen compounds. The exchange can often be carried out by the direct action of free halogen on the hydrocarbon. For example, if methane CH_4 and chlorine are heated together to 300°C in the presence of a catalyst or under ultraviolet light, CCl_4 is formed (*tetrachloromethane, carbon tetrachloride*; colorless, noninflammable liquid with b.p. 77°C). Many such halogen compounds have extensive application as solvents. Carbon tetrachloride is the principal cleaning agent of the dry-cleaning industry.

Recently fluoro derivatives have acquired great technical importance, among other things in the form of high-polymers (for example *Teflon*). The fluorine atoms become less and less reactive the more of them there are bound to the same carbon atom. Thus, *dichlorodifluoromethane* CCl_2F_2 is nontoxic, noninflammable, and does not attack metals. Because of its favorable thermal properties (b.p. $-29°C$) it is used as a cooling medium in refrigerators (*Freon*; however, this name is also used for other fluorochloro derivatives of methane and ethane, used for the same purpose).

f. Carbon monoxide. The structure of *carbon monoxide* CO was discussed in 5-3a, d. Thus, its structure is analogous to the structure of the isoelectronic nitrogen molecule N_2, and it also has nearly the same molecular weight. Some of its important properties are the following:

Melting point	Boiling point	Crit. temp.	Crit. press	ΔH^0	ΔG^0
$-205.1°C$	$-191.6°C$	$-140.2°C$	35.0 bar	-110.56 kJ	-137.32 kJ

The correspondence with nitrogen in melting point, boiling point and critical data is striking. The solubility in water is also of the same order as for nitrogen, and thus slight. Carbon monoxide is colorless and odorless.

The following equilibrium holds for the oxidation of carbon monoxide to carbon dioxide (thermodynamic data for 25°C):

$$CO(g) + \tfrac{1}{2}O_2(g) \rightleftharpoons CO_2(g); \quad \Delta G = -257.2 \text{ kJ}; \quad \Delta H = -283.1 \text{ kJ} \quad (23.1)$$

The very negative ΔG value shows that this equilibrium at room temperature is strongly displaced to the right, and this holds up to over 2000°C. Carbon monoxide therefore burns in air (blue flame) and is an important fuel (water gas, 18d; producer gas, see below). Also, its reducing power is of great technical importance, especially in the reduction of metal oxides to metals (21-5b). At ordinary temperature a catalyst is required in order that the oxidation shall proceed at an appreciable rate. Catalysis is also involved when noble metal ions in water solution are reduced by carbon monoxide at room temperature (Pd^{2+} + $H_2O + CO(g) \rightarrow Pd(s) + CO_2 + 2H^+$; the finely divided metal formed darkens the solution, and can be used as a test for carbon monoxide).

With hemoglobin carbon monoxide forms a compound, carbon monoxide hemoglobin, which cannot take up oxygen. Carbon monoxide is therefore a very dangerous poison. With a carbon monoxide content of 1% in the air death can occur in 10 minutes; with 0.1% headache and malaise set in after 1 hour; with 0.01%, daily exposure should not exceed 8 hours. Carbon monoxide is hardly adsorbed by active carbon, so that ordinary gas masks provide no protection. As a protective air filter against carbon monoxide metal oxides that catalyze the oxidation to carbon dioxide are used.

Besides oxygen, carbon monoxide readily combines also with many other nonmetals. Carbon monoxide mixed with chlorine, on irradiation or in the presence of active carbon at about 125°C, gives *carbonyl chloride, phosgene* $COCl_2$ (lung-damaging gas with no warning properties). On heating sulfur vapor with carbon monoxide *carbonyl sulfide* OCS is formed. Technically very important is the conversion of carbon monoxide to hydrocarbons with hydrogen gas in the presence of a catalyst (*Fischer-Tropsch process*, 14-4).

Carbon monoxide is most simply prepared in the laboratory by dropping conc. formic acid into conc. sulfuric acid at about 100°C: $HCOOH \rightarrow H_2O + CO(g)$. Commercially it is made by oxidation of carbon, either with water vapor (*water gas*, 18d) or with atmospheric oxygen (*air gas*). Air gas is made in a so-called gas producer, hence the resulting gas mixture is more often called *producer gas*. In the gas producer air is passed up from below through a heated layer of carbon (generally coke). Here the *producer gas equilibrium* is decisive:

$$CO_2(g) + C(s) \rightleftharpoons 2CO(g) \qquad (23.2)$$

At a total pressure of 1 atm it gives the following gas proportions:

	450	500	600	700	800	900	950	°C
CO_2	98	95	77	42	10	3	1.5	vol %
CO	2	5	23	58	90	97	98.5	vol %

As the air penetrates the lower parts of the carbon layer where the temperature is still low, carbon dioxide is thus predominantly formed according to $C(s) + O_2(g) \rightarrow CO_2(g)$. However, the carbon combustion according to this equation develops a large amount of heat so that the temperature higher up in the carbon layer rises to over 1000°C, and here the carbon dioxide is almost completely converted to carbon monoxide according to (23.2). Thus, producer gas finally consists mainly of carbon monoxide ($\sim 25\%$) and the accompanying atmospheric nitrogen ($\sim 70\%$).

In 2-3b it was shown that carbon monoxide at ordinary temperature is unstable with respect to decomposition into carbon dioxide and carbon, and this is supported by the data given above (the equilibrium (23.2) is strongly displaced to the left). However, the decomposition proceeds at an inconsequential rate, but it can be catalyzed at 300°C by platinum or nickel.

When there is so large an *excess of oxygen* that all carbon is burned, the equilibrium (23.2) no longer applies, but instead the dissociation equilibrum for carbon dioxide (23.1) determines the composition of the gas mixture. This equilibrium even at high temperature is so strongly shifted to the right that mostly carbon dioxide is formed. Even at 2000°C pure carbon dioxide at 1 atm is only 2% dissociated into CO and O_2.

Many transition metals coordinate carbon monoxide as a ligand to form *carbonylmetals* (formerly "metal carbonyls"). These are formed especially by the metals in groups 8, 9 and 10, but also by most of the metals in the two preceding groups

7 and 6 and also by V in group 5. The simplest carbonylmetals are mononuclear and are formed by metals with even atomic number. Among these we may mention *tetracarbonylnickel* $Ni(CO)_4$ (colorless, m.p. $-25°C$, b.p. $43°C$), *pentacarbonyliron* $Fe(CO)_5$ (yellow, m.p. $-20°C$, b.p. $103°C$) and *hexacarbonylchromium* $Cr(CO)_6$ (colorless, easily sublimable crystals). They are thus quite volatile. Like many other carbonylmetals they are formed by direct reaction between very finely divided metal and carbon monoxide. Since carbon monoxide is absorbed in the reaction, the process is favored by increase of pressure, and in addition the reaction rate is increased by gentle heating. However tetracarbonylnickel can be formed at atmospheric pressure and ordinary temperature. With stronger heating (above about 200°C) the carbonylmetals decompose into metal and carbon monoxide. Carbonylmetal synthesis and subsequent decomposition can be used for the preparation of pure metals, and tetracarbonylnickel has become technically very important in this connection (Mond process, 34-4).

Carbon monoxide is bonded to the metal atom by its carbon atom. The existence of the types mentioned above is customarily explained by the observation that the electron cloud of the metal atom, after acquiring one electron pair from each ligand, has the same configuration as the next-following noble gas; in the type examples given 36Kr (Ni, $28 + 4 \times 2$; Fe, $26 + 5 \times 2$; Cr, $24 + 6 \times 2$).

By the action of bases on carbonylmetals, carbonylmetallate ions are formed: $Fe(CO)_5 + 4OH^- \rightarrow Fe(CO)_4{}^{2-} + CO_3{}^{2-} + 2H_2O$. After acidification of the solution a corresponding, very weak acid can be distilled off; in the case of the example thus *dihydrogen tetracarbonylferrate(−II)* $H_2Fe(CO)_4$. $HCo(CO)_4$, for example, is also known. Here also the central atom may have a noble-gas configuration.

Metals of odd atomic number always give carbonylmetals with an even number of metal atoms in the molecule. The total number of electrons thus also becomes even. These polynuclear carbonylmetals are somewhat less fusible and less volatile than the mononuclear ones. One example is $Co_2(CO)_8$ with m.p. $51°C$.

The unusual oxidation numbers that the metal atoms formally take on in the carbonyl complexes have been discussed in 5-6c.

Other atoms or atomic groups can also be bound covalently to transition metal atoms together with carbon monoxide. For example, *halocarbonylmetals* such as $Fe(CO)_5X_2$ and $Cu(CO)X$ are known. The formation of the latter type of compound is made use of when a copper(I) halide solution is used to absorb carbon monoxide. By the action of nitrogen oxide on carbonyl metals *carbonylnitrosylmetals* can be obtained, containing CO and NO groups, which are probably analogously bound. However, for this bonding the NO molecule must give up its unpaired electron to the metal atom so that it is converted to the nitrosyl cation NO^+ (22-9b) analogous in structure to CO. That this does occur is shown by the existence of the compounds $Co(CO)_3(NO)$ and $Fe(CO)_2(NO)_2$, which are very similar to $Ni(CO)_4$, and in which the electron cloud of the metal atoms then acquires 36 electrons (Co, $27 + 1 + 4 \times 2$; Fe, $26 + 2 + 4 \times 2$).

g. Carbon dioxide and carbonic acid. The structure of *carbon dioxide* has been discussed in 5-3e. Some important data are the following:

Triple point	Sublimation point	Crit. temp.	Crit. press	ΔH^0	ΔG^0
$-56.6°C$, 5.18 bar	$-78.45°C$	$31.04°C$	73.80 bar	-393.6 kJ	-394.5 kJ

Concerning the formation of solid carbon dioxide (*carbon-dioxide snow, Dry Ice*) and its sublimation, see 13-4a. Carbon dioxide is colorless in all its forms. The molecular weight is approximately 50 % greater than for nitrogen and oxygen, so that carbon dioxide produced in the air (before diffusion is effective) collects in the lower air layers. The occurrence of carbon dioxide in nature is treated in 23-2i.

At 20°C and total pressure of 1 atm 100 g of water dissolves 0.17 g CO_2, corresponding to 0.93 times the volume of the water. The solution reacts weakly acid (carbon dioxide therefore has an acid odor and taste). In the solution carbon dioxide is dissolved to a large extent as CO_2 molecules, but these participate in the hydration equilibrium $CO_2 + H_2O \rightleftharpoons H_2CO_3$. *Carbonic acid* H_2CO_3 is then protolyzed in the usual way. The manner in which hydration influences the strength of carbonic acid and the retardation effect it leads to has been discussed in 12-2e.

When one attempts to isolate carbonic acid by removing water, the proportion of carbon dioxide in the solution is increased until its solubility is exceeded and it is given off as a gas. Pure H_2CO_3 is therefore unknown.

We have previously discussed the equilibrium between carbon dioxide and the elements (2-3b), and the dissociation equilibrium of carbon dioxide into carbon monoxide and oxygen (23-2f). It is apparent from the ΔG values given that it is very stable at room temperature. At higher temperature it can give up oxygen more easily and can then oxidize strong reducing agents (for example, carbon in the production of producer gas, 23-2f).

Carbon dioxide is obtained in a few places from natural carbon dioxide wells, but far more often as a by-product of the combustion of carboniferous fuels, on calcining natural carbonates (especially "lime burning": $CaCO_3(s) \rightarrow CaO(s) + CO_2(g)$) and fermentation.

If the proportion of carbon dioxide in the original gas mixture is low it must first be combined chemically, for example in an alkali carbonate solution: $CO_3^{2-} + H_2O + CO_2 \rightleftharpoons 2HCO_3^-$. At low temperature the process goes to the right, and when the conversion to hydrogen carbonate ion is practically complete the carbon dioxide is driven off by heating the solution.

Carbon dioxide is purified and condensed by cooling and compression. It is transported and marketed in steel cylinders, but nowadays more than half is transformed by expansion to carbon-dioxide snow (13-4a), which is then compressed into *Dry Ice*.

Gaseous carbon dioxide, obtained mostly from the liquid, is used in the manufacture of carbonated drinks and in fire extinguishers. Dry Ice is used mainly for cooling. Liquid carbon dioxide is used as a cooling medium in refrigerators where it is desired to avoid the risk of poisoning from leakage.

The monamide of carbon dioxide, *amidocarbonic acid* (older name "carbamic acid") NH_2COOH is known only in its salts and esters. *Diamidocarbonic acid* is *urea* $(NH_2)_2CO$. The structural formulas may be written as follows:

HO\C=O HO/	H₂N\C=O HO/	H₂N\C=O H₂N/
carbonic acid	amidocarbonic acid	urea

The ammonium salt of amidocarbonic acid, *ammonium amidocarbonate* NH_2COONH_4 is formed by reaction between ammonia and carbon dioxide (22-6c). By heating this salt under pressure urea is obtained: $NH_2COONH_4 \rightarrow (NH_2)_2CO + H_2O$. Urea is colorless, easily soluble in water and melts at 133°C. It is a very important nitrogen fertilizer, and is also the starting material for the manufacture of carbamide plastics. It is produced mainly by the method just described.

Tricarbon dioxide ("carbon suboxide") C_3O_2 is also known. The molecule is linear and probably has the structure

$$\underline{\overline{O}}=C=C=C=\underline{\overline{O}} \leftrightarrow |O\equiv C-C\equiv C-\overline{\underline{O}}| \leftrightarrow |\overline{\underline{O}}-C\equiv C-C\equiv O|$$

and thus shows great similarity to carbon dioxide. It can be obtained by extracting water from malonic acid: $HOOC-CH_2-COOH \rightarrow O-C-C-C-O + 2H_2O$. Tricarbon dioxide is a gas at room temperature (b.p. 7°C), which is irritating to the mucous membranes and poisonous. The type of mesomerism that characterizes carbon dioxide and tricarbon dioxide is possible only if the molecule has an odd number of carbon atoms. With an even number of carbon atoms no triple bonds can be obtained. Since delocalization of orbitals leads to increased stability (5-3e), this explains why C_2O_2 and C_4O_2 are unknown (C_5O_2 may possibly exist).

h. Carbonates. The salts of carbonic acid are called *carbonates* and contain the carbonate ion CO_3^{2-}, whose structure was described in 5-3e. Carbonate ion of the type CO_4^{4-} does not exist, probably because the carbon atom is too small (cf. the analogous situation with the nitrate ion, 22–10a). Solid *hydrogen carbonates* with hydrogen carbonate ion HCO_3^- are known with certainty only for the alkali metals except lithium, and for ammonium (cf. 12-1c). For certain other metals, for example, Cu(II) and Hg(II), only oxide or hydroxide carbonates are known; for Al(III), Cr(III) and Fe(III) no carbonates at all are known.

Carbonate ions are colorless. Silver carbonate Ag_2CO_3, however, is light yellow because of the strong polarizing effect of the Ag^+ ion (7-3b). With the exception of lithium carbonate, the carbonates of the alkali metals and ammonium are easily soluble in water, and the hydrogen carbonates are somewhat less soluble than the corresponding carbonates. Thallium(I) carbonate is moderately soluble, lithium carbonate is rather difficultly soluble, other carbonates are poorly soluble. However, all carbonates are decomposed by sufficiently acid solutions with the evolution of carbon dioxide.

The basic properties of the carbonate ion (chap. 12) make all fairly strong solutions of alkali and thallium(I) carbonates strongly basic.

A large number of carbonates of the formula type $M^{II}CO_3$ at atmospheric pressure have the same crystal structure as *calcite*, that is, the stable, trigonal (rhombohedral) modification of $CaCO_3$. Certain other carbonates of the same formula type, on the other hand, have the same structure as *aragonite*, the orthorhombic modification of $CaCO_3$, metastable at atmospheric pressure. If all these carbonates are arranged in order of increasing cationic radius, we obtain the series:

	Mg^{2+}	Co^{2+}	Zn^{2+}	Fe^{2+}	Mn^{2+}	Cd^{2+}	Ca^{2+}	Sr^{2+}	Pb^{2+}	Ba^{2+}
ionic radius, Å	0.65	0.72	0.74	0.75	0.80	0.97	0.99	1.13	1.21	1.35
			calcite structure						aragonite structure	

The aragonite structure, in which the cation is coordinated to eight oxygen atoms (in calcite only to six), appears when the cationic radius exceeds a certain value, and the radius of Ca^{2+} evidently lies so close to this value that the aragonite structure can be formed under certain conditions. At 100°C the stability region of aragonite is reached when the pressure exceeds about 4.5 kbar.

A similar transition is found in the alkali nitrates (the nitrate and carbonate ions have analogous structures), where $LiNO_3$ and $NaNO_3$ have the calcite structure but KNO_3, for which the cation is larger, has the aragonite structure (for $RbNO_3$ and $CsNO_3$ the cations are too large even for the aragonite structure).

Carbonates establish equilibria with metal oxide and carbon dioxide according to, for example, $M^{II}CO_3 \rightleftharpoons M^{II}O + CO_2(g)$. For $CaCO_3$ this equilibrium was discussed in 13-5e. The carbonate is more stable with respect to decomposition the larger the cationic radius is, and the smaller its charge. The alkali carbonates are most stable (fusible without appreciable decomposition) and their stability increases with increasing atomic number of the metal. The carbonates of the alkaline-earth metals are altogether less stable than the alkali carbonates, but also here a corresponding trend is noticeable. A carbon dioxide pressure of 1 atm is thus reached at the following temperatures: for $MgCO_3$, 353; $CaCO_3$, 898; $SrCO_3$, 1260; $BaCO_3$, 1430°C.

Carbonates of other metals as a rule are easily decomposed by heating. So also are all hydrogen carbonates, for example $2NaHCO_3(s) \rightarrow Na_2CO_3(s) + H_2O(g) + CO_2(g)$. The development of carbon dioxide on heating hydrogen carbonate is made use of when sodium or potassium hydrogen carbonate is used as a fire-extinguishing powder. Sodium hydrogen carbonate is used as baking powder, either alone ("baking soda") or mixed with acid-reacting substances (usually hydrogen phosphates), to facilitate the evolution of carbon dioxide.

i. The carbon cycle in nature. This is shown schematically in fig. 23.2. The great carbon reservoir directly available to life on the earth is the carbon dioxide in the atmosphere and the hydrosphere. To a large extent an equilibrium exists between the carbon dioxide in the atmosphere and in the hydrosphere, and in this equilibrium the solid carbonates on the ocean floors also participate. This equilibrium is very sluggishly established because of long transport routes. Large local variations can also occur, for example because of variations in temperature, salinity of the sea, etc. The proportion of carbon dioxide in the atmosphere on the average is 0.031 vol% = 0.046 wt%, corresponding to a total weight of approximately

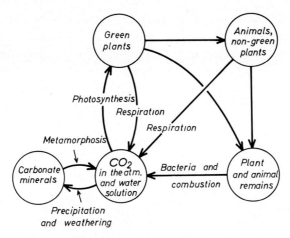

Fig. 23.2. The carbon cycle in nature.

2.3×10^{15} kg CO_2. The average for the hydrosphere is about 0.01 wt % (calculated as CO_2 but present as CO_2, H_2CO_3, HCO_3^- and CO_3^{2-}), which corresponds to a total weight of 141×10^{15} kg.

By far the most important conversion of carbon dioxide in the atmosphere and the hydrosphere takes place through living organisms and their remains. Photosynthesis on the earth yearly consumes about 5×10^{14} kg CO_2 (1.6×10^7 kg s^{-1}), which is thus brought to living organisms. Approximately the same amount of CO_2 is formed by respiration and putrefaction processes. But this may be compared with the fact that man by combustion of fuel (mostly carbon and oil) adds to the atmosphere about 0.1×10^{14} kg CO_2 yearly. If none of this additional CO_2 is dissolved in the hydrosphere, and other consumption and production of CO_2 are balanced out, the proportion of CO_2 in the atmosphere would be doubled in a little more than 300 years. As a matter of fact, a continual increase in the average CO_2 content can be detected.

Since the CO_2 molecule has strong infrared absorption, the CO_2 content of the atmosphere serves to decrease radiation of heat from the earth. A doubling of the CO_2 content would, barring other influencing factors, cause an increase in the mean temperature of the atmosphere of the whole earth of 3.6°C. It is therefore clear that the increase in the proportion of CO_2 just referred to can have climatic significance.

j. Sulfur compounds. The most important sulfide of carbon is *carbon disulfide* CS_2, whose structure may be considered to be analogous to that of carbon dioxide. At ordinary temperature it consists of a colorless, heavy, strongly light-refractive, volatile liquid (m.p. -112°C, b.p. 46. 3°C). In quite pure form it has hardly any odor, but the impurities that are commonly present give it an obnoxious smell. It is very poisonous and particularly easily flammable.

Carbon disulfide is produced commercially by reaction between charcoal and sulfur vapor at about 850°C ($C(s) + 2S \rightleftharpoons CS_2$), or between hydrocarbons and sulfur vapor ($CH_4 + 4S \rightleftharpoons CS_2 + 2H_2S$). The major portion of the carbon disulfide output is used in making cellulose xanthate, which, dissolved in lye to give *"viscose"*, is reconverted to cellulose in the form of fiber (rayon, that is, artificial silk and cellulose wool) or thin film (cellophane). Carbon disulfide is also used as a solvent for sulfur in the cold vulcanization of rubber, for the manufacture of xanthates used for flotation (15-2), and for combating injurious animals (insects and rodents).

Carbon disulfide with sulfide ion gives the *trithiocarbonate ion* CS_3^{2-}, analogous to the carbonate ion: $CS_2 + S^{2-} \rightleftharpoons CS_3^{2-}$. By treating a thiocarbonate with acid free *trithiocarbonic acid* H_2CS_3 can be obtained, which is an unstable, oily, red liquid. *Mono-* and *dithiocarbonate ions*, CSO_2^{2-} and CS_2O^{2-} respectively, are also known.

Carbonyl sulfide OCS was mentioned in 23-2f. At room temperature it consists of a flammable gas, which is colorless and odorless.

k. Nitrogen compounds. All the carbon-nitrogen compounds considered here are pseudohalide ions, or their corresponding acids or dimeric pseudohalogens. For a survey of this group of compounds reference should be made to 20-6. There the electronic structure of the ions is discussed, which for the diatomic cyanide ion CN^- is analogous to that of carbon monoxide (5-3a, d) and for the triatomic ions is analogous to that of carbon dioxide (5-3e).

1. *Cyanide ion* CN^-. The most commonly used cyanide is *sodium cyanide* NaCN (*potassium cyanide* KCN has become of less commercial importance). In the most important commercial method of production, ammonia is blown over a heated mixture of sodium and charcoal. The following series of reactions take place: $2NH_3 + 2Na \rightarrow 2NaNH_2$(sodium amide)$ + H_2$; $2NaNH_2 + C \rightarrow Na_2CN_2$ (sodium cyanamide, see below 23-2k6)$ + 2H_2$; $Na_2CN_2 + C \rightarrow 2NaCN$. In 23-2k6 a method for preparing calcium cyanide is also described.

Most cyanides are decomposed by acids with the formation of *hydrogen cyanide* HCN, which is generally given off as a colorless, flammable gas (m.p. $-13.4°C$, b.p. $25.5°C$). In hydrogen cyanide there is probably an easily shifted equilibrium between the acids H—C≡N and H—N≡C, which, however, cannot be separated from each other. As esters, on the other hand, *nitriles* R—C≡N as well as *isonitriles* R—N≡C are known (cf. the similar situation for sulfurous and nitrous acids described in 21-11a and 22-10c).

Hydrogen cyanide is very poisonous but is notable more for its extraordinarily fast action than for its minimum lethal dose (about 0.05 g). It is used to combat injurious animals. In high dilution it smells like bitter almond (it is present in several plant substances), in higher proportions it has a very unpleasant odor and taste. However, many people are quite insensitive to its odor. In water solution hydrogen cyanide is an acid, which in contrast to the hydrogen halides is very weak ($pK_a = 9.40$).

The cyanides of the alkali metals and the alkaline-earth metals are easily soluble in water, and because of the low acid strength of hydrogen cyanide the proportion of HCN in the solution may be so high that the solution smells strongly of it.

The cyanide ion easily forms *cyano complexes* with metal ions, especially of transition metals (5-6c; the small space requirements of the cyano ligands was discussed in 5-3i; see also 5-3j). The complexes of the transition metals beginning with group 6 and going to the right in the periodic table are especially stable. Thus, the *dicyanoargentate ion* $Ag(CN)_2^-$ is so stable that AgI in spite of its low solubility product cannot be precipitated from a solution containing this complex ion. The *hexacyanoferrate*(II) *ion* $Fe^{II}(CN)_6^{4-}$ ("ferrocyanide") and the *hexacyanoferrate*(III) *ion* $Fe^{III}(CN)_6^{3-}$ ("ferricyanide") are so stable that they are decomposed only slowly by strong acids (therefore, the former is practically non-poisonous). Because of the formation of cyano complexes, most poorly soluble heavy-metal cyanides dissolve in water when cyanide ion is added in excess (13-1e). The toxicity of the cyanides probably results from the formation of a cyano complex with metals in the respiratory enzymes, so that they are disabled.

Solutions of cyano complexes are very important in electrolytic metal plating (16f). Cyanides (including hydrogen cyanide) are much used in preparative organic chemistry.

2. *Cyanogen* $(CN)_2$. The dimeric cyanogen as a pseudohalogen corresponds to the dimeric halogen molecules (20-6). The formula $|N{\equiv}C{-}C{\equiv}N|$ represents the main aspect of its structure, which is linear. Cyanogen is a colorless, poisonous, flammable gas with a penetrating odor (m.p. $-34.4°C$, b.p. $-21.2°C$). It is formed by heating heavy metal cyanides ($Hg(CN)_2 \rightarrow Hg + (CN)_2$), but is most easily prepared by reaction between copper(II) ion and cyanide ion in rather concentrated solution: $Cu^{2+} + 2CN^- \rightarrow CuCN(s) + \frac{1}{2}(CN)_2(g)$. Gay-Lussac, who prepared cyanogen in 1814, named it "cyanogène" (Gk. κύανος, dark blue; γεννάω, I make), since he believed it to be the color-producing constituent of Prussian blue (34-5b; for the same reason hydrogen cyanide was formerly called "Prussic acid").

Cyanogen dissolves easily in water, ethanol and ether, but rapidly decomposes in solution to give various carbon- and nitrogen-containing substances. When heated to about 400°C it polymerizes into a solid white substance, *paracyanogen* $(CN)_n$.

The flame produced on combustion of cyanogen has a maximum temperature of 4,580°C, and is thus the hottest flame that can be achieved by purely chemical means. The reason for the high temperature lies in the formation of two very stable molecules, carbon monoxide and nitrogen, according to: $(CN)_2(g) + O_2(g) \rightarrow 2CO(g) + N_2(g)$; $\Delta H = -530$ kJ.

3. *Cyanate ion* OCN^-. *Cyanates* can be prepared by oxidation of cyanides: $CN^- + \frac{1}{2}O_2 \rightarrow OCN^-$. The cyanates are easily decomposed in water solution: $OCN^- + 2H_2O \rightarrow NH_3 + HCO_3^-$. Free cyanic acid is also very unstable. In this acid there probably exists an equilibrium between *cyanic acid* HOCN and *isocyanic*

acid HNCO, which, however, cannot be separated from each other. Esters are known only for isocyanic acid, which thus have the general formula RNCO.

4. *Fulminate ion* ONC$^-$. This ion, which is isomeric with the cyanate ion, is best known in *mercury*(II) *fulminate*, which precipitates out as a white, crystalline powder when a nitric acid solution of mercury(II) nitrate is poured into ethanol. The salt, like other fulminates (Lat. fulmen, lightning), detonates with shock or strong heating, and has been much used as a detonating primer under the name "fulminating mercury". It is now usually replaced by other substances (among others, lead azide, 22-6f). The acid HONC, *hydrogen fulminate* has also been called "fulminic acid".

5. *Thiocyanate ion* SCN$^-$. *Thiocyanates* can be prepared by fusing alkali cyanides with sulfur: CN$^-$ + S → SCN$^-$ (cf. preparation of cyanates). The thiocyanates are quite stable in water solution and in their reactions and solubilities particularly resemble bromides and iodides. A typical and very sensitive reaction for the thiocyanate ion is the red coloration that is produced with Fe(III) ion, which results mainly from the formation of the complexes FeSCN^{2+} and Fe(SCN)$_2^+$ (this color reaction led to the earlier designation of thiocyanates as "rhodanides" after the Gk. ῥόδεος, rose-red).

Thiocyanic acid is very unstable and shows the same type of isomerism as cyanic acid (see above). Free *thiocyanogen* (SCN)$_2$ ("rhodane") is also known.

6. *Cyanamide ion* NCN^{2-}. The ion is usually written CN$_2^{2-}$. The most important salt is *calcium cyanamide* CaCN$_2$. It is prepared by passing nitrogen gas over calcium carbide, which is locally heated to about 1000°C: CaC$_2$ + N$_2$ → CaCN$_2$ + C. Heat is released by the reaction so that continued heating is unnecessary. Pure calcium cyanamide is colorless, but the carbon formed colors the product gray-black. This product under the name *Cyanamid* is an important fertilizer. In the earth the calcium cyanamide is decomposed by water through several intermediate stages, the overall process, however, being: CaCN$_2$ + 3H$_2$O → CaCO$_3$ + 2NH$_3$. The weed-killing action of cyanamid is also found useful.

The carbon-calcium cyanamide mixture sets up an equilibrium CaCN$_2$ + C ⇌ Ca(CN)$_2$, which above 1000°C begins to be appreciably displaced to the right. This is used in the production of *calcium cyanide*. Addition of sodium chloride favors the shift in equilibrium, and decomposition of the calcium cyanide formed is prevented by rapid cooling. The method is an important source of cyanide.

Free *cyanamide* H$_2$NCN, which may be considered as the amide of cyanic acid, is formed by moderate acidification of cyanamide salt solutions: NCN^{2-} + 2H$^+$ → H$_2$NCN. Cyanamide (m.p.45°C) is colorless and easily soluble in water. It is easily dimerized to *dicyandiamide* (H$_2$NCN)$_2$, especially in weakly basic solution and on heating. Both in strongly acid or basic solution it is hydrolyzed to urea (23-2g): H$_2$NCN + H$_2$O → (NH$_2$)$_2$CO.

23-3. Silicon

a. Occurrence and history. Silicon is one of the most common elements in the universe (seventh in rank in tab. 17.1) and the next most abundant after oxygen in the accessible portion of the earth (tab. 17.2). It never occurs free and is almost always bound to oxygen in silicon dioxide SiO_2 and in various silicates. It generally plays a small role in living organisms. In plants silicon is found especially in diatoms, horsetails and grasses; in the animal world it is found especially in feathers, and in lesser amounts in connective tissue.

Torbern Bergman in the 1770's considered silicon dioxide to be an *earth*, as the class of refractory and difficultly reducible oxides was then designated. It was first established as an element in 1823 by Berzelius, who gave it the Swedish name *kisel* (Ger. Kiesel, flint). The name *silicon* is derived from the Latin word silex, flint.

b. Properties. As mentioned earlier in 23-1, silicon at ordinary pressure is an intrinsic semiconductor, and the pure element forms hard but brittle, lustrous, opaque crystals of dark gray color. In the visible spectral region its absorption is high (hence the opacity), but in the infrared range, on the other hand, very low. The density is 2.33 g cm^{-3}. The crystal structure is the same as that of diamond. The melting point is 1420°C, the boiling point 2477°C. Several other modifications of silicon have been found at high pressure. One of these, existing at a pressure above 120 kbar, has the same structure as white tin (23-5b) and shows metallic conductivity.

Silicon is chemically very resistant and is attacked at ordinary temperature only by fluorine and hydrogen fluoride. When heated in air it is oxidized only above 1000°C.

c. Production and uses. Silicon compounds as such, primarily silicates, have enormous technical importance (23-3i). However, the element is nowadays also important as a source material, and therefore its production plays a considerable role.

The most important source of silicon, silicon dioxide, is difficult to reduce (21-5b). The largest amount of commercial silicon is made by reduction of silicon dioxide as quartz by carbon in an electric arc furnace: $SiO_2 + 2C \rightarrow Si + 2CO(g)$. Iron is often added and then an iron-containing product is obtained, *ferrosilicon*, which can be used directly if the silicon is intended for steel production. Purer silicon is obtained by reduction of silicon tetrachloride, for example with hydrogen or zinc vapor at about 1000°C: $SiCl_4(g) + 2Zn(g) \rightarrow Si(s) + 2ZnCl_2(g)$; or by thermal dissociation of SiI_4.

Silicon is much used in metallurgy, partly as a reducing agent in metal production and in the deoxidation (21-5b) of metal melts, and partly as an alloy constituent (silicon-alloy steels, 34-3e). To an increasingly great extent silicon is used as a semiconductor in rectifiers and transistors (6-4d). It is then highly purified by zone melting (13-5f2). Because of its low infrared absorption, silicon is also used in infrared optics.

d. Silicides. According to the appropriate rules of nomenclature (5-2a, 5-3f) compounds of silicon with metals or boron are called silicides. The silicides of the transition elements have metallic character (6-4g). The silicides of other not too electropositive metals have metallic conductivity with moderate silicon proportions, and are semiconductors with higher silicon content. The composition of the metallic silicides is not determined by ordinary principles of valency, and is often variable. Many silicides are technically important because of their hardness, chemical resistivity and high melting points. They can generally be prepared by heating the metal and silicon together. Many silicides react with hydrogen ions (acids; the silicides of the more electropositive metals even with water) with the formation of silicon hydrides (23-3 e): $Mg_2Si(s) + 4H^+ \rightarrow 2Mg^{2+} + SiH_4(g)$.

e. Hydrides. Silicon forms a series of silicon hydrides ("silanes"), which are formally and structurally analogous to the saturated aliphatic hydrocarbons. The general formula is thus Si_nH_{2n+2}. Pure silicon hydrides up to Si_5H_{12} are known. Silicon hydrides arise on decomposition of silicides with acids (23-3 d), but in general several different hydrides are formed simultaneously. The reaction between silicon halides and lithium tetrahydridoaluminate (24-3 d) in ether solution gives better yields and more discrete products: $SiCl_4 + LiAlH_4 \rightarrow SiH_4(g) + LiAlCl_4$; $2Si_2Cl_6 + 3LiAlH_4 \rightarrow 2Si_2H_6(g) + 3LiAlCl_4$.

The physical properties of the silicon hydrides are in most cases reminiscent of those of the corresponding hydrocarbons, but their chemical properties are totally different. They are less stable with respect to a number of processes. The decomposition temperatures are lower, their oxidizability greater, and they are decomposed by traces of alkali (for example, from glass). They are all poisonous.

f. Halogen compounds. Silicon combines directly with halogens with the formation of *tetrahalides*: $Si(s) + 2Cl_2(g) \rightarrow SiCl_4(g)$. With fluorine the reaction goes spontaneously at room temperature, but as the atomic number of the halogen increases, higher and higher temperatures are required. With iodine red heat is necessary. The melting and boiling points of the tetrahalides increase with increasing molecular weight (at room temperature SiF_4 is a gas, $SiCl_4$ and $SiBr_4$ are liquids, and SiI_4 is solid).

By heating the tetrahalides with silicon, halides richer in silicon can be obtained. These contain silicon chains to which the halogen atoms are bound, and they therefore have the general formula Si_nX_{2n+2}.

The silicon halides react violently with water (hydrolysis of halides, 20-5a), generally with the formation of silicon dioxide hydrates: $SiX_4 + (2+n)H_2O \rightarrow SiO_2 \cdot nH_2O + 4HX$. They therefore fume in moist air.

Silicon tetrafluoride SiF_4 is obtained more conveniently than through synthesis from the elements by the action of hydrogen fluoride on silicon dioxide. Here there occurs the equilibrium $SiO_2 + 4HF \rightleftharpoons SiF_4 + 2H_2O$, and a requirement for the formation of SiF_4 is that H_2O be tied up. Usually one starts with calcium fluoride (fluorite), quartz sand and conc. sulfuric acid. The last reagent gives HF with CaF_2 ($CaF_2 + H_2SO_4 \rightarrow CaSO_4(s) + 2HF$) and ties up the water. Hydrogen fluoride also attacks silicates with the formation of SiF_4. In this property lies its ability to *etch glass*.

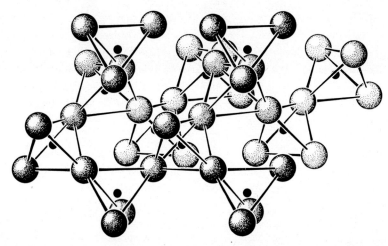

Fig. 23.3. The structure of high quartz. Small, black spheres = Si, large spheres = O.

Silicon tetrafluoride under ordinary conditions is a colorless gas with a pungent odor. Its hydrolysis corresponds to a reaction to the left in the above-mentioned equilibrium equation. In this hydrolysis the hydrogen fluoride formed reacts with the silicon tetrafluoride still unconverted to give *hexafluorosilicate ion* SiF_6^{2-}: $2HF + SiF_4 \rightleftharpoons 2H^+ + SiF_6^{2-}$ (cf. 23-1). The existence of an acid H_2SiF_6 has not been demonstrated (12-2f). If water is removed from the solution the reaction goes to the left. The hexafluorosilicate ion is poisonous.

g. Oxides. In silicon dioxide SiO_2 and oxosilicates silicon has the oxidation number IV and is almost always surrounded tetrahedrally by four oxygen atoms. The overriding tendency toward the coordination number four with respect to oxygen forms the basis of the structural chemistry of silicon. Only in the rather unimportant monoxide SiO does silicon have the oxidation number II.

Silicon monoxide SiO can be obtained as a gas by strong heating of a mixture of silicon and silicon dioxide. The equilibrium $Si + SiO_2 \rightleftharpoons 2SiO(g)$ is shifted appreciably to the right only at very high temperatures. However, by quenching the gas it can be condensed into metastable, solid SiO, a brownish-black substance, which probably has a very disordered crystal structure.

Silicon dioxide SiO_2 exists in several modifications with different crystal structures. In all except the high-pressure phase stishovite, mentioned below, each silicon atom coordinates four oxygen atoms by allowing the SiO_4 tetrahedra to have common corners. In this way molecules are formed with infinite extent in three dimensions, *framework molecules*, as in fig. 23.3. At ordinary pressure the following phases are stable:

(low)quartz[1] $\xrightleftharpoons{573°C}$ high quartz[1] $\xrightleftharpoons{870°C}$ tridymite $\xrightleftharpoons{1470°C}$ cristobalite $\xrightleftharpoons{1720°C}$ melt

---

[1] Quartz and high quartz are also called α and β quartz, respectively. However, it should be noted that these designations are sometimes reversed.

In quartz and high quartz the SiO_4 tetrahedra are connected together in the same way and the two phases are distinguished only by small differences in the relative positions of the tetrahedra (slightly different Si–O–Si angles). The phase transition between quartz and high quartz therefore goes very rapidly and reversibly. The other transitions, on the other hand, involve changes in the linkage of the tetrahedra and therefore take place very sluggishly. Both cristobalite and tridymite can thus easily be supercooled. Also, the melt especially easily supercools and at lower temperatures forms *silicon dioxide glass* ("fused quartz"; improperly called "quarz glass"). The reason is that the melt also contains linked SiO_4 tetrahedra and therefore a large number of strong bonds must be broken in order to build up an ordered crystal structure (6-3d).

For both supercooled cristobalite and tridymite rapid phase transitions occur of the same type as that between quartz and high quartz.

In both quartz and high quartz the structures lack mirror planes and centers of symmetry, so that they form optically active, left- and right-handed crystals (fig. 6.15).

Recently it has been questioned whether tridymite actually exists in the pure SiO_2 system. Its existence may be the result of impurities which need be present in only very small amounts.

At high pressures, other, denser SiO_2 modifications are stable. *Coesite* is known (stable at pressures > 20 kbar), which is built up of SiO_4 tetrahedra in a more compact linkage than in the above modifications, and *stishovite* (stable at pressures > 130 kbar), which has the rutile structure (29f) and is thus built up of SiO_6 octahedra. Transitions to and from these high-pressure modifications are also sluggish. In order for the phases to be formed at an appreciable rate from quartz, for example, temperatures above at least 500°C are required. Because of the sluggishness of the reaction, the high temperature phases remain after release of pressure at room temperature for an indefinite time. The rare occurrences of these two phases in nature has been found always to be associated with meteoric impacts, and their presence in a rock is used as evidence of such an event at that location in the geologic past.

If quartz powder is heated under pressure together with a very weak NaOH solution (380–585°C, 350–1250 bar) small crystals are slowly formed of still another SiO_2 modification. This modification, which has been named *keatite*, is built up of SiO_4 tetrahedra in a somewhat less dense arrangement than in quartz. The proportion of NaOH is critical; if it is too high quartz is formed (cf. below, the preparation of "synthetic" quartz), if it is too low cristobalite is produced.

Especially in the form of quartz, silicon dioxide occurs abundantly in the earth's crust. Larger, water-clear quartz crystals, *rock crystal*, are used as gemstones, as are colored quartz crystals, for example, *amethyst* (violet), *citrine* (yellow), *rose quartz* (light red) and *smoky quartz* (brown, sometimes erroneously called "smoky topaz"; the color is caused by F centers of the type described in 6-3a). Hydrated silicon dioxide is also common, with more or less amorphous structure; it is discussed in 23-3h6.

Silicon dioxide is an acid oxide and is therefore little attacked by acids, except hydrogen fluoride. On the other hand it is attacked by basic solutions and basic oxides. The formation of silicates in the latter case is much used in metallurgy (21-5a). Quartz is the most important starting material for the production of silicon and silicon compounds. Its hardness (7 on Mohs' scale, 6-5c) makes it useful also in grinding (quartz sand).

Silicon dioxide glass is very important as a material for apparatus that must be resistant to acids or high temperatures. Its small coefficient of thermal expansion permits it to tolerate sudden temperature changes without breaking. Fine filaments of silicon dioxide glass can easily be drawn, and because of its favorable mechanical properties (high strength and small elastic hysteresis effects) these are frequently used as suspension filaments in instruments. Ordinary silicon dioxide glass is prepared by heating quartz to just above the melting point, and then quenching. The glass then contains large amounts of small gas bubbles, which make it milky. Clear glass is obtained by allowing the product to collect in an oxyhydrogen flame, fed with a gas mixture carrying quartz powder or silicon tetrachloride (which gives silicon dioxide in the flame). Air occlusion then does not occur. At temperatures above about 1100°C (lower in the presence of certain impurities) silicon dioxide glass crystallizes quite rapidly to cristobalite and is then unusable.

Nowadays a less pure but very useful silicon dioxide glass is produced in a quite special way (trade name *Vycor*). By proper heat treatment, alkali borosilicate glass (23-3i2) can be transformed into two finely divided and intimately mixed glass phases. One of these contains about 96% SiO_2 and 3% B_2O_3, the other mainly alkali borate. The object made of these two phases is treated with warm acid, so that the latter phase is dissolved out and the former phase remains in a porous but coherent form. This is sintered at 1200°C, forming a homogeneous glass phase. On sintering strong shrinkage occurs, which must be taken into consideration in the design of the original form.

The low ultraviolet light absorption of both quartz and silicon dioxide glass is utilized in ultraviolet optics. This property together with its insensitivity to thermal shock makes silicon dioxide glass useful also in "quartz lamps".

The crystal structure of quartz is such that on a quartz crystal, when it is compressed or expanded, certain areas become positively charged and opposite areas become negatively charged (*piezoelectric effect*). Compression gives charges opposite to those produced by expansion. Conversely, a quartz crystal placed in a varying electric field will contract or expand in response to changes in field strength. Because of these properties quartz is much used for transforming mechanical to electrical oscillations and vice versa, for example in frequency control, under-water signalling, echo sounding and locating submarines. The scarcity of good, natural quartz crystals for such purposes has led to the production of "synthetic" quartz crystals. In a closed cylinder filled with NaOH solution the pressure is held at 1500 bar and the temperature at 350–400°C. The bottom of the cylinder is maintained at a somewhat higher temperature than the upper part. Quartz powder is placed in the bottom, and this slowly dissolves in the liquid. The solution rises upward and when its temperature thus drops quartz crystallizes out on seeds suspended in the upper region. A quartz crystal of 0.5 kg can be grown in less than one month.

h. Silicates and silicic acids. The major portion of the earth's crust consists of silicate phases (and silicon dioxide), which occur in very great variety. The

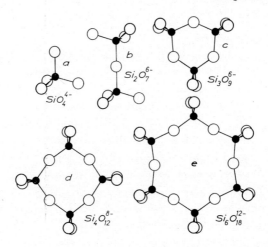

Fig. 23.4. Isolated silicate ions. Solid circles = Si, open circles = O.

structural principles of the silicates were unknown until their structures were determined by X-ray diffraction, especially by W. L. Bragg and his coworkers beginning in 1926. It has been said that this work "transformed a chemical riddle into a system of simple and elegant architecture". Among the scientists who have contributed to subsequent developments in this field we should especially mention Pauling and Belov.

As stated earlier (5-6b, 21-5a), the silicates are characterized especially by the fact that every silicon atom coordinates tetrahedrally four oxygen atoms. Therefore they invariably contain SiO_4 tetrahedra. If the silicate has less than four oxygen atoms per silicon atom (the number of oxygen atoms per silicon atom is hereafter written O/Si) the latter take care of the oxygen atoms by coordinating them in common. In this way SiO_4 tetrahedra are linked together by common corners so that complex, polynuclear polysilicate ions arise (cf. the wholly analogous formation of polyborate ions described in 5-6b). In the following section the silicates are classified according to the degree of coupling of the SiO_4 tetrahedra.

Although the ionic-bond contribution to the Si-O bond is at most 50 %, and free Si^{4+} ions are unknown, the silicon atoms are often assumed to be Si^{4+} ions and the oxygen atoms to be oxide ions O^{2-}. In discussions of the space requirements of the various kinds of atoms in silicate structures it is thus customary to use ionic radii throughout. Such is the case, for example, in accounting for the function of aluminum in silicates. Aluminum is the most common element next to silicon and oxygen in silicate minerals. The ionic radius of Al^{3+} is so little larger than the radius of Si^{4+} (fig. 5.5) that Al^{3+} can replace Si^{4+} as the central atom in SiO_4 tetrahedra. Thus, *aluminosilicates* are obtained. However, the value of its radius is sufficiently large to allow Al^{3+}, like other larger cations, also to occupy positions

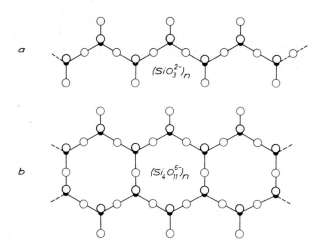

Fig. 23.5. Infinite chain and band ions in silicates. Solid circles = Si, open circles = O.

outside the tetrahedra in which more than four oxygen atoms can be coordinated (almost always six; cf. ionic size and coordination number, 5-2d).

Silicate minerals are very often solid solutions (6-3a). Minerals with different compositions and in many cases different names may belong to the same phase, and may be considered to be related to each other by different solution mechanisms (both substitution and interstitial solution occur). Sometimes this relationship involves several simultaneous solution processes of different kinds. For reasons of space only ideal formulas are given in the following paragraphs, but the actual compositions may depart significantly from these formulas.

1. *Isolated, mononuclear silicate ion* (*nesosilicates*, from Gk. νῆσος, island). If O/Si ⩾ 4, isolated SiO_4 groups can be formed. The characteristic building unit then becomes the ion SiO_4^{4-}, which is usually called *orthosilicate ion* (fig. 23.4a). Orthosilicates with the general formula $(M^{II})_2SiO_4$ belong to the *olivine group*. The most common olivines can be formulated as $(Mg, Fe^{II})_2SiO_4$. The cations Mg^{2+} and Fe^{2+} can thus substitute for each other, but two cations M^{2+} are always associated with one SiO_4^{4-} ion. Other orthosilicates are represented by *zircon* $ZrSiO_4$ and the *garnet group* with the general formula $(M^{II})_3(M^{III})_2(SiO_4)_3$ (usually, M^{II} = Ca, Mg, Fe, Mn; M^{III} = Al, Fe, Cr; garnets, usually red in color, are used as gemstones, and powdered garnet is used as an abrasive.)

A number of orthosilicates also contain other anions, for example oxide ions in *andalusite*, *sillimanite* and *kyanite*, which all have the formula Al_2OSiO_4, and fluoride ions in *topaz* $Al_2F_2SiO_4$ (as a gemstone, usually yellow). In Al_2OSiO_4, O/Si = 5, and the fifth oxygen atom is not coordinated to silicon (cf. 21-5a).

2. *Isolated, polynuclear silicate ions* (*sorosilicates*, from Gk. σωρός group). If O/Si < 4 the oxygen atoms are not sufficient to form isolated SiO_4^{4-} ions. However, tetrahedral coordination can be maintained if the oxygen atoms are shared in

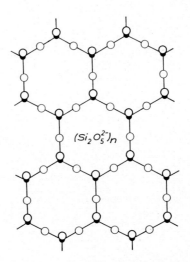

Fig. 23.6. Infinite layer ion in silicates. Solid circles = Si, open circles = O.

common by two tetrahedra. In this way polynuclear, but still isolated silicate ions can occur. The simplest of these are illustrated in fig. 23.4 b–e. The rings shown have the general formula $(SiO_3^{2-})_n$. The best known representative is *beryl* $Be_3Al_2Si_6O_{18}$ (the ion e). The highly prized gemstones *emerald* (colored green by a small amount of Cr^{3+} replacing Al^{3+}) and *aquamarine* (light blue-green) are colored beryls.

3. *Infinite chain and band structures.* (*inosilicates*, from Gk. ἰνός,, fiber). The SiO_4 tetrahedra may also be linked into straight, infinite chains (fig. 23.5a) with the formula $(SiO_3^{2-})_n$, which are characteristic of the large group of silicate minerals called *pyroxenes*. To this group belong, for example, the *augites* which can be derived from *diopside* $CaMg(SiO_3)_2$ by substitution especially of Ca^{2+} or Mg^{2+} by Fe^{2+}, and of $Mg^{2+}+Si^{4+}$ by $2Al^{3+}$ or by $Al^{3+}+Fe^{3+}$. Al can replace Mg as well as Si. In the latter case Al goes in as the central atom in some of the tetrahedra. Formerly all silicates with ions of the overall formula SiO_3^{2-} were called *metasilicates*. This name does not uniquely indicate the structure and should be avoided.

If two chains of the above-mentioned type are joined into an infinite band as in fig. 23.5b, still more oxygen atoms are saved. The overall formula is $(Si_4O_{11}^{6-})_n$ with $O/Si = 2.75$. Bands of this type characterize another large silicate group, the *amphiboles*. In these OH^- almost always enters in as another type of ion. To the amphiboles belong all *hornblendes*.

These chain or band ions are always oriented in parallel in the crystal and the other ions lie between them. The cohesive forces in the crystal are greatest along the length of the infinite ions and therefore such crystals have two pronounced cleavage planes (6-5b) parallel to this direction.

$$\text{tetr} \left\{ \begin{bmatrix} O_3 \\ Si_2 \\ O_2 \end{bmatrix}^{2-} \right. \qquad \text{oct} \left\{ \begin{matrix} (OH)_3 \\ Al_2 \\ (OH)_3 \end{matrix} \right. \qquad \text{oct} \left\{ \begin{matrix} (OH)_3 \\ Mg_3 \\ (OH)_3 \end{matrix} \right. \qquad \text{oct} \left\{ \begin{bmatrix} (OH)_3 \\ Mg_2Al \\ (OH)_3 \end{bmatrix}^{+} \right.$$

a. $Si_2O_5^{2-}$ b. $Al(OH)_3$ c. $Mg(OH)_2$ d. $Mg_2Al(OH)_6^{+}$

Fig. 23.7. Units of single tetrahedral and octahedral layers. Each layer may be considered as being infinite in extent in a plane that is normal to the picture plane and parallel to the lines of text.

4. *Infinite layers* (*phyllosilicates*, from Gk. φύλλον, leaf). If the ratio O/Si falls lower, a still more extensive linking of the SiO_4 tetrahedra is required. If the coupling of single SiO_3^{2-} chains that gave the $Si_4O_{11}^{6-}$ band is continually repeated in the same plane, an infinite layer ion with the overall formula $(Si_2O_5^{2-})_n$ is formed (fig. 23.6). In this layer, thus, O/Si = 2.5. These layers are found in a large number of silicates, including especially the *mica*, *chlorite* and *clay minerals*, but they then occur in combination with another type of layer in which Al or Mg atoms are the central atoms. The structural principles of this group have been elucidated especially by Pauling.

Fig. 23.6 shows that, for 1 Si, $\tfrac{3}{2}$O are directed toward one side of the layer and 1 O to the other side of the layer. Oxygen atoms of the latter type are each bound to only one silicon atom. One unit of the layer can be written schematically as in fig. 23.7a, where the coordination of O around Si is also indicated.

The pure hydroxides $Al(OH)_3$ and $Mg(OH)_2$ crystallize in layer structures; these layers have structures as shown in fig. 6.18b and 6.18a, respectively. Analogously to the method used to formulate the $Si_2O_5^{2-}$ layer, the units of the $Al(OH)_3$ layer can be written as in fig. 23.7b, and of the $Mg(OH)_2$ layer as in fig. 23.7c. The singly bound oxygen atoms in the $Si_2O_5^{2-}$ layer have practically the same relative positions as the oxygen atoms in $\tfrac{2}{3}$ of the OH groups on each side of these hydroxide layers. If one imagines a $Si_2O_5^{2-}$ layer and an $Al(OH)_3$ layer to be placed against each other with the coincident oxygen atoms in common (corresponding OH groups in the $Al(OH)_3$ layer are thus removed) a composite, electrically neutral layer is obtained, whose unit is shown in fig. 23.8a. Such double layers placed parallel build up the mineral *kaolinite*, which thus has the formula $Al_2(OH)_4Si_2O_5$. The white porcelain clay *kaolin* consists for the most part of kaolinite.

$$\text{tetr} \left\{ \begin{matrix} O_3 \\ Si_2 \\ O_2 \; OH \end{matrix} \right. \\ \text{oct} \left\{ \begin{matrix} Al_2 \\ (OH)_3 \end{matrix} \right.$$

a. Kaolinite

$$\text{tetr} \left\{ \begin{matrix} O_3 \\ Si_2 \\ O_2 \; OH \end{matrix} \right. \\ \text{oct} \left\{ \begin{matrix} Al_2 \\ O_2 \; OH \end{matrix} \right. \\ \text{tetr} \left\{ \begin{matrix} Si_2 \\ O_3 \end{matrix} \right.$$

b. Pyrophyllite

$$\text{tetr} \left\{ \begin{matrix} O_3 \\ Si_2 \\ O_2 \; OH \end{matrix} \right. \\ \text{oct} \left\{ \begin{matrix} Mg_3 \\ (OH)_3 \end{matrix} \right.$$

c. Chrysotile

$$\text{tetr} \left\{ \begin{matrix} O_3 \\ Si_2 \\ O_2 \; OH \end{matrix} \right. \\ \text{oct} \left\{ \begin{matrix} Mg_3 \\ O_2 \; OH \end{matrix} \right. \\ \text{tetr} \left\{ \begin{matrix} Si_2 \\ O_3 \end{matrix} \right.$$

d. Talc

Fig. 23.8. Units of composite, uncharged silicate layers.

$$\text{tetr} \begin{Bmatrix} O_3 \\ Si_{1\frac{1}{2}}Al_{\frac{1}{2}} \\ O_2 \quad OH \end{Bmatrix}^{-} \quad \text{tetr} \begin{Bmatrix} O_3 \\ SiAl \\ O_2 \quad OH \end{Bmatrix}^{2-} \quad \text{tetr} \begin{Bmatrix} O_3 \\ Si_{1\frac{1}{2}}Al_{\frac{1}{2}} \\ O_2 \quad OH \end{Bmatrix}^{-} \quad \text{tetr} \begin{Bmatrix} O_3 \\ Si_2 \\ O_2 \quad OH \end{Bmatrix}^{\frac{1}{3}-}$$

$$\text{oct} \{ Al_2 \quad\quad \text{oct} \{ Al_2 \quad\quad \text{oct} \{ Mg_3 \quad\quad \text{oct} \{ Al_{1\frac{2}{3}}Mg_{\frac{1}{3}}$$

$$\text{tetr} \begin{Bmatrix} O_2 \quad OH \\ Si_{1\frac{1}{2}}Al_{\frac{1}{2}} \\ O_3 \end{Bmatrix} \quad \text{tetr} \begin{Bmatrix} O_2 \quad OH \\ SiAl \\ O_3 \end{Bmatrix} \quad \text{tetr} \begin{Bmatrix} O_2 \quad OH \\ Si_{1\frac{1}{2}}Al_{\frac{1}{2}} \\ O_3 \end{Bmatrix} \quad \text{tetr} \begin{Bmatrix} O_2 \quad OH \\ Si_2 \\ O_3 \end{Bmatrix}$$

$$\quad\quad a \quad\quad\quad\quad\quad b \quad\quad\quad\quad\quad c \quad\quad\quad\quad\quad d$$

Fig. 23.9. Units of composite, charged silicate layers.

One $Al(OH)_3$ layer meanwhile has two equivalent sides, and an $Si_2O_5^{2-}$ layer can therefore be combined also with the other side. The result is a triple layer whose unit is shown in fig. 23.8b and has the formula $Al_2(OH)_2(Si_2O_5)_2$. The mineral *pyrophyllite* is made up of such layers.

If a $Mg(OH)_2$ layer is combined with an $Si_2O_5^{2-}$ layer, the double layer in fig. 23.8c is obtained with the unit $Mg_3(OH)_4Si_2O_5$. Such layers make up the mineral *chrysotile*, which constitutes the most important type of *asbestos*. Because of the poor fit between the component single layers, however, strains are produced which with crystal growth cause the layers to roll together into very small parallel tubes. These can be seen in the electron microscope.

The more symmetrical structure obtained when the $Mg(OH)_2$ layer is joined to a $Si_2O_5^{2-}$ layer on each side shows no sign of strains. The unit of this triple layer is shown in fig. 23.8d and has the composition $Mg_3(OH)_2(Si_2O_5)_2$. The mineral *talc* consists of such layers.

In the triple layers of the pyrophyllite and talc types, however, substitutions can occur that lead to negative charge. This is then compensated by corresponding positive charges of cations placed between the layers. Often Si^{4+} in the tetrahedra is partly replaced by Al^{3+}, but also the central atoms of the octahedra may be replaced (for example Al^{3+} partly by Mg^{2+}). Units that arise with partial substitution of Si^{4+} by Al^{3+} are shown in fig. 23.9a, b, c. These are incorporated in the *mica minerals*. The unit $a + K^+$ ions thus form *muscovite* with the formula $KAl_2(OH)_2(AlSi_3O_{10})$; the unit $b + Ca^{2+}$ ions form *margarite*, $CaAl_2(OH)_2(Al_2Si_2O_{10})$. The unit $c + K^+$ ions, derived from talc, form *phlogopite*, $KMg_3(OH)_2(AlSi_3O_{10})$. The common mineral *biotite* is phlogopite with Mg^{2+} partly replaced by Fe^{2+}.

The *chlorite minerals* are built up alternately of *anionic layers* derived from talc (for example, fig. 23.9c), and *cationic layers* of the type of fig. 23.7d.

The *clay minerals* include certain minerals with neutral layers such as kaolinite and pyrophyllite, but especially a large number of minerals which contain anionic layers with lower negative charge per layer unit than the anionic layers mentioned above. A typical unit is shown in fig. 23.9d, which together with $\frac{1}{3}Na^+$ is contained in *montmorillonite*. Here the metal cations are often hydrated, so that hydrogen bonding is also present. The clay minerals are formed from other silicates by hydrothermal processes (action of water under heat and pressure) and weathering. The crystals are often incompletely developed and have low order.

Transitions from mica minerals to montmorillonite-type minerals are common and lead to important clay minerals. They have lower proportions of alkali and alkaline-earth metals than the micas but higher water content because of a certain amount of hydration of the metal cations. They are commonly called *hydromicas* or *illites*.

The layer structure of all the silicates mentioned above leads to a pronounced cleavage parallel to the layers. The forces between the layers are weakest when the layer is electrically neutral (fig. 23.8). In pyrophyllite and talc the layers glide easily over each other, so that these minerals feel greasy and have lubricating properties. In kaolinite the lubricating power is much less, probably because one free side of the layer contains OH groups, whose hydrogen atoms form hydrogen bonds between the layers. In the minerals with somewhat more highly charged layers as in fig. 23.9 a–c the layers are bound together more strongly by the cations between them. Mica and chlorite minerals are therefore harder than silicate minerals with neutral layers and the layers do not glide over each other so easily. When muscovite and margarite are compared, for example, it is found that the higher negative charge per unit in the latter case leads to greater hardness and brittleness (margarite has been called a "brittle mica"). The clay minerals with low negative charge per unit (fig. 23.9 d) are similar in mechanical properties to silicates with neutral layers. In the clay minerals the contents of the interlayer region can also be very variable. The water content can be changed continuously (*zeolitic* water, 13-5e2), cations exchanged (*ion exchanger*, 15-3) and even organic substances may be absorbed between the layers, sometimes with significant increases in the interlayer distance. These properties are of the greatest importance to the behavior of soils, and are also made use of commercially when clay minerals are used as absorption media (especially *bentonite* with montmorillonite as the chief constituent).

5. *Frameworks* (*tectosilicates* from Gk. τεκτονική, carpentry). In the $Si_2O_5^{2-}$ sheet in fig. 23.6 three oxygen atoms in each tetrahedron are common to two tetrahedra. If also the fourth oxygen atom is shared by two tetrahedra, the greatest economy of oxygen atoms is achieved that is found in silicon-oxygen structures (two SiO_4 tetrahedra, in fact, never have more than *one* corner in common). This occurs in the silicon dioxide modifications (23-3g), for example in the high-quartz structure in fig. 23.3, where $O/Si = 2$. Such three-dimensionally linked tetrahedra can form the basis of silicate structure if part of the Si^{4+} ions are replaced by Al^{3+}. In this way there arises an anion infinite in all three dimensions, a framework anion, into whose atomic interstices cations can be inserted so that electrical neutrality is maintained. For example, if $\frac{1}{4}$ of the Si^{4+} ions in SiO_2 are replaced by Al^{3+}, the overall formula of the framework ion becomes $AlSi_3O_8^-$. In *orthoclase* one such ionic unit binds one K^+ ion to give $KAlSi_3O_8$.

Orthoclase is a *feldspar* and other feldspars have similar structures. Important examples are *albite* $NaAlSi_3O_8$ and *anorthite* $CaAl_2Si_2O_8$. In the latter, half of the Si^{4+} ions have been replaced by Al^{3+} and proper charge compensation achieved by introducing Ca^{2+} in place of Na^+. *Plagioclases* (with a number of special mineral names) have compositions between albite and anorthite. The plagioclases nearest each of the end members albite and anorthite are solid solutions, but in between the structures show distortions so that we cannot speak of an unbroken solid solution series.

To the framework silicates belong the *zeolites*, in which, however, the framework ion has a less compact structure than in the feldspars. The anion skeleton is penetrated by channels, which are large enough to permit exchange of certain ions and molecules (of the latter especially water, cf. 13-5e2) with the surroundings without breaking down the skeleton. Zeolites are therefore used as *ion exchangers* and *molecular sieves* (15-3). An example is *natrolite* $Na_2Al_2Si_3O_{10} \cdot 2H_2O$, whose ion-exchange properties were discussed in 15-3. Allied with the zeolites is the splendid blue mineral *lapis lazuli*, which corresponds closely to the formula $Na_8S_n(AlSiO_4)_6$. The color is produced by the polysulfide ion S_n^{2-} (21-8; here for the most part probably $n=2$). Powdered lapis lazuli has been an important pigment under the name *ultramarine*. Nowadays ultramarine is produced by igniting a mixture of kaolin, sodium carbonate and sulfur in the absence of air.

6. *Formation and properties of silicates.* Many anhydrous silicates can be prepared by reaction at high temperature (often in the melt) between silicon dioxide and the corresponding metal oxide. Such reactions have been discussed n 21-5a. Both from that discussion and the descriptions of the structures of the silicates given above it is apparent how they change with the proportion of oxide-ion content. At the same time the properties of both the solid silicates and the corresponding melts change in a way that is clearly related to the structure. In very acid silicates, which because of the high proportion of SiO_2 have low oxide-ion activity, framework anions arise, which lead to high viscosity of the melt and great tendency to glass formation (6-3d). When the proportion of metal oxide is increased the oxide-ion activity is increased, the large anions are broken up more and more, and thus the viscosity and the tendency for glass formation is decreased. To a large extent, there is also a drop in the transformation range (softening temperature) of the glass and the melting points of the corresponding crystalline silicates.

Most silicates are very poorly soluble in water, which can be explained in many cases by the presence of infinite ions. Only the silicates of the alkali metals, with the exception of lithium, are easily soluble. On the other hand, many silicates dissolve, although slowly, in alkaline solutions. However, in order to obtain quickly a water-soluble product the silicate is often melted with alkali hydroxide or carbonate ("fusion"). Silicates are also decomposed by hydrogen fluoride so that solution can take place.

The constitution of water solutions of alkali silicates is poorly understood. Infinite ions present in the solid silicates are, of course, broken down on solution (cf. 5-6b). In clear solutions quite small, isolated ions occur. The *silicic acid* $Si(OH)_4$ ("orthosilicic acid") and its first two ions $SiO(OH)_3^-$ and $SiO_2(OH)_2^{2-}$ are clearly present in dilute solutions (the two ions are probably also present in highly hydrated silicate crystals). Polynuclear silicate ions are also definitely present although the nuclear multiplicity is probably not very large (perhaps as high as 4). On acidification, even by the action of atmospheric carbon dioxide, larger colloidal and undefined particles are formed. At the same time the solution becomes turbid. On increasing the acidity the particle size increases and finally a colorless *silicic acid gel* precipitates. Such gels are also formed by hydrolysis of silicon halides. It was once thought that definite silicon oxide hydrates ("silicic acids") could be identified in silicic acid gel but such conclusions have been shown to be erroneous. Silicic acid gel can be dried to solid products with lower water content ("silica gel"), which have very high porosity and adsorptive ability. Silica gel is therefore used as an adsorbing medium for several different types of substances, especially water (drying agent for both gases and liquids) and as a catalyst support (14-4).

Hard silica gel constitutes the mineral *opal* (as a gemstone often milky white with beautiful color play). A lower water content and the beginnings of ordered structure are shown by *chalcedony* with many varieties such as *agate* (often banded), *carnelean* (yellow to red), *onyx* (black), *jasper* and *flint*. All are perhaps best written $SiO_2 \cdot nH_2O$. They are much used in jewelry and ornamentation. Agate because of its hardness is used for mortars, plates in balance suspensions, etc. *Kieselguhr* consists of the remains of opalized skeletons of diatoms.

The silicate sediments on the ocean floors play a great role in fixing the composition and the pH value of the ocean water. A shift in pH in one direction is compensated by the solution of certain silicate minerals and the precipitation of others. If the pH value is shifted in the opposite direction these processes are reversed.

i. Silicates in technology. Many silicate minerals are used directly without any particular alteration, for example various clay minerals (adsorbing media), asbestos (thermal insulation), mica (electrical insulation), agate (hard surfaces), gemstones and ornamental stones.

1. *Alkali silicates.* Sodium silicate is most widely used, potassium silicate in smaller amounts. Anhydrous sodium silicate is produced by melting quartz sand with sodium carbonate or sodium sulfate and carbon. Only at higher temperature and pressure does the melt dissolve at all rapidly in water (*water glass*). Sometimes the solution is produced directly by the treatment of quartz sand with sodium hydroxide solution. In "normal water glass" the ratio $SiO_2/Na_2O = 3.2–3.5$, but more alkaline glasses with lower values are also made. Solutions sold commercially may have very variable content of solid glass. Usually they have syrupy consistency at ordinary temperature. In addition, solid, rapidly dissolving sodium silicate hydrates are sold, particularly one which is probably $Na_2[SiO_2(OH)_2] \cdot 8H_2O$ (earlier thought to be $Na_2SiO_3 \cdot 9H_2O$ and therefore called "sodium metasilicate").

Sodium silicate is used especially as an adhesive, for example for pasting paper and bonding paper pulp, in asbestos cement tiles and in fireproof putty, sometimes together with asbestos. It is a constituent also of fireproof paints and is used in the manufacture of silica gel. The use of water glass for preserving eggs depends on a protective precipitate that fills the pores of the eggshell as a result of its reaction with the silicate solution. The "metasilicate" mentioned above is used as a detergent, (the water must not be hard, however —see 26-5c—since then insoluble calcium silicates precipitate in the fabrics).

2. *Glass.* Glass as a technical material in its own right has been known at

least since 3000 B.C. and very likely was first obtained in connection with pottery or metal making. Concerning the general properties of glass, see 6-3d.

Silicon dioxide glass (23-3g) has many excellent properties but requires too high temperatures for its production and treatment to be economical as an ordinary glass material. These temperatures can be reduced by going to more basic silicate melts (23-3h). *Silicate glasses* are then obtained. However, the addition of basic oxides is generally not so great that frameworks of SiO_4 tetrahedra do not still play a considerable role (in many common types of glass the ratio $O/Si = \sim 2.5$). The basic oxides may not be only alkali oxide, for then the glass would be water soluble. Therefore, in addition to alkali oxides, CaO, MgO, BaO, PbO, ZnO, Al_2O_3, among others, are used. In glasses containing Al_2O_3, Al^{3+} can partly replace Si^{4+} in SiO_4 tetrahedra, and partly occur outside the tetrahedral skeleton (cf. 23-3h). B_2O_3 is incorporated into many glasses, in which the glass-forming anion skeleton contains both B^{3+} and Si^{4+} as central atoms (borosilicate glass). Special glasses, and among these especially optical glasses, as would be expected, vary widely in composition. Some special glasses are completely free of silicon dioxide; certain glasses with low ultraviolet absorption are based on a phosphate-ion skeleton.

In the making of glass, so-called incidental glass additives are introduced. Among these are the *fining agents*, that is, substances that give off gases in the melt in large bubbles which sweep out the small bubbles that are always formed, and which otherwise would only slowly rise to the surface. The most common are As_2O_3 and alkali nitrates. When colorless glass is desired, *decolorizing agents* are added to eliminate impurities in the starting material that cause color. Discoloration usually results from iron. Fe^{2+} alone gives a light-blue color and Fe^{3+} alone a yellow to brown color. Most often Fe^{2+} and Fe^{3+} are present with a green color. Through complete oxidation to Fe^{3+} (with As_2O_3, alkali nitrates, manganese dioxide or cerium dioxide) the color becomes less perceptible. Coloring agents act in most cases by introducing either colored ions (for example, Co^{2+} which gives a blue color as in cobalt glass; lanthanoid ions) or particles of colloidal dimensions (metallic particles of Au or Cu give a ruby-red color, particles of the solid solution Cd(S,Se) give an orange-red color). To obtain *opal glass* a *clouding agent* is used, especially fluorite CaF_2.

The silicate glasses are classified according to the most important constituent oxides. *Calcium-alkali silicate glass* makes up the ordinary glass used for window glass, glass vessels and glass fiber (glass wool). The starting materials are alkali carbonate, calcium carbonate and quartz. The alkali metal is usually sodium (*soda glass*) but also potassium (*potash glass*). In *lead-alkali silicate glass* (*lead glass*) the major portion of the calcium oxide is replaced by lead oxide. It has a high refractive index and dispersion in the visible spectral region and therefore is extensively used as optical glass and in finer ornamental glasses. *Aluminum borosilicate glass* contains only small amounts of alkali oxides and calcium oxide. The melting point of the silicon dioxide is reduced mainly by the presence of boron trioxide B_2O_3 and to some extent by a certain amount of aluminum oxide. However, the softening and working temperatures are considerably higher than for the previously mentioned types of glass. Its advantages are high resistance to thermal shock (low coefficient of expansion) and high chemical stability (because of its low alkali content). Such glasses are therefore much used in laboratory apparatus and in chemical industry. One well-known commercial glass of this type is *Pyrex*. Concerning *Pyroceram*, which in finished form is no longer a glass, see 14-2a.

3. *Ceramic materials*. By ceramic materials is meant nowadays the material in an object that is made of inorganic substances that are shaped at ordinary temperature and then "fired" at higher temperature. The constituent substances may be of various kinds, for example minerals and other oxides or carbides. The firing serves to give the product the necessary strength, either by sintering together the crystallites in the paste (14-3b), or by partly melting the paste and binding them on setting. With melting, solid phases must remain in sufficient amounts so that the shape is not disturbed too much.

Here we consider only ceramic materials with clay (Gk. κέραμος, clay) as the plastic raw material essential for shaping and for the firing processes. The melt then consists of silicates and on cooling often gives a glass (6-3d).

Depending on the composition and firing temperature the main bulk of the product, the so-called body, is given varying porosity, which is used to determine its quality. If considerable melting occurs, the body does not absorb liquids; when immersed in water it does not gain weight, and it is not colored by fuchsin in methanol solution under pressure, thus constituting an *impervious* or *fuchsin-fast material*. Less dense material, according to the percent of absorbable water, is said to be *dense-fired* or *dense-sintered* ($<1\%$), *vitreous* ($1-3\%$), *semivitreous* ($3-10\%$) and *nonvitreous* ($>10\%$).

On heating the clay mineral kaolinite (23-3h4), for which we may write the "oxide formula" $Al_2O_3 \cdot 2SiO_2 \cdot 2H_2O$, the water is first given off at $500°-600°C$ with the formation of anhydrous kaolinite ("metakaolinite") with the oxide formula $Al_2O_3 \cdot 2SiO_2$. However, at higher temperatures the only stable, intermediate phase in the system $Al_2O_3-SiO_2$ is *mullite*, which has a somewhat variable composition (solid solution) close to the formula $3Al_2O_3 \cdot 2SiO_2$ (fig. 23.10). When equilibrium is more rapidly established on further raising the temperature, therefore, mullite (becoming noticeable at about $950°C$) and silicon dioxide are formed. The figure shows that mullite has a melting point maximum at $1850°C$ with eutectics at $1840°C$ on the Al_2O_3 side and $1595°C$ on the SiO_2 side. Therefore, preparations with SiO_2 content lower than mullite are completely solid up to $1840°C$ but in preparations with SiO_2 content greater than mullite liquid phase appears above $1595°C$. However, if the SiO_2 content is not too high, at the start the proportion of solid phases is so high and also the melt is so viscous that only a slow softening results. Thus, fired kaolinite is completely solid up to $1595°C$ and thereafter softens slowly. Ceramic materials that remain solid up to at least about $1580°C$ are designated as *fireproof*. Such materials are thus obtained from, among other things, kaolinite and clays that in addition to H_2O contain practically speaking only SiO_2 and Al_2O_3 (*fire clay*). The fired material is often called *chamotte*. If

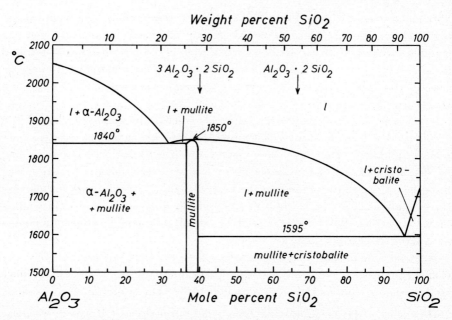

Fig. 23.10. Phase diagram for the system Al_2O_3–SiO_2.

enough Al_2O_3 is added (mostly in the form of bauxite AlO(OH)) so that the SiO_2 content becomes more or less the same as that of mullite, the softening temperature rises to 1840°C. One then approaches *refractory materials*, which, however, as a rule are not silicates, but usually consist of pure oxides, nitrides or carbides.

For practical and economic reasons less pure clays are often used, and the system becomes more complicated. Usually the softening and thus the firing temperature is lowered (the clay is no longer "fireproof"). Departures from the pure system Al_2O_3–SiO_2 are also caused by additions of nonplastic raw materials, especially quartz (sand, flint) and feldspar (various kinds). These "thin out" the mass and facilitate drying of the goods before firing, and decrease changes in shape on drying and firing. They also lower the firing temperature ("fluxes"). In most of these cases mullite still appears as an important constituent of the fired product.

The *color* of fired ceramic material is commonly caused by iron, which comes mainly from the clay material. With iron-free clay, for example kaolin, generally no color appears ("white-burning" clay). When iron is present, this occurs as Fe(III) after firing in an oxidizing atmosphere. In 14-3b it was stated that Fe_2O_3 that has not been heated too high can be a beautiful red color, but that the oxide on heating to higher temperatures (above $\sim 700°C$) becomes progessively darker (finally dark violet). When iron-containing clays are fired in an oxidizing atmosphere, however, the color remains red with firing temperatures up to about 1100°C ("red-burning" clay; for example, red brick), and this is caused by the formation of a solid solution of Fe_2O_3 with Al_2O_3, that is, the iron oxide enters into a phase $(Fe^{III}, Al)_2O_3$. If the firing is done in reducing atmosphere, weakly colored Fe(II) compounds are formed, which with low iron content hardly color the product.

Clays with high proportions of CaO (the proportion of SiO_2 being correspondingly less) on firing in oxidizing atmosphere take a yellowish color ("buff-burning" clay; for example, yellow brick).

In most cases (brick and tile are the most important exceptions) the ceramic material is covered with a glassy coating, a *glaze*. Usually oxide mixtures are used that have approximately eutectic composition so as to be easily fusible. The goods are dipped into a water suspension of this mixture. Simpler materials are glazed after drying with a single firing (smooth firing). Otherwise a raw-goods firing is first done followed by glazing and then smooth firing. Pigments, usually oxides of heavy metals, are applied either before glazing (*underglaze color*) or to the already glazed and fired object (*overglaze color*), for which a special firing of the color ("decorating firing") at 700–900°C is required.

4. *Cement.* Cement usually denotes a substance that after mixing with water hardens to a stonelike mass. In contrast to ordinary plaster (26-5a) the hardening even proceeds under water and the hardened product is very stable toward water. In ancient times natural cements were known (for example, slaked lime mixed with volcanic tuff called pozzuolana) but only at the beginning of the 19th century were methods developed in England for making really practical cement. Because of the similarity of the final product to a familiar building stone, Portland stone, it was called *Portland cement*. All similar types of cement are still referred to collectively by this name.

In the manufacture of Portland cement, clay, finely ground limestone and water are blended into a paste, which is dried and "burned" in a rotary kiln (about 150 m long) at a final temperature of 1400–1500°C. The material is thus sintered into *cement clinker*, which is then finely ground together with a few percent of calcium sulfate (gypsum). The product constitutes cement, and contains several solid phases. The chief constituent is a well-crystallized calcium oxide silicate Ca_3OSiO_4, but in addition other calcium silicates, some double and triple oxides (21-5a) of Ca, Al and Fe^{III} and glassy constituents are present. On addition of water, these phases are to a large extent decomposed; certain substances go into solution and the remainder consists largely of rather disordered, hydrated silicates and aluminosilicates. These react with each other and with the solution producing an array of solid phases which are complicated and in several cases ill-defined, but which can be described as hydrated silicates, aluminosilicates and aluminosulfates of calcium. The cement in this way hardens or sets (the proportion of water should not be greater than just the amount taken up by the solid phases) and because the new phases adhere effectively to the silicate crystals in the admixed sand and stone (as well as reinforcing iron) a mass with excellent properties of solidity and durability is obtained (*concrete*). The setting process continues over a very long time.

Cement with a much higher proportion of aluminum oxide and the least possible amount of silicon dioxide is made by melting bauxite (24-3a) with limestone. The main constituent is calcium aluminum oxide $CaAl_2O_4$, which after addition of water gives on setting hydrated calcium aluminates (*high-alumina cement*, also *fused cement* because of the method of fabrication). High-alumina cement sets much faster than Portland cement, but is considerably more expensive.

j. Silicones. The silicones belong to the class of carbon compounds, but are discussed here because their structure is based on silicon-oxygen skeletons, analogous to those of the silicates. If in the silicate ions in fig. 23.4–6 (with the exception of fig. 23.4a) the singly bonded oxygen atoms are replaced by organic residues R, silicones are obtained. In the SiO_2 structures one may imagine that a number of silicon atoms are removed and those oxygen atoms thus left with one bond are replaced by R. The most important silicones are high polymers and according to this viewpoint may be said to be based on silicon-oxygen skeletons as in fig. 23.5 (chains or bands), fig. 23.6 (layers) or SiO_2 structure (three-dimensional frameworks). Various types of skeletons and degrees of polymerization contribute to a multitude of products with various and valuable properties. Especially important are their chemical resistance and great stability toward heat, which are unusual in organic compounds. There are *oils*, which have very constant viscosity over a large temperature range, *resins* for insulation, impregnation and laquers, and *rubberlike* products. If on a silicone-coated surface hydrophobic (15-2) organic groups face outward, the surface becomes water-repellant, a property that has wide use.

k. Carbon, nitrogen and sulfur compounds. An excess of carbon in the reduction of silicon dioxide with carbon (23-3 c) leads to the formation of *silicon carbide* SiC: $SiO_2 + 3C \rightarrow SiC + 2CO(g)$. Silicon carbide exists in several crystal structures, all of which are types with tetrahedral coordination throughout, as discussed in 6-2c. The relationship to diamond with respect to both bond and structure type shows itself in very great hardness (harder than corundum with hardness number 9; cf. 6-5c). Pure silicon carbide is colorless but the commercial product is colored by small amounts of impurities. Depending on the conditions of manufacture the color varies from yellow to green and blue to almost black. The foreign atoms also lead to impurity semiconduction (6-4d). Silicon carbide with precisely controlled additions of foreign atoms have begun to be used in transistors, especially where high operating temperature is required.

Silicon carbide is a very important abrasive material and as such is commonly called *Carborundum* (the name signified a carbon compound with a hardness of the order of corundum; other names are also used, for example Crystallon). It is very stable at high temperature and is little affected then even by oxygen because of the formation of a protective layer of silicon dioxide. These properties toghether with its high resistivity make it a suitable material for electrical resistances and elements in resistance ovens. Here the carbide is found also under names such as *Silite, Silundum, Globar*. It is also used as a highly refractory material in ovens.

Silicon nitride Si_3N_4 is used as a newer, refractory and very hard ceramic material with high resistance to thermal shock and chemical attack (here also a protective layer of silicon dioxide is formed on oxidation), and good electrical-insulation properties even at high temperature. The desired object cannot be obtained by sintering ready-made nitride, since it decomposes at temperatures below the sintering temperature.

Compressed silicon powder is therefore used to start with, which is nitrided by heating (final temperature about 1400°C) in a stream of nitrogen or ammonia gas. The reaction is accompanied by sintering (*reaction sintering*).

Silicon sulfide SiS_2 is produced by melting together the elements at red heat. It crystallizes in colorless, sublimable needles, built up of SiS_4 tetrahedra joined into chains through two common corners (thus common edges). The formula can therefore be written schematically as

$$\begin{array}{c}\diagdown_S\diagdown\quad\diagup^S\diagdown\quad\diagup^S\diagdown\\ Si\qquad Si\qquad Si\\ \diagup\diagdown_S\diagup\quad\diagdown_S\diagup\quad\diagdown_S\diagup\end{array}$$

. In SiO_2 with analogous composition tetrahedral coordination around the silicon atoms is achieved only by three-dimensional linking since two SiO_4 tetrahedra never have more than *one* common corner.

23-4. Germanium

In ionic radius and chemical properties germanium resembles silicon and therefore is widely distributed, although in very small amounts, replacing silicon in silicate minerals. It occurs rarely in more concentrated form in sulfide minerals such as *argyrodite* Ag_8GeS_6. In argyrodite from Saxony Winkler discovered germanium in 1886 and named the new element after the country of origin. However, as early as 1871 Mendeleev predicted both the existence of this element (to which he gave the provisional name ekasilicon) and many of its properties (4-2a). The enrichment of germanium in coal ash was mentioned in 17c. Coal ash is used as a source of germanium as well also as the even richer deposits in the flues of coal-fired furnaces.

As mentioned in 23-1, germanium at ordinary pressure is an intrinsic semiconductor (6-4d) and, like silicon, crystallizes with the diamond structure. Its color is gray-white, and the melting point is 959°C. At high pressures several other modifications of germanium have been found. One of these, obtained at a pressure above 120 kbar, has the same structure as white tin (23-5b) and metallic conductivity.

The chemistry of germanium as stated above is quite like that of silicon. The principal oxidation number is IV, but in addition Ge(II) compounds are known, which, however, are easily oxidized. *Germanium dioxide* GeO_2 crystallizes in one modification with the low-quartz structure. It has a predominantly acid character and with basic oxides and in basic solutions gives *germanates*, analogous to silicates.

Elementary germanium has recently become of great importance as a semiconducting material, especially in rectifiers and transistors (6-4d).

23-5. Tin

a. Occurrence and history. As can be seen in tab. 17.3 the average proportion of tin in the earth's crust is low. However in a few places it has been concentrated into mineral deposits. The only tin mineral of real importance as ore is *cassiterite* SnO_2.

The most important tin deposits are in southeast Asia (Indonesia with the islands Banka and Billiton, the Malay peninsula, Thailand), Bolivia and Katanga.

Tin was much used formerly both pure and alloyed with copper in bronze (36-5a). In ancient times it was mined in Cornwall but these deposits are no longer important. The Latin name is *stannum*, from which the symbol Sn is derived.

b. Properties. Tin has a transition point at 13°C (an older value of 18°C is incorrect). Below this temperature α tin with the cubic diamond structure is stable, and is usually a gray powder (*gray tin*). α tin is an intrinsic semiconductor (6-4d). Above 13°C β tin is stable, a malleable, lustrous, silver-white metal (*white tin*) with a tetragonal structure. On bending a bar of white tin a squeaking sound is heard (the tin "cries"), which results from the rubbing of the crystals against one another. The α-tin structure is less compact than the β-tin structure; in the vicinity of the transition temperature the density of α tin is 5.75 and of β tin 7.31 g cm^{-3}.

The β-tin structure is stable up to the melting point at 231.9°C, but at 160°C certain changes in properties occur. Among other things the metal is brittle above this temperature.

Since β tin is formed by cooling molten tin, this modification is the one usually met with. The transition β tin → α tin as a rule takes place very slowly, even far below the transition point, because α nuclei are only slowly formed. If α tin is added, or the rearrangement of tin atoms in the outer layers facilitated by being able to occur through a solution, the transition goes more rapidly. The possibility of seed formation increases with falling temperature, but at the same time the reaction rate decreases so that the rate of α-tin formation has a maximum at about −50°C (cf. 14-2a). Here α tin forms quite rapidly. The presence of foreign metals generally lowers the rate of transition (however, aluminum increases the rate).

On transition to α tin, gray areas appear on the β-tin surface, and since the former modification has a greater specific volume, these areas become raised. Finally complete decomposition to a gray powder takes place. In order to be sure to avoid this *tin disease*, tin objects should not be stored at temperatures below 13°C.

At ordinary temperature tin is hardly attacked by air or water. However, a thin oxide layer is formed on the surface, and on heating in air this layer grows in thickness. At high temperature combustion to SnO_2 may occur. On heating in the absence of air tin boils at about 2600°C.

Tin is slowly attacked by dilute acids with the formation of tin(II) salts and hydrogen. Higher acid strengths and heating enhances the attack, but with oxidizing acids other reactions take place. Conc. nitric acid produces fine-grained, white SnO_2 with varying amounts of adsorbed water. Warm conc. sulfuric acid attacks tin more sluggishly with reduction to sulfur dioxide.

On boiling with alkali hydroxide solution hexahydroxostannate(IV) ion is obtained: $Sn(s) + 2OH^- + 4H_2O \rightarrow Sn(OH)_6^{2-} + 2H_2(g)$.

c. Production and uses. After enrichment and possibly roasting to remove sulfur and arsenic, tin is recovered from cassiterite by reduction with carbon in a reverberatory or blast furnace: $SnO_2 + 2C \rightarrow Sn + 2CO(g)$. The crude tin is purified, among other ways, by melting. Many impurities are, in fact, only slightly soluble in liquid tin immediately above its melting point. Rather pure tin can therefore be drawn off at this temperature.

The scarcity of tin makes it relatively expensive and the location of its deposits made it difficult to obtain during World War II. Therefore, serious attempts have been made in various ways to decrease consumption, but in many cases tin is irreplaceable.

The largest amounts of tin are used as a corrosion-protective coating on metals (*tinning*; for protective action, see 16i). Sheet iron is mainly tinned, and the product (*tin plate*) is used especially for the tin can industry. The coating is nowadays often done by electrolytic means, which conserves tin in relation to tin dipping which predominated earlier. In the latter, use is made of the ability of molten tin to adhere to and flow over ("wet") clean surfaces of many metals. The same applies to the application of tin for *soldering*. In both cases, ordinarily tin-lead alloys are used (*soft solder*) since these are harder, more easily melted and cheaper than pure tin (however, lead must be avoided in contact with foodstuffs). Most easily melted is the eutectic with 38 wt % lead (fig. 23.11). However, higher proportions of lead are often used (up to 60 wt % lead), partly for economic reasons, and partly because this alloy within a certain temperature interval corresponding to the two-phase region Pb phase $+1$ (13-5f) has a pasty consistency, which can be advantageous for certain work.

For tinning and soldering the metal surface must be completely clean and free of oxides. Previous to tinning the oxides are dissolved by acid treatment, *pickling*, often with dilute sulfuric or hydrochloric acid. During the tinning process itself, to remove the remaining oxides and prevent new oxidation, the presence of a *flux* is necessary. This protects the metal surface by flowing out over it but at the same time is displaced by the soldering tin. In addition it should be able to dissolve oxides. For tinning and soldering with tin or tin alloys (soft soldering), among other things NH_4Cl is used (22-6c), but more often a water solution of $ZnCl_2$ and NH_4Cl. From this the salt $(NH_4)_2ZnCl_4$ (37-5) crystallizes, which is also used in preparing soldering solution. For soft soldering more easily damaged parts occasionally organic substances are used, among others, rosins.

Tin is incorporated into many alloys, of which we mention here only those in which it is the main constituent (the others, for example bronze, are referred to under the primary metal). *Soft solder* has already been mentioned. *Tin-base bearing metals* (*white metal, babbitt*) besides tin contain 7–20 % antimony and a few percent of copper. Here hard crystals of, among other things, SnSb are formed embedded in a fine-grained, softer but nevertheless stable matrix (eutectic). Recently tin has been largely replaced by lead (lead-base bearing metals). Formerly tin-lead alloys (these include English pewter) were used for household and ornamental objects. Higher luster and nontoxicity is obtained by alloying the tin with 4–8 % antimony and about 2 % copper (brittania metal). Tin for tubes and foil has now been nearly completely replaced by aluminum.

Tin in tin-plate scrap (tin cans) must be removed before remelting the scrap, since tin is detrimental to steel. In this way also large amounts of tin can be recovered.

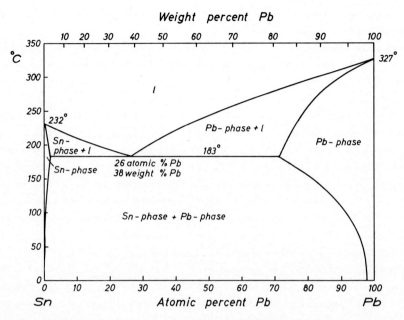

Fig. 23.11. Phase diagram for the system tin–lead.

Formerly the scrap was treated with chlorine. In the absence of water and at temperatures below 50°C the iron is not attacked but SnCl$_4$ is produced. Nowadays a hot NaOH solution together with an oxidizing agent (nitrate or nitrite) is used, with which the tin gives hexahydroxostannate(IV) ion (Sn + 6OH$^-$ → Sn(OH)$_6^{2-}$ + 4e$^-$) but the iron remains behind. Metallic tin can be obtained from the stannate solution by electrolysis.

d. Tin hydride. Only *tin tetrahydride* SnH$_4$ is known, which is produced in gaseous form on decomposing MgSn$_2$ with hydrochloric acid, and by the action of directly formed hydrogen on tin(II) salt solutions (magnesium powder in the acidified solution or cathodically developed hydrogen). A better yield is obtained by reaction between tin tetrachloride and lithium tetrahydridoaluminate (24-3d) in ether solution at −30°C: SnCl$_4$ + 4LiAlH$_4$ → SnH$_4$ + 4LiCl + 4AlH$_3$. The hydride decomposes slowly at room temperature, rapidly on heating to about 150°C (in a glass tube a "tin mirror" is formed). It can be condensed to a liquid with the boiling point −52°C.

e. Halogen compounds. *Tin*(II) *chloride* (*tin dichloride*) SnCl$_2$ is obtained anhydrous as a white mass (m.p. 247°C, b.p. 605°C) by heating tin in a stream of hydrogen chloride. From a solution of tin in hydrochloric acid the dihydrate SnCl$_2$·2H$_2$O crystallizes ("tin salt"), which is easily soluble in water. In water solution strong hydrolysis sets in, on which among other things poorly soluble tin(II) hydroxide chloride Sn(OH)Cl is formed. The hydrolysis prevents the preparation of anhydrous SnCl$_2$ by directly heating the dihydrate (20-5a). The heating must then be done in a stream of hydrogen chloride.

Tin(II) chloride in hydrochloric acid solution is a strong reducing agent and is used as such both in analysis and industry.

Tin(IV) chloride (tin tetrachloride) $SnCl_4$, obtained by treatment of tin with chlorine, is a colorless liquid (b.p. 114.1°C), which fumes in moist air. It is strongly hydrolyzed by water with appreciable development of heat and formation of tin-dioxide gel with adsorbed water. The overall reaction can be written $SnCl_4 + 2H_2O \rightleftharpoons SnO_2(s) + 4H^+ + 4Cl^-$. However, from very concentrated solution the hydrate $SnCl_4 \cdot 5H_2O$ crystallizes. $SnCl_4$ also enters into the equilibrium $SnCl_4 + 2Cl^- \rightleftharpoons SnCl_6^{2-}$. The formation of *hexachlorostannate*(IV) *ion* is favored by high concentration and addition of HCl, so that the hydrolysis is suppressed and the latter equilibrium shifted to the right. On passing HCl into a conc. solution of $SnCl_4$, $H_2SnCl_6 \cdot 6H_2O$ crystallizes (cf. 12-2f). Alkali and ammonium salts are obtained by mixing a conc. solution of $SnCl_4$ with the corresponding chlorides.

f. Oxygen compounds. On adding a small amount of alkali hydroxide to a solution of a Sn^{2+} salt a white, poorly soluble precipitate of *tin*(II) *oxide hydroxide* is formed, in which units $Sn_6O_4(OH)_4$ have been found. No solid hydroxide $Sn(OH)_2$ is known. The precipitate has amphoteric character (12-4d); it dissolves in acids with the formation of Sn^{2+} ions, and in strong bases with the formation of *hydroxostannate*(II) *ions* (older name, stannite ions), particularly $Sn(OH)_3^-$. Probably the coordination number six is achieved in the oxide hydroxide as well as in $Sn(OH)_3^-$ through the formation of polynuclear complexes (cf. 12-1b). Hydroxostannate(II) ions are strongly reducing and are thus converted to *hydroxostannate*(IV) *ions*: $Sn(OH)_3^- + 3OH^- \rightarrow Sn(OH)_6^{2-} + 2e^-$; $e^0 = -0.93$ V. This property is utilized in many industrial processes.

When the oxide hydroxide is heated in the absence of air, water is given off with the formation of *tin*(II) *oxide* SnO, which forms a blue-black powder. On heating in air oxidation proceeds to SnO_2.

Tin(IV) oxide, tin dioxide SnO_2 constitutes the mineral *cassiterite*, which is found as tetragonal crystals, often more or less brownish (the color is probably caused by the presence of iron or manganese). The dioxide is formed by combustion of tin in a stream of air as a white powder poorly soluble in water, acids and alkali solutions. It also often remains unchanged in glass melts to which it gives a milk-white color, so that it is used for making white glazes and enamels. On the other hand it is dissolved by molten alkali oxides and hydroxides, with the formation of double oxides (21-5a), for example Na_2SnO_3 (formerly erroneously called a "stannate"). From a concentrated water solution of such a fused product easily soluble *alkali hexahydroxostannate*(IV) crystallizes out, for example, $Na_2Sn(OH)_6$ (a salt such as this was considered formerly to be $Na_2SnO_3 \cdot 3H_2O$). If acids are added to the salt solution a gel of SnO_2 with adsorbed water is formed, which is similar to the gel formed by hydrolysis of $SnCl_4$ (23-5e). In freshly precipitated form ("α stannic acid") these gels dissolve easily in both strong bases (on which the hexahydroxostannate(IV) is regenerated) and in acids. When allowed to stand, they age (12-4d, 14-2d), a process that is hastened by heating. The grain size and order increase, the water content decreases and a product is finally obtained that is insoluble in both bases and acids ("β stannic acid", earlier also called "metastannic acid") This product is also obtained by oxidation of tin with conc. nitric acid (23-5b). Hydrated SnO_2 in its various forms has high adsorptivity. Tin compounds such as $SnCl_4$, $(NH_4)_2SnCl_6$ and $Na_2Sn(OH)_6$ were formerly used as mordants in textile dying. Here hydrated SnO_2 precipitates on the textile fibers and seals in the coloring substance.

g. Sulfur compounds. *Tin*(II) *sulfide* SnS is obtained as a blue-gray mass of metallic luster by melting together tin and sulfur. It appears as a dark-brown precipitate on passing hydrogen sulfide into solutions of Sn(II) salts. The precipitate is poorly soluble in dilute, strong acids and in alkali and ammonium sulfide solutions. On the other hand it dissolves in polysulfide solutions (21-8) with oxidation to thiostannate(IV) ion: $SnS(s) + S_2^{2-} \rightarrow SnS_3^{2-}$.

Tin(IV) *sulfide* SnS_2 since early times has been made by heating together tin (or better tin amalgam), sulfur, and ammonium chloride (the process is obscure) and is then obtained as golden-yellow crystal flakes. In this form it is used under the name mosaic gold as a pigment in bronze paint ("tin bronze"). With Sn(IV) solutions hydrogen sulfide gives a yellow precipitate of SnS_2. It is decomposed by moderately strong, hot hydrochloric acid with the evolution of H_2S, and dissolves in alkali and ammonium sulfide solutions with the formation of *thiostannate*(IV) *ion*: $SnS_2(s) + S^{2-} \rightleftharpoons SnS_3^{2-}$. Alkali hydroxide solutions dissolve SnS_2 according to $3SnS_2(s) + 6OH^- \rightarrow 2SnS_3^{2-} + Sn(OH)_6^{2-}$.

23-6. Lead

a. Occurrence and history. The most common lead mineral is *galena* PbS, which is also the predominant ore of lead. Some lead is also mined as *cerussite* $PbCO_3$, which is a weathering product of galena. In addition a number of lead salts $Pb^{II}A^{VI}O_4$ (A = S, Cr, Mo, W) are known as minerals. Galena very often contains silver as argentite Ag_2S precipitated in finely divided form (probably the silver was dissolved in the galena at higher temperature). Also, the ore often contains zinc as admixed sphalerite ZnS.

Various isotopes of lead are embodied in each of the natural, radiactive series (9-2a).

The world's most important deposits of galena occur in the United States, Canada (British Columbia), Mexico and Australia. Sweden is also an important source of lead ore, especially from Laisvall and Vassbo since the 1950's, where galena-impregnated sandstone is mined. The Laisvall deposit is considered to be one of the largest known lead-ore bodies in the world.

Lead was already known in ancient Egypt. Greek and Phoenician colonists mined galena in Spain and the Romans later took over these mines, which are still worked today. The Romans used lead for, among other things, water pipes, whence the term "plumbing", from Lat. *plumbum*, lead.

b. Properties. In 9-3c, d, we discussed the isotope composition of lead and its application to radioactive age determination. "Ordinary" lead has the isotope composition 1.4 atom% ^{204}Pb, 25.2 atom% ^{206}Pb, 21.7 atom% ^{207}Pb, 51.7 atom% ^{208}Pb.

Lead is a very soft metal. A freshly cut surface is lustrous blue-white, but in moist air is quickly coated with an oxide skin with a dull, blue-gray color. Solid lead has a cubic-close-packed structure up to the melting point (327.4°C). Molten lead boils at 1740°C. Lead of the isotope composition given above has the density 11.34 g cm^{-3} at room temperature. Lead from uranium ores, which contains mostly ^{206}Pb, has the density 11.27 g cm^{-3}.

The above-mentioned oxide skin protects lead from further oxidation in air at ordinary temperature. However, above the melting point the oxidation proceeds, on which first PbO and then Pb_3O_4 is formed. Distilled and air-free water does not attack lead. In the presence of oxygen, on the other hand, a slow reaction takes place: $Pb(s) + H_2O + \frac{1}{2}O_2 \rightarrow Pb^{2+} + 2OH^-$. Oxide hydroxide (23-6e) then precipitates, but since the solubility product of this is relatively high the water carries an appreciable amount of Pb^{2+} ions. Because of the toxicity of lead compounds water carried through lead pipes may be harmful to the health. However, with hard water the risk is slight. Hard water contains hydrogen carbonate and sulfate ions (26-5c) so that the lead surface acquires a protective layer of very insoluble hydroxide carbonate and sulfate respectively: $3Pb^{2+} + 4OH^- + 2HCO_3^- \rightarrow Pb_3(OH)_2(CO_3)_2(s) + 2H_2O$; $Pb^{2+} + SO_4^{2-} \rightarrow PbSO_4(s)$.

With acids lead gives Pb(II) salts. However, sulfuric acid produces no continuing attack because of the formation of lead(II) sulfate. Lead has thus been important as a material resistant even to conc. sulfuric acid. In a similar way a lead(II) chloride layer prevents continued attack by hydrochloric acid. In other cases lead is attacked by acids; oxidizing acids such as nitric acid act quickly and directly, nonoxidizing acids such as acetic acid act only in the presence of atmospheric oxygen.

All lead compounds are very toxic, and wherever lead or lead compounds are used one must always guard against the risk of lead poisoning. The more severe cases produce, among other things, stomach and intestinal symptoms. In 5-6e the complex formation with EDTA was mentioned as a remedial agent. The detection of lead in the blood or urine provides a positive diagnosis.

c. Production and uses. The raw material with few exceptions is galena. This is crushed and concentrated, nowadays usually by flotation (15-2). The concentrate is mixed with limestone as a slag former, and then roasted and sintered simultaneously (sinter-roasting). The roasting is continued for various lengths of time according to the method of recovery to be used.

Rich ores, which consist predominantly of galena, are often treated by the *roast-reaction process*. In this case the ore is roasted until approximately two thirds of the sulfide is oxidized to PbO and $PbSO_4$ according to $2PbS + 3O_2(g) \rightarrow 2PbO + 2SO_2(g)$ and $PbS + 2O_2(g) \rightarrow PbSO_4$. The sinter is smelted in an electric furnace with the addition of limestone and smaller amounts of coke. The most important reactions in the smelting are: $PbS + 2PbO \rightarrow 3Pb + SO_2(g)$, and $PbS + PbSO_4 \rightarrow 2Pb + 2SO_2(g)$. The main purpose of the coke is to lower the amount of lead in the slag. The melt drawn off still contains sulfide, which by a subsequent air blast in a converter is reduced to Pb: $PbS + O_2(g) \rightarrow Pb + SO_2(g)$. The "hearth smelting" formerly used, where the roasting and the smelting were done one after the other in the same hearth, is beginning to fall into disuse.

When larger amounts of other sulfides and SiO_2 are present, however, the roast-reaction process is not applicable. PbO and $PbSO_4$ react with SiO_2 according to $2PbO + SiO_2 \rightarrow Pb_2SiO_4$ and $2PbSO_4 + SiO_2 \rightarrow Pb_2SiO_4 + 2SO_3(g)$, and the lead silicate formed is not reduced. In such cases the *roast-reduction process* is used. Here the concentrate is roasted until oxidation by the above reactions is practically complete. The SiO_2 present reacts at the same time to give Pb_2SiO_4. Smelting then is carried out in a cylindrical furnace which is charged with sinter and coke and functions roughly like a blast furnace. The CaO already formed from the limestone in sintering serves in the usual way as a slag former, but now also has the special purpose of taking care of the silicic acid in Pb_2SiO_4. The processes in the blast furnace can be written schematically as: $PbO + CO(g) \rightarrow Pb + CO_2(g)$; $PbSO_4 + 2CO(g) \rightarrow Pb + SO_2(g) + 2CO_2(g)$; $Pb_2SiO_4 + 2CaO + 2CO(g) \rightarrow 2Pb + Ca_2SiO_4 + 2CO_2(g)$.

The crude lead (base bullion) obtained by these methods contains many impurities, especially a number of dissolved metals, and must be refined. The recovery of the silver, which is almost always present and is therefore of especial economic importance, is described in 36-4b.

Metallic lead is much used for its chemical resistivity and its easy fusibility and softness, which make it easy to cast and shape. When greater hardness is needed, lead is alloyed, particularly with antimony, and is then called *hard lead*. The chemical industry uses a considerable amount of lead for vessels and conduits, but a considerably larger amount is used for electric-cable sheething. However, it is likely that in both these cases plastics will come to replace lead to a large extent. The high density of lead is utilized in lead counterbalance weights and similar applications, and lead radiation shielding. The density and ease of molding make lead useful for filling sheethed projectiles for small-arms weapons, and for lead shot. Lead shot is given increased hardness by alloying with about 0.3 % arsenic. Other lead alloys that may be mentioned are *soft solder* (23-5c) and *lead-base bearing metals* (23-5c). *Type metal* usually contains 70–85 % lead and in addition antimony and a few percent of tin. The molten alloy expands on solidifying

Group 14, The Carbon Group

(cf. bismuth, 22-3e) and therefore fills the matrices well on casting. Large amounts of lead are used for the manufacture of lead storage batteries (16h).

d. Lead hydride. Only *lead tetrahydride* PbH_4 is known. It is formed by reactions similar to those that give tin hydride (23-5d) but with much smaller yields. It is also considerably less stable.

As analogs of PbH_4 we may include the more stable compounds PbR_4 where hydrogen is replaced by alkyl groups. Of these *tetramethyl-* and *tetraethyllead* in spite of their toxicity are extensively used as an antiknock agent in gasoline (8g). They are produced industrially by the reaction of alkyl halides with a lead-sodium alloy.

e. Oxygen compounds. The lowest lead oxide known with certainty is *lead*(II) *oxide* PbO. It constitutes the most important commercial lead compound and is usually made by oxidation of molten lead with air. After melting (m.p. 886°C) it solidifies to a red-yellow, platy, crystalline mass (*litharge*). It exists in two modifications: PbO (tetr., red) $\xrightleftharpoons{491°C}$ PbO (orthorh., yellow). Inasmuch as the rate of transition is low, the yellow modification can be obtained metastable below the transition point, for example by careful heating of lead(II) carbonate or nitrate. PbO, which is a predominantly basic oxide, is the most common starting material for making other lead compounds. However, the largest amounts are used in the manufacture of lead-storage-battery plates (16h).

PbO since early times has been used as a "dryer", that is, added to boiling linseed oil to make it more easily drying. Here lead salts of the fatty acids ("lead soap") are formed, whose Pb(II) ions catalyze the oxidation of the oil and thus its drying. Certain other metal ions, for example Mn(II) and Co(II), have a similar effect.

On oxidation of finely divided PbO in air at about 500°C, yellow-red *lead*(II,IV) *oxide, red lead* Pb_3O_4 is formed. Red lead is not, as sometimes suggested, a lead(II) plumbate(IV), but may be said to be a double oxide (21-5a) of Pb(II) and Pb(IV). Red lead with linseed oil as a binder makes an effective rust-protective paint, whose effect depends partly on the good mechanical and adhesive properties of the lead soaps formed, partly on the weakly basic medium produced by the red lead, and finally on the fact that the red lead in some way renders the iron surface passive. Only recently have other equally effective rust-protective paints been developed, especially zinc chromate (31-5c).

Strong oxidation of Pb(II) compounds gives *lead*(IV) *oxide* PbO_2 (the name *lead dioxide* is also correct, but not the earlier name lead peroxide since peroxide ions are not present). The oxidation is usually carried out anodically ($Pb^{2+} + 2H_2O \rightarrow PbO_2 + 4H^+ + 2e^-$; this process also takes place at the positive plates on charging a lead storage cell—see 16h) or with chlorine ($Pb^{2+} + 2H_2O + Cl_2 \rightarrow PbO_2 + 4H^+ + 2Cl^-$). The oxide forms a dark-brown, strongly oxidizing powder. It has amphoteric properties but the basic character is weak. Its solubility in acids is thus rather slight. Its acid character is more pronounced. When it is dissolved in alkali hydroxides *hexahydroxoplumbates*(IV) are formed, analogous to the corresponding tin compounds (23-5e): $PbO_2 + 2OH^- + 2H_2O \rightarrow Pb(OH)_6^{2-}$.

Several *unstable lead*(II,IV) *oxides* are also known, with compositions close to the formula Pb_2O_3. One of these, which is usually designated as Pb_2O_3, can only be obtained from water solutions. The others are obtained by oxidation of PbO or decomposition of PbO_2 with heat.

On the addition of alkali hydroxide to solutions of Pb(II) salts a white precipitate of *lead*(II) *oxide hydroxide* is formed, in which hexanuclear units $Pb_6O_4(OH)_4$ have been found. No hydroxide $Pb(OH)_2$ is known in solid form. The oxide hydroxide is somewhat soluble in water with a basic reaction. In acids it dissolves easily with formation of Pb^{2+} ion. However, it also has weakly acid properties and is dissolved by alkali hydroxide solution with the formation of *hydroxoplumbate*(II) *ions* (older name, plumbite ions), especially $Pb(OH)_3^-$. The analogy with the corresponding tin compounds (23-5f) is striking.

f. Other compounds. In *lead*(II) *salts* the ionic-bond contribution is appreciable. However, most of these salts are poorly soluble, and of these we should especially mention the *halides*, the *azide* (22-6f), *sulfate*, *carbonate* and *chromate*. The most important soluble Pb(II) salts are *lead*(II) *nitrate* and *lead*(II) *acetate* (called "sugar of lead" because of its sweet taste). However, when these salts are dissolved in water hydrolysis takes place so that hydroxo complexes are formed and hydroxide salts may precipitate. The hydrolysis is prevented by addition of acid.

The Pb(II) ion is colorless. Color may be produced by the anion and several such Pb(II) compounds are used as pigments (31-5c).

Lead(II) *hydroxide carbonate* $Pb_3(OH)_2(CO_3)_2$ constitutes an important white paint pigment, *white lead*. This is produced by several methods, which as a rule involve the conversion of lead to lead(II) acetate with acetic acid, followed by reaction with carbon dioxide. White lead has excellent adhering and covering ability and is therefore still much used in spite of its toxicity and sensitivity to hydrogen sulfide (brown coloration of lead(II) sulfide).

Lead(II) *sulfide* PbS constitutes the mineral *galena*, which crystallizes in lead-gray crystals with metallic luster and cubic symmetry (sodium-chloride structure). It is formed as a black precipitate on passing hydrogen sulfide into lead(II) salt solutions. Its low solubility product (13-1 d, e, tab. 13.1) makes the precipitation practically complete provided the pH of the solution does not fall below 1. For this reason, and because of the black color of the sulfide, hydrogen sulfide is a sensitive test reagent for Pb^{2+} ions.

As mentioned earlier the ionic character is less pronounced for Pb(IV) compounds. These are easily reduced to Pb(II) compounds. *Lead*(IV) *chloride* or *lead tetrachloride* $PbCl_4$ is an unstable yellow liquid which is readily hydrolyzed and therefore fumes in moist air. If chlorine is passed into a suspension of $PbCl_2$ in hydrochloric acid at 10–15°C, a solution is obtained that contains chloro complexes of lead(IV), possibly uncharged $PbCl_4$, and presumably hexachloroplumbate ion $PbCl_6^{2-}$. The high HCl concentration prevents precipitation of PbO_2, for example according to $PbCl_2 + Cl_2 + 2H_2O \rightleftharpoons PbO_2(s) + 4H^+ + 4Cl^-$. If conc. NH_4Cl solution is added to this solution, yellow crystals of *ammonium hexachloroplumbate*(IV) $(NH_4)_2PbCl_6$ separate out. $PbCl_4$ can be prepared from these by immersing the crystals into chilled conc. sulfuric acid: $(NH_4)_2PbCl_6(s) + H_2SO_4 \rightarrow (NH_4)_2SO_4 + PbCl_4(l) + 2HCl$. $PbCl_4$ is insoluble in conc. sulfuric acid and therefore appears as a separate liquid phase.

CHAPTER 24

Group 13, The Boron Group

24-1. Survey

The most common oxidation number in the boron group is III. The oxidation number I appears very rarely with aluminum, but with increasing atomic number its relative stability increases so that with thallium it gives the most stable compounds. With the oxidation number I the "inert" electron pair ns^2 is retained, but with the number III this also has been given off (5-2f; cf. the analogous situation in group 14 and to a certain degree also in 15). The oxidation number II also exists, although it is seldom encountered under ordinary conditions.

Boron in many respects shows strong analogies with silicon, although the different oxidation numbers impose formal differences. Its quite high electronegativity makes the covalent contribution strong even in bonds with very electronegative elements. With rising atomic number within the boron group the electronegativity and consequently the covalent-bond contribution decreases; already with aluminum the ionic contribution is considerably more marked than with boron. The Tl^+ ion in many respects is like an alkali-metal ion. Parallel to these changes the metallic properties increase. Boron is an insulator or a semiconductor (24-2b); the other elements of the group are metals. The basicity of the oxides and hydroxides increases with the ionic-bond contribution. The boron oxide B_2O_3 is an acid oxide, and aluminum oxide Al_2O_3 is intermediate. $B(OH)_3$ is an acid and is therefore also written H_3BO_3. $Al(OH)_3$, $Ga(OH)_3$ and $In(OH)_3$ are amphoteric, $TlOH$ gives strongly basic solutions.

In boron the number of orbitals available for bonding is four, but often not all of these can be used because of lack of space around the small boron atom. Thus four fluorine atoms are very readily coordinated but four larger atoms less easily. With oxygen both threefold and fourfold coordination occur. In aluminum d orbitals can also be used for bonding. Here coordination number six is possible, and is also permitted by the size of the aluminum atom. With oxygen the number six is very common, but four also often occurs, for example in aluminosilicates (23-3h).

The hydrogen compounds (hydrides) are built with covalent bonds and their stability decreases with rising atomic number.

24-2. Boron

a. Occurrence and history. Boron is present on the average in rather small proportion in the earth's crust, where it is always combined with oxygen in boric acid, borates or borosilicates. The largest amounts occur in *tourmaline*, a very complicated borosilicate of alkali metals and magnesium, but this mineral does not form minable deposits. Sea water contains an average of 5 mg B per l, which probably came from solution of rocks and, in the form of boric acid, from the interior of the earth by way of volcanic gases (see below). Evaporation of water in inland seas and undrained lakes has probably been the origin of the greater portion of the minable borate concentrations. The most important borate mineral in these is *kernite* $Na_2B_4O_7 \cdot 4H_2O$. Other important borate minerals are *borax* $Na_2B_4O_7 \cdot 10H_2O$, and several other borates of Na, Ca and Mg. In volcanic regions *boric acid* $B(OH)_3$ is obtained (as the mineral *sassolite*), which is brought up with ejected, superheated water vapor.

Kernite occurs in very large deposits in southern California (Kern County). The United States is consequently by far the largest producer of boron.

Plants generally contain boron in small amounts, which, however, is of essential importance. A number of boron-deficiency diseases are known in plants, but not in animals, on the other hand. Land plants probably get their boron from the weathering products of tourmaline.

In 1785 Torbern Bergman established that a substance that had long been known as the product of treating borax with an acid, was itself an acid. Bergman proposed that this acid should be called boracic acid, but the name with usage became boric acid. By reducing boric acid with potassium, Gay-Lussac and Thénard in 1808 prepared what they presumed to be the "radical" of boric acid and called it in French *bore*. However, a pure preparation of the element was not obtained (cf. 24-2c).

b. Properties. Boron has the isotopic composition $\sim 20\%$ ^{10}B, $\sim 80\%$ ^{11}B. Only in the 1940's was boron first obtained in pure form. It exists in at least three solid modifications, all with complicated crystal structures. Characteristic building units in these modifications are B_{12} groups in which the boron atoms are situated at the corners of an icosahedron (fig. 24.1a). The crystals are harder than silicon carbide, nearly opaque and dark red to black. The conductivity is very low and increases with heating. It is still uncertain whether absolutely pure boron is a semiconductor or whether it is an insulator and the conductivity observed is a result of impurity semiconduction (6-4d). The melting point is about 2300°C.

At room temperature boron is rather inert, but on heating reactions readily occur. Especially notable then is its oxidizability, which makes hot boron an effective reducing agent. Finely divided boron ignites in air at about 700°C and burns to B_2O_3. Hot oxidizing acids also attack boron.

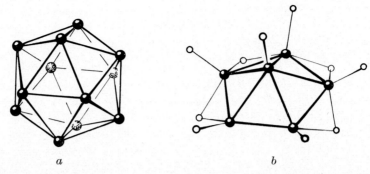

Fig. 24.1. (a) Icosahedral group B_{12} appearing in crystallized boron, boron carbide and certain borides and boron hydrides. (b) The boron hydride B_6H_{10}. Large spheres = B, small spheres = H.

The boron isotope ^{10}B has a very high absorption cross section for neutrons (9-2 b), involving the nuclear reaction ^{10}B(n,α)^{7}Li. Boron must therefore be absent from nuclear reactor materials. On the other hand the high neutron absorption of boron is utilized for neutron shielding, control elements in nuclear reactors (9-4a) and for the detection of neutrons.

c. Preparation and uses. By the reaction of B_2O_3 with metals (alkali metals, magnesium, aluminum) a brown powder is obtained which contains up to 98 % B and is called "amorphous" boron. However, it probably contains no elementary boron but rather metal borides and oxygen-containing compounds with a high proportion of boron. The purest preparations are nowadays obtained from boron tribromide BBr_3 or boron triiodide BI_3, which is decomposed in the gaseous form at a surface heated to at least 800°C (often electrically heated tungsten or tantalum wires). The reaction can be facilitated by mixing with hydrogen gas.

Pure boron has practically no applications. Steel is sometimes alloyed (34-3e) with small amounts of boron, which, however, is then added in the form of an iron-boron alloy, *ferroboron*. This is made by melting iron with B_2O_3 in the presence of a reducing metal (for example, aluminum). The most important commercial starting materials for the preparation of other boron compounds are borax and boric acid.

d. Borides. The borides are much like the silicides (23-3d) in their properties, and some of them are similarly used for their hardness, chemical resistivity and high melting point.

e. Hydrides. The majority of well-defined boron hydrides ("boranes") belong to two groups, B_nH_{n+4} (well characterized for $n=2, 5, 6, 8, 10, 18$) and B_nH_{n+6} (well characterized for $n=4, 5, 9, 10$). A number of hydrides with $n>10$ certainly exist in both groups. At and above about $n=10$ the boron hydrides are solid at room temperature. The lighter molecules form easily volatile liquids except B_2H_6 (*diborane*), which is gaseous. The hydrides of the second group are appreciably less stable than those of the first.

It has not been possible to isolate the simple boron hydride BH_3. It appears as a transitory intermediate, however, in many processes and combines easily with electron-pair donors (5-3c). When B_2H_6 is heated to about 100°C or slightly higher is splits at first into BH_3, which reacts extremely rapidly with remaining B_2H_6. The following reactions, among others, may take place: $BH_3 + B_2H_6 \rightarrow B_3H_9$; $B_3H_9 \rightarrow B_3H_7 + H_2$; $B_3H_7 + B_2H_6 \rightarrow B_5H_{11} + H_2$. Higher hydrides are thus formed together with hydrogen. High-molecular-weight and undefined solid hydrides may also be found among the products. At red heat all boron hydrides decompose into boron and hydrogen.

All boron hydrides are hydrolyzed by water, generally more slowly the higher the molecular weight is. B_2H_6 is hydrolyzed rapidly even in cold water: $B_2H_6 + 6H_2O \rightarrow 2B(OH)_3 + 6H_2(g)$. In air oxidation takes place, often with spontaneous ignition. The volatile hydrides are foul-smelling and poisonous.

The boron hydride molecules in the groups mentioned above are *electron-deficient molecules* with three-atom bonds which give hydrogen bridges of the type $B{\overset{\cdot H \cdot}{}}B$. See 5-6b for a discussion of these bonds and also a schematic structural formula for B_2H_6. Fig. 24.1b shows B_6H_{10} with four hydrogen bridges. However, in 1963 by thermal decomposition of $B_{10}H_{14}$ an unusually hydrogen-poor hydride $B_{20}H_{16}$ was obtained (m.p. 196°C with sublimation), in which each hydrogen atom is bonded to only one boron atom. In several boron hydrides the molecule has a boron skeleton that may be considered as a smaller or larger fragment of a B_{12} icosahedron. Such is the case, for example, in B_6H_{10} (cf. fig. 24.1*a* and *b*).

Boron hydrides are formed by reaction of borides (for example, magnesium boride) with acids. However, the yield of B_2H_6 is slight because of the rapid hydrolysis of this hydride and it is therefore most easily prepared by reaction between boron trichloride and lithium tetrahydridoaluminate (24-3d) in ether solution: $4BCl_3 + 3LiAlH_4 \rightarrow 2B_2H_6(g) + 3LiCl + 3AlCl_3$. The others are best obtained by thermal decomposition of B_2H_6, sometimes with an excess of hydrogen gas, in which the chief products depend on the conditions of reaction (pressure, temperature). This is followed by fractional distillation.

The deficiency of electrons causes boron hydrides to take up electrons from, for example, alkali metals so that anions with a smaller or no electron deficiency are formed: $2K + B_2H_6 \rightarrow (K^+)_2(B_2H_6)^{2-}$. The anion $B_2H_6^{2-}$ is isoelectronic with ethane C_2H_6. The reaction of B_2H_6 with lithium hydride LiH in ether solution gives $2LiH + B_2H_6 \rightarrow 2Li^+(BH_4)^-$. The tetrahydridoborate ion BH_4^- has a normal structure and is isoelectronic with methane CH_4 and the ammonium ion NH_4^+ (5-6b). Similar compounds are known with other alkali metals, but for other metals the bonding becomes more and more covalent the more electronegative the metal is. More covalent bonds may be considered to arise through the boron hydride molecule functioning as an electron-pair acceptor (5-3c).

In many reactions between boron hydrides and electron-pair donors BH_3 is split off, which immediately combines with the donor. For example, if the electron-pair donor is A, we may obtain $2A + B_2H_6 \rightarrow 2ABH_3$.

With an electron-pair donor such as triethylamine $(C_2H_5)_3N$, ions $B_nH_n^{2-}$ without electron deficiency are formed together with hydrogen. From theoretical considerations ions representing all n values from 5 to 12 are expected to exist, and ions with $n=6$, 9, 10, 11, and 12 have so far been found. The boron framework in $B_{12}H_{12}^{2-}$ is a complete icosahedron (fig. 24.1 a). If diethyl sulphide $(C_2H_5)_2S$ is used as the electron-pair donor in the presence of acetylene, neutral molecules (*carboranes*) are formed which can be derived from these ions by replacing two boron atoms with carbon atoms. For example, $B_{12}H_{12}^{2-}$ gives the isoelectronic carborane $B_{10}C_2H_{12}$ in which the boron-carbon framework forms an icosahedron. From $B_{10}C_2H_{12}$ the carborane ions $B_9C_2H_{12}^{-}$ and $B_9C_2H_{11}^{2-}$ can be obtained.

In the large, highly symmetrical molecules and ions of boron hydrides and carboranes the bonding electrons are strongly delocalized, which often leads to "aromatic" character (5-3e). The ion $B_9C_2H_{11}^{2-}$ forms sandwich complexes (5-6c) with ions of transition elements.

The boron hydrides have been tested as rocket fuels (they have large energy content per unit of mass and volume, and are relatively easily handled) and have therefore been produced on a large scale. For this purpose mostly the relatively stable B_5H_9 (b.p. 48°C), $B_{10}H_{14}$ (m.p. 99.7°C), and also alkyl-substituted boron hydrides have been used. The last have also begun to be used as additives to engine fuels. Alkali tetrahydridoborates are used as reducing agents in organic syntheses.

f. Halogen compounds. The most important halides of boron have the general formula BX_3. They can all be prepared by direct combination of the elements. Even with X = F the covalent-bond contribution is predominant, and this increases as the halogen becomes heavier and less electronegative. The discrete molecules lead to low melting points and boiling points (BF_3, -128, -100; BCl_3, -107, 12; BBr_3, -46, 91; BI_3, 50, 210; all in °C).

As described in 5-3e for BF_3 the boron atom is sp^2-hybridized and thus lies in the same plane as the ligands. The third p orbital gives a delocalized π bond.

The structure can thus be written
$$\begin{array}{c} |X| \\ \| \\ B \\ / \ \backslash \\ |\underline{X}| \quad |\underline{X}| \end{array}$$
with the double bond delocalized.

The boron halides are electron-pair acceptors and the trifluoride is one of the strongest such known. As with SO_3 (21-10b) the state where the central atom fulfills the octet rule by bonding four ligands is clearly more stable than the state with double bonding in the free molecule. Formally the absorption of an electron pair can be written, for example, as

$$\begin{array}{c} |\underline{F}| \\ | \\ |\underline{F}|-B \\ | \\ |\underline{F}| \end{array} + \left[|\underline{F}|\right]^- \rightarrow \left[\begin{array}{c} |\underline{F}| \\ | \\ |\underline{F}|-B-\underline{F}| \\ | \\ |\underline{F}| \end{array}\right]^-$$

where a tetrahedral *(tetra)fluoroborate ion* BF_4^- is formed. This ion appears when boron trifluoride is passed into a water solution of hydrogen fluoride: $HF+BF_3 \rightleftharpoons H^+ + BF_4^-$. No acid HBF_4 has been shown to exist (12-2f). If water is removed from the solution the reaction goes to the left. Fluoroborate can also be prepared by the action of hydrogen fluoride in water solution on borate: $B(OH)_4^- + 4HF \rightleftharpoons BF_4^- + 4H_2O$.

Boron trifluoride BF_3 can be formed by the reaction between boron and fluorine. An old method, which, however, gives a low yield, consists of heating diboron trioxide with calcium fluoride and conc. sulfuric acid. Here hydrogen fluoride is formed first, which then reacts with the oxide: $B_2O_3 + 6HF \rightarrow 2BF_3 + 3H_2O$. On an industrial scale it is now manufactured by heating a fluoroborate and diboron trioxide with fuming sulfuric acid: $6BF_4^- + B_2O_3 + 6H^+ \rightarrow 8BF_3 + 3H_2O$ (in practice usually a mixture effectively of fluoroborate and diboron trioxide is made by allowing hydrogen fluoride to react with borax; see above). Boron trifluoride is marketed compressed in steel cylinders. Because of its properties as an electron-pair acceptor and its related effect as a catalyst in many organic reactions it has found wide use. Fluoroborate solutions are currently used in electrolytic metal plating (16f).

All boron halides are hydrolyzed by water and the volatile halides therefore fume in moist air.

g. Oxygen compounds. Compounds containing boron–oxygen bonds exist in great number, and boron occurs in nature only in such compounds. In these the boron atom shows the coordination numbers three and four. With respect to coordination with oxygen boron thus lies on the boundary between atoms with threefold coordination, for example carbon(IV) and nitrogen(V), and atoms with fourfold coordination, for example silicon(IV), phosphorus(V), sulfur(VI) and chlorine(VII).

The most important boron oxide is *diboron trioxide* B_2O_3, which can be made by combustion of boron, but is usually prepared by igniting boric acid: $2B(OH)_3 \rightleftharpoons B_2O_3 + 3H_2O$. It easily takes up water to regenerate boric acid. Diboron trioxide is usually formed as a colorless glass, which can be caused to crystallize only with great difficulty. In both the crystals (m.p. 450°C) and the melt the structural principle is a network infinite in three dimensions, held together by B–O–B bonds. The melt is therefore easily supercooled (6-3d).

Diboron trioxide is an acid oxide (21-5a). When melted together with basic oxides it forms *borates* of widely varying compositions and structures. In 5-6b it was shown how the occurrence of *polyborate ions* can be explained by the tendency of the borate ion to coordinate at least three oxygen atoms. In addition to the mononuclear *orthoborate ion* BO_3^{3-} (fig. 5.29a), the dinuclear *diborate ion* $B_2O_5^{4-}$ (fig. 5.29b), the trinuclear *triborate ion* $B_3O_6^{3-}$ (fig. 5.29c) and the infinite-chain *metaborate ion* $(BO_2^-)_n$ (fig. 5.29d) were described. Thus the borate ions, like the silicate ions (23-3h), become larger and more extended the higher the proportion of the acid oxide is.

Among the crystal structures of borates formed from water solution, in addition to the tetrahedral ion $B(OH)_4^-$, a number of other polyborate ions have been found, both isolated and infinite in one or two dimensions. One such borate is *borax*, a *sodium tetraborate*, which has commonly been written $Na_2B_4O_7 \cdot 10H_2O$.

In all structures of borates formed from water solution so far determined, boron coordinates either OH groups that are not shared with other boron atoms, or oxygen atoms that are shared with another boron atom. The coordination of oxygen atoms around the boron atoms is either threefold (plane) or fourfold (tetrahedral). Very often B_3O_3 rings appear similar to the triborate ion (fig. 5.29c). Examples are:

$$\begin{bmatrix} & & \text{OH} & & \\ & & | & & \\ & O\!-\!B\!-\!O & & \\ & / & | & \backslash & \\ HO\!-\!B & & O & & B\!-\!OH \\ & \backslash & | & / & \\ & & O\!-\!B\!-\!O & & \\ & & | & & \\ & & \text{OH} & & \end{bmatrix}^{2-} \quad \begin{bmatrix} HO & & & & OH \\ \backslash & & & & / \\ B\!-\!O & & O\!-\!B \\ / & \backslash & / & \backslash \\ O & & B & & O \\ \backslash & / & \backslash & / \\ B\!-\!O & & O\!-\!B \\ / & & & & \backslash \\ HO & & & & OH \end{bmatrix}^-$$

tetraborate ion $B_4O_5(OH)_4^{2-}$ pentaborate ion $B_5O_6(OH)_4^-$
(the ring system is not plane)   (the two ring planes lie at right angles to each other)

Borax contains tetraborate ions, which are bound together into infinite chains by hydrogen bonds. These are interleaved with infinite cation chains consisting of $Na(H_2O)_6^+$ octahedra joined together by common edges, so that the overall composition of a chain is $Na(H_2O)_4^+$. Borax is thus properly represented by the formula $[Na(H_2O)_4]_2[B_4O_5(OH)_4]$.

Borate melts with large borate ions have a great tendency to supercooling and therefore glass formation (6-3d). Since the melting points of borates are generally rather low, borate melts are much used in making enamels and glazes. As mentioned in 23-3i2 diboron trioxide is also incorporated into certain types of glasses, in which then a borosilicate network is formed.

The reactions of borates in oxide melts correspond to what one would expect from the structural features described and the acid character of diboron trioxide, according to 21-5a. Borax is much used as a flux for brazing. After the water is driven off with heating an acid melt is obtained with the overall formula $Na_2O \cdot 2B_2O_3$ and this easily dissolves basic oxides on the metal surface with the formation of easily fusible, more basic borates. The same types of reactions are involved when basic oxides are dissolved in a "borax bead" that is obtained by heating a few grains of borax. The melt often solidifies to a glass that shows a color characteristic of the included metal.

Alkali borates are easily soluble in water, other borates poorly soluble.

The most important boric acid is *(ortho)boric acid* $B(OH)_3$, which crystallizes out of acidified water solutions of alkali borates. It forms colorless, leaflike crystals, built up of infinite parallel-oriented layers. The layers consist of $B(OH)_3$ molecules bound together by hydrogen bonds (fig. 24.2). The bond forces between the layers are weak, leading to good cleavage and an easy sliding of the layers

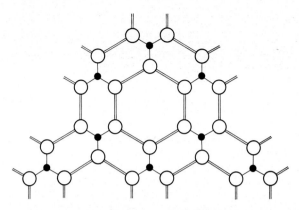

Fig. 24.2. A portion of the infinite layer in boric acid B(OH)$_3$. Solid circles = B, open circles = O. The double lines indicate hydrogen bonds, and a hydrogen atom thus lies in each.

over one another (boric acid crystals therefore feel "greasy" and have a certain lubricating ability). On heating, orthoboric acid transforms to more water-poor, condensed acids, among others *metaboric acid* (several modifications) with the overall formula HBO$_2$. Finally, B$_2$O$_3$ is obtained. If the evolution of water is prevented by the presence of water vapor at high enough pressure, B(OH)$_3$ is vaporized (cf. 21-5d). Therefore, boric acid may be present in volcanic gases (24-2a).

Boric acid is a very weak acid. It is protolyzed in water solution, probably mainly according to B(OH)$_3$ + H$_2$O ⇌ B(OH)$_4^-$ + H$^+$ and for this equilibrium at 25°C, pK_a = 9.00. However, in borate solutions several other types of ions occur, among others, the tetraborate ion described above. With organic molecules that contain several OH groups properly spaced with respect to each other (for example fructose, mannite), boric acid complexes are formed according to the following schematic equation:

$$\begin{matrix} \diagdown \\ C-OH \\ | \\ C-OH \\ \diagup \end{matrix} + \begin{matrix} HO \diagdown \\ B \\ HO \diagup \end{matrix} \begin{matrix} OH \\ \\ OH \end{matrix} + \begin{matrix} HO-C \diagup \\ | \\ HO-C \diagdown \end{matrix} \rightleftharpoons \left[\begin{matrix} \diagdown \\ C-O \\ | \\ C-O \\ \diagup \end{matrix} \begin{matrix} \\ B \\ \end{matrix} \begin{matrix} O-C \diagup \\ | \\ O-C \diagdown \end{matrix} \right]^- + H^+ + 3H_2O$$

If a strong base is added to a solution of such a complex, the reaction goes to the right and the titration curve obtained corresponds approximately to that of an acid of the strength of acetic acid.

Methyl and ethyl esters of boric acid (obtained by mixing together boric acid, alcohol and conc. sulfuric acid) are volatile and burn with a green flame (spectral bands, probably caused by boron oxides), which can be used as a test for boron.

Boric acid is used as a weak antiseptic, especially in treatment of the eyes (it should not be used internally). The boric acid system is often used as a buffer to control pH.

With hydrogen peroxide alkali borate solutions give *alkali peroxoborates*. Ordinary commercial *"sodium perborate"* contains the ion

$$\begin{bmatrix} HO & O-O & OH \\ & B & B & \\ HO & O-O & OH \end{bmatrix}^{2-}$$

The peroxoborates are widely used as washing and bleaching agents.

h. Nitrides and carbides. A large number of compounds containing an equal number of atoms of boron and either nitrogen or phosphorus show many similarities to carbon compounds. This is especially true of boron-nitrogen compounds, since $\frac{1}{2}(B+N)$ is isoelectronic with C. Thus, by heating boron in nitrogen gas or diboron trioxide in ammonia *boron nitride* BN is produced, which has a structure consisting of parallel layers of the same type as in graphite (fig. 5.21) but with alternating boron and nitrogen atoms (in boron nitride the sixfold rings in the sheets lie directly over each other). Boron nitride like graphite has lubricating properties, but is colorless and an insulator. Such extensive delocalization of the π orbitals as in graphite (5-3e) therefore does not occur. It is of great value as a ceramic material (in inert atmosphere it is stable up to the melting point at about 3000°C) and as a lubricating agent. In 1957 the diamond analog of boron nitride (trade name Borazon) was successfully prepared at high pressure and high temperature (about 55 kbar, 1650°C). This has the hardness of diamond and, in contrast to diamond, will tolerate heating up to 1900°C. Its greater brittleness has so far prevented it from gaining any technical importance.

By heating B_2H_6 and ammonia to 250–300°C the benzene analog *borazole*

$$\begin{array}{c} H \\ B \\ HN \quad NH \\ | \quad \| \\ HB \quad BH \\ N \\ H \end{array}$$

is obtained, whose physical properties are very much like those of benzene. The chemical properties, on the other hand, are quite different.

Boron carbide is obtained as black crystals by reduction of B_2O_3 with carbon at 2500°C. The crystals contain B_{12} icosahedra (fig. 24.1) and in the interstices between these up to three carbon atoms per B_{12} group can be situated. The formula then is $B_{12}C_3$, but has often been written B_4C. But the carbide phase remains homogeneous even if at least one of these three carbon atoms is replaced by boron ($B_{12}BC_2$). Boron carbide, like boron, is harder than silicon carbide, melts only at 2350°C, and is chemically exceedingly inert. It is used as a durable construction material and in metallurgy as a reducing agent. In nuclear energy technology it is used as a neutron absorber when high abrasion resistance is also required.

24-3. Aluminum

a. Occurrence and history. As can be seen in tab. 17.2, aluminum is one of the most common elements in the accessible parts of the earth. It never occurs free and is almost always combined with oxygen, especially in aluminosilicates (23-3h). The most important of these are the *feldspars* (23-3h5), the *micas* and *clay minerals*

(23-3h4). One important oxygen-free aluminum mineral is *cryolite*, sodium hexafluoroaluminate Na_3AlF_6. The amount of aluminum in living organisms is very small and its presence has not been proven to be essential.

The most important source of aluminum is supplied by weathering products, which are given the collective name *bauxite*. Bauxite consists to a large extent of aluminum hydroxide AlO(OH), but a large proportion of impurities is commonly present, especially silicon dioxide and iron(III) hydroxide.

The most important European bauxite deposits are in France (among other places, Les Baux, whence the name), Hungary, Italy, Yugoslavia and the Soviet Union. In North America it is mined in the United States (especially Arkansas) and Jamaica; elsewhere important reserves are located in the Guianas of South America, the countries around the Gulf of Guinea, and India.

In ancient times *alum* (Latin name *alumen*) $KAl(SO_4)_2 \cdot 12H_2O$ was used in medicine and in color dying. In the middle of the 18th century the oxide Al_2O_3 was prepared from alum. It was designated as an earth (23-3a) and called *terra aluminis*. The metal presumed to be contained in terra aluminis was called *aluminum* by Davy in 1808, which gradually in most languages became *aluminium* (in America the earlier name has been retained). The metal was prepared in 1825 by Ørsted by reduction of aluminum chloride with potassium amalgam.

The metal aluminum first became of practical importance when Héroult in France and the American chemist Hall in 1886, simultaneously but independently of each other, developed the method for its preparation by electrolysis of melts.

b. Properties. Aluminum is a silver-white metal with low density (2.70 g cm^{-3}). Up to the melting point at 660°C the structure is cubic close packed. The boiling point is about 2300°C. Pure aluminum can be molded plastically both cold and hot. The thermal and electrical conductivities are high (the latter approximately two thirds that of copper).

Aluminum is a quite nonnoble metal ($e^0 = -1.66$ V), but as mentioned in 16e its surface in many cases is protected against chemical attack by the rapid formation of an oxide layer of a thickness of about 0.01 μm both in water and in air. The layer is transparent but very dense, hard and clinging. The presence of mercury or its salts destroys the passivity so that both water and oxygen rapidly react. Through intensified oxidation the layer can be made thicker (of the order of 10 μm) and thereby made to give still greater protective action. For this, especially anodic oxidation (anodization) is used. The thick oxide layer makes the formerly bright surface dull and can be colored with both inorganic and organic dyes. Impurities or alloying elements generally decrease the chemical resistivity of aluminum.

The oxide layer is also protective when aluminum is heated in air. For this reason and because of the high thermal conductivity, combustion with a flame occurs only at high temperature or if the metal is finely divided. The heat of combustion is very high ($2Al(s) + \frac{3}{2}O_2(g) \rightarrow Al_2O_3(s)$; $\Delta H^0 = -1674$ kJ) and the flame temperature is therefore very high. Mixtures of aluminum powder and air are explosive. In oxygen gas aluminum foil or wire will burn with strong evolution of light (flash bulbs).

Basic and many sufficiently strongly acid solutions dissolve the spontaneously formed, thin oxide layer and then attack the metal with the evolution of hydrogen. The basic solutions give hydroxoaluminate ions and the acid solutions aluminum ion. The processes can be represented by the following very schematic equations (among other things hydration—12-4d—is not shown): in basic solution, $Al_2O_3 + 2OH^- + 3H_2O \to 2Al(OH)_4^-$ followed by $2Al + 2OH^- + 6H_2O \to 2Al(OH)_4^- + 3H_2(g)$; in acid solution, $Al_2O_3 + 6H^+ \to 2Al^{3+} + 3H_2O$ followed by $2Al + 6H^+ \to 2Al^{3+} + 3H_2(g)$. However, strongly oxidizing acids, by strengthening the oxide layer, may become quite inactive; conc. nitric acid does not dissolve aluminum at room temperature. Here also the passivating layer is destroyed in the presence of mercury or its salts. The thick layers obtained by anodization greatly increase the resistance to both basic and acid solutions.

For aluminum the oxidation number III is wholly predominant and is the only one considered in the following sections. The bond character and coordination number have been discussed in 24-1.

c. Production and uses. As stated in 16f aluminum cannot be deposited electrolytically from water solution, but on the other hand, it can be from anhydrous melts. The most readily available, pure aluminum compound, namely the oxide Al_2O_3, however, cannot be electrolyzed directly because of its high melting point (2050°C). On the other hand a suitable electrolyte is obtained by dissolving the oxide in cryolite Na_3AlF_6, which melts at 1000°C. The cryolite solution further has a quite high conductivity. Some aluminum fluoride AlF_3 is also added to the melt. The electrolyte usually contains 5–10 wt % aluminum oxide and is held at about 1000°C by the heat of the electric current.

The aluminum oxide is extracted from bauxite and must be very pure. Various processes have been used for purification but the most common nowadays is the so-called *Bayer process*. In this process the dried and ground bauxite is leached at about 170°C (under pressure) with strong sodium hydroxide solution: $AlO(OH) + OH^- + H_2O \to Al(OH)_4^-$. Among the impurities many of the oxides (especially of iron) are insoluble, and silicon dioxide forms a poorly soluble sodium aluminosilicate. These solid phases formed by the impurities are filtered off. The sodium aluminate solution is cooled and seeded with crystals of aluminum hydroxide $Al(OH)_3$, following which hydroxide slowly precipitates: $Al(OH)_4^- \to Al(OH)_3 + OH^-$. The hydroxide is filtered off (the remaining sodium hydroxide solution is concentrated and used again for leaching), washed, dried and ignited so that the oxide is formed: $2Al(OH)_3 \to Al_2O_3 + 3H_2O$.

The cost of the process depends highly on the amount of silicon dioxide in the bauxite. The higher this is, the larger are the amounts of aluminum and sodium (and thus usable sodium hydroxide) lost in the form of the sodium aluminosilicate mentioned above. For higher contents of silicon dioxide the *Pedersen process* is applied. The bauxite is first melted with limestone, so that calcium silicate and calcium aluminum oxide are formed. On leaching with sodium carbonate solution only the latter is attacked, which gives sodium aluminate solution and insoluble calcium carbonate. The aluminate solution is treated approximately as described above.

Cryolite was formerly obtained exclusively as a mineral but as the only deposit (in Greenland) is exhausted it is now made synthetically, usually by treating a sodium aluminate solution with hydrogen fluoride: $Al(OH)_4^- + 6HF \rightarrow AlF_6^{3-} + 2H^+ + 4H_2O$. Na_3AlF_6 is poorly soluble in water and therefore precipitates out.

The vessel that holds the melt is lined with carbon, which serves as the cathode for the electrolysis. The molten metal formed collects at the bottom of the vessel and can be drawn off. The anodes, also made of carbon, are lowered into the melt.

The conditions within the melt are poorly understood, but a number of both simple and complex ions are present. Concerning the electrode processes only the final products are known and therefore only the overall process can be stated. Neither can it be assumed that the main portion of the current is carried by the ions that are discharged at the electrodes (cf. 16f). At the *cathode* a small quantity of sodium ions is discharged ($Na^+ + e^- \rightarrow Na$) in an undesirable side reaction (the sodium metal formed goes off as a gas or burns when tapped off with the aluminum), but the most important overall process is $Al^{3+} + 3e^- \rightarrow Al$. The overall process at the *anode* can be written $2O^{2-} \rightarrow O_2(g) + 4e^-$. The oxygen produced oxidizes the carbon anode to CO_2, which with the remaining electrode carbon takes part in the equilibrium $CO_2(g) + C(s) \rightleftharpoons 2CO(g)$. The anodes are thus strongly attacked and evolve a mixture of CO_2 and CO.

If we overlook the processes involving the anode carbon, the main process is a decomposition of aluminum oxide according to $Al_2O_3 \rightarrow 2Al + \frac{3}{2}O_2(g)$. The consumption of cryolite is therefore slight. For the production of 1000 kg aluminum approximately 2000 kg aluminum oxide, 60 kg cryolite and aluminum fluoride, and 500 kg of electrode material are used up. The energy consumed is approximately 20 000 kWh.

The tapped metal is refined, among other ways, by remelting, which gives a purity of 99.3–99.7 % Al (commercial aluminum). Nowadays by repeated melt electrolysis a degree of purity of 99.99 % can be achieved on a commercial scale. Such pure material is used for special purposes and as starting material for aluminum alloys where precisely controlled composition is required.

Aluminum is one of the most important metals of our time. This results especially from its low density, good stability toward corrosion and the high strength that can be obtained by alloying the metal.

The addition of foreign elements to a metal generally decreases its corrosion stability (16i). Even the impurities in commercial aluminum have this effect, but such "unalloyed" aluminum is still much used when fairly good corrosion stability is required. Examples are decorative fixtures, cooking vessels, equipment for the chemical and foodstuffs industries, storage tanks, packaging (food cans, tubes, foil) and electrical conductors.

Alloys are especially used when greater strength is needed, but additives generally lower the stability toward corrosion. The most common alloy elements are copper, magnesium, silicon and manganese. A distinction is made between alloys in which an increase in strength can only be obtained by cold working (14-3c) and alloys in which great strength and hardness can be obtained by precipitation hardening (13-5f2). The former contain a lower proportion of alloying elements (an example is 1.2 % Mn in addition to the "normal" impurities) and thus still have a relatively good stability toward corrosion. This can also be enhanced by anodization. Besides the applications mentioned above there are structural applications where strength is required in more corrosive environment, for example ship construction materials. Alloys for marine use often have magnesium as the principal additive. Larger proportions of magnesium (for example, *magnalium* with 10–30 % Mg) are sometimes used when especially low density is required.

For the greatest hardness and strength, such as in airplane structures, stressed ship and motor parts, precipitation-hardened alloys are used. The proportions of alloying elements here are higher, and copper is often included. A typical alloy (*Duralumin type*) contains 4.5 % Cu, 0.6 % Mg, 0.7 % Si, 0.8 % Mn. Alloys suitable for casting often contain Cu or Si as the chief alloying element; many such alloys can be hardened by precipitation.

Alloyed aluminum is sometimes coated (plated) with pure aluminum so that maximum stability toward corrosion can be combined with other desired properties.

The high reflectivity of aluminum even in the ultraviolet (7-3a), together with its stable surface, make it suitable in many cases for mirrors. Astronomical mirrors nowadays consist usually of an aluminum layer vaporized on glass. Finely powdered aluminum is used as a pigment, partly in "bronze paints", and partly in asphalt paints that provide corrosion-protective coatings under water. Aluminum also is included in alloys with other metals as the main component, for example with copper (aluminum bronze) and magnesium.

The very high bond energy of aluminum oxide makes aluminum an effective reducing agent at high temperature. It is thus used in metallurgy for deoxidation of metal melts and for production of metals from difficultly reducible oxides (*"aluminothermic"* reduction according to H. Goldschmidt, 21-5b). Liquid iron for sealing cracks can be readily obtained by igniting a mixture of iron oxide and aluminum powder (*Thermit*): $2Al + Fe_2O_3 \rightarrow 2Fe(l) + Al_2O_3(s)$. Aluminum powder is more and more used as an oxidizable substance in solid rocket fuels and explosives with ammonium perchlorate and ammonium nitrate as oxidizing agents (20-8e, 22-6c).

d. Hydrides. The *aluminum hydride* AlH_3 is known, but is unstable and easily oxidized and is of little interest. More important, on the other hand, is the *tetrahydridoaluminate ion* AlH_4^- (concerning its constitution, see 5-6b). *Lithium tetrahydridoaluminate* can be prepared by reaction between lithium hydride and aluminum chloride in ether solution: $4LiH + AlCl_3 \rightarrow LiAlH_4 + 3LiCl$. It consists of a colorless, solid substance, which can be stored in dry air, and is much used in the synthesis of boron hydrides

(24-2e) and as a selective reducing agent in organic chemistry. The sodium compound has begun to be used for the same purposes. *Alkylaluminums* AlR_3, (R = alkyl), with structures analogous to AlH_3, are incorporated in catalysts for organic polymerization processes.

e. Halogen compounds. *Aluminum fluoride* AlF_3 is produced by the action of hydrogen fluoride on hot aluminum, aluminum oxide or aluminum hydroxide, and forms a colorless powder (sublimation point 1260°C) insoluble in water, acids and bases. The aluminum atom has a pronounced tendency to coordinate six fluorine atoms octahedrally. In AlF_3 this is achieved through a network of AlF_6 octahedra linked so that every F atom is common to two octahedra. This structure accounts for its poor solubility and stability to heat (13-4b). On heating with other fluorides, fluoride ions are added and therefore, according to the definition in 12-1e, aluminum fluoride is an acid. With increasing proportion of fluoride ion the network structure is successively broken down. In the simplest examples, octahedral layers are obtained with $AlF_3 + F^- \rightarrow AlF_4^-$, octahedral chains with $AlF_3 + 2F^- \rightarrow AlF_5^{2-}$ and isolated octahedra with $AlF_3 + 3F^- \rightarrow AlF_6^{3-}$. In *cryolite* Na_3AlF_6 the anions consist of such isolated octahedra. In many cases, however, the structures are more complicated. The analogy with silicon dioxide and silicates (23-3h) is nevertheless clear.

Anhydrous *aluminum chloride* $AlCl_3$ is produced commercially usually by heating aluminum scrap in a stream of chlorine or hydrogen chloride. It is then obtained as a colorless mass subliming at 180°C. Like BF_3 (24-2f) the molecule can act as an electron-pair acceptor, but at the same time, by means of the free electron pairs on the chlorine atoms it can act as an electron-pair donor. In this way dimeric molecules Al_2Cl_6 are formed, consisting of two $AlCl_4$ tetrahedra with two common chlorine atoms, thus (schematically):

$$\begin{array}{c} Cl \\ Cl \end{array} Al \begin{array}{c} Cl \\ Cl \end{array} Al \begin{array}{c} Cl \\ Cl \end{array}$$

. The equilibrium $2AlCl_3 \rightleftharpoons Al_2Cl_6$ in the gas phase at the sublimation point is very strongly shifted to the right, but is displaced to the left with increasing temperature. At 800°C the dissociation to $AlCl_3$ is practically complete. Its ability to form addition compounds with electron-pair donors makes aluminum chloride a useful catalyst, especially in organic chemistry (Friedel-Crafts reactions).

A solution in hydrochloric acid of aluminum, its oxides (however, the α oxide —24-3f—is insoluble) or hydroxides, on evaporation gives $AlCl_3 \cdot 6H_2O$. Here the cation is $Al(H_2O)_6^{3+}$, and a more exact formulation is therefore $[Al(H_2O)_6]Cl_3$. This cation takes part in the equilibria summarized in equation (12.71). On heating aluminum chloride hydrate finally HCl is given off (cf. 20-5a) and therefore anhydrous $AlCl_3$ is not obtained but rather the hydroxide (with stronger heating the oxide). When the chloride is dissolved in water it is hydrolyzed. The anhydrous chloride fumes in moist air as a result of the formation of hydrogen chloride by the hydrolysis.

f. Oxygen compounds. The coordination number of aluminum with respect to oxygen is somewhat variable. The presence of d orbitals usable for bonding permits the number six (in contrast to boron, cf. 24-1) and the aluminum atom is also large enough for this number (in contrast to silicon). This number is thus found in the oxides, hydroxides and most double oxides with aluminum as cation. As mentioned in 23-3h, however, the radius is not so large but that aluminum in the aluminosilicates can partly replace silicon in the SiO_4 tetrahedra and thus assume the coordination number four. But here (for example, in micas and clay minerals) aluminum may also be found outside the tetrahedra and then almost always with coordination number six with oxygen.

The only *aluminum oxide* of importance is Al_2O_3. It is formed by oxidation of the metal (24-3b) and by heating the hydroxide. Beginning at about 300–400°C a number of metastable oxide phases appear, whose structures depend on the starting material. These are poorly soluble in water but soluble in acids and bases and are hygroscopic. They have high adsorptivity and are used as adsorption media and catalyst supports (14-4). Above 1000°C the metastable oxide phases are rapidly transformed to the stable phase $\alpha\, Al_2O_3$, which is very hard, insoluble in acids and not hygroscopic. If aluminum oxide is to be weighed in quantitative work it should therefore be heated above 1000°C. The melting point is high, 2050°C. The structure is trigonal and can be described as a hexagonal close packing (6-1f, 6-2c) of oxygen atoms, within whose interstices aluminum atoms are located so that they have octahedral oxygen coordination. Industrially $\alpha\, Al_2O_3$ is produced in large amounts from bauxite (24-3c). $\alpha\, Al_2O_3$ constitutes the mineral *corundum* (hardness 9, see 6-5c). Pure corundum is colorless, and in gem quality corundum crystals are used for precious stones (white sapphire). Poorer material is used as an abrasive, such as *emery*, which consists of impure corundum (major portion), together with magnetite, hematite and quartz. $\alpha\, Al_2O_3$ colored by certain transition metal oxides is highly prized for gem stones, especially *ruby* (colored red by the replacement of a small amount of Al^{3+} ions by Cr^{3+} ions) and *sapphire* (colored blue by partial replacement of Al^{3+} by Ti^{3+}).

By melting aluminum oxide in an electric oven, well-crystallized corundum is obtained, which is used as an abrasive and as refractory ceramic material. Large corundum crystals can be made by allowing powdered Al_2O_3 to fall through an oxyhydrogen flame and collecting the droplets formed on an Al_2O_3 pedestal. The droplets crystallize on the pedestal and the crystal formed in the first drop grows in the accumulating drops so that a large, single crystal (*boule*) is formed. With a small amount of Cr_2O_3 added to the powder ruby is obtained, and with a small amount of added $TiO_2 + Fe_2O_3$ sapphire is obtained. From these artificial gem stones are made, which can only be distinguished from natural stones by their greater freedom from defects. The products are also used as bearing stones in timepieces and electrical measuring instruments ("sapphire" used for phonograph needles is made from corundum synthesized in this way). Quite recently large crystals of corundum have also been successfully grown from solutions of aluminum oxide in molten lead fluoride PbF_2 (m.p. 824°C).

In 12-1b and 12-4d the properties and reactions of *aluminum hydroxide* as an amphoteric hydroxide were discussed. The structure of the stable form of $Al(OH)_3$, known as the mineral *gibbsite* (also called *hydrargillite*) was shown in fig. 6.18b. On heating, aluminum hydroxide loses water and finally goes over to the oxide. *Aluminum oxide hydroxide* $AlO(OH)$ may be formed as an intermediate product. Two different modifications of this, *diaspore* ($\alpha\ AlO(OH)$), and *boehmite* ($\gamma\ AlO(OH)$), are known as minerals and are present among others in bauxite.

Aluminum hydroxide is used as an adsorbing medium and for water purification. When the gelatinous hydroxide precipitates out on hydrolysis of a dissolved aluminum compound (usually aluminum sulfate is used) it carries with it suspended solid particles and high-molecular-weight, dissolved impurities. The method makes it possible to purify water more rapidly than by ordinary filtration.

When aluminum hydroxide is dissolved in a base, according to 12-4d the formation of the *(hydroxo)aluminate ion* $Al(OH)_4^-$ may be assumed to be the final stage. The salts that crystallize from these solutions may possibly contain such ions and thus be true aluminates. On the other hand, if they are melted or if aluminum oxide is melted with other metal oxides, in general aluminates are not obtained but rather *double oxides* (21-5a), for example with the formulas M^IAlO_2 and $M^{II}Al_2O_4$. The latter are usually of the spinel type (21-5a). The mineral *spinel* in the original sense is $MgAl_2O_4$. Many spinels are used as gem stones, and synthetic gems can be made in the same way as aluminum oxide gem stones. The mineral *chrysoberyl* $BeAl_2O_4$ is also a double oxide but does not have the spinel structure.

g. Other compounds. The most important aluminum compounds of primarily salt-like character contain sulfate ion as the anion. *Aluminum sulfate* $Al_2(SO_4)_3$ crystallizes out of water solution as colorless crystals with $18H_2O$. Commercially aluminum sulfate solution is made by dissolving the hydroxide or bauxite in hot conc. sulfuric acid. Usually the products are marketed with smaller amounts of water than in the hydrate mentioned above. The water solution of the sulfate is acid (12-1b). When the solution is made nearly neutral the hydroxide precipitates, and this is used as mentioned in 24-3f for water purification (the proper pH value is obtained by adding lime). In a similar way aluminum hydroxide is precipitated in textile fibers, which then become more easily dyed (mordant). The sulfate is also used for sizing paper, in which the aluminum salts of resin acids are precipitated among the paper fibers.

The double salt *potassium aluminum sulfate* $KAl(SO_4)_2 \cdot 12H_2O$ since early times has been called *alum (potassium alum)* and is the parent type for the alums (21-11b3). Potassium alum is nowadays prepared by crystallization from a solution of potassium sulfate and aluminum sulfate. It is used for the so-called tawing of leather and in pharmacology as an astringent. Earlier alum was much used as a mordant and for paper sizing but has been more and more replaced by aluminum sulfate.

Aluminum sulfide Al_2S_3 can be made from the elements at red heat. It is decomposed by water into aluminum hydroxide and hydrogen sulfide and therefore cannot be obtained by precipitation from solution. Concerning *aluminum nitride* AlN see 22-5.

24-4. Gallium, indium and thallium

These three metals occur widespread in the earth's crust in very small average amounts and are not believed to be concentrated in any local deposits. Gallium always follows aluminum (the radius of Ga^{3+} is slightly larger than that of Al^{3+}) and often zinc. Indium especially follows zinc in sphalerite. Thallium is often found in pyrite and sphalerite and can then be recovered from the flue dusts produced by roasting these minerals. Thallium isotopes are included in each of the natural radioactive series (9-2a).

All three elements were discovered spectroscopically, thallium in 1861, indium in 1863 and gallium in 1875. In 1871, however, Mendeleev predicted the existence of an element ("eka-aluminum") with approximately the properties of gallium (4-2a). Indium and thallium were named for the colors of their most characteristic spectral lines (indigo blue and green, respectively, the latter from Gk. ϑαλλός, green twig), gallium for France (Lat. Gallia) where it was discovered.

The metals are all soft, gallium and indium quite like tin in appearance, thallium more like lead. The melting points are: Ga, 29.8; In, 156.2; Tl, 304°C. Gallium is thus solid at ordinary room temperature, but melts in the hand. For gallium at the melting point the solid phase has a greater specific volume than the liquid phase (1-4h). Therefore, if gallium is stored in a glass vessel the glass may easily be broken if the temperature varies around the melting point. The boiling point of gallium is 2237°C and consequently at atmospheric pressure the metal is liquid over an unusually wide temperature range. Gallium and indium are not altered by dry air, and thallium becomes coated with a gray-black oxide film, which protects it from further attack. All three metals are definitely more noble than aluminum ($Ga \rightleftharpoons Ga^{3+} + 3e^-$, $e^0 = -0.53$ V; $In \rightleftharpoons In^{3+} + 3e^-$, $e^0 = -0.34$ V; $Tl \rightleftharpoons Tl^+ + e^-$, $e^0 = -0.34$ V) but are quite easily soluble in acids. They can be prepared by electrolysis of water solution of their salts.

The oxidation numbers were discussed in 24-1. The numbers I and III occur for all three elements but III decreases and I increases in importance as the atomic number increases. For gallium and indium III is still the most important, for thallium I has become predominant. In many cases which correspond formally to the number II, a mixture of an equal number of atoms with the numbers I and III is present.

Compounds with the oxidation number III for all three elements show a strong analogy to the corresponding Al(III) compounds, although the effect of the increasing basicity of the oxides and hydroxides with increasing atomic number is noticeable. Tl(III) is easily reduced to Tl(I). Tl(I) compounds show many similarities in some ways to compounds of the intermediate alkali metals (for example, hydroxides, carbonates, sulfates), and in other ways to silver compounds (for example, oxides, sulfides, halides). It should especially be noted that Tl(OH) is easily soluble and gives a strongly basic solution.

Thallium compounds are poisonous, like the compounds of its neighbors in the periodic system, mercury and lead.

Liquid gallium will not wet silicon dioxide and therefore can be used as a thermometric liquid in thermometers of silicon dioxide glass, which will withstand temperatures up to about 1200°C. GaP and GaAs are semiconductors and used in electronics. Thallium compounds (especially the sulfate) have been used for eradicating rats and injurious insects. Tl(I) halides have a low infrared absorption and are therefore used in infrared optics.

CHAPTER 25

Group 1, Alkali Metals

25-1. Survey

The alkali metals show so great similarities to each other that they are properly treated together. Among them 87 Fr, francium, is so little known that it is often excluded in comparisons made in the following sections.

In the alkali metals also many properties are affected by the increase in atomic or ionic radius with atomic number. The radii (fig. 5.2 and 5.5) show quite large increases in the steps Li–Na–K but smaller increases in the steps K–Rb–Cs. K, Rb and Cs (as well as Fr) therefore constitute in many respects a group separate from Li and Na. The two latter elements are also less similar to each other, and especially lithium assumes a special position apart from all the other alkali metals. As a matter of fact lithium in many ways resembles magnesium, which has nearly the same atomic and ionic radii.

Many of the typical properties of the alkali metals depend upon the fact that the valence electron in their s shells is easily given off (4-2b), which leads to a very low ionization energy (5-2b, tab. 5.1 and fig. 5.4). This in turn leads to a very low electronegativity (fig. 5.23), which decreases with increasing atomic number. Potassium, rubidium, cesium and francium are less electronegative than any of the elements in other groups. In chemical processes the neutral atom never gives up more than one electron, and never takes up any electrons. Therefore, the only oxidation number besides zero is I.

The small electronegativity of the alkali metals causes the bonds in their compounds with nonmetals always to have high ionic-bond character. The compounds therefore very often constitute typical salts. Since the oxidation number is always I, in these cases the presence of ions M^+ can be assumed.

The alkali metal ions generally form very weak complexes. However, they are hydrated in water solution. The tendency for hydration decreases with increasing ionic size (5-2h). It is also found that water of hydration less and less commonly is carried along from the water solution on crystallization as the alkali metal ion gets bigger (it will be noted, however, that water molecules bound to the anions may appear).

Typical alkali metal salts are colorless as long as the anion is not colored. As a rule they are easily soluble. With alkali ions of increasing size, some anions give a quite sharply increasing solubility, and other anions a quite sharply decreasing solubility. The first type gives a number of rather poorly soluble salts of lithium (occasionally also of sodium). Among these are LiF (very high lattice energy because both kinds of ions are very small; 6-2d, e), Li_2CO_3, Li_3PO_4, $LiSb(OH)_6$ and $NaSb(OH)_6$ (22-13c). The second type gives rather poorly soluble salts generally of all the metals in the group consisting of potassium, rubidium, and cesium. These include the perchlorates $MClO_4$ and the hexachloroplatinates(IV) M_2PtCl_6 of these metals.

25-2. Occurrence and history

a. Occurrence. The low electronegativity of the alkali metals makes them very nonnoble and they therefore do not occur in free form on the earth.

The average proportion of *lithium* in the earth's crust is quite low and specific lithium minerals are few. Among these may be mentioned the currently most important lithium mineral *spodumene*, which is a lithium pyroxene (23-3h3); *lepidolite*, which is a lithium mica (23-3h4); a less common lithium aluminosilicate called *petalite* $LiAlSi_4O_{10}$; and *amblygonite* $(Li,Na)Al(OH,F)PO_4$.

The largest deposits of spodumene are found in the United States in South Dakota and North Carolina. In the United States lithium is also obtained along with potash recovery from salt lakes in the western states (see below). The lepidolite deposits in Southern Rhodesia are also important. Lithium also occurs in a number of plants, among others tobacco, but watering with lithium-containing well water has been shown to injure plants.

Sodium is one of the most common elements in the accessible parts of the earth (tab. 17.2). In the solid earth's crust it is found especially in *feldspars* (albite and sodium-rich plagioclases; see 23-3h5). Sodium ions have been dissolved from minerals by flowing water and carried to the sea, where its average content has risen by constant addition from river water and evaporation to the present-day value of 0.46 M Na (corresponding to 2.7 wt % sodium chloride). The most important vehicle of sodium in commerce and industry is *sodium chloride*.

In enclosed drainage areas where the evaporation is greater than the precipitation, the salt content increases in the salt lakes and solid salt may even crystallize out. Lakes with insufficient or no drainage (Dead Sea, Great Salt Lake) may in this way become useful for the recovery of sodium chloride as well as other salts. In certain such lakes (in Egypt but especially in California and Kenya) large amounts of sodium carbonate are also deposited (soda lakes). Large beds of solid sodium chloride (rock salt) have been formed in a similar way during earlier geologic periods. In connection with such salt beds brines are often found that can be exploited. In warm and dry climates sea water is evaporated in enclosed "salt pans".

Sodium and potassium ions are the most common cations in the body fluids and soft tissue of animals. Sodium predominates in the intercellular fluids, while potassium predominates in the cell fluid itself. Plants generally require only small amounts of sodium. Appreciable amounts of sodium are found only in marine and shore plants.

Potassium occurs in the accessible parts of the earth (tab. 17.2) in hardly less amounts than sodium. It occurs especially in the potassium feldspars *orthoclase* and *microcline* (23-3h5) and the mica minerals *muscovite* and *biotite* (23-3h4), which, however, cannot be economically worked for potassium. Much less potassium than sodium from the weathering of minerals is carried to the sea, since potassium ions are adsorbed and retained by sediments, especially clay minerals (23-3h4), much more strongly than sodium ions. In sea water, therefore, the ratio Na/K = 28 by weight and 48 by number of atoms. However, through appropriate crystallization conditions potassium salts have accumulated in such large amounts in a number of salt lakes, and also in previously formed solid beds, that recovery is possible. The most important potassium minerals in these deposits are *sylvite* KCl and the double salts *carnallite* $KMgCl_3 \cdot 6H_2O$ and *kainite* $KMgClSO_4 \cdot 3H_2O$.

Important salt lakes for potash recovery are, among others, the Dead Sea and Searles Lake in California. Solid salt beds in Germany, for example at Stassfurt, were formed from a Permian inland sea. Other important beds are found in Elsass, Poland, the Soviet Union, and New Mexico in the United States.

The occurrence of potassium in animal organisms was mentioned above in connection with sodium. In plants potassium is an important constituent and is taken up from the earth in the form of potassium ions. Deficiency of potassium is fatal to plants and must be compensated by fertilization with potassium salts. For this purpose the above-mentioned salt minerals are widely used, and also potassium nitrate (25-5b).

The average amount of *rubidium* in the earth's crust is not especially low (tab. 17.3), but it does not occur concentrated in any minable minerals. The atomic radius is not much larger than that of potassium so that it follows potassium and is recovered along with potassium salts. *Cesium* is much rarer. It also follows potassium to a certain extent, but also occurs in a rare silicate mineral *pollucite* $Cs_4Al_4Si_9O_{26} \cdot H_2O$, the largest known occurrences being in Varuträsk in Sweden and in Manitoba, Canada.

Cesium occurs in nature only as ^{133}Cs. However, from ^{137}Xe formed by fission of heavy atomic nuclei (9-2c) the processes $^{137}Xe \xrightarrow[3.9 \text{ min}]{\beta^-} {}^{137}Cs \xrightarrow[30 \text{ y}]{\beta^-} {}^{137}Ba$ take place. The isotope ^{137}Cs therefore arises from nuclear fission and thereby, among other things, in the fallout from the explosion of nuclear weapons. It poses a threat to health because its half-life leads to a relatively high activity over a quite long time, and it is taken up by plants in place of potassium.

Francium occurs in nature only as the radioactive nuclide ^{223}Fr in a side branch of the actinium series (9-2a), which can be written $^{227}Ac \xrightarrow[21 \text{ y}]{\alpha} {}^{223}Fr \xrightarrow[22 \text{ min}]{\beta^-} {}^{223}Ra$. Since the half-life of francium is so short and the side branch accounts for only 1.2 % of the decaying ^{227}Ac, the amount of francium is very low. The properties of francium have therefore mainly been studied on preparations produced by artificial nuclear reactions (9-2b). Other francium isotopes have even shorter half-lives than ^{223}Fr.

b. History. In ancient times, for washing among other things, sodium carbonate (soda) obtained from Egyptian salt lakes was used, as well as potassium carbonate (potash) made from plant ash. The two salts could not be distinguished and they were both called in Hebrew neter, in Greek νίτρον and in Latin nitrum. From these was derived the Arabic alchemical name *natron*, while nitrum came gradually to refer to potassium nitrate (22-2b). At the end of the Middle Ages the term *alcali* or *alkali* (probably after Arabic al-qali, plant ash, which from land plants contains potassium carbonate) began to be used in the same sense as natron. In the 18th century it was learned that sodium and potassium compounds were distinct from each other. The base (12-1d) of the former was then called *alcali minerale* or *natron*, the base of the latter *alcali vegetabile* or *kali*. By electrolysis of the molten hydroxides Davy in 1807 prepared the two metals, which he called *sodium* and *potassium*. Later all the metals in group 1 were given the collective name *alkali metals*.

In 1817 Arfvedson discovered in petalite a new alkali, for which Berzelius proposed the name lithion (gk. λίθεος, from stone) "since this alkali was first found in the mineral kingdom". The corresponding metal *lithium* was prepared in 1818 by Davy by electrolysis of the molten oxide.

Bunsen and Kirchhoff discovered spectroscopically *cesium* in 1860 and *rubidium* in 1861. The names were given for the colors of typical spectral lines (Lat. caesius, blue, and rubidus, dark red).

Francium was detected in 1939 in the actinium series by Mlle. Perey and named for her homeland. The chemical characteristics of francium have been ascertained by carrier techniques (9-5a).

25-3. Properties of the alkali metals

The heaviest alkali metal, francium, has only unstable isotopes. Among the isotopes of the others, ^{40}K and ^{87}Rb are radioactive and are incorporated in the natural isotope mixtures to the extent of 0.01 % and 27.85 % respectively (tab. 1.3). ^{40}K decays by two routes (9-2a), 89 % of the nuclei according to ^{40}K $\xrightarrow{\beta^-}$ ^{40}Ca and 11 % according to ^{40}K $\xrightarrow{EC}$ ^{40}Ar, which together give an average half-life of 1.25×10^9 years. ^{87}Rb decomposes only according to ^{87}Rb $\xrightarrow[5 \times 10^{10} \text{ y}]{\beta^-}$ ^{87}Sr. Because of the long half-lifes the activities are low. However, the decay of ^{40}K probably plays a role in the heat balance of the earth (17b).

The densities, melting points and boiling points of the alkali metals are summarized in tab. 25.1. From these it is apparent that at room temperature they form solid phases of low density. Lithium has the lowest density of all elements that are solid at room temperature. For the most part the density increases with rising atomic number, while the melting and boiling points decrease. On vaporization mainly single atoms are produced, which, however, are in equilibrium with a small amount of diatomic molecules (5-3a). The solid alkali metals all have body-centered cubic crystal structures (6-2c, fig. 6.12c) and are soft and ductile; the hardness decreases in the sequence Li > K > Na > Rb > Cs, but even lithium is softer than lead. The clean metal surfaces are silver-white.

The low ionization energies make ionization possible even in a flame, so that simple line spectra arise in the visible region. These are very useful for the detec-

Table 25.1. *Physical properties of the alkali metals.*

	3Li	11Na	19K	37Rb	55Cs
Density at 20°C, g cm^{-3}	0.53	0.96	0.86	1.59	1.99
Melting point, °C	180	97.8	63.3	38.9	28.7
Boiling point, °C	1331	890	766	701	685

tion and analysis of the alkali metals and also produce typical flame colors (for Li, carmine red; Na, yellow; K, Rb and Cs, red violet).

The colors and wavelengths (in Å) of the strongest lines, arranged according to decreasing intensity, are as follows: Li, red 6708, orange 6104; Na, yellow, close doublet 5896/5890; K, red doublet 7699/7655, violet, close doublet 4047/4044; Rb, violet doublet 4216/4202, red doublet 7948/7800; Cs, blue doublet 4593/4555, orange 6213.

All alkali metals have high conductivity. The mechanism of conduction was discussed (with lithium as an example) in 6-4b, c.

The low electronegativities cause the alkali metals to react readily with a large number of nonmetals and also with many metals. The oxidizability and thus the reducing power is high. With reactions in water solution the equilibria are affected also by the hydration of the alkali ions. This was discussed in 5-2h, i, where it was shown how the hydration enthalpy influences the ionization of the metals in water solution and thus the standard potentials (16c, tab. 16.1). For the most part, however, the reactivity increases with increasing atomic number of the alkali metal. All alkali metals decompose water with the evolution of hydrogen ($M + H_2O \rightarrow M^+ + OH^- + \frac{1}{2}H_2(g)$). For lithium the heat developed in this reaction is less than for the other alkali metals and the reaction goes much less energetically. With large pieces of sodium, or finely divided sodium which reacts very rapidly, the heat of reaction causes spontaneous ignition (of the hydrogen gas produced as well as the undecomposed metal). With smaller pieces of sodium ignition is prevented by the cooling brought about by the floating about of the pieces on the water surface. With potassium, rubidium and cesium, the heat of reaction is so high that ignition always takes place.

In absolutely dry air lithium and sodium are not oxidized, and probably not potassium either, but on the other hand rubidium and cesium are oxidized and spontaneously ignite in air. In moist air all alkali metals form hydroxides, which are rapidly converted to carbonates.

Worthy of note is the solubility of the alkali metals in liquid ammonia, which was discussed in 22-6a.

25-4. Production and uses of the alkali metals

The alkali metals can, of course, be prepared from their compounds by chemical reduction processes (in which the volatility of the metal favors the shift in the equilibrium toward the metal) but this method is used technically only for rubidium and cesium. For example, hydroxides, carbonates or chlorides are reduced with a metal such as magnesium, calcium, barium or lithium. The temperature is maintained high enough so that the metal distills off (RbOH + Mg → MgO(s) + Rb(g) + $\frac{1}{2}$H$_2$(g)).

The other alkali metals are made by electrolysis, either of a hydroxide or a chloride melt. The hydroxide melts more easily than the chloride (NaOH m.p. 318°C, NaCl m.p. 801°C) but in chloride electrolysis the melting point can be lowered by adding another chloride or a fluoride (eutectic or melting point minimum, see 13-5f, the latter illustrated by fig. 13.11b). Chloride electrolysis has the advantage over hydroxide electrolysis of using directly chloride raw material, but the disadvantage of greater corrosion of the apparatus. For the production of sodium, which is by far the most important process, both hydroxide and chloride electrolysis are used. The cathode is generally made of iron, the anode n hydroxide electrolysis of nickel, in chloride electrolysis of carbon. The electrode regions are separated from each other by a wire metal screen and some sort of bell-shaped arrangement. The alkali metal is lighter than the melt and floats up to the surface in the cathode chamber.

Sodium metal is used in the manufacture of sodium compounds such as sodium peroxide (21-6a, 25-5a) and sodium cyanide (23-2k1) and in organic syntheses, especially in the manufacture of tetramethyl- and tetraethyllead (23-6d). Sodium is also used as a reducing agent in the production of metals such as titanium and zirconium (29e) and in the deoxidation of metal melts (21-5b). Molten sodium is used for heat exchange, for example to dissipate the heat in engine valves that have partly sodium-filled hollow spaces, and to cool nuclear reactors (9-4a). In the latter case sodium-potassium alloys have also been used, which are liquid even at room temperature (the system Na–K has a eutectic at −12.6°C and 78 wt% potassium). Sodium is also used in vapor-discharge lamps (sodium lamps).

Nowadays the most used alkali metal after sodium is lithium. The largest amounts are used in metallurgy as additives in small quantities, and it then acts most often as a deoxidation agent. The use of lithium in thermonuclear reactions was discussed in 9-4b. The manufacture of lithium hydride (18f) is also important, which is used to a large extent in the production of lithium tetrahydridoaluminate (24-3d), lithium tetrahydridoborate (24-2e), and alkyllithiums for use in organic syntheses in processes of the Grignard type.

For the remaining alkali metals we should mention the use that is made in photocells and electronics of their ionization on irradiation or heating (K → K$^+$ + e$^-$). Here mostly potassium is used, which is relatively cheap, and cesium when especially high sensitivity is required (lower ionization energy but higher cost).

25-5. Compounds of the alkali metals

a. Oxygen compounds. Lithium and sodium form *oxides* M_2O (21-5a, b) and *peroxides* M_2O_2 (21-6a), the other alkali metals also *hyperoxides* MO_2 (21-6c). On oxidation of alkali metals with air or oxygen generally a mixture of these oxides is formed. With an excess of oxygen, however, lithium gives mainly Li_2O, sodium Na_2O_2 and the other alkali metals MO_2. The tendency to combine with oxygen thus increases with increasing metal-atom radius. An oxygen-poor compound can be obtained by thermal dissociation of an oxygen-rich compound, by heating the latter with metal or by oxidation of the metal with a limited oxygen supply. Concerning their properties reference should be made to the sections cited above. The most important of all these compounds is *sodium peroxide* Na_2O_2, which is used as a bleaching agent and in the preparation of peroxy compounds.

Alkali-metal *hydroxides* MOH (21-5d) are easily soluble in water and give strongly basic solutions (since early times called "lye"). *Sodium hydroxide* NaOH and *potassium hydroxide* KOH are very important industrial products, which nowadays are manufactured by *chlor-alkali electrolysis* (20-4b). The old method of "causticizing" carbonate is still used to some extent, however, when the cost of electric power is high. In this method burnt lime is added to a carbonate solution. The lime first reacts according to $CaO(s) + H_2O \rightarrow Ca^{2+} + 2OH^-$, following which the equilibrium $2Na^+ + CO_3^{2-} + Ca^{2+} + 2OH^- \rightleftharpoons 2Na^+ + 2OH^- + CaCO_3(s)$ is set up. Because of the poor solubility of calcium carbonate the equilibrium is shifted strongly to the right. Sodium and potassium hydroxide form colorless, hard, crystalline masses of fibrous texture, which easily take up water and carbon dioxide with the formation at first of hydroxide hydrates and carbonates. In laboratory work especially potassium hydroxide is used as a drying agent and absorption agent for carbon dioxide.

Sodium hydroxide is most important industrially. Large amounts are used in the cellulose industry (kraft process 21-7a; rayon manufacture 23-2j), in the manufacture of soaps (sodium salts of fatty acids) and other sodium compounds. The organic-chemical industry (among others, oil refining) is a large consumer. Potassium hydroxide is used for making soft soaps (potassium salts of fatty acids).

b. Other compounds. Much information concerning compounds of the alkali metals has already been given, for the ionic compounds, among other places under the corresponding anion. Generally this information will not be repeated here, and also those compounds will not be mentioned whose existence and properties are apparent from what has already been said. The following sections will serve to amplify certain points.

Sodium chloride, table salt, NaCl. (M.p. 801°C; b.p. 1440°C). Its solubility behavior and hydrates were discussed in 13-5f3. Anhydrous NaCl usually crystallizes in cubes with the sodium-chloride-type structure (fig. 6.16a). If the crystallization occurs from water solution the crystals easily enclose water inclusions, which on heating burst out with a crackling sound—the salt "decrepitates". The mineral *halite* (rock salt) is sometimes blue in color, because of the presence of F centers (6-3a), which probably arose from radiation from ^{40}K (25-3) enclosed in the crystal. Pure sodium chloride is not hygroscopic, but commercial salt often contains hygroscopic magnesium salts as impurities.

Sodium chloride has always been of great commercial importance, and is now one of the most important industrial raw materials. It is the basic raw material for practically all sodium and chlorine compounds and is also used in large amounts as a condiment and as a preservative.

Sodium sulfate Na_2SO_4 is mainly obtained as a by-product of the manufacture of hydrogen chloride (20-5d). A saturated water solution of the salt is in equilibrium with one of two stable solid phases, $Na_2SO_4 \cdot 10H_2O$ (*Glauber's salt*) below a peritectic temperature (13-5f2) of 32.38°C (very accurately determined and used for thermometer calibration), and Na_2SO_4 above this temperature. Sodium sulfate is used especially in the manufacture of kraft pulp (21-7a).

Sodium nitrate (*Chile saltpeter*) $NaNO_3$ crystallizes with the calcite structure (23-2h). Concerning its occurrence, see 22-2a. Sodium nitrate has been of great importance as a source of nitrogen and as a nitrogen fertilizer, but has been largely replaced by products of atmospheric nitrogen fixation (22-6b, c).

Potassium nitrate KNO_3 since early times has been called saltpeter (Lat. sal petrae, stone salt). It has the aragonite structure (23-2h). By fermentation of nitrogen-rich organic material with the formation of nitrate (22-16) in potassium-rich strata, and favorable conditions of rainfall, potassium-nitrate-bearing beds may be formed. This is the case in Bengal where formerly the largest amounts of potassium nitrate were recovered by leaching such deposits. Nowadays potassium nitrate is produced synthetically. For this purpose mostly the so-called *conversion process* is used, which involves the interaction between KCl and $NaNO_3$. Of the four salts that can be formed from the ions present, NaCl has the lowest solubility at 100°C and KNO_3 the lowest at room temperature (the change in solubility with temperature for these two salts is shown in fig. 13.1). Therefore, on mixing concentrated, hot solutions of KCl and $NaNO_3$ in equivalent amounts NaCl crystallizes, which is filtered off while hot. The filtrate contains mainly K^+ and NO_3^- and on cooling KNO_3 crystallizes. By washing with cold water and recrystallization the KNO_3 is freed of the remaining NaCl.

Potassium nitrate constitutes the oxidizing agent in *gunpowder* (*black powder*), which consists of an intimate mixture of potassium nitrate, charcoal and sulfur. Sodium nitrate cannot be used because, in contrast to potassium nitrate, it is hygroscopic. Potassium nitrate for the manufacture of gunpowder was imported from Bengal, but it was also made by allowing nitrogen-rich organic material to ferment in the presence of wood ash (potassium carbonate). Potassium nitrate is also used as a fertilizer (often incorporated into mixed fertilizers), supplying plants with both potassium and nitrogen.

Sodium carbonate, soda, washing soda Na_2CO_3 forms three stable hydrates with 10, 7 and 1 molecules of H_2O. From a water solution of sodium carbonate at room temperature the decahydrate (*"crystal soda"*) crystallizes, which on heating to a temperature of 32.0°C transforms peritectically (13-5f2) to heptahydrate + solution. The heptahydrate transforms peritectically in turn at 35.3°C to monohydrate + solution. The monohydrate, finally, transforms peritectically at about 112°C to anhydrous Na_2CO_3 + solution; the latter then is above its boiling point and therefore rapidly gives off water with continued deposition of anhydrous salt. The vapor pressure of the decahydrate and heptahydrate is high enough at room temperature (at 20°C about 16 mbar) so that in relatively dry air they effloresce (13-5e 1) to the monohydrate.

Anhydrous Na_2CO_3 ("soda ash") melts at 860°C with very slow decomposition. Rapid decomposition occurs only at appreciably higher temperature (23-2h). The anhydrous salt dissolves in water with the development of considerable heat.

The basic properties of the carbonate ion make water solutions of sodium carbonate quite strongly basic, and this property is often exploited. For example, soda has been used for washing since early times; the carbonate ion, by forming poorly soluble carbonates with polyvalent cations, also softens hard water (26-5c).

Sodium carbonate occurs in nature especially in the soda lakes mentioned in 25-2a. Formerly it was also obtained by leaching the ash of marine plants (seaweed). However, the great demand forced attempts to develop an economical method for producing sodium carbonate from sodium chloride, and in this Leblanc succeeded in 1791.

The first step of the *Leblanc process* was the preparation of sodium sulfate: $2NaCl + H_2SO_4 \rightarrow Na_2SO_4 + 2HCl(g)$. The sulfate was then fused with limestone and carbon: $Na_2SO_4(l) + 2C(s) \rightarrow Na_2S(s) + 2CO_2(g)$; $Na_2S(s) + CaCO_3(s) \rightarrow Na_2CO_3(l) + CaS(s)$. The sodium carbonate was leached from the solidified melt with water, and the calcium sulfide remained behind. The manufacture of sodium carbonate by this method was the first chemical industry.

Nowadays the Leblanc process has been replaced by the *Solvay* or *ammonia-soda process*, developed about 1870. It is based on the fact that sodium hydrogen carbonate $NaHCO_3$ has rather low solubility and therefore precipitates out of a sulution with sufficiently high concentrations of Na^+ and HCO_3^-. Into a saturated solution of sodium chloride ammonia is first passed, and then carbon dioxide (made by roasting limestone, according to equation (1) below). In the first step, then, the process (2) takes place, and when the proportion of HCO_3^- becomes high enough, sodium hydrogen carbonate precipitates out (3). The precipitate is filtered off, dried and heated to about 200°C, forming sodium carbonate according to (4). The carbon dioxide given off in the last stage is reused. The calcium oxide formed in roasting the limestone is slaked to the hydroxide (5), which with the ammonium ions remaining in the filtrate gives ammonia (6), which is reused.

$$CaCO_3(s) \rightarrow CaO(s) + CO_2(g) \qquad (1)$$
$$2NH_3(g) + 2CO_2(g) + 2H_2O(l) \rightarrow 2NH_4^+ + 2HCO_3^- \qquad (2)$$
$$2Na^+ + 2HCO_3^- \rightarrow 2NaHCO_3(s) \qquad (3)$$
$$2NaHCO_3(s) \rightarrow Na_2CO_3(s) + H_2O(g) + CO_2(g) \qquad (4)$$
$$CaO(s) + H_2O(l) \rightarrow Ca^{2+} + 2OH^- \qquad (5)$$
$$2NH_4^+ + 2OH^- \rightarrow 2NH_3(g) + 2H_2O(l) \qquad (6)$$
$$\overline{CaCO_3(s) + 2Na^+ \rightarrow Na_2CO_3(s) + Ca^{2+} \qquad (7)}$$

The total process is (7) or, since chloride ions are present in the solution throughout, $CaCO_3 + 2NaCl \rightarrow Na_2CO_3 + CaCl_2$. Thus, primarily limestone and sodium chloride are consumed. Unfortunately, the by-product calcium chloride cannot be entirely utilized, although increasing amounts are now used to salt roadways (26-5b).

The manufacture of sodium carbonate by reaction of sodium hydroxide and carbon dioxide ("carbonizing") has been attempted, but because of the high cost of sodium hydroxide has not been found to be profitable.

Sodium carbonate is used especially in the glass industry (23-3i2), and in the manufacture of soap, detergents and other sodium compounds.

Sodium hydrogen carbonate, baking soda $NaHCO_3$ (obsolete name sodium bicarbonate, in colloquial speech "bicarbonate"; cf. 12-1c) can most simply be prepared by passing carbon dioxide into a cold, saturated solution of sodium carbonate: $2Na^+ + CO_3^{2-} + H_2O + CO_2(g) \rightarrow 2NaHCO_3(s)$. The salt precipitates because of its quite low solubility. Industrially it is obtained as an intermediate product in the manufacture of Solvay soda (see above). In nature it is found deposited with sodium carbonate in soda lakes (*nahcolite*). Its decomposition on heating and its application because of this property as a fire extinguisher and as baking powder has been mentioned in 23-2h. Sodium hydrogen carbonate gives weakly basic water solutions (pH ≈ 8) and has been much used to counteract excess stomach acidity.

Potassium carbonate, potash K_2CO_3. At room temperature the hydrate $K_2CO_3 \cdot 1.5H_2O$ crystallizes out of water solution. The potassium carbonate of commerce is usually the anhydrous salt, which, however, is hygroscopic and easily deliquesces (13-5e1). Potassium carbonate was obtained formerly by leaching wood ash (in clay pots, hence the name potash). For technical production an analog of the Solvay process cannot be used because potassium hydrogen carbonate is too soluble. The largest amounts are probably made now from potassium hydroxide and carbon dioxide. The major portion is used in the manufacture of potash glass (23-3i2) and soap.

CHAPTER 26

Group 2, Alkaline-Earth Metals

26-1. Survey

The alkaline-earth metals in many ways are similar to the alkali metals. Here also we can distinguish a group of closely similar elements Ca, Sr, Ba and Ra corresponding to K, Rb, Cs and Fr among the alkali metals. The two lightest elements Be and Mg stand apart from the four others, and are also less similar to each other. Beryllium, whose atomic and ionic radii are very small, is most unlike the others and in certain respects is similar to aluminum.

In 5-2f it was explained how the alkaline-earth metals mainly form noble-gas ions M^{2+} by giving up the s^2 pair outside the noble-gas shell. Only in exceptional cases have M^+ ions been observed, and the oxidation number is almost always II. Ionic bonding occurs very easily with the heavier metals, and these have low electronegativities (fig. 5.23). The standard potentials are also very low, and for calcium, strontium and barium lie between the potentials for sodium and cesium (tab. 16.1). The electronegativity and standard potential increases as the atom becomes lighter. The electronegativity of beryllium is approximately the same as that of aluminum. Beryllium therefore does not form typically ionic bonds. On the other hand, in bonds with nonmetals the ionic bond contribution is high for calcium and the subsequent heavier metals. Magnesium has intermediate character. The large covalent contribution in bonds with beryllium can also be explained by the small size of the beryllium atom, which makes its polarizing effect large (5-2e). Beryllium also has an appreciably greater tendency for complex formation than the other metals of the group.

In BeO the bonding is sufficiently covalent for the oxide to have intermediate character (21-5a), but thereafter the ionic contribution and basicity increase with each stage toward higher atomic number. MgO is already a basic oxide. The same trend is found for the hydroxides. The hydroxides of calcium, strontium and barium give strongly basic solutions, but because of the lower solubilities (cf. 21-5d), they are less strong than solutions of the alkali-metal hydroxides.

As stated in 5-2h the alkaline-earth-metal ions are more hydrated than the alkali-metal ions. Salts crystallized from water solution generally contain water of crystallization, and salts prepared without water of crystallization (or with low water content) usually absorb water. Among these are found many often used drying agents, for example, $CaCl_2$, $CaSO_4$ (Drierite), $Mg(NO_3)_2$, $Mg(ClO_4)_2$ (Anhydrone), $Mg(ClO_4)_2 \cdot 3H_2O$ (Dehydrite), $Ba(ClO_4)_2$.

As ions the alkaline-earth metals have no color, and also the more covalent beryllium and magnesium compounds are often colorless. Beryllium and magnesium hydroxides are very poorly soluble in water, but with increasing atomic number of the metal the solubility increases; barium hydroxide is rather easily soluble in cold water. The solubilities of the sulfates change in the opposite direction, and thus magnesium sulfate is easily soluble and barium sulfate very poorly soluble. All chlorides and nitrates are easily soluble. Salts of anions of medium-strong or weak acids are most often poorly soluble; for example, fluorides (except beryllium fluoride), phosphates (however the dihydrogen phosphates are easily soluble or rather soluble), carbonates and oxalates. Sulfides, cyanides, thiocyanates and acetates are easily soluble and thus constitute exceptions to this rule.

The covalent contribution in beryllium compounds becomes apparent in that these compounds often form molecular crystals (6-2a) and then generally are rather easily fusible and volatile (13-4b). Molten beryllium halides also have very low conductivity. However, if the crystal does not contain finite molecules it may, of course, be more difficultly fusible and less volatile. An example is BeO, which crystallizes with the wurtzite-type structure (6-2c), that is, a tetrahedral lattice related to that of diamond. Here, like diamond, we have a giant molecule with high melting point (2530°C; cf. 13-4b).

26-2. Occurrence and history

a. Occurrence. Free alkaline-earth metals are not found in nature.

Beryllium is present in very low proportion in the earth's crust, but occurs concentrated in a number of minerals, of which only *beryl* (23-3h2) is of practical importance. The largest amounts of beryl are mined in Argentina, Brazil and India.

Beryllium and its compounds are highly poisonous (often with a prolonged effect). This property is probably related to its complex-forming ability.

Magnesium is a common element on the earth (17b, tab. 17.2). The largest amount is found in silicates, for example the minerals *olivine, augite, hornblende, talc* and *biotite* mentioned in 23-3h. However, silicates are not important for the production of magnesium. As the carbonate magnesium is found in the form of the double salt *dolomite* $CaMg(CO_3)_2$, which can make up entire rock formations, and *magnesite* $MgCO_3$, the latter occurring in large deposits over the whole world. Very often magnesium is also dissolved in calcium carbonate, which may then be written $(Ca,Mg)CO_3$ (as a rock somewhat improperly called dolomitic limestone). In the salt deposits referred to in 25-2a magnesium is found in, among other things, *kieserite* $MgSO_4 \cdot H_2O$, *carnallite* $KMgCl_3 \cdot 6H_2O$, and *kainite* $KMgClSO_4 \cdot 3H_2O$. Sea water contains a quite large amount of magnesium ions (on the average 0.05 M Mg, that is, 0.1 of the Na molarity). For the production of magnesium dolomite, magnesite and the above-mentioned mineral salts are all exploited, as well as sea water. In recent times sea water has become the most important source of magnesium.

Magnesium is present in both plant and animal organisms. In the former it is incorporated into chlorophyll. The abundance of magnesium minerals generally makes the soil sufficiently rich in this element so that magnesium fertilization is unnecessary.

Calcium is more common in the earth's crust than magnesium (tab. 17.2). The most abundant and most important of the calcium minerals is *calcite* $CaCO_3$, contained in *limestone* (pure and well-crystallized in *marble*) and *chalk* (soft, white, consisting mainly of shell remains). Large amounts of calcium also occur in *apatite, phosphorite* (22-2a), *fluorite* (20-2a) and in silicates (cf. 23-3h). The sulfate as *gypsum* $CaSO_4 \cdot 2H_2O$ (pure, fine-grained gypsum is called *alabaster*), and *anhydrite* $CaSO_4$, can form extensive beds. Flowing water may contain a high proportion of calcium ions dissolved from minerals (cf. 26-5b, c), but in the ocean these are to a large extent precipitated as calcium carbonate, so that the mean content of the water is only 0.01 M Ca.

Calcium is present in both plant and animal organisms, in the latter most especially as carbonate and phosphate (apatite, see 22-2a) in shell and skeletal material. In general the soil contains sufficient amounts of calcium to supply the needs of plants. However, in many cases burnt lime, that is, calcium oxide CaO, or ground limestone is added, supposedly to act as a soil conditioner.

Strontium and *barium* are each present in the earth's crust in approximately one hundredth the amount by weight of calcium (tab. 17.3). The ionic radii (fig. 5.5) show why strontium preferably follows calcium and potassium. The ionic radius of barium is too large for this metal to be able to follow calcium to any great extent, but, on the other hand, it can follow potassium. The most important minerals of both strontium and barium are the carbonates and the sulfates, of which we may mention *strontianite* $SrCO_3$, *celestite* $SrSO_4$ and *barite* $BaSO_4$.

Through fission of heavy atomic nuclei (9-2c) there is produced, among other things, ^{90}Kr, which decays according to $^{90}Kr \xrightarrow[33\ s]{\beta^-} {}^{90}Rb \xrightarrow[2.7\ min]{\beta^-} {}^{90}Sr \xrightarrow[28\ y]{\beta^-} {}^{90}Y$. In the fallout following nuclear weapon explosions, therefore, ^{90}Sr is present, whose half-life leads to rather high activity over a quite long time. Strontium is readily taken up by organisms in place of calcium, and in vertebrate animals it is stored up especially in the skeleton. The presence of ^{90}Sr is therefore very dangerous. The danger can be reduced by simultaneously supplying generous amounts of calcium, so that the ^{90}Sr is, so to say, diluted.

All isotopes of *radium* are unstable. Radium isotopes form a part of each of the natural radioactive series (9-2a), but only ^{226}Ra included in the uranium series has a long enough half-life (1622 years) for it to be present in amounts that make possible its complete isolation by ordinary chemical methods. Such amounts are found in uranium minerals.

b. History. As mentioned earlier, difficultly reduced, melted and volatilized oxides were formerly designated *earths*. These were divided into *true earths*, which were considered to be insoluble in water, and *alkaline earths*, which give water solutions resembling those of the alkalis. In 1817 Berzelius included with the latter talc, lime, strontian and baryte earths, that is, magnesium, calcium, strontium and barium oxide, respectively. The two latter elements were named after the minerals strontianite (found in Strontian, Scotland) and barite (Gk. βᾰρύς, heavy). Talc earth was also called magnesia after the carbonate, which, like pyrolusite MnO_2 and magnetite Fe_3O_4, since ancient times had been called *magnesia* or *magnes* after a locality in Magnesia in Asia Minor. By electrolysis of wet hydroxides with mercury cathodes, Davy in 1808 obtained amalgams of magnesium, calcium, strontium and barium, from which by distilling off the mercury he obtained the metals (not completely pure; somewhat earlier Berzelius in a similar way made amalgams of calcium and barium). Davy also named the new metals (magnesium he called "magnium").

The beryl earth obtained from beryl, that is, beryllium oxide, was counted by Berzelius as a true earth. The fact that beryllium is now said to be an alkaline-earth metal depends on the structure of its electron cloud. The metal was prepared in 1828 by Wöhler and Bussy independently of each other by reduction of the chloride with potassium, and the name was given by Wöhler (in French beryllium has long been called "glucinium" because of the sweet taste of its salts).

Radium was discovered in 1898 in the uranium mineral pitchblende by Pierre and Marie Curie, and was given its name for its radiation.

26-3. Properties of the alkaline-earth metals

The data given in tab. 26.1 show a rather irregular pattern. Beryllium is a very hard and, at ordinary temperature, brittle metal (at higher temperatures it can be plastically shaped, to a maximum degree at about 400°C). For the metals after beryllium the hardness tends to decrease with increasing atomic number. The softest alkaline earth metals are harder than the hardest alkali metal, lithium.

Compounds of calcium and the subsequent, heavier metals give typical flame colors: calcium, brick red; strontium, carmine red; barium, green; radium, carmine red. On vaporization in a flame these compounds are dissociated to a lesser degree than alkali-metal compounds, and therefore give in addition to atomic spectra (line spectra) also molecular spectra (band spectra; see 3d). In flame spectra the oxide bands are almost always present, and it is mainly these that give the flame colors. Only at higher temperatures (electric arc, spark) is the degree of dissociation so great that atomic spectra predominate. Beryllium and magnesium give no flame spectra, but on the other hand, do give arc and spark spectra. On heating in air the oxide is first formed, and its volatility in these cases is too low to produce a high enough gas concentration in a flame.

The chemical properties resemble those of the alkali metals but the reactivity is generally lower. However, it increases in most types of reactions with increasing atomic number of the metal. In absolutely dry air strontium and barium are oxidized, but not the lighter metals. Oxidation on heating in air also takes place more easily the heavier the metal is. At high enough temperatures combustion

Table 26.1. *Physical properties of the alkaline-earth metals.*

	4Be	12Mg	20Ca	38Sr	56Ba	88Ra
Density at 20°C, g cm^{-3}	1.85	1.74	1.54	2.62	3.74	—
Melting point, °C	1283	650	850	770	710	700
Boiling point, °C	2480	1120	1490	1370	1640	—

with a flame will occur, but the ignition temperature depends of the fineness of division. The temperature of combustion is high (magnesium flare). Water and moist air do not attack beryllium, but on the other hand, they do attack the other metals. Magnesium is slowly attacked (here a hydroxide film gives some protection but is easily loosened), the others more rapidly the heavier the metal is. In acids (but not in strongly oxidizing acids) all alkaline-earth metals dissolve with the evolution of hydrogen. Beryllium dissolves in strong bases with the formation of beryllate ions (26-5a).

Calcium, strontium and barium, like the alkali metals, dissolve in liquid ammonia (22-6a).

Soluble barium compounds are poisonous.

26-4. Production and uses

The alkaline-earth metals are produced technically either by electrolysis of a halide melt or by thermal reduction of the oxide or halide.

In the production of *beryllium*, beryllium hydroxide is the starting material, but this is obtained from the raw material beryl in various ways. One currently important method involves the leaching with sulfuric acid of beryl that has been made more easily attacked by fusing and quenching. The hydroxide is precipitated from the resulting sulfate solution with sodium hydroxide solution at boiling temperature (see 26-5a). The hydroxide is converted either to the chloride, which is electrolyzed, or the fluoride, which is reduced with magnesium. The metal is purified by melting in the absence of air. The difficulty of working the metal because of its brittleness makes it preferable to shape the metal by sintering (14-3b) beryllium powder under pressure at about 1000°C.

The beryllium atom has a low neutron absorption, but because of its lightness it effectively slows neutrons. Beryllium metal or oxide is therefore used in nuclear reactors as a moderator or reflector material (9-4a). The metal has also begun to be used to encapsulate reactor fuel elements. Because of its low absorption of X rays the metal is much used as windows in X-ray tubes. Its low density, hardness, rather high melting point, and good corrosion properties would make the metal more useful were it not for its high cost, brittleness and toxicity. The largest amounts of beryllium are now used for the production of *beryllium copper* (36-5a).

Magnesium is produced for the most part by electrolysis of molten, anhydrous magnesium chloride. The most important raw material at present is sea water, from which magnesium hydroxide is precipitated according to $Mg^{2+} + 2OH^- \rightarrow Mg(OH)_2(s)$. The requisite hydroxide ion concentration is obtained by adding burnt lime: $CaO + H_2O \rightarrow Ca^{2+} + 2OH^-$. Sometimes burnt dolomite $(CaO + MgO)$ is used, so that the magnesium in the dolomite is also converted to the hydroxide. The conversion of the hydroxide to chloride for melt electrolysis is accomplished by various methods, which are designed to arrive at the anhydrous chloride without too great losses. If the chloride is obtained with hydrochloric acid, all of the water cannot be removed simply by heating without decomposing the chloride (20-5a). After it was found that the water is rapidly given off without appreciable hydrolysis if a still slightly hydrated chloride is added slowly and continuously to the melt during the course of the electrolysis, this method has been used to a large extent. To give the melt low viscosity and high conductivity other chlorides (for example, sodium and calcium chlorides) are also mixed in. The cathode is iron, the anode, graphite. The metal formed floats to the surface of the salt melt and is collected there in an inverted trough.

To a lesser extent magnesium is also made by thermal reduction of the oxide (reducing agents are carbon, calcium carbide or silicon).

Magnesium metal is used as a reducing agent in metal production (especially titanium and uranium, but also zirconium and beryllium) and deoxidation of metal melts (21-5b), as electrodes for corrosion protection (16i), in fireworks (magnesium flare) and in the preparation of Grignard reagents for organic syntheses. However, the major portion goes to light metal alloys, in which aluminum is the most important other alloy constituent. Light alloys with magnesium as the main metal and up to 10 wt % aluminum are particularly important as construction material for aircraft. Alloys with aluminum as the main metal were discussed in 24-3c.

Calcium is produced by electrolysis of molten calcium chloride or by thermal reduction of calcium oxide with aluminum. The metal is used as a reducing agent in the preparation of certain metals, for deoxidation, and to some extent as an alloy element. On heating, calcium combines with hydrogen, oxygen, nitrogen, and water vapor, and is therefore used as a scavenger ("getter") for these gases in high-vacuum tubes and for the purification of noble gases.

26-5. Compounds of the alkaline-earth metals

a. Oxygen compounds. The small size of the beryllium atom gives it the coordination number four with respect to oxygen, in contrast to the number six or higher for the other alkaline-earth metals.

Of the oxides MO, magnesium and calcum oxide are the most important. Both are made by heating (*burning*) the carbonates (cf. below); for example, *calcium oxide* (*burnt lime, quicklime*) in lime kilns at about 1000°C: $CaCO_3(s) \rightarrow CaO(s) +$

$CO_2(g)$. Calcium oxide is much used for making plaster, and as a soil conditioner, although its value in the latter case is in dispute. It is extremely important as a cheap source of base, both for water solutions (for example, for the making of hydrogen sulfite solutions in the sulfite-cellulose process), and oxide melts (for example, as a slag-former in the smelting of silica-rich ores, in glass and cement manufacture). In the former case it is primarily the hydroxide that is active; in the latter case calcium oxide is usually added in the form of the carbonate (limestone). *Magnesium oxide* MgO sinters at very high temperature (for pure material about 2000°C, for less pure material somewhat lower) to a mass that does not react appreciably with water and is used as a refractory ceramic material (m.p. 2800°C).

Beryllium hydroxide $Be(OH)_2$ is amphoteric (12-4d). When freshly precipitated, it gives with acids hydrated beryllium ions, primarily $Be(H_2O)_4^{2+}$, and with bases *(hydroxo)beryllate ions*, often written as $Be(OH)_4^{2-}$. Aging of the hydroxide, which is greatly hastened by heating, results in a transformation to a more difficultly soluble form. This form also precipitates from a boiling beryllate solution.

The low solubility of *magnesium hydroxide* $Mg(OH)_2$ has been used for analytic separation. Solutions of alkali hydroxides precipitate $Mg(OH)_2$ rather completely, but the solubility product of the hydroxide is high enough (tab. 13.1) so that complete precipitation can hardly be obtained with ammonia solution, in which the pH value is not so high. In addition the increase of pH is counteracted by the NH_4^+ formed. If NH_4^+ has been added in sufficient amount beforehand, the $[OH^-]$ value required to initiate precipitation of $Mg(OH)_2$ cannot be reached at all.

Calcium hydroxide $Ca(OH)_2$ is produced by "slaking" calcium oxide with water: $CaO(s) + H_2O(l) \rightarrow Ca(OH)_2(s)$; $\Delta H = -65.3$ kJ. The ΔH value indicates a strong evolution of heat. With sufficient excess of water a hydroxide solution, *lime water*, is obtained. If the amount of water is not sufficient for complete solution, a suspension is formed of hydroxide in saturated solution, *milk of lime*. An aggregate of slaked lime, sand and water constitutes *mortar* or *lime plaster*, which hardens as it absorbs carbon dioxide from the air: $Ca(OH)_2 + CO_2 \rightarrow CaCO_3(s) + H_2O$. In order to hasten the hardening of plaster inside a house, carbon dioxide is sometimes supplied by an open coke fire, which also favors the removal of water by raising the temperature. In thick layers, however, the hardening of mortar or plaster continues over many years. Calcium hydroxide is used in several processes of sugar refining. Among other things, sugar is precipitated as calcium saccharate, which can be isolated and then decomposed by passing in carbon dioxide.

b. Other compounds. *Magnesium chloride* $MgCl_2$ is very hygroscopic, and when it is present as an impurity in table salt the latter becomes moist and forms lumps. When a sufficient amount of magnesium oxide is added to a concentrated solution of magnesium chloride a hard mass is formed of magnesium hydroxide chloride. This is the bonding constituent in *magnesia cement*, which is used, among other things, for flooring. *Calcium chloride* forms several hydrates, of which the hexahydrate $CaCl_2 \cdot 6H_2O$ crystallizes from water solution at room temperature. This forms a eutectic with ice at $-55°C$ and 59 wt% $CaCl_2 \cdot 6H_2O$ (30 wt% $CaCl_2$). This temperature can therefore be obtained in a freezing mixture (13-5f3) consisting of the hexahydrate and ice. By careful heating (20-5a) anhydrous calcium chloride can be prepared, which very readily takes up water (drying agent). Calcium chloride with a water content corresponding approximately to $CaCl_2 \cdot 2H_2O$ is sufficiently deliquescent to be used to hold down dust on roadways (cf. 25-5b). A calcium chloride solution of the eutectic composition mentioned above, because of its low freezing point, is used as a heat-transfer medium ("brine") in refrigeration equipment.

Magnesium sulfate $MgSO_4$, among other things, forms a monohydrate (the mineral *kieserite*) and a heptahydrate (*Epsom salt*, cf. 21-11b3). *Calcium sulfate* $CaSO_4$ occurs in nature as the dihydrate *gypsum* $CaSO_4 \cdot 2H_2O$, and anhydrous as *anhydrite*. Gypsum crystallizes from water solution at room temperature. On heating, gypsum is dehydrated, but it can easily be rehydrated if it is not heated too strongly. The processes are complicated and not well understood, but recent studies suggest that gentle heating forms a "subhydrate" with variable water content, $CaSO_4 \cdot (0-\frac{2}{3})H_2O$. In this form the water is zeolitically bound (13-5e2) and can be given off and taken up fairly reversibly. Just above 100°C a water content of about $\frac{1}{2}H_2O$ per $CaSO_4$ is quickly reached, and this product has earlier been referred to as a hemihydrate. On heating to about 200°C the subhydrate phase rapidly becomes anhydrous, but it can still take up water ("*soluble anhydrite*").

$CaSO_4$ used as a drying agent (26-1) consists of soluble anhydrite, and after use can be regenerated by heating to about 200°C. *Plaster of Paris* has been heated to between 120 and 200°C and therefore consists of subhydrate with a water content between about $\frac{1}{2}$ and 0 H_2O per $CaSO_4$. When plaster of Paris is treated with the proper amount of water, the water is taken up with the formation of a solid mass of the dihydrate. This property is the basis of the wide use made of gypsum for casting and structural purposes. However, if gypsum is heated above about 200°C, a new anhydrous phase is formed, identical with the mineral anhydrite, which does not reabsorb water at a noticeable rate ("*insoluble anhydrite*"). Gypsum is sufficiently soluble in water (0.2 g per 100 g water at 20°C) that objects made of gypsum (alabaster) cannot be kept out of doors. Natural water also often contains dissolved calcium sulfate.

Barium sulfate $BaSO_4$ is used to increase contrast in X-ray photographs of the stomach and intestinal tract, and because of its low solubility (0.00028 g per 100 g water at 25°C) produces no toxic effects as long as it is free of other barium salts.

Strontium and *barium nitrate* are used in fireworks to give red and green flames, respectively.

Concerning the structures and decomposition temperatures of the *carbonates* of the alkaline-earth metals, reference should be made to 23-2h. For calcium carbonate the stable (at atmospheric pressure) modification *calcite* and the metastable (13-3a) *aragonite* were mentioned in that section. At ordinary temperatures the transition of aragonite to calcite does not take place at a measurable rate. Only calcite is a rock-forming mineral and is incorporated in limestone and chalk. When calcium carbonate is precipitated from water solution at room temperature calcite is formed, but from boiling solutions predominantly aragonite, even though it is metastable (aragonite is therefore often found in hot-spring deposits). Limestone is an exceedingly important material in chemical industry and metallurgy. Here it is converted, either before use or simultaneously with use, into calcium oxide or hydroxide (26-5a).

On adding carbonate ion to solutions of salts of the alkaline-earth metals the carbonates MCO_3 of Ca, Sr, Ba and Ra are precipitated, but usually hydroxide carbonates of Be and Mg of varying compositions are produced. *Magnesium hydroxide carbonate* since early times has been called "magnesia alba".

When a poorly soluble carbonate is in solubility equilibrium with a water solution, this equilibrium is connected in the first place with the acid-base equilibrium $CO_3^{2-} + H^+ \rightleftharpoons HCO_3^-$ in the manner described in 13-1e. When the acidity is increased, the acid-base equilibrium is shifted to the right and the solubility of the carbonate increases. When the acidity is lowered, the equilibrium is shifted to the left and the solubility of the carbonate is lowered. In nature the acidity of water often rises through solution of carbon dioxide (mainly formed by organic decay), and such water dissolves appreciable amounts of carbonate material. Water that is in equilibrium at 25°C with calcite contains 0.52 mg per 100 g water if it is free of carbon dioxide, but if it is saturated with carbon dioxide ($p_{CO_2} = 1$ atm) it contains 38 mg per 100 g. Calcium carbonate is here most commonly encountered because of its prevalence, but magnesium and iron(II) carbonates behave in a similar manner. The significance of this behavior in connection with the hardness of water is discussed in 26-5c. If the carbonate solutions are made basic, carbonates precipitate. Precipitation is also caused if carbon dioxide is given off on boiling or if the water evaporates (stalactites). It is often said that the increased solubility in carbon-dioxide-containing water depends on the easier solubility of the hydrogen carbonates. Since the solid hydrogen carbonates are unknown, this explanation is meaningless.

Calcium carbide CaC_2 ("carbide") is made industrially by heating calcium oxide with carbon in an electric arc. The process is controlled mainly by the equilibrium $CaO + 3C \rightleftharpoons CaC_2 + CO(g)$, which has a favorable balance at the working temperature, about 2200°C. However, other equilibria are involved, and furthermore the carbide is formed in a melt in which the oxide dissolves. The melt is drawn off and the product contains usually about 15% calcium oxide and a few percent of carbon (the carbon gives the product a gray-black color; pure calcium carbide is colorless).

Calcium carbide is used especially in the production of acetylene (23-2c), and is therefore a very important starting material in the organic-chemical industry. It is also converted to *calcium cyanamide* (Cyanamid, cf. 23-2k6).

Many compounds of alkaline-earth metals are used in finely ground or precipitated form as white *pigment* or as *filler* in paper, rubber and paint. This applies especially to calcium carbonate (chalk), gypsum, insoluble anhydrite, magnesium hydroxide carbonate (magnesia alba) and barium sulfate. As a paint pigment with high covering ability and stability, *lithopone* is important, a mixture of barium sulfate and zinc sulfide, which is obtained by precipitation from barium sulfide solution with zinc sulfate solution: $Ba^{2+} + S^{2-} + Zn^{2+} + SO_4^{2-} \rightarrow BaSO_4(s) + ZnS(s)$.

c. Hardness of water. The hardness of water depends on the presence of polyvalent cations, especially Ca^{2+}, Mg^{2+} and in certain types of water, Fe^{2+}. In general Ca^{2+} plays the biggest role. These cations may be present in quite high proportions if the water does not contain too great amounts of anions with which the cations can form poorly soluble compounds. If the carbonic-acid system is present this should then not contain CO_3^{2-} in too high proportion. Usually the anions HCO_3^-, SO_4^{2-} and Cl^- may be expected in hard water.

The ion HCO_3^- is found in nature especially when carbon-dioxide-containing water has dissolved carbonates of Ca^{2+}, Mg^{2+} or Fe^{2+} (26-5b). The combination $Ca^{2+} + 2HCO_3^-$ is the most common cause of hard water. HCO_3^- participates, among other things, in the equilibria $HCO_3^- + H^+ \rightleftharpoons H_2CO_3 \rightleftharpoons H_2O + CO_2$. If such water is heated CO_2 is given off (most effectively by boiling), the water becomes more basic, and the corresponding amount of carbonates of the cations mentioned precipitates out. What is left is mainly the amounts of these ions corresponding to other anions. Carbonate precipitation on boiling causes the water to lose its *carbonate hardness* or *temporary hardness* while a *permanent hardness* remains. The sum of these two hardness quantities constitutes the *total hardness*.

Water hardness is often expressed in "German degrees", °dH. 1 °dH corresponds to 1 mg CaO (= 0.71 mg Ca) per 100 ml water, or the equivalent amounts, 0.72 mg MgO (= 0.43 mg Mg) and 1.28 mg FeO (= 1.00 mg Fe) per 100 ml water. Water with a total hardness in °dH of 0–2 is said to be very soft; 2–5, soft; 5–10, medium hard; 10–20, hard; >20, very hard. In general, lakes and streams give soft water, ground waters in carbonate-containing rock beds give hard water.

Hard water is unsuitable for washing, since the cations mentioned form poorly soluble salts with the fatty acids in soap ("lime soap", etc.) and thus greatly increase the consumption of soap. For other household uses also, for feed water and for chemical industry it is unsuitable because of the precipitation of poorly soluble substances, for example on boiling and evaporation.

Hard water can, of course, be completely deionized by some of the methods mentioned in 21-5c. Often, however, only the polyvalent cations are removed by exchange with Na^+ in an ion exchanger (15-3). In many cases, often in laundering, precipitation or complex formation is used. To precipitate carbonates it is sufficient to add a base (cf. 26-5b). On a larger scale, calcium hydroxide is used for this purpose, and also sodium hydroxide. However, the permanent hardness still remains. A more complete softening is obtained with anions that precipitate the polyvalent cations, for example with sodium carbonate or various phosphates. The polyphosphates are especially effective because of their ability to form strong complexes with polyvalent cations (22-12e4). Most of these substances also give basic solutions, but the relatively expensive polyphosphate is conserved if the water is first made basic with a cheaper material. In laundering the water should also be basic in order that the fat can be dissolved through soap formation and the acid in the soiled clothes can be neutralized; the most suitable pH value is said to lie between 10 and 11.

CHAPTER 27

The Transition Elements

The remaining chapters of this book are devoted to groups 3 to 12, taken in that order (cf. tab. 4.3). According to 4-2b the elements in these groups except group 12 are called *transition elements* or, since all are metals, *transition metals*.

As an introduction to the special treatment of the transition elements a review is recommended of the following sections of Part I where they are considered more generally:

Electron configuration and position in the periodic system 4-2b, c (fig. 4.7).
Atomic radii 5-1d (fig. 5.2).
Ionization energies 5-2b (tab. 5.1, fig. 5.4), and ion formation 5-2f (fig. 5.8). Ionic radii 5-2c (fig. 5.5). Polarization effect of ions 5-2e.
The structures of transition-metal complexes were discussed in 5-3i (influence of d orbitals), 5-6c (general on metal complexes), 5-3j (ligand-field theory).
Metallic bonding in the elements 6-4f, among other things illustrated by melting point variations (fig. 6.26). Crystal structures of the transition elements 6-2c, 6-4b. Metallic phases containing transition elements and nonmetals 6-4g.
Color 7-3. Magnetic properties 7-4. Role in catalysts 8h and 14-4.

The properties of the transition elements change relatively little in going from element to element in the same period ("horizontal similarity"). This is related among other things to the fact that adjacent ions with the same charge have very similar radii. Especially small are the differences among the lanthanoids and actinoids where changes in the electron cloud take place in its interior ($4f$ or $5f$) while the outer portions remain undisturbed. The differences are quite small also among the three elements within the same period of groups 8, 9 and 10. This is the reason that Mendeleev collected these three columns of elements into one group (collectively designated by him as group 8—cf. 4-2c; at the beginning Mendeleev also included copper, silver and gold in group 8).

If instead we go downward from period 4 to period 5 within one group the changes in properties become considerably more marked. This can be explained to a large extent by the quite large increase in atomic radius with this step. Going further downward to period 6 further changes become small ("vertical similarity") since the radii change only slightly because of the "lanthanoid contraction". For the transition elements also, the first element in a group thus has properties more divergent from those of the other elements (for the nontransition elements this is true also for other reasons; see beginning of 5-6). The similarities between elements lying above and below each other in periods 5 and 6 are greatest immediately after the lanthanoids, that is, in group 4 (Zr and Hf). Going from here to the right in the system they become gradually less striking, but are still apparent in groups 8–10. In these three groups the three elements in period 4 (Fe, Co, Ni) are customarily called collectively the *iron metals* and the six elements of these groups in periods 5 and 6 (Ru, Rh, Pd, Os, Ir, Pt) are called *platinum metals*.

In the following discussion of the oxidation numbers of the transition elements we do not consider the lanthanoids and actinoids, for which this question is more extensively treated in 28-1. Fig. 5.8 shows that nearly every one of the other transition elements occurs in several different oxidation states. The highest, and at least at the beginning of the periods, most important oxidation numbers involve the formation of ions with either a noble-gas shell, or a noble-gas shell together with a completely filled *f* group in the interior of the electron cloud. However, the oxidation number VIII is never exceeded. The elements after group 8 therefore have maximum oxidation numbers that are smaller than those required to give shells of the type mentioned.

In compounds the transition elements normally do not have lower oxidation numbers than I, and this number is important only in group 11 (cf. 5-2f). As stated in 5-6c, however, oxidation numbers 0, −I and −II exist in complexes with certain ligands, but it may be pointed out that these low oxidation numbers then are considered to have only formal significance.

As stated earlier in 5-2f, when one goes toward heavier elements in one group of transition elements, the ionic stability shifts toward higher oxidation numbers. High oxidation numbers thus are particularly represented among the heavier elements of the group, low oxidation numbers among the lighter elements.

When ions with the same charge within one group are compared, it is found that the smallest have the greatest polarizing ability and therefore form the most covalent bonds (5-2e). The covalent-bond contribution is thus greatest with the ion in period 4 and decreases downward in the group. For one element the polarizing ability increases with positive ionization, which implies a smaller radius and greater charge. Thus, the covalent-bond contribution increases. Since for the transition elements the high oxidation numbers become more prominent as we go downward in the group, the strongest covalent bonds within the group are generally found with its heaviest elements. In other groups of the periodic system usually the opposite behavior is found. In groups 1 and 2 this results entirely from the fact that each group has only one important oxidation number. For the elements in groups 13–17 nonmetals occur uppermost in the groups, and on going downward the metallic properties increase and a stronger and stronger shift toward lower oxidation numbers takes place.

For the transition elements the covalent-bond forces at higher oxidation numbers lead among other things to the appearance of *oxo ions* (end of 5-6c), which play a large role in the chemistry of many transition elements.

As described in 6-4g the transition metals are notable in that they often form metallic phases with certain nonmetals. Many of these phases are of practical importance. Hydrides, (18f), in which hydrogen is incorporated in atomic form and which often have variable composition (interstitial solid solutions, 6-3a), certainly are involved when transition elements in metallic form serve as catalysts (14-4) in hydrogenation and dehydrogenation processes. A number of carbides (23-2c) are of fundamental importance to the properties of steel (34-3).

In all groups of transition elements except group 3 the lightest element of the group (belonging to period 4) is more common than the heavier elements. This probably is a result of the minimum in the nuclear energy and hence the maximum in nuclear stability in the vicinity of iron (fig. 9.2).

CHAPTER 28

Group 3, The Scandium Group with Lanthanoids and Actinoids

28-1. Survey

With group 3 are included, besides scandium and yttrium, also the *lanthanoids* 57 La–71 Lu, and the *actinoids* 89 Ac–103 Lr. The elements with atomic numbers higher than 92 U are often called *transuranium elements*. The name *uranoids* for the elements 92 U–103 Lr may also be used.

Concerning the nuclear properties, especially of the actinoids, reference should be made generally to chap. 9. There it was stated among other things, that all nuclides heavier than $^{209}_{83}$Bi have unstable atomic nuclei. For group 3 this implies instability for all nuclides beginning with actinium. Some of these have rather long half-lives (28-2a) but for the most part the stability decreases rapidly with increasing atomic number (cf. 9-1b). For the heaviest elements, because of the short lifetime and the very small amounts produced, only the nuclear properties and possibly a few general chemical properties can be determined. Of the lighter elements of the group promethium has only unstable isotopes. Among the other lanthanoids six unstable nuclides occur in nature, mentioned in 9-2a.

In 4-2b it was stated that the lanthanoids show a pronounced $4f$ character and in 5-2f (fig. 5.8) that the oxidation number III predominates. The oxidation numbers II and IV do occur, however, especially when they lead to a noble-gas shell as in Ce(IV); a filled $4f$ group as in Yb(II); or a half-filled $4f$ group as in Eu(II), Tb(IV). In addition Pr(IV), Sm(II) and Tm(II) exist, which indicates a tendency to approach the types mentioned. For the lanthanoids the oxidation stage III has the greatest stability with respect to oxidation and reduction, from which it follows that compounds in the stage II are reducing and in stage IV are oxidizing.

The tendency to attain a filled or half-filled $4f$ group is also apparent from the fact that in europium and ytterbium in the metallic state (oxidation number 0) one $5d$ electron has been transferred to the $4f$ group (5-1d). In this way the number of electrons participating in the metallic bond is reduced from three to two, so that the bond becomes weaker. The consequences this has for the properties of europium and ytterbium in the metallic state were discussed in 5-1d and 6-4f.

The actinoids at the beginning of the series, on the other hand, vacillate between $6d$ and $5f$ character. As can be seen from Fig. 5.8 the oxidation numbers

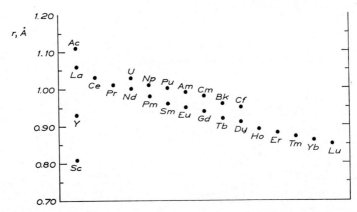

Fig. 28.1. Radii of M^{3+} ions in group 3. Analogous ions are placed in the same vertical column.

vary between III and VI (while they do not exceed that corresponding to a noble-gas shell). From approximately 95 Am onward the $5f$ and thus the actinoid character is quite definite. At the same time the oxidation number III becomes predominant and the properties much like those of actinium and the lanthanoids in the same oxidation state. The $6d$ character at the beginning of the series confers such chemical properties on these elements (among others the value of the highest oxidation number, which in fact predominates) that when the periodic system was first set forth the then known 90 Th and 92 U were quite naturally placed in the titanium group (4) and chromium group (6), respectively. When later 91 Pa was discovered, this element was placed in the vanadium group (5).

In the end of the actinoid series the oxidation number II appears in Md(II) and No(II) showing the tendency to attain a filled $5f$ group.

The oxidation numbers II–VI for the elements in group 3 may be schematically represented by the cations M^{2+}, M^{3+}, M^{4+}, $M^{V}O_2^+$, $M^{VI}O_2^{2+}$. We write these cations in their simplest forms (without hydration) and ignore the fact that on hydrolysis (12-1b) many other types of cations are formed from these. In a solid phase MO_2^+ and MO_2^{2+} may also be represented by infinite ions with these overall compositions. However, the series given above shows the increasing tendency to combine with oxygen ligands with increasing oxidation number (5-6c, last part). The oxidation numbers V and VI appear only among the actinoids 91 Pa–95 Am, and consequently the two types of oxo cations can occur only with these elements (for example, the neptunyl(V) ion NpO_2^+ and the uranyl(VI) ion UO_2^{2+}). For the oxidation number VI a number of compounds are known, for example with the alkali or alkaline-earth metals, which from their formulas would be assumed to contain oxo anions such as UO_4^{2-} ("uranate ion") or $U_2O_7^{2-}$ ("diuranate ion"). However, in all the structures so far studied the oxo anions appear to be infinite in one or two dimensions. This explains why even alkali uranates and diuranates are poorly soluble in water (cf., for example, the easily soluble alkali chromates and dichromates, where CrO_4 tetrahedra make up the finite ions CrO_4^{2-} and $Cr_2O_7^{2-}$).

In group 3 many analogous ions are of nearly equal sizes. For M^{3+} ions this is shown by fig. 28.1. The slow change from element to element in the lanthanoid and actinoid series is apparent. M^{3+} ions of neighboring elements, as well as the corresponding compounds, therefore have very similar properties. This applies, of course, also to each of the other types of ions, but examples are most numerous among M^{3+} ions because these are so common.

Of the M^{3+} ions Sc^{3+} is the smallest and therefore forms the most covalent bonds (chap. 27) among these ions, and has the greatest complex-forming ability. All the M(III) hydroxides give basic water solutions, but the basic character is weakest for scandium and increases with increasing radius of the M^{3+} ion (cf. again fig. 28.1). The basic character is quite strong for lanthanum and the lighter lanthanoids in the M(III) state. These hydroxides therefore behave chemically more like calcium hydroxide than like the formally analogous aluminum hydroxide (the ionic radius for Ca^{2+} is 0.94 Å, but for Al^{3+} only 0.45 Å). However, like calcium hydroxide (26-1), the basic strengths of the hydroxide solutions are limited by the quite low solubilities.

The M^{3+} ions with noble-gas shells (Sc^{3+}, Y^{3+}, La^{3+}, Ac^{3+}; cf. fig. 5.8) show no absorption in the visible spectrum and are thus colorless. The same is true of M^{3+} ions with half-filled (Gd^{3+}, Cm^{3+}) or filled (Lu^{3+}) f groups. A number of ions which in fig. 5.8 lie adjacent to those mentioned are also colorless (Ce^{3+}, Yb^{3+}), but stronger and stronger color appears the farther away from these we go. Thus, Eu^{3+} and Tb^{3+} are very weakly colored, Nd^{3+} and Ho^{3+} strongly colored. Ce^{4+}, Th^{4+}, Pa^{5+} and U^{6+} would be considered as noble-gas ions, but the high charges cause them always to form quite strongly covalent bonds (5-6c). Therefore color may appear (7-3b), as in the case of Ce^{4+} and U^{6+} (for example, in the uranyl(VI) ion UO_2^{2+}, which is yellow in color). Th^{4+} and Pa^{5+}, on the other hand are colorless. The insensitivity of the absorption bands caused by the incomplete f groups to changes in bonding conditions was discussed in 7-3b. The absorption bands of the lanthanoids and actinoids are valuable in analysis.

The incomplete f groups in lanthanoids and actinoids often lead to the existence of unpaired electrons, the more so because the external shielding has the effect that the f electrons hardly participate in any pair formation in chemical bonding. The large number of f orbitals (7) also means that the number of unpaired electrons can be unusually high. Paramagnetism (7-4) is therefore common and often attains high strengths.

28-2. Occurrence and history

a. Occurrence. The most common elements of group 3 in the earth's crust are yttrium and cerium, each having an average of about 40 p.p.m. Next follow lanthanum and neodymium with about 20 p.p.m. These elements are more common than many well-known elements (for example lead with 15 p.p.m.), and the old designation "rare-earth metals" (originally including scandium, yttrium, and the lanthanoids) is not entirely appropriate. The proportions of the lanthanoids follow Harkins' rule very beautifully (17a, 9-1b), as would be expected, since here no great difference in chemical properties would give rise to deviations depending on different natural occurrences. All isotopes of promethium are unstable (^{145}Pm has the longest half-life, 18 years), but since the element is continually being formed by spontaneous fission of uranium (9-2c) it is found in extremely small amounts in uranium ores.

Scandium, yttrium, and most lanthanoids occur in nature predominantly in the oxidation state III, which together with the small variations in the radii of the M^{3+} ions (fig. 28.1) causes them to be able to replace each other within wide limits in minerals (solid solutions). Usually a distinction can be made among minerals containing (a) mainly the larger M^{3+} ions (La to about Eu), (b) mainly the smaller M^{3+} ions (Y, approximately Gd to Lu and to some extent Sc) and (c) in rather rare cases, both the larger and smaller M^{3+} ions with no appreciable dominance of either category. Within the first two classes most often the most common element in the class predominates, thus in a Ce and in b Y. As early as the first half of the 19th century the known oxides of the elements of class a were therefore called *cerium earths* and of the elements of class b *yttrium earths*. In minerals of class a thorium is also often present. The ion Th^{4+} has the radius 0.99 Å and therefore readily replaces atoms of the cerium-earth metals in solid solutions (while the difference in charge is compensated as described in 6-3a).

If the environment in which the mineral was formed was not sufficiently strongly oxidizing, those lanthanoids which can also exist in the oxidation state II (fig. 5.8) may be reduced to this state. Eu^{3+} is most easily reduced and therefore as Eu^{2+} usually follows the divalent ions Ca^{2+}, Sr^{2+} and Pb^{2+}, and even K^+ (ionic radii, fig. 5.5).

The most important mineral for the lanthanoids is *monazite*, which can be formulated $CePO_4$, where a quite large proportion of the Ce^{3+} ions is replaced by M^{3+} ions of other lanthanoids, especially the cerium-earth elements, and Th^{4+}. Monazite is also the most important source of thorium. The mineral is mainly found weathered and concentrated in *monazite sand*.

For the recovery of lanthanum the monazite deposits in Brazil, India and South Africa play the largest role. Of some importance also is *bastnäsite*, cerium(III) fluoride carbonate $CeFCO_3$, where cerium is partly replaced in the same way as in monazite. However, the thorium content is low. Bastnäsite was first found at Bastnäs in central Sweden, but the most important deposits are at Mountain Pass, California. *Cerite*, cerium(III) silicate is also found at Bastnäs and at Mountain Pass. *Gadolinite* is most nearly yttrium iron(II) silicate, where Y^{3+} is replaced more or less by M^{3+} ions of yttrium-earth elements. Gadolinite is used as a source of these elements. The mineral was first recognized in pegmatite from a feldspar mine in Ytterby in east-central Sweden and named for the Finnish chemist Gadolin who studied it.

Larger amounts of actinoids are found only for thorium and uranium, in which ^{232}Th and ^{238}U have the long half-lives 1.4×10^{10} and 4.5×10^9 y, respectively. The only transuranium elements that can be detected in nature are ^{237}Np and ^{239}Pu, which are formed in very small amounts in uranium minerals through the absorption of neutrons by ^{238}U (9-2c, last part). Actinium and protoactinium are included in the natural radioactive decay series (9-2a), but all of their isotopes have such short half-lives (longest for ^{231}Pa with 3.2×10^4 y) that they occur in nature only in very small amounts (and then only in uranium minerals).

For thorium, as already mentioned, monazite is the most important mineral. The uranium minerals are of great importance because of the fundamental role played by uranium in the exploitation of nuclear energy (9-4a). The proportion or uranium in the earth's crust is on the average 4 p.p.m., that is, greater than, for example, silver. Many uranium minerals are known, but only a few have any great significance. The most important are *pitchblende* and *uraninite*, which are both oxide phases with extended homogeneity ranges. This range for pitchblende lies around $UO_{2.6}$ and for uraninite around UO_2. In *thucholite* thorium and uranium oxides are mixed with asphaltlike substances. *Carnotite*, potassium uranyl vanadate, and *tyuyamunite*, the corresponding calcium compound, should be mentioned. In some places uranium-bearing sedimentary rocks are found, especially phosphorites and slates. The proportion of uranium is low and not connected with any particular mineral, but often represents large total amounts of uranium. The uranium here has probably been adsorbed from sea water in connection with the formation of the sediments.

It is difficult to give a geographical survey of uranium mining, partly because of vigorous prospecting which constantly turns up new deposits, and partly because production is often kept more or less secret. A classic occurrence of uranium is Joachimstal in Czechoslovakia (pitchblende) but beyond this we have no reliable information from the Eastern bloc. In Africa a rich deposit was found in Katanga (pitchblende), but this is now believed to be practically exhausted. Recently a significant production has begun in South Africa, especially Witwatersrand. Gold-bearing quartz here also contains uranium minerals (uraninite and thucholite) and the combination with gold mining makes for reasonable costs in spite of the low uranium content (about 200 p.p.m.). In Canada very large pitchblende and uraninite deposits occur at Great Bear Lake, Lake Athabaska and Blind River (about 1000 p.p.m.). The United States has quite large deposits in Colorado, Utah and Arizona (uraninite, pitchblende and carnotite, 2000–5000 p.p.m.).

b. History. In 1794 Gadolin found an "earth", that is, oxide (26-2b), in a mineral from Ytterby, which was later named gadolinite (28-2a). The new oxide was studied by Ekeberg among others, who gave it the name *yttrium earth* after the locality. Ten years later Berzelius and Hisinger discovered another new oxide on studying a mineral from Bastnäs. They called the corresponding metal *cerium* and the mineral cerite after the asteroid Ceres which was discovered a few years earlier. The same new oxide was identified simultaneously and independently by Klaproth.

For a long time yttrium earth and cerium earth were assumed to be oxides of their corresponding pure metals, but in the years 1839–43 Mosander discovered that both consisted of several metal oxides with very similar properties. Thereafter and on into this century there followed a more and more complete separation into new oxides until in 1907 scandium, yttrium and all the lanthanoids except element 61 were known. This separation involved very difficult and time-consuming work, carried out by investigators such as Cleve, L. F. Nilson, de Boisbeaudran, Urbain, de Marignac and Auer von Welsbach. During this development the names of the elements were often changed and displaced. It is notable that yttrium, terbium, erbium (names given by Mosander, although the last two were used for what were later shown to be mixtures) and ytterbium are all named after Ytterby. Scandium was discovered by Nilson in 1879; however, in 1871 Mendeleev had already predicted its existence and properties (4-2a). Holmium, thulium and the element that finally took the name erbium, were discovered by Cleve in 1879 (holmium, which was named after Stockholm, had been independently identified in 1878 by Soret). The missing lanthanoid 61 was first found in 1947 among the fission products from a uranium reactor, and was named promethium after Prometheus of Greek mythology, who gave the fire of the gods to man.

In 1789 Klaproth found a new element in pitchblende, to which he gave the name *uranium* after the planet Uranus discovered a few years earlier. *Thorium* was discovered in the form of its dioxide in 1828 by Berzelius during a study of a mineral from Norway, and was named after the god Thor. Actinium was discovered in 1899 and protoactinium in 1917, both in uranium preparations.

By then the actinoids 89 Ac, 90 Th, 91 Pa, and 92 U were thus known. In 1940 McMillan and Abelson found that an isotope of element 93 appeared by β^- decay of ^{239}U which was formed by neutron capture by ^{238}U (9-2b). This transuranium element was called *neptunium* since it lay outside uranium, as the planet Neptune lies outside Uranus. The formation processes were

$$^{238}_{92}U(n,\gamma)^{239}_{92}U \xrightarrow[23.5 \text{ min}]{\beta^-} {}^{239}_{93}Np \left(\xrightarrow[2.3 \text{ d}]{\beta^-} {}^{239}_{94}Pu \right)$$

It was immediately suspected that ^{239}Np decays to an isotope of the transuranium element 94 according to the process given in parentheses above, but this could not be proved. This element, which was given the name *plutonium* after the planet Pluto lying outside Neptune, was identified with certainty in 1941 by Seaborg, McMillan and coworkers after it was produced according to

$$^{238}_{92}U(d,2n)^{238}_{93}Np \xrightarrow[2.1 \text{ d}]{\beta^-} {}^{238}_{94}Pu$$

After this, other nuclear reactions (9-2b) were developed and the transuranium elements 95–103 were identified and characterized. The major portion of this work has been carried out at the University of California at Berkeley under the leadership of Seaborg. These transuranium elements have been named 95 *americium*, 96 *curium* (after Curie), 97 *berkelium* (after Berkeley), 98 *californium*, 99 *einsteinium*, 100 *fermium*, (after Fermi; 9-4a, last part), 101 *mendelevium* (after Mendeleev), 102 *nobelium* (after Alfred Nobel), 103 *lawrencium* (after the inventor of the cyclotron, E. O. Lawrence).

28-3. Properties of the group 3 elements

Of the three last actinoids 101 mendelevium, 102 nobelium and 103 lawrencium, of which all isotopes have very short half-lives (cf. tab. 1.3), only extremely small

Table 28.1. *Densities and melting points of the group 3 elements.*

	Density 25°C, g cm⁻³	Melting pt. °C		Density 25°C, g cm⁻³	Melting pt. °C
21 Sc	2.99	1539			
39 Y	4.48	1509			
57 La	6.17	920	89 Ac	10.06	~1100
58 Ce	6.77	795	90 Th	11.72	1750
59 Pr	6.78	935	91 Pa	15.37	<1873
60 Nd	7.00	1024	92 U	19.04	1132
61 Pm	—	—	93 Np	20.45	640
62 Sm	7.54	1072	94 Pu	19.74	639.5
63 Eu	5.26	826	95 Am	13.67	1176
64 Gd	7.90	1312	96 Cm	13.51	1340
65 Tb	8.27	1356	97 Bk		
66 Dy	8.54	1407	98 Cf		
67 Ho	8.80	1461	99 Es		
68 Er	9.05	1497	100 Fm		
69 Tm	9.33	1545	101 Md		
70 Yb	6.98	824	102 No		
71 Lu	9.84	1652	103 Lr		

quantities have been prepared. As a consequence it has been possible only to determine their nuclear properties, and, for 101 mendelevium and 102 nobelium using carrier techniques (9-5a) and ion-exchange chromatography (15-4), to draw general conclusions regarding the chemical properties. Only up to 100 fermium has it been possible to obtain large enough quantities of the element or its compounds in a condensed state to enable the determination of properties associated with such states (for example, crystal structure, density, melting point).

As stated earlier (6-4b) nearly all metallic elements have either close-packed or body-centered cubic crystal structures. This is also true of the structures known so far of the group 3 elements with the important exceptions of uranium, neptunium and plutonium, which each have several crystalline modifications with more complicated crystal structures. For these metals the physical properties are also strongly dependent on the particular structure involved.

Tab. 28.1 gives densities and melting points for the group 3 elements. The melting points of the lighter of these elements are shown also in fig. 6.26.

In 28-1 and earlier sections the special position of europium and ytterbium with regard to properties of the metallic state was mentioned. The weaker metallic bonding in these two elements is evident from the density and melting-point values, which are appreciably lower than those of neighboring elements (although the density is dependent not only on interatomic distances and atomic weight, but also on structure type).

The metals are silver-white, gray or slightly yellowish, and not particularly hard. Europium and ytterbium, because of the weaker metallic bonding, are softer than their neighbors (nearly as soft as lead).

Their quite strongly electropositive properties make the metals, in many reactions, similar to the alkaline-earth metals (cf. positions of the standard potentials for the pairs La–La^{3+}, Lu–Lu^{3+}, Sc–Sc^{3+} in tab. 16.1). It may be said that generally they react with oxygen, moist air, water and acids approximately as calcium does (26-3). However, variations do appear. Uranium is one of the most reactive of the group 3 elements, and requires great care in its practical applications (for example, the careful encapsulation of fuel elements in nuclear reactors).

28-4. Production and uses of group 3 elements

In group 3, with the exception of 61 Pm, the elements up to uranium can be obtained from natural material. On the other hand, all transuranium elements must be prepared by nuclear reactions. Many of them, especially ^{239}Pu, are obtained in appreciable quantities in nuclear reactors (9-4a). The principles of preparation of the heavier transuranic elements by bombardment of lighter nuclei have been outlined in 9-2b (see especially fig. 9.4). In this way the elements up to 100 Fm have been prepared in weighable amounts, but 101 Md and 102 No have so far been obtained in such small amounts that they can only be studied by carrier techniques and ion-exchange chromatography. Of the element 103 Lr only a few scores of atoms have been obtained in each experiment. Since in addition the half-life of its most stable isotope is only 45 seconds, chemical identification has so far not been possible.

In the following section only the most important chemical problems in the preparation of pure elements from the mineral or from spent fuel elements (9-4a) are considered. Isotope separation in the preparation of enriched fuel-element material is also described.

Since the group 3 elements often occur together and exhibit great chemical similarity, the preparation of pure elements must generally be carried out by careful separation processes. This is true on the one hand for scandium, yttrium, and the lanthanoids, which follow one another in minerals (28-2a), and on the other hand the transuranium elements, which are obtained mixed together in spent fuel elements.

If certain of the elements present can change oxidation number with respect to the others, a separation can easily be made between the groups with different oxidation numbers. Thus, for the lanthanoids, the oxidation numbers II and IV as distinct from III can be used, and for the actinoids up to plutonium a number of oxidation states > III (fig. 5.8) can be used. But in many cases where the principal oxidation number for neighboring elements is the same (this applies especially to the number III) the problem becomes more difficult. In the classical work on lanthanoids use was made particularly of solubility differences among salts (fractional crystallization and fractional hydroxide precipitation), and also, for example, differences in the decomposition temperatures of the nitrates. In spite of the generally extremely large number of steps in the process, and consequently very time-consuming work, the separations were rarely entirely complete. Here the situation has completely changed since with *ion-exchange chromatography* (15-4) and elution with complex-formers (for example, citrate solution or EDTA) or *solvent extraction* (13-1f) it is now possible to get both a rapid and complete separation.

For the recovery of *uranium* from ore, after grinding and concentration the ore is leached, often with sulfuric acid (this dissolves uranium(VI); if the ore contains lower oxidation states of uranium it must be oxidized at the same time). The solution contains uranyl(VI) ion UO_2^{2+}, which forms complexes with the anions of the leaching agent. Thus, with sulfuric acid, according to the sulfate concentration, UO_2SO_4, $UO_2(SO_4)_2^{2-}$ or $UO_2(SO_4)_3^{4-}$ are obtained. The anion complex can be absorbed in a nitrate-saturated *anion exchanger* and eluted with acid nitrate solution, by which process the ion exchanger is regenerated. *Extraction* (13-1f) is also used, but with a nonaqueous liquid. This liquid should either itself or through a substance dissolved in it bind the complex more strongly than it is bound in water. As a binding substance for sulfato complexes alkyl phosphates or alkyl amines are used, usually dissolved in heavier hydrocarbons. The complex goes over to the nonaqueous phase and is then extracted again into the water phase. Both after ion exchange and extraction the final product obtained is practically always uranium oxides.

Group 3 elements in metallic form are produced by reduction of oxides or halides with alkali metals (sodium), alkaline-earth metals (magnesium, calcium) or hydrides of the latter (calcium hydride CaH_2). Reduction of halides is preferred since the by-product—the slag—then is more easily melted and collected. Electrolysis of halide melts is also used. The most important method for producing metallic uranium from uranium oxides consists of reduction with hydrogen at 600°C to uranium dioxide UO_2, followed by conversion to uranium tetrafluoride UF_4 at 500° according to $UO_2(s) + 4HF(g) \rightarrow UF_4(s) + 2H_2O(g)$. The uranium tetrafluoride is reduced with molten magnesium according to $UF_4 + 2Mg \rightarrow U + 2MgF_2$. The heat of reaction is so high that the products are melted.

Enrichment of the fissionable isotope ^{235}U is accomplished by gaseous diffusion of uranium hexafluoride UF_6 (9-3b), which is made at 250°C according to $UF_4(s) + F_2(g) \rightarrow UF_6(g)$. After the desired separation into $^{235}UF_6$ and $^{238}UF_6$ is achieved these are processed separately (for example, in preparing the metal, by reconversion to tetrafluoride and reduction).

The purpose of *processing spent fuel elements of a nuclear reactor* is apparent from 9-4a. Here it is necessary to separate the plutonium formed (^{239}Pu) and the remaining uranium from other transuranium elements and fission products. Several methods have been used, but solvent extraction (13-1f) predominates in newer installations. Use is made of the fact that ions M^{4+} and $M^{VI}O_2^{2+}$ have significantly greater tendency than other cations to form nitrato complexes, and these complexes can be extracted from water solution of high ionic strength with organic solvents that contain oxygen atoms (alcohols, ethers, ketones, esters lately often tributyl phosphate). In a nitrate solution with high enough oxidation potential, plutonium and uranium are easily obtained in the oxidation states IV and VI, that is, as the ions M^{4+} and $M^{VI}O_2^{2+}$ (cf. 28-1). These then form nitrato complexes which can be extracted. The few fission products with the oxidation number IV that may be present (especially cerium and zirconium) may follow along, and these are removed by special methods. Most of the other cations remain in the water solution. The organic solution is subsequently treated with water together with reduction so that plutonium goes over to Pu(III), but uranium remains as U(IV) and U(VI). The Pu(III) nitrate then cannot remain in the organic phase but goes over to the water phase. The uranium stays in the organic phase. Thus, the separation problem is solved. Each extraction process is carried out in practice in many steps, connected according to the cascade principle (9-3b), which makes possible continuous operation.

The actinoids that in metallic form have been most used in peaceful and military applications are, besides "natural" uranium, uranium enriched in ^{235}U, ^{239}Pu and to a lesser degree, thorium. The principles have been described in 9-4. As mentioned there, fuel elements consisting of compounds, for example uranium dioxide UO_2 and uranium carbide UC, are also currently in use.

An alloy is made directly from monazite, consisting of the constituent metals, that is, cerium earth elements (mostly cerium and lanthanum) and some iron. It is commonly called "Misch metal" and is used in metallurgy as an alloy additive. When Misch metal is filed, particles are formed that are *pyrophoric*, that is, ignite spontaneously in air. Addition of more iron (up to about 30 wt %) increases the hardness and production of sparks and gives the *flint* much used in cigarette lighters, etc.

28-5. Compounds of the group 3 elements

Here we give only some brief information on compounds of uranium and a few applications not previously mentioned of compounds of the group 3 elements.

a. Uranium compounds. The system *uranium–oxygen* is very complicated and still not completely understood. At room temperature the brown *uranium dioxide* UO_2 exists, which at higher temperature can dissolve additional oxygen. Another oxide phase (dark brown to black) at room temperature has a composition varying between $UO_{2.60}$ and $UO_{2.67}$ and can be designated as a U_3O_8 *phase*. Between these phases, however, there is at least one additional oxide phase. By careful heating of uranyl nitrate *uranium trioxide* UO_3 is obtained as an orange-red powder: $UO_2(NO_3)_2 \rightarrow UO_3(s) + 2NO_2(g) + \frac{1}{2}O_2(g)$. With stronger heating in air all other oxides go over to the U_3O_8 phase.

All uranium oxides dissolve in nitric acid and from the solution *uranyl*(VI) *nitrate* $UO_2(NO_3)_2$ crystallizes as the hexahydrate in the form of lemon-yellow crystals with yellow-green fluorescence.

Uranyl(VI) solutions give with hydroxide ion a yellow precipitate, at first of *uranyl*(VI) *hydroxide* $UO_2(OH)_2$, which with an excess of the precipitating agent goes over to *"polyuranates"*, which also are yellow and difficultly soluble. One schematic equation is $2UO_2(OH)_2 + 2Na^+ + 2OH^- \rightarrow Na_2U_2O_7(s) + 3H_2O$, but the product contains water and several different phases are believed to be formed. *Uranates*(VI) of the type Na_2UO_4 can be obtained from melts, but not from water solution. The uranyl(VI) ion easily forms *carbonato, sulfato* and *nitrato complexes* of which the two latter are important in the solvent extraction of uranium (28-4).

Uranium tetrafluoride UF_4 (green, m.p. 960°C) is used in the production of metallic uranium (28-4). *Uranium hexafluoride* UF_6 (colorless crystals, which sublime at 56°C and 1 atm) is one of the few volatile uranium compounds and is used in isotope separation by gaseous diffusion (9-3 b).

b. Some further applications. Oxides of the group 3 elements are high-melting, and this fact is used in some cases in making refractory materials. Thus, from *thorium dioxide* ThO_2 (m.p. 3050°C) objects are made for laboratory use at high temperatures (crucibles). The sulfides are also stable and high-melting; *cerium sulfide* CeS can be heated up to the melting point (2450°C) without decomposition.

For *gas incandescent lamps* (Auer von Welsbach) a solid solution of 1 wt % cerium dioxide CeO_2 in thorium dioxide ThO_2 is used. A "gas mantle" is made by immersing a net fabric in a conc. nitrate solution of the corresponding composition, then drying and igniting it. The fabric burns and the nitrates decompose to dioxides, which hold the form of the fabric. In a gas lamp this oxide mixture emits an intense, white light.

The radiation from pure heated ThO_2 (like many other colorless oxides) is rather small in the visible spectrum, very small in the near infrared, and is appreciable only in the far infrared. The heat radiation from the oxide is therefore fairly low, which with a good supply of heat leads to high temperature.

Pure CeO_2, on the other hand, at high temperature has a strong radiation in the visible spectrum (especially in the blue) and quite strong radiation also in the near infrared. When heated in the same flame as ThO_2, therefore, CeO_2 radiates much more heat and never reaches the same temperature. It turns out, however, that a solid solution of a small proportion of CeO_2 in ThO_2 has a radiating ability in the visible spectrum approximately as great as that of pure CeO_2 and in the near infrared nearly as low as that of pure ThO_2. Here the advantages of high radiating ability in the visible spectrum and the feasibility of high temperature are thus combined. A large proportion of the radiation of the gas mantle is therefore visible light. With higher proportions of CeO_2 the radiating ability in the near infrared increases and the temperature is lowered.

Rare-earth oxides are incorporated in carbon arcs to increase light intensity as much as tenfold. Recently, europium-activated yttrium oxide and yttrium vanadium oxide have shown great promise as red phosphors for color television tubes.

The colored lanthanoid ions are used in many ways in the *glass industry* (23-3 i2). Finer glasses (special glasses, optical glasses and also sometimes enamels and glazes) are thus colored by adding lanthanoid oxides to the glass melt (cerium dioxide is also used as an oxidizing agent to decolorize glass). By varying the proportions of praseodymium and neodymium oxides, neutral gray or blue optical glasses can be obtained. The strong absorption bands in yellow of such glasses are also used in glass-blowing goggles to filter out sodium light.

A ceramic material (trade name Yttralox), consisting of yttrium oxide Y_2O_3 with about 10% thorium dioxide ThO_2 in solid solution, is after firing at about 2200°C transparent as glass in spite of its polycrystalline nature. This is caused by its being made up of a single phase with cubic symmetry, which is, therefore, optically isotropic (6-1e) and the fact that it can be obtained quite free from pores. Yttralox is used for heat-resistant optical articles, and its low absorption for visible and infrared light is an additional advantage.

Cerium dioxide, pure or mixed with other lanthanoid oxides, is nowadays an important polishing agent for optical glass.

Acid solutions of Ce(IV) are strong oxidizing agents. They contain complex cerate(IV) ions, formed by adding the anions of the acid to Ce^{4+}. The redox pair produced is dependent on the type of anion and also on the anion concentration, since the number of ligands in the complex changes with the concentration. Each such redox pair has its own standard potential (16c). Especially high redox potentials are obtained in perchloric acid solutions of cerium(IV) perchlorate $Ce(ClO_4)_4$. This is perhaps the most strongly oxidizing solution that is stable enough for redox titration(16d). The sulfate and nitrate are also used for redox titration.

CHAPTER 29

Group 4, The Titanium Group

a. Survey. The familiar elements in group 4 are titanium, zirconium, and hafnium. In this group we may include (see fig. 4.7 and tab. 4.3) the element 104, of which during 1964 Russian scientists synthesized an isotope with the mass number 260. This decays by spontaneous fission with a half-life of 0.3 s. The short half-life and the tiny amounts that can be obtained greatly hinders the study of the other properties of the element.

Titanium, zirconium, and hafnium exist preferably in the oxidation state IV, which for Ti^{4+} and Zr^{4+} gives a noble-gas shell and for Hf^{4+} a xenon shell plus a completely filled $4f$ group in the interior of the electron cloud (cf. fig. 5.8). Oxidation numbers II and III are well known only for titanium (cf. chap. 27). The tendency to go over to the state IV makes these states strongly reducing. When the ions Ti^{4+}, Zr^{4+}, and Hf^{4+} form compounds with not too strongly polarizable ions they give no color (7-3b; on the other hand, TiI_4, for example, is red). In the lower oxidation states color arises because of the incomplete d groups. Thus, Ti(III) compounds are violet.

It has been stated several times that, because of the lanthanoid contraction, zirconium and hafnium have nearly the same atomic and ionic radii in corresponding states (figs. 5.2 and 5.5). They therefore have extremely similar chemical properties and can completely replace each other in crystals. The radius values for titanium, on the other hand, are somewhat smaller.

In comparing compounds of the same type it is found that titanium is bonded with the most covalent bonds (chap. 27). Titanium dioxide TiO_2 has intermediate character, but is predominately acid. Zirconium and hafnium dioxides are more basic. With the oxidation number IV the high charge favors the formation of oxo cations, which is evident especially for titanium, where at the same time the radius is least. In solid titanium(IV) salts the most common cation thus has the net composition TiO^{2+} (oxotitanium(IV) ion, earlier called titanyl ion), although this appears to exist only as infinite chains $(TiO^{2+})_n$.

b. Occurrence. *Titanium* has a quite high average abundance in the earth's crust (tab. 17.2 and 17.3) and is generally very widespread. The most common mineral and the greatest source of titanium is *ilmenite*, which consists of the double oxide $Fe^{II}Ti^{IV}O_3$. Ilmenite often occurs also as lamellae in magnetite Fe_3O_4 (*titanomagnetite*), and in appreciable proportions the recovery of titanium may be profitable after the crushed magnetite is removed magnetically. *Rutile* TiO_2 is also an important source of titanium.

The most important deposits of titanium nowadays are in India (ilmenite concentrated in beach sand) and Florida (ilmenite in prehistoric beach sand). Rutile sand is found in Australia. However, currently attempts are being made to exploit vein deposits, which contain mainly ilmenite (for example, in New York state and Quebec) and titanomagnetite. Ilmenite and quite rich titanomagnetite are mined in Norway.

Zirconium occurs in moderate abundance in the earth's crust and is more common than, for example, nickel, zinc and copper (tab. 17.3). The most widespread and technically the most important mineral is *zircon*, zirconium silicate $ZrSiO_4$ (23-3h1), but *baddeleyite* ZrO_2 may also be mentioned.

Zircon is obtained especially concentrated in beach sand in Australia. Baddeleyite is found particularly in southern Brazil.

Hafnium does not occur in any particular mineral, but universally follows zirconium. All zirconium minerals contain hafnium and its average proportion in the elementary mixture is usually 1–5 wt %.

The three titanium group elements are not considered to play any active role in plant or animal organisms, but on the other hand, they do not produce any toxic effects.

c. History. In 1791 Gregor prepared from an English titanium mineral an oxide, which he assumed to contain a new metal. Independently Klaproth in 1795 found that rutile was the oxide of a new metal, which he called *titanium*. The metal was first prepared, although not pure, by Berzelius (1825).

Already in 1787 Klaproth had isolated from zircon the oxide of another new metal, which he called *zirconium* after the mineral. This metal was also prepared in impure form by Berzelius (1824).

Through study of the X-ray spectra of the elements and application of Moseley's law (4-3b), it was soon found that element 72 was not represented among the known elements. It was quite generally believed that this element, like the elements 57–71, would be a "rare-earth metal". A search was made for its X-ray spectrum in minerals of such metals, but without success. After Bohr in 1922 on the basis of his atomic theory proposed the element to be a zirconium homolog (the 4f group must in fact already be filled at 71 Lu), Coster and de Hevesy in Bohr's institute the same year studied the X-ray spectra of zirconium minerals and immediately found lines corresponding to the atomic number 72. The new element was given the name *hafnium* (Hafnia, Lat. for Copenhagen). Its great similarity to zirconium had frustrated all earlier attempts to find it by chemical means.

In 1964 element 104 (still not officially named) was synthesized in the Soviet Union by bombardment of plutonium with neon nuclei according to

$$^{242}_{94}Pu + ^{22}_{10}Ne \rightarrow ^{260}_{104}X + 4^1_0n$$

d. Properties of the group 4 elements. Some physical properties are given in tab. 29.1. The pure metals have a steel-gray color and are malleable. However, even

Table 29.1. *Physical properties of the elements of the titanium group.*

	22Ti	40Zr	72Hf
Density at 20°C, g cm^{-3}	4.50	6.50	13.07
Melting point, °C	1680	1852	1977
Boiling point, °C	3280	4380	~5200

very small amounts of nonmetals (for example, oxygen, nitrogen, carbon, hydrogen) make them brittle.

In compact form and at moderate temperatures the metals to a large extent are chemically very resistant. In oxidizing medium, this results from the formation of an oxide film, which prevents further attack. This protection is especially effective in titanium. In a nonoxidizing medium, on the other hand, zirconium and hafnium are more resistant than titanium. At about 450°C all three metals begin to react appreciably with a large number of substances and with further increase in temperature these reactions become rapidly more vigorous. At temperatures above about 800°C the reactivity is very great and makes metallurgical work and processing difficult. Especially serious is the absorption of oxygen, nitrogen, carbon and hydrogen. In small amounts these elements form solid solutions in the metal phase, but with larger amounts new phases are formed (oxides, nitrides, carbides, hydrides). Thus, with strong oxidation the dioxide MO_2 is obtained as the final product. The reaction $M(s) + O_2(g) \rightarrow MO_2(s)$ takes place with strong evolution of heat (ΔH^0 for TiO_2 is -931 kJ, and for ZrO_2, -1086 kJ).

Strong nitric acid and alkali solutions hardly react, nor do oxidizing chloride solutions (sea water). Hydrofluoric acid, on the other hand attacks even in high dilution. Titanium tolerates only dilute solutions (<about 3%) of hydrochloric, sulfuric and phosphoric acid but zirconium and hafnium resist appreciably stronger solutions.

e. Production and uses of the group 4 elements. The oxides are reduced with difficulty (cf. tab. 21.1), and hydrogen or carbon cannot be used as reducing agents because of reactions with these elements mentioned above. All processes at higher temperature must also be carried out in the absence of nitrogen. The most important industrial method is the reduction of tetrachloride with magnesium or sodium.

The tetrachlorides are made by chlorination at about 600°C of a mixture of ore (that is, more or less pure oxide) and carbon: for example, $TiO_2 + 2C + 2Cl_2 \rightarrow TiCl_4 + 2CO$. The tetrachlorides are purified, among other ways, by fractional distillation. This may be followed by a separation of zirconium from hafnium. The mixture of zirconium and hafnium tetrachlorides is converted to suitable salts and the separation usually carried out by solvent extraction (13-1f) or ion-exchange chromatography (15-4). From the separated solutions $ZrO_2 \cdot nH_2O$ and $HfO_2 \cdot nH_2O$ respectively are precipitated by hydrolysis, which are ignited to the dioxides and again chlorinated. The reduction of the tetrachlorides is carried out at about 800°C in steel vessels under an inert gas (He, Ar), for example according to $TiCl_4 + 2Mg \rightarrow Ti + 2MgCl_2$. The salt formed in the reaction ($MgCl_2$ or NaCl) is removed mostly by melting or leaching, and then by vaporization through heating. The porous metal remains ("titanium sponge", "zirconium sponge"). Before further processing these are melted, and new melting methods have made possible the modern applications of these metals.

The main difficulty in melting has been reaction with the crucible material, and this has now been completely avoided by a special type of electric arc smelting. The metal sponge (sometimes with alloy metals added) is compressed into a rod, which is lowered from above into a closed furnace. This is either evacuated or filled with argon at rather low pressure. The arc is struck between the lower end of the rod and pure metal that is held in a water-cooled copper crucible. The metal in the crucible is melted at the surface but the remainder stays solid and does not react with the crucible. From the end of the rod molten metal drops into the crucible ("consumable electrode"), and while this takes place the crucible with its added contents is lowered at the same time that the rod is advanced.

Very pure metal in small quantities can be obtained by dissociation of tetraiodide (all the tetrahalides are dissociated to metal and halogen on heating, but the iodides are least stable and require the lowest temperature) on a wire of the metal in question heated electrically to about 1400°C.

Titanium particularly, but also zirconium, have found greatly increased application since World War II. *Titanium* metal is used as such or in alloys, mainly those in which titanium is the chief metal (alloying elements—very often aluminum—seldom exceed 10 wt% altogether), but also in steel. The former are appreciably lighter than steel but have better mechanical properties than light metals. In contrast to these the titanium alloys can be used in the temperature range 200–450°C. Stability toward chemicals is good in many cases. Disadvantages are high price and difficulties of working. Titanium alloys have up to now been most used in the aircraft industry, especially in jet engine parts. Applications of both pure titanium metal and titanium alloys are increasing as corrosion-resistant materials in chemical industry. *Zirconium*, because of its low neutron absorption and high resistance to corrosion, is used for encapsulation of fuel elements in nuclear reactors (among metals only Be and Mg have lower neutron absorption than Zr, but Be is brittle and Mg has a low melting point and is easily corroded). It must then be totally free of hafnium since the natural hafnium isotopes have high neutron absorption. To further increase the stability toward corrosion of the encapsulating material the zirconium is generally alloyed with about 1.5 wt% tin and smaller amounts of iron, nickel and chromium (Zircaloy). Zirconium is also used as the oxidizable metal in photoflash bulbs.

Hafnium has up to now been used almost exclusively as a neutron absorber in control elements of nuclear reactors.

f. Compounds. The most important oxides, the *dioxides*, have already been mentioned in 29b and d. *Titanium dioxide* TiO_2 exists in three modifications, known as the minerals *rutile* (tetragonal), *anatase* (tetragonal) and *brookite* (orthorhombic). At ordinary pressure only rutile is stable, while the other two are metastable (13-3a). All three are built up of TiO_6 octahedra, in which each oxygen atom is common to three octahedra, but they differ in the mutual arrangement of the octahedra.

On melting with basic oxides titanium dioxide gives double oxides, which formerly were considered to be titanates. Here belong also the double oxides that occur as the minerals ilmenite (29b) and perovskite (21-5a).

Titanium dioxide is much used as a durable and well-covering paint pigment (*titanium white*), often mixed, for example, with calcium sulfate. It is also used to render porcelain enamels opaque white (cf. also 14-2a). Synthetic rutile, a highly prized gem stone because of its unusually high refractive index, is made from titanium dioxide by a method analogous to that used for the Al_2O_3 gem stones (24-3f).

Titanium dioxide is produced usually by solution of ilmenite in conc. sulfuric acid, in which sulfates of Fe(II) and Fe(III) are formed together with oxotitanium sulfate $TiOSO_4$. Fe(III) is reduced with iron scrap to Fe(II), the greater part of the $FeSO_4$ is removed for the most part by freezing, following which the $TiOSO_4$ is hydrolyzed by heating the solution, sometimes after addition of alkali: $TiO^{2+} + 2OH^- \rightarrow TiO_2 + H_2O$. Mainly hydrated titanium dioxide is formed (older name "titanic acid"), which is ignited until TiO_2 remains. Some TiO_2 has also begun to be produced by oxidation of gaseous $TiCl_4$.

Zirconium dioxide ZrO_2 has a high melting point (2700°C) and since it also is chemically resistant it is used as a refractory material in crucibles, etc.

A rod of 85 wt% ZrO_2 + 15 wt% Y_2O_3 after preheating becomes electrically conducting and can then be strongly heated by the passage of current. The oxide mixture behaves analogously to that used in the gas mantle (28-5b) and the result is high temperature and high emission of visible light (*Nernst lamp*).

The most important of the *halogen compounds* are the *tetrahalides* MX_4, and the *hexahalometallate ions* MX_6^{2-} which they form by addition of halide ions. The tetrahalides can be formed by reaction between the elements (an industrial method for the preparation of the tetrachlorides was described in 29e). The metal-halogen bond is strongest in the fluorine compounds, which are therefore the most stable.

Titanium tetrachloride $TiCl_4$ is a colorless liquid (b.p. 136.5°C; the gas consists of $TiCl_4$ molecules), which is easily hydrolyzed ($TiCl_4 + 2H_2O \rightarrow TiO_2(s) + 4HCl$) and therefore has a sharp odor and fumes in moist air. In conc. hydrochloric acid the hydrolysis is suppressed, and if such a solution is reduced cathodically or with zinc it becomes violet and on crystallization gives *titanium trichloride hexahydrate* $TiCl_3 \cdot 6H_2O$. The solution is strongly reducing and is used occasionally for redox titration (color change violet→nearly colorless with the transition Ti(III)→Ti(IV)). Anhydrous *titanium trichloride* (violet powder) can be obtained by reduction of tetrachloride with hydrogen. It is used as a polymerization catalyst in plastics manufacture.

Zirconium and *hafnium tetrachlorides* at ordinary temperature are solid, colorless substances.

The *metallic nitrides*, *carbides* and *borides* of the titanium metals are notable for their hardness, high melting points and chemical resistivity. Of these *titanium carbide* TiC (m.p. 3150°C) is used as the hard constituent in hard metals (14-3b) and, like *titanium diboride* TiB$_2$ (m.p. 2920°C), as a high-temperature material. Concerning *hafnium carbide* ∼HfC, see 30f.

CHAPTER 30

Group 5, The Vanadium Group

a. Survey. Vanadium, niobium and tantalum occur primarily with the oxidation number V, which gives a noble-gas or noble-gas-like shell (see fig. 5.8). The oxidation numbers II, III and IV are also known but are of significance only for vanadium (cf. chap. 27). Compounds in the state V are colorless or slightly colored, compounds in the lower states are colored (incomplete d groups).

Vanadium has a smaller atomic radius than niobium and tantalum in corresponding states (fig. 5.2). The radii of the latter on the other hand are nearly the same because of the lanthanoid contraction. Niobium and tantalum thus acquire very similar chemical properties although the similarity is not as great as that between zirconium and hafnium.

In accordance with chap. 27 it is found on comparison of analogous bonds that the highest covalent contribution occurs with vanadium. For the same element the covalent contribution increases with the oxidation number. Thus, for the vanadium oxides, VO and V_2O_3 are basic, VO_2 is intermediate and V_2O_5 predominantly acid. Vanadium(II) and vanadium(III) occur as the ions V^{2+} and V^{3+} in fairly typical salts, for example in the sulfates. With the oxidation numbers IV and V the appearance of oxo ions is favored; thus in strongly acid solutions the ions $V^{IV}O^{2+}$ (oxovanadium(IV) ion) and $V^{V}O_2^{+}$ (dioxovanadium(V) ion) respectively exist.

Nb_2O_5 and Ta_2O_5 have intermediate character but are predominantly acid.

b. Occurrence. Among the elements of this group, as would be expected (chap. 27), vanadium has the highest average abundance in the earth's crust (somewhat higher than nickel; see tab. 17.3). All three elements are quite evenly distributed with no great local concentrations.

The most important *vanadium* mineral is *patronite* VS_4. *Vanadinite* is also of importance, and is an analog of apatite (22-2a) with the formula $Pb_5Cl(VO_4)_3$. Vanadium is also obtained in connection with the recovery of uranium from *carnotite*, potassium uranyl vanadate. Very large amounts of vanadium occur in iron ores where vanadium in low concentrations follows titanium and phosphorus, and also in mica (*roscoelite*) and clay silicate minerals.

The largest single occurrence of vanadium is in Peru, and consists mainly of patronite. Vanadium and uranium are recovered together from oxide and carnotite ores in Colorado, Utah and Arizona. Vanadiferous iron ores are found in Scandinavia.

Table 30.1. *Physical properties of the elements of the vanadium group.*

	23V	41Nb	73Ta
Density at 20°C, g cm^{-3}	6.11	8.58	16.7
Melting point, °C	1920	2500	3000
Boiling point, °C	3380	~4900	~5400

Surprisingly enough, vanadium is the active transition metal in a respiratory pigment in an order of lower marine animals, the tunicates (cf. 8h).

Niobium and *tantalum* follow each other in nature and are obtained mainly from the double oxide *columbite* $Fe^{II}(Nb^V, Ta^V)_2O_6$. The mineral is sometimes called *niobite* or *tantalite* according to the metal that predominates. Nigeria and Katanga produce the largest amounts of these minerals.

c. History. In 1801 Hatchett prepared from columbite a new metal oxide and called the metal *columbium* after the mineral. The following year Ekeberg found in several minerals, of which one was tantalite, an oxide with new properties. Its insolubility in acids was striking and Ekeberg gave the name *tantalum* after Tantalus of Greek mythology, "in allusion to its [the oxide's] inability, in the midst of a flood of acid, to take any for itself and be satisfied". It was believed for a long time that columbium and tantalum were identical, but in 1844 H. Rose showed that columbium consisted to a large extent of another metal than Ekeberg's tantalum. He called it *niobium* (Niobe was the daughter of Tantalus). The name columbium was formerly much used in the United States, but should be avoided.

In 1831 Sefström discovered *vanadium* in iron made from an ore from Taberg, Sweden. The name was given after the Scandinavian goddess Vanadis.

d. Properties of the group 5 elements. The physical properties are given in tab. 30.1. The metals are very difficultly melted. They have steel-gray color, are hard, malleable in the pure state but like the metals of the titanium group become brittle with very small amounts of nonmetals. Mainly because of passivation by a protective oxide film they possess great resistance to attack by a number of substances at moderate temperature and the resistivity increases in the order $V < Nb < Ta$. Among the acids, niobium and tantalum are attacked only by hydrofluoric acid and conc. sulfuric acid, but vanadium is also attacked by strongly oxidizing acids in general. Only very strong alkali solutions react. At temperatures above about 300°C all three metals become quite reactive.

e. Production and uses of the group 5 elements. The metal(V) oxides M_2O_5 contained in ores or their roasting products are fused with alkali, forming water soluble compounds (30f). These are leached out with water and from the solution the metal(V) oxide hydrates are precipitated by acidification. Drying of the hydrates gives metal(V) oxides.

Vanadium is seldom used in pure form, but has long been produced in an alloy with about 50% iron (*ferrovanadium*) by reduction of vanadium(V) oxide with carbon, silicon (in an electric furnace) or aluminum (aluminothermic reduction, 24-3c) in the presence of iron. Ferrovanadium is used in the production of vanadium-alloy steels (34-3e).

Metal(V) oxides of *niobium* and *tantalum* are usually obtained together and must be separated (see below). The metals are now made mostly by heating in vacuum a mixture of metal(V) oxide and carbide (obtained by reduction of metal(V) oxide with carbon) in exact proportions: $M_2O_5 + 5MC \rightarrow 7M + 5CO(g)$. Final melting is done in a manner analogous to that of titanium and zirconium (29e).

For the separation of niobium and tantalum the metal(V) oxides are dissolved in hydrofluoric acid: $M_2O_5 + 10HF \rightarrow 2MF_5 + 5H_2O$; fluorometallate ions are also formed, especially according to $MF_5 + 2F^- \rightarrow MF_7^{2-}$ (notice the unusual coordination number; see 5-3i). This is followed by solvent extraction (13-1f) with methyl isobutyl ketone, which preferentially takes up niobium fluorides from water solutions with high acidity, and tantalum fluorides from water solutions with low acidity. From the ketone solutions the fluorides are absorbed by water from which the metal(V) oxide hydrates are precipitated.

Niobium may be considered as a by-product of the more important tantalum production. The metal has begun to be used in the manufacture of steel and for encapsulating fuel elements in nuclear reactors. Some metallic niobium compounds of the general formula Nb_3X (X among other elements = Sn, Al, Ge) are superconducting (6-4a) up to unusually high temperatures (max. $\sim 21°K$), which makes them of practical importance.

Tantalum is used mainly in pure form, and is valuable because of its good corrosion properties. It is used as an acid-stable material in chemical industry, in surgery for permanent reconstructions (not attacked by body fluids and compatible with the tissues) and in electron-tube parts.

f. Compounds. *Vanadium(V) oxide (divanadium pentoxide)* V_2O_5 forms yellow-red crystals (m.p. 648°C). In the crystals vanadium coordinates five oxygen atoms, which form a square pyramid, and such pyramids are linked together by common oxygen atoms. However, an oxygen atom belonging to another pyramid other than the nearest linked neighbors comes near enough so that we may speak of the coordination number six with a very irregular coordination octahedron. Vanadium(V) oxide is used as a catalyst in redox processes, especially in the contact process for sulfuric acid (21-10b). The activity results from a change in the oxidation number of vanadium, which is favored by the small energy differences between the different states.

Vanadium(V) oxide disolves slightly in water, the solution becoming acid. If the solution is made strongly basic, solution easily takes place with the formation of numerous *oxo-* and *hydroxovanadate(V) ions*, many of which are polynuclear. It also dissolves in strong acids, with the formation mainly of the ion $V^VO_2^+$.

From melts of vanadium(V) oxide with basic oxides solid *oxovanadates*(V) crystallize; for example, strongly basic $3M_2O + V_2O_5 \rightarrow 2M_3VO_4$ (commonly called orthovanadates), or less strongly basic $M_2O + V_2O_5 \rightarrow 2MVO_3$ (commonly called metavanadates). In M_3VO_4 the ions are VO_4^{3-} tetrahedra; in MVO_3 they are infinite tetrahedral chains $(VO_3^-)_n$, analogous to the polyphosphate chains $(PO_3^-)_n$, and geometrically similar to the chains $(SiO_3^{2-})_n$ in fig. 23.5a. Vanadium thus here coordinates four oxygen atoms. Other, both finite and infinite vanadate ions are found in other solid vanadates. All these give water solutions of the same type as the basic water solutions mentioned above. The ion VO_4^{3-} can also exist in strongly basic water solution (pH $\geqslant 12$). At lower pH values the solutions contain a number of *polyvanadate ions*, which with pH < 6 are orange or red in color; in these the orange-colored *decavanadate ion* $V_{10}O_{28}^{6-}$ is prominent, in which vanadium approaches octahedral coordination.

Solutions of niobium(V) and tantalum(V) oxides in strong bases also contain oxo- and hydroxometallate(V) ions. However, these oxides are so weakly acid that melting with basic oxides does not give niobates or tantalates but double oxides (21-5a). The sizes of the niobium(V) and tantalum(V) atoms causes them always to coordinate six oxygen atoms.

On adding acid, solutions of vanadium(V), niobium(V) and tantalum(V) precipitate hydrated metal(V) oxides (formerly called vanadic, niobic and tantalic acids, respectively).

Tantalum carbide $\sim$ TiC, like *hafnium carbide* $\sim$ HfC, is one of the most high-melting substances known (melting points probably about 3800°C). The highest melting point, about 3950°C, appears to be attained by a solid solution of HfC in TaC.

CHAPTER 31

Group 6, The Chromium Group

31-1. Survey

For chromium, molybdenum and tungsten the oxidation numbers II–VI normally occur (cf. chap. 27). For chromium the chromium(III) compounds are the most stable under ordinary conditions and the next-most-important chromium(VI) compounds are therefore strongly oxidizing. For molybdenum and tungsten on the other hand the metal(VI) compounds are the most stable (cf. chap. 27). The ions Cr^{6+}, Mo^{6+} and W^{6+} may be thought of as noble-gas-like ions, but their high charges make them so strongly polarizing that they often form colored compounds. This is especially true of Cr^{6+}, which has the smallest radius.

The ion Cr^{3+} is little larger than Al^{3+} (fig. 5.5), which causes Cr(III) to resemble Al(III) in many respects.

Molybdenum and tungsten have larger radii than chromium in corresponding states (fig. 5.2) but are rather similar to each other in size because of the lanthanoid contraction. Their chemical properties thus in many cases are similar, but also show many differences.

The Cr^{6+} ion because of its small size coordinates oxygen tetrahedrally (for example, in the chromate(VI) ion CrO_4^{2-}), but otherwise octahedral coordination is predominant. Tetrahedral coordination of oxygen also occurs with Mo^{6+} and W^{6+}, however, especially in the important ions MoO_4^{2-} and WO_4^{2-}. The cyano complexes of the type $Mo^{IV}(CN)_8^{4-}$ and $Mo^V(CN)_8^{3-}$ described in 5-3i should also be noted.

Oxides of the same element become more acid the higher the oxidation number is. Cr_2O_3 has intermediate character and CrO_3 is acid. MoO_3 and WO_3 are more weakly acid than CrO_3.

With the oxidation number VI oxo anions are very common and occur both in solid phases and in liquid solutions. Thus, solid *chromates*(VI), *molybdates*(VI) and *tungstates*(VI) also exist. All three metals can form polynuclear oxo anions, and the tendency for this behavior is especially strong with molybdenum and tungsten. *Polymolybdates* and *polytungstates* represent important branches of the chemistry of molybdenum and tungsten.

31-2. Occurrence and history

a. Occurrence. *Chromium* occurs in moderate abundance in the earth's crust. The element is widespread, but concentrations in minable deposits are relatively rare. *Chromite* $Fe^{II}(Cr^{III})_2O_4$, a double oxide of the spinel type (21-5a) is by far the most important ore.

The most important chromite deposits are in South Africa, Asia Minor, the Urals and New Caledonia.

Molybdenum and tungsten are rare metals with about equal abundances (tab. 17.3). The most important molybdenum mineral is *molybdenite*, molybdenum disulfide MoS_2 (closely resembling graphite but with higher density, 5.06 g cm^{-3}).

The largest molybdenite deposits are found in the United States, especially in Colorado.

The particular ore-forming tungsten minerals are *scheelite*, calcium tungstate $CaWO_4$, and *wolframite*, a solid solution of iron(II) and manganese(II) tungstates $(Fe^{II}Mn^{II})WO_4$.

The most important deposits are in China, but deposits in Burma, the Malay peninsula and in Alaska, Canada and western United States are also important.

In 22-16 it was mentioned that nitrogen fixation by bacteria is dependent on the presence of molybdenum in the soil. Probably a supply of molybdenum is necessary for all plants, although in such small amounts that deficiencies seldom occur. The vegetation on molybdenum-rich soils may take up so much molybdenum that it becomes dangerous to domestic animals. Molybdenum is also incorporated in certain enzymes in the animal body.

b. History. *Chromium* was discovered in 1797 by Vauquelin in a lead chromate mineral. The name was chosen because of the colors of the compounds of the new element (Gk. χρῶμα, color). Relatively pure metal was first prepared in 1894 by H. Goldschmidt by aluminothermy (24-3c).

Until the 18th century various substances were designated by the Latin name "molybdaena" (Gk. μόλυβδος, lead) because like lead they gave a black streak. This was especially true of the closely similar minerals graphite and molybdenite. The difference in properties between these two minerals was first demonstrated by Qvist in 1754. Scheele in 1778 prepared from molybdenite a new oxide (MoO_3), which he called *molybdenum earth*. By reduction of this oxide with carbon Hjelm in 1781 prepared carboniferous *molybdenum*.

The term *wolfram* (Ger. Wolf +Rahm, froth) was used as early as the 16th century as a name for the mineral wolframite, which, when it occurred together with cassiterite, made difficult the recovery of tin ("ate up the tin"). The mineral now called scheelite was earlier called *tungsten* (heavy stone) in Sweden, and from this in 1781 Scheele prepared a new oxide (WO_3), which he called tungstic acid. In the same year Torbern Bergman pointed out that the quite weak acid properties of tungstic acid indicated that it should be possible to reduce it to the metal. This was done in 1783 by two Spanish chemists, the brothers J. J. and F. de Elhuyar, of whom the former had worked with Bergman. The reduction was carried out on oxide obtained from wolframite, and the metal was thus given the European name wolfram. The names tungsten and tungstène are still preferred for the metal in English and French, respectively.

Table 31.1. *Physical properties of the elements of the chromium group.*

	24Cr	42Mo	74W
Density at 20°C, g cm^{-3}	7.23	10.2	19.3
Melting point, °C	1903	2617	3380
Boiling point, °C	2640	~4800	~5500

31-3. Properties of the group 6 elements

Some physical properties of these elements are given in tab. 31.1. The melting points are high and tungsten is the most difficultly melting of all metals. The metals have a light, steel-like color and are hard. In the pure state they are malleable but easily become brittle with improper working and heat treatment. For chromium up to 1840°C and for molybdenum and tungsten up to their melting points the stable crystal structures are body-centered cubic.

In compact form all three metals are stable in air and water at ordinary temperature. Chromium is the most resistant to heating in air. Oxidation to Cr_2O_3 occurs then but this is nonvolatile and protects against further attack. Molybdenum and tungsten begin to be oxidized at 400–500°C and at higher temperature the reaction goes rapidly because the trioxides formed are vaporized.

The standard potential for the process $Cr \rightleftharpoons Cr^{3+} + 3e^-$ is $e^0 = -0.74$ V. Chromium is thus less noble than iron and lies just below zinc (cf. tab. 16.1). It also dissolves easily in nonoxidizing acids. Strongly oxidizing acids (nitric acid, aqua regia) or anodic oxidation, however, cause *passivation* through the formation of a stable but invisible oxide film, which prevents further dissolution. The oxide film can for a time also prevent subsequent attack by nonoxidizing acids, but it disappears quite rapidly with reduction, so that attack begins. Alkali hydroxide solutions slowly dissolve chromium with the formation of hydrogen and hydroxochromate(III) ions of various compositions. The processes may be written schematically (hydration is not shown): $2Cr(s) + 6H^+ \rightarrow 2Cr^{3+} + 3H_2(g)$ and $2Cr(s) + 2OH^- + 6H_2O \rightarrow 2Cr(OH)_4^- + 3H_2(g)$.

Molybdenum and tungsten are passivated less easily than chromium and are therefore more easily attacked by oxidizing acids than the latter. On the other hand they are much more stable than chromium toward nonoxidizing acids. They are little attacked by alkali hydroxide solutions.

31-4. Production and uses of the group 6 elements

a. Production. In the production of pure *chromium* from chromite, the iron must first be removed. Chromite powder mixed with sodium carbonate and calcium oxide (which keeps the sodium carbonate from melting together and interfering with access to air) is heated in a furnace in an oxidizing atmosphere. The oxidation $Cr(III) \rightarrow Cr(VI)$ converts the chromium in the ore to water-soluble sodium chromate, and the reaction may be written $Cr_2O_3 + 2Na_2CO_3 + \frac{3}{2}O_2(g) \rightarrow 2Na_2CrO_4 + 2CO_2(g)$. The sodium chromate is leached out and converted by acidification to sodium dichromate $(2CrO_4^{2-} + 2H^+ \rightarrow Cr_2O_7^{2-} + H_2O)$, which is crystallized out and reduced with carbon: $Na_2Cr_2O_7 + 2C \rightarrow Cr_2O_3 + Na_2CO_3 + CO(g)$.

Cr_2O_3 can be reduced with hydrogen only with difficulty (21-5b, tab. 21.1), and on reduction with carbon or carbon monoxide, chromium carbides are formed Formerly aluminum, but now usually silicon is used as the reducing agent. In the latter case heat must be supplied in the reduction.

Electrolytic *chromium plating* is an excellent and very common protection against corrosion. As an electrolyte a water solution of chromium trioxide is used with added sulfuric acid or chromium(III) sulfate (sulfate ions facilitate the deposition of the metal). The anode is lead.

Molybdenum and *tungsten* are obtained quite easily by reduction of the trioxides with hydrogen (21.1). The terminal temperatures on reduction usually are about 1100°C for MoO_3 and 800°C for WO_3. The metals are then obtained in powdered form. Because of their high melting points they are often not melted but rather sintered. Rods pressed from the powder are strongly heated by passage of current and then further shaped by hot working. Molybdenum trioxide is prepared by roasting molybdenite: $MoS_2 + \frac{7}{2}O_2(g) \rightarrow MoO_3 + 2SO_2(g)$. The trioxide is purified, either by sublimation or by leaching the product with ammonia solution, following which molybdenum trioxide is obtained by heating the ammonium molybdate formed. Tungsten trioxide is produced by oxidation smelting of the mineral with sodium carbonate, so that sodium tungstate is formed (cf. above for chromium), which is leached out with water. Trioxide hydrate precipitates from the solution on acidification.

All three metals are extensively used as alloying elements in steels, and in that case added in the form of iron alloys, *ferrochromium*, *ferromolybdenum* and *ferrotungsten*. Certain grades of ferrochromium are obtained by direct reduction of the mineral chromite with carbon in an electric arc furnace: $FeCr_2O_4 + 4C \rightarrow Fe + 2Cr + 4CO(g)$.

b. Uses. Chromium is one of the most important alloying elements in steel and is used especially to increase its hardness and its stability against corrosion (*stainless steel*) and oxidation at higher temperatures. Alloys of chromium with cobalt and molybdenum or tungsten are very hard and are used in cutting tools. Certain alloys (*Nichrome*, *Nichrothal*) with nickel (nickel is usually the major constituent), chromium and greater or lesser amounts of iron have low conductivity and good resistance to oxidation up to about 1200°C. They are therefore much used as resistance elements for electrical heating. The Swedish alloy *Kanthal*, containing 20–30 wt % chromium, about 5 wt % aluminum, 2 wt % cobalt and the remainder iron, resists oxidation up to 1375°C.

Molybdenum is used as an alloying element in steel (34-3e) and in pure form in large amounts for metal parts in incandescent lamps and electron tubes.

Tungsten is also an important alloying element in steel (34-3e). Pure tungsten has become enormously important as a filament material in incandescent lamps and electron tubes. It is also used for electrical contacts and spark plug electrodes where high temperatures must be tolerated.

31-5. Chromium compounds

Here only the oxidation number II, III and VI are considered.

a. Chromium(II) compounds. These are obtained by reduction of Cr(III) with strong reducing agents. Thus, for example, reduction of a hydrochloric acid solution of $CrCl_3$ with zinc in the absence of air gives a blue solution of $CrCl_2$. The blue color is caused by the hydrated Cr^{2+} ion. For the pair $Cr^{2+} \rightleftharpoons Cr^{3+} + e^-$, $e^0 = -0.41$ V (tab. 16.1), which shows that the Cr^{2+} ion is strongly reducing. The solution just mentioned is therefore rapidly oxidized by atmospheric oxygen, and also slowly by hydrogen ions in the solution: $2Cr^{2+} + 2H^+ \rightarrow 2Cr^{3+} + H_2(g)$. Poorly soluble chromium(II) compounds, for example *chromium(II) acetate*, which can be obtained readily in solid form by precipitation, after drying are relatively stable.

b. Chromium(III) compounds. As mentioned earlier, chromium(III) compounds show many similarities to aluminum compounds.

Anhydrous *chromium(III) chloride* $CrCl_3$ is formed by heating chromium, or chromium(III) oxide plus carbon, in a stream of chlorine gas. It sublimes and is condensed in the form of red-violet crystal plates. It is very difficultly soluble in water, but in the presence of small amounts of Cr(II) it dissolves easily with strong evolution of heat. The solution is dark green at first, and contains mainly the cation $CrCl_2(H_2O)_4^+$ and Cl^-. Gradually the color changes through light green to violet as a result of the replacement of the chlorine ligands in the complex cation by water in two stages. On heating, the violet solution changes in the opposite direction, and after cooling reverts slowly to violet. Therefore we may write

$$CrCl_2(H_2O)_4^+ + Cl^- + 2H_2O \rightleftharpoons CrCl(H_2O)_5^{2+} + 2Cl^- + H_2O \rightleftharpoons Cr(H_2O)_6^{3+} + 3Cl^-$$
$$\text{dark green} \qquad\qquad\qquad \text{light green} \qquad\qquad\qquad \text{violet}$$

From the solutions the solid hydrates $[CrCl_2(H_2O)_4]Cl \cdot 2H_2O$, $[CrCl(H_2O)_5]Cl_2 \cdot H_2O$ and $[Cr(H_2O)_6]Cl_3$ can be obtained. Since the total composition is constant, three isomers are obtained (*hydrate isomerism*). In the three hydrates $\frac{1}{3}$, $\frac{2}{3}$, and $\frac{3}{3}$, respectively, of the total chlorine is bound by ionic bonds, and only these portions can be precipitated out of the solutions with Ag^+ as AgCl.

The ions mentioned above show the tendency of chromium(III) to form complexes with octahedral coordination. A large number of such complexes are known, and the ammine complexes are very typical. Examples of these are *hexa-*

minechromium(III) ion $Cr(NH_3)_6^{3+}$, aquopentamminechromium(III) ion $Cr(H_2O)$ $(NH_3)_5^{3+}$, chloropentamminechromium(III) ion $CrCl(NH_3)_5^{2+}$, trichlorotriamminechromium(III) $CrCl_3(NH_3)_3$, which is not an electrolyte, and tetrachlorodiamminechromate(III) ion $CrCl_4(NH_3)_2^-$. Many polynuclear complexes are also known.

The *hexaquochromium*(III) *ion* $Cr(H_2O)_6^{3+}$ participates in acid-base equilibria that are analogous to those for aluminum and are discussed in 12-1b and 12-4d. If a solution of a chromium(III) salt is not made too strongly basic (for example, with ammonia) *chromium*(III) *hydroxide* $Cr(OH)_3$ comes out as a pale grayish-green, flocculent precipitate of variable water content. In the freshly precipitated state this precipitate is soluble in both acids and bases and is thus amphoteric. The solubility minimum lies at pH ≈ 8.5, which should therefore not be exceeded on adding base. When dissolved in bases, the hydroxide forms *hydroxochromate*(III) *ions*, as a final product probably $Cr(OH)_4^-$ (earlier called chromite ion). If the precipitate is allowed to stand it becomes difficultly soluble in both acids and bases. Above about 70°C *chromium*(III) *oxide hydroxide* $CrO(OH)$ is produced.

Chromium(III) *oxide* Cr_2O_3 is obtained by igniting $Cr(OH)_3$ or chromium trioxide CrO_3. The oxide is isomorphous with $\alpha\ Al_2O_3$ (24-3f). Because of its green color and stability it is used as a paint pigment, *chrome green*. It is often made industrially by igniting alkali chromates or dichromates with sulfur: $Na_2Cr_2O_7 + S \rightarrow Cr_2O_3 + Na_2SO_4$ (the sodium sulfate is dissolved out with water). Addition of Cr_2O_3 to a glass melt gives the glass a green color.

The most important chromium(III) compounds with saltlike character are the sulfates. A solution of chromium(III) hydroxide in sulfuric acid gives violet crystals of *chromium*(III) *sulfate* with the constitution $[Cr(H_2O)_6]_2(SO_4)_3 \cdot nH_2O$ ($n = 2 - 6$). The color of the water solution on heating goes from violet to green as water in the complex ion is replaced with sulfate groups. The dodecahydrate of *potassium chromium*(III) *sulfate* $KCr(SO_4)_2 \cdot 12H_2O$ is an alum (21-11 b3), *chrome alum*. It forms violet crystals, whose solution behaves in the same way as the solution of the simple sulfate. Solutions of chromium(III) sulfate are quite strongly acid owing to hydrolysis (12-1b). Solutions brought to a pH value of about 3 by adding base are much used in a very important leather-tanning process (*chrome tanning*). Here complex formation probably occurs between Cr^{3+} ions and carboxyl groups in the peptide chains that make up the protein in the skins, so that the chains are bound more tightly together.

c. Chromium(VI) compounds. The tetrahedral *tetraoxochromate*(VI) *ion* CrO_4^{2-} has long been called *chromate ion*. The production of chromate from chromium(III) oxide was discussed in 31-4a. In the laboratory the fusion can be made more effectively with potassium nitrate.

The chromate ion gives a yellow color. In general, the solubilities of chromates resemble those of sulfates and many chromates are also isomorphous with the corresponding sulfates. Alkali chromates and ammonium chromate are easily soluble in water, magnesium chromate is similarly so, but for the other alkaline-earth metals the solubility of the chromate decreases rapidly with increasing atomic number. Barium chromate is thus very poorly soluble. Most other metal chromates are poorly soluble, and this is true especially of the chromates of lead(II), silver and mercury(I) and (II). *Lead(II) chromate* $PbCrO_4$ is used as a paint pigment, *chrome yellow*, and produces various hues by being mixed with lead sulfate. Another chromate pigment is *lead(II) oxide chromate* Pb_2OCrO_4, *chrome red*.

It has not been possible to isolate *chromic acid* H_2CrO_4, corresponding to the chromates. In its first protolytic stage it is a strong acid, in its second a weak acid ($pK_{a2} = 6.50$). Therefore, alkali chromate solutions react basic: $CrO_4^{2-} + H_2O \rightleftharpoons HCrO_4^- + OH^-$. Solid hydrogen chromates $MHCrO_4$ are unknown.

However, the hydrogen chromate ion takes part also in other equilibria, the most important being with the *dichromate ion* $Cr_2O_7^{2-}$. We may write

$$2CrO_4^{2-} + 2H^+ \rightleftharpoons 2HCrO_4^- \rightleftharpoons Cr_2O_7^{2-} + H_2O \qquad (31.1)$$

In strongly acid solutions (pH < 0) other products also appear. With high chromium contents probably *polychromate ions* with greater nuclearities are here formed. Solid *dichromates* $(M^I)_2Cr_2O_7$ (yellow-red), *trichromates* $(M^I)_2Cr_3O_{10}$ (deep red) and *tetrachromates* $(M^I)_2Cr_4O_{13}$ (dark red) are all known. In solution the trichromate ion, for example, can be formed according to: $Cr_2O_7^{2-} + CrO_4^{2-} + 2H^+ \rightleftharpoons Cr_3O_{10}^{2-} + H_2O$. The structures of the dichromate and trichromate ions have been determined, and they consist of CrO_4 tetrahedra linked together with one corner common to two tetrahedra. We may thus write the schematic formulas

$$\begin{bmatrix} & O & & O & \\ O & Cr & O & Cr & O \\ & O & & O & \end{bmatrix}^{2-} \qquad \begin{bmatrix} & O & & O & & O & \\ O & Cr & O & Cr & O & Cr & O \\ & O & & O & & O & \end{bmatrix}^{2-}$$

dichromate ion $Cr_2O_7^{2-}$ trichromate ion $Cr_3O_{10}^{2-}$

Probably the tetrachromate ion is a similar chain of four tetrahedra.

Dichromates of alkali and alkaline-earth metals crystallize from acidified chromate solutions. They are all easily soluble in water. If a dichromate solution is added to a solution containing other metal ions, generally the dichromate of the metal does not precipitate, but rather its chromate. The latter's solubility product is usually so small that it is exceeded by the amount of chromate ion allowed by the equilibria (31.1). However, precipitation can occur of some poorly soluble dichromates such as *silver dichromate* $Ag_2Cr_2O_7$ (dark red).

Water solutions of di- and polychromates are acid. For the dichromate this results from a process displacing the equilibria (31.1) to the left (cf. 12-1a).

From a very strongly acidified dichromate solution dark red, needle-shaped crystals of *chromium trioxide* CrO_3 precipitate, consisting of infinite, parallel-oriented chains of the same type as in the trichromate ion. Chromium trioxide may therefore be considered as probably being the end member in the polychromate series (cf. the series sulfate–polysulfate–sulfur trioxide, 21-11a). Chromium trioxide melts at 196°C without decomposition, but gives off oxygen at about 250°C: $2CrO_3(l) \rightarrow Cr_2O_3(s) + \frac{3}{2}O_2(g)$. It is easily soluble in water and deliquesces easily. A dilute solution contains primarily hydrogen chromate ion, a concentrated solution dichromate and probably polychromate ions. Large amounts of chromium trioxide are used for chromium plating (31-4a).

Chromium(VI) compounds are strong oxidizing agents and are reduced in the process generally to chromium(III) compounds. Especially strongly oxidizing are di- and polychromates (for the dichromate ion see tab. 16.1) and chromium trioxide. Their technical applications generally depend primarily on their oxidizing power, as, for example, in organic synthesis. Acid dichromate solution is used as an oxidizing agent in redox titration (16d). Strong sulfuric acid dichromate solution has long been employed for cleaning laboratory glassware. Many metals, when treated with chromate or dichromate solution, become less subject to corrosion. This probably results from a chromate layer formed on the surface, which helps to maintain a protective metal-oxide layer. Recently developed anticorrosion paints with pigments (*zinc yellow*) consisting of potassium zinc hydroxide chromate or zinc hydroxide chromate probably act in a similar manner. For land construction and marine superstructures these are now the most common anticorrosion paints and they have largely replaced red lead. The solubility of zinc yellow is high enough so that it cannot be used under water, and a covering paint coat is always necessary.

The oxidation of certain organic substances with dichromate ion only takes place with irradiation. If gelatine containing ammonium dichromate ("bichromated gelatine") is illuminated the Cr^{3+} ions formed make the gelatine insoluble (cf. chrome tanning, 31-5b). This fact is made use of in several ways in photographic reproduction and preparation of photoengravings.

Technically the most important chromium(VI) compounds are *sodium chromate* $NaCrO_4$ and *sodium dichromate* $Na_2Cr_2O_7 \cdot 2H_2O$. Concerning their production, see 31-4a. Because of its lower price and greater solubility, sodium dichromate has to a large extent replaced the formerly more common *potassium dichromate* $K_2Cr_2O_7$. These chromates and dichromates are much used as oxidizing agents and also in the manufacture of paint pigments.

When *ammonium dichromate* $(NH_4)_2Cr_2O_7$ is heated to about 200°C the ammonium ion is oxidized by the dichromate ion: $(NH_4)_2Cr_2O_7 \rightarrow N_2(g) + Cr_2O_3 + 4H_2O$. Here the evolution of heat is considerable so that once started the reaction continues throughout the whole mass of dichromate.

With strong hydrochloric acid dichromate ion gives *chromyl chloride* CrO_2Cl_2 according to: $Cr_2O_7^{2-} + 6H^+ + 4Cl^- \rightleftharpoons 2CrO_2Cl_2 + 3H_2O$. Chromyl chloride is easily hydrolyzed (reaction to the left) and therefore, to obtain a good yield, the water must be tied up. This can be done with conc. sulfuric acid, and the preparation is conveniently carried out by heating together anhydrous sodium chromate or dichromate, sodium chloride and conc. sulfuric acid. Chromyl chloride is distilled off and condensed as a dark red liquid (b.p. 116.7°C). In chromyl chloride the coordination is fourfold as in the chromate and dichromate ions.

Addition of 30% hydrogen peroxide solution to a dichromate solution with cooling gives a blue complex, which is assumed to be a *peroxochromate*(VI) *ion* $CrO(O_2)_2OH^-$, with the possible structure *a*.

$$\begin{bmatrix} O-O \\ \vee \\ O=Cr-OH \\ \wedge \\ O-O \end{bmatrix}^- \qquad \begin{bmatrix} O-O \\ O \vee O \\ | > Cr < | \\ O \wedge O \\ O-O \end{bmatrix}^{3-}$$

a. Peroxochromate(VI) ion $CrO(O_2)_2OH^-$

b. Peroxochromate(V) ion $Cr(O_2)_4{}^{3-}$

Peroxochromate(VI) ion decomposes easily. However, with many organic substances more stable addition compounds are formed. When the water solution is shaken with ether the ether phase is colored blue by such a compound (sensitive reaction for Cr(VI)).

If instead hydrogen peroxide is added to an alkaline chromate solution ($CrO_4{}^{2-}$ ion), red *peroxochromate*(V) *ion* $Cr(O_2)_4{}^{3-}$ is formed (structure *b* above), according to $2CrO_4{}^{2-} + 2OH^- + 7H_2O_2 \rightarrow 2Cr(O_2)_4{}^{3-} + 8H_2O$.

Chromium(VI) compounds are very poisonous, both because they precipitate protein and because of their strong oxidizing effect. Chromium trioxide is used in medicine to cauterize body tissues.

31-6. Molybdenum and tungsten compounds

a. Oxygen compounds of molybdenum(VI) and tungsten(VI). When the metals and many compounds (molybdenum disulfide, ammonium molybdates and tungstates, trioxide hydrates; see below), are heated in air they give the trioxides. *Molybdenum trioxide* MoO_3 is obtained as a white, *tungsten trioxide* WO_3 as a lemon-yellow, crystalline powder. WO_3 is best thought of as a framework molecule, built up by linking together WO_6 octahedra. MoO_3 has a different structure. Here MoO_4 tetrahedra are more easily recognized, linked together in infinite chains. Both are acid oxides and with basic melts or solutions give molybdates and tungstates, respectively (see below). They are poorly soluble in water.

The terms *molybdate* and *tungstate* generally refer to oxomolybdate(VI) and oxotungstate(VI), respectively. As mentioned earlier the mononuclear and tetrahedral ions $MoO_4{}^{2-}$ and $WO_4{}^{2-}$ exist. If the ratio O/M < 4, polynuclear *polyanions* appear, *polymolybdate* and *polytungstate ions*. A large number are known and they consist usually of MO_6 octahedra linked together by common corners or edges. Their geometric properties are therefore completely different from those of the chromates, in which the oxygen coordination is always tetrahedral. Such polyanions can be obtained from mononuclear molybdates or tungstates if melts of these salts have trioxide added to them, or if trioxide or a strong acid is added to

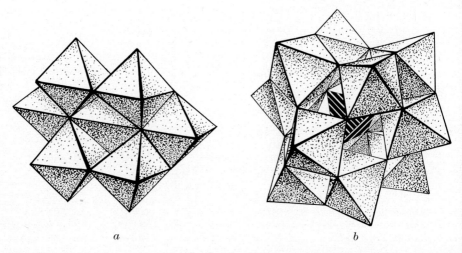

Fig. 31.1. Polyanions with the coordination polyhedra shown as solids. The MoO_6 octahedra are not quite regular. (a) The heptamolybdate ion $Mo_7O_{24}^{6-}$. (b) The dodecamolybdophosphate ion $PMo_{12}O_{40}^{3-}$. This ion consists of a central PO_4 tetrahedron (striped), surrounded by twelve MoO_6 octahedra. Each corner (oxygen atom) of the PO_4 tetrahedron forms at the same time corners of three MoO_6 octahedra.

a water solution. However, the composition and structures of ions formed from the melt and from solution are usually different (in the former case MO_4 tetrahedra also sometimes appear). When acid is added to a water solution of MoO_4^{2-} ions a *heptamolybdate ion* $Mo_7O_{24}^{6-}$ (formerly called paramolybdate) may be formed according to $7MoO_4^{2-} + 8H^+ \rightleftharpoons Mo_7O_{24}^{6-} + 4H_2O$. The structure of the heptamolybdate ion is shown in fig. 31.1a. Further increase in the acidity produces first several (not fully defined) intermediate products, and then a precipitate of molybdenum trioxide hydrate.

When many oxoanions, such as borate, silicate, phosphate and arsenate ions, are added to acid molybdate and tungstate solutions, anions are formed in which the central atom of the oxo anion surrounds itself with oxygen atoms belonging to MoO_6 or WO_6 octahedra respectively. In these ions there are thus several different kinds of central atoms and they are therefore called *heteropoly anions* (in contrast an ion such as $Mo_7O_{24}^{6-}$ is an *isopoly anion*). Fig. 31.1b shows the structure of the *dodecamolybdophosphate ion* $PMo_{12}O_{40}^{3-}$. Ammonium dodecamolybdophosphate $(NH_4)_3PMo_{12}O_{40}$ is poorly soluble in water and precipitates (lemon yellow) on addition of acid ammonium molybdate solution to a phosphate solution. This is used as a test for phosphoric acid as well as for its quantitative determination.

b. Oxygen compounds with lower oxidation number than VI. By reduction of the trioxides, most easily by heating in vacuum with appropriate amounts of the respective metal, there are obtained the dioxides MoO_2 and WO_2 (oxidation number IV; both give lustrous, violet-brown crystals) as well as several oxides with average oxidation number between V and VI (the oxides M_2O_5 are not believed to exist). The latter are strongly colored (dark blue or violet), probably because both M(V) and M(VI) are present simultaneously (7-3b). In these lower oxides also mainly MO_6 octahedra are found (occasionally MO_4 tetrahedra) to be the structural units. The structures of several (V,VI) oxides have been determined by X-ray diffraction, for which very complicated formulas such as Mo_8O_{23} and $W_{18}O_{49}$ have been established.

Partial reduction of molybdate(VI) or tungstate(VI) ions in solution gives intensely blue, colloidal products, *molybdenum blue* and *tungsten blue*. They are probably hydroxo complexes with metal atoms in the oxidation states V and VI. They are easily adsorbed by textile fibers, and molybdenum blue is used as a textile dye.

Among the substances with average oxidation numbers between V and VI we may mention the so-called *tungsten bronzes* whose composition may be expressed as M_xWO_3 (M = alkali metal, $x < 1$). The tungsten bronzes are semiconductors with high conductivity, metallic luster and splendid colors. They can be prepared by reducing a mixture of alkali tungstate and tungsten trioxide in the appropriate ratio by heating in a stream of hydrogen gas or with metallic tungsten. For M = Na and $0.93 \geqslant x \geqslant 0.32$ the structure is based on a cubic perovskite structure (21-5a) with the ideal composition NaW^VO_3. Even at the most sodium-rich limit $Na_{0.93}WO_3$ there are vacancies in the Na^+ lattice. Meanwhile electroneutrality is maintained by the presence of W(VI). The homogeneity range of the tungsten bronzes extends to $Na_{0.32}WO_3$ by an increase in the number of Na^+ vacancies simultaneously with an increase in the number of W(VI) atoms (see the general discussion of such compensation in 6-3a). At the same time the color changes from golden yellow through red and violet to dark blue. The tungsten bronzes are chemically extremely resistant and are used to some extent as paint pigments.

c. Other compounds. *Molybdenum disulfide* MoS_2 has a layer structure with MX_2 layers as in fig. 6.18a and with relatively weak forces between the layers (in 31-2a it was mentioned that as the mineral *molybdenite* it greatly resembles graphite). The layers therefore easily glide over each other, producing a greasy effect. MoS_2 is used as a lubricant, especially at high working temperature (up to 370°C). *Molybdenum disilicide* $MoSi_2$ even at high temperature is very resistant to atmospheric oxidation because of a protective layer of silicon dioxide formed on the surface. Since it has a suitable conductivity it is used as a resistance material for electrical heating. *Tungsten carbide* WC is the most common hard constituent of hard metals (14-3b).

CHAPTER 32

Group 7, The Manganese Group

32-1. Survey

Among the elements of this group, technetium has only unstable isotopes, although three of them (^{97}Tc, ^{98}Tc, ^{99}Tc) have long half-lives. The decay processes for these are

$$^{97}_{43}\text{Tc} \xrightarrow[2.6\times10^6\,\text{y}]{\text{EC}} {}^{97}_{42}\text{Mo}, \quad ^{98}_{43}\text{Tc} \xrightarrow[1.5\times10^6\,\text{y}]{\beta^-} {}^{98}_{44}\text{Ru} \quad \text{and} \quad ^{99}_{43}\text{Tc} \xrightarrow[2.1\times10^5\,\text{y}]{\beta^-} {}^{99}_{44}\text{Ru}.$$

Technetium is formed by spontaneous fission of uranium (9-2c) and is therefore found in very small amounts in uranium ores. Technetium has also been detected in certain red-giant stars. All technetium that has been studied chemically has been produced in nuclear reactors.

For manganese normally all the oxidation numbers II–VII exist. The most important are II, IV and VII, of which II is the most stable. For technetium and rhenium also probably all these oxidation numbers occur, although representatives of some of them have not yet been identified. Here the shift of stability toward higher oxidation numbers is clearly apparent (chap. 27), inasmuch as VII is the most stable. While Mn(VII) is strongly oxidizing, Tc(VII) is weakly and Re(VII) very weakly oxidizing. Sometimes the oxidation number −I has been assumed for rhenium, but several recent studies indicate that the compounds in question contain rhenium with a positive oxidation number, bound in a complex with hydride ions, H$^-$. All compounds of elements of the manganese group with oxidation number lower than VII are colored. Manganese(VII) compounds are strongly colored (cf. MnO$_4^-$); technetium(VII) and rhenium(VII) compounds as a rule are more weakly colored (the ions TcO$_4^-$ and ReO$_4^-$ are colorless).

Technetium and rhenium have quite similar properties and are clearly distinct from manganese (chap. 27). On the other hand, these two metals show many similarities to molybdenum and tungsten ("horizontal similarity"; chap. 27).

In compounds where the oxidation numbers are I–IV, octahedral coordination is the most common. The oxidation numbers V–VII are mainly represented by oxides and the oxo anions M^{V}O$_4^{3-}$, M^{VI}O$_4^{2-}$ and M^{VII}O$_4^-$. These oxo anions are tetrahedral.

Polyanions are not common, wherein the manganese group is distinctly different from the two preceeding groups.

The oxides have larger covalent-bond contribution and become more acid, the higher the oxidation number is. MnO has an ionic lattice (sodium-chloride type) and can be considered as a basic oxide, Mn_2O_3 has intermediate character, and Mn_2O_7 is an acid oxide with finite molecules (a volatile liquid at 0°C). Tc_2O_7 and Re_2O_7 also form finite molecules and are acid.

32-2. Occurrence and history

a. Occurrence. *Manganese* is rather abundant in the earth's crust (tab. 17.3). The most important ores are oxides and oxide hydroxides, especially *pyrolusite* MnO_2, but also *manganite* $MnO(OH)$ and *hausmannite* Mn_3O_4. Among other manganese minerals we may mention the carbonate *rhodochrosite* $MnCO_3$ and the oxide silicate *braunite* $3Mn_2O_3 \cdot MnSiO_3$. Manganese ores commonly occur with iron ores, and also manganese is often dissolved in iron minerals (for the most part substitution of Fe^{2+} by the nearly equal-sized ion Mn^{2+}). Great amounts of manganese are present in deep-sea sediments as concretions ("manganese nodules") containing high proportions of oxidic Mn(IV).

The largest deposits of manganese lie in the Soviet Union (Ukraine and the Caucasus), Africa (Union of South Africa, Gabon, Ghana, Morocco), Brazil and India.

Manganese in small amounts is believed to be essential to the formation of chlorophyll in green plants, although it is not incorporated in chlorophyll. Manganese is taken up by plants in the form of Mn(II). As is shown later (32-5a), Mn(II) is more easily oxidized the more basic the medium is. Therefore, in basic soils the plants show evidences of manganese deficiency (often gray or light spots on the leaves) in the presence of much higher amounts of manganese than in more acid soils. In treating basic soils with Mn(II) in order to eliminate manganese deficiency the oxidation must be prevented by acidification or by combining Mn(II) into difficultly oxidized complexes.

Manganese in small amounts enters into certain animal enzyme systems, but its role is still little understood. Manganese-deficiency disease is known to occur in domestic animals. Larger amounts of manganese produce toxic effects.

Concerning the occurrence of *technetium*, see 32-1.

The average abundance of *rhenium* in the earth's crust is very low. No specific rhenium minerals are known, and generally rhenium follows molybdenum in very small amounts. It is recovered mainly in connection with the production of molybdenum from molybdenite.

b. History. As mentioned in 26-2b in ancient times the name *magnesia* or *magnes* was given to several different minerals, among others pyrolusite (also called in particular magnesia nigra). The mineral was used as a decolorizing agent for glass (23-3i2), which is probably the reason that its name, by association with the Greek verb μαγγα-νίζειν = purify, later sometimes was modified to *manganese*. Torbern Bergman and Scheele showed that pyrolusite must constitute the oxide of a previously unknown metal, and in 1774 Gahn succeeded by reduction with carbon in obtaining this metal, although most probably with a very high proportion of carbon. Gahn called the new

Table 32.1. *Physical properties of the manganese group elements.*

	25Mn	43Tc	75Re
Density at 20°C, g cm^{-3}	7.50 (α Mn)	11.5	21.0
Melting point, °C	1244	2200	3180
Boiling point, °C	2041	∼4600	∼5600

metal magnesium, but after Mg was prepared the name was changed to *manganesium*, *manganium* or *manganese*.

Technetium was first detected in 1937 in molybdenum that had been bombarded with deuterons: $_{42}$Mo(d,n)$_{43}$Tc. The name was given in allusion to the fact that it could only be obtained artificially.

Rhenium was discovered in 1925 by Noddack and Ida Tacke in gadolinite (28-2a) and columbite (30b). The name refers to the Rhine River.

32-3. Properties of the group 7 elements

Physical properties of the group 7 elements are given in tab. 32.1. Next to tungsten, rhenium is the most difficultly melting of all metals. All three metals are light steel-gray. Manganese exists in four different crystal structures:

$$\alpha\,\text{Mn} \xrightleftharpoons{727°C} \beta\,\text{Mn} \xrightleftharpoons{1095°C} \gamma\,\text{Mn} \xrightleftharpoons{1133°C} \delta\,\text{Mn} \xrightleftharpoons{1244°C} \text{melt}$$

| cubic, 58 atoms per unit cell | cubic, 20 atoms per unit cell | face-centered cubic | body-centered cubic |

α and β Mn have complicated structures and are brittle (β Mn can be brought to room temperature by quenching). Technetium and rhenium crystallize in hexagonal-close-packed structures and are both malleable.

In compact form manganese is somewhat tarnished in moist air, but thereafter is attacked only slowly. In finely powdered form it is easily oxidized and generally shows great reactivity. It easily dissolves in acids, including oxidizing acids.

Technetium and rhenium dissolve easily in concentrated, oxidizing acids. On the other hand, neither technetium nor rhenium will dissolve in the absence of air in hydrogen halide acids, nor will rhenium dissolve in dilute oxidizing acids.

32-4. Production and uses of the group 7 elements

a. Production. *Manganese* can be produced by reduction of oxides aluminothermically (24-3c) or with silicon in an electric furnace. However, the products are quite impure. Pure manganese is made nowadays mainly by electrolysis of Mn(II) sulfate solution. Reduction of manganese oxides with carbon in a blast furnace gives products with a high carbon content, but this constitutes the most important metallurgical manganese preparation. As stated in 32-2a, manganese ore usually contains iron. One product of the reduction of rich manganese ore is *ferromanganese* with 20–80 wt % Mn, 6–8 wt % C, and the rest iron. Poorer manganese ores (manganiferous iron ore) give *spiegeleisen* (the fracture shows large, mirrorlike crystal blades of manganiferous cementite; see 34-3a), with 5–20 wt % Mn, 4–5 wt % C, and the rest iron.

Technetium is prepared nowadays from fission products from nuclear reactors, in which the element is represented mainly by ^{99}Tc. Separation is achieved by ion exchange (15-3) and solvent extraction (13-1f) and the end product then is ammonium pertechnetate NH_4TcO_4. From this the metal can be obtained by reduction with hydrogen.

Rhenium is produced primarily by processing the dust which deposits in the smoke flues on roasting molybdenite (31-4a), and which contains rhenium(VII) oxide Re_2O_7. The dust is leached with water forming perrhenic acid $HReO_4$, which is converted with ammonia to ammonium perrhenate NH_4ReO_4. From the latter the metal can be obtained by reduction with hydrogen.

b. Uses. *Manganese* is of great importance in the manufacture of steel. Manganese added to a steel melt combines partly with oxygen (deoxidation, 21-5b) and partly with sulfur. In addition manganese makes up one of the most important alloying elements in steel (34-3e). When the requirements for the carbon content of the steel allow the use of ferromanganese or spiegeleisen these are employed, but if low carbon content is needed, lower-carbon manganese preparations, for example electrolytic manganese, may be necessary. Electrolytic manganese is also used in the production of nonferrous manganese alloys.

Rhenium, because of its high melting point, has begun to be used in combination with tungsten in thermocouples for measurement of temperatures up to 2500°C. It is also used to some extent for construction details in electron tubes.

32-5. Manganese compounds

a. Manganese(II) compounds. Mn^{2+} occurs as a fairly typical ion. Its charge and radius (fig. 5.5) cause the salts in many respects (for example, solubility) to resemble corresponding Mg^{2+} and Fe^{2+} salts. Thus, the chloride, sulfate and nitrate are easily soluble, the fluoride, carbonate and phosphate poorly soluble. The hydrated Mn^{2+} ion gives a light red color to the water solutions and hydrates of its salts.

In relatively strong acid solution Mn(II) compounds are rather stable toward oxidation. However, they are oxidized by strong oxidizing agents, especially to Mn(IV) in the form of MnO_2 according to $Mn^{2+} + 2H_2O \rightleftharpoons MnO_2(s) + 4H^+ + 2e^-$; $e^0 = 1.23$ V. The equilibrium is thus shifted toward MnO_2 with rise in pH. In basic medium Mn(II) occurs as $Mn(OH)_2$ and the most important redox pairs are: $Mn(OH)_2 + OH^- \rightleftharpoons MnO(OH) + H_2O + e^-$, $e^0 = 0.1$ V; and $Mn(OH)_2 + 2OH^- \rightleftharpoons MnO_2(s) + 2H_2O + 2e^-$; $e^0 = -0.05$ V. The low values of the standard potentials in the two latter cases show that here oxidation occurs much more easily than in acid medium. Here also the equilibrium is shifted toward the oxidized form with rise in pH.

Manganese(II) *oxide* MnO is obtained as a gray-green powder by reduction with hydrogen of all other manganese oxides. MnO can be considered as a basic oxide and crystallizes with the sodium-chloride structure. Without losing its homogeneity the oxide can take up somewhat more oxygen than is implied by the formula MnO and should then be written $\sim$ MnO. Here vacancies occur in the manganese positions while at the same time some of the remaining manganese atoms attain higher oxidation numbers than II (cf. 6-3 a).

When alkali hydroxide is added to solutions of Mn(II) salts, white *manganese*(II) *hydroxide* $Mn(OH)_2$ precipitates. Its solubility product is too high (tab. 13.1) to make possible complete precipitation with ammonia solution, at least if the amount of ammonium ion present is great. The precipitation behavior of $Mn(OH)_2$ is thus similar to that of $Mg(OH)_2$ (26-5 a), but in the presence of the ammonia system probably the formation of amminemanganese(II) complexes contributes to the suppression of the precipitation of the hydroxide. $Mn(OH)_2$ is rapidly oxidized by atmospheric oxygen (see above) while turning brown in color.

Anhydrous *manganese*(II) *chloride* $MnCl_2$ can be obtained by combustion of manganese in chlorine, or by heating manganese or manganese(II) oxide in dry hydrogen chloride. The *tetrahydrate* $MnCl_2 \cdot 4H_2O$ crystallizes from a water solution at room temperature. The water solution can also be obtained by heating manganese oxides with hydrochloric acid, for example as in the preparation of chlorine from pyrolusite and hydrochloric acid (20-4 b). *Manganese*(II) *sulfate* $MnSO_4$ is obtained as a white mass on heating all manganese oxides with sulfuric acid. From water solutions of this material several rose-colored hydrates can be obtained, of which the metastable *tetrahydrate* $MnSO_4 \cdot 4H_2O$ constitutes the most common commercial variety and is one of the most important of manganese compounds. Quite large amounts of this product are used as manganese fertilizer (32-2 a).

The solubility product of *manganese*(II) *sulfide* MnS is so high (tab. 13.1) that it is precipitated completely by sulphide ion only from basic solution. When freshly precipitated it is light red, but on aging in the absence of air it goes over to a more stable, green form, which is also obtained by melting manganese with sulfur. In air it turns brown through oxidation.

Light red manganese(II) sulfide can be obtained in two forms, one with the sphalerite structure, the other with wurtzite structure (6-2 c). Thus, in both the coordination is tetrahedral. The green sulfide has the sodium-chloride structure and thus octahedral coordination. The bonding conditions are clearly entirely different, causing the difference in color.

b. Manganese(III) compounds. In water solution Mn(III) is reasonably stable only in complexes. Free Mn³⁺ ions are easily disproportionated (16d) according to $2Mn^{3+} + 2H_2O \rightleftharpoons Mn^{2+} + MnO_2(s) + 4H^+$. *Manganese*(III) *oxide* Mn_2O_3 is obtained as a brown-black powder when MnO_2 is heated in air below about 900°C. At higher temperature further oxygen is given off with the formation of *manganese*(II,III) *oxide* Mn_3O_4 ($Mn^{II}(Mn^{III})_2O_4$). The latter oxide constitutes the mineral *hausmannite* (brown-black). *Manganese*(III) *oxide hydroxide* MnO(OH) is well-defined, occurring as the mineral *manganite* (steel-gray to brown-black).

c. Manganese(IV) compounds. *Manganese*(IV) *oxide* (*manganese dioxide, pyrolusite*) MnO_2 is the most stable oxide. The composition is somewhat variable and the oxide should thus properly be written $\sim MnO_2$. The dioxide is black or gray-black. It is used as an oxidizing agent, especially in Leclanché cells (16g). On oxidation of Mn(II) salts or reduction of manganate(VI) or permanganate in basic solution, black-brown precipitates of manganese dioxide with varying water contents are obtained.

In addition we should mention the rather unstable *hexahalomanganates*(IV) with ions MnX_6^{2-}, especially MnF_6^{2-}.

d. Manganese(V) compounds. On oxidation of Mn(IV) compounds or reduction of Mn(VI) or Mn(VII) compounds, *manganates*(V), containing the unstable ion MnO_4^{3-}, may appear. If permanganate dissolved in ice-cold conc. sodium hydroxide solution is reduced with, for example, SO_3^{2-} ion, a blue solution is obtained, from which $Na_3MnO_4 \cdot 10H_2O$ crystallizes.

e. Manganese(VI) compounds. This oxidation state is represented chiefly by *manganates*(VI), often called *manganates*, and containing the tetrahedral ion MnO_4^{2-}. These are obtained by melting manganese dioxide with alkali hydroxide with oxidation either by atmospheric oxygen or an oxidizing agent such as alkali nitrate or chlorate: $MnO_2(s) + 2KOH(l) + \frac{1}{2}O_2(g) \rightarrow K_2MnO_4 + H_2O(g)$. The melt is colored green by the manganate ion. The alkali manganates dissolve easily with a green color in dilute alkali hydroxide solution. If the solution is neutralized or acidified, rapid disproportionation takes place into manganese dioxide and permanganate, the latter coloring the solution violet: $3MnO_4^{2-} + 4H^+ \rightleftharpoons MnO_2(s) + 2MnO_4^- + 2H_2O$.

f. Manganese(VII) compounds. The chief representatives of this group are *manganates*(VII), since early times called *permanganates*, and containing the tetrahedral ion MnO_4^-. Alkali permanganates were made formerly by oxidation of a manganate(VI) solution with chlorine or ozone. Nowadays anodic oxidation is used exclusively: $MnO_4^{2-} \rightarrow MnO_4^- + e^-$. The most important compound is *potassium permanganate* $KMnO_4$. This salt forms dark violet, iridescent crystals, which are rather easily soluble in water. The solution is violet with a strong color even in high dilution. In acid solution the permanganate ion is reduced according to $MnO_4^- + 8H^+ + 5e^- \rightleftharpoons Mn^{2+} + 4H_2O$; $e^0 = 1.51$ V. The high standard potential shows that the permanganate ion is a very powerful oxidizing agent. It can be seen from the equation, and it was shown in more detail in 16d, that the oxidizing power of the ion increases rapidly with increase in the acidity of the solution. In neutral or weakly basic solutions the reduction takes place according to $MnO_4^- + 2H_2O + 3e^- \rightleftharpoons MnO_2(s) + 4OH^-$; $e^0 = 0.588$ V. In strongly basic solution the permanganate ion is reduced according to $MnO_4^- + e^- \rightleftharpoons MnO_4^{2-}$; $e^0 = 0.564$ V. Here hydroxide ions are oxidized according to $4OH^- \rightleftharpoons O_2(g) + 2H_2O + 4e^-$; $e^0 = 0.401$ V. The overall reaction thus is $4MnO_4^- + 4OH^- \rightleftharpoons 4MnO_4^{2-} + O_2(g) + 2H_2O$. Potassium permanganate is much used as an oxidizing agent. This is especially true in organic syntheses, where, however, it has recently begun to be replaced by more specific oxidizing agents. Its oxidizing action is utilized also for bleaching and disinfection. It is used as an oxidizing agent in redox titrations. These are usually done in strongly acid solutions. Since the Mn^{2+} ion then formed is very weakly colored and the MnO_4^- ion very strongly colored, the endpoint is apparent without the use of an indicator.

The strong acid *permanganic acid* $HMnO_4$ has not been isolated, but *manganese*(VII) *oxide* (*dimanganese heptoxide*) Mn_2O_7 can be obtained free by reaction between potassium permanganate and cold ($< -5°C$) conc. sulfuric acid. The oxide forms a volatile, oily, dark green-brown liquid, which easily explodes.

32-6. Technetium and rhenium compounds

Here the reader should refer to the survey in 32-1. The most important oxidation state for both technetium and rhenium is VII. The oxides Tc_2O_7 (m.p. 120°C, b.p. 311°C) and Re_2O_7 (m.p. 304°C, b.p. 350°C) are unexpectedly stable and are the final products obtained on heating the metals in air. Both give acid water solutions. From a solution of Tc_2O_7 dark red crystals of the strong acid *pertechnetic acid* $HTcO_4$ crystallize. Solid perrhenic acid $HReO_4$ has been prepared by adding a limited amount of water to Re_2O_7. With bases *pertechnetates* and *perrhenates* are obtained, containing the ions TcO_4^- and ReO_4^-, respectively. Both these ions are colorless, but TcO_4^- has strong ultraviolet absorption very close to the visible spectral region. The color of $HTcO_4$ results from an alteration of the TcO_4^- group sufficient to shift the absorption into the visible region.

CHAPTER 33

Survey of Groups 8, 9 and 10

As mentioned in chap. 27 the similarities between neighboring elements in the same period ("horizontal similarities") are great for groups 8, 9 and 10, and these three columns of elements were, therefore, placed by Mendeleev in one group. In chap. 27 it was also stated that the great "vertical similarities" between the elements in periods 5 and 6, a consequence of the lanthanoid contraction, led to a division into *iron metals* and *platinum metals*. The latter are also often divided into "light" (period 5: Ru, Rh, Pd) and "heavy" (period 6: Os, Ir, Pt) platinum metals. However, notable horizontal similarities extend beyond the limits of groups 8, 9 and 10. Thus, the elements of group 8, Fe, Ru, Os, show many similarities to the elements of the previous column, Mn, Tc, Re in group 7. In the same way many similarities are found between the elements of group 10, Ni, Pd, Pt, and respectively Cu, Ag, Au in the subsequent group 11.

As usual both the highest oxidation number, and the most stable oxidation number relative to other numbers, increase as one goes downward in each group. On the other hand, as one goes to the right in these three groups, in each period a corresponding decrease in these numbers occurs. This is shown by fig. 5.8, but the following array, in which the known oxidation numbers are differently indicated according to their stability relative to the other oxidation numbers of the element, shows this more clearly. The low oxidation numbers $(0, -I, -II)$, which appear in bonds with certain ligands (for example, CO and CN^-), and which in 5-6c were stated to have essentially only formal significance, have not been indicated.

	Group 8								Group 9						Group 10					
	I	II	III	IV	V	VI	VII	VIII	I	II	III	IV	V	VI	I	II	III	IV	V	VI
Fe	(+)	+	+	(+)		+														
Ru	(+)	+	+	+	(+)	+	+	+												
Os	(+)	(+)	+	+	(+)	+	(+)	+												
Co									(+)	+	+	+								
Rh									(+)	(+)	+	(+)	+	(+)						
Ir									(+)	(+)	+	+	(+)	(+)						
Ni															(+)	+	(+)	+		
Pd															(+)	+	(+)	+		(+)
Pt																+	(+)	+		(+)

Of the oxidation numbers in this table, V has little importance, and the numbers VII and VIII occur only for ruthenium and osmium. For the most part, the highest numbers are attained only in bonding with strongly electronegative ligands (5-3g, 5-6c): $Fe^{VI}O_4^{2-}$, $Ru^{VII}O_4^-$, $Os^{VIII}O_4$, $Ru^{VI}F_6$, $Ir^{VI}F_6$.

The nobility of the metals increases sharply in going in one group from the iron metal to the light platinum metal (period 5). In going in the same group to the heavy platinum metal (period 6), a considerably smaller increase is found. Within each period the nobility increases to the right. The platinum metals therefore form a group with high nobility, increasing to the right. As a consequence they are chemically very resistant, especially against acids. Through passivation, however, the iron metals can often become quite resistant in oxidizing environments.

The affinity for oxygen decreases to the right in each period and downward in each group; the affinity for sulfur, on the other hand, increases for the most part in these directions. Here we notice a transition to group 11 where the tendency to combine with sulfur is particularly great in relation to the tendency to combine with oxygen.

Relatively stable *oxo anions*, analogous to those formed in the previous groups (for example, CrO_4^{2-}, MnO_4^{2-}, MnO_4^-) are formed only by the platinum metals in group 8 ($Ru^{VI}O_4^{2-}$, $Ru^{VII}O_4^-$, $Os^{VI}O_4^{2-}$). On the other hand the ion $Fe^{VI}O_4^{2-}$ is very unstable (strongly oxidizing).

Oxides with the oxidation numbers II and III are basic or intermediate. With increasing oxidation number the oxides generally take on more acid character. However, the remarkable, volatile tetroxides RuO_4 and OsO_4 show only weak acid properties.

The *ability to form complexes* is great throughout all three groups. For most ligands it increases on going to the right in a period. Within each group it is less for the iron metal than for the platinum metals, and the former therefore occur more frequently as "free" ions than the latter. This leads to quite large similarities between "free" ions of the same type among the iron metals, but when complex compounds are compared, great vertical similarities are found. For example, Fe, Ru and Os form stable hexacyanometallate(II) ions $M^{II}(CN)_6^{4-}$, and still greater similarity is found between complexes of Co, Rh and Ir, especially in the oxidation state III (for example, ammine complexes such as $M^{III}(NH_3)_6^{3+}$, hexacyanometallate(III) ions $M^{III}(CN)_6^{3-}$ and hexanitrometallate(III) ions $M^{III}(NO_2)_6^{3-}$). Octahedral coordination is the most common. Square coordination (5-3i, j) occurs especially with Ni(II), Pd(II) and Pt(II), all with eight d electrons. Tetrahedral coordination is found in the previously mentioned oxo anions and the tetroxides in group 8, but also, for example, in $Co^{II}Cl_4^{2-}$ and $Ni(CO)_4$. *Ammine complexes* are very numerous, but especially typical for these groups are complexes with ligands CO, NO^+ (*carbonyl* and *nitrosyl complexes*, 23-2f, 22-9b) and CN^- (*cyano complexes*, 23-2k1).

The elements of groups 8, 9 and 10 are often good *contact catalysts*. In processes involving hydrogen this may result from the fact that all the metals to a greater or lesser degree dissolve hydrogen. The hydrogen occurs in the metallic phase in the atomic state and is therefore very reactive when it is released.

CHAPTER 34

The Iron Metals Iron, Cobalt and Nickel

34-1. Occurrence and history

a. Occurrence. As was shown in 17b *iron* is one of the most common elements in the body of the earth as a whole. In the accessible parts of the earth (tab. 17.2) the proportion of iron is also high. Iron occurs combined in a wide variety of minerals but in the earth's crust is rarely found in native form. Iron meteorites consist mainly of iron but usually they have a quite high nickel content (17b). The most important ore minerals are oxides, but carbonates and sulfides are also important. The oxides are *hematite* Fe_2O_3, and *magnetite* Fe_3O_4. *Limonite* may be considered to be an amorphous iron(III) oxide hydroxide $FeO(OH)$ with variable water content. The crystalline minerals *goethite* α $FeO(OH)$ and *lepidocrocite* γ $FeO(OH)$ have the same composition but different crystal structures. *Bog ore* consists mainly of limonite. *Siderite* $FeCO_3$ makes up important deposits in some places. *Pyrite* FeS_2 (21-2a, 21-8) is used mainly as a sulfur ore, but the roasting residues formed in the recovery of sulfur (21-4b) are used among other things to produce iron.

Minable iron ore deposits are widespread over the earth. The Soviet Union since the end of the 1950's has been the largest producer of iron ore, with the heaviest output in the Ukraine, Kursk between the Ukraine and Moscow, and the Urals. In 1966 the United States was the next largest producer, mostly from Lake Superior (Masabi Range), followed by Canada (Lake Superior and Labrador), China (Manchuria), France (limonite with high phosphorus content, so-called minette ore, in Lothringen), Sweden and India. The ore body in Kiruna in northern Sweden is one of the world's largest.

Iron is an indispensible element in living organisms. In these it functions as an electron-transfer intermediary (cf. the importance of the transition elements as homogeneous catalysts, 8h) in respiratory processes. The cytochromes, which act as respiratory enzymes in both plants and animals, thus contain iron. It is also incorporated in the respiratory pigments hemoglobin and myoglobin (the latter provides an oxygen reservoir in the muscle tissues) in vertebrate animals. An iron reserve is held by the high-iron-containing protein ferritin, which is stored in the spleen and intestinal walls. Iron is involved in the formation of chlorophyll in plants, although it is not incorporated in the chlorophyll itself.

Cobalt and *nickel* have rather low abundance in the earth's crust (tab. 17.3); nickel is about three times more common than cobalt. The proportion of nickel in the earth's core is probably high (17b). The nearly equal atomic radii of iron, cobalt and nickel causes them easily to replace each other in solid solution and therefore they often follow each other in solid phases. For the same reason cobalt and nickel also follow copper. The most important cobalt and nickel minerals are sulfides and arsenides. Some of these belong to the nickel-arsenide type (6-2c), for example *niccolite* NiAs. Others belong to the pyrite type (21-2a, 21-8), in which about half of the sulfur is replaced by arsenic or antimony, and in which iron, cobalt and nickel may replace each other. The most important of these minerals is *cobaltite* (Co,Fe)AsS. Several arsenide minerals have the same structure as *skutterudite* $CoAs_3$, in which the cobalt may be partly replaced by nickel and partly also by iron, and the arsenic content may go as low as about 2As per metal atom. The most important of these is *smaltite*, often written as $CoAs_2$. Relatively pure sulfides are *millerite* NiS (NiS is polymorphous and millerite corresponds to the modification called γ NiS), *pentlandite* $(Ni,Fe)_9S_8$ (predominantly nickel) and *linnaeite* Co_3S_4 (spinel type, 21-5a). Nickel is also obtained from *garnierite*, which may be considered as the silicate chrysotile (23-3h4) with magnesium partly replaced by nickel.

Over half of the world's production of *cobalt* comes from ore from Katanga. Here is mined chiefly hydrated cobalt-copper oxides of rather indefinite composition, which are formed secondarily from sulfides. Other cobalt mines are operating especially in Northern Rhodesia, Morocco, Canada (Ontario, probably the world's largest deposits; arsenides and arsenide sulfides, sometimes together with silver minerals), and the United States. The most important *nickel* mines are situated in Canada (Ontario, close to the cobalt deposits; mostly pentlandite). Lesser deposits are also worked in Cuba and in the Soviet Union (Petjenga, Kola Peninsula).

Cobalt has a very important biological function as a constituent of vitamin B_{12}. This may be formed by certain bacteria and fungi that, among other things, may be active in the alimentary canal of animals. Of course, the food must then contain cobalt. The production of B_{12} is particularly prominent in the ruminants, which seem to have especially great requirements for it. In most other cases, for example in humans, the production of B_{12} is insignificant, and here the food must contain the extremely small amounts of B_{12} required to sustain life. Pernicious anemia is caused by a deficiency of vitamin B_{12}. A deficiency of cobalt in the soil or in domestic animal feed is quite rare and mostly affects ruminants.

b. History. *Iron* has been known and used since prehistoric times. In Greek it was called σίδηρος and in Latin, *ferrum*.

In the Middle Ages German miners gave to certain minerals, which in spite of their promising appearance did not yield good metal, the deprecatory name "Kobalt", probably in reference to the evil mining spirits "Kobolde". A number of these minerals proved to be useful for coloring glass and glazes, and these in particular were later

Table 34.1. *Physical properties of the iron metals.*

	26Fe	27Co	28Ni
Density at 20°C, g cm^{-3}	7.87 (α Fe)	8.89	8.90
Melting point, °C	1539	1495	1455
Boiling point, °C	2890	2880	2840
Curie point, °C	768	1150	353

named "Kobalt". From one such mineral Brandt in Stockholm at least before 1738 prepared what he designated as a new semimetal (cf. arsenic, 22-2b) and called it *kobalt*.

A mineral was also known (niccolite NiAs) that, because of its coppery red color, was thought to be a copper ore, but which yielded no copper. This was called "Kupfernickel", which may be said to signify "false copper". By reduction of kupfernickel, Cronstedt in 1751 obtained a new "semimetal", which he somewhat later called *nickel* after the mineral name.

34-2. Properties of the iron metals

Some important physical properties are collected in tab. 34.1. The pure metals are malleable and have rather low hardness. They are almost silver-white, with a slight tinge of yellow in cobalt and nickel. All three metals are ferromagnetic up to the Curie points (6-3b, 7-4a).

For nickel there is only one, face-centered cubic (=cubic-close-packed, 6-1f) structure, while iron and cobalt are polymorphous. The iron structures are:

$$\alpha \text{ Fe} \underset{}{\overset{906°C}{\rightleftarrows}} \gamma \text{ Fe} \underset{}{\overset{1401°C}{\rightleftarrows}} \delta \text{ Fe} \underset{}{\overset{1539°C}{\rightleftarrows}} \text{melt}$$

body-centered cubic / face-centered cubic / body-centered cubic

Formerly it was believed that the Curie point for iron at 768°C involved a phase transition and the paramagnetic phase between 768 and 906°C was then called β Fe. Notice that α and δ Fe have the same structure.

It has recently been found that a new, nonferromagnetic, hexagonal-close-packed iron phase is stable at pressures above 130 kbar.

The cobalt structures are:

$$\alpha \text{ Co} \underset{}{\overset{417°C}{\rightleftarrows}} \beta \text{ Co} \underset{}{\overset{1495°C}{\rightleftarrows}} \text{melt}$$

hexagonal close packed / cubic close packed = face-centered cubic

The two close-packed structures have very similar energies, which causes the transition between them to go slowly as long as the temperature is not high enough to make the atomic motions large. Even at the transition temperature the atomic mobility is too small for rapid transition and on cooling β Co the metal does not go completely over to α Co. The structure is always close-packed, but the layer sequence (6-1f) is not $ABABAB$ ... throughout as in hexagonal close packing, but here and there "mistakes" appear, which may be said to be remains of the cubic sequence $ABCABC$ ...

It was stated in chap. 33 that the nobility of the iron metals increases with increasing atomic number. In compact form and dry air iron is oxidized only at about 200°C, and the oxidation becomes apparent in *tarnish colors* caused by light interferences in the thin oxide layers. In moist air, and more rapidly in oxygen-containing water, *rusting* takes place. For rusting in moist air an adsorbed water layer is necessary, and this is formed especially easily on impurities in the iron surface. When the relative humidity is less than about 60% iron does not rust. If the surface is not completely clean, rusting takes place when the humidity exceeds 60%, but an absolutely clean and bright surface may remain unchanged at considerably higher humidity. Rusting may be considered to occur in two steps. First, the water oxidizes the iron to Fe(II): $Fe(s) + 2H_2O \rightarrow Fe(OH)_2(s) + H_2(g)$. For absolutely pure iron this reaction is strongly inhibited by the hydrogen overvoltage (16e), but this effect decreases in the presence of foreign substances in the metal surface. The reaction also is favored by an increase in the acidity, which can be seen, among other things, in the strong attack in air containing combustion gases (CO_2 and especially SO_2). In the second step oxygen oxidizes the iron further to Fe(III): $2Fe(OH)_2(s) + \frac{1}{2}O_2 \rightarrow 2FeO(OH)(s) + H_2O$. *Rust* may be considered as consisting of iron(III) oxide hydroxide $FeO(OH)$ with variable water content.

Iron is generally attacked by acid and alkaline solutions. In dilute acids first Fe(II) is formed with the evolution of hydrogen, but at higher redox potential oxidation to Fe(III) takes place. In strongly oxidizing media (for example, in conc. nitric acid or with anodic oxidation) passivation (31-3) sets in, which, however, can be easily broken down, for example by reduction, scratching, striking or magnetic action.

Compact cobalt and nickel are hardly attacked by water or oxygen gas at ordinary temperature. Cobalt is attacked by acids and alkalis, although less easily than iron. Nickel is attacked still less and is especially resistant to alkalis. Strongly basic substances, for example alkali hydroxides, can be fused in nickel vessels. Both cobalt and nickel can be passivated.

In very finely divided form, iron, cobalt and nickel are all pyrophoric (14-3a).

34-3. Metallurgy and metallography of iron and iron alloys

The extreme importance of iron and iron alloys makes it worthwhile to treat their metallurgy in considerable detail.

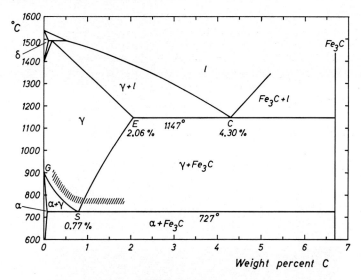

Fig. 34.1. Phase diagram for the system iron–carbon.

α = α Fe and interstitial solid solutions of C therein, *ferrite* (max. 0.02 wt% C)
γ = γ Fe and interstitial solid solutions of C therein, *austenite* (max. 2.06 wt% C)
δ = δ Fe and interstitial solid solutions of C therein (max. 0.10 wt% C)
Fe_3C = the iron carbide *cementite* (cf. 23-2 c)
l = melt.

Pure iron is prepared on a laboratory scale by reduction of pure iron oxides with hydrogen, but electrolysis is technically more important, usually of iron(II) chloride in water solution (*electrolytic iron*). Thermal decomposition of pentacarbonyl iron $Fe(CO)_5$ (23-2f) is also used (the product is called *carbonyl iron*). However, the major production of *technical iron* is done by reduction of ore with carbon monoxide. The products then contain carbon, and both because of this and because of its special features the system iron–carbon is fundamental for the understanding of the metallurgy of iron and for important properties of many iron alloys in practical use.

a. The system iron–carbon. The phase diagram for proportions of carbon that are of practical significance is shown in fig. 34.1. Of the phases shown cementite is not stable with respect to decomposition into iron and graphite. Often this decomposition takes place at an undetectable, slow rate (this is always so at room temperature), and cementite can thus be designated as metastable. Fig. 34.1 therefore shows the *metastable system* Fe–C. The *stable system* Fe–C contains *graphite* instead of cementite and along with this substitution the lines of the phase diagram that determine the metastable equilibria in which cementite takes part are shifted. However, the displacements are so small that fig. 34.1 may be considered as a diagram also for the stable system with graphite merely replacing cementite.

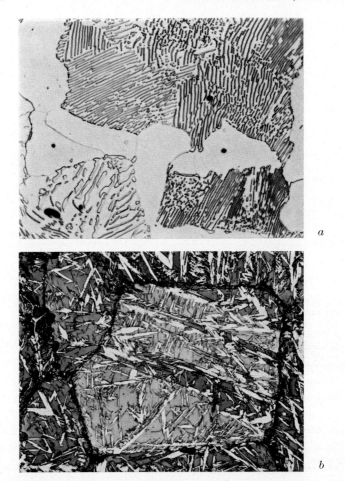

Fig. 34.2. Photomicrographs of carbon steel that has been flat-ground, polished and etched. (a) 0.5 wt% C, slowly cooled from 1100°C. Magnification 1300 ×. Ferrite (white) and pearlite, consisting of lamellae of ferrite (white) and cementite (gray). (b) 1.4 wt% C, quenched from 1100°C. Magnification 180 ×. White martensite lamellae in still-remaining austenite (gray). The original austenite grain boundaries are visible. (S. Modin.)

In the upper left part of the diagram there is a peritectic transformation (13-5f2) in which the δ phase takes part. When a melt solidifies and passes this region, the high temperature causes the equilibria to be reached very rapidly. Any iron-carbon alloy with less than 2 % carbon that comes down into the austenite region therefore consists of austenite, and the preceding transitions can be disregarded. With sufficiently rapid quenching of austenite (not so rapidly that the austenite is supercooled—see below—and not so slowly that the cementite is decomposed with the formation of graphite), depending on the carbon content, either α phase (ferrite) separates along *GS*, or cementite along *SE*. When the eutectoidal point S is reached, the remaining austenite is converted to a eutectoid consisting of alternating lamellae of ferrite and cementite, called *pearlite* (pearlite is thus not the name of one phase but a mixture of two phases). At high enough magnification the pearlite is strongly reminiscent of a eutectic (fig. 34.2a).

By quenching austenite (for example, by immersion in water) the eutectoid temperature can be passed through with no transition taking place. By further rapid cooling the austenite can remain well below the eutectoid temperature, but at rather low temperature the major portion of it goes over to a metastable phase, *martensite*. The temperature of formation of martensite depends on the carbon content of the martensite, and for eutectoidal austenite (0.77 % C) is approximately 200°C. The martensite still contains all of the carbon in solid solution (it thus has the same carbon content as the parent austenite), but the iron lattice, in whose interstices the carbon atoms lie, has changed to a body-centered tetragonal lattice, which can be said to be a deformed α-iron lattice. Martensite may be thought of as a supersaturated and therefore metastable ferrite, whose α-iron lattice is distorted because of the excess carbon. The transition austenite→martensite takes place very rapidly, inasmuch as it does not require any material transport. In fig. 34.2b the martensite forms plates (in a section most of them look like needles) embedded in untransformed austenite. With somewhat slower cooling or reheating (as low as about 100°C) the martensite decomposes through complicated intermediate stages into cementite and stable, low-carbon ferrite.

Martensite is hard and brittle, and the hardness increases with the carbon content up to about 1 % C, after which it remains roughly constant. The conversion of a suitable iron-carbon alloy (steel, see below) to austenite by heating, followed by conversion of soft austenite to hard martensite by quenching, is the oldest known *hardening* process. However, a steel that is merely hardened is generally too hard and brittle for most purposes. The steel is therefore reheated (*annealed, tempered*) to decompose the martensite to a certain extent, so that the steel becomes somewhat softer and tougher. By annealing, strains that arose in quenching are also evened out. Since early times the annealing temperature was monitored by observing the tarnish colors (34-2) of the surface. A very tough but less hard material is obtained by annealing at high temperature, 400–650°C.

Rather effective hardening generally occurs after quenching from the shaded region in fig. 34.1, which extends from approximately 0.2 % to 2 % C. The position of this region is dependent mainly on the conditions that there must be a sufficient amount of austenite to start with, and that the steel must not be heated to too high a temperature before quenching since the austenite grains will then be able to grow too large (14-2d, 14-3a), making the mechanical properties poorer. However, with the lowest carbon contents the hardness after hardening is not so great (among other things, because of the lower hardness of low-carbon martensite). All iron-carbon alloys with less than 2 % C are now called *steel* (if otherwise unalloyed, *carbon steel*). Depending on whether the carbon content is less than, equal to or greater than the eutectoid carbon content (0.77 % C), the steel is said to be *hypoeutectoid, eutectoid* or *hypereutectoid*, respectively. We speak also of *mild (soft)* = low-carbon and *hard* = high-carbon steels. Even without hardening the hardness and strength increase with increasing carbon content. Steel is ductile and can be rolled. The especially easily forged and rolled, but not hardenable, steel with less than 0.2 % C constitutes "commercial iron", making up the largest proportion of steel production.

It is apparent from the above discussion that the properties of carbon steel depend to a high degree on its *heat treatment*, that is, such procedures as holding at certain temperatures, cooling, often at particular cooling rates, and reheating.

Besides iron and carbon, carbon steel always contains other substances, mainly silicon and manganese. The proportions of these are generally too small to alter significantly the iron–carbon diagram. For steel with other elements purposely added (alloy steel, 34-3e), on the other hand, the phase diagram may have a notably different character. For alloy steels also, the heat treatment is very important.

Iron-carbon alloys with over 2 % C are called *cast iron*, because they have such low melting points (fig. 34.1) that by older methods they could be melted and poured. In *white cast iron* the carbon is present combined in cementite, and in *gray cast iron* it is present at least partly as graphite. The presence of silicon favors the formation of graphite nuclei and with amounts over 1.5 % Si generally the iron may be expected to solidify as gray cast iron. Cast iron with lower silicon content, on the other hand, solidifies in practice as white cast iron, provided the cooling is not so slow that the cementite is allowed to decompose. Manganese stabilizes the cementite and thus promotes the formation of white cast iron (cf. spiegeleisen, 32-4a). The high proportion of graphite or cementite in cast iron makes it brittle. Both this and the fact that cast iron as low as 1147°C contains liquid phase, mean that it cannot be forged. It is clear from the previous discussion that it cannot be hardened.

Both in the metastable (with cementite, white cast iron) and the stable (with graphite, gray cast iron) system a eutectic C appears. In the first case this consists primarily of cementite and austenite (which later decomposes as mentioned earlier) and is called *ledeburite*. According to whether the carbon content is less than, equal to or greater than the carbon content of the eutectic, the cast iron is called *hypoeutectic, eutectic* or *hypereutectic*, respectively. For casting mostly hypoeutectic gray cast iron (containing silicon) is used. White cast iron is more brittle than gray but has greater resistance to wear and tear (cementite is very hard), so that it is preferred for certain purposes. However, the product is more difficult to machine.

b. Reduction of ore. By far the most important method for reducing ore is the *blast-furnace process*, which is carried out in an almost cylindrical furnace (fig. 34.3), usually lined with chamotte brick (23-3i3). The blast furnace operates on the counter-current principle. It is charged from above with ore, reducing agent in the form of coke, and slag-forming substances, usually limestone. The amount

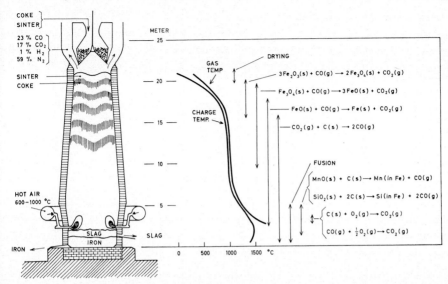

Fig. 34.3. Schematic vertical section of a blast furnace. The chart at the right shows the temperature distribution and the chief processes that take place in different zones in the furnace.

of the last varies according to the acid-base character (21-5a) of the gangue of the ore, which often contains an excess of SiO_2-containing minerals. Nowadays an increasingly large proportion of the ore supplied consists of finely divided concentrates and, therefore, it is more and more usual that these concentrates are sintered together with the required amount of limestone and powdered coke. The blast furnace charge thus consists of this type of sinter together with additional coke. When hot air is blown through the blast furnace from below, carbon monoxide is formed somewhat as in a gas producer (23-2f), and this reduces the ore (21-5b). The iron formed melts and dissolves carbon from the coke to the extent of about 3 to 4 %. A small portion of the *pig iron* produced is used after remelting for casting, but the major portion goes to the production of steel, mainly in liquid form (34-3c).

The increased use of sinter has had great significance in the economy of the blast-furnace process, since it makes possible a faster reaction rate and thus decreased loss of coke and increased production. More recently coke has begun to be replaced by other fuels such as oil, natural gas or powdered coal, which are blown in from below together with the air blast. Mixtures of hydrogen and carbon monoxide as well as pure hydrogen have also been tried. In *electric pig-iron furnaces* the heating is done electrically (the current is carried partly by the charge and partly by arcing through the interstices of the charge), and the function of the carbon is confined to reduction. The method presupposes a low cost of power in relation to the price of coal. Modern blast furnaces are sometimes built as very large units. The largest have diameters of about 10 meters and a production of 3000 tons per day.

Besides carbon, pig iron contains other substances. Part of the silicon and manganese of the ore goes into the iron during the process. These substances, which are always contained in small amounts in nearly all types of steel, and which besides are oxidized ahead of the iron in the steel-making processes (34-3c), are harmless. The amount of phosphorus in the pig iron depends on the amount contained in the ore. Usually a low phosphorus content in pig iron is sought for, but if only a phosphorus-rich ore is available (French minette ores, Swedish apatite iron ores, 34-1a), necessity has been turned to advantage by devising steel-making processes that use the phosphorus as fuel (Thomas and Kaldo processes, 34-3c). Sulfur enters into the pig iron as impurity from the coke. Both phosphorus and sulfur are generally detrimental to steel products that are subsequently produced from the pig iron. A phosphorus content of over $\sim 0.05\%$ leads to poor tensile strength at ordinary temperature ("cold brittleness"), as a result mainly of the great inhomogeneity of the material. Sulfur contents above $\sim 0.04\%$ cause brittleness at red heat and thus under hot working ("red shortness"). This brittleness results from precipitation in the grain boundaries (14-2c) of iron sulfide FeS with low strength.

c. Steel melting. In order to prepare from pig iron a material that can be hot worked (14-3c) and hardened, the pig iron must be transformed into *steel*, in which process primarily the carbon content of the pig iron is reduced. The process is an oxidizing melting in which the carbon, silicon, manganese and phosphorus contained in the charge material are for the most part oxidized and either go off as a gas or are carried over to the slag. Minor amounts of iron are, of course, oxidized at the same time, and the iron oxides also go over to the slag. The oxidizing agent is air, oxygen gas or oxygen contained in scrap iron or iron ore. Toward the end of the process, when the desired carbon content in the melt is reached and at the same time its oxygen content has increased, a deoxidation of the melt is carried out (21-5b) by adding to it alloys of Si, Mn or Al. In this way the oxygen content is brought down to a low value, and the amount of slag inclusions in the solidified steel can be minimized. In the manufacture of alloy steel (34-3e) the alloy substances are usually added at the end of the process. The process is terminated by pouring the melt into cast iron molds (ingot molds), forming ingots (*ingot steel*). The ingot is then worked by rolling or forging. Smaller amounts of molten steel are poured directly into sand molds to give finished products, often of complex shapes (cast steel products). In former times the pig iron was oxidized at temperatures below the melting point of steel. These so-called wrought-iron processes are now entirely obsolete.

The steel-melting processes are classified as *acid* and *basic*, according to the acid-base character (21-5a) of the blast furnace lining. In the acid processes quartz is used for the lining, and in the basic processes burnt dolomite. The processes are carried out so that the slag has the same acid-base character as the furnace lining, in order that the slag will not react too much with the latter. In the basic processes limestone or burnt lime are added as slag former. Since high-SiO_2-containing slags cannot take up either phosphorus or sulfur, the acid processes require a charge that is nearly free of these substances. These processes, which historically are the earliest, have therefore declined in importance. The basic methods permit greater latitude in the quality of the charge, since especially the phosphorus but also the sulfur can be carried over to the slag. The phosphorus, which is oxidized to P_2O_5, with the basic slags give calcium phosphate silicate. If the charge is so rich in phosphorus that the P_2O_5 content of the slag reaches about 15%, the latter is used under the name *Thomas phosphate* (first obtained in the Thomas process mentioned below), or *basic slag*, as an important phosphate fertilizer (22-12 e4).

The most important of the steel-producing methods is the *Siemens-Martin* or *open-hearth process*, which is carried out in a trough-shaped reverberatory furnace, usually having a basic lining. The process provides great latitude in the proportions of scrap iron and pig iron in the charge; the latter can be introduced advantageously in liquid form. Oil, preheated producer gas (23-2f) or coke-oven gas (18d) is burned over the surface of the melt with hot air, which is sometimes charged with oxygen. Oxygen for the oxidation is supplied by the flame, by the oxides in the iron ore and scrap iron, and nowadays to a considerable extent by blowing pure oxygen into the bath (21-4a).

In the *electrosteel processes* the heating is done electrically, either in an arc furnace or an induction furnace. The charge consists of scrap iron, pig iron and iron ore. The melting down and oxidation take place in the arc furnace in the same way as in the open-hearth furnace, but with iron ore or oxygen as oxidizing agent. After the oxidizing slag is removed, the process can be terminated by treating the steel bath under a reducing slag, so that the amounts of oxygen and sulfur dissolved in the steel are reduced to low values. Such treatment is important in the manufacture of quality steel, especially alloy steels with high requirements for freedom from impurities and slag inclusions. For this reason, the major portion of the world production of quality steel is carried out in arc furnaces.

The *pneumatic steel processes* consist of a special group, whose main characteristic is that the starting material consists of liquid pig iron. These methods are currently being rapidly developed, both with respect to process technology and commercial application. The oldest pneumatic process and at the same time the oldest of all ingot-steel methods is the *Bessemer process*, which was patented by Bessemer in 1855. However, it was first carried out with satisfactory results in Sweden in 1858 (G. F. Göransson). The process is carried out in an acid-lined, pear-shaped, tiltable furnace, called a converter. After the liquid pig iron is poured into it, air is blown through the melt from holes in the bottom of the converter. Thus, the silicon and manganese in the pig iron is burned with an especially high heat of oxidation, and the temperature rises rapidly. This leads to stronger oxidation of the carbon, which quickly decreases in amount. The entire process takes 15 to 20 min. Its course can be followed by observing the color (or better the spectrum) of the flame produced by the waste gases. Since phosphorus is not removed in the acid Bessemer process, the latter is now of little importance. The basic counterpart of the method is called the *Thomas process*. It is used to a considerable extent in countries where steel manufacture is based on phosphorus-rich ores (Belgium, France, West Germany). The pig iron used has a P content of 1.5 to 2%, which serves as fuel in the process and goes over to the basic CaO-rich slag (Thomas phosphate). The Thomas process cannot convert any large amount of scrap iron, which, of course, is a disadvantage.

In these older converter methods the air blown through the melt causes the melt to take up small amounts of nitrogen, which already in concentrations above 0.01% give the steel a special kind of brittleness. This can be avoided if oxygen is substituted for air. At the same time the heat losses are reduced. Large amounts of heat leave the converter with the waste gases, and if the nitrogen of the air can be left out, the waste gas volume for a given amount of oxygen is considerably reduced. The heat gain obtained in this way makes it possible to add large quantities of iron scrap or ore, which otherwise would cool the charge too much. But if oxygen were to be blown through the melt from below as in the Bessemer and Thomas converters the bottom of the furnace would melt rapidly. Therefore the oxygen is blown from above against the upper surface of the melt. In order to obtain an effective reaction either strong convection currents are set up in the melt (*LD process*), or the melt is mechanically stirred by rolling the converter (*Kaldo process*). The advantages of these new converter methods using oxygen are so obvious that they may be expected to replace the older steel-melting processes quite rapidly.

d. Case hardening, nitriding, malleableizing. If an iron object with low carbon content is heated in a carbon-containing medium the surface layer may take up an amount of carbon that is suitable for hardening by quenching. In this way a preformed object can be given a hard surface, while its interior remains soft and tough. The carbonation itself is called *carburizing* and the whole process *case hardening*. The carburizing medium may be gaseous (carbon monoxide, hydrocarbons), liquid (salt melts containing cyanides) or solid (charcoal powder with activating additives, for example carbonate; however, the carbon is transferred as carbon monoxide). The temperature is generally 800–950°C.

Nitriding is another way of obtaining high surface hardness. A steel is then used that is alloyed with substances which combine strongly with nitrogen, especially aluminum, chromium and vanadium (nitriding steel). The previously worked and heat-treated object is heated in a stream of ammonia at 500–550°C, usually for 25–100 hours. The ammonia is decomposed at the surface of the object and the nitrogen formed diffuses inward. In the surface layer there are thus formed nitrides of aluminum, chromium, etc., which precipitate in finely divided form and lead to a particularly great hardness (cf. precipitation hardening, 13-5f 2). No subsequent hardening by quenching is necessary.

To obtain a softer, relatively tough material from the easily cast but brittle cast iron, *malleableizing* is used. The method is suitable for smaller objects cast from cast iron that solidifies white (34-3a). Packed in sand, sometimes mixed with iron oxide (iron-ore concentrate), they are heated for several days in the temperature range of 700–900°C. The cementite is then decomposed and the graphite formed is oxidized at the surface of the object. The decrease in proportion of carbon at the surface leads to a diffusion of carbon from the interior. After cooling a carbon content decreasing toward the surface is then obtained. The regions which thus have a suitable carbon content can then also be hardened.

e. Alloy steels. It was previously stated that especially manganese, silicon and aluminum are often added in steel melting for deoxidation (21-5b). When these elements are added for other purposes in amounts over and above those needed for deoxidation, they are considered as special alloying elements and the steel is designated as *alloy steel*. Other alloying elements in steel are very often transition elements (like manganese). Vanadium, chromium, molybdenum, tungsten, cobalt and nickel are common. Of the nonmetals boron is also sometimes used in addition to silicon. A distinction is made between *high-alloy steel* and *low-alloy steel*, and the boundary between these is usually drawn at a total content of the special alloying substances of about 5 wt %. If this total content exceeds about 50 wt %, we no longer speak of steel but of an alloy with this or that "base" (=the main metal).

The properties of alloy steels may be varied within wide limits by proper choice of alloying substances and heat treatment. In this way may be obtained good mechanical properties such as high tensile strength, hardness and abrasion resistance, increased stability toward oxidation at higher temperature ("heat-resistant steel") and corrosion (stainless steel), appropriate magnetic properties, etc.

An important effect of the alloying elements is alteration of the α–γ and γ–δ transition temperatures. The breadth of the γ (austenite) region may in this way be changed appreciably in relation to the γ region of the pure carbon steel (fig. 34.1). It may be said for the most part that the γ region is broadened by those elements that themselves have a face-centered-cubic modification (Mn, Ni, Co), but is narrowed by those elements that lack such a modification (Si, V, Cr, Mo, W). The decrease of the γ region in the latter case causes the α and δ regions to coalesce (α and δ Fe have the same structure; see 34-2) with sufficient amounts of alloying elements. In these cases other carbides than cementite also often appear.

The presence of new elements, which because of their larger atoms diffuse much more slowly than carbon, cause the processes in the solid state that depend on diffusion to go more slowly in alloy steel than in carbon steel (but not, on the other hand, the transition austenite → martensite, which is diffusionless; 34-3a).

If alloying elements are dissolved in austenite, their diffusion-rate-decreasing effect often is very important. The austenite can then be supercooled more easily than in a carbon steel. A reduction of the eutectoid temperature (broadened γ region) may also contribute to this effect. At the same time the formation temperature of martensite is lowered and may go below room temperature. In such cases the austenite at room temperature exists as a metastable phase (with increased γ region and very high alloy content, possibly as a stable phase). The steel is then said to be an *austenitic steel* and does not become hard through formation of martensite. On the other hand, if the martensite formation temperature is not lowered below room temperature, martensite is formed. The steel can then be hardened, and because the austenite is more easily supercooled, the cooling in the process need not be so sharp (sometimes cooling in air suffices). For the same reason the hardening effect extends to areas lying far below the surface where the cooling rate otherwise would not be fast enough. The steel acquires greater *hardenability*.

If the γ region is squeezed out, a steel may exist in the $\alpha(=\delta)$ state at all temperatures up to the melting point. Such a *ferritic steel*, of course, cannot be hardened.

Carbide formation can cause an increase in hardness, partly because the carbide itself is hard, and partly through precipitation hardening (13-5f 2) in connection with beginning carbide crystallization. Precipitated carbides may also hinder growth of the crystal grains of the matrix and thus prevent harmful coarse granularity.

Low-alloy steels are heat treated for the most part in the same way as carbon steels, while the strongly altered properties of high-alloy steels often lead to the need for completely different heat treatment.

In the following a number of important alloy steels are mentioned in connection with their areas of application. All proportions are in weight per cent.

In common *structural steel* and similar types, up to 1% Mn is used to increase tensile strength. Steels with 12–15% Mn and about 1% C (austenitic steel) are tough with very great abrasion resistance and are therefore used in stone crushers, mills, ditch diggers and track switches.

In steel for *machine parts*, which is case-hardened (34-3d), great hardenability is required, which is achieved by alloying with nickel, chromium or molybdenum, either alone or in combination. Low-alloy *tool steels* often contain chromium, nickel, molybdenum, tungsten and vanadium. For boring and cutting steels, especially in automatic machines, a high working speed is desirable, with which the edge becomes very hot. Retention of hardness at the working temperature (high heat hardness) is then necessary and can be obtained in high-alloy steels with molybdenum, tungsten, vanadium and cobalt. *High-speed steel*, which contains as the most important alloying elements 16–20% W and 3–6% Cr, can operate at about 600°C. Here the hardness results from an incipient precipitation of the tungsten carbide W_2C. *Spring steel* is often alloyed with silicon and manganese, sometimes also with chromium. In this way the hardness throughout is increased and the tendency to fatigue decreased.

Corrosion resistance is obtained especially by alloying with chromium. The resistance to corrosion of the steel increases with chromium content (note the propensity of chromium for passivation, 31-3) and with proportions of chromium above about 13% we come into the region of the *stainless steels*. These often also contain nickel, the most common stainless steel containing 18% Cr, 8% Ni and less than 0.1% C ("18-8 steel"; austenitic). Following improper heat treatment chromium carbide may be deposited in the grain boundaries of such a steel. With the formation of chromium carbide chromium is drawn out of the adjacent austenite regions, which then lose their resistance to corrosion. The result is corrosion in the grain boundaries and breakdown of the steel. This can be prevented, either by decreasing the carbon content, or by adding titanium or niobium with which the carbon then forms carbides instead of with chromium. It must also always be remembered that "stainless steel" can rust in reducing media, which destroy the passivation. Chromium-nickel-alloy steels, similar to the stainless steels, are used where good tensile quality is required at high temperature. Addition of cobalt may further improve the high-temperature strength (cf. also 34-4b). High chromium content confers stability toward oxidation at higher temperature (heat-resistant steel; see also 31-4b).

In transformer cores and motor pole pieces a *magnetically soft material* (7-4a) is required. Pure iron may be used for this purpose, but smaller losses are obtained if practically pure iron is alloyed with up to about 5% Si. To conserve weight high magnetic permeability is also needed, which can be obtained through high proportions of cobalt or nickel (for example alloys of the Permalloy type, where the highest permeability is attained with about 80% Ni). Permanent magnets, on the other hand, must be made of *magnetically hard material*, and for this nowadays alloys are used that contain high proportions of at least three of the metals cobalt, nickel, aluminum and copper (the iron content is often <50%, sometimes even 0). Austenitic steel is nonferromagnetic.

Concerning *nitriding steel*, see 34-3d. Finally we may mention *invar* with 36% Ni (the remainder iron with altogether <1% of other elements), which has an extremely low thermal expansion coefficient and is therefore used in length-measuring instruments, time pieces, etc.

34-4. Production and uses of cobalt and nickel

a. Production. The methods of production are quite complicated, among other things because the ores as a rule contain a number of different metals. Especially in cobalt metallurgy many details are also kept as trade secrets. The description is therefore necessarily very schematic. Very often cobalt and nickel are obtained in connection with the manufacture of copper, and portions of the production processes are, in fact, very similar for the three metals. The description in 36-4a of the metallurgy of copper should thus be referred to at this point.

The most important cobalt and nickel ores are sulfide and arsenide ores, which simultaneously contain iron, cobalt and nickel, usually also copper and sometimes silver. In a melt of sulfides and arsenides of these metals there appear over a wide composition range one sulfide phase (*matte*) and one arsenide phase (*speiss*). The system thus has a miscibility gap in which it is two-phase (13-5a). The sulfide phase can dissolve some arsenic and the arsenic phase some sulfur. With low amounts of arsenic, therefore, only the sulfide phase appears, and with low amounts of sulfur, only the arsenide phase. In practice there is always also an oxide (silicate) phase, slag. Iron, cobalt and nickel are bound more strongly to arsenic than to sulfur and are thus drawn to the arsenide phase if one is present. On the other hand, copper and silver prefer to stay in the sulfide phase.

Here we describe first the production of *nickel*, this being the most important. The sulfide and arsenide ores are concentrated, and if possible an attempt is made here by flotation to separate the ores into copper-rich and nickel-rich fractions. This is generally followed by a partial oxidation roasting (21-5b) to reduce the sulfur content to a level suitable for subsequent smelting processes. The arsenic content of the arsenide ores is also reduced by roasting. The roasting product, which for the most part consists of oxides and sulfides of copper, nickel and iron, and silicates, is melted as in copper smelting to matte, which is blast-smelted into a more concentrated matte. However, the smelting is not continued to the metal, but the matte is tapped off, containing 75–80% (Ni+Cu) and about 20% S. This is roasted to oxide, after which the copper oxide is leached out with sulfuric acid (the copper is recovered from the solution electrolytically). The remainder, which is mainly nickel oxide, is reduced. In the so-called *Mond process* a very fine powder is desired, and the reduction is done with water gas at 350–400°C. The powder is then treated with carbon monoxide at about 60°C to form tetracarbonylnickel (23-2f): $Ni(s) + 4CO(g) \rightleftharpoons Ni(CO)_4(g)$. No other metal carbonyls are formed under these conditions, so that all impurities are left behind when the tetracarbonylnickel is distilled off. This is heated to about 180°C, upon which it decomposes and very pure nickel falls out (reaction direction to the left). If the Mond process is not appropriate, the reduction is carried out instead with carbon in an electric furnace with added slag-formers, after which the impure metal is cast into anodes. Then follows electrolytic purification, although this does not remove the cobalt, which is generally carried along from the ore.

Nickel silicate ore (garnierite) is mixed with gypsum and smelted with coke and slag-formers. The mixture is thereby reduced to a sulfide phase (matte), which is treated as above.

The production of *cobalt* is complicated among other things by the greater difficulty of separating cobalt from other metals (a process analogous to the Mond process cannot be applied). The final processing is done by wet methods, and to obtain the material in soluble form various methods are utilized. Oxidic ores are often leached directly with sulfuric acid, but sulfide ores must first be converted to soluble chlorides and sulfates by means of chlorinating and sulfating roasts (36-4a). Smelting methods are also used. In these, ores with a high percentage of cobalt are roasted until they are free of sulfur, after which the sintered product is smelted in an electric furnace together with coke and a slag former. A Co-Cu-Fe alloy is thus obtained, which separates into two layers, one rich in cobalt and poor in copper, the other poor in cobalt and rich in copper. The former is dissolved in sulfuric acid for cobalt processing, while copper is recovered from the latter. Ores poor in cobalt are roasted incompletely and then smelted. Depending on the sulfur and arsenic contents either a sulfide phase (matte) or an arsenide phase (speiss) or both are then formed. The matte and speiss are roasted and leached.

From the solutions a large portion of the copper is deposited electrolytically, after which purification is carried out by means of various precipitation operations. Finally cobalt(III) oxide hydroxide is precipitated with slaked lime in the presence of an oxidizing agent. The hydroxide may be dissolved in sulfuric acid to form sulfate, which is electrolyzed, or also ignited to Co_3O_4 (cf. 34-6b), which is reduced.

b. Uses. *Cobalt* is an important alloy metal in steels (34-3e). It is also used in various alloys for permanent magnets, among other things, together with nickel. Cobalt is also of great importance as a binder in hard metals with the tungsten carbide WC as the hard substance (14-3b).

The cobalt isotope ^{60}Co, which by β^- decay (half-life 5.2 y) gives an intense γ radiation of high energy, is used as a γ-ray source, among other things for radiation therapy ("cobalt gun"). ^{60}Co is prepared by neutron irradiation of ^{59}Co, the only natural cobalt isotope, in a nuclear reactor (9-2b).

Nickel is used in pure form for apparatus for chemical work because of its stability toward corrosion. The largest quantities of metallic nickel go for the production of alloy steels (34-3e), but it is included also in many other types of alloys.

Among those with nickel as the main metal we may mention high-temperature alloys with low conductivity such as *Nichrome* and *Nichrothal* (31-4b), alloys of nickel-molybdenum-iron and nickel-molybdenum-chromium-iron with high resistance to acids, and alloys with copper, which have good mechanical properties (can also be precipitation hardened, 13-5f2) and quite high chemical resistivity. Of the latter the best known is *Monel metal* with about 70 % Ni and 30 % Cu (Monel metal is produced directly from ores containing approximately this ratio of nickel to copper by roasting concentrated matte to an oxide mixture, which is reduced). A series of alloys containing among other things nickel, cobalt and chromium, and with either nickel or chromium as a base, is used in gas turbines, jet engines and rockets because of their great hot tensile strengths. Nickel alloys with various magnetic properties have been mentioned in 34-3e. Large quantities of nickel anodes are used for nickel plating. Finely divided nickel, obtained by reduction of nickel(II) oxide with hydrogen, is an important catalyst in hydrogenation reactions.

34-5. Iron compounds

a. Iron(II) compounds. In many iron(II) compounds the Fe^{2+} ion, long called *ferrous ion*, appears as a building unit without appreciable covalent character. Earlier it was stated that the charge and radius of the Fe^{2+} ion (fig. 5.5) cause its salts to show many similarities to corresponding Mg^{2+} and Mn^{2+} salts. For the same reason there are also great similarities between Fe^{2+} salts and the corresponding Co^{2+} and Ni^{2+} salts. In all these cases the chlorides, sulfates and nitrates are easily soluble, but fluorides, carbonates and phosphates are more or less difficultly soluble. The hydrated Fe^{2+} ion gives a light blue-green color to its water solutions and hydrates. If other colors appear not originating in other atoms or atomic groups, more or less covalent bonding is present between Fe(II) and its ligands.

In water solution Fe(II) is oxidized to Fe(III) more easily the more basic the solution is, and the reason for this was described in 16d. Even in neutral solution the oxidation takes place easily. On the other hand, in acid solution Fe(II) is quite stable toward oxidation. This leads to the result that even oxidizing acids dissolve iron with the formation of Fe(II) and not Fe(III), as long as the acid is dilute and cold.

Iron(II) *oxide* ~FeO has a wide range of homogeneity and an iron content that is generally lower than that corresponding to the formula FeO. It then also contains some Fe(III) (6-3a). This oxide is easily obtained as a black, pyrophoric powder by heating iron(II) oxalate in the absence of air: $FeC_2O_4(s) \rightarrow FeO(s) + CO(g) + CO_2(g)$.

Iron(II) *hydroxide* $Fe(OH)_2$ is obtained as a white precipitate by adding alkali solution to iron(II) salt solutions in the absence of air. The precipitate is easily oxidized in air (see above), on which the color turns to a dirty green, becoming darker until it is finally red-brown (iron(III) oxide hydroxide). The dark intermediate color stages result from the simultaneous presence of Fe(II) and Fe(III) (7-3b). Precipitation of $Fe(OH)_2$ with ammonia solution and the associated effect of ammonium ions is entirely analogous to that of $Mn(OH)_2$ (32-5a).

Iron(II) *chloride* $FeCl_2$ is obtained anhydrous as a white mass on heating iron in dry hydrogen chloride gas. If iron is dissolved in hydrochloric acid in the absence of air, a green solution of the chloride is obtained from which green crystals of the *tetrahydrate* $FeCl_2 \cdot 4H_2O$ are formed. The most important iron(II) salt is *iron*(II) *sulfate*, which from water solution at ordinary temperature is obtained as a green *heptahydrate* (*iron vitriol*) $FeSO_4 \cdot 7H_2O$ (21-11b3). In dry air the heptahydrate loses water and also is oxidizied on the surface to a yellow-brown iron(III) hydroxide sulfate. The double salt $(NH_4)_2Fe(SO_4)_2 \cdot 6H_2O$ (*Mohr's salt*) is appreciably more stable toward air and is therefore used in the preparation of Fe(II) solutions of prescribed compositions. Iron(II) sulfate is produced by dissolving iron in sulfuric acid, the largest quantities being obtained from baths in which iron is pickled (21-11b2, 23-5c) with sulfuric acid. An old method consists of allowing moist pyrite to be oxidized by the air: $2FeS_2 + 7O_2 + 2H_2O \rightarrow 2FeSO_4 + 2H_2SO_4$. On heating iron(II) sulfate heptahydrate $6H_2O$ are easily driven off and the seventh water molecule with more difficulty. The anhydrous salt is a white powder, which decomposes above 500°C with the evolution of SO_2 and SO_3. Fe_2O_3 remains at the end: $2FeSO_4(s) \rightarrow Fe_2O_3(s) + SO_2(g) + SO_3(g)$. Formerly the gas mixture obtained in this way was passed into water to produce sulfuric acid, "oil of vitriol". Iron(II) sulfate is used for the preparation of other iron compounds, in dyeing, as a reducing agent, as a wood preservative and a weed killer. Blue ink is made from iron(II) sulfate and gallic acid. This solution is only weakly colored, but on the page oxidation takes place and a blue-black precipitate of an iron(III) gallate complex is formed. To give the ink an initial color a blue dye is used.

Iron(II) *carbonate* $FeCO_3$ is formed by adding alkali carbonate to Fe(II) salt solutions in the absence of air. The white carbonate is oxidized in air to iron(III) oxide hydroxide with the evolution of carbon dioxide: $2FeCO_3(s) + \frac{1}{2}O_2 + H_2O \rightarrow 2FeO(OH)(s) + 2CO_2(g)$. The carbonate constitutes the mineral *siderite*. Like calcium carbonate, it dissolves in water containing carbon dioxide (26-5b, c).

Alkali cyanides with an iron(II) salt solution give a red-brown precipitate of *iron*(II) *cyanide* $Fe(CN)_2$, which dissolves with an excess of cyanide ion to a yellow solution containing the octahedral *hexacyanoferrate*(II) *ion* $Fe^{II}(CN)_6^{4-}$. If the cations are potassium ions, large sulfur-yellow crystals of $K_4[Fe(CN)_6] \cdot 3H_2O$ crystallize out (formerly called yellow prussiate of potash). The ion $Fe(CN)_6^{4-}$ is a very strong complex (23-2k1) and its water solution shows none of the ordinary reactions of Fe^{2+} and CN^-. Neither are its salts poisonous. As stated earlier (5-3i) the ligands are bonded to the central atom with a strong covalent contribution. Hexacyanoferrates(II) of alkali metals and alkaline-earth metals except barium are easily soluble in water; salts of other metals are poorly soluble. By adding conc. hydrochloric acid to a conc. solution of an easily soluble hexacyanoferrate(II), *hydrogen hexacyanoferrate*(II) (*hexacyanoferric*(II) *acid*) $H_4[Fe(CN)_6]$ is obtained as a white precipitate, which is easily oxidized with the formation of a blue color.

The *iron sulfides* $\sim$ FeS and FeS_2 are usually considered formally to be *iron*(II) *sulfide* and *iron*(II) *disulfide* respectively, but both are semiconductors with very weak ionic-bond contribution. The former is precipitated completely only from basic solution (for example, with ammonium hydrogen sulfide solution, 21-7a) because of its relatively high solubility product. Technically the sulfide is made by melting iron with sulfur or pyrite ($Fe + FeS_2 \rightarrow 2FeS$). The crystal structure is of the nickel-arsenide type (6-2c); however, the phase has an extended homogeneity range from the ideal composition FeS to at least the composition $Fe_{0.82}S$. The decrease in iron content is attained by vacancies in the iron lattice (6-3a), but these vacancies result in structural changes such that we do not have here a straightforward case of solid solution. Simultaneously the sulfide becomes ferrimagnetic (7-4a) and changes color from almost black to bronze-colored. The mineral *pyrrhotite* consists of such a sulfide phase with iron content less than that corresponding to the formula FeS. The disulfide FeS_2 usually occurs as the mineral *pyrite* and has been described in 21-2a, 21-8 and 34-1a. It is brass-yellow and has metallic luster.

Previously mentioned Fe(II) compounds are the ion $Fe(NO)^{2+}$, the *nitroprusside ion* $Fe(CN)_5NO^{2-}$ (both in 22-9b) and *ferrocene* $Fe(C_5H_5)_2$ (5-6c).

b. Iron(III) compounds. In these the covalent character is greater than in the Fe(II) compounds. Iron(III) compounds have great similarities to the corresponding aluminum and chromium(III) compounds. The ion Fe^{3+} has long been called *ferric ion*. The hydrated ion $Fe(H_2O)_6^{3+}$ is colorless, but it easily gives off protons (hydrolyzed, 12-1b), and the hydroxo complexes so formed are yellow to red-brown. Solutions of iron(III) salts that do not contain excess acid thus take on this color. If they are made strongly acid, the hydrolysis is suppressed. Often complexes are then formed with the anions of the acid as ligands (acido complexes), and if these complexes are colored the solution takes on their color. Thus, solutions of iron(III) chloride acidified with hydrochloric acid are yellow.

Iron(III) *oxide* Fe_2O_3 is formed on heating iron(III) salts of volatile acids, iron(III) oxide hydroxide, and by burning iron in a stream of oxygen. However, the temperature must here not go so high that Fe_2O_3 gives off oxygen with the formation of Fe_3O_4 (see below). Otherwise, this process gives stable α Fe_2O_3, which has the same crystal structure as α Al_2O_3 (24-3f) and constitutes the mineral *hematite*. Concerning metastable γ Fe_2O_3, see below. The manner in which the color and reactivity of iron(III) oxide depends on the degree of order in its structure, and thus the method of formation, was described in 14-3b. Various paint pigments under different names (Venetian red, English red) consist of iron(III) oxide. Ignited oxide is hard and is used as polishing powder ("rouge", "crocus").

Hematite has a blue-gray to black-gray metallic luster, sometimes with a red tinge. Powdered hematite is dark red, so that the mineral on scratching shows a red "streak" (the names hematite and bloodstone refer to this property).

At high temperature iron(III) oxide gives off oxygen with the formation of *iron*(II,III) *oxide* $\sim Fe_3O_4$ ($Fe^{II}(Fe^{III})_2O_4$). From tab. 21.1 we find that for the equilibrium $6Fe_2O_3(s) \rightleftharpoons 4Fe_3O_4(s) + O_2(g)$ at 1000°C, $\log K_p = -11.36$. The equilibrium at this temperature thus implies an extremely low oxygen pressure, and if this pressure is raised above this level the reaction goes to the left. At 1150°C the oxygen pressure is 0.5 mbar, so that oxygen can be easily pumped away from Fe_2O_3 (reaction to the right). At still higher temperature Fe_2O_3 rapidly gives off oxygen even in air at atmospheric pressure. The oxide Fe_3O_4 is a black (two valence states, 7-3b) semiconductor with metallic luster. Its magnetism was early noticed in the mineral *magnetite* (for the origin of this name, see 26-2b), then also called *loadstone*. The crystal structure is the spinel type (21-5a). When iron is oxidized during heat-treatment and hot-working processes a *scale* is formed, which consists of Fe_3O_4. Fe_3O_4 is most easily prepared by passing water vapor over iron at about 600°C (18d).

As stated in 7-4a, Fe_3O_4 is ferrimagnetic. The same is true of many other phases of the type $M^{II}(Fe^{III})_2O_4$, where M^{II} is an element in the divalent state and of approximately the same ionic radius as Fe^{2+}, or a mixture of several such elements. Most important are Mn, Fe, Co, Ni, Cu, Zn, Cd, Mg. Most of these phases have the spinel structure. Like all spinels they are double oxides (21-5a), but are improperly called *ferrites*. Because they often have a high total magnetic moment but appreciably lower conductivity than the metals, they have been widely used as (sintered) magnetic core material for high-frequency applications (for example, Ferroxcube).

If very finely divided Fe_3O_4 is oxidized at temperatures below about 350°C, a metastable Fe_2O_3 modification $\gamma\ Fe_2O_3$ is formed. In this the oxygen positions of the spinel structure are preserved, but vacancies appear in the iron lattice while Fe(II) goes over to Fe(III) (see 6-3a). The color changes to a reddish Fe_2O_3 color but the ferrimagnetism remains. The magnetic coating on red-brown magnetic recording tape consists of $\gamma\ Fe_2O_3$.

The manner in which the hydrated Fe^{3+} ion reacts has been described in 12-1b. What was said there and in 12-4d concerning $Al(OH)_3$ and its reactions with acid and basic solutions can also in principle be applied under corresponding conditions to iron(III). When alkali hydroxides or ammonia is added to an iron(III) salt solution a red-brown gelationous precipitate is formed, which on aging slowly forms *iron*(III) *oxide hydroxide* FeO(OH). When freshly precipitated it probably consists of FeO(OH) with adsorbed water in indefinite amount. In this form the precipitate easily dissolves in acids but on the other hand only to a slight extent in bases. The oxide hydroxide exists in two different modifications, α and γ FeO(OH), which also occur as minerals (34-1a). The structures are the same as α and γ AlO(OH) respectively. Iron(III) hydroxide $Fe(OH)_3$ is unknown.

Anhydrous *iron*(III) *chloride* $FeCl_3$ is most easily obtained by heating iron in a stream of chlorine gas. The chloride sublimes and deposits as strongly hygroscopic crystal plates with dark green surface color and dark red transmission color (cf. 7-3a). Analogously with $AlCl_3$ (24-3e) the equilibrium $2FeCl_3 \rightleftharpoons Fe_2Cl_6$ prevails in the gaseous phase. A water solution of iron(III) chloride is produced industrially by dissolving iron in hydrochloric acid followed by oxidation with chlorine. There are a number of different hydrates, but under ordinary conditions the yellow *hexahydrate* $FeCl_3 \cdot 6H_2O$ crystallizes, the most common commercial product. Its strongly acid solution is used for photoengraving.

Iron(III) *sulfate* $Fe_2(SO_4)_3$ is obtained as a white powder on evaporating the water solution. From solutions also containing sulfates of monovalent cations, especially K^+ and NH_4^+, *iron alums* $M^I Fe^{III}(SO_4)_2 \cdot 12H_2O$ (12-11 b3) are obtained, which easily form large crystals. Completely pure iron alum crystals are colorless, but usually traces of Mn(III) give them a pale violet color. The same is true of the color of *iron*(III) *nitrate hexahydrate* $Fe(NO_3)_3 \cdot 6H_2O$.

Iron(III) readily forms complexes with halide and pseudohalide ions (20-6). Typical complexes are $Fe^{III}F_6^{3-}$, $Fe^{III}Cl_4^-$, and $Fe^{III}(CN)_6^{3-}$. The last, octahedral *hexacyanoferrate*(III) *ion*, is contained in $K_3Fe(CN)_6$ ("red prussiate of potash") which is obtained by oxidation, for example with chlorine, of a solution of the yellow prussiate acidified with hydrochloric acid. It is less stable than the hexacyanoferrate(II) ion and is therefore poisonous. On mixing solutions of hexacyanoferrate(II) and iron(III) salts or of hexacyanoferrate(III) and iron(II) salts, in both cases a dark blue precipitate is formed. Formerly the first was called "Prussian blue" and the second "Turnbull's blue", but it is now known that they are the same substance and are both called *Prussian blue*. The composition depends on the conditions of the experiment, but a typical formulation is $M^I Fe^{II} Fe^{III}(CN)_6$. In the crystals probably part of the iron atoms continually exchange electrons and the intense color is related to this property (7-3b). The formation of Prussian blue is useful for the detection of iron.

Blueprint paper is made by treating paper in darkness with a solution of ammonium hexacyanoferrate(III) and iron(III) citrate. On illumination the iron(III) of the citrate is reduced by the citrate ions, and Prussian blue is formed. Afterward it is fixed by dissolving out the unreacted salts with water.

Still more sensitive than Prussian blue formation as a test for iron are the intensely red *thiocyanatoiron*(III) *ions*, $FeSCN^{2+}$ and $Fe(SCN)_2^+$ discussed in 23-2k5.

Iron(III) sulfide probably does not exist, although the mixed iron(II,III) sulfide Fe_3S_4 corresponding to Fe_3O_4 has recently been found as a mineral. In basic solutions Fe(III) salts with sulfide ion gives $2FeS(s) + S(s)$.

c. Iron(VI) compounds. These are represented mainly by *tetraoxoferrates*(VI) with the ion $Fe^{VI}O_4^{2-}$, often called merely *ferrates*. Oxidation of iron powder in a potassium nitrate melt followed by leaching with water, or oxidation of iron(III) oxide hydroxide in conc. alkali hydroxide solution with chlorine or bromine, gives a red-violet solution containing FeO_4^{2-}. Barium ions precipitate from this solution dark red $BaFeO_4$, which is relatively stable. The ferrates are strong oxidizing agents.

d. Other oxidation states of iron. Here we will refer only to the *carbonyliron compounds*, which were described in 23-2f.

34-6. Cobalt compounds

a. Cobalt(II) compounds. A sufficient number of water ligands around a cobalt(II) ion gives it a red color in solid hydrates as well as in solution. Such is the case with the *hexaquocobalt*(II) *ion* $Co(H_2O)_6^{2+}$ which occurs in hydrates and water solutions of most salts with oxo anions (nitrate, sulfate, chlorate, etc.) Other anions, such as halide and hydroxide ions, may force out the water ligands and then the color will be different, very often blue.

In complexes cobalt(II) usually assumes the coordination numbers 4 or 6. The former gives tetrahedral complexes such as $CoCl_4^{2-}$, the latter octahedral complexes such as $Co(H_2O)_6^{2+}$ and $Co(NH_3)_6^{2+}$. However, the coordination number 5 does occur. The highest cyano complex is probably $Co(CN)_5^{3-}$ and not $Co(CN)_6^{4-}$. As stated in 34-6b many cobalt(II) complexes are easily oxidized to cobalt(III) complexes.

Cobalt(II) *oxide* $\sim CoO$ is formed by heating in the absence of air cobalt(II) salts of volatile acids and of cobalt(II) hydroxide. *Cobalt*(II) *hydroxide* $Co(OH)_2$ is precipitated from cobalt(II) salt solutions by alkali hydroxide. The precipitate is blue at first, but afterward light red. It is easily oxidized to brown *cobalt*(III) *oxide hydroxide* $CoO(OH)$ with variable water content. If the precipitation is carried out with ammonia solution, $Co(OH)_2$ precipitates provided the proportion of NH_3 is small. As the amount of NH_3 increases, however, more and more *amminecobalt*(II) *ions* are formed. In this way the amount of Co^{2+} is lowered so much that the precipitate goes into solution again.

If cobalt(II) oxide is melted with quartz and potassium carbonate, a blue glass of *potassium cobalt*(II) *silicate* is obtained. Under the name *smalt* this is used to give a blue color to glass, glazes and enamels. A "borax bead" (24-2g) is colored blue by cobalt(II) compounds. Many metallic oxides give characteristic colors on igniting with cobalt(II) oxide (the metal oxide is often moistened with cobalt(II) nitrate solution which on ignition leaves the oxide). Aluminum oxide gives a blue color caused by the spinel $CoAl_2O_4$ ("Thénard's blue"), zinc oxide gives a green color ("Rinman's green") and magnesium oxide a rose color.

Cobalt(II) *chloride* $CoCl_2$ gives a light-red water solution, in which the major portion of the cobalt(II) is present as the hexaquocobalt(II) ion $Co(H_2O)_6^{2+}$. The *hexahydrate* crystallizes from the solution at room temperature. This is made up of *dichlorotetraquocobalt*(II) groups $CoCl_2(H_2O)_4$ among which a further $2H_2O$ per Co are packed. The chloride has several lower hydrates and in these more and more

Co–Cl bonds appear as the water content becomes smaller. At the same time the color becomes more and more blue. At 20°C and a water vapor pressure above 7.2 mbar the hexahydrate is the only solid phase but at lower vapor pressure and the same temperature water is given off under the formation first of red-violet *dihydrate* and then blue *monohydrate*. These color changes are used to indicate changes in humidity, for example to warn when a drying agent for gases is exhausted. Even on slight warming of the hexahydrate water is given off and the color turns blue, becoming red again on cooling in sufficiently moist air. Writing made with a dilute water solution is nearly invisible but appears with a blue color when warmed ("sympathetic ink"). In water solution also the color turns blue when the Co^{2+} ion acquires other ligands than water. On heating the solution chloride ions replace more and more of the water molecules as ligands, and this more easily the more concentrated the solution is. Addition of hydrochloric acid or chlorides of alkali or alkaline-earth metals increases both the number of chloro-complexes and their chlorine content, leading to a change of color toward blue even at room temperature.

Sulfide ions precipitate black *cobalt(II) sulfide* CoS from basic cobalt(II) salt solutions. Freshly precipitated sulfide dissolves in dilute acid, and the sulfide will not form in acid solution. However, if the sulfide is allowed to stand it soon becomes difficultly soluble in dilute strong acids. This has been said to result from a transition to a more poorly soluble modification or from oxidation.

b. Cobalt(III) compounds. According to 16c the ion Co^{3+} is a very strong oxidizing agent and therefore extremely unstable toward reduction to Co^{2+}. As stated in 16e it even decomposes water with the evolution of oxygen. However, it was shown in 16c that the *hexaminecobalt*(III) *ion* $Co(NH_3)_6^{3+}$ is only a weak oxidizing agent. The same is true of a very large number of strong cobalt(III) complexes, which are thus stable toward reduction (cf. also 5-6c). For the most part cobalt(III) occurs mostly in complexes, which, in fact, are known in extremely large numbers. Because of their minimal oxidizing action, they are easily formed by oxidizing a cobalt(II) solution containing the desired ligands. Other complexes may then be formed by exchange of ligands. When air is passed into an ammoniacal cobalt(II) salt solution, $Co(NH_3)_6^{3+}$ is formed. If carbonates are present, $Co(NH_3)_4CO_3^+$ is formed, in which the carbonate group is bound as a chelate (5-6c). With conc. hydrochloric acid this ion gives $Co(NH_3)_4Cl_2^+$. In all these ions the coordination is octahedral as in most cobalt(III) complexes.

Potassium hexanitrocobaltate(III) $K_3[Co(NO_2)_6]$ is poorly soluble (yellow) and the easily soluble sodium salt is therefore used as a test reagent for potassium.

It was largely through the study of cobalt(III) complexes that Werner during the 1890's was able to carry out his investigations on the structures of inorganic molecules (cf. 7-5).

Concerning *cobalt*(III) *oxide hydroxide*, see 34-6a. In cases where the heating of cobalt(III) compounds would be expected to give cobalt(III) oxide Co_2O_3, black cobalt(II,III) *oxide* Co_3O_4 (spinel type) is always formed. On heating this oxide in air to about 900°C it gives off oxygen with the formation of cobalt(II) oxide CoO.

34-7. Nickel compounds

a. Nickel(II) compounds. For nickel the oxidation number II is wholly predominant. Nickel(II) compounds show great similarities to the corresponding cobalt(II) as well as copper(II) compounds. The most important nickel(II) complexes are either octahedral and then usually green, blue or violet; or square (5-3i, j) and then most often yellow to red. $Ni(H_2O)_6^{2+}$ (green) and $Ni(NH_3)_6^{2+}$ (blue) belong to the former group, $Ni(CN)_4^{2-}$ (yellow) to the latter. The *hexaquonickel*(II) *ion* $Ni(H_2O)_6^{2+}$ appears in hydrates and water solutions of nickel(II) salts, which are then colored green. *Nickel*(II) *oxide* ~NiO (greenish gray), *nickel*(II) *hydroxide* $Ni(OH)_2$ (apple green, not oxidized by air) and *nickel*(II) *sulfide* NiS (black) are formed in a manner quite analogous to the cobalt(II) compounds. The decrease in solubility of the sulfide after precipitation here results from a transition to a more difficultly soluble modification.

The most common commercial compounds are *nickel*(II) *chloride hexahydrate* $NiCl_2 \cdot 6H_2O$ and *nickel*(II) *sulfate* $NiSO_4$, especially the *heptahydrate* ("nickel vitriol") which crystallizes out of water solution below 31.5°C. Both hydrates are much used to prepare electrolytes for nickel plating.

b. Other oxidation states of nickel. In 23-2f *tetracarbonylnickel* $Ni(CO)_4$ was described and its technical application discussed in 34-4a. The molecule is tetrahedral (concerning the oxidation number of nickel in this compound, see 5-6c).

Ni(III) and Ni(IV) occur in NiO(OH) and NiO_2, which are obtained with varying water content by strong oxidation (chlorine or anodic oxidation) of nickel(II) salt solutions in the presence of alkali. These compounds take part in processes in the alkaline nickel storage batteries (16h).

CHAPTER 35

The Platinum Metals

a. Occurrence. The platinum metals usually follow each other in nature because their atoms have nearly the same size. Palladium and platinum are the most abundant. The nobility of the platinum metals and their relative affinity for oxygen and sulfur (chap. 33) indicate that they may be expected most of all in metallic phases, then in sulfide phases and in smallest amounts in oxide phases. They should therefore occur in larger amounts in the interior of the earth than in the earth's crust, where their abundance is very low (17b, tab. 17.3). This explains also their relatively high proportion in iron meteorites (on the average, 10 p.p.m. each). In the earth's crust the platinum metals occur mostly in metallic form and as sulfides (platinum also as arsenides and sulfide arsenides). In metallic form they are most often alloyed with each other and with iron, sometimes also with gold. The largest amounts seem to be associated with certain rocks very rich in olivine (23-3h1), in which they occur both in metallic form and as sulfides (arsenides), very often in association with chromite (31-2a). Not infrequently, nickel sulfide minerals (especially pentlandite, 34-1a) containing platinum metals in solid solution also occur in these rocks. Metal grains released on weathering of platinum-metal-bearing rocks and concentrated because of their high density in certain sedimentary deposits (for example, alluvial sand) originally constituted the most important source of the platinum metals.

The largest deposits of the platinum metals in the world so far known are in the Union of South Africa (Transvaal) and consist of sulfides, arsenides and metal in olivine-rich rocks. Deposits in Canada (Ontario) consisting of sulfides and arsenides are also of great importance. Here the platinum metals are recovered entirely in connection with the processing of simultaneously occurring nickel ores (34-1a). Production in the Soviet Union is also large, coming from sedimentary as well as primary deposits, mainly in the Urals and Siberia. The classical sedimentary occurrences in Columbia and similar deposits in Alaska also are still of some importance.

b. History. Platinum metals in the form of grains from alluvial sand were noticed very early in both the Old and New Worlds and were used sporadically alloyed with gold in jewelry and art objects. In South America the Spanish gradually learned to use the metals. In the 1740's Wood and Brownrigg studied a sample from Colombia, which they proposed to be a new semimetal (they were unable to melt it, and considered fusibility to be a criterion of a true metal). For this they adopted the name platina (diminutive of plata, silver) already used by the Spanish. In 1803–1804 Wollaston found

Table 35.1. *Physical properties of the platinum metals.*

	44Ru	45Rh	46Pd
Crystal structure	hexagonal-close-packed	face-centered cubic	face-centered cubic
Density at 20°C, g cm^{-3}	12.4	12.4	12.0
Melting point, °C	2310	1960	1550

	76Os	77Ir	78Pt
Crystal structure	hexagonal-close-packed	face-centered cubic	face-centered cubic
Density at 20°C, g cm^{-3}	22.6	22.55	21.45
Melting point, °C	3050	2450	1770

two new metals in crude platinum and named them *palladium* (after the just-discovered asteroid Pallas) and *rhodium* (after Gk. ῥόδον, rose, because of its rose-colored salt solutions). Almost simultaneously Tennant discovered two further platinum metals, which he called *osmium* (after Gk. ὀσμή, smell, because of the strong smell of the tetroxide), and *iridium* (after Gk. ἶρις, rainbow, because of the many different colors of its compounds). The sixth platinum metal was discovered by Klaus and was given the name *ruthenium* (after New Latin Ruthenia = Russia).

c. Properties. As shown in tab. 35.1 the melting points of the platinum metals fall to the right in each period but increase on going down in each group. The hardness of the metals varies in a similar way. Therefore, osmium is the most difficultly melted (of all metals only tungsten and rhenium have higher melting points) and the hardest (hardness on Mohs' scale = 7.5; cf. 6-5c), while palladium is the most easily melted and the softest. The three hardest metals, iridium, ruthenium and osmium exhibit an increasing brittleness and decreasing malleability in that order. The other metals are easily malleable. Osmium is gray-blue, platinum gray-white, the others more silver-white.

Because of the nobility of the platinum metals their compounds with volatile substances on igniting give the metals as residues. These then take a spongy form (*platinum sponge*, etc.), which often has a very high surface area and therefore much greater reactivity than the compact metals. Reduction of platinum metal compounds in solution (for example, cathodically) may deposit a still more finely divided black metal (*platinum black*, etc.). The platinum metals are easily obtained in *colloidally dispersed* form (15-4), usually as hydrosols, by dispersal in an electric arc.

In compact form at room temperature the platinum metals are not noticeably affected by air or water. On heating in air ruthenium and osmium form volatile tetroxides RuO_4 and OsO_4. Rhodium, iridium and palladium are somewhat tarnished at red heat in air or oxygen, but the oxides formed (Rh_2O_3, IrO_2 and

PdO) give off oxygen again at temperatures over about 1100°C. Platinum is not oxidized at temperatures below about 1000°C. On the other hand, platinum, like rhodium, iridium and palladium, slowly decreases in weight on heating in air or oxygen above about 1000°C. At this temperature the decrease in weight probably results mostly from the formation of volatile oxides and only at about 1500°C do the vapor pressures of the metals become high enough to have any importance.

On heating platinum metals with halogens and with nonmetals that are solid at room temperature, a reaction generally takes place. Metal melts usually react with the platinum metals to form alloys. Easily melted metals may therefore attack the platinum metals at quite low temperature. Those substances that give metals easily on heating are therefore also harmful.

Very finely divided platinum metals are sometimes attacked by acids, while in compact form, on the other hand, they are extremely resistant. In chap. 33 it was stated that the nobility of the platinum metals increases to the right in the periodic system. But all the platinum metals have great complex-forming tendencies and the stability of most of their complexes increases the further to the right the metals lie. This results in the fact (cf. 16c) that attack by acids containing strong complex-formers (the anion of the acid or other ligands) increases to the right. Palladium and platinum are therefore attacked by aqua regia, which is resisted by the other metals in compact form. Palladium is also attacked by oxidizing acids, for example nitric acid. Rhodium and iridium in compact form are practically completely resistant to all acids, while ruthenium and osmium are affected somewhat by oxidizing acids. Alkali melts and other salt melts often attack the metals. However, platinum resists nonoxidizing alkali melts in the absence of air. At ordinary temperature alkali hydroxide solutions mainly attack only ruthenium and osmium.

In a finely divided state all platinum metals easily take up hydrogen, which at low temperature is probably for the most part adsorbed on the metal surface. At higher temperature solution of hydrogen in atomic form (chap. 27) also occurs. The solubility of hydrogen gives rise to a rapid diffusion of hydrogen through compact metal at red heat. Palladium absorbs hydrogen appreciably more easily than the other platinum metals and also dissolves much larger amounts. 1 vol. of compact palladium at room temperature dissolves about 850 vol. of hydrogen and in finely divided form considerably more. The permeability of palladium to hydrogen is high already at 200°C and this is also the case for certain alloys rich in palladium. If a platinum metal is exposed to atomic hydrogen, for example as a cathode at which hydrogen is formed on electrolysis, it absorbs hydrogen at ordinary temperature even in compact form. The low hydrogen overvoltage on platinum metals (16f) is doubtless related to this effect.

The great ability of the platinum metals to adsorb and dissolve hydrogen certainly contributes to their effectiveness as contact catalysts in hydrogenation and dehydrogenation processes. But they also catalyze processes in which hydrogen is not involved, for example oxidation of sulfur dioxide (21-10b). A classic example of platinum as a contact catalyst where hydrogen is involved is the ignition of oxyhydrogen with platinum sponge.

d. Production. Concentration of platinum metals is carried out mainly by panning (metal phases) and flotation (sulfide and arsenide minerals). The subsequent concentration of platinum metals from sulfide and arsenide minerals is carried on in Canada in connection with the recovery of nickel (34-4a). The platinum metals are then obtained either from the residues left by the Mond process after treatment with carbon monoxide, or from the anode slime formed in the electrolytic refinement of nickel (16f). The South African ores are more and more worked only for platinum metals but the methods are mostly the same, although the Mond process is not involved here. The platinum metals in the concentrates recovered in these various ways are separated from each other and from other metals by several different methods, which are very complicated and to some extent not made public. Here we can only note that the concentrate is generally first treated with aqua regia. This dissolves especially platinum and palladium and foreign metals. The other platinum metals remain largely undissolved. The solution is carefully reduced, leaving mainly platinum in the tetravalent state (as $PtCl_6^{-2}$ ion). Addition of ammonium chloride precipitates ammonium hexachloroplatinate(IV) $(NH_4)_2PtCl_6$, which on heating to red heat is decomposed leaving platinum as platinum sponge.

e. Uses. The platinum metals are used almost exclusively in metallic form, and then it is particularly their chemical and thermal stability and their catalytic properties that are utilized. Among the platinum metals platinum has been most used up to now for vessels for chemical processes, both in the laboratory (especially crucibles) and in industry. Platinum is also used as resistance material in laboratory furnaces and as electrode material. The jewelry industry consumes a large part of the platinum production, but here palladium, the cheapest of the platinum metals, has also begun to be used. Since both palladium and platinum are very soft, they are then usually alloyed with some of the other platinum metals to make them harder. Such alloys have also been used in other cases where both hardness and chemical or thermal stability are desired (for example, fountain pen nibs, electrical contacts, spinnerettes for spinning rayon). Gold alloyed with platinum or palladium is much used in dental practice. Palladium has also begun to be used for many other purposes as an alloying substance together with nickel, copper, silver and gold. Platinum metals are also incorporated into vacuum-tube parts. Electrolytically deposited rhodium plating is used where hardness, resistance to corrosion, high reflectivity and low contact resistance are required; for example, for precision weights, electrical contacts, and searchlight reflectors. Platinum metals are incorporated in several important thermocouple combinations, the most common consisting of pure platinum welded to an alloy 10–13 wt% rhodium in platinum.

Nowadays, all of the platinum metals, either pure or in alloys, are used for various catalytic applications.

f. Compounds. The great ability of the platinum metals to form complexes means that they are hardly ever encountered as typical ions but rather mainly in complexes (cf. chap. 33). Octahedral coordination is the most common, but palladium(II) and platinum(II) form a very large number of complexes with coordination number 4 and square coordination. In other cases four-coordinated complexes are most often tetrahedral.

The most important platinum metal compounds are *halides* and *halometallate ions*. Many halides can be obtained by reaction of the metal with the halogen. The fluorides have the highest oxidation numbers, and *hexafluorides* MF_6 of all platinum metals except palladium are known. *Osmium heptafluoride* OsF_7 appears to exist, but it has not been possible to prove the existence of the earlier proposed OsF_8. Of the chlorides MCl_2, MCl_3 and MCl_4 are known for most of the metals (however, $RhCl_4$, $PdCl_3$ and $PdCl_4$ have not been proved). The most important of the *halometallate ions* are $M^{II}X_4^{2-}$ (only for Pd^{II} and Pt^{II}; square coordination) and the very common octahedral $M^{III}X_6^{3-}$ and $M^{IV}X_6^{2-}$.

When platinum is dissolved in aqua regia *hexachloroplatinate(IV) ion* $PtCl_6^{2-}$ is formed, and after the nitric acid is removed by heating with excess of hydrochloric acid yellow crystals of $H_2PtX_6 \cdot 6H_2O$ can be obtained from the solution. This hydrate has been called "chloroplatinic acid", but probably no acid H_2PtCl_6 exists (12-2f). Silver and cesium hexachloroplatinates are poorly soluble in water, the rubidium, potassium and ammonium salts are rather poorly soluble but the sodium and lithium salts are easily soluble. *Ammonium hexachloroplatinate(IV)* is very important in the production of platinum because it is easily precipitated out of a solution containing $PtCl_6^{2-}$ ion and then converted to platinum sponge at low red heat (35d). When $H_2PtCl_6 \cdot 6H_2O$ is heated brown, water-insoluble *platinum(II) chloride* $PtCl_2$ is formed. When this is dissolved in hydrochloric acid, square-coordinated *tetrachloroplatinate(II) ion* $PtCl_4^{2-}$ is obtained, of which several salts are known. This ion can also be obtained by reduction of $PtCl_6^{2-}$, with sulfur dioxide, for example. The above reactions are also exhibited to a large extent by palladium.

Among the other square complexes of platinum(II) and palladium(II) we should especially note the numerous *ammine complexes* (for example, $M^{II}(NH_3)_4^{2+}$ and $M^{II}(NH_3)_2X_2$, the latter in two forms, cis and trans; see 7-5).

Among the *oxides*, *ruthenium* and *osmium tetroxides* RuO_4 and OsO_4 are of especial interest, partly because of their high oxidation number VIII and partly because of their high volatility. RuO_4 is formed in small amounts by heating the metal above 600°C in oxygen, but it is best prepared by oxidation of a conc. potassium ruthenate(VI) solution with chlorine ($RuO_4^{2-} \rightarrow RuO_4 + 2e^-$). At room temperature it forms rather volatile, yellow crystals, which melt at 26°C to an orange-red, volatile liquid. The boiling point cannot be reached at atmospheric pressure, since the liquid decomposes first explosively into $RuO_2 + O_2$. OsO_4 is always formed when osmium or an osmium compound is oxidized with oxygen and is most simply prepared by heating osmium powder in a stream of oxygen gas. At room temperature it forms rather volatile, pale yellow crystals, which melt at 40°C. The liquid boils without decomposition at 130°C and the gas tolerates temperatures up to 1500°C. Both the tetroxides have unpleasant odors and irritate the mucous membranes and eyes. They dissolve in water to form practically neutral solutions. The water solution of OsO_4 is reduced by many organic substances giving a black color resulting from the formation of metal or lower oxides. Unsaturated fatty acids are especially effective reducing agents, and the solution is used to stain fats in microscopic preparations.

Ruthenium powder in an oxidizing alkali hydroxide melt (for example, $KOH + KNO_3$) gives tetrahedral *ruthenate*(VI) *ion* $Ru^{VI}O_4^{2-}$. If chlorine is passed into a ruthenate(VI) solution its color is changed from red to green because of the formation of similarly tetrahedral *ruthenate*(VII) *ion* $Ru^{VII}O_4^-$ (older name "perruthenate ion"; note the similarity to the metals in group 7). Crystals of alkali ruthenate(VII) are black. On the further addition of chlorine RuO_4 is formed as mentioned above. In an analogous manner *osmate*(VI) is obtained with the ion $Os^{VI}O_4^{2-}$.

Platinum(II) *sulfide* PtS and *platinum*(IV) *sulfide* PtS_2 are formed as black precipitates when hydrogen sulfide is passed into solutions containing $PtCl_4^{2-}$ and $PtCl_6^{2-}$, respectively. Both sulfides are poorly soluble in acids and alkali solutions; PtS hardly dissolves at all in aqua regia, while freshly precipitated PtS_2 dissolves in hot, conc. nitric acid and aqua regia. Alkali sulfide solutions dissolve PtS with difficulty and PtS_2 only when freshly precipitated or when present in solid solution in other sulfides soluble in these reagents. In ammonium polysulfide solution PtS_2 dissolves more easily, with the formation of *thioplatinates*(IV).

CHAPTER 36

Group 11, The Copper Group or the Coinage Metals

36-1. Survey

The elements of this group can all exist with the oxidation numbers I, II and III. The most common oxidation numbers for copper are I and II, for silver I, for gold I and III. The less common of the three oxidation numbers are nearly always found only in strong complexes. The number II is extremely rare for gold.

With the oxidation number I a filled $(n-1)d$ group (d^{10}) is obtained and the result is formation of ions with 18-shells (5-2f; fig. 5.8). Ag(I) can form rather saltlike and easily water-soluble compounds, especially AgF, $AgNO_3$ and $AgClO_4$, but most other Ag(I), compounds are, like Cu(I) and Au(I), notably covalent in structure and form poorly water-soluble substances or strong complexes. The 18-shell ions do not contribute any color as long as they are not combined with easily polarizable atoms or atomic groups. They are thus colorless in the hydrated state. Tetrahedral coordination is the most common for Cu(I) and Ag(I). However, they also occur in linear or nearly linear coordination, thus with the coordination number 2, and for Au(I) this is predominant (cf. 5-3j).

Oxidation numbers greater than I imply that the $(n-1)d$ group is incomplete —d^9 for II, d^8 for III—and we have typical transition metal ions. These are colored in most all cases; the most notable exceptions are CuF_2 and $CuSO_4$ in the anhydrous states. Of Cu(II) and Au(III) the latter forms the strongest complexes. The d^8 and d^9 groups provide the necessary conditions for square coordination (5-3i, j) which here is the most common. Often, however, and especially for Au(III), there are two further ligands at greater distance, so that an augmentation to an irregular coordination octahedron occurs.

The metals become more noble the heavier the atom is (tab. 16.1). Silver and gold are very noble. Silver is not oxidized in air or water at ordinary temperature, nor is gold under any conditions. Gold compounds are very easily reduced to the metal. Copper and silver have a great affinity for sulfur.

36-2. Occurrence and history

a. Occurrence. As shown in tab. 17.3 the mean abundances of silver and gold in the earth's crust are appreciably less than that of copper. The mean abundance of gold is very small. However, silver and especially gold probably are present in considerably higher proportion in the interior of the earth (17b). The slight affinity of the metals for oxygen means that they are seldom contained in oxidic minerals. On the other hand their high affinity for sulfur leads them often to form sulfide minerals (here besides sulfides we may include arsenides, antimonides and tellurides). All three metals can occur in native form, but this is especially true of gold, which is found mostly in this form.

Of the world's known *copper* resources about 90 % consist of sulfide ores, about 9 % of oxidic ores, while native copper (mainly at Lake Superior in the United States) accounts for less than 1 %. The most important ore-forming sulfide minerals are *chalcocite* Cu_2S, *chalcopyrite* $CuFeS_2$ (the most common copper mineral), and *bornite* Cu_5FeS_4. Oxidic copper minerals are formed from sulfide ores by the action of air and water and substances dissolved in the water. Their occurrences therefore lie near the surface. The most important are *cuprite* Cu_2O, *malachite* $Cu_2(OH)_2CO_3$ (green) and *azurite* $Cu_3(OH)_2(CO_3)_2$ (blue).

The United States is the largest producer of copper with the principal deposits located at the southern end of the Rocky Mountains. In Canada there are rich deposits, especially around Hudson Bay, and in South America especially in Chile. In Africa very large amounts of copper ore are mined in Zambia and Katanga. Beside these, only the Soviet Union, with ores mainly from the Urals, approaches a production comparable with those regions mentioned above. The chalcopyrite ore of Falun, Sweden, which was first worked in the 13th century, for a long time made Sweden the world's largest producer of copper. Now the largest Scandinavian copper production comes from Outokumpo, Finland, followed by that from the Skellefteå field (Here the oldest mine, in Boliden, is now exhausted) in Sweden.

Copper is found in both plant and animal organisms, but its physiological function in many cases is not completely understood. It is incorporated into certain enzyme-type proteins. A copper-containing, blue globulin, ceruloplasmin, occurs in the blood plasma of all mammals. Hemocyanin, which serves the same purpose for crustaceans and molluscs that hemoglobin does for vertebrates, contains copper as the active transition metal. An insufficient supply of copper causes deficiency diseases in both plants and animals, which may be serious in raising livestock as well as food crops; ruminant animals are especially susceptible to copper deficiency. However, larger amounts of copper have toxic effects, and especially lower organisms are sensitive to it. For humans copper and its compounds are moderately toxic. Serious poisoning is rare, among other things because vomiting easily sets in.

The most common *silver* mineral is *argentite* Ag_2S, which is analogous in structure to chalcocite Cu_2S. Minerals consisting of solid solutions between these two are also known. The thioarsenites and thioantimonites of silver (22-14) should also be mentioned, especially *pyrargyrite* Ag_3SbS_3 (less important is *proustite* Ag_3AsS_3; see 22-2a). *Cerargyrite* is AgCl. However, the most important source of silver is contained in other sulfides, especially galena in which it is believed mostly to be included as argentite (up to 1 wt % Ag), but also in chalcocite in which the silver is in solid solution. Silver is therefore mostly recovered in connection with lead and copper production. At present the yield of silver does not cover the demand.

About 70 % of the world's production of silver comes from North and South America with large deposits in the Cordilleran mountain chain from Alaska in the north to Chile in the south.

Gold as stated earlier occurs mostly in native form, but is then usually alloyed with other metals, especially silver. The most important ore-forming gold compounds are tellurides mostly of the type $(Au,Ag)Te_2$. However, these are easily weathered and the metal released. When the residues of completely weathered gold-bearing rocks are carried away by water the heavy native gold is often concentrated by washing in certain parts of the sediments formed (gold sand).

The classical gold deposits in, for example, California, Australia, Alaska, and Russia consisted of alluvial sediments, while the gold in the Union of South Africa, which accounts for nearly half of the world's production, is contained almost entirely in Algonquin rocks. The most important producers of gold besides the Union of South Africa now are the Soviet Union and Canada.

Soluble silver and gold compounds are poisonous.

b. History. The rather common occurrence of metallic gold in river sands presumably is the reason why gold was the first metal that mankind became acquainted with. Since gold particles could readily be forged together by hammering, it was easy to make objects from it. Copper probably came next in the history of discovery, but all three of the coinage metals have been known since prehistoric times. The Greeks called copper χαλκός (cf. 4-2c), and this root appears in the names of many copper minerals. The Romans used the name *cuprum* after Cyprus where important copper mines were located. They called silver *argentum* (from Gk. ἀργύρεος) and gold *aurum*. The three Latin names are used in deriving chemical names from names of the elements.

36-3. Properties of the coinage metals

Some physical properties of the coinage metals are summarized in tab. 36.1. All three of the metals have face-centered cubic (= cubic-close-packed, 6-1f) structures. They are extremely good conductors of both electricity and heat. Silver has the highest conductivity of all the elements, and next comes copper. The red surface color of copper, the white color of silver and the yellow color of gold are familiar to everyone. The hardness decreases with increasing atomic number and at the same time the malleability, which is already high in copper, increases. Gold is extraordinarily soft and malleable and can be rolled and hammered out to extremely thin foil (*gold leaf* is usually about 0.1 μm = 1000 Å thick). In sufficiently thin sheets copper and gold have green to blue-green transmission colors (7-3a), silver has blue to violet.

Table 36.1. *Physical properties of the coinage metals.*

	29Cu	47Ag	79Au
Density at 20°C, g cm^{-3}	8.94	10.49	19.31
Melting point, °C	1083	960.5[a]	1063[a]
Boiling point, °C	~2580	~2180	~2710

[a] Calibration point in the International Practical Temperature Scale.

Compact copper is not oxidized in dry air at room temperature, and even in moist but otherwise pure air or pure water it is attacked very slowly. On heating in air oxidation begins to be quite rapid at about 200°C and the surface then first becomes yellowish red (Cu_2O) and then black (CuO). Concerning the mechanism of oxidation, see 6-3c. When copper is attacked in moist air or water at ordinary temperature forming a green surface coating (*patina*), other substances are active. A copper roof is generally first blackened by copper sulfides. Through oxidation of these gradually copper(II) hydroxide sulfates (for example, $Cu_4(OH)_6SO_4$) appear, which are also formed by the action of air containing sulfur dioxide. In inland regions hydroxide sulfates make up most of the patina. In addition it usually contains copper(II) hydroxide carbonates (for example, $Cu_2(OH)_2CO_3$, the mineral malachite). The proportion of these is small in country areas, larger in the cities, but there also less than that of the hydroxide sulfates. On the seacoast hydroxide chlorides, for example $Cu_2(OH)_3Cl$, may reach high proportions. On objects of copper or copper alloys that have lain in the ground, the patina may also have taken up other ions, for example phosphate ions.

Silver is not attacked by oxygen at atmospheric pressure but in oxygen at a pressure >15 bar and at 300°C it forms silver oxide Ag_2O. Molten silver dissolves oxygen from the air (about 20 volume units per volume unit of silver) but gives off the greater portion of it on solidifying. This solubility of oxygen causes difficulties in melting and casting pure silver, but is practically absent with common silver alloys. Gold does not at all react with oxygen.

At ordinary temperature fluorine attacks copper with the formation of a protective coating, but on the other hand does not attack silver or gold. None of the three metals are attacked at ordinary temperature by the other halogens if they are completely dry. In the presence of water vapor or at higher temperature, on the other hand, reaction does occur. Sulfur and many sulfur compounds attack copper and silver even at ordinary temperature while gold in most cases is unaffected. When copper and silver tarnish in dry air, this results mainly from the formation of sulfide by the action of H_2S. So-called "oxidized" silver has been blackened by sulfide by treatment with alkali sulfide solution.

The behavior of the coinage metals in acids follows from the descriptions of standard potentials given previously (end of 16c) and of sulfuric acid (21-11b1), nitric acid and aqua regia (both in 22-10d1).

Copper is quite resistant to alkali hydroxide solutions in the absence of air; silver and gold are also resistant to alkali melts as long as they are not oxidizing. Copper dissolves in conc. alkali cyanide solutions with the evolution of hydrogen: $Cu(s) + 2CN^- + H_2O \rightarrow Cu(CN)_2^- + OH^- + \frac{1}{2}H_2(g)$. To dissolve gold and silver in cyanide solution oxygen must also be introduced, and then no hydrogen is evolved (see 36-4 b).

All three metals easily combine with mercury to give amalgams.

Gold and silver particularly can be easily precipitated as metals by reduction of their solutions. In this way, but also by dispersal in an electric arc (15-4), the metals can also be obtained in colloidal form. Copper sols are also known, but in hydrosols of copper the particles are often oxidized. Copper and gold ruby glass (23-3i2) acquire their red colors from copper and gold particles of colloidal dimensions.

36-4. Production of the coinage metals

a. Copper. The iron content of chalcopyrite, in addition to the fact that pyrite often is present in sulfide copper ores, means that iron plays an important role in the metallurgy of copper. As a rule, copper ores also contain silver and gold, which must be recovered.

Here we discuss only the treatment of sulfide ores, by far the most important copper ores. The copper content is usually low and therefore the copper minerals are concentrated by flotation (15-2). This is followed either by smelting ("dry") or solution ("wet") processes. The latter are used mainly for low-grade ores, which are not worthwhile smelting. In the *dry processes*, the copper/sulfur ratio is first adjusted either by roasting in which the sulfur is partly burned off, or by mixing in copper scrap. Roasting also reduces the amount of substances with volatile oxides, especially arsenic, antimony and selenium. The roasted material consists mainly of oxides and sulfides of copper and iron, and silicates. They are smelted in a reverberatory furnace or electric furnace with added slag formers, usually quartz. The result depends primarily on the fact that compared to iron, copper has a greater affinity for sulfur and less for oxygen. The sulfur therefore combines first with copper, and then, according to the supply, with iron. The excess iron forms iron(II) oxide, while all iron(III) is reduced by sulfide sulfur. The iron(II) oxide enters into a silicate phase, consisting chiefly of iron(II) silicate, which forms slag. Under the slag a heavier, liquid sulfide phase (*matte*, cf. 34-4a) is formed, consisting mostly of $Cu_2S + FeS$.

The matte is tapped off and air blown through it in a *converter*. Here the iron sulfide reacts first: $FeS(l) + \frac{3}{2}O_2(g) \rightarrow FeO + SO_2(g)$. The iron(II) oxide combines with the added quartz to give iron(II) silicate, which forms slag (this stage of blowing is therefore called the "slagging period"). The slag is tapped off, and there remains a sulfide phase with higher copper content than before, a more concentrated matte. Blowing is resumed and now the reaction $Cu_2S(l) + O_2(g) \rightarrow$

$2Cu(l) + SO_2(g)$ takes place ("blister-forming period"). The molten copper sinks and forms a layer under the sulfide (the copper converter is shaped like a horizontal cylinder and the air is blown from the sides above the copper layer into the sulfide layer). As long as sulfide remains, Cu_2O cannot be formed. When practically all of the sulfur is oxidized the blowing is halted and the copper, which is still quite impure, is tapped off.

The molten copper may be transferred directly to a *refining furnace*, in which air is blown through the melt. In other cases the copper is first cast as "blister copper" (this is very blistery mainly because of its sulfur content), which is then melted down in the refining furnace. Here most of the impurities are oxidized and those that do not go off in gaseous form are removed in a slag. Some sulfur remains, but also the copper is oxidized to some extent. The sulfur and oxygen are removed by reaction with the dry-distillation products from green tree logs (poles, hence the term *poling*) that are immersed in the melt. A more recent method is to blow a reducing gas (mixture of CO and H_2 obtained according to 18d, or NH_3) through the melt. The copper from the refining furnace is cast directly into anodes (the refining furnace is therefore also called "anode furnace") for *electrolytic refining* (16f). The electrolyte is a sulfuric acid-copper sulfate solution, and *electrolytic copper* is deposited on thin cathode sheets of pure copper. Most of the impurities, among others silver, gold and selenium, form an anode slime, which is further processed. In general the electrolytic copper is melted, and by air blowing and then poling or gas reduction the traces of sulfur carried from the electrolyte are eliminated. Finally the copper is cast.

In the *wet process* the copper is leached out of the ore. Oxidic ores can generally be leached directly with acid, but the sulfide ores may not dissolve easily. The copper is then converted to a more easily soluble form by special roasting processes. In *chloridizing roasting* sodium chloride is added, and by complicated and not completely understood processes the copper and iron content of the ore is converted mainly to $CuSO_4$, $CuCl_2$ and Fe_2O_3. *Sulfatizing roasting* is done without additives; here by precise temperature control $CuSO_4$ and Fe_2O_3 are formed. The leaching is carried out mostly with sulfuric acid and the solution then allowed to flow over iron scrap. The copper precipitates in the form of so-called *cement copper* ($Cu^{2+} + Fe(s) \rightarrow Cu(s) + Fe^{2+}$); sometimes accompanying silver and gold are also precipitated. The cement copper is refined as above.

b. Silver and gold. Silver minerals are concentrated by flotation. For gold minerals (generally metallic gold, usually alloyed with silver), because of their high density, washing can also be used. Then the metals are dissolved, for which the principal method is *cyanide leaching*. For gold this is still usually preceded by *almalgamation*. The concentrate after flotation is treated with mercury, which takes up the gold and the silver dissolved in the gold. The mercury is removed from the amalgam formed by distillation in iron retorts, and is condensed and reused. Rarely more than 60% of the gold can be recovered by amalgamation, but after the cyanide leaching the total recovery is about 95%. Cyanide leaching is based on the fact that gold and silver as well as silver sulfide and silver chloride are dissolved by alkali cyanides in the presence of air according to

$$2Au(s) + 4CN^- + H_2O + \tfrac{1}{2}O_2 \rightarrow 2Au(CN)_2^- + 2OH^- \quad \text{(same for Ag)}$$
$$Ag_2S(s) + 4CN^- + 2O_2 \rightarrow 2Ag(CN)_2^- + SO_4^{2-}$$
$$AgCl(s) + 2CN^- \rightarrow Ag(CN)_2^- + Cl^-$$

Usually sodium cyanide solution is used through which air is blown. The noble metals are precipitated from the solution with zinc dust: $2Au(CN)_2^- + Zn(s) \rightarrow Zn(CN)_4^{2-} + 2Au(s)$. Any excess of zinc is dissolved in sulfuric acid and the remaining noble metals melted. The separation of silver and gold in the raw metal obtained is nowadays accomplished mostly electrolytically (16f). If the amount of other metals present is low, the metal can also be leached with sulfuric acid, which dissolves silver but not gold (end of 16c). For such "parting" nitric acid was formerly used, but this is more expensive and also cannot be used for all proportions of silver and gold.

Large amounts of silver and gold are obtained as by-products of the production of other metals. The major portion of the world's production of silver is accounted for in this way, most coming from silver-bearing lead ores.

Silver (and sometimes accompanying gold) is extracted from argentiferous crude lead ("base bullion"; 23-6c) with zinc by the *Parkes process* (13-1f). The zinc is distilled off and "rich bullion" with about 10% silver is left behind. From this the silver is recovered by *cupellation*. The lead is melted and oxidized along with other base metals by an air stream blown on the surface of the melt. Finally a melt of fairly pure silver (sometimes containing gold) lies on the bottom of the furnace.

Silver and gold by-products of the production of copper are collected in the anode slime during copper electrolysis (36-4a). The processing of this slime is quite complicated and cannot be described here.

36-5. Uses of the coinage metals

a. Copper. Unalloyed *copper* is used especially for its high electrical conductivity, its high thermal conductivity and its resistance to corrosion. The electrical conductivity is sharply decreased by substances dissolved in the solid copper phase, for example phosphorus, arsenic and nickel. Other substances that dissolve in molten copper but to a lesser degree in solid copper and therefore precipitate in the grain boundaries when the melt solidifies (14-2c), reduce the tensile strength. Among these is, for example, bismuth. The electrical industry, which is one of the largest consumers of copper, therefore requires material that is at least 99.9 wt% Cu. In general, 0.03–0.06 wt% oxygen is then present. Unalloyed copper is also used for chemical-industrial apparatus, building material (especially roofing sheet), water pipes, boiler tubes, etc.

Copper alloys are of extreme technical importance. Their production places lower demands on the quality of the copper. Copper is often contained in alloys with other metals as base and such alloys have been described previously. Here we consider only alloys with copper as base.

Copper-base alloys with less than 98 wt % copper and with zinc as the most important alloying element are called *brass*. The proportion of zinc usually lies between about 5 and 45 wt %. These alloys generally have good ductility both cold and hot. With low zinc content the color is red but as the proportion of zinc increases becomes lighter, "brass yellow". Other alloying metals are often added to give special properties. Addition of lead (*leaded brass*) imparts favorable machining properties. *Nickel silver (German silver)* has a quite white color and good corrosion properties. It is often silver plated. A common nickel-silver alloy contains 60 wt % copper, 22 wt % zinc, 18 wt % nickel. Brass alloys of various compositions are also used as brazing alloys.

Other copper-base alloys with less than 98 wt % copper are called *bronzes*. *Tin bronze* may contain 5–30, but nowadays rarely over 10 wt % tin. Its great importance in ancient times depended on its being quite hard and at the same time easily melted and cast. *Phosphor bronze* contains in addition to tin a small amount of phosphorus (usually <0.25 wt %), which improves the mechanical properties to a large degree.

Aluminum bronze, containing about 10 wt % aluminum and also often other alloying elements in small amounts, has high tensile strength and resistance against wear. *Silicon bronze* contains up to 3 wt % silicon. Among the bronzes we may include *cupro nickel* with 10–30 wt % nickel, which has excellent corrosion properties. *Constantan* has still higher nickel content (40–45 wt %), and has low conductivity with low temperature dependence.

Among the copper alloys with about 98 wt % copper we may mention *beryllium copper* (also called *beryllium bronze*), which generally contains up to 2 wt % beryllium. It can be precipitation hardened (13-5f2) and is used when a hard and flexible but nonmagnetic and nonsparking material is required. Its high conductivity may also be important.

b. Silver. Pure silver is used where high conductivity or resistance to corrosion is required. However, its softness makes it necessary to alloy it even for moderate tensile strength. For most applications such as coinage, table silver and silverware, silver is alloyed with copper. Silver-base alloys, generally containing copper and zinc, are used as brazing alloys where a strong joint is required. Silver is also used in dental amalgam.

Large quantities of silver are consumed in the production of silver halides for photographic purposes (36-8).

Silver plating of metal objects is nowadays done electrolytically. Silver-plated mirrors are most often made by precipitating the silver from an ammoniacal silver nitrate solution (36-7a).

c. Gold. Like silver, gold is alloyed whenever any hardness or strength is required. In goldsmith work gold is mostly alloyed with copper, but alloys with silver, nickel and zinc are also used ("white" gold gets its light color from nickel or palladium, but contains other elements also). Gold coinage always contains copper, sometimes also silver. Much gold is used in dentistry. Here also it is alloyed mainly with copper, but alloys with silver, nickel, zinc and platinum metals are also used.

Gold plating of metal objects is usually carried out electrolytically. When porcelain or glass are gilded for decorative purposes, a gold compound painted on the object is decomposed by heating. Wood, gypsum, leather and similar materials are gilded by gluing on gold leaf. Thin gold sheet or gold leaf was used in ancient times for gilding. Later, gilding was done by applying a coat of gold amalgam, which was then heated to evaporate the mercury (fire gilding).

36-6. Copper compounds

a. Copper(I) compounds. The disproportionation of Cu^+ in equilibrium with water according to $2Cu^+ \rightleftharpoons Cu + Cu^{2+}$ becomes negligible only if the amount of Cu^+ ion is very small because it is strongly tied up (16d). In equilibrium with water, therefore, the oxidation number Cu(I) is found only in poorly soluble compounds or strong complexes.

Copper(I) *oxide* $\sim Cu_2O$ exists as the red mineral *cuprite*. As stated in 6-3a the composition of the oxide can vary at least from Cu_2O to $Cu_{1.996}O$. With departure from the ideal composition a small amount of Cu(II) appears along with a corresponding number of cation vacancies. The semiconducting properties of the oxide were discussed in 6-4d. The oxide is formed on heating copper in air (cf. also 6-3c), but then becomes easily contaminated by copper(II) oxide CuO. As long as metal remains, however, the layer immediately outside it consists of Cu_2O. The equilibrium $2CuO \rightleftharpoons Cu_2O + \frac{1}{2}O_2(g)$ is displaced to the right on raising the temperature, and above 1050°C almost exclusively Cu_2O is formed. To prevent the reformation of CuO the cooling must be done very quickly or in a protective atmosphere. Copper(I) oxide is practically insoluble in water and is therefore obtained by addition of alkali hydroxide to copper(I) salt solutions ($2Cu^+ + 2OH^- \rightarrow Cu_2O(s) + H_2O$) or by reduction of alkaline copper(II) salt solutions. The precipitate is yellow to begin with, but becomes red on boiling or drying, probably as a result of increase in grain size (7-3a). Already at 135°C hydrogen begins appreciably to reduce copper(I) oxide to the metal.

Technically, copper(I) oxide is mostly produced from copper scrap by electrolytic means. The scrap serves as anode and sodium chloride solution as the electrolyte. At the anode $CuCl_2^-$ ions are formed ($Cu + 2Cl^- \rightarrow CuCl_2^- + e^-$) and these are converted by the OH^- ions formed at the cathode to Cu_2O ($2CuCl_2^- + OH^- \rightarrow Cu_2O(s) + H^+ + 4Cl^-$). The oxide is widely used as a pigment in ship-bottom paint, where its toxic action prevents marine animal and plant growth.

Copper(I) *halides* (except the fluoride, which is not known with certainty) at room temperature are colorless and poorly soluble in water. At this temperature they all have the sphalerite structure (6-2c, fig. 6.16d), which indicates that the bonds are strongly covalent. They are formed by reduction of Cu(II) salt solutions containing the corresponding halide ion. The iodide $\sim$CuI is formed just by adding iodide ion ($Cu^{2+} + 2I^- \rightarrow CuI(s) + \frac{1}{2}I_2(s)$), as a result partly of the fact that iodide ion is a stronger reducing agent than the other halide ions (tab. 16.1), and partly that the iodide is the least soluble copper(I) halide (tab. 13.1). For the preparation of the chloride and bromide, sulfur dioxide is a suitable reducing agent. The chloride is also obtained by reduction with metallic copper in strong hydrochloric acid solution. Concerning the accompanying color changes, see 7-3b. Analogously with the iodide, *copper*(I) *cyanide* CuCN is formed on addition of cyanide ion to a copper(II) salt solution: $Cu^{2+} + 2CN^- \rightarrow CuCN(s) + \frac{1}{2}(CN)_2(g)$ (cf. 23-2k2).

Complex, colorless *halocuprate*(I) *ions* are easily formed, and therefore copper(I) halides go into solution with excess of halide ion: $CuX(s) + X^- \rightarrow CuX_2^-$. With strong excess of cyanide ion, the cyanide gives the colorless, tetrahedral *tetracyanocuprate*(I) *ion* $Cu(CN)_4^{3-}$. Here the copper is so strongly bound that no copper sulfide precipitate occurs with hydrogen sulfide. At lower concentrations of cyanide the number of cyanide ligands is less than four and the complexes are decomposed by hydrogen sulfide. With ammonia solution colorless *amminecopper*(I) *ions* are obtained, especially according to $CuX(s) + 2NH_3 \rightarrow Cu(NH_3)_2^+ + X^-$.

Solutions of copper(I) compounds absorb carbon monoxide with the formation of *halocarbonylcopper*(I) Cu(CO)X (23-2f).

Copper(I) *sulfide* $\sim Cu_2S$ has a very low solubility product (tab. 13.1) and is therefore precipitated by hydrogen sulfide even from strongly acid copper(I) solutions (cf. 13-1e). The precipitate is black. When the sulfide is prepared by heating copper with sulfur, it is metallic lead gray like the mineral *chalcocite*.

b. Copper(II) compounds. Numerous salts of copper(II) are known, of which many are soluble in water and pratically completely dissociated in water solution. In salt solutions and in certain salt hydrates the blue hydrated *tetraquocopper*(II) *ion* $Cu(H_2O)_4^{2+}$ exists. The four ligands are square-coordinated (5-3i, j), but two additional ligands may lie at a greater distance so that an irregular coordination octahedron results. The ion gives off protons (cf. 12-1b) so that copper(II) solutions are weakly acid. It probably participates in the following equilibrium, among others:

$$2Cu(H_2O)_4^{2+} \rightleftharpoons \left[(H_2O)_2Cu \begin{matrix} H \\ O \\ \diagup \diagdown \\ \diagdown \diagup \\ O \\ H \end{matrix} Cu(H_2O)_2 \right]^{2+} + 2H^+ + 2H_2O$$

The square coordination is maintained in the binuclear complex ion.

Water solutions with high proportions of certain copper(II) salts have other colors than blue because the Cu^{2+} ion to a greater or lesser extent also binds other ligands than water. Such is the case with halides, where halo complexes are formed. Copper(II) salts with ammonia give *amminecopper(II) ions,* and the stepwise addition of ammonia has been discussed in 10c. The appearance of copper(II)–ammonia bonds is indicated by an increasingly dark blue color. The most important complex is the *tetramminecopper(II) ion* $Cu(NH_3)_4^{2+}$, with square coordination.

Copper(II) oxide CuO is formed as a black powder by complete oxidation in air of heated copper (cf. Cu_2O, 36-6a). It is easily reduced and can therefore be used with heat as an oxidizing agent, for example in organic elemental analysis.

When a copper(II) salt solution is made basic, a pale-blue, flocculent and voluminous precipitate is formed, which aside from its variable water content may be considered as *copper(II) hydroxide* $Cu(OH)_2$. The precipitate at ordinary temperature darkens slowly, with boiling more rapidly, through the formation of CuO. When freshly precipitated it dissolves appreciably in concentrated alkali hydroxide solution with a deep-blue color, with the formation, among other things, of *tetrahydroxocuprate(II) ion* $Cu(OH)_4^{2-}$. The oxide and hydroxide easily dissolve in ammonia to form tetramminecopper(II) ion.

Such a tetramminecopper(II) solution is used as a solvent for cellulose (*Schweizer's reagent*). On addition of acid the cellulose reprecipitates and can then be formed into fibers (rayon; cf. 23-2j).

Certain organic molecules containing OH groups, for example the ion of tartaric acid, tartrate ion, are bound through the O atoms of these groups so strongly to Cu^{2+} ions that copper(II) hydroxide does not precipitate when the solution is made basic. However, if copper(II) is reduced to copper(I) the bonds to the organic molecules become weaker or are ruptured, and Cu_2O precipitates in the basic solution. This property is made use of when a basic solution containing tartratocuprate(II) ions (*Fehling's solution*; cf. 13-1e) is used as a reagent for organic reducing agents, for example aldehydes and monosaccharides.

From a solution of copper(II) oxide, hydroxide or carbonate in hydrochloric acid the easily soluble *copper(II) chloride dihydrate* $CuCl_2 \cdot 2H_2O$ may be crystallized, which in pure form is blue but, probably because of excess water on the surface, usually has a green color. The crystals are built up of $CuCl_2(H_2O)_2$ groups with square coordination. However, these are packed so that two additional chlorine atoms in neighboring groups are coordinated to the Cu^{2+} ion at a greater distance resulting in an irregular octahedral coordination (cf. 5-3i, j). Anhydrous *copper(II) chloride* $CuCl_2$ is obtained as a brown mass when the dihydrate is heated in a stream of hydrogen chloride gas. The structure of the anhydrous chloride and its behavior when dissolved in water has been discussed in 5-6b. A very concentrated solution is brown, but when diluted the color changes first to green and then to blue, as a result of the progressive breakdown of the complexes in which more and more of the copper–chlorine bonds are replaced by copper–water bonds.

Copper(II) *sulfate* $CuSO_4$ is best known as the *pentahydrate* $CuSO_4 \cdot 5H_2O$ (*copper vitriol*; cf. 21-11 b3). The system $CuSO_4$–H_2O has been treated in 13-5e1. Crystals of the pentahydrate are built up of tetraquocopper(II) ions and sulfate ions. The fifth water molecule is not coordinated by copper and the formula may therefore be written $[Cu(H_2O)_4]SO_4 \cdot H_2O$. The structure explains the blue color of the crystals (large crystals have a strong blue color, the crystalline powder is light blue; see 7-3a). The blue color becomes lighter as the water content of the hydrate and thus the number of copper–water bonds decreases. The anhydrous sulfate is colorless. Copper(II) sulfate is easily soluble in water and the solutions always show a blue color.

Copper(II) sulfate may be prepared by dissolving copper in hot conc. sulfuric acid: $Cu(s) + 2H_2SO_4 \rightarrow Cu^{2+} + SO_4^{2-} + 2H_2O + SO_2(g)$. Industrially, copper scrap is dissolved in hot dilute sulfuric acid with aeration: $Cu(s) + H_2SO_4 + \frac{1}{2}O_2(g) \rightarrow Cu^{2+} + SO_4^{2-} + H_2O$. The sulfate is the most important commercial copper salt. It is used as a starting material for other copper compounds, and to a large extent for combating fungus parasites on plants.

From solutions of copper in nitric acid at room temperature light-blue crystals of *copper*(II) *nitrate* crystallize out in the form of the *hexahydrate* $Cu(NO_3)_2 \cdot 6H_2O$. The hydrate dissolves very readily in water to give solutions of light-blue color. It melts peritectically (13-5f3) already at 26°C.

Copper(II) salt solutions with alkali carbonate give green to blue precipitates of *copper*(II) *hydroxide carbonate*. From such precipitates various crystalline hydroxide carbonates have been obtained, among others $Cu_2(OH)_2CO_3$ and $Cu_3(OH)_2(CO_3)_2$, known also as the minerals *malachite* and *azurite*, respectively. Patina (36-3) often contains minor amounts of copper(II) hydroxide carbonate, but consists usually mainly of hydroxide sulfates. Normal copper(II) carbonate has not been successfully prepared.

Copper(II) *sulfide* $\sim CuS$ can be obtained by heating copper and sulfur together or by precipitation from copper(II) salt solutions with hydrogen sulfide. Its very low solubility product (tab. 13.1) makes the precipitation practically complete even from strongly acid solution (cf. 13-1e). The precipitated sulfide is black. In large crystals it is dark blue with metallic luster, and is found as the mineral *covellite*.

36-7. Silver compounds

a. Silver(I) compounds. In all its most stable and important compounds, silver has the oxidation number I. Silver(I) compounds are chemically very similar to copper(I) compounds, aside from the instability of the latter toward disproportionation (36-6a). Most silver(I) compounds are poorly soluble in water. The most important exceptions are the fluoride AgF, the nitrate $AgNO_3$, the chlorate $AgClO_3$

and perchlorate $AgClO_4$. The sulfate Ag_2SO_4 and the acetate are rather poorly soluble. All these compounds are colorless. The silver(I) ion Ag^+ is hydrated and the hydrate behaves through proton release as a very weak acid.

Many anions give such strong silver(I) complexes, *argentate*(I) *ions*, that their presence strongly increases the solubility of silver(I) salts that are otherwise poorly soluble in pure water. They may thus bring them into solution, or prevent their precipitation (13-1e). The strongest of these complexes is the *dicyanoargentate*(I) *ion* $Ag(CN)_2^-$ and consequently cyanide ion is an excellent medium for the solution of poorly soluble silver compounds (cf. 36-4b). With ammonia the quite strong complex *diamminesilver*(I) *ion* $Ag(NH_3)_2^+$ is obtained. Many poorly soluble silver(I) salts therefore dissolve in ammonia solution.

Silver is reduced to the metal from most compounds by electromagnetic radiation; in many cases, however, the radiation must be ultraviolet or of still shorter wavelength.

Silver(I) *oxide* Ag_2O is formed as a dark-brown precipitate on addition of alkali hydroxide to silver(I) salt solutions: $2Ag^+ + 2OH^- \rightleftharpoons Ag_2O(s) + H_2O$. The equilibrium is strongly shifted to the right because the oxide is very poorly soluble. Nevertheless, water in equilibrium with the oxide reacts distinctly basic. Silver(I) salt solutions with ammonia solution give at first an oxide precipitate. However, the major portion of the silver remains in solution as $Ag(NH_3)_2^+$ and excess ammonia also dissolves the precipitated oxide. Solid silver(I) hydroxide AgOH is unknown, although a complex with this formula exists in small amounts in water solution. When the solubility product of AgOH is (improperly) spoken of, reference is made to the product $[Ag^+][OH^-]$ in a solution in equilibrium with solid Ag_2O (tab. 13.1). Chlorides, bromides and iodides can be converted into hydroxides by treatment with a suspension of silver oxide in water, since the corresponding silver halides are less soluble than the oxide: $2Cl^- + Ag_2O(s) + H_2O \rightarrow 2AgCl(s) + 2OH^-$. The oxide is easily reduced to metal and decomposes into the elements even with gentle heating.

In the *silver*(I) *halides* the ionic bond contribution is relatively great only in the fluoride AgF. As would be expected it decreases as one goes to more and more electropositive halogens. The solubility in polar solvents (water) also decreases as the halogen becomes more electropositive. The fluoride is easily soluble in water (and forms hydrates, which is unusual for silver compounds), but already the chloride is very poorly soluble (tab. 13.1). The fluoride and chloride are colorless, the bromide yellowish-white and the iodide yellow (cf. 7-3b). The chloride is less stable than the fluoride but more stable than the bromide and iodide (6-2e). This is indicated also by their sensitivity to light. On irradiation with visible light the fluoride is unaltered, while the others are darkened by the precipitation of silver: $2AgCl(s) + h\nu \rightarrow 2Ag(s) + Cl_2(g)$. The sensitivity to light increases in the sequence chloride < iodide < bromide. More energetic radiation than visible light (ultraviolet and shorter wavelengths) decomposes all silver halides.

Because of complex formation the solubility of the silver halides increases with sufficient increase in halide ion concentration, as mentioned in 13-1e for the chloride. Thus the mononuclear complexes AgX, AgX_2^-, AgX_3^{2-}, AgX_4^{3-} are formed, as well as polynuclear complexes with the bromide and still more so with the iodide ion. The greater the solubility product of the halide is, the more easily the ion product can fall below it, with increased solubility as a result. Further, the solubility becomes greater, the stronger the complex formed. In ammonia solution, where $Ag(NH_3)_2^+$ is formed, the solubility of the chloride is high, that of the bromide less, and of the iodide very low. Thiosulfates give stronger complexes, especially $Ag(S_2O_3)_2^{3-}$ and $Ag(S_2O_3)_3^{5-}$ (cf. 21-11f1), and lead to quite high solubility even of the iodide. Cyanide ions, which give primarily the very strong complex ion $Ag(CN)_2^-$, dissolve all silver halides very easily.

The silver halides are much used in the production of photographic emulsions (36-8).

Finely divided metallic silver with a silver fluoride solution gives bronze-colored crystals of *disilver fluoride* Ag_2F. These are built up of parallel layers of silver and fluorine atoms in the sequence ... AgAgFAgAgF .... The shortest Ag–Ag distance is nearly the same as in silver metal and the distance Ag–F nearly the same as in silver fluoride. The conductivity is high, at least along the plane of the layers. It has therefore been said that disilver fluoride is a hybrid between a metal and a salt.

Silver(I) *pseudohalides* (20-6) show many similarities to the halides, among other things in that they are all poorly soluble in water. We should mention *silver cyanide* AgCN and *silver thiocyanate* AgSCN, which both dissolve in excess of the anion with the formation of strong *cyano-* and *thiocyanatoargentate*(I) *complexes*, respectively. In both cases the mononuclear complexes AgX, AgX_2^-, AgX_3^{2-} and AgX_4^{3-} have been identified, and in addition polynuclear thiocyanato complexes. Alkali salts of the *dicyanoargentate*(I) *ion* $Ag(CN)_2^-$ are commonly used in the electrolyte for silver plating (cf. 16f).

The low solubility of the silver halides and pseudohalides and their generally well-defined composition make them very useful in analysis.

Silver(I) *nitrate* $AgNO_3$ is obtained by solution of silver in nitric acid, and on evaporation of the solution forms beautiful, colorless crystals. They are very easily soluble in water. The melting point, 212°C, is remarkably low. Many organic reducing agents (for example aldehydes and monosaccharides) reduce silver nitrate to the metal. When a silver nitrate solution is reduced in this way, the silver is deposited on glass as a mirror. A large number of organic compounds have a reducing action under illumination, while completely pure silver nitrate is not affected by light. Proteins are coagulated by silver nitrate, which therefore can be used to cauterize organic tissue ("lapis infernalis", "lunar caustic") When applied to the skin, the skin is blackened in the light. Silver nitrate is the most important silver salt and is the starting material for the production of most other silver compounds.

Silver(I) *sulfide* Ag_2S is formed in a multitude of reactions between silver or silver compounds and sulfur-bearing substances. As a precipitate the sulfide is black, in larger crystals gray-black with metallic luster. It has an extremely low solubility product (tab. 13.1), so that precipitation with hydrogen sulfide is very complete, even from strongly acid solutions.

The sulfide is precipitated even from solutions containing cyanoargentate ion. This is surprising since according to 36-4b the sulfide is dissolved by cyanide ion. In the former case we are concerned with the equilibrium $2Ag(CN)_2^- + S^{2-} \rightleftharpoons Ag_2S(s) + 4CN^-$, which because of the low solubility of the sulfide goes to the right. In the latter case, the reaction given in 36-4b for the solution of the sulfide involves the oxidation of the sulfide ions in this equilibrium to sulfate ions by oxygen. The sulfide ions are so completely consumed in this way that the equilibrium is displaced to the left.

b. Higher oxidation states of silver. Free Ag^{2+} ion is strongly oxidizing and not stable in water solution. In the solid form, *silver*(II) *fluoride* AgF_2 is known; all other silver(II) compounds are complex. In most of these the coordination around silver(II) is square and very often the ligand atoms are nitrogen atoms.

Silver and many silver compounds react with ozone to give a gray-black substance, which has been stated to be silver(II) oxide AgO. However it is considered nowadays to be $Ag^IAg^{III}O_2$.

36-8. The basic photographic processes

The previously mentioned decomposition of the silver halides on irradiation requires large amounts of light, but was formerly used when copies were made without development on so-called printing-out paper. However, with much smaller quantities of light invisible changes already set in, which with subsequent reduction cause a visible silver precipitation (*development*). These invisible changes imply the presence of a so-called *latent image*. In modern photographic methods amounts of light greater than are required only to produce a latent image are almost never used.

The light-sensitive photographic material consists generally of very small silver halide crystals suspended in gelatin (called "emulsion" although "suspension" would be the proper term; cf. 15-4), which is spread on a support of glass, cellulose acetate or paper. For negative material mostly silver bromide with a small iodide content is used (solid solution Ag(Br,I)), and for positive material silver chloride or bromide is used, either alone or in solid solution with each other. In the following the halide is always indicated as the bromide.

It has long been known that some of the silver halide crystals in an emulsion become developable after exposure, while others do not. The greater the amount of light that has reached the emulsion, the greater is the number of developable crystals. On development, the visible silver precipitate begins at one or a few points in the crystal (*development nuclei, sensitive specks*) and spreads from these throughout its volume. The different blackening in a picture is not caused by varying degrees of silver precipitation in all crystals, but rather by a varying number of crystals that are completely or nearly completely blackened. It has also been found that the appearance of a latent image is dependent to a large degree on the fact that silver halide crystals contain Ag_2S as an impurity. Sulfur is always taken from the proteins in the gelatin, which thus increase the photosensitivity of the silver halide. Ag_2S occurs adsorbed on the surface of the crystal and especially along the edges formed by the steps in its growth spirals (14-2b). The increase in sensitivity (*sensitization*), which results when the emulsion is allowed to stand while warm and "ripen" before pouring, is caused by the fact that more Ag_2S is thus formed. Further sensitization can be achieved among other ways by treatment with sulfur compounds, for example thiosulfate ions.

The nature of the latent image was little understood for a long time, but has been explained especially through the work of Gurney and Mott in 1938, and Mitchell in 1957. Their hypotheses have been based both on experimental observation and energy calculations. On irradiation a halide ion absorbs energy and gives off an electron: $Br^- + h\nu \rightarrow Br + e^-$. The released electron becomes a conduction electron (6-4d). The bromine atom forms a positive electron hole (6-4d) in the crystal and this can move by absorbing an electron from a neighboring bromide ion, which then becomes a new positive hole, and so on. The migration of the hole continues until it encounters another atom or group of atoms whose electron cloud can easily donate an electron to the hole. The hole is filled and the atom or atomic group is said to be a "trap" for the hole. A study of the energy conditions shows that the most important trap for the positive hole is the trace of Ag_2S in the crystal mentioned above. When Ag_2S gives electrons to the holes AgS molecules appear, and also Ag^+ which occurs as an interstitial ion: $Ag_2S - e^- \rightarrow Ag_2S^+ \rightarrow AgS + Ag^+$. The Ag^+ ions wander (still in interstitial sites), and if they reach distorted lattice regions at the edges of steps on the crystal surface and at edge dislocations (6-5a), the energetic possibility arises for their combination with conduction electrons according to $Ag^+ + e^- \rightarrow Ag$. By repetition of the process a group of silver atoms lying adjacent to each other is obtained. When this group becomes large enough to continue growing under the action of a developer, a latent image has been produced. The sensitive specks thus lie at lattice distortions and it is only here that development nuclei occur, in spite of the fact that the light has been absorbed randomly over the whole crystal.

Even without a latent image a certain amount of reduction of the silver halide takes place on development. This causes an even blackening (chemical fog), which becomes stronger the longer the development time is. The fogging increases also with the age of the emulsion.

Pure silver bromide emulsion has a very low absorption and consequently low sensitivity for light with wavelengths greater than 4900 Å (blue-green). Therefore, dye substances with strong absorption at longer wavelengths (mostly cyanines) are adsorbed on the surface of the silver bromide crystals. The energy absorbed by the dye is transferred to the silver bromide crystal with the same effect as though it were irradiated with shorter wavelength light. This dye sensitization is used in the production of "orthochromatic" and "panchromatic" emulsions.

The *development* involves a reduction of silver halide, initiated at a development nucleus: $AgBr(s) + e^- \rightarrow Ag(s) + Br^-$. Extensive formation of silver now takes place. The particular developing substance in the development solution usually consists of an organic reducing agent which can function in basic solution. An appropriate and relatively stable pH value is obtained by using carbonate ion (sometimes also borate ion) in sufficient amount for adequate buffering action. Alkali sulfite is also added. The sulfite ions are oxidized by the air more easily than the developing substance and thus protects the latter against unnecessary oxidation ("preservative").

Inasmuch as the sulfite ions to some extent dissolve silver bromide with the formation of sulfitoargentate complexes, they also break incidental contacts between different silver grains and thus decrease the possibility for the spreading of silver precipitation from grain to grain. The formation of larger silver aggregates is inhibited and the silver precipitation becomes more "fine-grained."

Addition of bromide ion decreases the rate of development, especially for grains without development nuclei. In this way chemical fogging is suppressed relative to the image, which thus becomes clearer.

Following development the unreduced silver halide must be dissolved out in order that it will not later become blackened. This *fixing* is usually done with a solution of sodium thiosulfate, which gives soluble thiosulfatoargentate complexes (36-7a).

36-9. Gold compounds

a. Gold(I) compounds. In equilibrium with water relatively stable gold(I) compounds appear only when the gold atom is so strongly tied up that the amount of Au^+ in the solution is very small. Disproportionation according to the equilibrium $3Au^+ \rightleftharpoons 2Au(0) + Au(III)$ then becomes negligible. This is the case with poorly soluble compounds or strong complexes (cf. copper(I), 16d).

Gold(I) *halides* AuX (except the fluoride) can be obtained as an intermediate stage between gold and gold(III) halides AuX_3, when gold reacts with halogens or when gold(III) halide is decomposed by heat: $Au + \frac{3}{2}Cl_2 \rightleftharpoons AuCl + Cl_2 \rightleftharpoons AuCl_3$. All three halides are rather unstable, both with respect to disproportionation and dissociation. Disproportionation is particularly prominent in the chloride, especially under the influence of water: $3AuCl \rightarrow 2Au + AuCl_3$. The very poorly soluble iodide disproportionates less easily, but dissociates, slowly at room temperature, rapidly with gentle heating: $2AuI \rightarrow 2Au + I_2$. The bromide has an intermediate position.

The *dihaloaurate*(I) *ions* AuX_2^- are remarkably stable. When gold is dissolved in cyanide solution in the presence of air, as mentioned in 36-4b, *dicyanoaurate*(I) *ion* $Au(CN)_2^-$ is formed. Its alkali salts are used in the preparation of electrolytes for gold plating (16f).

b. Gold(III) compounds. Gold(III) salt solutions with alkali hydroxide or carbonate give a yellow-brown precipitate, which probably can be considered as *gold*(III) *oxide hydroxide* AuO(OH). It has amphoteric properties, but the acid character predominates (it has been called "auric acid"), and it dissolves easily in an excess of the precipitating agent. From the solution *hydroxoaurates*(III) crystallize, for example, K[Au(OH)$_4$]. On heating AuO(OH) water goes off, but already at about 150°C oxygen begins to split off so that finally only gold remains.

Gold(III) *chloride* AuCl$_3$ is formed when chlorine is passed over finely divided gold at about 200°C. The chloride sublimes as Au$_2$Cl$_6$ molecules and on condensation forms red, needle-shaped crystals. Decomposition according to 36-9a is noticeable already at 100°C and becomes rapid at 250°C. The chloride dissolves in hydrochloric acid and from the solution yellow, needle-shaped crystals of HAuCl$_4$·4H$_2$O can be obtained. This substance is also formed if a solution of gold in aqua regia is evaporated together with hydrochloric acid, and has been called both "chloroauric acid" (cf. 12-2f) and "auric chloride." Many salts are known containing the square *tetrachloroaurate*(III) *ion* AuCl$_4^-$. Sodium and potassium salts, which occur as light-yellow crystals, constitute the most important commercial gold salts.

Some compounds, earlier assumed to contain gold(II), have been shown to be gold (I,III) compounds. One such compound is CsAuCl$_3$, whose crystals contain an equal number of linear Au^{I}Cl$_2^-$ and square Au^{III}Cl$_4^-$ ions. The formula Cs$_2$[Au^{I}Cl$_2$][Au^{III}Cl$_4$] expresses this structure. The black color of the compound has been discussed previously (7-3b).

On passing hydrogen sulfide into solutions of gold(III) salts a black precipitate is formed which is poorly soluble in acids. It is possible that this consists primarily of the sulfide Au$_2$S$_3$, which, however, quickly decomposes to a large extent to give gold.

CHAPTER 37

Group 12, The Zinc Group

37-1. Survey

The most important oxidation number for all three elements in the group is II, which implies ions with 18-shells (5-2f, fig. 5.8) and thus outermost a filled $(n-1)d$ group (d^{10}). Mercury also often exhibits the oxidation number I, which is probably always associated with the dimeric ion Hg_2^{2+}. Even if the number I should exist for zinc and cadmium it has no significance, at least in water solutions. Oxidation numbers higher than II, which would make the $(n-1)d$ group incompletely filled and thus confer transition-element character, are not known with certainty. The difficulty of breaking into the 18-shell of the M^{2+} ions is shown by the high values for the ionization energies I_3 (given in fig. 5.4 for Zn^{2+} and Cd^{2+}).

In spite of the fact that the effect of the lanthanoid contraction on atomic size is still apparent within this group, mercury is rather distinct in its properties from the mutually often similar zinc and cadmium. The standard potentials in tab. 16.1 show that zinc and cadmium are quite base metals, while mercury, on the other hand, is approximately as noble as silver. Zinc and cadmium are also more easily oxidized by oxygen than mercury. The oxides of the former are therefore more difficultly reduced than mercury(II) oxide HgO, which is easily decomposed by heating. Both zinc and cadmium form stable hydroxides, while mercury does not. Zinc and cadmium show many similarities to magnesium, while mercury is similar to the coinage metals.

Within the group the covalent-bond contribution generally increases with increasing atomic number. Comparing the sulfides, for example, the solubility decreases and the color deepens as we go downward in the group. Hg(II) also is incorporated in a multitude of complexes. Halo complexes are very strong with mercury, somewhat weaker with cadmium and zinc.

All three of the metals, but especially mercury, have a high affinity for sulfur.

The boiling points of the metals and the heats of vaporization are lower than for any other metals with the exception of the alkali metals.

In the oxidation state II all three metals often show tetrahedral coordination. For Zn(II) and Cd(II), however, in many cases the maximum coordination number may increase to 6 with octahedral coordination. For Hg(II) the type of twofold linear or nearly linear coordination that also occurs with Cu(I), Ag(I) and Au(I) is very common (5-3j).

37-2. Occurrence and history

a. Occurrence. The abundance of *zinc* in the earth's crust is rather low, approximately the same as for nickel (tab. 17.3). The most important ore mineral is *sphalerite* (*zinc blende*), cubic zinc sulfide ZnS. Sphalerite often occurs together with other sulfides, one particularly important associated mineral being galena. Other zinc minerals are considered to have been formed by oxidation of primary sphalerite. Among these we may mention *zincite* ZnO, *smithsonite* $ZnCO_3$ and *hemimorphite* $Zn_4(OH)_2Si_2O_7 \cdot H_2O$.

The largest zinc ore operations are in the United States. Deposits are found in many states, the most important currently are in Montana and Idaho. Canada, Australia, Mexico, and the Soviet Union are also large producers.

Zinc in very small amounts is essential to both plants and animals. Zinc deficiencies in vegetation are rare, but occur mostly in neutral soils when the zinc compounds in the soil are most difficult to absorb. Zinc deficiency causes, among other things, chlorophyll defects. In mammals zinc is incorporated in a variety of enzymes, and in insulin. Zinc deficiency in animals is unknown with natural fodder.

Soluble zinc compounds in larger amounts can cause severe poisoning because of their corrosive action on the tissues.

The abundance of *cadmium* in the earth's crust is significantly less than that of zinc (tab. 17.3). Cadmium occurs mostly as a follower of zinc, dissolved in zinc minerals in an amount that is usually less than 0.5 wt %. Specific cadmium minerals are rare.

Cadmium is therefore recovered only in connection with zinc, and the United States is here also the largest producer.

Cadmium compounds are definitely more poisonous than zinc compounds. Cadmium-plated objects therefore should not come in contact with foodstuffs. Severe cadmium poisioning, sometimes with fatal results, has occurred on breathing cadmium vapor and cadmium oxide smoke, formed on heating cadmium or cadmium-plated material, for example in welding.

Mercury is also a rare metal, although its abundance in the earth's crust is somewhat higher than that of cadmium. The only important mercury mineral is *cinnabar* HgS, together with which drops of metallic mercury are sometimes found. Occasionally amalgams, especially with silver, are also found.

The old mercury mines in Spain (Almadén) and Italy (Idria) still provide over half of the world's production of mercury. Then follow Jugoslavia, the United States (especially California and Nevada), Mexico and the Soviet Union.

Mercury and mercury compounds are very poisonous, mainly because Hg(II) combines strongly with sulfur atoms of the proteins. The metal is volatile enough even at room temperature so that staying in a room where there are exposed mercury surfaces, for example from spilled mercury, will easily result in poisoning. Metallic mercury can be absorbed through the unbroken skin. Acute mercury poisoning produces severe stomach and intestinal symptoms and later serious kidney damage. In chronic poisoning similar but less severe effects appear, often accompanied by nervous disturbances. For acute poisoning protein substances can be taken as first aid, for example egg white or milk. The toxic effects of mercury are utilized in the fungicidal treatment of seed. Mercury preparations are also used in medicine.

b. History. As early as the time of Christ, long before metallic *zinc* was known, its ores, especially smithsonite, were used in making brass. The smithsonite was heated with carbon and copper, so that the reduced zinc was directly alloyed with the copper. The Greeks called smithsonite καδμεία, the Romans called it cadmia. At least as early as the 14th century zinc was produced in China and India by reduction of smithsonite with organic substances in a closed vessel. The metal was brought from the Orient to Europe, but it was probably first made there in the 17th century. Approximately at that time it began to be called *zinc*, a name of obscure middle-European origin.

In 1817 Stromeyer, in studying a preparation of zinc carbonate, which on ignition gave a yellow-colored oxide, found that it contained a new metal. He called it *cadmium* after the ancient name for zinc carbonate.

The ease with which *mercury* is prepared from cinnabar caused the metal to be known even in ancient times. The Romans, who for their production worked the mines in Almadén, used the Greek name *hydrargyrum* (water silver), from which the symbol Hg is derived. The alchemists believed that mercury was an essential constituent of all metals. They gave it the same symbol as that of the planet Mercury, from which the name comes.

37-3. Properties of the group 12 elements

Some physical properties are collected in tab. 37.1. Zinc and cadmium crystallize in a somewhat deformed hexagonal close packing (6-1f). The atoms form close-packed layers as in fig. 6.13a, but the distance between the layers is approximately 15 % greater than in an ideal close packing. Solid mercury formed at ordinary pressure always has a trigonal structure. A tetragonal structure with a denser packing than the trigonal one is stable below $-194°C$ but is not obtained at ordinary pressure because the rate of transition is too slow. At high pressure the transition to this phase takes place readily.

Zinc is a blue-white, and cadmium a silver-white metal. However, both are easily attacked in moist air (see below) and then become dull. Mercury is silver-white. Mercury is the only metal that is liquid at ordinary room temperature. The remarkably low boiling points were mentioned in 37-1. At 20°C the vapor pressure of mercury is 0.0016 mbar, but this is enough to make air saturated with mercury vapor very dangerous to the health on long exposure. Zinc and cadmium have moderate, and mercury low conductivity.

Table 37.1. *Physical properties of the zinc-group elements.*

	30Zn	48Cd	80Hg
Density at 20°C, g cm^{-3}	7.13	8.64	13.55
Melting point, °C	419.505^a	321	−38.86
Boiling point, °C	908	765	356.73

^{a} Calibration point in the International Practical Temperature Scale.

At room temperature pure zinc is ductile, but because of impurities, especially iron, commercial zinc is rather brittle. However, at 100°C commercial zinc also becomes ductile. The tensile properties of zinc and cadmium are moderate.

Zinc and cadmium are quite similar in their behavior to chemical attack, but cadmium is somewhat more resistant. In compact form both remain unaffected by dry air at room temperature, but are attacked by moist air, cadmium less than zinc. Then a thin layer of hydroxide is formed, and in the presence of carbon dioxide also carbonate or hydroxide carbonate, which to a certain extent protects the metal from further attack. On heating oxidation soon becomes appreciable. Above about 250°C it becomes rapid, and on strong heating the metals may burn with a flame. Completely pure mercury at room temperature remains intact even in moist and carbon-dioxide-containing air. However, impurities in the form of dissolved, less noble metals, even in small amounts, leads to the formation of a thin oxide coating in moist air. On heating mercury in air mercury(II) oxide HgO is formed. The oxidation proceeds quite rapidly at temperatures above 300°C, and here also is enhanced by dissolved impurities. If the temperature rises above about 500°C, however, the oxide begins to decompose into mercury and oxygen.

Zinc and cadmium are noticeably attacked by water only in the presence of oxygen. Both metals dissolve easily in acids. When dissolved in nonoxidizing acids hydrogen is evolved. In oxidizing acids, hydrogen may be given off, especially if the acid is dilute, but in addition the anion of the acid may be reduced (in nitric acid, among other things, to ammonia). Alkaline solutions attack zinc, but cadmium hardly at all. Mercury because of its high standard potential (tab. 16.1) is stable toward nonoxidizing acids. On the other hand mercury is dissolved by nitric acid and hot conc. sulfuric acid. With excess of mercury, mercury(I) salts are then formed, otherwise mercury(II) salts.

Because of its liquid state, mercury even at room temperature alloys with a large number of metals. The products are called *amalgams* (37-7e).

37-4. Production and uses of the group 12 elements

a. Production. As mentioned earlier in 37-2a *zinc* in its ores is followed by *cadmium* in small amounts. Cadmium is therefore obtained as a by-product in the preparation of zinc, which here sets the pattern.

In bonding hydroxide ions to Zn^{2+} ions in water solutions the coordination number probably does not exceed four.

Zinc forms quite strong ammine complexes. The *tetramminezinc ion* $Zn(NH_3)_4^{2+}$ has the highest known number of ammine ligands in water solution. By the action of ammonia gas on solid zinc salts solid ammines have been prepared containing *hexamminezinc ion* $Zn(NH_3)_6^{2+}$. Through the formation of amminezinc ions many poorly soluble zinc compounds will dissolve in ammonia solution.

Zinc oxide ZnO is found in nature as the mineral *zincite*. As stated in 6-3a zinc oxide may have a very small excess of zinc above the formula ZnO, which is believed to result from a few Zn^{2+} ions having gone over to Zn^+ ions. At ordinary temperature ZnO is colorless but becomes more and more yellow on heating. On cooling the yellow color disappears (*heat color*, 7-3a). Zinc oxide is practically insoluble in water. It is produced technically by oxidizing zinc vapor with air. The vapor is obtained either by boiling the metal, or directly from roasted zinc ore by reduction with carbon. Zinc oxide is the most important commercial zinc compound. It is used especially for paint pigment (*zinc white*), but also to introduce zinc ions into glasses, enamels and glazes for ceramics. Addition of zinc oxide improves the quality of rubber in many respects. Zinc oxide as a catalyst has been discussed in 14-4. Finally zinc oxide is an important ingredient in many medical powder and ointment preparations where it acts as a weak astringent and antiseptic.

Zinc hydroxide $Zn(OH)_2$ is obtained as a white precipitate on addition of alkali hydroxide to zinc salt solutions. This has variable water content and consists of very small particles with quite disordered structure. However, several hydroxides can be obtained, for example by very slow precipitation, all with the composition $Zn(OH)_2$, but with different, ordered crystal structures. In these the hydroxide ions are coordinated by the zinc ions either tetrahedrally or octahedrally. The hydroxide dissolves both in excess of the precipitation reagent and in acids. From solutions in alkali hydroxides many *hydroxozincates* can be crystallized, for example, $Na_2Zn(OH)_4$. Hydroxozincate ions are also formed by dissolving zinc in alkali hydroxide solution, for example according to $Zn(s) + 2OH^- + 2H_2O \rightarrow Zn(OH)_4^{2-} + H_2(g)$. When ammonia is added to a zinc salt solution hydroxide is precipitated at first, but with excess of ammonia the precipitate is dissolved with the formation of amminezinc ions.

Zinc chloride $ZnCl_2$ can be obtained by heating zinc in a stream of chlorine gas. Hydrates crystallize at room temperature from solutions of zinc in hydrochloric acid. The anhydrous chloride is very hygroscopic and is therefore used as a water-absorbing agent in organic syntheses. Zinc chloride is also used as a mordant in dying and printing textiles and in the making of organic dyestuffs. From mixtures of zinc chloride solution with zinc oxide *zinc hydroxide chlorides* crystallize, among others, $Zn(OH)Cl$.

Zinc chloride solutions with high proportions of chloride ion contain chloro complexes such as $ZnCl^+$, $ZnCl_2$, $ZnCl_3^-$ and $ZnCl_4^{2-}$. In the solid salts that can be crystallized from these solutions, the *tetrachlorozincate ion* $ZnCl_4^{2-}$ is present. This occurs in the salt $(NH_4)_2ZnCl_4$ (23-5c), which crystallizes from a solution of $ZnCl_2$ and NH_4Cl, and is used as a flux for soldering.

Zinc cyanide $Zn(CN)_2$ is poorly soluble in water and therefore precipitates on adding cyanide ion to a zinc salt solution. It is easily soluble in an excess of the precipitating reagent with the formation of *cyanozincate ions*, among others, *tetracyanozincate ion* $Zn(CN)_4^{2-}$.

Zinc sulfate $ZnSO_4$ is prepared commercially by dissolving zinc scrap in sulfuric acid or by leaching roasted zinc ore with sulfuric acid. At room temperature the *heptahydrate* $ZnSO_4 \cdot 7H_2O$ ("zinc vitriol", 21-11 b 3) crystallizes out of the water solution, and is isomorphous with the corresponding compounds of magnesium and nickel(II). Zinc sulfate is used in the manufacture of rayon, as zinc fertilizer, for combating plant parasites, as a flotation reagent, as a mordant in dying textiles and in the preparation of electrolytes for zinc plating. In very dilute water solution it is sometimes used as an external antiseptic.

Alkali carbonate solutions with zinc salt solutions usually give precipitates of *zinc hydroxide carbonate*. Only if the mixture is more acid than about $pH = 5$, which is best obtained with hydrogen carbonate solution under compressed carbon dioxide, does *zinc carbonate* $ZnCO_3$ precipitate. This constitutes the mineral *smithsonite*.

Zinc sulfide ZnS is found in nature as *sphalerite* and sometimes as the less common *wurtzite*. Both forms have tetrahedral structures, sphalerite with cubic and wurtzite with hexagonal symmetry (6-2c). In pure form the sulfide is colorless, but the minerals are more or less dark-colored, primarily because of iron in solid solution. From a zinc salt solution zinc sulfide is precipitated by sulfide ions practically completely if $pH \geqslant 3$. The freshly precipitated sulfide is easily soluble in dilute, strong acids. When allowed to stand, however, the solubility decreases. Zinc sulfide, sometimes mixed with cadmium sulfide, is much used as a luminescent material (6-4e).

37-6. Cadmium compounds

Cadmium compounds are in most cases very similar to the corresponding zinc compounds, and thus only some of the differences will be discussed here.

The cadmium ion generally has a greater tendency toward complex formation than the zinc ion. This is perhaps best shown by the *chloride, bromide*, and especially the *iodide*. These are all easily soluble in water, but in water solutions high proportions of halo complexes are present, such as CdX^+, CdX_2, CdX_3^-, and CdX_4^{2-}. When halide ion is added, the average number of ligands shifts toward higher and higher values and finally CdX_4^{2-} ions predominate.

Cadmium oxide CdO is brown in color in contrast to the colorless zinc oxide. *Cadmium hydroxide* $Cd(OH)_2$ dissolves somewhat in a strong excess of alkali hydrox-

Production is carried out thermally or electrolytically. In both cases oxidic material is the starting point. After concentration the ore as a rule is roasted, which for sphalerite produces the reaction $2ZnS(s) + 3O_2(g) \rightarrow 2ZnO(s) + 2SO_2(g)$, and for other ores decreases the amount of water and carbon dioxide (in smithsonite).

In the thermal process the zinc oxide is reduced with carbon, and carbon monoxide becomes the immediate reducing agent. The equilibrium $ZnO(s) + CO(g) \rightleftharpoons Zn(g) + CO_2(g)$ is displaced to the right with increasing temperature, but adequate conversion is attained only above the boiling point of zinc. The zinc is thus evolved in gaseous form, and if the mixture of zinc vapor and carbon dioxide is allowed to leave the furnace, reoxidation takes place when the temperature drops. Therefore an excess of carbon is used so that the carbon dioxide is reduced to carbon monoxide, which will not oxidize zinc. The process is often carried out in retorts with carbon as fuel, or in an electric furnace. The zinc vapor is condensed, but the condensate usually also contains other easily volatile metals, especially lead and cadmium. However, the boiling points of lead (1740°C), zinc (908°C) and cadmium (765°C) are so different that the three metals can easily be separated from each other by fractional distillation in a column (13-5c).

For the electrolytic process the roasting product is leached with sulfuric acid. Other metals are separated from the impure zinc sulfate solution in various ways. Since cadmium is more noble than zinc (tab. 16.1), it can be precipitated by adding zinc dust: $Cd^{2+} + Zn(s) \rightarrow Cd(s) + Zn^{2+}$. The cadmium concentrate is dissolved in sulfuric acid and cadmium deposited from it by electrolysis. The purified zinc sulfate solution is also electrolyzed, the zinc being deposited on aluminum cathodes.

Mercury is obtained from cinnabar quite simply by heating in a stream of air ($HgS(s) + O_2(g) \rightarrow Hg(g) + SO_2(g)$) and condensing the vapor. It is marketed in steel flasks.

The mercury so obtained is fairly pure. To obtain still higher purity it can be washed with dilute nitric acid (which dissolves out the base metals and metal oxides) and then distilled in vacuum. These procedures become more effective if previously air is blown through the metal heated to about 250°C, so that the major portion of the impurities is oxidized.

b. Uses. *Zinc* is used mainly as a protection against corrosion for iron objects, and in alloys. Formerly zinc sheet was much used also for roofing, sink surfaces, kitchen vessels, etc. Zinc plating for protection against corrosion can be done electrolytically (*galvanizing* in its true sense). When thicker layers are needed, especially for outdoor protection, the plating is usually done by dipping the object previously pickled in hydrochloric or sulfuric acid (20-5d, 21-11b2) into molten zinc (*hot-dipped zinc coating*, improperly called "hot galvanizing"). The object may also be coated with zinc dust and heated to just below the melting point of zinc, or it may be sprayed with molten zinc. Zinc dust is also used as a protective pigment in antirust paint. The electrochemical action of zinc in preventing corrosion was discussed in 16i.

Large amounts of zinc are used in making brass (36-5a). Recently, alloys with over 90 wt % zinc plus aluminum and small amounts of copper and magnesium have been used more and more for die (pressure) casting.

In electrical primary cells zinc is most often used for the minus pole (16g). In the form of zinc dust it is a much used reducing agent. In this function, it is also used for the precipitation of nobler metals, for example gold and silver after cyanide leaching (36-4b).

In the manufacture of zinc compounds such as paint pigment and zinc salts the metal is usually the starting material.

Most *cadmium* goes for corrosion-protective coatings, which are mostly applied electrolytically (*cadmium plating*). Cadmium protects somewhat better than zinc, so that a cadmium coating can be made thinner than a zinc coating. However, cadmium is much more expensive than zinc. Lesser amounts of cadmium are used in bearing metal for higher temperatures, with cadmium as the base metal alloyed with either nickel or with silver together with copper, all in rather small amounts. Cadmium is also used in soldering metal, in certain low-melting alloys, for example Wood's metal (13-6), and in Jungner storage batteries (16h). Its high neutron absorption makes it useful for control elements in nuclear reactors (9-4a).

Mercury has very special applications because it is a liquid metal at ordinary temperature. In addition, its quite high nobility is a great advantage. The largest amounts are involved in applications of the metal itself in electrical apparatus (circuit breakers and relays with liquid contacts, electrochemical apparatus, discharge lamps), in other instruments (thermometers, manometers, barometers, barriers for gases) and in chloralkali electrolysis (20-4b). The metal is also used directly in the preparation of dental amalgam (37-7e) and in the amalgamation processes for the recovery of gold and silver (36-4b). Large amounts of mercury are used in the treatment of seed, in drugs, for catalysts in the organic chemical industry, and for mercuric oxide cells (16g).

37-5. Zinc compounds

Only the oxidation number II is of any importance here. All zinc compounds, in which color is not caused by other atoms or groups, are colorless. This is true even of the sulfide and the oxide at ordinary temperature.

Among the simple zinc salts the halides except the fluoride, and also the nitrate, sulfate, and acetate, are easily soluble in water. Of the poorly soluble zinc salts we may mention the fluoride, cyanide, carbonate, phosphate, and silicates.

The hydrated Zn^{2+} ion can function as an acid and then converts to hydroxo complexes (12-1b). Dilute water solutions of zinc salts are rather weakly acid.

ide and, like zinc hydroxide, in ammonia solution with the formation of *amminecadmium ions*. From cadmium salt solutions *cadmium sulfide* CdS is practically completely precipitated by hydrogen sulfide if $\text{pH} \geqslant 1$. Depending on the conditions of formation the color of the sulfide is light yellow to dark orange. Even *cyanocadmiate* complexes are decomposed by sulfide ion with the formation of cadmium sulfide. Freshly precipitated cadmium sulfide is hardly soluble in dilute acids. Hot concentrated acids will dissolve even aged sulfide. Cadmium sulfide is an important paint pigment, *cadmium yellow*. We may also mention *cadmium red*, solid solutions Cd(S,Se), which become darker red with increasing selenium content.

Metallic cadmium dissolves in molten cadmium chloride (m.p. 558°C), which is considered to indicate the formation of Cd_2^{2+} ions (20-5a). Cadmium is then in the oxidation state I and is analogous to mercury.

37-7. Mercury compounds

a. General. The stability of mercury in its various oxidation states is best discussed with the standard potentials in tab. 16.1 as a starting point. We find there that for the redox pair $2Hg-Hg_2^{2+}$, $e_{01}^0 = 0.793$ V; for the pair $Hg-Hg^{2+}$, $e_{02}^0 = 0.850$ V; and for the pair $Hg_2^{2+}-2Hg^{2+}$, $e_{12}^0 = 0.920$ V. The standard potentials e_{01}^0 and e_{02}^0 show that strong oxidizing agents are needed to oxidize metallic mercury, but that most oxidizing agents that can oxidize the metal to Hg_2^{2+} are also strong enough to oxidize to Hg^{2+}. The value of e_{12}^0 shows that reduction of Hg^{2+} to Hg_2^{2+} proceeds with weak reducing agents. Excess of such agents reduce Hg_2^{2+} further to Hg. The presence of substances that form complex or poorly soluble compounds whose strength or solubility (respectively) are different for Hg(I) and Hg(II) may alter these conditions (16c, d).

For the equilibrium $Hg^{2+} + Hg \rightleftharpoons Hg_2^{2+}$ we calculate according to 16d from the standard potentials e_{01}^0 and e_{02}^0 the equilibrium constant $\{Hg_2^{2+}\}/\{Hg^{2+}\} = 90$ (25°C). The ion Hg_2^{2+} is therefore stable with respect to disproportionation into Hg^{2+} and Hg, and this implies also that Hg^{2+} is reduced by Hg to Hg_2^{2+}. However, the equilibrium constant is so close to 1 that when a mercury(II) compound is appreciably less soluble or is a much stronger complex than the corresponding mercury(I) compound, the major portion of Hg_2^{2+} will be disproportionated, going according to the reaction given above to the left. Thus we have, for example, the reactions: $Hg_2^{2+} + S^{2-} \to HgS(s) + Hg(l)$; $Hg_2^{2+} + 2OH^- \to HgO(s) + Hg(l) + H_2O$; and $Hg_2^{2+} + 2CN^- \to Hg(CN)_2 + Hg(l)$. The reaction products of mercury(II) with ammonia (37-7d) are so poorly soluble that all mercury(I) compounds are decomposed by ammonia with the formation of such products and mercury. In all these reactions the mercury precipitates out in very finely divided form with a black color.

b. Mercury(I) compounds. Mercury(I) compounds have many similarities to silver(I) compounds, in spite of the fact that in all states of aggregation they all contain the dimeric ion Hg_2^{2+}. Strong analogies are found, for example, in their water solubilities. Most mercury(I) compounds are thus poorly soluble. The nitrate, chlorate and perchlorate are fairly soluble. These are practically completely dissociated in water solution. The hydrated ion Hg_2^{2+} is a weak acid, and the acid reaction of its solution results for the most part from the presence of Hg^{2+}, which in hydrated form is a rather strong acid (12-1b). Generally speaking no strong complexes of Hg_2^{2+} are known, in striking contrast to the strong complex-forming tendency of Hg^{2+}.

Analogously with many silver compounds mercury(I) compounds are often affected by light and darken with the precipitation of metallic mercury.

Mercury(I) chloride Hg_2Cl_2 is usually prepared by precipitation from mercury(I) nitrate solution with chloride ion or by heating mercury with a restricted amount of chlorine. In the latter case mercury(I) and mercury(II) chloride are formed simultaneously, which are both sublimed and condensed. The former is very poorly soluble and the latter rather easily soluble in water, so that they can easily be separated. The precipitated mercury(I) chloride is a white powder, and the sublimed chloride forms a lustrous crystalline mass. Even with gentle heating a yellow heat color appears. The crystals are built up of linear molecules Cl–Hg–Hg–Cl and these are also present in the gas if it is completely dry. In the presence of water vapor disproportionation takes place to $HgCl_2$ and Hg. In the light mercury(I) chloride is darkened on the surface by mercury. If ammonia solution is poured over the chloride, as stated in 37-7a mercury(II) compounds are formed (37-7d) together with mercury which blackens the preparation. It is probably this reaction that led to the old name *calomel* (Gk. καλός μέλας, beautiful black) for mercury(I) chloride. The low solubility of the chloride renders it relatively nonpoisonous and it is used to some extent in medicine both externally and internally.

Mercury(I) nitrate $Hg_2(NO_3)_2$ is obtained by the action of dilute, cold nitric acid on an excess of mercury. It is formed also by reduction of mercury(II) nitrate in water solution with mercury. From the solution colorless crystals of the dihydrate are obtained, which are easily soluble in dilute nitric acid. In pure water the dihydrate gives yellow hydroxide nitrates. The solutions are easily oxidized by air, but this can be prevented if they are allowed to stand in contact with mercury.

Mercury(I) sulfide Hg_2S may possibly be formed initially on precipitation of a mercury(I) salt solution with sulfide ion, but as stated in 37-7a it immediately disproportionates.

c. Mercury(II) compounds. Some mercury(II) salts, for example the nitrate and perchlorate, dissociate in water solution approximately like normal salts. The rather strong acid properties of the Hg^{2+} ion (12-1b) leads to considerable hydrolysis, however, which as a rule causes the formation of poorly soluble oxide or hydroxide salts. If the anion is that of a strong acid, the hydrolysis can be suppressed and a clear solution obtained with an excess of the acid. In such strongly acid solutions the nitrate and perchlorate are rather easily soluble.

In general, however, mercury(II) has a pronounced tendency to form complexes. In solutions of the chloride, bromide, iodide and cyanide the Hg^{2+} ion exists mainly in complexes with the anions and it has been possible to identify HgX^+, HgX_2, HgX_3^- and HgX_4^{2-}. Of the Hg^{2+} salts the chloride and the cyanide are rather easily soluble, the bromide rather poorly soluble and the iodide very poorly soluble. The strong complex formation here decreases the hydrolysis and the appearance of oxide and hydroxide salts.

Among the poorly soluble mercury(II) salts we may also mention the carbonate and phosphate. The sulfide is very insoluble.

With ammonia Hg^{2+} gives ammine complexes, of which the *diamminemercury*(II) *ion* $Hg(NH_3)_2^{2+}$ is especially prominent. The ion is linear, $[H_3N-Hg-NH_3]^{2+}$. With very high ammonia concentration the tetrahedral *tetramminemercury*(II) *ion* $Hg(NH_3)_4^{2+}$ is also formed. As described in 37-7d other processes also occur with ammonia.

Mercury(II) salts with anions of strong acids are colorless. The poorly soluble oxide and hydroxide salts are usually yellow, as are the salts with anions of weak acids. The oxide and sulfide are colored (see below).

Mercury(II) *oxide* HgO, as stated in 37-3, is formed on heating mercury in air but decomposes if the temperature rises above about 500°C. It is obtained more easily by mild heating of mercury(II) nitrate: $Hg(NO_3)_2 \rightarrow HgO(s) + 2NO_2(g) + \frac{1}{2}O_2(g)$. In these cases it is red. It is obtained as a yellow precipitate on addition of alkali hydroxide in excess to a mercury(II) solution: $Hg^{2+} + 2OH^- \rightarrow HgO(s) + H_2O$. On heating the yellow oxide the color changes to red, which remains after cooling. The yellow oxide is more reactive than the red. It has been suggested that the yellow form has smaller crystal size or a more distorted structure than the red (7-3a), but it may be that the two are distinct in some other way. The crystals are built up of infinite chains $\diagdown_{Hg}\diagup^O\diagdown_{Hg}\diagup^O\diagdown_{Hg}\diagup\diagdown_O\diagup$. The twofold coordination around mercury is also found here. If alkali hydroxide is added to a mercury(I) solution, a black precipitate of $HgO + Hg$ is obtained, since Hg_2O is not stable (37-7a).

Mercury(II) *chloride* $HgCl_2$ is made commercially mainly by heating mercury with an excess of chlorine in fused silica retorts. The chloride is vaporized and condensed in lead chambers as colorless crystals. An old method of heating mercury(II) sulfate with sodium chloride ($HgSO_4 + 2NaCl \rightarrow Na_2SO_4 + HgCl_2(g)$) is hardly used today. The temperature can be kept below the melting point of mercury(II) chloride (280°C) since its vapor pressure is already high enough so that active sublimation takes place. For this reason, and because of its caustic action on body tissues, mercury(II) chloride has since early times been called *corrosive sublimate*. In the gas phase nearly linear molecules Cl–Hg–Cl are present and these can also be distinguished in the crystals. In the crystals, each mercury atom coordinates two diametrically opposed chlorine atoms at short distances, but four further chlorine atoms in neighboring molecules lie close enough so that we can also speak of an irregular octahedral coordination (5-3j). Mercury(II) chloride is rather easily soluble in water (100 g water dissolves 7.4 g at 20°C, 55 g at 100°C), but as stated earlier most of the dissolved chloride is present as complexes, mainly as $HgCl_2$.

Mercury(II) chloride is a very powerful poison, but because of its antiseptic action has been much used externally in medicine.

Mercury(II) *iodide* HgI_2 exists in two forms with different crystal structures and different colors: HgI_2(tetragonal, red) $\xrightleftharpoons{127°C}$ HgI_2(orthorhombic, yellow). If mercury is ground with iodine at ordinary temperature the red form is obtained, but if the reaction takes place above 127°C, for example between the vapors, the yellow form is produced. Because of its low solubility the iodide is precipitated on adding iodide ion to a mercury(II) salt solution. In that case the yellow form appears first, which rapidly transforms to the red (13-3a). The iodide dissolves easily in excess of iodide ion with the formation of *iodomercurate*(II) *ions*. A light-yellow dihydrate of *potassium tetraiodomercurate*(II) $K_2HgI_4 \cdot 2H_2O$ crystallizes out of a solution containing potassium iodide.

Mercury(II) *cyanide* $Hg(CN)_2$ constitutes such a strong complex that it arises when nearly all mercury(II) compounds are treated with cyanides. It is even formed when mercury(II) oxide is heated with a water suspension of Prussian blue (34-5b). In spite of the strong complex formation the cyanide is rather easily soluble in water. However, the solution shows none of the usual mercury reactions except that mercury(II) sulfide can be precipitated because of its extremely low solubility product (see below).

Mercury(II) *nitrate* $Hg(NO_3)_2$ and *mercury*(II) *sulfate* $HgSO_4$ are obtained by heating mercury in an excess of nitric acid and sulfuric acid, respectively. As mentioned earlier mercury(II) nitrate is rather easily soluble in water containing excess nitric acid. On the other hand, the sulfate is quite insoluble. From the acid solutions solid salts can be obtained, which are easily hydrolyzed by water with the formation of oxide and hydroxide salts.

Mercury(II) *sulfide* HgS is found in nature as the red mineral *cinnabar*, whose trigonal crystals are built up of chains Hg–S–Hg–S–Hg–S . A black modification of HgS with the sphalerite structure (cubic, 6-2c, fig. 6.16d) forms the rare mineral *metacinnabarite*. Cinnabar is the stable modification up to about 380°C, but in spite of this metacinnabarite is often formed as a primary phase. Thus, sulfide ions generally give a black mercury(II) sulfide precipitate. Both modifications are extremely insoluble in water (tab. 13.1 gives an approximate pK_s value, applicable to both). Precipitation with sulfide ion therefore takes place even in strongly acid solution, and the precipitate does not dissolve in acids provided they are not hot and concentrated. Aqua regia easily dissolves the sulfide with the separation of sulfur, probably because of the formation of complexes such as $HgCl_4^{2-}$. The sulfide does not dissolve in solutions of alkali hydroxides or ammonium sulfide, but it will dissolve in concentrated solutions of alkali sulfides. *Thiomercurate(II) ions* such as HgS_2^{2-} are then formed. On dilution or acidification of such solutions the sulfide reprecipitates: $HgS_2^{2-} + H_2O \rightleftharpoons HgS(s) + OH^- + SH^-$.

Especially formerly, cinnabar has been used as a brilliant red paint pigment. However, it slowly darkens in the light (the process is not well understood), which has damaged many older works of art, and now cadmium red (37-6) is usually used instead. Cinnabar is usually made by heating the black sulfide with an amount of concentrated alkali sulfide solution small enough so that only a little of the sulfide is dissolved. The black sulfide is then transformed through the solution quite rapidly to the stable and somewhat less soluble red form.

d. Compounds with mercury-nitrogen bonds. As stated in 22-6a many compounds of metals with molecules and ions of the ammonia system are known. Aside from ammine complexes, however, these are usually decomposed by water and therefore they are stable only in its absence (for example, in liquid ammonia). But many mercury compounds of this type, among others the halides, are attacked so slowly by water that they are also formed in water solution. These compounds because of their structures constitute an interesting group. The structures are determined by the fact that Hg always coordinates 2N linearly or nearly linearly, while N always coordinates Hg and H tetrahedrally.

The relationship among the most important cations is shown by the following scheme:

$$2Hg^{2+} + 4NH_3 \rightleftharpoons 2Hg(NH_3)_2^{2+} \xrightleftharpoons[+2NH_4^+]{-2NH_4^+} 2HgNH_2^+ \xrightleftharpoons[+NH_4^+]{-NH_4^+} Hg_2N^+$$

If a $HgCl_2$ solution containing much ammonium chloride is treated with ammonia solution, a white precipitate of *diamminemercury(II) chloride* $Hg(NH_3)_2Cl_2$ is formed according to the first step above. Since this precipitate can be melted (although with decomposition) it was formerly often called "fusible white precipitate." The cation in these crystals is the linear ion $[H_3N-Hg-NH_3]^{2+}$.

If the $HgCl_2$ solution contains little or no ammonium chloride, the process proceeds to the next step, in which a more nitrogen-poor white precipitate of *amidomercury(II) chloride* $HgNH_2Cl$ appears. On heating this decomposes completely without melting and has therefore been called "infusible white precipitate." Its crystals contain infinite cation chains, which can be written $(HgNH_2^+)_n$ and have the structure

$$\begin{array}{ccccc} & NH_2 & & NH_2 & \\ & / \backslash & & / \backslash & \\ & Hg & Hg & Hg & \\ \backslash & / & \backslash & / & \\ & NH_2 & & NH_2 & \end{array}$$

The compound is also formed by the action of liquid ammonia on solid $HgCl_2$: $HgCl_2(s) + 2NH_3 \rightarrow HgNH_2Cl(s) + NH_4Cl$.

If Hg_2Cl_2 is treated with ammonia solution, there is formed according to circumstances one of the compounds just mentioned and mercury, which gives a black color (37-7a, b).

The last and most nitrogen-poor step in the above scheme is reached more easily with more dissociated mercury(II) salts than the halides. If a solution of mercury(II) nitrate is treated with ammonia solution, a yellow-white precipitate of $Hg_2NNO_3 \cdot H_2O$ is thus obtained. If ammonia solution is poured over yellow and thus relatively reactive mercury(II) oxide, the dihydrate of the corresponding hydroxide $Hg_2NOH \cdot 2H_2O$ is obtained as a light-yellow powder. This hydroxide is called *Millon's base*. Besides the nitrate many other salts Hg_2NX are known, among others the halides. The iodide is obtained as a brown precipitate when an alkaline solution containing the ion HgI_4^{2-} (37-7c) is treated with ammonia: $2HgI_4^{2-} + 3OH^- + NH_3 \rightleftharpoons Hg_2NI(s) + 7I^- + 3H_2O$. The formation of this iodide is used as an extremely sensitive test reaction for ammonia, for example in drinking water. For this an alkaline solution of $K_2HgI_4 \cdot 2H_2O$ (*Nessler's reagent*) is used. With very small amounts of ammonia the formation of Hg_2NI becomes apparent as a yellow coloration.

According to the reaction scheme given above, treatment of Hg_2NX compounds with NH_4^+ ion leads to reactions to the left and thus formation of the previously mentioned types.

The composition of the Hg_2N^+ ion means that the above-mentioned coordination conditions can be fulfilled only if the ion forms a three-dimensional framework. This is brought about by structures that from a purely geometrical viewpoint are analogous to one of the structures of the silicon dioxide modifications cristobalite or tridymite (23-3g). Si, which tetrahedrally coordinates 4O, corresponds here to N, which tetrahedrally coordinates 4Hg. O, which lies between two Si, corresponds to Hg, which lies between two N (fig. 23.3 will illustrate the principle of coordination although it shows a different silicon dioxide modification than those mentioned here). Thus, a framework cation $(Hg_2N^+)_n$ is obtained and in its interstices lie the anions and incidental water molecules. It is also apparent that these can easily be exchanged (cf. the zeolites, 23-3h 5), while the framework cation remains unchanged.

e. Amalgams. Since early times the alloys of mercury have been called *amalgams*. The name is applied regardless of whether the alloy is single or multiphase, or whether the phases are solid or liquid. With high mercury content or high temperature the amalgam may constitute a single liquid phase, consisting of a solution of one or several metals in mercury. If the amalgam consists of both solid and liquid phases, it may have a more or less doughy consistency. Mercury forms amalgams with a large number of metals, and because of its liquid state the amalgamation can often be brought about merely by contact at ordinary temperature. However, with the transition metals such amalgamation generally occurs sluggishly or not at all. To get an amalgam in these cases more effective methods may be necessary, for example higher temperature or electrolytic deposition of the metal on a mercury cathode. For chromium, iron and cobalt, it has not been possible with such methods to obtain anything but a suspension of very finely divided metal in the mercury phase.

Amalgams have great practical importance—see for example 20-4b and 36-4b. In organic reductions a sodium amalgam is often used. Amalgams used for dental fillings have various compositions, but nowadays often contain mainly silver and tin as other metals. They are prepared from metal powder and mercury immediately before insertion and then have a doughy consistency. They rapidly convert to solid phases and through further processes in the solid system the hardness gradually increases with time. Ammonium amalgam was discussed in 22-6c.

APPENDIX 1

Symbols

Letter symbols

Quantities are set in italics (for example, V); units, chemical symbols and symbols for substances are in Roman (for example, cm, Hg, X). Quantities are indicated as follows (with volume as an example): V, applies to a whole system; v_i applies to one mole of the substance "i" (occasionally partial molar volume, see 1-5e). Atomic and molecular species in general have often been indicated by A, B, C, L, M, X, Y. Symbols often have an index attached. A number of such symbols are given in the list, as well as the meaning of certain indexes. In other cases the meaning of the index should be clear. Symbols for elementary particles will be found in tab. 1.2.

A	ampere.	e	base of natural logarithms; index for vapor formation (evaporation), 13-5b.
Å	ångström (10^{-10} m).		
A	mass number, 1-3b; Helmholtz free energy, 2-3b.	e	unit of charge, 1-3a; electrode potential, 16c.
a	acid, 12-1a, 12-1e.	e^-	electron (compl. symbol $_{-1}^{0}$e), 1-3e.
a	activity, 10a.	e^+	positron (compl. symbol $_{+1}^{0}$e), 1-3e.
aq	water (aqua).	EC	electron capture, 9-2a.
B	bond energy, 5-1c.	eV	electron volt, 1-2c.
b	base, 12-1a, 12-1e.	F	faraday, 1-4d.
C	coulomb.	f	index for fusion, 13-5b; symbol for function.
C	total concentration, 1-4e; molarity, 1-4e; heat capacity C_p at const. press., C_v at const. vol., 2-2b.	f	activity factor, 10a; number of degrees of freedom, 13-2a; symbol for quantum number $l=3$, 4-1b.
c	velocity of light, 1-2a; concentration, 1-4e; molarity, 1-4e; molar heat c_p at const. press., c_v at const. vol. 2-2b.	G	Gibbs free energy, 2-3b.
		g	gram; gas.
D	enthalpy or energy of dissociation, 6-2d.	H	enthalpy, 2-1d.
		h	Planck's constant, 1-2b.
d	deuteron (compl. symbol $_1^2$d), 1-3e; day.	Hz	hertz.
		I	ionization energy, 5-2b; ionic strength, 11d; moment of inertia, 3c.
d	symbol for quantum number $l=2$, 4-1b.	i	index for substance or order number.
∂	symbol for partial differentiation, 1-5e.	J	joule.
E	energy, 1-2a.	j	rotation quantum number, 3c.
	E_A activation energy, 8d; attraction energy, , 5-1c.	K	equilibrium constant, 10b; with index, in particular
	E_P potential energy, 5-1c.		K_a acidic constant, 12-2a
	E_R repulsion energy, 5-1c		K_b basic constant, 12-2a
	electron affinity, 5-2b; emf for cell, 16b.		K_c with activities expressed in concentrations, 10b

Appendix I. Symbols

K_p	with gas activities in pressures, 10b	q	quantity of heat added, 2-1c;
K_s	solubility product, 13-1d		q_e heat of vaporization (e = evaporation), 13-5b.
K_w	acidic constant of water, 12-2d		q_f heat of fusion, 13-5b.
K_x	with activities expressed in mole fractions, 10b;	R	gas constant, 1-4f.
	K-electron shell, 4-1b.	r	radius; distance.
k	kilo- (10^3).	S	entropy, 2-3a; shielding constant, 4-2d.
k	equilibrium constant with concentrations instead of activities, 10b; mixed or incomplete constant, 12-2e; with index, in particular:	s	solid substance (solidus); second.
		s	symbol for quantum number $l=0$, 4-1b; spin quantum number, 4-1b; solubility, 13-1b.
	k_a acidic constant, 12-2e		
	k_b basic constant, 12-2e	T	absolute temperature.
	k_s solubility product, 13-1d		T_c Curie temperature, 7-4a.
	k_w acidic constant of water, 12-2e; further other constants, for example:	t	temperature in °C; time; $t_{\frac{1}{2}}$ half-life, 8a.
		U	internal energy, 2-1c.
		u	unified atomic mass unit = $1/12$ of atomic mass of ^{12}C, 1-3f.
	k_R Rydberg's constant, 4-3a; rate constant, 8a; force constant, 3c; number of components, 13-2a.	u	ionic mobility, 11b.
		V	volt.
L	sublimation enthalpy, 5-2i; L-electron shell, 4-1b.	V	volume.
		v	velocity; mole volume, 1-5e; vibration quantum number, 3c.
l	liter; liquid (liquidus).		
l	length; azimuthal quantum number, 4-1b.	w	quantity of work added, 2-1c.
		X	lattice energy, 6-2b.
M	mol l^{-1}, molar, 1-4e; metal atom; mega- (10^6).	x	mole fraction, 1-4e.
			x_a acid fraction, 12-3b.
M	molecular weight, 1-4d; M-electron shell, 4-1b.		x_b base fraction, 12-3b. electronegativity, 5-3f.
m	meter; milli- (10^{-3}).	Y	hydration enthalpy, 5-2i.
m	mass; m_e rest mass of the electron, 1-3a; mol kg^{-1} solvent, molal, 1-4e; magnetic quantum number, 4-1b.	y	year.
		Z	atomic number, nuclear charge, 1-3b.
		z	charge, 1-3f.
N	newton, App. 2.	α	degree of dissociation, 10c; α particle, 9-2a.
N	number of nuclear neutrons, 1-3b. N-electron shell, 4-1b.	β^-	electron, 1-3e.
N_A	Avogadro's number, 1-4c.	β^+	positron, 1-3e.
n	neutron (compl. symbol ^1_0n), 1-3e.	Δ	increase of a quantity in a process, 1-5d.
n	number; transference number (also n_+ and n_-), 6-3c; principal quantum number, 4-1b.	λ	wavelength.
		μ	chemical potential, 2-3d; dipole moment, 5-2g; micro- (10^{-6}).
P	pressure.	ν	frequency.
p	operator symbol for $-\log$, 1-4e; proton (compl. symbol ^1_1p), 1-3e.	Π	osmotic pressure, 13-1g.
		π	π bond, 5-3a.
p	momentum, 1-2b; symbol for quantum number $l=1$, 4-1b; number of phases, 13-2a	ρ	density.
		Σ	sum.
		σ	σ bond, 5-3a.
p_i	partial pressure, 1-4f.	ψ	wave function, orbital, 3c, 4-1b.

Other symbols

⁰   as a right upper index denotes a quantity corresponding to a standard state, 2-2a, 2-3b.

*   radioactive (labeled) atom 9-3e; excited atom, 4-1d; antibonding orbital, 5-3a.

[ ]   concentration, 1-4e.
{ }   activity, 10a.
| contact between two phases, 16b.
‖   salt bridge, 16b.

Greek alphabet

Α	α	alpha	Ξ	ξ	xi
Β	β	beta	Ο	ο	omicrón
Γ	γ	gamma	Π	π	pi
Δ	δ	delta	Ρ	ρ	rho
Ε	ε	epsilón	Σ	σ	sigma
Ζ	ζ	zeta		ς	sigma at end of word
Η	η	eta			
Θ	ϑ	theta	Τ	τ	tau
Ι	ι	iota	Υ	υ	upsilón
Κ	κ	kappa	Φ	φ	phi
Λ	λ	lambda	Χ	χ	chi
Μ	μ	mu	Ψ	ψ	psi
Ν	ν	nu	Ω	ω	oméga

APPENDIX 2

Constants and Conversion Factors

The constants where relevant refer to the atomic weight scale based on $^{12}C = 12$, and most of them have been taken from the compilation by J. W. M. DuMond and E. R. Cohen in the Technical News Bulletin of the U.S. National Bureau of Standards for October, 1963.

(handwritten: mol vol = 22.4 l 0°C ; 24.0 20°C)

$c = 2.997925 \times 10^8$ m s^{-1}
$h = 6.6256 \times 10^{-34}$ J s
$N_A = 6.02252 \times 10^{23}$ formula units mol^{-1}
$e = 1.60210 \times 10^{-19}$ C
$F = N_A e = 96\,487$ C equiv^{-1}
$m_e = 9.1091 \times 10^{-31}$ kg
$R = 8.3143$ J deg^{-1} mol^{-1}
$ = 8.3143 \times 10^{-5}$ m^3 bar deg^{-1} mol^{-1}
$ = 8.2055 \times 10^{-5}$ m^3 atm deg^{-1} mol^{-1}
$ = 1.9865$ cal (15°C) deg^{-1} mol^{-1}
$273.15\,R = 2.2711$ kJ mol^{-1}
$ = 2.2711 \times 10^{-2}$ m^3 bar mol^{-1}
$ = 2.2414 \times 10^{-2}$ m^3 atm mol^{-1}
$298.15\,R = 2.4789$ kJ mol^{-1}
$ = 2.4789 \times 10^{-2}$ m^3 bar mol^{-1}
$ = 2.4465 \times 10^{-2}$ m^3 atm mol^{-1}
$R(\ln 10)/F = 0.198414$ mV deg^{-1}
$298.15\,R(\ln 10)/F = 59.16$ mV
$k = R/N_A = 1.3805 \times 10^{-23}$ J deg^{-1}
$1\,\text{Å} = 10^{-10}$ m

1 l (older) = 1.000028×10^{-3} m^3
1 l (after 1964) = 10^{-3} m^3
1 u = $1/N_A$ g = 1.66043×10^{-27} kg
1 N (newton) = 10^5 dyn
1 bar = 10^5 N m^{-2}
1 atm (normal atm) = 1.01325 bar (definition)
1 at (techn. atm) = 0.980665 bar (definition)
1 torr = 1/760 atm (definition)
$\phantom{1 \text{torr}} = 1.333224$ mbar
1 J (joule) = 1 N m = 1 W s = 10^7 erg
1 eV = 1.60210×10^{-19} J
1 cal (15°C) = 4.1855 J
1 cal (NBS) = 4.1840 J (definition)
1 cal (Steam Tables) = 4.1868 J
1 m^3 atm = 1.01325×10^5 J
0°C = 273.15° K

T (tera)	10^{12}	m (milli)	10^{-3}
G (giga)	10^9	μ (micro)	10^{-6}
M (mega)	10^6	n (nano)	10^{-9}
k (kilo)	10^3	p (pico)	10^{-12}

$e = 2.71828$ $\quad$ $\log e = 0.434294$ $\quad$ $\ln 10 = 2.302585$

(handwritten: 1 liter atm = 101.33 J.)

Relative values of atomic weights and masses in various atomic weight scales

Scale:	"physical"	"chemical"	new
Standard:	$^{16}O = 16$	"natural" oxygen = 16	$^{12}C = 12$
	1	0.999725	0.9996822
	1.000275	1	0.999957
	1.0003179	1.000043	1

INDEX OF NAMES
AND BIOGRAPHICAL DATA

Abbreviations: Austral., Australia; Aust., Austria; Bel., Belgium; Can., Canada; Den., Denmark; Eng., England; Fin., Finland; Fr., France; Ger., Germany; Gr., Greece; Hung., Hungary; Ind., India; Ire., Ireland; It., Italy; Neth., Netherlands; Nor., Norway; Pol., Poland; Rus., Russia; Sp., Spain; Swed., Sweden; Switz., Switzerland; U.S., United States.

Abelson, P. H. 1913–, U.S.: 676
Acheson, E. G. 1856–1931, U.S.: 589
Ampère, André-Marie 1775–1836, Fr.: 460
Arfvedson, Johan August 1792–1841, Swed.: 650
Arrhenius, Svante 1859–1927, Swed.: 254, 314, 323
Aston, Francis W. 1877–1945, Eng.: 27
Auer von Welsbach, Carl 1858–1929, Aust.: 679, 681
Avogadro, Amadeo 1776–1856, It.: 26, 36, 40

Balard, A. J. 1802–76, Fr.: 460
Balmer, J. J. 1825–98, Switz.: 82, 94
Bartlett, Neill 1930–, U.S. (b. Eng.): 460
Bayer, K. J., late 19th cent., Ger.: 640
Becquerel, A. H. 1852–1908, Fr.: 265
Belov, Nikolaj Vasilevich 1891–, Rus.: 606
Berggren-Pasch, G. E. 1788–1862, Swed.: 581
Bergman, Torbern 1735–84, Swed.: 15, 601, 631, 694, 705
Berthollet, Claude Louis 1748–1822, Fr.: 43
Berzelius, Jöns Jacob 1779–1848, Swed.: 7, 24, 93, 98 f, 492, 601, 650, 660, 675, 676, 684
Bessemer, Henry 1813–98, Eng.: 723
Birkeland, Kristian 1867–1917, Nor.: 549, 560
Bjerrum, Jannik 1909–, Den.: 304
Bjerrum Niels, 1879–1958, Den.: 315
Black, Joseph 1728–99, Eng.: 15
Bohr, Niels 1885–1962, Den.: 75, 99, 684
de Boisbaudran, P.-É. Lecoq 1838–1919, Fr.: 676
Boltzmann, Ludwig 1844–1906, Aust.: 40
Born, Max 1882–, Eng. (b. Ger.): 66
Bosch, Carl 1874–1940, Ger.: 281, 307, 398, 399, 455, 549, 568
Boyle, Robert 1627–91, Eng.: 15, 24, 322, 491

Bragg, William Henry 1862–1942, Eng. (also Austral.): 238
Bragg, William Lawrence 1890–, Eng. (b. Austral.): 233, 238, 606
Brand, Hennig, late 17th cent., Germ.: 538
Brandt, Georg 1694–1768, Swed.: 538, 715
de Broglie, Louis 1892–, Fr.: 66
Brownrigg, William 1711–1800, Eng.: 737
Brønsted, Johannes Nikolaus 1879–1947, Den.: 317, 323, 511
Bunsen, Robert W. 1811–99, Ger.: 650
Bussy, A. A. B. 1794–1882, Fr.: 660

Caro, Heinrich 1834–1910, Ger.: 532
Cavendish, Henry 1731–1810, Eng.: 451, 458, 507
Chadwick, James 1881–, Eng.: 20
Chalmers, T. A. Eng.: 295
de Chancourtois, A.-É. B. 1819–86, Fr.: 85
Chaptal, J. A. 1756–1832, Fr.: 538
Clapeyron, B. P. É. 1799–1864, Fr.: 361
Cleve, Per Teodor 1840–1905, Swed.: 24, 676
Clusius, Klaus 1903–63, Ger., Switz. (b. Ger.): 282
Coster, Dirk 1889–, Neth.: 684
Courtois, Bernard 1777–1838, Fr.: 460
Crafts, James Mason 1839–1917, U.S.: 643
Cronstedt, Axel Fredrik 1722–65, Swed.: 715
Crookes, William 1832–1919, Eng.: 458
Curie, Marie Sklodowska 1867–1934, Fr. (b. Pol.): 266, 492, 660
Curie, Pierre 1859–1906, Fr.: 266, 492, 660

Dalton, John 1766–1844, Eng.: 19, 41, 43
Davy, Humphrey 1778–1829, Eng.: 323, 460, 639, 650, 660
Deacon, Henry 1822–76, Eng.: 469

Index of names

Debye, Peter 1884–1966, Switz., Ger., U.S. (b. Neth.): 313, 315
Dirac, P. A. M. 1902–, Eng.: 66
Dulong, P. L. 1785–1838, Fr.: 59

Edison, Thomas Alva 1847–1931, U.S.: 437
Einstein, Albert 1879–1955, Ger., Switz., U.S. (b. Ger.): 15, 51, 52, 676
Ekeberg, Anders Gustaf 1767–1813, Swed.: 323, 675, 690
de Elhuyar, Fausto 1755–1833, Sp.: 694
de Elhuyar, Juan José 1754–96, Sp.: 694
Enskog, David 1884–1947, Swed.: 282
Eyde, Samuel 1866–1940, Nor.: 549, 560

Faraday, Michael 1791–1867, Eng.: 37, 309
von Fehling, Hermann 1811–85, Ger.: 753
Fermi, Enrico 1901–54, It., U.S. (b. It.): 292, 676
Fischer, Franz 1877–1948, Ger.: 399, 592
Flood, Håkon 1905–, Nor.: 324, 501
Frank, F. C. 1911–, Eng.: 391
Frankland, Edward 1825–99, Eng.: 458
Frasch, Hermann 1851–1914, Ger., U.S. (b. Ger.): 499
Friedel, Charles 1832–99, Fr.: 643

Gadolin, Johan 1760–1852, Fin.: 674, 675
Gahn, Johan Gottlieb 1745–1818, Swed.: 705
Gay-Lussac, J. L. 1778–1850, Fr.: 40, 529, 599, 631
Gibbs, J. Willard 1839–1903, U.S.: 51, 60, 61, 360
Glauber, J. R. 1604–68, Ger., Aust., Neth. (b. Ger.): 654
Glover, John 1817–1902, Eng.: 529
Goldschmidt, Hans 1861–1923, Ger.: 507, 642, 692
Goldschmidt, Victor Moritz 1888–1947, Ger., Nor. (b. Switz.): 202
Goldstein, Eugen 1850–1930, Ger.: 19
Göransson, Göran Fredrik 1819–1900, Swed.: 723
Graham, Thomas 1805–69, Eng.: 408
Gregor, William 1761–1817, Eng.: 684
Guldberg, Cato M. 1836–1902, Nor.: 302
Gurney, Ronald W. 1899–1953, Eng., U.S. (b. Eng.): 758
Gutmann, Viktor 1921–, Aust.: 324, 501, 524

Haber, Fritz 1868–1934, Ger.: 281, 307, 398, 399, 455, 549, 568

Hahn, Otto 1879–1968, Ger.: 276
Hall, Charles M. 1863–1914, U.S.: 639
Harkins, William D. 1873–1951, U.S.: 265, 440, 674
Hatchett, Charles 1765–1847, Eng.: 690
Hedvall, J. Arvid 1888–, Swed.: 389, 396
Heisenberg, Werner 1901–, Ger.: 66 f.
Heitler, Walter 1904–, Ger., Ire., Switz. (b. Ger.): 99, 130
van Helmont, J. B. 1577–1644, Bel.: 39
Henry, William 1774–1836, Eng.: 349
Héroult, Paul 1863–1914, Fr.: 639
Hess, Germain Henri 1802–50, Rus. (b. Switz.): 56
de Hevesy, Georges 1885–1966, Ger., Den., Swed. (b. Hung.): 285, 684
Hisinger, Wilhelm 1766–1852, Swed.: 675
Hjelm, Peter Jacob 1746–1813, Swed.: 694
van't Hoff, Jacobus Henricus 1852–1911, Neth., Ger. (b. Neth.): 359, 373
Hooke, Robert 1635–1703, Eng.: 491
Hückel, Erich 1896–, Ger.: 313, 315
Hume-Rothery, William 1899–1968, Eng.: 231
Hund, Friedrich 1896–, Ger.: 84, 129

Joliot, Frédéric 1900–58, Fr.: 266, 272
Joliot-Curie, Irène 1897–1956, Fr.: 266, 272
Joule, James Prescott 1818–89, Eng.: 16, 364
Jungner, E. W. 1869–1924, Swed.: 437, 766

Kayser, Heinrich 1853–1940, Ger.: 458
Kelvin, Lord (William Thomson) 1824–1907, Eng.: 346
Kirchhoff, Gustaf Robert 1824–87, Ger.: 650
Klaproth, M. H. 1743–1817, Ger.: 675, 676, 684
Klaus, Karl Karlovich 1796–1864, Rus.: 738
Kossel, Walter 1888–1956, Ger.: 99
von Laue, Max 1879–1960, Ger.: 186, 238

Lavoisier, Antoine Laurent 1743–94, Fr.: 15, 24, 323, 451, 492, 508, 538, 586
Lawrence, Ernest O. 1901–58, U.S.: 676
Leblanc, Nicolas 1742–1806, Fr.: 655
Le Chatelier, Henry 1850–1936, Fr.: 307
Leclanché, Georges 1839–87, Fr.: 434
Lewis, Gilbert Newton 1875–1946, U.S.: 99, 155, 297, 323 f, 523
Libby, Willard F. 1908–, U.S.: 284
Lindqvist, Ingvar 1921–, Swed.: 324, 501, 524

Lockyer, Norman 1836–1920, Eng.: 458
London, Fritz 1900–54, Ger., Switz., Eng., Fr., U.S. (b. Ger.): 99, 130
Lowry, Thomas M. 1874–1936, Eng.: 316
Lux, Hermann 1904–, Ger.: 324, 501
Lyman, Theodore 1874–1954, U.S.: 82, 94

McMillan, Edwin M. 1907–, U.S.: 676
de Marignac, J.-Ch. G. 1817–94, Switz.: 676
Marsh, James 1790–1846, Eng.: 556
Martin, Pierre Émile 1824–1915, Fr.: 723
Maxwell, James Clark 1831–79, Eng.: 52
Mayer, Julius Robert 1814–78, Ger.: 16
Mayow, John 1643–79, Eng.: 491
Mendeleev, Dmitri Ivanovich 1834–1907, Rus.: 86, 619, 646, 668, 676, 711
Meyer, Lothar 1830–95, Ger.: 86
Millon, Eugène 1812–67, Fr.: 774
Mitchell, J. W. 1913–, Eng.: 758
Mohs, Friedrich 1773–1839, Ger., Aust. (b. Ger.): 236
Moissan, Henri 1852–1907, Fr.: 466
Mond, Ludwig 1837–1909, Eng. (b. Ger.): 593, 727
Mosander, Carl Gustaf 1797–1858, Swed.: 676
Moseley, H. G. J. 1887–1915.: 86, 95, 684
Mott, N. F. 1905–, Eng.: 758
Müller von Reichenstein, Franz Joseph 1740–1825, Aust.: 492
Mulliken, Robert S. 1896–, U.S.: 99, 150

Nernst, Walther 1864–1941, Ger.: 687
Nessler, Julius 1827–1905, Ger.: 774
Newlands, J. A. R. 1838–98, Eng.: 85
Nier, Alfred O. 1911–, U.S.: 27
Nilson, Lars Fredrik 1840–99, Swed.: 676
Nobel, Alfred Bernhard 1833–96, Swed.: 676
Noddack, Walter 1893–1960, Ger.: 706

Ölander, Arne 1902–, Swed.: 27
Ørsted, Hans Christian 1777–1851, Den.: 639
Ostwald, Wilhelm 1853–1932, Ger.: 548, 560, 563, 568

Paneth, Fritz 1887–1958, Aust., Eng. (b. Aust.): 285
Parkes, Alexander 1813–90, Eng.: 357, 749
Paschen, Friedrich 1865–1947, Ger.: 82, 94
Pauli, Wolfgang 1900–58, Ger., Switz., (b. Aust.): 77, 135
Pauling, Linus 1901–, U.S.: 99, 150, 154, 606, 609

Pedersen, Harald C. 1888–, Nor.: 640
Perey, Marguerite 1909–, Fr.: 650
Petit, A. T. 1791–1820, Fr.: 59
Phragmén, Gösta 1898–1944, Swed.: 231
Planck, Max 1858–1947, Ger.: 16
Priestley, Joseph 1733–1804, Eng.: 492
Proust, J. L. 1754–1826, Fr.: 43
Prout, William 1785–1850, Eng.: 32

Qvist, Bengt 1726–99, Swed.: 694

Raman, Venkata 1888–, Ind.: 240
Ramsay, William 1852–1916, Eng.: 458
Raoult, François-Marie 1830–1901, Fr.: 345, 371, 373
Rayleigh, Lord (John William Strutt) 1842–1919, Eng.: 458
Rinman, Sven 1720–92, Swed.: 734
Rose, Heinrich 1795–1864, Ger.: 690
Rose, Valentin (the elder) 1736–71, Ger.: 386
Rouelle, G. F. 1703–70, Fr.: 323
Rutherford, Ernst 1871–1937, Eng. (b. New Zealand): 19, 75, 266, 271, 283, 458
Rydberg, Janne 1854–1919, Swed.: 94

Scheele, Carl Wilhelm 1742–86, Swed.: 466, 492, 538, 694, 705
Schrödinger, Erwin 1887–1961, Aust., Ger., Switz., Ire. (b. Aust.): 66, 69,
Schweizer, Eduard, middle of 19th cent., Switz.: 753
Seaborg, Glenn T. 1912–, U.S.: 676
Sefström, Nils Gabriel 1787–1845, Swed.: 690
Sillén, Lars Gunnar 1916–, Swed.: 319, 505
Slater, John C. 1900–, U.S.: 99
Soddy, Frederick 1877–1956, Eng.: 458
Solvay, Ernest 1838–1922, Bel.: 551
Soret, J.-Louis 1827–90, Switz.: 676
Stahl, Georg Ernst 1660–1743, Ger.: 491
Stark, Johannes 1874–1957, Ger.: 78
Stensen, Niels (Nicolaus Steno) 1638–86, Den.: 192
Strassman, Fritz 1902–, Ger.: 276
Stromeyer, Friedrich 1776–1835, Ger.: 763
Suess, Hans 1909–, Ger., U.S. (b. Ger.): 440
Svedberg, The 1884–, Swed.: 408
Szilard, Leo 1898–, U.S. (b. Hung.) :295

Tacke-Noddack, Ida 1896–, Ger.: 706
Tennant, Smithson 1761–1815, Eng.: 738
Thénard, L. J. 1777–1857, Fr.: 631, 734
Theophrastus, ca. 370–285 B.C., Gr.: 538

Index of names

Thomas, Sidney 1850–85, Eng.: 722, 723, 724
Thomson, J. J. 1856–1940, Eng.: 19, 27
Thomson, William (Lord Kelvin) 1824–1907, Eng.: 346
Travers, Morris W. 1872–1961, Eng.: 458
Tropsch, Hans 1889–1935, Ger.: 399, 592
Trouton, F. Th. 1863–1922, Eng.: 64
Tyndall, John 1820–93, Eng.: 408

Urbain, Georges 1872–1938, Fr.: 676
Urey, Harold C. 1893–, U.S.: 440, 451

Vauquelin, Louis Nicolas 1763–1829, Fr.: 694
Volta, Allessandro 1745–1827, It.: 433

Waage, Peter 1835–1900, Nor.: 302
van der Waals, J. D. 1837–1923, Neth.: 98, 165 f
Wackenroder, H. W. F. 1798–1854, Ger.: 534
Werner, Alfred 1866–1919, Switz.: 250, 735
Westgren, Arne 1889–, Swed.: 231
Winkler, Clemens 1838–1904, Ger.: 619
Wöhler, Friedrich 1800–82, Ger.: 660
Wollaston, W. H. 1766–1828, Eng.: 737
Wood, B. 19th cent., U. S.: 386
Wood, Charles 18th cent., Eng.: 737

Zeeman, Pieter 1865–1943, Neth.: 78

SUBJECT INDEX

A

Absorption, optical 72; selective optical 73; nuclear resonance 274
  spectrum 73
Accelerator 273
Accumulator, see storage battery
Acetic acid–water protolytic system 318 f, 337 f
Acetylene 143, 590, 665
Acheson graphite 589
Acid 300 f, 501; according to Brønsted 316; according to Gutmann-Lindqvist 324; according to Lewis 323 f; divalent (=diprotic) 316; polyvalent (=polyprotic) 316; trivalent (=triprotic) 316
  anhydrides 501
  -base concept, development of 322
  -base equilibria 316 f; calculations 340
  -base pair 316
  -base process 317
  -base system 316
  fraction 338
Acidic constant 325, 329; apparent 330; incomplete (=mixed) 329; of water 329 f; table 329
Acidity 326
Acid reaction 329
Actinium 31, 670 f; see actinoids
Actinium series 270
Actinoids 89, 671 f; history 676; occurrence 674 f; production and uses 678 f; properties 676 f
Actinoid contraction 113
Actinon 458
Activated complex 255
Activation
  by catalysts 398
  analysis 296
  energy 255
Activator 228
Activity 297 f; 312 f
  factor 299, 312
Addition reactions 302, 304
Adsorbed water 510
Adsorbent 400
Adsorption 400 f; negative 400; positive 400
  chromatography 406
  isotherm 401
Aerosols 409
Agate 613
Age determination, radioactive 283 f
Aging of precipitates 345, 395, 511
Air gas 592
Alabaster 659, 664

Albite 212, 611, 648
Alcali 650
  minerale 650
  vegetabile 650
Alcosols 409
Algaroth powder 558
Alkali 650
  hexachloroplatinates(IV) 648
  hexahydroxostannates(IV) 623
  hydridoborates 633
  hydroxides 653
  hyperoxides 516, 653
  metals 93, 647 f; history 650; occurrence 648 f; production and uses 651 f; properties 650 f
Alkaline-earth carbonates 596, 665
  elements 93, 657 f; history 660; occurrence 658 f; production and uses 661 f; properties 660 f
  hydroxides 657, 663
Alkaline earths 660
Alkaline reaction 329
Alkali-nickel storage battery 437
Alkali
  nitrates 596
  oxides 653
  perchlorates 648
  peroxides 513, 653
  peroxoborates 638
  silicates 613
  tetrahydridoborates 633
Alkyl
  -aluminums 643
  -lithiums 652
  sulfinate 526, 534
  -sulfinic acid 526
  sulfonate 526
  -sulfonic acid 526
Allotropism 198
Alloy 230
Alloy steels 725
Alpha decay 267
Alpha particle 265, 267
Alum 322, 531, 639, 645
Alumen 639
Aluminate ion 645
Aluminosilicates 606, 630
Aluminothermic reduction 506, 642
Aluminum 28, 630, 638 f; history 639; occurrence 638 f; production 640 f; properties 676 f; uses 641 f
  alloys 641 f
  borosilicate glass 614
  bronze 642, 750
  chloride 474, 643

fluoride 368, 643
hydride 642
hydroxide 204, 321, 645; amphoterism of 345 f
ion, hydrated 643; protolysis of 321
nitride 205, 544
oxide 644, 734; α Al_2O_3 644
oxide hydroxide 546; α AlO(OH) 645; γAlO(OH) 645
oxide–silicon dioxide system 615, 616
phosphorus oxide 504
sulfate 645
sulfide 646
Alum-type double salts 531
Amalgamation (for silver and gold) 748, 766
Amalgams 764, 774 f
Amblygonite 648
Americium 31; see actinoids
Amethyst 604
Amide ion 174, 546, 547
Amido
 -carbonic acid 594 f
 -mercury(II) chloride 773
 -mercury(II) ion 773
Amino group 546
Ammine 126
 -cadmium ions 769
 -cobalt(II) ions 734
 complexes 126, 304 f, 547, 712
 -copper(I) ions 752
 -copper(II) ions 304 f, 753
 -zinc ions 767
Ammonia 545 f; crystal structure 168; formation of 307; production and uses 549 f
 molecule 124, 140, 144, 174
 protolytic system 332
 -soda process 551, 655
 -water protolytic system 318 f, 339 f
Ammonium 551
 amalgam 551
 amidocarbonate 552, 595
 carbonate 552
 chloride 366, 551, 621
 compounds 550 f
 dichromate 700
 dihydrogen phosphate 552, 576
 dodecamolybdophosphate 702
 fluoride 508; crystal structure of 205
 hexachloroplatinate(IV) 740, 741
 hexachloroplumbate(IV) 629
 hydrogen carbonate 552
 hydrogen sulfide 517, 520
 hydroxide 548
 ion 216, 546, 547, 550; bonding in 140, 144, 174
 iron(II) sulfate hexahydrate 730
 nitrate 551 f, 568
 perchlorate 487, 551
 perrhenate 707
 pertechnetate 707
 sulfate 531, 551
 tetrachlorozincate 621
Amorphous solid phases 42, 186, 218 f
Amphiboles 608

Amphiprotic properties 317
Ampholyte 317, 342 f, 356, 511; colloidal 411
Amphoteric properties 317, 502
Analysis, phase 386; thermal 387
Analytical chemistry 10
Anatase 686
Andalusite 502, 607
Anhydrite 491, 659, 664; soluble 664; insoluble 664, 666
Anhydrone 658
Anion
Anion exchanger 405
Anisotropy 186
Annealing 397, 719
Annihilation radiation 21, 266
Anode furnace (for copper) 748
Anode slime 433, 499, 740, 748, 749
Anodization 639
Anorthite 611
Anthracite 589
Antichlor 534
Antielectron (= positron) 20, 21
Antiferromagnetism 247, 248
Antifluorite structure type 203
Anti-isomorphism 203
Antimonic acid 579
Antimonides 544 f
Antimonites 579
Antimony 30, 536; black 542; explosive 542; history 538; metallic 172, 542; occurrence 538; production and uses 544; properties 542; structure 172
 –bismuth system 378, 379, 383
 butter 558
 -(III) hydroxide salts 579
Antimonyl ion 579
Antimony mirror 556
 -(III) oxide salts 579
 pentachloride 558
 tetroxide 579
 trichloride 558
 trihydride 556
Antineutron 20, 21
Antiparticle 20, 21
Antipode 198
Antiproton 20, 21
Apatite 228, 236, 465, 537, 575, 659
 iron ore 722
Aqua regia 425, 567
Aquamarine 608
Aquo complex ions 178, 509 f, 512
Aquopentamminechromium(III) ion 698
Aragonite 362, 596, 665; crystal structure of 596
Area, centered 189; doubly primitive 188; primitive 188
Argentate(I) ions 755, 756
Argentite 624, 744, 745
Argentum 29, 745
Argon 28, 457 f; history 458; occurrence 457; properties 458 f; recovery 459 f; uses 460
 arc welding 460
Argyrodite 619
Aromatic character 149

Arrested reactions 45, 62, 256, 428 f
Arrhenius's equation 254
Arsenates 578
Arsenic 29, 366, 536 f; black 172, 542; history 538; metallic (=gray) 172, 541; occurrence 537 f; production and uses 544; properties 541 f; structure 172; "white" 538, 577; yellow 172, 541
   acid 578
   mirror 542, 556, 578
   sulfides 580
   trichloride 324, 558
   trihydride 556
Arsenides 544 f
Arsenious acid 578
Arsenites 578, 581
Arsenopyrite 538, 544
Arsine 556
Asbestos 610
Association reactions 302
Astatine 31, 439; history 441; occurrence 440
Atmosphere 445
Atom 19, 33; central 99; characteristic 100; isoelectronic 109; labelled 285; many-electron 83 f, 95; one-electron 82 f, 94
Atomic
   arc welding 453
   heat 58
   ion 105; formation of 117 f
   mass 26; absolute 26; relative 26 f
   nucleus 21 f, 259 f
   number 22
   orbital 77
   percent 38
   radius, covalent 163 f; effective 102 f
   spectrum 73, 94 f
   weight 26 f; scale 27; table 28 f
Atomization
   energy 56, 102
   enthalpy 56
Attraction, energy of 101
Augite 608, 658
Auric chloride 760
Aurum 31, 745
Austenite 717 f, 725
Autocatalysis 258
Autoionization 155, 324
   equilibrium 324
Autoprotolysis 327, 343
Avogadro's law 40
Avogadro's number 36
Azeotropic liquid 375
Azide ion 150, 479, 554
Azides 554
Azimuthal quantum number 76
Azote 28, 538
Azurite 744, 754

B

Babbitt 621
Baddeleyite 684
Baking powder 577, 596
Baking soda 596, 656

Balmar series 82, 94
Band spectrum 73
Barite 659, 660
Barium 30, 657 f; history 660; occurrence 659; properties 660 f
   carbonate 596
   chromate(VI) 699
   ferrate(VI) 734
   nitrate 664
   perchlorate 658
   sulfate 530, 531, 664, 666; solution equilibrium 354 f
   sulfide 228
Barrier layer 228
Baryon 20, 21
Base 301 f, 501; according to Brønsted 316; according to Gutmann-Lindqvist 324; according to Lewis 323 f; divalent (=diprotic) 316; polyvalent (=polyprotic) 316; trivalent (=triprotic) 316
Base bullion 626, 749
Base fraction 338
Base metal 424
Basic constant 326, 330
Basic reaction 329
Basic slag 576, 723
Bastnäsite 674
Bauxite 616, 639, 640
Bayer process 640
Bearing metals, lead-base 621, 626; tin-base 621
Bentonite 611
Benzene 124, 164, 590; bonding in 147 f
Berkelium 31; see actinoids
Berthollide 44
Beryl 195, 608, 658, 660, 661
Berryllate ions 661, 663
Beryllium 28, 32, 657 f; history 660; occurrence 658; production and uses 661; properties 660 f
   bronze 750
   copper 661, 750
   halides 658
   hydroxide 663
   oxide 657, 658, 660
Bessemer process 723
Beta decay 266
Beta particle 266
Bicarbonate 656
Bichromated gelatine 700
Binary systems 359, 369 f
Binder 397
Biochemistry 9
Biosphere 445
Biotite 610, 649, 658
Birkeland-Eyde process 549, 560
Bismuth 31, 536; history 538; occurrence 538; production and uses 544; properties 542 f; structure 172
   -antimony system 378, 379, 383
   -(III) fluoride–lead(II) fluoride system 214
   halides 558
   hydroxide nitrate 580
Bismuthinite 538, 582

Bismuth
  mirror 556
  oxide chloride 580
  oxide hydrate 580
  oxide nitrate 580
  sulfide 582
  trichloride 579
  trihydride 556
  triiodide 204
  trinitrate 579
Bis(thiocyanato)iron(III) ion 600, 733
Bis(thiosulfato)argentate(I) ion 534, 756
Bituminous coal 589
Black liquor 517
Black powder 654
Blast furnace 720, 721
Blast-furnace process 720 f
Bleaching powder 469, 484
Blister copper 748
Blister-forming period (in copper smelting) 748
Blueprint paper 733
Blue vitriol 531
Body-centered cubic lattice 196
Boehmite 645
Bog ore 713
Bohr atom 75
Boiling point 365
  curve 373
  diagram 373 f; see also phase diagram
  elevation 271 f; molal 372
Boliden impregnation 578
Boltzmann's constant 40
Bomb, fission 292; thermonuclear 292
Bomb calorimeter 54
Bond, chemical 97 f; covalent 98, 128 f; dipole 98, 124 f; dipole-dipole 125; double 137, 139; electrostatic 98, 105 f; hydrogen 167 f; ion-dipole 98, 124 f; ionic 98, 105 f; metallic 98, 229 f; one-electron 132, 139; single 137, 139; three-center 175; three-electron 133, 137 f, 139; triple 136 f; unsaturated 163; valence 139 f; van der Waals 165; π 134; σ 134
  angle 184, 185
  axis 129
  dissociation energy 102, 162
  energy 100 f, 161 f
  length 99, 164
  number 138 f; 163
  order 138
Boranes 632 f
Borate ion 147, 175, 176, 635
Borates 635 f; structures of 176 f
Borax 631, 636
  bead 636, 734
Borazole 638
Borazon 638
Boric acid 631, 636 f; protolysis of 319, 637
  esters 637
Borides 632
Born-Haber cycle 206
Bornite 744
Boron 28, 32, 630 f; amorphous 632; history 631; occurrence 631; production and uses 632; properties 631 f
  carbide 638
  -group elements 630 f
  hydrides 175, 632 f
  nitride 638
  tribromide 632, 634
  trichloride 124, 634
  trifluoride 147, 634, 635
  triiodide 632, 634
Borosilicate glass 614
Boule 644
Branched-chain reaction 258, 288
Brass 750, 766; β 216, 231
Braunite 705
Breeder reactor 292
Brick 616 f
Bridging ligand 100
Brittania metal 621
Brittle micas 611
Bromate ion 481
Bromic acid 481, 485
Bromine 29, 286, 464 f; history 466; occurrence 465; production and uses 470; properties 471 f; reaction with water 468
  dioxide 480, 481
  trifluoride 477 f
  water 468
Bronzes 750
Brookite 686
Buffer 341, 576
  action 341
  solution 341
Building unit 200
Burnt lime 659, 662

C

Cadmium 29, 761 f; history 763; occurrence 762; production 765; properties 763 f; uses 766
  bromide 204, 768
  chloride 768
  compounds 768 f
  hydroxide 768 f
  iodide 204, 768
  -(I) ion 769
  -(II) ion 768; halo complexes 768
  oxide 768
  plating 766
  red 769, 773
  sulfide 228, 769
  yellow 769
Cage structure 215
Calcined phosphate 576
Calcite 195, 236, 362, 595, 659, 665; crystal structure of 596
Calcium 28, 657 f; history 660; occurrence 659; production and uses 662; properties 660 f
  arsenate 578

carbide 590, 665
carbonate 595 f, 665, 666; dissociation 375
chloride 474, 658, 664
cyanamide 600, 665
-cyanamide process 549
cyanide 600
fluoride 195, 207, 208, 209; crystal structure of 203
hydride 456
hydrogen phosphate 530
hydroxide 204, 663
hypochlorite 469, 484
metaborate 176, 177
metaphosphate 576
oxide 659, 662 f
phosphide 544
sulfate 531, 658
sulfide 228
tungstate 694
Californium 31, 264; see actinoids
Calomel 770
Calorie 18
Capture, electron 266; particle 272
Carbamic acid 594
Carbamide plastics 595
Carbide ion 122
Carbides 590
Carbo 586
Carbon 584 f; active 589; amorphous 588; heat of combustion of 57, 63; history 586; isotopes of 27, 28, 32, 283 f, 586; occurrence 585 f; properties and uses 586 f; P, T diagram 586
Carbonate hardness 666
Carbonate ion 147, 175, 595
Carbonates 595 f
Carbonatotetramminecobalt(III) ion 735
Carbon
black 588
cycle in nature 596
dioxide 124, 149, 330, 366, 593 f; in nature 596 f
-dioxide snow 366, 594
disulfide 150, 597 f
Carbone 586
Carbon glass 589
Carbon-group elements 584 f
Carbonic acid 330, 594
—water protolytic system 330, 594
Carbonizing (for Na_2CO_3) 656
Carbon
-14 method 284 f
monoxide 135, 137, 143 f, 398, 591 f
monoxide hemoglobin 591
steel 719
suboxide 595
tetrachloride 124, 141, 591
Carbonyl
chloride 592
complexes 179, 183, 592 f
iron 717
-metals 592
-nitrosylmetals 593
sulfide 150, 592, 598

Carboranes 634
Carborundum 618
Carboxylic acids 169
Carburizing 724
Carnallite 649, 658
Carnelian 613
Carnotite 675, 689
Caro's acid 532
Carrier 294; isotopic 294, 403; nonisotopic 294, 403
technique 273, 295
Cascade 279
Case hardening 724
Cassiterite 619, 623, 694
Cast iron 720; eutectic 720; gray 720; hypereutectic 720; hypoeutectic 720; white 720
Catalysis 45; heterogeneous (=contact) 397 f; homogeneous 258; selective 399
Catalyst 45, 62, 397 f; mixed 398, 399
support 399, 613
Cation 23
exchanger 405
Causticizing 653
Celestite 659
Cell, electrochemical 414 f; electrolytic 414, 417; irreversible 417; reversible 417
process 414 f
Cellophane 598
Cellulose 598, 753
xanthate 598
Cement 617; high-alumina (=fused) 617 f
Cement clinker 617
Cement copper 748
Cementite 590, 717 f
Center of symmetry 190
Central atom 99, 181
Central ion 99
Centrifugation 15
Ceramics materials 615 f; dense-fired (dense-sintered) 615; fireproof 615; impervious (=fuchsin-fast) 615; nonvitreous 615; refractory 616; semivitreous 615; vitreous 615
Cerargyrite 744
Cerite 674, 675
Cerium 30, 121, 671; see lanthanoids
dioxide 681, 682
earths 674, 676
-(IV) perchlorate 682
sulfide 681
Cerrusite 624
Ceruloplasmin 744
Cesium 30, 647; history 650; occurrence 649; production and uses 651 f; properties 650 f
bromide 203, 207
chloride 202, 207
-chloride structure type 202, 207
dichloroaurate(I) 245
gold(I,III) chloride 245, 760
iodide 203, 207, 208, 226
octafluoroxenate(VI) 463
tetrachloroaurate(III) 245
Chain reaction 257 f, 288; branched 258, 288

Chalcedony 613
Chalcocite 744, 745
Chalcogens 93, 489 f; history 491; occurrence 490
Chalcopyrite 744, 747
Chalk 659, 665, 666
Chamber
  acid 529
  crystals 529
Chamotte 615
Charge, electrical unit (=elementary) 19; formal 154; specific 409
Charge compensation 212
Chelate 179, 180
  effect 180
Chemical bond 97 f
Chemical potential 64, 297
Chemiluminescence 74
Chemistry, analytical 10; general 9; inorganic 9; organic 9; physical 10; theoretical 10
Chile salpeter 537, 549, 654
Chlor-alkali electrolysis 453, 469 f, 653, 766
Chloramine 553, 556
Chlorates 485 f
Chlorate ion 175, 287, 295, 481, 482; structure of 145
Chloric acid 335, 481, 482, 485
Chlorine 28, 110, 257, 464 f; history 466; occurrence 465; production 469 f; properties 471 f; reaction with water 427, 428; uses 470
  dioxide 245, 287, 480, 481
  molecule 166, 167
  water 468
Chlorite ion 287, 481, 482, 485; structure of 145
Chlorite minerals 609, 610
Chlorites 485
Chloroauric acid 760
Chloropentamminechromium(III) ion 698
Chlorophyll 181, 659, 705, 713
Chloroplatinic acid 741
Chlorosulfuric acid 521
Chlorous acid 335, 481, 482, 483, 485
Chromate(VI) ion 691, 693, 698
Chromates(VI) 693, 698 f
Chromatography 406 f
Chrome
  alum 698
  green 698
  red 699
  tanning 698
  yellow 699
Chromic acid 699
Chromite 694, 695, 696
Chromite ion 698
Chromium 29, 693 f; history 694; occurrence 693 f; production 695 f; properties 695; uses 696
  -(II) acetate 697
  -(III) arsenate 578
  -(II) chloride 697
  -(III) chloride 204, 697
  -(II) compounds 697

  -(III) compounds 693, 697 f
  -(IV) compounds 693, 698 f
  -group elements 693 f; history 694; occurrence 693 f; production 695 f; properties 695; uses 696 f
  -(III) hydroxide 698
  -(III) oxide 248, 693, 696, 698
  -(IV) oxide 693, 696, 699 f, 701
  -(III) oxide hydroxide 698
  plating 696, 700
  -(III) sulfate 696, 698
  trioxide 693, 696, 699 f, 701
Chromyl chloride 700 f
Chrysoberyl 645
Chrysotile 610
Cinnabar 762, 765, 772 f
Cis form 249, 250
Citrine 604
Clapeyron equation 361
Clathrate compounds 214 f
Clay, buff-burning 617; red-burning 617; white-burning 617
Clay minerals 406, 609, 610, 613, 638
Cleavage 235; nuclear 275
  plane 235
Close packing 196 f; cubic 197; hexagonal 197
Clouding agent (for glass) 614
Coagulation 409
Coal 585, 589
Coal tar 589
Coarsely dispersed system 408
Cobalt 29, 248, 711 f; history 714 f; occurrence 714; production 728; properties 715, 716; uses 728; α Co 715; β Co 715
  alloys 728
  -(II) chloride 734 f
  -(II) chloride dihydrate 735
  -(II) chloride hexahydrate 734
  -(II) chloride monohydrate 735
  -(II)-cobalt(III) redox pair 425
  -(II) compounds 734 f
  -(III) compounds 735 f
  gun 728
  -(II) hydroxide 734
Cobaltite 714
Cobalt
  -(II) oxide 734, 736
  -(II, III) oxie 736
  -(III) oxide hydroxide 728, 734
  -(II) sulfide 735
Coercive force 247
Coesite 604
Coinage metals 93, 231; see copper-group elements
Coke 589
  plant 589
Cold brittleness 722
Cold working 397
Collector 404
Collision complex 255
Collision number 254 f
Colloidally dispersed system 408
Colloids 408 f; hydrophilic 410; hydrophobic

410; lyophilic 410; lyophobic 410; protective 410
Color 241 f; ion 243; surface 242; tarnish (=iridescent) 243, 716; transmission 241 f
Color center 214
Columbite 690, 706
Column 374, 406
Combustion 253, 412; heat of 55; spontaneous 254
Commercial iron 720
Complex 34; activated 255; collision 255; metal 178 f; polynuclear 175 f
  constant 302
  formation 178 f, 302, 304, 358
  ion 105
Component 35, 359
Compound, binary 173 f; chemical 43
Compound nucleus 272
Concentration 37 f; total 38
Concentration polarization 418
Concrete 617
Condensation 410
Conductivity water 509
Conductor, electronic 218, 221; ionic 217; metallic 224
Conservation
  of energy, law of 16, 50
  of mass, law of 16
  of matter, law of 15
Constantan 750
Consumable electrode 686
Contact catalysis 397
Contact catalysts 712
Contact process (for sulfuric acid) 524, 529
Control element 290
Conversion (of gas) 589
Conversion process 654
Converter 723, 747
Cooking liquor 517
Coordination 99, 114 f
  configuration 113, 114 f
  number 99; characteristic 135; maximum 115
Copper 29, 728, 743 f; history 745; occurrence 743 f; oxidation of metal 218; production 747 f; properties 745 f; uses 749 f
  alloys 216, 231, 750
  -(II) arsenate 758
  -(I) bromide 205, 752
  -(II) chelate complexes 179
  -(I) chloride 205, 752
  -(II) chloride 245, 753; structure of 177
  -(II) chloride dihydrate 753
  -(II) chloride–potassium chloride system 382, 385
  -(I) compounds 751 f
  -(II) compounds 752 f
  —copper(I) redox pais 427
  -(I)–copper(II) redox pair 427
  -(I) cyanide 752
  -(II) fluoride 743
  -gold alloys 210
  -group elements 743 f; history 745; occurrence 743 f; production 747 f; properties 745 f; uses 749 f
  -(I) halides 752
  hydride 573
  -(II) hydroxide 753
  -(II) hydroxide carbonate 744, 746, 754
  -(II) hydroxide sulfate 746, 754
  -(II) hydroxide chloride 746
  -(I) iodide 205, 213, 752
  -(I) ion, disproportionation 427, 751
  -(II) nitrate 568
  -(II) nitrate hexahydrate 754
  -(I) oxide 213, 218, 228, 744, 746, 751
  -(II) oxide 746, 753
  –silver system 379, 380, 383
  -(II) sulfate 742, 754
  -(II) sulfate hydrates 376 f, 754
  -(II) sulfate pentahydrate 754
  -(I) sulfide 744, 752
  -(II) sulfide 754
  vitriol 531, 754
Coprecipitation 286, 402 f
Coring 379
Corrosion resistance 726
Corrosive sublimate 711
Corundum 236, 644
Coulomb energy 102
Coulomb force 102
Coulomb's law 102
Counter ion 401
Covalence 139
Covalent
  bond 98, 128 f, 236
  radii 103, 163 f
Covellite 754
Cracking gas 589
Cristobalite 603, 604, 616
Critical
  point 364
  pressure 364
  size 289
  temperature 364
Crocus 731
Crust of the earth 441
Cryolite 465, 639, 640, 641, 643
Crystal 34, 43, 186 f; ideal 42, 186; distorted 193
  aggregate 194
  elements 190
  energy 200
  nucleation and growth 106 f, 389 f
  planes 192
  soda 655
  structure 34, 201 f
  symmetry 190 f
  system 195
Crystalline phase 42
Crystallization, fractional 383; water of 126, 510
Crystallography, geometrical 186 f
Crystallon 618
Cubic system 195
Cupellation 749
Cuprite 744, 751
Cupro nickel 750

Subject index

Cuprum 29, 745
Curie
　point 248
　temperature 248
　transition 216
Curium 31; see actinoids
Cyanamid 600, 665
Cyanamide ion 150, 479, 600
Cyanamides 479, 600
Cyanate ion 150, 479, 599
Cyanic acid 479, 599
Cyanide ion 135, 137, 479, 598
Cyanide leaching (for silver and gold) 748 f, 766
Cyanides 598
Cyano
　-argentate(I) complexes 756
　-cadmiate complexes 769
　-cobalt(II) ions 734
　complexes 157, 178, 183, 433, 599, 712
Cyanogen 479, 599
　flame 599
Cyanozincate ion 768
Cyclopentadiene ion 182
Cyclotron 273
Cytochromes 713

D

Daltonide 44
Dalton's law of partial pressures 41
Dark reaction 257
Daughter
　nucleus 266
　nuclide 266
D-D process 275, 293, 294
Deacon process 469
Decavanadate ion 692
Decay, alpha 267; beta ($=\beta^-$) 266; isobaric 266; positron ($=\beta^+$) 266
Decay constant 251
Decolorizing agents (for glass) 614
Decomposition voltage 430
Decrepitation 654
Defects, crystal 215
Definitions 15 f
Deformation, crystal 217, 231 f; elastic 231; plastic 232
Deformation energy 231
Degeneracy 78
Dehydrite 658
Deliquescence 376
Delocalized electrons 98, 145 f, 223
Dendrite 392
Denitrification 583
Dense-fired (=dense-sintered) material 615
Density, from crystal unit cell 189
Dental amalgam 750, 766, 774
Deoxidation 504, 601, 662, 722
Dephlogisticated air 492
Depolarizer 433
Derivate, partial 49
Desalination of sea water 510

Deuterium 25, 280 f, 293, 431, 432, 451, 452
Deuteron 25, 272, 453
Development, photographic 757 f
Development nucleus 757 f
Dialkyl sulfite 526
　sulfone 526
　sulfoxalate 526
Dialysis 388
Diamagnetism 246
Diamidocarbonic acid 594 f
Diammine-mercury(II) chloride 773
　-mercury(II) ion 771, 773
　-platinum(II) chloride 250
　-silver(I) ion 126, 356, 755, 756
Diammonium hydrogen phosphate 552, 576
Diamond 236, 362, 585 f; bonding and structure 142, 205
Diantimony
　pentasulfide 581, 582
　pentoxide 579
　trioxide 578 f
　trisulfide 581
Diaphragm process 469
Diarsenic
　pentoxide 578
　pentasulfide 580 f
　trioxide 220, 577 f
　trisulfide 580 f
Diarsine 556
Diaspore 645
Dibismuth trioxide 579
　trisulfide 582
Diborane 175, 633
Diborate ion 176, 635
Diboron trioxide 220, 635
Dibromine oxide 479, 481
Dicarbonyldinitrosyliron 593
Dichlorine heptoxide 481
　hexaoxide 481
　oxide 479 f, 481
　trioxide 481
Dichloro
　-aurate(I) ion 245, 760
　-cuprate(I) ion 751
　-diammineplatinum(II) 250
　-difluoromethane 591
　-tetramminecobalt(III) ion 250, 735
　-tetraquocobalt(II) 734
Dichromate ion 699; protolysis of 319
Dichromates 699
Dicyandiamide 600
Dicyano
　-argentate(I) ion 357, 432, 599, 749, 755 f,
　-aurate(I) ion 749, 759
　-cuprate(I) ion 747
Dielectric constant 123
Diffraction 237; electron 238; neutron 238; X-ray 238
　methods 237 f
　pattern 237
Diffusion, in liquids 41, 388 f; in solids 42, 217, 389
Diffusion layer 388
Dihaloaurate(I) ions 759

Dihydrogen
 hexachloroplatinate(IV) hexahydrate 741
 hexachlorostannate(IV) hexahydrate 623
 selenide 519
 sulfide 518
 telluride 519
 tetracarbonylferrate(−II) 593
Diiodine pentoxide 481
Diiodotyrosine 465
Dimanganese heptoxide 710
Dimer 34
Dimorphism 197
Dinitrogen oxide 150, 154, 559
 pentoxide 559, 563; decomposition of 252
 tetroxide 559 f, 562
 trioxide 559, 562
Diopside 608
Dioxonitrate(II) ion 564
Dioxivanadium(V) ion 689, 691
Dioxygen difluoride 479
Diphosphate ion 571, 573
Diphosphine 555
Diphosphoric acid 571, 573, 575
 system 336 f, 344, 345
Diphosphorys pentoxide 569 f
 tetrahydride 555
 trioxide 569
Dipole 98, 101; electric 123; induced 123; permanent 123
 bond 98, 124 f
 -dipole bond 125, 369
 moment 123
Disilver fluoride 756
Dislocation 232; edge 232; screw 391
Disorder 60, 215 f
 phenomena 210 f
Dispersion medium 14, 408 f
Disproportionation 427
Dissociation 56, 302 f, 310 f; electrolytic 308;
 degree of 302
 constant 302
 energy 56, 240
 enthalpy 56
Distillation, fractional 374; dry 589
Distribution, ordered 211; statistical 212
 coefficient (=factor, ratio) 357
 equilibrium 348 f, 357 f
 function, radial 78, 79, 85
Disulfate ion 178, 526, 532
Disulfates 532
Disulfide ion 519
Disulfite ion 527, 533
Disulfites 533
Disulfur dichloride 474, 520
Disulfuric acid 526, 532
Disulfurous acid 533
Dithiocarbonate ion 598
Dithionate ion 527, 532, 534
Dithionic acid 527, 532, 534
Dithionite ion 527, 533
Dithionous acid 527
Diuranate ion 672
Divanadium pentoxide 230, 689, 691
Division, mechanical 234 f

Dodecamolybdophosphate ion 702
Dolomite 658
Dolomitic limestone 658
Domain, magnetic 247 f
Double bond 137
Double oxide 503
Double salt 322
Doubly primitive area 188
Drierite 658
Dry cell 434
Dry Ice 366, 594
Dryer 627
Dulong and Petit, rule of 59
Duralumin 642
Dye sensitization 758
Dysprosium 30; see lanthanoids

E

Earths (oxides) 601, 639, 660
Earth's crust 441
Edge dislocation 232
Edison storage cell 437
EDTA (=ethylenediamine tetraacetate) 180, 625
Effective radius 102 f
Efflorescence 376
Effusion 278
Einsteinium 31; see actinoids
Einstein's equation 15
Ekaaluminium 646
Ekasilicon 619
Electroanalysis 432
Electrochemistry 412 f
Electrode potential 415, 419 f
Electrode process 414 f
Electrodialysis 510
Electrolysis 308, 309 f, 430 f
Electrolyte 308 f; strong 311; weak 311
Electrolytic
 cell 414, 417
 copper 748
 dissociation 308; Arrhenius's theory of 314
 iron 717
 manganese 606
 refining 433, 748
Electromotive force 415
Electron 19, 20; delocalized 98, 145 f, 223;
 localized 98; odd (=unpaired) 84, 137 f;
 245, 249; solvated 547 f; valence 85
 affinity 106 f, 110 f
 capture 266
 cloud 19, 22 f, 75 f, 86 f
 compound 231
 -deficit conduction 227
 -deficient molecule 175, 633
 density 76, 78, 79 f
 diffraction 238
 distribution 83 f, 159 f, 184 f
Electronegative element 111
Electronegativity 111, 150 f
Electron-excess conduction 227

Electron hole 225
Electronic
  conduction 221, 224 f
  energy 51, 239
  transition 82
Electron
  neutrino 20
  pair 77, 108; bonding (=shared) 132; inert 108; lone (=unshared) 136
  -pair acceptor 140, 324
  -pair bond 98, 132
  -pair donor 140, 324
  shell 77
  spin 76; antiparallel 76; parallel 76
  spin resonance 249
Electrophoresis 409
Electroplating 432
Electropositive element 111
Electropositivity 111
Electrostatic bond 98, 105 f
Electrosteel processes 723
Electrovalence 111
Elements 22, 24 f; building up of 86 f; mean abundance in earth's crust 445 f; occurrence 440 f; structure of 170 f; table of isotopes of 28 f
Elementary area 188
Elution 406
Emanation 271, 458
Emerald 608
Emery 644
Emf (=electromotive force) 415
Emission 72
Emission spectrum 73
Emulsion 408; photographic 757 f
Enantiomorphism 198 f, 249, 250
Endogenic process 445
Endothermic process 55
Energy 15 f, 50 f; binding (nuclear) 264; chemical 52; forms of 51 f; free 61; internal 52 f; quantization of 70; total 51
  band 223
  exchange 72 f
  levels 70, 75, 82 f, 129 f, 157 f
  principle 16, 50
  production (nuclear) 287 f
  threshold (=energy barrier) 62, 256
English red 731
Enriched fuel 291
Enriching section 279, 280
Enthalpy 54
Entropy 60
Enzyme 258
Epsom salt 664
Equilibrium 45 f, 300 f; chemical 59 f; dynamic 46; heterogeneous 302; homogeneous 302; ionic 308 f; metastable 361, 363, 366; partial 46; radioactive 270; static 46
  constant 300
  diagram (=phase diagram) 363 f
  equation 300 f
  pressure 361
  temperature 361
Equivalence, symmetry 190

Equivalent 37; electrochemical 37
  conductivity 311
Erbium 30; see lanthanoids
ESR (=electron spin resonance) 249
Eternit 613
Ethane 166
Ethanol–benzene boiling-point diagram 374
Ethanol–water protolytic system 318 f
Ethanol–water azeotrope 375
Ethylene 143, 163, 590
  bromide 470
  -diamine 180, 250
  -diamine tetraacetate 180
Europium 30, 104; see lanthanoids
Eutectic 380
  composition 380
  line 380
  point 380
  reaction 381
  temperature 380
Eutectoid 382
Exchange reaction 281, 286, 305
Excitation energy 83
Excited state 70, 83
Exogenic process 445
Exothermic process 55
Explosives 567, 568
Extraction 357 f
Extrinsic semiconduction 226

F

Face-centered cubic lattice 196
Faraday 37
Faraday's electrolytic laws 309
F center 214
Fehling's solution 753
Feldspars 611, 616, 638, 648, 649
Fermium 31; see actinoids
Ferrate(VI) ion 161, 733 f
Ferrates 733 f
Feeric ion (=iron(III) ion) 731
Ferricyanide ion 599
Ferrimagnetism 247, 248
Ferrite 717, 718 f
Ferrites 732
Ferritin 713
Ferroboron 632
Ferrocene 182
Ferrochromium 696
Ferrocyanide ion 599
Ferromagnetism 247 f
Ferromanganese 707
Ferromolybdenum 696
Ferrosilicon 601
Ferrovanadium 691
Ferrotungsten 696
Ferrous ion (=iron(II) ion) 729
Ferroxcube 732
Ferrum 29, 714
Final nucleus 272
Fining 578, 614
Fire air 492

Fire clay 615
Fire gilding 751
Firing (ceramics) 615 f; smooth 617; decorative 617
Fischer-Tropsch process 399, 455, 592
Fission, of atomic nuclei 275 f, 288 f; spontaneous 277
  bomb 292
  neutron 288 f
Fixing, photographic 534, 759
Flint 613
Flint (lighter) 680
Flotation 403 f
Fluorapatite 537
Fluorescence 74
Fluorine 28, 32, 110, 464 f; history 466; occurrence 465; production and uses 468 f; properties 471 f
  molecule 135, 138, 139, 170
Fluorite 195, 236, 465, 659
  structure type 203
Fluoroborate ion 635
Fluorspar 465
Flux 466, 616; soldering 503, 551, 621
Fogging, photographic 758
Force constant 71
Formation, energy of 56; enthalpy of 56, 57
Formula unit 34
Formula weight 36
Foul air 492, 538
Fractional crystallization 383
Fractional distillation 374
Fracture, intercrystalline (= grain boundary) 235; conchoidal 235
Framework molecule 34, 200, 236
Francium 31, 647; history 650; occurrence 649
  -(I) ion 117
Frasch process 499
Freedom, degree of 48, 360
Free radical 183 f, 245
Free rotation 137
Freezing mixture 385, 664
Freezing-point depression 371 f; molal 372
Freon 591
Frequency 18
Friedel-Crafts reactions 643
Frother 403
Fuel cell 433
Fuel element 289 f, 678; processing 291, 679; encapsulation 661, 678, 686, 691
Fulminate ion 150, 479, 600
Fulminating mercury 600
Fulminic acid 479, 600
Furnace process (for phosphoric acid) 574
Fused cement 617 f
Fused quartz 604
Fusion 503, 532, 612; nuclear 275, 292 f
Fusion bomb 292 f

G

Gadolinite 674, 706
Gadolinium 30, 248; see lanthanoids
Galena 243, 624, 628, 744
Gallium 29, 630, 646
  arsenide 646
  -(III) hydroxide 430
  phosphide 646
Galvanic cell 414, 416
Galvanized iron 439, 765
Galvanizing 765
Gamma rays 267
Gangue 404, 503
Garnet group 607
Garnierite 714, 728
Gas 39 f; artificial (=illuminating) 589; ideal 39; liquifaction of 364 f; natural 589, 590
  chromatography 407
  constant 40
  curve 373
  density 40
Gaseous diffusion 278 f
Gas
  explosion 257
  hydrate 215
  mantle lamp 681
  mixture, ideal 40 f, 297, 299
  producer 592
  -works 589
Gay-Lussac's volume law 40
Gay-Lussac tower 529
Gel 410, 511
Germanates 619
German degrees 666
Germanium 29, 205, 584, 619; semiconduction in 226, 619
  dioxide 220, 619
German silver 750
Getter 662
Giant molecule 33, 200
Gibbs free energy 61
Gibbsite 645
Gibbs phase rule 360
Gilding 751
Glass 42, 219 f, 613 f
  electrode 326
  etching 476, 602
Glauber's salt 377, 654
Glaze 617
Gliding plane, in crystal deformation 232
Gliding 232
Globar 618
Glover acid 530
Glover tower 529
Glucinium 660
Goethite 713
Gold 31, 743 f; history 745; occurrence 743 f, 745; production 748 f; properties 745 f; uses 751; "white" 751
  -(III) chloride 760
  coinage 751
  -(I) compounds 759
  -(III) compounds 759 f
  -(I) halides 759
  -(III) halides 759, 760
  -(I) iodide 759
  leaf 745, 751

-(III) oxide hydroxide 759 f
  plating 751, 759
  silver tellurides 745
  sulfide 760
Graham's salt 577
Grain boundary 393
Grain-boundary fracture 235, 394
Grain-boundary migration 235
Graphite 63, 362, 585 f; crystal structure and bonding 148, 149, 164; in iron metallurgy 717, 720; pyrolytic 588, 590
Graphite oxide 588
Graphitization 588
Green vitriol 531
Ground state 70
Ground water 511
Group, in electron cloud 77; in periodic system 90
Growth spiral 392, 393
Guest molecules 215
Guldberg-Waage law 302
Gunpowder 654
Gypsum 236, 659, 664, 666

H

Haber-Bosch process 281, 307, 398, 399, 455 549 f, 568
Habit, crystal 193
Hafnium 30, 683; history 684; occurrence 684; production and uses 685 f; properties 685
  carbide 692
  tetrachloride 685
Halates 485
Half cell 415
Half-life 252, 266
Halic acids 485
Halides 471; formation of 474; general properties 471 f; hydrolysis 472 f; solution of metals in molten 474
Halite 654
Halo
  -argentate(I) complexes 756
  -carbonylcopper(I) 593, 752
  -carbonylmetals 593
  -cuprate(I) ions 752
Halogen dioxides 480
  oxides 479 f
Halogens 93, 464 f; history 466; occurrence 465 f; production and uses 468 f; properties 466 f
Haloids 478
Halopentacarbonyliron 593
Hardenability 726
Hardening (steel) 719, 724
Hard lead 626
Hard magnetic material 247, 727
Hard metal 396
Hardness 235 f
Hard substance 396
Hard water 511, 666 f
Harkins' rule 265, 440, 446, 674

Hartshorn salt 552
Hausmannite 705, 709
Heart smelting (lead) 626
Heat 54, 55; specific 58
  capacity 58; mean 58
  color 242
  content 54
  energy 52
  explosion 254
  motion 35, 39, 42, 52
  of combustion 55
  of formation 55
  of reaction 55
  of solution 55
  -resistant steel 696, 726
  treatment (steel) 720
Heavy water 280, 289
Helium 28, 32, 267, 457 f; I 459; II 459; history 458; liquid 365, 459; occurrence 457; properties 458 f; recovery 460; uses 460
Helium molecule ion 133
Helmholtz free energy 61
Hematite 713, 731
Hemimorphite 762
Hemocyanin 744
Hemoglobin 181, 713
Henry's law 349
Heptafluoro
  -niobate(V) ion 157, 691
  -tantalate(V) ion 157, 691
  -zirconate(IV) ion 157
Heptamolybdate ion 702
Hess's law 56
Heteropoly anions 177, 702
Hexacarbonylchromium 593
Hexachloro
  -antimonate(V) ion 558
  -platinate(IV) ion 204, 740, 741
  -plumbate(IV) ion 628
  -stannate(IV) ion 623
  -titanate(IV) ion 161
Hexacyano
  -ferrate(II) ion 156, 183, 287, 599, 730
  -ferrate(III) ion 159, 160, 178, 183, 287, 599, 733
  -ferrates(II) 730, 733
  -ferric(II) acid 731
Hexafluoro
  -ferrate(III) ion 156, 159, 160, 161
  -manganate(IV) ion 709
  -silicate ion 603
  -zirconate(IV) ion 181
Hexagonal system 195
Hexahalomanganates(IV) 709
Hexahydroxo
  -antimonate(V) ion 579
  -plumbate(IV) ion 627
  -stannate(IV) ion 620, 622
Hexammine
  -chromium(III) ion 161, 698, 735
  -cobalt(II)–hexamminecobalt(III)   redox pair 425
  -cobalt(II) ion 425, 734
  -cobalt(III) ion 159, 161, 182, 425, 735

-nickel(II) ion 204, 245, 736
-nickel(II) chloride 204
-zinc ion 161, 767
Hexaquo
 -aluminium ion 643
 -chromium(III) ion 156, 698
 -cobalt(II) ion 734
 -iron(III) ion 159, 731
 -nickel(II) ion 161, 243, 736
 -magnesium ion 509
Hexaoxoiodic(VII) acid 481, 487 f
Hexathionate ion 527, 534
Hexathionic acid 527, 534
High-alumina cement 617 f
High polymer 34
High-speed steel 726
High spin 159
Hole conduction 226
Holmium 30; see lanthanoids
Homogeneity, region of 35, 44, 210
Horizontal similarity 668, 711
Hornblende 608, 658
Hot atom 296
Hot-dipped zinc coating 765
Hot galvanizing 765
Hot working 397
Hund's rule 84, 129, 135
Hybrid orbital 141
Hybridization 140 f, 155 f; dsp^2 156; d^2sp^3 156; sp 143; sp^2 143; sp^3 141 f, 144 f; sp^3d 155
Hydrargillite 645
Hydrargyrum 763
Hydrate 126, 510
 isomerism 697
Hydration 125 f, 509 f; energy of 127; enthalpy of 127
 constant 330
 equilibrium 330
Hydrazine 546, 552 f
Hydrazinium(+) ion 546, 553
Hydrazinium (2+) ion 546, 553
Hydrazinium (2+) sulfate 553
Hydrazoic acid 554; see hydrogen azide
Hydride ion 122, 456
Hydrides 455 f; boiling points 369; covalent 456; melting points 367; metallic 456; saltlike 456
Hydrocarbons 590
Hydrochloric acid 332, 475, 476
Hydrofluoric acid 475, 476
Hydrogen 28, 32, 451 f; atomic 453; atomic structure of 78, 79 f; heavy 452; history 451; isotope separation 280 f; light 452; liquid 365; occurrence 451; ortho- 452; para- 452; production 453 f; properties 451 f; uses 455
 azide 479, 546, 554
 bomb 293
 bond 98, 167 f
  bridge 175
 bromide, decomposition 257; production 477; properties 475
 bromide molecule 146, 151, 166
 carbonate ion 322, 330, 595

 carbonates 595
 cell 433, 435
 chloride 474 f; formation 257; production 476; properties 475; uses 476
 chloride hydrates 318, 475
 chloride molecule 124, 146, 151, 166, 216
 chloride–water azeotrope 375
 compounds 455 f
 cyanide 479, 598
 difluoride ion 168, 322
 electrode 326, 415
 fluoride 174, 347, 474 f; production and uses 476; properties 475
 structure 168 f
 fulminate 600
 hexacyanoferrate(II) 731
 halides 474 f; acid strengths 333 f; production and uses 476 f
 iodide, preparation 475; properties 477
 iodide molecule 166
 ion 310, 327 f
 ion activity 326
 molecule 132, 139
 molecule ion 130 f
 orthotellurates 535
 peroxide 429 f, 514 f; production and uses 515
 peroxide ion 513, 514
 polysulfides 520
 sulfate ion 528
 sulfates 531
 sulfide 517, 518
 sulfide ion 517
 sulfide molecule 144
 sulfide system 517
 sulfide water 518
 sulfite ion 533
 sulfites 533
 tetracarbonylcobaltate 593
 tetrachloroaurate(III) tetrahydrate 760
Hydrolysis 319
Hydrolysis of metal ions 319 f
Hydromicas 611
Hydronium ion 318 (footnote)
Hydrophilic colloid 410
Hydrophilic phase 403
Hydrophobic colloid 410
Hydrophobic phase 403
Hydrosols 409
Hydrosphere 445
Hydrothermal process 610
Hydroxide carbonates 595
Hydroxide ion 174, 310, 327 f, 511 f
Hydroxides 511 f; amphoteric 345 f, 511
Hydroxide salts 512
Hydroxo
 -aluminate ion 645
 -aurates(III) 759 f
 -beryllate ions 663
 -chromate(III) ions 698
 complex 512
Hydroxonium ion 318 (footnote)
Hydroxo
 -iron(II) complexes 427

-iron(III) complexes 427
-plumbate(II) ions 628
-stannate(II) ions 623
-stannate(IV) ions 623
-vanadate ions 691
-zincate ions 434 f, 767
Hydroxyapatite 537
Hydroxylamine 546, 553 f
Hydroxylammonium ion 546, 553
Hyperon 20, 21
Hyperoxide ion 245, 340, 515; structure and bonding 135, 138, 153
Hyperoxides 515 f
Hypo 534
Hypobromite ion 481, 484, 485
Hypobromous acid 335, 468, 481, 484, 485
Hypochlorite ion 145, 481, 482, 485
Hypochlorites 484 f
Hypochlorous acid 335, 427, 468, 482, 483, 484 f
Hypohalites 484 f
Hypohalous acid 484 f
Hypoiodite ion 481, 484, 485
Hypoiodous acid 335, 481, 485
Hyponitrite ion 564
Hyponitrites 564
Hyponitrous acid 564
Hypophosphate ion 572
Hypophosphates 573
Hypophosphoric acid 572, 573
Hypophosphite ion 571
Hypophosphites 572 f
Hypophosphorous acid 571, 573
Hysteresis 247

I

Ice 168, 508 f; amorphous 508; cubic 508·hexagonal (=trigonal) 508; structure of 168, 169, 205
Ideal crystal 42, 186
Ideal gas 39
Ideal solution 43, 297
Identity (lattice) 187
Identity distance 187
Illites 611
Illuminating gas 589
Ilmenite 681, 687
Imide ion 174, 546, 547
Imino group 546
Impurity level 226
Impurity semiconduction 226
Indicator 326, 342
    paper 342
Indium 30, 630, 646
    antimonide 205
    -(III) hydroxide 630
    -(III) ion, protolysis of 320
    oxide chloride, structure 321
Induction period 258
Inductive effect 336
Industrial diamonds 585, 587
Inert pair 108

Infinitely nuclear molecule 100
Infrared 17; far 239; near 239, 240
    spectrum 239
Ingot steel 722
Inhibitor 45, 258
Initial nucleus 272
Ink 730; sympathetic 735
Inner core of the earth 441, 442
Inner sphere 125
Inorganic chemistry 9
Inorganic rubber 582
Inosilicates 608
Insulator 221, 225
Intensive quantity 14
Intercrystalline fracture 235, 394
Interference 18, 237
Interference pattern 237
Interhalogen compounds 477 f
Interstitial atom 210
Interstitial solid solution 210
Intrinsic semiconductor 225, 226
Invar 727
Inversion, molecular 144; symmetry 190
Iodate ion 481, 485
Iodates 486
Iodic acid 481, 485, 486
Iodide ion 244
Iodimetry 534
Iodine 30, 226, 464 f; history 466; occurrence 465; production and uses 470 f; properties 243, 245, 471 f
-(III) acetate 465
-(I) chloride 366, 473
-(III) phosphate 465
Iodomercurate(II) ions 772
Ion 22 f, 33; atomic 105, 117 f; complex 105; finite 176; hydrated 125; infinite 176; isoelectronic 109; net charge on 25
    cluster 105, 118
    color 243
    -dipole bond 98, 124 f
    exchangers 214, 611, 612
    -exchange chromatography 273, 407
Ionic bond 98, 105 f, 236
    conduction 217 f
    crystal 105, 200
    equilibrium 308 f
    exchanger 214, 404 f; organic 405
    migration 309 f
    mobility 309 f
    radii 103, 111 f
    reaction 256
    size 111 f
    strength 312 f
Ion-ion bond 105
Ionization energy 83, 94, 106 f
Ionization stages 119
Ionolysis 324
Ion pair 105, 118
Ion transfer 324
Iridescent color 243
Iridium 31, 711 f; see platinum metals
    -(IV) oxide 738
Iron 29, 248, 711 f; carbonyl 717; commercial

720 electrolytic 717; history 714; occurrence 713; metallography 716 f; metallurgy 716 f; properties 715, 716; pure 717; technical 717; α Fe 715, 717, 718 f, 725 f; β Fe 715; γ Fe 715, 717, 718 f, 725 f; δ Fe 715, 717, 718, 725 f
alloys 725
alums 733
-(III) bromide 204
–carbon system 717 f
-(II) carbonate 730
-(II) chloride 204, 730
-(III) chloride 731, 732 f
-(III) chloride hexahydrate 733
-(II) chloride tetrahydrate 730
-(III) citrate 733
-(II) compounds 729 f
-(III) compounds 731 f
-(IV) compounds 733 f
-(II) cyanide 730
-(II) disulfide 731
-(III) gallate complex 730
-(II) hydroxide 437, 729 f
-(II)–iron(III) redox pair 415, 425 f, 427
metals 93, 669, 711 f, 713 f; history 714 f; occurrence 713 f; properties 715 f
-(III) nitrate hexahydrate 733
-(II) oxalate 729
-(II) oxide 203, 213, 248, 729
-(II, III) oxide 213, 245, 248, 732
-(III) oxide 242, 731; formation 396; α Fe_2O_3 731; γ Fe_2O_3 213, 732
-(III) oxide hydroxide 403, 732; α FeO(OH) 732; γ FeO(OH) 732
-(II) sulfate 730
-(III) sulfate 733
-(II) sulfate heptahydrate 510, 730
-(II) sulfide 213, 731
vitriol 531, 730
Isobar 22, 266
Isocyanic acid 599
Isoelectric point 344
Isoelectric atoms and ions 109
Isomer 34
Isomeric transformation 267
Isomerism 34; hydrate 697; nuclear 259, 267
Isometric system 195
Isomorphism 197
Isonitriles 598
Isopoly anion 177, 702
Isostructural substances 197
Isotope 22: 277 f; table 28 f, 32
analysis 278
effect 277
mixture, natural 24, 277 f, 282 f
Isotope separation 278 f: electromagnetic 278
by centrifugation 282
by electrolysis 280 f
by exchange reactions 280 f
by fractional distillation 278 f
by gaseous diffusion 278 f
by ionic migration 282
by thermal diffusion 282
Isotype 197

J

Jasper 613
Joule-Thomson effect 364
Jungner storage cell 437, 766

K

K meson 20
Kainite 649, 658
Kaldo process 722, 724
Kalium 28
Kanthal 696
Kali 650
Kaolin 609, 616
Kaolinite 609, 610, 611, 615
Keatite 604
Kernite 631
Kieselguhr 613
Kieserite 658, 664
Kraft process 517, 653, 654
Krypton 29, 457: history 458; occurrence 457; properties 458 f; recovery 459 f; uses 460
compounds 461, 463
difluoride 463
tetrafluoride 463
Kupfernickel 715
Kyanite 607

L

Lambda hyperon 20
Lamp back 588
Lanthanoid contraction 104, 113, 668
Lanthanoid ions 121, 673; color 673
Lanthanoids 88, 104, 671 f; history 675 f; occurrence 673 f; production and uses 678 f; properties 230, 673, 676 f
Lanthanum 30; see lanthanoids
Lapis infernalis 756
Lapis lazuli 612
Laser 73
Latent image 757
Lattice 187; body-centered cubic 196; face-centered cubic 196; simple cubic 196
energy 200 f, 206 f; table 207
Laughing gas 559
Laundering 667
Lawrencium 31; see actinoids
Layer molecule 200, 204
LCAO (=linear combination of atomic orbitals) 129
LD process 724
Lead 584 f, 624 f; hard 626; history 625; isotopes 31, 283, 284, 286; occurrence 624; production 625 f; properties 625; uses 626 f
-(II) acetate 628
alloys 626
arsenate 578
azide 554, 628
-base bearing metal 621, 626
-chamber process 529

-(II) carbonate 627, 628
-(IV) chloride 628 f
-(II) chromate 628, 699
dioxide 627
-(II) fluoride–bismuth(III) fluoride system 214
glass 614
-(II) halides 628
-(II) hydroxide carbonate 628
-(II) iodide 204
-(II) ion 627, 628
-(II) nitrate 563, 627, 628
-(II) oxide 627
-(II, IV) oxide 627
-(IV) oxide 436, 627
-(II) oxide chromate 699
-(II) oxide hydroxide 628
-(II, IV) oxides, unstable 628
poisoning 180, 625
-(II) salts 628
soap 627
storage cell 436
-(II) sulfate 436, 530, 531, 625, 628
-(II) sulfide 203, 628
tetrachloride 628 f
tetrahydride 627
–tin system 621, 622
Leblanc process 655
Le Chatelier's principle 307
Leclanché cell 434
Ledeburite 720
Left-handed quartz 199, 604
Lepidocrocite 713
Lepidolite 648
Lepton 20, 21
Levelling effect 332
Lever rule 371
Leweis acid 324
Ligand 99, 179; bidentate 180; bridging 180; chelate 180; sexadentate 180; terdentate 180
atom 99, 179
-field theory 157 f
Lime soap 666
Limestone 659, 665
Lime water 663
Limonite 713
Linear combination of atomic orbitals 129
Line spectrum 73
Linnaeite 714
Linseed oil 627
Liquid 41; structure of 34, 41
Liquid air 459 f, 498
Liquid ammonia 366; solution equilibria in 547; metal solutions in 547 f
Liquid curve 373
Liquid phosphorus hydride 555
Liquid potential 415
Liquidus curve 378
Liquidus surface 378
Litharge 627
Lithium 28, 647 f; history 650; in fusion bombs 293; metallic bonding in 222, 223 f; occurrence 648; production and uses 652; properties 650 f
bismuthide 231
bromide 207, 208
carbonate 595, 648
chloride 207, 208, 209
chloride–magnesium chloride system 213
deuteride 293
fluoride 207, 208, 209, 648
hexahydroxyantimonate(V) 579, 648
hydride 203, 456
iodide 151, 207, 208
molecule 135 f, 171, 222
nitride 544
nitrate 596
oxide 653
phosphate 648
tetrahydridoaluminate 602, 622, 642
Lithopone 666
Loadstone 732
Local element 438
Localized electrons 98
Long-range order 41, 186, 216
Lonsdaleite 587
Low quartz 603 f
Low spin 159
Lox 498
Luminescence 74, 228
Lunar caustic 756
Lutetium 30; see lanthanoids
Lye 653
Lyman series 82, 94
Lyophilic colloid 410
Lyophilic phase 403
Lyophobic colloid 410
Lyophobic phase 403

M

Magic numbers 265
Magnalium 642
Magnes 660, 705
Magnesia 660, 705
  alba 665, 666
  cement 663
  mixture 574
Magnesite 658
Magnesium 28, 395, 657 f; history 660; occurrence 658 f; production and uses 662; properties 660 f
-air cell 435
ammonium phosphate 574
antimonide 231
carbonate 596
chloride 473, 662, 663
chloride–lithium chloride system 213
chromate 699
diborate 176
flare 661, 662
fluoride 368, 474
hydroxide 204, 662, 663
hydroxide carbonate 665, 666
hydroxide chloride 663
nitrate 658

nitride 544, 545
orthoborate 176
oxide 207, 208, 657, 663, 734
perchlorate trihydrate 658
silicide 231
sulfate 664; aqueous solution 47, 48
Magnetic
  domains 247 f
  moment 246; orbital 246; spin 246; permanent 246; spontaneous 247
  material, hard 247, 727; soft 247, 727
  quantum number 76
  saturation 247
Magnetism 246 f
Magnetite 660, 683, 713, 732; color 245; magnetic properties 248
Main series (radioactive) 270
Malachite 744, 746, 754
Malleableizing 724
Manganates 709
Manganate(V) ion 704, 709
Manganates(V) 709
Manganate(VI) ion 704, 709
Manganates(VI) 709
Manganate(VII) ion 704, 709 f
Manganates(VII) 709 f
Manganese 29, 704 f; history 705 f; occurrence 705; production and uses 707; properties 706; α Mn 706; β Mn 706; γ Mn 706; δ Mn 706
 -(II) compounds 707 f
 -(III) compounds 709
 -(IV) compounds 709
 -(V) compounds 709
 -(VI) compounds 709
 -(VII) compounds 709 f
 dioxide 434, 709
 -(II) chloride 708
 -(II) chloride tetrahydrate 708
 -(II) fluoride 248
 -group elements 704 f; history 705 f; occurrence 705; production 706 f; properties 706; uses 707
 -(II) hydroxide 707, 708
 nodules 705
 -(II) oxide 248, 704, 708
 -(II, III) oxide 709
 -(III) oxide 705, 709
 -(IV) oxide 248, 434, 469, 709
 -(VII) oxide 121, 705, 710
 -(III) oxide hydroxide 434, 709
 -(IV) oxide–permanganate redox pair 423, 710
 (II)–permanganate redox pair 426, 710
 -(II) sulfate 708
 -(II) sulfate tetrahydrate 708
 -(II) sulfide 707
Manganite 705, 709
Mantle of the earth 441 f
Marble 659
Marcasite 362
Margarite 610, 611
Marsch test 556
Martensite 717, 719

Martin process 723
Mass-action law 302
Mass number 21, 25; table 28 f
Mass spectrograph 26, 278
Matches 544, 580, 581
  Matte 727, 728, 747
Matter 15 f; forms of 33 f
  waves 66
Maximum work 61
Maxwell distribution law 52
Mean bond energy 162
Mean heat capacity 58
Mean ionic activity 314
Melting 365, 366 f; congruent 385; incongruent 385
 curve 364
 point 365
Mendelevium 31; see actinoids
Mercuric oxide cell 434, 766
Mercury 31, 761 f; history 763; occurrence 762 f; production 765; properties 763 f; uses 766
 -(II) bromide 771
 -(II) carbonate 771
 cathode process 469
 -(I) chlorate 770
 -(I) chloride 770
 -(II) chloride 311, 322, 771 f
 -(I) chromate 699
 -(II) chromate 699
 -(I) compounds 770
 -(II) compounds 770 f
 -(II) cyanide 769, 771, 772
 -(II) fulminate 600
 -(I) hydroxide nitrates 770
 -(II) iodide 362, 771, 772
 -(I) ion 249, 761, 769, 770
 -(II) ion 320, 769, 771; halo complexes 771; ammine complexes 771
 -(I) nitrate 770
 -(II) nitrate 772
 -(I) nitrate dihydrate 770
 -nitrogen compounds 773 f
 -(II) oxide 434, 761, 769, 771
 -(I) perchlorate 770
 -(II) phosphate 771
 poisoning 763
 -(II) sulfate 531, 772
 -(I) sulfide 770
 -(II) sulfide 205, 762, 769, 771, 772 f
Mesomerism 146
Meson 20, 21
Metaborates 176
Metaborate ion 176, 635
Metaboric acid 637
Metacinnabarite 772
Matakaolinite 615
Metal 92, 221, 222; oxidation of 218
 -air cell 435
 carbonyls 592
 complexes 178 f
 corrosion 438
 working 397
Metallic

bond 98, 229 f
crystal 200
luster 242
phase 230 f
radii 103
Metalloid 93
Metaphosphates 572
Metaphosphate composition 572
Metaphosphate ions 572, 573
Metaphosphoric acids 572, 573
Metasilicates 608
Metastannic acid 624
Metavanadates 692
Methane 124, 141, 174, 216, 590; heat of formation 57; bond energy 162
Mica minerals 609, 610, 638
Microcline 649
Microelement 285 (footnote), 447
Microwave spectrum 239
Milk of lime 484, 663
Millerite 714
Millon's base 774
Minette ore 713, 722
Mirror image 198
Mirror plane 190
Misch metal 680
Miscibility gap 370, 375, 380
Mixed catalyst 398
Mixed crystal 211
Mixture 35
MO (= molecular orbital) 129
MO method 140
Moderator 289
Modification 197
Mohr's salt 730
Mohs' hardness scale 236
Molality 38
Molar heat 58 f
Molarity 38; volume 38; weight 38
Mole 36 f
   fraction 38
   percent 38
   volume 48; partial 49
Molecular
   bond 98
   crystal 200
   mass 36
   orbital 129 f; in solids 222 f
   orbital method 140
   sieve 406, 612
   spectrum 73, 239 f
   structure 34
   weight 36
Molecule 33 f, 39; activated 255; complex 34; dimeric 34; dinuclear 100; electrophilic 140; energy-rich 183; framework 34; high-polymer 34; infinite 33 f; 100; nucleophilic 140; odd 137; polynuclear 100; trimeric 34
Molybdates(VI) 693, 701
Molybdenite 694, 705
Molybdenum 29, 693 f; history 694; occurrence 694; production 696; properties 695; uses 697
   blue 703

compounds 701 f
dioxide 702
disilicide 703
disulfide 694, 701, 703
earth 694
trioxide 693, 696
Moment, dipole 123; magnetic 246 f
Momentum 18
Monatomic ions 117 f
Monatomic layer 400
Monazite 674, 675, 680
   sand 674
Mond process 593, 727 f
Monel metal 729
Monoclinic system 195
Monomolecular layer 400
Monothiocarbonate ion 598
Montmorillonite 610, 611
Mordant 624, 645, 767
Mortar 663
Mosaic gold 624
Moseley's law 95, 684
Mother nucleus 266
Mother nuclide 266
Mullite 615, 616
Muriatic acid 476
Muon 20
   neutrino 20
Muscovite 610, 611, 649
Myoglobin 713

N

Nahcolite 656
Natrium 28
Natrolite 404, 612
Natron 650
Natural radioactive series 269
Neodymium 30; see lanthanoids
Neon 28, 32, 170, 456 f; history 458; occurrence 457; properties 458 f; recovery 459 f; uses 460
   molecule ion 135, 138
Neptunium 31; see actinoids
   series 270
Neptunyl(V) ion 672
Nernst lamp 687
Nesosilicates 607
Nessler's regent 774
Neutral reaction 329
Neutrino 20
Neutron 20, 273; fast 273; intermediate 273; slow 273; thermal 273
   bombardment 273 f
   diffraction 238, 248
Niccolite 714, 715
Nichrome 696, 729
Nichrothal 696, 729
Nickel 29, 248, 711 f; history 715; occurrence 714; production 727 f; properties 716 f; uses 728 f
   alloys 728 f
   -arsenide structure type 205
   -(II) chloride hexahydrate 736

compounds 736
-(II) hydroxide 437, 736
-(II) oxide 736; catalytic action 398
-(IV) oxide 736
-(III) oxide hydroxide 437, 736
plating 729, 736
silver 750
-(II) sulfate 736
-(II) sulfate heptahydrate 736
-(II) sulfide 714, 736
vitriol 736
Nife storage cell 437
Niobic acid 692
Niobite 690
Niobium 29, 689 f; history 690; occurrence 690; production and uses 690 f; properties 690
-(V) oxide 689, 691
Niton 458
Nitrate ion 287, 563, 568; bonding in 147, 154, 164, 175
Nitrates 568
Nitric acid 347, 563, 566 f; red fuming 566
hydrates 566
−water system 382, 566
Nitride ion 122, 547
Nitrided steel 724
Nitrides 539, 544 f
Nitriding 724
Nitriles 598
Nitrites 565 f; organic 565
Nitrite ion 563, 565; bonding in 147
Nitrito complex 565
Nitro complex 565
Nitro compounds (organic) 565, 567
Nitrogen 28, 32, 208, 563 f; history 538; occurrence 536 f; preparation and uses 543; properties 538 f
cycle in nature 582 f
dioxide 245, 562 f, 566
fertilizers 550, 583, 595, 600
-group elements 536 f; history 538; occurrence 536 f; production and uses 543 f; properties 538 f
-hydrogen bond, compounds with 545 f
molecule 135, 136, 139, 154, 163, 170
oxide 245, 559, 560, 566; bonding in 135, 137
oxides 558 f
−oxygen phase diagram 372, 373 f
trichloride 155, 473, 556
trifluoride 473, 557
triiodide 557
Nitrogène 538
Nitrohydroxylaminic acid 564
Nitronium ion 563
Nitroprusside ion 561
Nitrose 529
Nitrosyl
cation 561; bonding in 135, 137
chloride 567
complexes 561, 593
group 561
halides 561

hydrogen sulfate 529, 561, 562
perchlorate 561
sulfuric acid 529, 561, 562
Nitrous acid 562, 563, 565
Nitrous oxide 150, 559
Nitroxylic acid 564
Nitrum 650
Nitryl
cation 347, 530, 563, 566; bonding in 150
nitrate 563
perchlorate 563
NMR (=nuclear magnetic resonance) 246
Nobelium 31; see actinoids
Noble gases 85, 93, 367, 369; history 458; occurrence 457; origin 443; properties 458 f; recovery 459 f; uses 460
Noble-gas
compounds 460 f
ions 120 f
shell 85
Noble metal 424
Node 70
Nomenclature 100, 106, 151, 152 f, 456
Nonmetals 92
n-type conductor 227
Nuclear
charge 22, 25; effective 93
chemistry 10
energy 51, 262 f
isomer 259
magnetic resonance 246
reaction 265 f; induced 271 f; spontaneous 265 f
reactor 288 f
spin 246, 452
Nucleation, crystal 105 f, 389 f
Nucleon 20
Nucleus, atomic 19, 21 f, 259 f; complex 99 f; compound 272; growth 389; target 272
Nuclide 22, 260 f; radiogenic 282; stable 259; table 32; unstable 259

O

Occlusion (crystal) 402
Octacarbonyldichromium 593
Octacyanomolybdate(IV) ion 157, 693
(V) ion 157, 693
Octafluoroxenates(VI) 463
Octet rule 154 f
Oil of vitriol 730
Oleum 528, 529
Olivine 502, 607, 658
group 607
Omega hyperon 20
One-component systems 359, 363 f
One-electron atoms 82 f
One-electron bond 132, 133, 139
One-phase
area 381, 387
line 381
region 363, 370

Subject index

-onium ion 456
Onyx 613
Opal 613
  glass 614
Open-hearth process 723
Optical activity 199
Optical isotropy 196
Orbit 76, 77
Orbital 77 f; antibonding 131; bonding 131; delocalized 138, 146 f; localized 138; non-bonding 135; octahedral 156; square 156; tetrahedral 141; unsymmetrical 145 f; π 134; σ 134
  magnetic moment 246
Order of rotation symmetry 190
Order-disorder transition 216, 217, 363
Ore concentrate 397
Organic chemistry 9
Orientation polarization 123
Orpiment 538, 580
Orthoborate ion 147, 175, 176, 635
Orthoboric acid 636 f
Orthochromatic emulsion 758
Orthoclase 236, 611, 649
Orthohydrogen 452
Orthonitric acid 566
Orthoperiodate ion 175, 481
Orthoperiodates 488
Orthoperiodic acid 481, 487 f
Orthophosphate ion 571, 573, 574
Orthophosphoric acid 571, 573, 574
Orthohombic system 195
Orthosilicate ion 145, 175, 606, 607
Orthosilicic acid 612
Orthotellurate ion 175
Orthotellurates 535
Orthotelluric acid 535
Ortovanadate(V) ion 183, 692
Orthovanadates 692
Oscillation 39, 71
Oscillator 72
Osmate(VI) ion 742
Osmium 31, 711 f; see platinum
  heptafluoride 741
  tetroxide 121, 738, 741 f
Osmosis, reverse 359, 510
Osmotic pressure 358 f
Ostwald process 548, 560, 563, 568
Outer core of the earth 441, 442
Overglaze color 617
Overlap, electron 97, 101
Overvoltage 431
Oxidation 412 f
  of metals 218
  number 119, 152 f, 412
  state 152
Oxide
  carbonates 595
  halides 521
  hydroxides 512
  ion 122, 174, 501 f
  ion activity 503
  —oxygen equilibrium constants, table 505 f
  salt 320, 512

Oxides 500 f; acid 324, 501; basic 324, 501; double 503; formation 504 f; intermediate 501; reduction of 504 f; triple 503
Oxidized form 413
Oxidizing acid 424
Oxidizing roasting 504
Oxo
  acid 334; acid strength 334 f
  complex 512
  ions 500
  -chlorate ions 145
  -dinitrate ions 564
  -diphosphate ion 571, 573
  -disulfate ion 526
  -halates 481 f
  -halic acids 481 f
  -molybdate(VI) 701
  -nitrates 563 f
  -nitrate(I) ion 561, 563
  -nitric(I) acid 563
  -nitric acids 563 f
Oxonium ion 144, 174, 310, 318, 487, 528; hydrated 318, 475, 487
Oxonium perchlorate 487
Oxo
  -phosphates 570 f
  -phosphoric acids 570 f
  -polyphosphate ions 571, 573
  -polysulfate ions 526
  -selenic acids 535
  -sulfate(0) ion 525
  -sulfates 525 f
  -sulfuric(0) acid 525
  -sulfuric acids 525 f
  -telluric acids 525 f, 535
  -thioarsenate ions 581
  -thio ions 516
  -titanium(IV) ion 681
  -titanium sulfate 687
  -tungstate(VI) 701
  -vanadate(V) ions 691
  -vanadates(V) 692
  -vanadium(IV) ion 182, 689
  -vanadium(V) ion 182, 689, 691
Oxygen 28, 32, 153, 489; atomic 493; history 491 f; occurrence 490; production and uses 498; properties 492 f
  cations 122
  cycle in nature 490
  difluoride 174, 479
  molecule 135, 138, 163, 170
  —nitrogen phase diagram 372, 373 f
Oxygène 492
Oxygenium 323
Oxyhydrogen 452
  flame 453
Ozone 147, 493

P

Packing fraction 264
Pair formation 21
Palladium 29, 711 f; hydrogen permeability 452, 739; see platinum metals
  -(II) oxide 739

Panchromatic emulsion 758
Paper chromatography 407
Paracyanogen 599
Parahydrogen 452
Paramagnetism 246
Paramolybdate ion 702
Paris green 578
Parkes process 357, 749
Partial derivative 49
Partial molar quantity 49
Partial molar volume 49
Partial pressure 40 f
Particle, elementary 19 f; energy 17; resonance 21; specific charge 409; table of elementary 20; α 265, 267; β^+ 266; β^- 266
  density 69, 71
  in a box 68, 69, 76
Parting 749
Partition chromatography 407
Paschen series 82, 94
Passivity 424, 429, 529, 567, 639, 640, 695
Patina 746, 754
Patronite 689
Pauli principle 77, 129, 132, 135
Pearlite 719
Pedersen process 640
Pellets 397
Pencil lead 588
Pentaborate ion 636
Pentacarbonyliron 593
Pentacyanocobaltate(II) ion 734
Pentathionate ion 527
Pentathionic acid 527
Pentlandite 714, 737
Peptization 410
Perbromic acid 481, 487
Perbromate ion 481
Perbromates 487
Perchlorate ion 216, 287, 481, 482, 487 f; bonding in 145, 181
Perchlorates 486 f, 648
Perchloric acid 331, 332, 335, 481, 482, 487
Perchloric acid hydrates 318
Perhalates 486 f
Perhalic acids 486
Period 90
Periodates 482, 488
Periodic acid 487 f
Periodic system 75 f, 85 f; table 91
Peritectic 381
  line 382
  reaction 382
  temperature 382
Peritectoid 382
Permalloy 727
Permanent hardness 666
Permanganates 709
Permanganate ion 161, 287, 426, 704
Permanganic acid 710
Permutit 404
Perovskite 504, 687
  structure type 504
Peroxide ion 513; bonding in 135, 138, 153
Peroxides 513 f

Peroxo
  acids 513
  ions 513
  -borates 638
  -chromate(V) ion 701
  -chromate(VI) ion 701
  -disulfate ion 431, 513, 527
  -disulfuric acid 513, 527, 532
  -sulfate ion 513, 527, 532
  -sulfuric acid 513, 527, 532
Perrhenate ion 704, 710
Perrhenates 710
Perrhenic acid 707, 710
Perruthenate ion 742
Persulfate ion 532
Pertechnetate ion 704, 710
Pertechnetates 710
Pertechnetic acid 710
Perxenate ion 462
Petalite 648, 650
Petroleum coke 589
Pewter 621
pH 326
Phase 14 f; condensed 15; dispersed 14, 408; intermediate 370; intermetallic 230; metallic 230 f; metastable 366
  analysis 386
Phase diagram 363 f; general properties 386 f
  for carbon 586
  for sulfur 497
  for acetone–chloroform system 373
  for Ag–Cu system 379
  for Ag–Pt system 379
  for Al_2O_3–SiO_2 system 616
  for Bi–Sb system 379
  for ethanol–benzene system 374
  for Fe–C system 717
  for H_2O–HNO_3 system 382
  for H_2O–NaCl system 382
  for KCl–CuCl system 382
  for N_2–O_2 system 372
  for Sn–Pb system 622
Phase
  growth 391 f
  region 363
  rule 360
  transformation 45, 361
pH indicator 342
Phlogiston 491
  theory 466, 491 f
Phlogopite 610
Phosgene 592
Phosphate
  fertilizers 552, 574, 576
  ion 145, 175, 287, 571, 573
  rock 537, 543, 574
Phosphates(V) 571, 573, 574 f; condensed 572, 573, 575 f; uses 576 f
Phosphating 577
Phosphatoiron(III) ions 574
Phosphide ion 122
Phosphides 544 f
Phosphine 555
Phosphite ion 571, 573

Phosphites 573
Phosphonium ion 555
Phosphor bronze 543, 750
Phosphor salt 575
Phosphorescence 74
Phosphoric(V) acids 571, 573, 574; condenced 572, 573, 575 f; uses 576
Phosphoric acid system 333, 335 f, 342, 343 f
Phosphorite 537, 659
Phosphorous acid 571, 573
Phosphors 228
Phosphorous 28, 536; black 172, 540 f; history 538; occurrence 537; production and uses 543 f; properties 539 f; red 539, 540; structure 172; violet 541; white 172, 539 f; yellow 539
  halides 155, 557 f
  hydride, solid 555; liquid 555
  hydrides 555
  necrosis 540
  -nitride dichlorides 582
  -(III) oxide 569
  -(III, V) oxide 569
  -(V) oxide 569 f
  pentabromide 155
  pentachloride 155, 324, 473, 557
  pentafluoride 155
  pentasulfide 580
  poisoning 540
  sulfides 580
  trichloride 153, 557, 573
  trihydride 555
Phosphoryl(V) chloride 473, 558
Photochemical equivalence law 256
Photochemistry 256 f
Photoflash 639, 646
Photographic process 757 f
Photon 17 f, 20, 21
Photoreaction 256
Photosemiconduction 226
Photosynthesis 597
Phyllosilicates 609 f
Physical chemistry 10
Pi meson 20
Pickling 476, 530, 621, 765
Piezoelectric effect 605
Pig iron 721 f
  furnace, electric 721
Pitchblende 492, 660, 675
Plagioclases 611, 648
Planck's constant 17, 67
Plasma 123, 184, 293
Plaster 663
  of Paris 664
Plate, in distillation column 374
Platinum 31; see platinum metals
  black 415, 738
  -(II) chloride 741
  metals 93, 669, 711 f, 737 f; heavy 93; history 737 f; light 93; occurrence 737; production 740; properties 738 f; uses 740 f
  metal halides 741
  metal hexafluorides 741
  –silver phase diagram 379
  sponge 738, 740
  -(II) sulfide 742
  -(IV) sulfide 742
Plumbite ions 628
Plumbum 31, 625
Plutonium 31, 291; see actinoids
Pneumatic steel processes 723
Poisoning of catalysts 399
Polarity, bond 145, 151
Polarizability 116, 123
Polarization, concentration 418; of electrodes, chemical 418; of ions 116 f; of molecules, induced 123; orientation 123
Polarizing effect 116
Pole sign 417
Poling 748
Pollucite 649
Polonium 31, 489; history 492; occurrence 491; properties 498
Polyanion 176
Polyborate ions 176, 635
Polychromate ions 699
Polycrystalline substance 194, 393
Polyhalide ions 478
Polyhalides 477 f
Polymer 34
Polymolybdate 701 f
Polymolybdates 693, 701 f
Polymorphism 197
Polynuclear molecules 100
Polyphosphate ions 571, 573
Polyphosphates 571 f, 667
Polyphosphoric acids 571, 573
Polysulfate ions 526, 532
Polysulfide groups 519
Polysulfide ions 519
Polysulfides 519 f
Polysulfuric acids 526, 532
Polythionate ions 527, 535
Polythionates 534 f
Polythionic acids 527, 534 f
Polytungstate ions 701
Polyuranates 680
Polyvanadate ions 692
Polytungstates 693, 701
Portland cement 617
Positron 20, 21, 266
Positron decay 266
Potash 650, 656
  glass 614, 656
Potassium 28, 647; history 650; occurrence 649; production and uses 652; properties 650 f
  alum 645
  aluminum sulfate 645
  antimony(III) oxide tartrate 519
  biiodate 486
  bromide 352
  carbonate 656
  chlorate 486
  chloride 207, 218
  chloride–copper(I) chloride system 382, 385
  chloride–sodium chloride system 212, 379
  chromium(III) sulfate 698

cobalt(II) silicate 734
cyanide 598
dichromate 700
disulfide 532
disulfite 534
hexachlorostannate(IV) 204
hexachloroplatinate(IV) 204
hexacyanoferrate(II) trihydrate 730
hexacyanoferrate(II, III) 245
hexanitrocobaltate(III) 735
hydrogen carbonate 596
hydroxide 340, 653
metabisulfite 533
nitrate 352, 568, 596, 654
perchlorate 352, 474, 487
permanganate 469, 709
silicate 613
tetrahydroxoaurate(III) 760
tetraiodomercurate(II) dihydrate 772, 774
triborate 176
zinc hydroxide chromate 700
Potential, chemical 64 f, 297; electrode 415, 419 f; redox 419 f
Powder metallurgy 396
Powder reaction 396
Power-producing reactors 292
Pozzuolana 617
Praseodymium 30, 121, 671; see lanthanoids
Precipitate 355, 394, 409; purifying 402; aging of 345, 395
Precipitation hardening 383, 724
Preservative, in developer 759
Pressure, critical 364; partial 40; radiation 17
Primary cell 433 f
Primitive area 188
Primitive cell 189
Principal quantum number 76
Probability 60, 70
Process 45; chemical 45; hydrothermal 610; isotope-separating 283
Producer gas 592
  equilibrium 592
Projectile particle 272
Promethium 30, 674; see lanthanoids
Promoter, catalytic 398
Property 13 f; accidental 13; anisotropic 186; chemical 14; essential 13; isotropic 186
Proportion 37
Protactinium 31; see actinoids
Protective colloid 410
Protium 452
Protolysis 316; degree of 338
  constant 327
  equilibrium 317
Protolyte 316; colloidal 411; polyvalent 342 f; strength of 325 f; table 333
Protolytic system 316
Proton 20
  acceptor 316
  donor 316
  exchange 316
  jump 310
  -proton chain 294
  transfer 316 f

Proustite 538, 744
Prout's hypothesis 32 f
Prussian blue 245, 599, 733
Prussiate of potash, yellow 730; red 733
Prussic acid 599
Pseudohalide ions 478 f, 598 f
Pseudohalogens 478 f, 599 f
Pseudosystem 362
P, T diagram 363 f, 370; see phase diagram
P, T, x diagram 370
p-type conductor 227
P, x diagram 370
Pyrargyrite 744
Pyrex 615
Pyrite 195, 362, 491, 520, 713, 730, 731, 747
Pyroceram 391, 615
Pyrolusite 660, 705
Pyrophoric substance 395
Pyrophosphoric acid 574
Pyrophyllite 610, 611
Pyrosulfates 532
Pyrosulfuric acid 532
Pyroxenes 608
Pyrrhotite 248, 731

Q

Quantity 13; extensive 13; intensive 14; molar 48; partial molar 49; specific 13
Quantization 70 f
Quantum 16
  mechanics 66 f, 75
  number 70, 76; azimuthal 76; magnetic 76; principal 76; spin 76
  yield 256
Quartz 236, 603 f, 616; crystal form 193, 199; high 603 f; left-handed 199, 604; low 603 f; right-handed 199, 604; α 603; β 603
Quartz-crystal synthesis 605
Quaternary system 359, 386
Quenching 383
Quicklime 662

R

Radial distribution function 78, 79, 85
Radiant energy 16 f
Radiation, annihilation 21, 266; cosmic 273; electromagnetic 16; electromagnetic, table of 17; α 267; β 19, 266; γ 266, 267
  chemistry 295
Radical, free 183 f, 245
Radioactive equilibrium 270
Radioactive infection 271
Radioactive series 267 f; natural 269
Radioactivity 259, 265 f; in the earth's core 442
Radium 31, 657; history 660; occurrence 659
Radius, covalent 163 f; ionic 111 f; metallic 103; tetrahedral 163
  ratio 115

Radon 31, 271; history 458; occurrence 457 f;
  properties 458 f
  compounds 461, 463
  fluorides 463
Raman spectrum 240 f
Raoult's law 349, 371
Rare-earth metals 674; see lanthanoids
Rate constant 251
Raw-goods firing 617
Rayon 598, 653, 753
Reaction, arrested 45, 62, 256, 428 f; bimolecular 253; chain 257 f; chemical 45; dark 257; driving force of 59; eutectic 381; first-order 251 f; fusion 275; heat of 55; heterogeneous 251; homogeneous 251; ionic 256; monomolecular 252; peritectic 382; second-order 253; thermonuclear 292; third-order 253; zero-order 253
  energy 55
  enthalpy 55
  kinetics, heterogeneous systems 388 f; homogeneous systems 251 f
  rate 45, 251 f, 254
  sintering 619
  temperature 395
Reactor 288 f; fast 292; heterogeneous 289; homogeneous 291; slow 289; thermal 289
  core 290
Realgar 195, 538, 580
Recrystallization 395
Rectifier 228
Red lead 627
Red ochre 296
Red shortness 722
Redox
  buffer 428
  electrode 415
  equilibrium 413, 425
  indicator 428
  pair 413
  potential 419 f
  process 413; equations for 414
  system 413
  titration 428
Reduced form 413
Reduction 412 f
Refining, electrolytic 433
Refining furnace (for copper) 748
Reflector 290
Refractory materials 616
Remanence 247
Reprecipitation 402
Repulsion, energy of 101
Reserve-type cell 435
Residual current 430
Resonance 146
  absorption 274
Rest mass 16
Rhenium 31, 704 f; history 706; occurrence 705; production 707; properties 706; uses 707
  compounds 704, 710
  -(VII) oxide 121, 705, 707
Rhodane 479, 600

Rhodanide ion 479, 600
Rhodium 29, 711 f; see platinum metals
  -(III) oxide 738
Rhodochrosite 705
Rhombic system 195
Rhombohedral system 195
Rich bullion 749
Right-handed quartz 199, 604
Rinman's green 734
Ripening 758
RM cell 434
Roasting, chloridizing 748; oxidizing 504; sulfatizing 748
Roast-reaction process for lead 626
Roast-reduction process for lead 626
Rock crystal 604
Rock salt 648, 654
Roscoelite 689
Rose quartz 604
Rose's metal 386
Rotation 39, 71; free 137
Rotation axis 190
Rotational
  energy 51, 71, 239, 278
  quantum number 72, 240
  spectrum 239
Rotator 72
Rouge 731
Rubidium 29, 647; history 650; occurrence 649; production and uses 651 f; properties 650 f
  iodide 207, 226
  octafluoroxenate(VI) 463
Ruby 644
  glass 747
Rusting 716
Ruthenate(VI) ion 742
Ruthenate(VII) ion 742
Ruthenium 29, 711 f; see platinum metals
  tetroxide 121, 738, 741 f
Rutile 684, 686
  structure type 203
Rydberg's constant 94

S

Safety matches 581
Sal ammoniac 551
Salt 321 f, 471; acid 322; double 322; normal 322; oxide (=basic) 320
  bridge 415, 417
  hydrates 376
Saltpeter 568
Samarium 30, 121, 671; see lanthanoids
Sandwich complex 182, 634
Sapphire 644
Sassolite 631
Saturation, magnetic 247
Scale 732
Scandium 28, 673; history 676; occurrence 674
  -group elements 671 f; history 675 f; occurrence 673 f; production and uses 678 f; properties 676 f

Scavenger (precipitate) 403
Scheele's green 578
Scheelite 694
Scheidewasser 567
Schönite 531
  -type double salts 531
Schrödinger equation 69
Schungite 589
Schweizer's reagent 753
Screening 93
  constant 93
Screw dislocation 391
Secondary cell 436 f
Sedimentation 408
Seeding 389
Selenates 535
Selenic acid 535
Selenides 519
Selenious acid 523, 535
Selenium 29, 226, 489; glassy 220 f, 497; history 492; monoclinic (=red) 498; occurrence 491; production 499; properties 497 f; structure of 171, 172; trigonal (=hexagonal, gray, metallic) 497; uses 500
  dioxide 523
  halides 521
Selenocyanate ion 479
Selenocyanogen 479
Self-diffusion 286
Self-discharge 434
Semiconductor 221, 225 f; intrinsic 225, 226; extrinsic (=impurity) 226
Semimetals 92
Semipermeable membrane 359
Sensitive specks 757
Sensitization, photographic 758
Separation, by flotation 397, 403 f; of isotopes 278 f; magnetic 397
Sets of lines 187, 188
Sets of planes 187
Shell, electron 77, 84; noble-gas 85; 8-shell 85; 18-shell 88; $(18+2)$-shell 09
Shell structure in atomic nucleus 259
Shield gas 460
Short-range order 41, 186, 216
Siderite 713, 730
Siemens-Martin process 723
Sigma hyperon 20
Silanes 602
Silica gel 613
Silicate glasses 614 f; calcium-alkali 614; lead-alkali 614
Silicate melts 502, 614
Silicates 605 f; formztion and properties 612 f; production and uses 613 f
Silicic acid gel 612 f
Silicic acid 612 f
Silicides 601 f
Silicon 28, 227, 584 f, 600 f; history 601; occurrence 600 f; production and uses 601; properties 601
  bronze 750
  carbide 227, 618; structure 205

  dioxide 220, 601, 603 f
  dioxide–aluminum dioxide system 615 f
  dioxide glass 604, 605, 614
  disulfide 619
Silicon
  halides 602 f
  hydrides 602
  monoxide 603
  nitride 618 f
  tetrafluoride 368, 473, 602 f
  tetrahalides 602 f
Silicones 618
Silite 618
Sillimanite 607
Silundum 618
Silver 29, 743 f; history 745; occurrence 643 f; "oxidized" 746; production 748 f; properties 745 f; uses 750
-(I) acetate 755
-(I) arsenate 578
-(I) bromide 218, 244, 755, 757 f; lattice energy 207, 208; structure 203
-(I) carbonate 244, 595
-(I) chlorate 754
-(I) chloride 244, 435, 744, 655, 757; lattice energy 207, 208; solution equilibrium 126, 356; structure 203
chloride cell 435
-(I) chromate 699
coinage 750
-(I) compounds 754 f
–copper system 379, 380, 383
-(I) cyanide 357, 756
-(I) dichromate 699
-(I) fluoride 743, 754, 755; lattice energy 207, 208; structure 203
-(II) fluoride 757
-(I) halides 755, 757 f
-(I) hyponitrite 564
-(I) iodide 218, 244, 755, 757; lattice energy 207, 208; precipitation of 401 f, 411; structure 205
-(I) ion 244, 755, 758; ammine complexes 755, 756; cyano complexes 755, 756; thiosulfato complexes 756
-(II) ion 757
-(I) nitrate 568, 743, 754, 756
-(I) nitrite 565
-(I) oxide 353, 746, 755
-(I, III) oxide 437 f, 757
-(I) perchlorate 743, 755
-(I) phosphate 244, 574, 575
plating 750, 756
–platinum phase diagram 379
-(I) pseudohalides 756
-(I) sulfate 531, 755
-(I) sulfide 744, 756 f, 758
-(I) thiocyanate 756
-zinc storage cell 437
Simple cubic lattice 196
Single crystal 394
Sintering 396
Size, critical 289; of atoms 23; of ions 111 f
Skutterudite 714

Slag formation 503
Slag inclusions (in steel) 722
Slagging period (in copper smelting) 747
Slaked lime 484
Smalt 734
Smaltite 714
Smithsonite 762, 763, 765, 768
Smoke acid 522
Smpky quartz 604
Smoky topaz (=smoky quartz) 604
Soap 653, 656, 666
Soda 650, 655
   ash 655
   glass 614
Sodium 28, 647 f; history 650; occurrence 648; production 651 f; properties 650 f; uses 652
   amalgam 471 f, 774
   arsenate 578
   bicarbonate 656
   bismuth(V) oxide 580
   carbonate 650, 655
   chlorate 352, 486
   chloride 648, 653 f; aqueous solution 151, 385; bonding in 151; energy of formation 118; enthalpy of formation 127; lattice energy 206, 207; structure 202
   chloride dihydrate 385
   chloride–potassium chloride system 212, 379
   -chloride structure type 202, 207
   chloride–water system 382, 385
   chlorite 485
   chromate 696, 700
   cyanide 598
   dichromate 696, 700
   fluoride 207, 368
   hexafluoroaluminate 465, 639, 640, 641, 643
   hexahydroxoantimonate(V) 579, 648
   hexahydroxostannate(IV) 623
   hexametaphosphate 577
   hydrogen carbonate 596, 656
   hydroxide 653
   hypochlorite 484
   hyposulfite 534
   iodate 465, 470, 486
   manganate(V) 709
   metasilicate 613
   nitrate 537, 568, 596, 654
   nitrite 565
   oxide 203, 653
   perborate 638
   peroxide 653
   pyroantimonate 579
   silicate 613
   sulfate 352, 377, 654
   sulfate decahydrate 352, 377, 654
   sulfide 203, 517 f
   sulfite 533
   tetraborate 636
   tetrahydroxozincate 767
   thiosulfate 534
   triphosphate 577
   xenate(VIII) 462
Soft magnetic material 247, 726

Softening of water 511, 577, 666 f
Sol 409
Solder, soft 621, 626
Soldering 621
Solid 41 f; amorphous 42; crystalline 42; structure of 34, 42
   phosphorus hydride 555
   solution 43, 210 f; addition 210; interstitial 210; substitution 210; subtraction 210
   state 186 f
Solidus curve 379
Solubility 349 f
Solubility product 353 f; table 354
Solute 35
Solution 35; acid 329; alkaline (=basic) 329; binary 35; ideal 43, 297: ideal dilute 43; neutral 329; saturated 350; solid 43, 210; superacid 340; superbasic 340; supersaturated 350; ternary 35; true 408
Solvate 127
Solvated electron 547 f
Solvated ion 127
Solvation 127; energy of 127; enthalpy of 127
Solvay process 551, 655
Solvent 35; aprotic (=nonprotolytic) 317; nonaqueous 346 f; polar 124; protolytic 317
Soot 588
Sorosilicates 607 f
Spallation 277
Specific heat 58 f
Spectra of many-electron atoms 95; of one-electron atoms 94
Spectral band 73
Spectral line 72
Spectroscopic methods 239 f
Spectrum 72 f, 94 f: absorption 73; atomic 73; band 73; emission 73; infrared 239 f; line 72, 73; microwave 239; molecular 73; Raman 240 f; rotational 239; vibrational-rotational 241; X-ray 95
Speiss 727, 728
Sphalerite 624, 762, 765, 768; structure 204 f
Spiegeleisen 707
Spin 20, 76; antiparallel 76; parallel 76
Spinel 503, 645
   structure type 503
Spin magnetic moment 246
Spin quantum number 76
Spodumene 648
Spring steel 726
Square antiprism 113, 114
Stability constant 302
Stainless steel 696, 726
Stalactite 665
Standard
   hydrogen electrode 419
   potential 419; table 421 f
   state 57, 297, 298
Stannic acid, α 623; β 624
Stannite ions 623
Stannum 30, 619
Starch 468, 534
Stark effect 78
State 44 f; electronic 76 f; function of 47;

general equation of (for gases) 39; mesomeric 146; metallic 222; metastable 46, 362; stationary 66, 75; variable of 47 diagram 363, see phase diagram
Steel 719 f, 722 f; alloy 725 f; austenitic 725, 726; carbon 719; eutectoid 720; ferritic 726; hard 720; heat-resistant 726; high-alloy 725; hypereutectoid 720; hypoeutectoid 720; low-alloy 725; mild 720; stainless 696; 726; 18-8 726; tempered 719
  melting 722 f; acid 722; basic 722
Stereochemical methods 250
Stereoisomerism 198
Steric factor 255
Stibine 556
Stibium 30, 538
Stibnite 538, 544, 581
Stishovite 604
Stoichiometry 33
Storage cell 436 f
Streak 242
Stripping section 279, 280
Strontianite 659, 660
Strontium 29, 657 f; history 660; occurrence 659; properties 660 f
  carbonate 596
  chloride 203
  nitrate 664
  sulfate 531
Structural steel 726
Structure 34, 170 f
  determination 237 f
Struvite 574
Sublimation 365 f; enthalpy of 128
  point 365
Subshell 77
Substance 13; amorphous 42; crystalline 42; polycrystalline 194, 393; pure 35; solid 41
Substitution reaction 305
Sugar of lead 628
Sulfate ion 287, 525; bonding in 145, 155, 175
Sulfates 530 f
Sulfide ion 122, 517
Sulfides 516 f; acid 516; basic 516
Sulfinyl ion 522
Sulfite cellulose 533
Sulfite ion 286 f, 522, 525, 533, 759
Sulfites 533
Sulfitoargentate(I) complexes 759
Sulfonyl chloride 521
Sulfoxylate ion 525, 534
Sulfoxylates 534
Sulfoxylic acid 525, 526, 534
Sulfur 28, 489; flowers of 496; history 492; monoclinic 56, 63, 495; occurrence 490 f; orthorhombic 56, 63, 195, 495; plastic 220, 496; production 499; properties 495 f; P,T diagram for 497; structure 171, 172; trigonal (=rhombohedral) 496; uses 499
  dichloride 520
  dioxide 147, 522 f
  halides 520
  hexafluoride 473, 520
  tetrachloride 473, 520
  trioxide 147, 366, 523 f
  trioxide–water system 528
Sulfuric acid 525, 528 f; acid equilibria 332, 340, 347; fuming 528, 529, 532; manufacture 529 f; uses 530
Sulfurous acid 522, 525, 526, 532 f
Sulfurol chloride 521
Sun 294
Superconductivity 221
Supercooling 219, 352, 365, 390
Superoxides (=hyperoxides) 514
Superphosphate 530, 576
Supersaturation 390
Surface
  chemistry 400 f
  color 242
  energy 234
  migration 192
Suspension 408
Sylvite 649
Symbol, chemical 24 f
Symmetry 190 f; external 191
  center 190
  element 190
Sympathetic ink 735
Synthesis gas 454
System 14; binary 359; dispersed 14; heterogeneous 14, 388 f; homogeneous 14, 251 f; quaternary 359; ternary 359; unary 359; 363 f
Szilard-Chalmers process 295

T

Table salt 653, 663
Talc 236, 610, 611, 658
Tall oil 403
Tantalic acid 692
Tantalite 690
Tantalum 31, 689 f; history 690; occurrence 690; production 690; properties 690; uses 691
  -(V) oxide 689, 691
  carbide 692
Target nucleus 272
Tarnish color 243, 716
Tartar emetic 579
Tartratocuprate(II) ion 753
T-D process 293, 294
Technetium 29, 704; history 706; occurrence 704; preparation 707; properties 706 compounds 710
  -(VII) oxide 705, 710
Tectosilicate 611 f
Teflon 591
Telluric acid 535
Telluride ion 122
Tellurides 519
Tellurites 523, 535
Tellurium 30, 489; glassy 220 f; history 492; occurrence 491; properties 498; structure 171, 172; uses 500

Subject index   811

halides 521
dioxide 523
Tellurous acid 535
Temperature, critical 364
Tempering (steel) 719
Temporary hardness 666
Terbium 30, 121, 671; see lanthanoids
Ternary system 359, 386
Terra aluminis 639
Tetraarsenic tetrasulfide 580, 581
Tetraborate ion 636
Tetracarbonyliron(II) ion 183
Tetracarbonylnickel 183, 593, 728
Tetrachloro
 -aurate(III) ion 245, 760
 -cobaltate(II) ion 734
 -diamminechromate(III) ion 698
 -ferrate(III) ion 161
 -methane 124, 141, 474, 591
 -platinate(II) ion 741
 -titanate(IV) ion 161
 -zincate ion 161, 768
Tetrachromate ion 699
Tetrachromates 699
Tetracyano
 -cadmium ion 178
 -cuprate(I) ion 752
 -nickelate(0) ion 183
 -nickelate(II) ion 245, 736
 -zincate ion 768
Tetraethyllead 627, 652
Tetrafluoroberyllate ion 145
Tetrafluoroborate ion 145, 331, 635
Tetragonal system 195
Tetrahedral framework structure 204 f
Tetrahedral radii 163
Tetrahedrite 195
Tetrahydridoaluminate ion 175, 179, 642
Tetrahydridoborate ion 174 f, 633
Tetrahydridoborates 634
Tetrahydroxo
 -antimonate(III) ion 579
 -borate ion 175, 637
 -cuprate(II) ion 753
 zincate ion 434
Tetraiodomercurate(II) ion 774
Tetrametaphosphate ion 572, 575
Tetrametaphosphoric acid 576
Tetramethylammonium hydroxide 551
Tetramethyllead 627, 652
Tetrammine
 -copper(II) ion 156, 179, 304 f, 753
 -mercury(II) ion 771
 -zinc ion 767
Tetranitrogen tetrasulfide 580
Tetraoxo
 -chromate(VI) ion 698
 -ferrates(VI) 733
 -iodic(VII) acid 487
 -phosphate(V) 574
 -sulfate(VI) 530
Tetraoxygen 494
Tetraquoberyllium ion 663
Tetraquocopper(II) ion 156, 752

Tetrathioarsenate ion 516, 581
Tetrathionate ion 527
Tetrathionate acid 527
Tetravanadium oxide 230
Thallium 31, 630, 646
 -(I) carbonate 595
 -(I) halides 646
 -(I) hydroxide 630, 646
 -(I)–thallium(III) redox pair 425 f
Thénard's blue 734
Theoretical chemistry 10
Thermal analysis 387
Thermal diffusion 282
Thermistor 228
Thermit 642
Thermochemistry 55 f
Thermodynamics 50; first law of 51
Thermonuclear bomb 292
Thermonuclear reaction 292
Thio ion 516
Thioarsenite ions 580, 581
Thiocyanates 600
Thiocyanate ion 150, 479, 600
Thiocyanatoargentate(I) complexes 756
Thiocyanatoiron(III) ions 600, 733
Thiocyanic acid 479, 600
Thiocyanogen 479, 600
Thiomercurate(II) ions 773
Thionyl chloride 521
Thionyl ion 522
Thioplatinates(IV) 742
Thiostannate(IV) ion 624
Thiosulfates 534
Thiosulfate ion 286 f, 516, 527, 534
Thiosulfatoargentate(I) complexes 534, 756, 759
Thiosulfuric acid 527, 534
Thomas phosphate 537, 576, 723
Thomas process 722, 724
Thorium 31, 674; see actinoids
 dioxide 203, 681
 series 270
Thoron 458
Three-center bond 175
Three-electron bond 133, 139
Three-phase line 380
Thucholite 675
Thulium 30, 671; see lanthanoids
Thyroxine 465
Tin 30, 584 f, 619 f; gray ($=\alpha$) 205, 620; occurrence and history 619; production 620 f; properties 620; white ($=\beta$) 620
 alloys 621
 -base bearing metals 621
 bronze 624, 750
 -(II) chloride 622
 -(IV) chloride 622, 623
 -(II) chloride dihydrate 622
 dichloride 622
 dioxide 619
 -dioxide gel 623
 disease 620
 -(II) hydroxide chloride 622
 –lead system 621, 622

mirror 622
Tinning 621
Tin(II) oxide 623
  -(IV) oxide 623
  -(II) oxide hydroxide 623
  plate 621
  -(II) sulfide 624
  -(IV) sulfide 204, 624
  tetrachloride 622, 623
  tetrahydride 622
Titanic acid 687
Titanium 29, 683 f; history 684; occurrence 683 f; production 685 f; properties 685; uses 686
  carbide 203, 688
  diboride 688
  dioxide 683, 685, 686, 687; structure 203
  -group elements 683 f; history 684; occurrence 683 f; production 685 f; properties 684 f; uses 686
  -(II) oxide 215
  sponge 686
  tetrachloride 324, 474, 687
  tetraiodide 683
  trichloride 687
  trichloride hexahydrate 687
  white 687
Titanomagnetite 683 f
Titanyl ion 683
Titration curve, acid-base 341; redox 428
Tool steel 726
Topaz 236, 607
Total concentration 38
Total hardness 666
Total pressure 40 f
Tourmaline 631
Trace atom 285
Trace element 285
Tracer 278, 285 f
Tracer methods 285 f, 294
Transfer reaction 305
Transference number 218
Trans form 248, 250
Transformation range 220
Transistor 228
Transition elements 89, 668 f
Transition metals 89, 668 f
Transition range, indicator 342
Translation 39
Translational energy 51
Transmission color 241 f
Transuranium elements 89, 671; preparation 274; see also actinoids
Triborate ion 176, 635
Tricarbon dioxide 595
Tricarbonylnitrosylcobalt 593
Trichlorotriamminechromium(III) 698
Trichromate ion 699
Trichromates 699
Triclinic system 195
Tridymite 603, 604
Trigonal bipyramid 113, 114, 115
Trigonal system 195
Trimer 34

Trimetaphosphate ion 572, 575
Trimetaphosphoric acid 576
Trioxidinitrate(II) ion 564
Trioxidinitric acid 564
Trioxonitrate(V) ion (= nitrate ion) 568
Triphosphate ion 571, 575
Triphosphoric acid 571, 575
Triple bond 136
Triple oxides 503
Triple phosphate 576
Triple point 364
Triple-point pressure 366
Tris(thiosulfato)argentate(I) ion 534, 756
Trisulfate ion 526
Trisulfuric acid 526, 532
Trithiocarbonate ion 598
Trithiocarbonic acid 516, 598
Trithionate ion 527
Trithionic acid 527
Tritium 25, 293, 452, 455
Trouton's rule 64
T-T process 293
Tungstates(VI) 693, 701
Tungsten 31, 693; history 694; occurrence 694; production 695; properties 694; uses 697
  blue 703
  bronzes 703
  carbide 703
  compounds 701 f
  dioxide 702
  trioxide 693, 694, 696
Tungstic acid 694
Turnbull's blue 733
Turquoise 575
Twinned crystals 194
Two-component systems 359, 369 f
Two-phase area 381
Two-phase region 364, 370
T, x diagram 370; see also phase diagram
Tyndall effect 408
Type metal 626
Tyuyamunite 675

U

Ultrabasic rocks 442
Ultracentrifugation 408
Ultramarine 612
Ultramicroscope 409
Ultraviolet spectrum 240
Unary systems 359, 363 f
Uncertainty principle 66 f
Underglaze color 617
Unified atomic mass unit 26
Unit area 188
Unit cell 189
Unit charge 19
Universal composition 440, 443
Universal indicator 342
U_3O_8 phase 680
Uranate(VI) ion 672

Uranates(VI) 680
Uraninite 675
Uranium 31, 672 f; fission 275 f; production 679; separation of isotopes 278 f; see also actinoids
  carbide 289, 680
  dioxide 289, 679, 680
  hexafluoride 278 f, 679, 680
  –oxygen system 680
  series 269
  tetrafluoride 679, 680
  trioxide 680
Uranoids 671
Uranyl(VI)
  hydroxide 680
  ion 672, 673, 679, 680
  nitrate 680
Urea 594 f

V

Valence 111
  bond 139
  -bond method 140
  electron 85
  -electron concentration 231
Vanadic acid 692
Vanadinite 689
Vanadium 29, 689 f; history 690; occurrence 689; production 690; properties 690; uses 691
  -group elements 689 f; history 690; occurrence 689 f; production 690 f; properties 690; uses 691
  -(III) chloride 474
  -(IV) chloride 474
  -(II) ion 182
  -(III) ion 182
  -(II) oxide 689
  -(III) oxide 689
  -(IV) oxide 689
  -(V) oxide 230, 689, 691
Vaporization 366 f
Vapor pressure 361
  curve 364
  depression 349, 371
  relations 348
VB (= valence-bond) method 140
Venetian red 731
Vertical similarity 668, 711
Vibration 39, 71
Vibrational energy 51, 71, 73, 239, 278
Vibrational quantum number 71, 240
Vibrational-rotational spectrum 240
Visible spectrum 240
Vitamin B_{12} 714
Vitriol, oil of 730
Vitriols 531
Voltaic pile 433
Volume molarity 38
Vycor 605
van der Waals forces 98, 165 f, 236, 367 f
van der Waals radii 103, 166 f

W

Wackenroder's solution 534
Walter turbine 515
Washing soda 655
Water 56, 168, 507 f; acidic constant 328, 329; adsorbed 510; autoprotolysis 327 f; decomposition 431; deionized 405, 510; dissociation energy 162; distillation 510; electrolysis 510; hard 511, 666 f; heavy 28, ionic product 328; soft 511, 666; softening of 511, 577, 666 f; zeolitic 212, 378, 510, 611
  gas 454, 592
  -gas equilibrium 302, 453 f, 589
  glass 613
  hardness 511
  molecule 124, 140, 144, 166, 174
  of crystallization 126, 510
  protolytic system 317 f, 332, 337 f
  vapor 508
Wave
  equation 69
  function 69; orbital 77
  -length 17, 18, 240
  mechanics 66
  number 18
Weight molarity 38
Wet process (for phosphoric acid) 574; (for copper) 748
Whiskers 234
White lead 628
White metal 621
Whie precipitate, fusible 773; infusible 773
White vitriol 531
Wolfram (= tungsten) 31, 694
Wolframite 694
Work 53; electrical 53; maximum 61; pressure-volume 53
  content 61
Wood's metal 386, 766
Wurtzite 768; structure 205

X

X-ray diffraction 238
X-rays 95 f
X-ray spectra 95
Xenate(VIII) ion 462
Xenic acid 461
Xenon 30, 457; history 458; occurrence 457; properties 458 f; recovery 459; uses 460
  compounds 461 f
  difluoride 461
  hexafluoride 461
  (VI) oxide 461
  (VIII) oxide 462
  (VI) oxide tetrafluoride 462
  tetrafluoride 461
  tetroxide 462
  trioxide 461
Xi hyperon 20

Y

Ytterbium 30, 104, 121, 671; see lanthanoids
Yttralox 681
Yttrium 29, 673; history 676; occurrence 673, 674
  earths 674, 675
  oxide 681
  vanadium oxide 681

Z

Zeeman effect 78
Zeolites 378, 404, 611 f
Zeolitic water 212, 378, 510, 611
Zero-point energy 68, 72
Zinc 29, 761 f; history 763; occurrence 762; production 764 f; properties 763 f; uses 765 f
  acetate 766
  -air cell 435
  blende 762
  carbonate 762, 766, 768
  chloride 621, 767; electrochemistry 416 f
  compounds 766 f
  cyanide 766, 768
  dust 766
  fluoride 766
  -group elements 761 f; history 763; occurrence 762 f; production 764 f; properties 673 f; uses 765 f
  hydrogen arsenate 578
  hydroxide 435, 437 f, 767
  hydroxide carbonate 768
  hydroxide chloride 767
  hydroxide chromate 700
  hydroxide silicate 762
  lime 491
  manganese(III) oxide 734
  nitrate 766
  orthosilicate 228
  oxide 228, 734, 765, 767
  phosphate 766
  plating 765, 768
  silicates 228, 762, 766
  sulfate 766, 768
  sulfate heptahydrate 768
  sulfide 228, 666, 762, 768; structure 204 f
  vitriol 768
  white 767
  yellow 700
  −zinc(II) redox pair 416 f, 433 f
Zincite 762, 767
Zircaloy 686
Zircon 195, 607, 684
Zirconium 29, 683 f; history 684; occurrence 684; production 685 f; properties 685; uses 686
  dioxide 684, 685, 687
  silicate 684
  sponge 686
  tetrachloride 685, 688
Zone melting 383
Zoning 379